# African Rain Forest Ecology and Conservation

Banyang-Mbo Wildlife Sanctuary

Korup National Park

Lobeke Reserve

Dzanga-Sangha Reserve Complex

Nouabale-Ndoki National Park

Monveda Apagibeti

Ituri Forest/Okapi Wildlife Reserve

Kibale Forest National Park

Budongo Forest Reserve

Bwindi Impenetrable Forest National Park

Ghana Sacred Groves and Forest Reserves

Makokou

Ogooué Middle Valley

Lopé Faunal Reserve

Virunga/Volcano National Parks

Udzungwa Mountains National Park

Masoala National Park

Kahuzi-Biega National Park

Nyungwe Forest Reserve

# African Rain Forest Ecology and Conservation

An Interdisciplinary Perspective

*Edited by*
William Weber
Lee J. T. White
Amy Vedder
*and* Lisa Naughton-Treves

YALE UNIVERSITY PRESS
NEW HAVEN AND LONDON

**WCS**
WILDLIFE CONSERVATION SOCIETY

Published with the support of a grant from the Wildlife Conservation Society.

Designed by Thomas Whitridge
Set in Ehrhardt type by Ink, Inc., New York
Printed in the United States of America by Sheridan Books, Chelsea, Michigan.

Library of Congress Cataloging-in-Publication Data
African rain forest ecology and conservation: an interdisciplinary perspective / edited by William Weber...[et al.].
p.   cm.
Includes bibliographical references.
ISBN 0-300-08433-1 (alk. paper)
1. Rain forest ecology—Africa. 2. Rain forest conservation—Africa. I. Weber, William, 1950–
QH194 .A38 2000
577.34'096—dc21      00-043678

A catalogue record for this book is available from the British Library.

The paper in this book meets the guidelines for permanence and durability of the Committee on Production Guidelines for Book Longevity of the Council on Library Resources.

*frontispiece:* The African rain forest (shaded), showing key sites mentioned in this book.

10   9   8   7   6   5   4   3   2   1

# Contents

**Part V.**
**Applied Research and Management**

**Part VI.**
**Conclusion**

# Preface

William Weber and Amy Vedder

*Going up that river was like traveling back to the earliest beginnings of the world, when vegetation rioted on the earth and the big trees were kings. An empty stream, a great silence, an impenetrable forest.*
—Joseph Conrad, *Heart of Darkness*

NEARLY ONE HUNDRED YEARS after Conrad's first florid descriptions in 1902, the African forest remains a source of powerful imagery—and gross inaccuracy—in both literature and film. Scientifically, the vast central African rain forest is the least known of the world's tropical forest regions, at a time when its great biological and cultural diversity is buffeted by an increasingly complex array of economic and political forces.

This book will be useful to a broad range of individuals with interests in the world of the African rain forest. Its authors view that varied world from a diversity of perspectives, which are nevertheless interrelated. They also share a desire for more information, tempered by a recognition of the urgent need for action: action to save as much of the forest and its exceptional biological wealth as possible, and action to preserve the cultural heritage and economic potential of the forest for generations to come. It is our deepest hope that many readers will be inspired to add to our collective understanding, and to act on that knowledge for the sake of the forest, its wildlife, and its people.

Twenty-eight years ago, the two of us began a complicated and extremely rewarding relationship with the world of forested Africa. The exceptional wildlife spectacles and diversity of that world surrounded us on the first of many visits to Kahuzi-Biega National Park in August 1973, in the eastern Democratic Republic of the Congo (throughout this book, the nation formerly known as Zaire is referred to as DR Congo, whereas the Congo Republic, with its capital in Brazzaville, is referred to as Congo). Semi-habituated gorillas greeted our arrival with a mix of aggressive displays and passive acceptance that seemed to reflect their ambivalence toward our presence, which they tolerated for no more than twenty minutes. Leaving the gorillas that first day, we walked back through a blur of plants, birds, beetles, and bugs—all new to us—until our Pygmy guide brought our focus back to another species. Following a bee's flight, he ignored repeated stings to tear off part of its hive, and soon our hands and faces were smeared with an unspeakably rich mix of honey, bits of honeycomb, and larvae. We were in an exotic new world full of biological and cultural diversity.

Our more quotidian experience as teachers in the village of Bagira, not far from the park, made it clear that we were also in a land of almost unfathomable poverty—a land where basic human needs placed people in direct conflict with wildlife over use of the same natural forest resource base. This juxtaposition of biological wealth and human poverty was a keystone of our early African experience that has informed and influenced all our subsequent work.

In 1977, we received a grant from the Wildlife Conservation Society (formerly the New York Zoological Society) to return to the east-central African highlands to study and address the causes of a drastic decline in the mountain gorilla population shared by Rwanda, DR Congo, and Uganda. It was an unusual study for its time because of its interdisciplinary nature. Amy concentrated her efforts on basic conservation biology: gorilla feeding ecology, habitat requirements, and demography. Bill worked primarily on social, economic, and political factors affecting the gorillas' decline. Together, our results composed a far more complete picture of the situation. Our findings also allowed us to construct a "gorilla conservation equation"—a relative weighting of factors working for and against conservation—from which we produced recommendations for action to improve the welfare of the animals, the forest, and the people.

The Mountain Gorilla Project, which grew out of our work and which we stayed on to help initiate in Rwanda, has been hailed as a conservation success story. It certainly helped to restore the gorilla population and reverse negative local attitudes toward the Volcanoes National Park, and the financial return from its tourism component has helped to protect the park and gorillas even throughout the brutal civil war that has buffeted the surrounding region. Whether this protection can continue in the face of the extreme socioeconomic and political pressures now confronting Rwanda remains to be seen.

We left the Mountain Gorilla Project in 1980, convinced that an interdisciplinary, applied approach could produce positive results for conservation. Over the past two decades this approach has been tested by ourselves and many different colleagues and local partners in the forested regions of Rwanda, Burundi, Cameroon, Central African Republic, Congo, DR Congo, Madagascar, and Uganda, among others. In each instance, the ability to base management plans and other actions on an expanded, integrated information base has unquestionably improved the ultimate conservation impact. At the same time, the growing number and interrelatedness of socioeconomic and biological factors to be considered have greatly increased the complexity of any conservation planning exercise. This is in stark contrast to the not-so-distant past when conservation was seen as little more than deciding where to draw the boundaries of a park, with often inadequate knowledge of the relevant ecosystem and little or no concern for local human interests.

In the early 1980s, field experience such as ours began to inform and support a paradigm shift in the field of conservation, away from an emphasis on just parks and protected areas and toward balance and connectivity between wildlife conservation and human development interests. As described in the *World Conservation Strategy* published in 1981 by the International Union for the Conservation of Nature, this linkage was, to many people, both obvious and essential: "For if the object of development is to provide for social and economic welfare, the object of conservation is to ensure Earth's capacity to sustain development and to support all life."

Over the past twenty years, this concept has evolved from an initial focus on "integrated conservation and development" to the more broadly defined "sustainable development." In the process, it has emerged as the dominant model in the field of conservation, although it is not without its critics. While some wonder if wildlife and nature remain central tenets in the new creed, others question whether any modern form of economic development is truly sustainable.

Our own experience underscores the importance of a more comprehensive view of conservation, in which human beliefs, behaviors, politics, and economies are seen as powerful forces shaping the forest ecosystem. Yet we are also convinced that current conservation theory has raced far ahead of real-world, on-the-ground experience. This is first and foremost because we lack critical information from all disciplines involved: biology, ecology, political science, economics, and anthropology, among others. Second, even where adequate information exists within various disciplines, the leap from an aggregated multidisciplinary information base to a truly integrated *interdisciplinary* foundation—analogous to the difference between a mixture and a true compound—is both daunting and difficult. Finally, and perhaps most frustrating, we rarely monitor the impacts of whatever actions we do take over time. Thus an integrated conservation and development project in Cameroon can carry out its agenda and change the way a particular forest is managed over a five-year period. Yet, at the end of that time, no one can say whether the forest, wildlife, or local people are faring better or worse as a result of our interventions, because there were no baseline data sets and no one monitored change over the life of the project.

By the early 1990s, conservation researchers and practitioners were increasingly frustrated by two concerns. First, the gap between theory and practice was expanding, as more and more money was made available for a growing number of projects based on the integrated conservation and development model, but very little support was given for related research and monitoring. And second, the limits to comprehensive understanding across disciplines were becoming increasingly apparent.

In May 1993 the Wildlife Conservation Society organized a three-day conference in Essex, Massachusetts, to address these concerns (with additional support from the John D. and Catherine T. MacArthur Foundation). Participants included more than two dozen Africans, Europeans, and Americans with extensive experience as field biologists, social scientists, and conservation practitioners. Much information was exchanged, personal contacts were broadened and deepened, some sound recommendations were produced, and several productive field collaborations ensued. But the major result of the meeting was excitement about the potential to build bridges across existing barriers: barriers between academic disciplines, especially the biological and social sciences, and between researchers and practitioners.

The idea for this book grew out of that meeting. Over the next several years, other professional responsibilities took precedence, but the idea never died. Several of the participants in the Essex conference have contributed chapters to this volume and thus helped to rekindle the spirit of our gathering. Many others have joined us since then, greatly enriching the final product through their contributions. The chapters that follow are aimed at researchers and students from a broad range of academic disciplines, conservationists from across the tropics and around the world, and all those who seek a more comprehensive understanding of the central African rain forest.

This book is ambitious in scope. The rain forest itself is an extremely complex community, maintained in a dynamic, ever-changing state through the combined effects of climate and the interactions among millions of species. Some species may have a disproportionate impact on the ecosystem, but discerning what those effects are, or even identifying such species, is devilishly difficult. In fact, we are far from understanding the basic ecology of some of the rain forest's most common large mammals. Conditions are no less complicated when we move into the human sphere. The central African forest region has been racked by civil war and political instability in recent years. This is partly a lingering effect of the continent's dark and divisive colonial legacy, but civil wars and ethnic conflicts are an all-too-common part of the long and difficult nation-building process across continents, cultures, and time. Development, too, is a complex, nonlinear process that moves forward in fits and spurts, propelling one region ahead and leaving another behind. For better and for worse, much of the African rain forest region has been left behind at the beginning of the twenty-first century.

Set against its ambitions, this book falls short. Despite its great breadth of geographic and topical coverage, the information gaps are numerous and deep. Despite its strong multidisciplinary flavor, true integration across disciplines is rare. Despite an almost missionary zeal for information-based action, there are currently few concrete examples to demonstrate that this is indeed the true path to effective conservation. Yet perhaps its shortcomings provide the best reason to produce this book. Because in its gaps and weaknesses, in its partial answers and imperfect solutions, this collection is an accurate reflection of the current state of understanding and conservation in the African rain forest. It is, in fact, a call for others to become informed about, concerned for, and ultimately involved with this most important, fascinating, and threatened region.

# Acknowledgments

This book would not be possible without the support of the Wildlife Conservation Society. Its commitment of human and financial resources toward this project has been indispensable, and its dedication to applied multidisciplinary research and conservation in dozens of projects across the African rain forest is a continued source of inspiration. The John T. and Catherine D. MacArthur Foundation also provided partial support for the conference that stimulated this book.

Hilary Simons Morland helped with early organization of the book, as did Angie Hodgson and Kristi MacDonald. Nathan Havill assisted with fine-scale editing. Sarah Ward took control of the final production phase and drove us to complete the task at hand with a combination of good humor, determination, and high standards. Jean Thomson Black contributed appropriate doses of wisdom, encouragement, and admonitions in her role as senior editor at Yale University Press, with steady support from her editorial assistant, Nush Powell. Manuscript Editor Jenya Weinreb's attention to detail greatly improved the final product.

We thank our distinguished colleagues who distilled their extensive field experiences into the diverse chapters that make up this book. Many contributors also reviewed and commented on chapters other than their own. Others who made important editorial suggestions include Kate Abernethy and Adrian Treves, both of whom read several versions of many chapters and offered prompt, constructive criticism.

Finally, we thank the countless local people who have helped in myriad ways to make the African rain forest more accessible and understandable to all who wish to know more about their special world.

**WCS**
WILDLIFE CONSERVATION SOCIETY

**I**

**The African Context**

# CHAPTER I

# The African Rain Forest

## Climate and Vegetation

Lee J. T. White

The African rain forest has long fired peoples' imaginations: the heart of Conrad's darkness; testing ground and meeting place of Stanley and Livingstone; home of du Chaillu's savage ape; deathbed of numerous intrepid explorers; and fetid breeding ground of deadly diseases, most recently the sinister ebola virus. With time, much of the darkness has given way to light, as epitomized by the new image of the silverback gorilla as a gentle family man. However, logistical difficulties, political disturbances, underdevelopment (in the modern sense), and the forest's great size mean that our knowledge of what it is and how it works remains embryonic. In this chapter I shall introduce the vegetation of the African rain forest system and highlight some of the issues (both intellectual and practical) facing ecologists and conservationists, many of which will be explored in greater detail in the chapters that follow.

In order to provide a feel for what constitutes the African rain forest, let me document the life-history of one of its most spectacular inhabitants. In the year A.D. 1000 a forest elephant defecated as it moved along a game trail that followed a ridgeline running through the central African rain forest. The steaming dung pile contained the seeds of several forest trees, whose fallen fruits the elephant had collected from the forest floor during the previous few days. There were several seeds of the large forest tree *Baillonella toxisperma*, known at the time by the local name Moabi. Many seeds germinated, but over the following months all but one of the Moabi seedlings perished, trampled underfoot by elephants, browsed by duikers, consumed by rodents, or attacked by foliverous insects. The surviving Moabi seedling gradually thickened and became stronger but barely increased in height for many years.

When the Moabi seedling was about 20 years old but still barely 1 m tall, there was a violent tropical storm that toppled a large emergent tree standing about 20 m away. As it fell to the ground, it pushed over several other trees, creating a large gap in the forest and allowing light to penetrate to the normally gloomy forest floor. This event initiated a race. The Moabi began to grow toward the sky, as did other plants in and around the gap. Forest species growing in the center of the opening were scorched by the sun and died. Other plants were broken by elephants feeding on the lush vegetation in the gap, or were bent into nests by families of gorillas, but the Moabi escaped these influences and continued to grow taller. It gradually rose toward the opening above, carrying with it a shroud of lianas, which had grown around its trunk and branches. Gradually the Moabi rose above the other trees growing around it and, spreading its branches, cast its competitors into the shade of its canopy.

For many years the Moabi did not flower or fruit, but its trunk and branches gradually thickened and its canopy grew wider. Epiphytes were established in crevices on the trunk and created microhabitats for numerous invertebrates. When the Moabi, aged about 100 years, reached a diameter of about 70 cm, it flowered and fruited for the first time. Only a few fruits developed, and these were destroyed by a group of black colobus monkeys, which ate the immature seeds. It was some years before it flowered again.

Three hundred years later the Moabi had developed into a large tree and was producing a large fruit crop every three years or so. The large fruits fell 45 m to the ground as soon as they ripened, and their yeasty smell attracted large numbers of mammals to the base of the tree, which was now scarred and swollen from elephants having repeatedly fed on the bark. Elephants, gorillas, chimpanzees, mandrills, duikers, and monkeys, as well as many species of birds and insects, feasted on successive crops. Humans from an Iron Age village located some way further along the ridge also visited in the fruiting season to collect seeds, with which they made highly prized and nutritious cooking oil. As the tree grew, its fruit crop increased and it became the center of a radiation of elephant trails. More and more epiphytes accumulated on its branches, and hornbills nested in a crevice where rot had set in after one of the branches fell during a storm.

The Iron Age villagers moved to a site closer to the Moabi and cleared the forest all around it to grow bananas. They cut some of the lianas that were supported by the Moabi, but left the tree standing and continued to collect its seeds, which were an important part of their diet.

Eventually the people moved away and the forest around the Moabi grew back, but it now stood just over 60 m tall and towered above the surrounding canopies. Late in its life, in 1999, loggers built a road along the Moabi's ridge, following the well-worn elephant trail. They felled the rain forest giant, by then almost 3 m in diameter and 65 m tall, just a year before it passed 1,000 years of age, creating a huge light gap where its crown had crashed to the ground. Its trunk was dragged away by a bulldozer, and the orchids covering its branches gradually yellowed and died.

This necessarily schematic representation of the life of a rain forest tree provides just a glimpse of the forest, which is ancient but dynamic, and made up of many interrelated components whose relationships we are just beginning to study. In order to understand what has shaped the African rain forest as we see it today, one needs to comprehend geological time, as well as both the history and the prehistory of the African continent. In a system where a single tree can be up to 1,000 years old (J. M. Fay, pers. comm.), what we see today may have been shaped by events that occurred many decades or centuries ago.

## The Origins of the African Rain Forest System

Maley (1996) provides an excellent review of the origins and history of the African rain forest (figure 1.1). The roots of the modern African forests can be traced to the late Jurassic or lower Cretaceous, 136 million years ago, when the continent of Gondwana was just beginning to split along the African and South American tectonic plates. At this time the dominant trees were Gymnosperms, adapted to the hot, dry conditions in continental Gondwana (Boltenhagen et al. 1985; Salard-Cheboldaeff and Dejax 1991; Salard-Cheboldaeff and Boltenhagen 1992). The small, rigid, sclerified leaves of the Gymnosperms (of which such modern African representatives as

*Podocarpus* and *Juniperus* survive in montane forests above about 1,800 m: see, e.g., Letouzey 1968, 1985) are able to condense atmospheric moisture (mist and low cloud). The Gymnosperms of the Jurassic may have obtained much of their water in this way.

Angiosperms began to be present by the Barremian of the lower Cretaceous. They were herbaceous plants of such families as the Chloranthaceae (see, e.g., Doyle 1978, and many other references cited in Maley 1996), found today in the moist understory of Southeast Asian rain forests (Blanc 1989). In the late Barremian and early Aptian, the pollen types *Afropollis* and *Walkeripollis* linked to the Winteraceae arose (Doyle et al. 1982). Today this family is not found in continental Africa but survives in the austral temperate forests of South America, Madagascar, and Australia and in cool, moist mountainous zones of tropical regions (Doyle et al. 1990a, 1990b) in areas where one also finds the surviving Gymnosperm *Podocarpus* (White 1981).

As the African and South American tectonic plates gradually moved apart during the Cretaceous, numerous lagoons formed and the marine influence gradually increased. By the Senonian, the Atlantic Ocean was opening up. The climate became wetter, favoring the radiation of Angiosperms and the decline of the Gymnosperms in the tropics; hence the archetype of the modern African rain forest appeared (Doyle 1978; Salard-Cheboldaeff and Dejax 1991), about 35 million years after the early Angiosperms began to radiate.

These early forests, which developed toward the end of the Cretaceous and covered an extensive area of South America and Africa, were characterized by a radiation of the palms (Herngreen and Chlonova 1981). The transition between the Cretaceous and the Tertiary was associated with global extinctions (see Upchurch and Wolfe 1987) and represents the end of the Gymnosperms in the lowland trop-

ics, with the exception of a few species in the Gnetaceae and others surviving in climatic anomalies such as on high mountains. This period was also marked by the radiation of mammals, whose origins were in the Triassic, about 200 million years ago. By the Eocene some modern genera of plants began to appear in the fossil record (wood and pollen). At this time, areas presently on the equator were approaching their modern positions as the African continental plate moved gradually southward (Guiraud and Maurin 1991). Forests with affinities to modern African rain forest stretched across northern Egypt, with such genera as *Pentaclethra* and *Calpocalyx* (Mimosaceae) (Guinet et al. 1987), while in Libya fossilized wood has been found of species that today occur in wooded savannas (see Maley 1996).

In the upper Eocene, about 45 million years ago, the floristic composition of southern Cameroon began to resemble its modern state, with the first appearance in the fossil record of many genera and some species still occurring today (Salard-Cheboldaeff 1981; see figure 1.1). At this time the rain forest may well have stretched as far as Egypt in an unbroken swathe (Maley 1996). In the Oligocene there was a decline in palm abundance and variety, while many new Angiosperm genera appeared for the first time (Guinet and Salard-Cheboldaeff 1975). This radiation continued into the Miocene, and in the upper Miocene the first evidence of *Rhizophora* mangroves is found along the Gulf of Guinea.

In the middle Miocene, about 10.5 million years ago, there was a major marine regression, corresponding to an extension of the Antarctic ice cap. This resulted in increased grass and *Podocarpus* pollen in sediments off the Niger delta (Poumot 1989), suggesting that vegetation became more open and that montane elements expanded. There was a corresponding sharp decline in palm pollen, suggesting that the Are-

| ERA | PERIOD | EPOCH | TIME (Ma) | |
|---|---|---|---|---|
| | | | 0 | |
| Use of fire as of 500,000 B.P. | QUATERNARY | HOLOCENE | 0.01 | Forest re-establishes by 8,000 B.P., but regresses 3,500-2,000 B.P. |
| | First documented humans in Gabon c. 400,000 B.P. | PLEISTOCENE | | Series of 21 world glacials, some of which probably resulted in cool dry climate and forest regression. From 70,000 B.P. last world glacial, with maximal forest fragmentation 20,000 - 15,000 B.P. |
| First record of Homo 2.4 Ma in East Africa | Glacial advance in N hemisphere c. 2.5Ma / Marine regression | | 1.8 | |
| | | PLIOCENE | 2.5 | Climate more seasonal and erosive, peak in Podocarpus and grass pollen. |
| CENOZOIC | NEOGENE | Upper MIOCENE | 5 | Sahara opens up, climate arid, grass pollen and burnt cuticles in sediments of river Niger, landscape generally more open. First hominids. |
| | Marine regression | Lower MIOCENE | 10.5 | Palms decline. Podocarpus up. — In wet climate of Miocene the forest spreads north and further diversifies, e.g., Fillaeopsis, Tetrapleura (Mimosaceae), Klaineanthus (Euphorbiaceae), Pentadesma (Guttiferae), and Rhizophora (i.e., mangroves). |
| | TERTIARY | OLIGOCENE | 26 | Egypt with vegetation of rain forest and semi-deciduous forest. Pollen includes that of Pemtaclethra. |
| | Palms decline | Upper EOCENE | 37 | Libya with vegetation intermediary between rain forest and Sudanian zone. |
| | PALAEOGENE | Lower EOCENE | 45 | Central forests begin to resemble modern species composition, e.g., Pentaclethra, Calpocalyx (Mimosaceae), Alchornea (Euphorbiaceae), Symphonia globulifera (Guttiferae), Petersianthus macrocarpus (Lecythidaceae), and Bombax buonopozense (Bombacaceae). |
| | Equator near current position (c. 5°N) | EOCENE | 53 | |
| | Extinctions -- mammal radiation | PALAEOCENE | 65 | Increased humidity (marine influence) as continents separate. Gymnosperms decline in archetype of the rain forest. Radiation of Araceae (palms). |
| Madagascar separates from continent | Equator along south of modern Sahara | MAASTRICHTIAN / SENONIAN / TURONIAN | | |
| | CRETACEOUS | CENOMANIAN | 100 | Proliferation of Angiosperms but Gymnosperms remain dominant. |
| | | ALBIAN | | |
| MESOZOIC | | APTIAN | 120 | Early Angiosperms characteristic of aquatic / humid conditions, e.g., Winteraceae such as Afropollis and Walkeripollis. |
| | Gondwanaland begins to split | BARREMIAN | 136 | Climate hot and dry, Gymnosperms dominate—many species, including Podocarpus. |
| | JURASSIC | First birds appear | | |

Figure 1.1. Summary of the history of the African rain forest (compiled from Maley 1996 and other sources; see Maley 1996 for further information).

caceae declined in the changing climate. In the upper Miocene and into the Pliocene there is evidence that the equatorial African landscape began to be more open in response to global cooling (Broecker and Denton 1989). Fossilized wood of xeric species dated to about 4–5 million years ago suggests that the Sahara desert opened up in northern Africa in the early Pliocene and that climatic zonation similar to that existing today became established (Maley 1980, 1996). Morley and Richards (1993) found evidence of burnt grass cuticles in sediments of the Niger delta, suggesting that fires were prevalent in the grasslands of the time

About 2.5 million years ago the first northern hemisphere glacial advance occurred, at the same time as an expansion of the Antarctic ice cap (Mercer 1983; Barker and Kennett 1988), causing a major marine regression with corresponding peaks in grass and *Podocarpus* pollen recorded in marine deposits (Poumot 1989). This highlights the relationship between tropical lowland aridification and global cooling (Maley 1996), and evidence from sediments suggests that the climate of the period was also more seasonal and hence more erosive over much of the Niger basin (Knaap 1971). Bonnefille (1983) presents evidence of a similar cool, dry climate in the east African highlands at about the same period.

After the first major northern hemisphere glacial advance began 2.5 million years ago, a series of cycles of expansion and contraction, each lasting 41,000 years, occurred up until 800,000 years ago, when 100,000-year cycles with increased amplitude began (Ruddiman et al. 1989). The most recent world glacial cycle began about 70,000 years ago, peaking between about 20,000 and 15,000 B.P. The resulting changes in climate and vegetation (discussed in Chapter 5) provide a model for the previous twenty glaciations, which have affected the African forest system during the course of the Pleistocene.

This brief discussion of the general history of African forests illustrates the conceptual leap in time that one has to make if one is to understand the complexity of the system we see today. For example, the rain forest tree *Pentaclethra* is currently represented by two species in Africa (and others in South America). The first pollen record for this genus is from northern Egypt. It dates back about 50 million years (Guinet et al. 1987) and seems to correspond to *P. macrophylla*, a large tree that is common in many parts of the African rain forest today and which currently occurs from Senegal to the Democratic Republic of the Congo (henceforth DR Congo), but not in Egypt. During the course of this species' history great changes occurred in the distribution and extent of the African forest. The Kalahari desert has expanded into Congo from the south and then been recolonized by forest (see Maley 1996), while in the north, parts of the previous range of *P. macrophylla* have given way to the Sahara. It has seen forest extend in a continuous belt from Cameroon to Egypt, and contract into relatively small, isolated refugia (see, e.g., Maley 1987, 1996). As a result, distribution patterns of *Pentaclethra*, and the species it grows in association with, are likely to be extremely complex and to reflect historical events that are only scantily documented at best (White 1990).

Humans are notoriously limited in their ability to look at long-term implications of their actions. Even the concept of a tree that is 1,000 years old, as are some of those cut by foresters in northern Congo (J. M. Fay, pers. comm.), is difficult to envisage. The attempt to understand patterns resulting from the cumulative effects of 50 million years or more of forest history is one of the great challenges faced by ecologists, particularly those working in areas that were not completely wiped clean by the passage of an ice sheet, such as the African tropics.

## Modern Patterns of Species Distribution in the African Forest System: The AETFAT Chorological Classification

It is somewhat self-evident to say that the rain forest of tropical Africa is botanically distinct from temperate Europe or America, even if certain plants, such as the ubiquitous bracken fern, *Pteridium aquilinium*, are common to all these areas. However, it is instructive from historical,

biogeographical, and conservation perspectives to attempt classification of relatively homogeneous units with characteristic plant species composition. In 1983, building on earlier works (e.g., Lebrun 1947, 1960, 1961; Aubréville 1950, 1962; Monod 1957; Keay 1959), the Association pour l'Etude Taxonomique de la Flore d'Afrique Tropicale (AETFAT) published a vegetation map of Africa and an accompanying

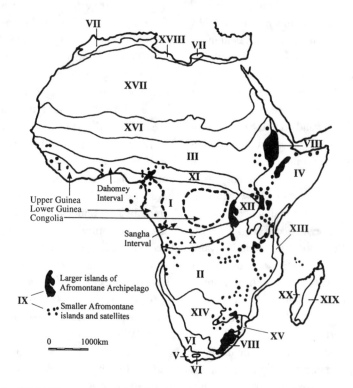

Figure 1.2. The AETFAT chorological classification of Africa and Madagascar, showing the subdivision of the Guineo-Congolian regional center of endemism: (I) Guineo-Congolian regional center of endemism; (II) Zambezian regional center of endemism; (III) Sudanian regional center of endemism; (IV) Somalia-Masai regional center of endemism; (V) Cape regional center of endemism; (VI) Karoo-Namib regional center of endemism; (VII) Mediterranean regional center of endemism; (VIII) Afromontane archipelago–like regional center of endemism, including (IX) Afroalpine archipelago–like region of extreme floristic impoverishment; (X) Guineo-Congolia/Zambezia regional transition zone; (XI) Guinea-Congolia/Sudania regional transition zone; (XII) Lake Victoria regional mosaic; (XIII) Zanzibar-Inhambane regional mosaic; (XIV) Kalahari-Highveld regional transition zone; (XV) Tongaland-Pondoland regional mosaic; (XVI) Sahel regional transition zone; (XVII) Sahara regional transition zone; (XVIII) Mediterranean/Sahara regional transition zone; (XIX) East Malagasy regional center of endemism; (XX) West Malagasy regional center of endemism. Adapted from White 1983a, © UNESCO. Reproduced by permission of UNESCO.

descriptive memoir (figure 1.2). It was based on the work of more than one hundred scientists with collective experience covering the whole of continental Africa and Madagascar (White 1983a, 1983b; see also White 1993a). The most original feature of the map is that vegetation types are depicted within a regional chorological framework. White 1993a gives a detailed account of the development of the map.

The phytochoria of the AETFAT map are based on plant species distributions and are defined by the number and proportion of species unique to them (a measure of their distinctness). These phytochoria fall into five categories:

1. Seven *regional centers of endemism* are distinguished on the African mainland (and two in Madagascar) (see figure 1.2), each with more than 50% of its phanerogamic

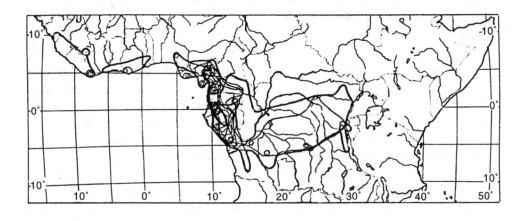

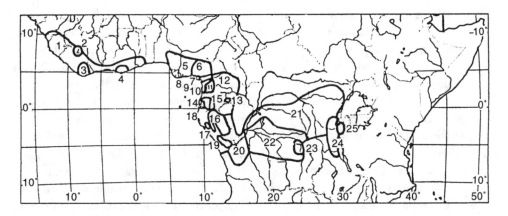

Figure 1.3. Ranges of all species and subspecies of the sections *Loasibegonia* and *Scutobegonia* (*top*) and the areas of endemism that can be defined from their distribution (*bottom*): (1) west tropical Africa; (2) Mount Nimba; (3) Cape Palmas; (4) Cape Three Points; (5) southeast Nigeria; (6) west Cameroon mountains; (7) Mount Cameroon; (8) Bioko; (9) north coastal Cameroon; (10) south coastal Cameroon; (11) Cameroon Plateau; (12) upland Cameroon/Gabon; (13) Bélinga; (14) west Crystal Mountains; (15) east Crystal Mountains; (16) Massif du Chaillu; (17) Doudou Mountains; (18) coastal Gabon; (19) Mayombe; (20) west Congo/DR Congo; (21) north central DR Congo; (22) south central DR Congo; (23) Sankuru; (24) east DR Congo; (25) Rwanda/Burundi. From Sosef 1994, reproduced by permission of the author and Wageningen Agricultural University.

species confined to it and a total of more than 1,000 endemic or near-endemic species. For example, the Guineo-Congolian regional center of endemism contains an estimated eight thousand species, of which 80% are endemic. About 25% of all genera, as well as nine families, are endemic.

2. There is one *archipelago-like regional center of endemism*, the Afromontane region, which has at least 4,000 species, of which about 75% are endemic or nearly so. It differs from other regional centers of endemism in its fragmented distribution (figure 1.3).

3. The 280 species that occur on the summits of the higher east African and Ethiopian mountains form the Afroalpine *archipelago-like region of extreme floristic impoverishment*, with virtually no endemic genera. Most species in this region are shared with the Afromontane region, and the two are generally united when considering classifications at a pan-African or worldwide scale.

4. Six *regional transition zones* separate regional centers of endemism. Floras of African transition zones have fewer than 1,000 endemic species, representing less than 50% of the total flora. For example, the Guinea-Congolia/Sudania regional transition zone, which surrounds the main forest block, probably contains fewer than 2,000 species, of which perhaps less than 50 are endemics.

5. Three small regions cannot be neatly categorized as regional centers of endemism or as transition zones and are named *regional mosaics*. They consist of areas with a mosaic of physiognomically and floristically distinct vegetation types and a small extent of intermingling of otherwise distinct floras. For example, all three regional mosaics contain disjunct populations of species characteristic of the Guineo-Congolian region.

Most of this book is naturally centered on White's (1983a) Guineo-Congolian regional center of endemism—including its extensions into the Lake Victoria regional mosaic (particularly in Uganda) and the Guinea-Congolia/Sudania regional transition zone—and, to a lesser extent, the Afromontane archipelago (see figure 1.2). In the following summary I shall highlight some of the salient features of these two zones as originally described in the AETFAT memoir.

### THE GUINEO-CONGOLIAN REGIONAL CENTER OF ENDEMISM

This region covers an area of 2.8 million km² and can conveniently be divided (at least physically) into two blocks, on either side of the dry Dahomey interval. Nearly everywhere the altitude is below 1,000 m. On the rare higher peaks, vegetation gradually acquires an Afromontane element.

Compared with rain forests of other continents, the Guineo-Congolian region is relatively dry, the majority receiving between 1,600 and 2,000 mm of rain per year. Two coastal areas, the first between the Guinea Republic and Liberia, and the second around Mount Cameroon, receive more than 3,000 mm of rain annually. Exceptionally, at the foot of Mount Cameroon annual precipitation exceeds 10,000 mm. Almost nowhere does mean monthly rainfall remain above 100 mm throughout the year. In equatorial parts of the DR Congo Basin, one or two months have only 50–100 mm of rain, and further from the equator and toward the Atlantic coast the length and severity of the dry period increase. At its dry extreme, a type of Guineo-Congolian rain forest survives where annual rainfall is as low as 1,230 mm, with a dry season of three months. Most places experience two peaks in rainfall, separated by one severe and one less-severe dry interval.

About 8,000 species of plants occur in the region, of which more than 80% are endemic.

The endemic families are Dioncophyllaceae, Hoplestigmataceae, Huaceae, Lepidobotryaceae, Medusandraceae, Octoknemaceae, Pandaceae, Pentadiplandraceae, and Scytopetalaceae. Characteristic large trees are commonly in the Leguminosaea (Caesalpinioideae and Mimosaceae), Chrysobalanaceae, Guttiferae, Irvingiaceae, Meliaceae, Moraceae, Myristicaceae, Sapotaceae, Sterculiaceae, and Ulmaceae. Smaller woody plants are abundantly represented by Anacardiaceae, Annonaceae, Apocynaceae, Celastraceae, Dichapetalaceae, Ebenaceae, Euphorbiaceae, Flacourtiaceae, Guttiferae, Icaciniaceae, Ochnaceae, Olacaceae, Rubiaceae, Sapindaceae, and Tiliaceae. The Caesalpinioideae family is particularly rich in endemic and near-endemic genera. The canopy of the Guineo-Congolian rain forest is usually at least 30 m high (and is often much taller), with emergent trees sometimes exceeding 60 m (e.g., Letouzey [1968] measured a *Baillonella toxisperma* [Sapotaceae] at 68 m) and attaining large diameters (e.g., van Rompaey [1993] measured an *Entandophragma candollei* [Meliaceae—a mahogany] of 6 m diameter). Most plants are woody. In Ghanaian forests, for example (Hall and Swaine 1981), 37% of plants are non-climbing phanerophytes (mostly trees), 31% are climbers (mostly woody species of liana, which can exceed 50 cm diameter [Reitsma 1988] and be more than 100 m long), 10% are epiphytic herbs, mainly orchids and ferns, and 22% are terrestrial herbs.

It is generally accepted that the forests of Africa are less diverse than those of South America and Asia (e.g., Richards 1973; Brenan 1978; White 1983a; Pannell and White 1988), although Hladik (1986), Reitsma (1988), and Gentry (1993) have argued that botanical plots in Africa may be similar in species richness to those in South America. This is true to some extent, but Phillips et al. (1994) cite a number of plots in South America and Asia with double the number of tree species of the most diverse of Reitsma's plots in Gabon. If one analyzes the

diversity of certain families found in humid forest undergrowth, such as the Lauraceae, Palmae (Arecaceae), Araceae, Piperaceae, Gesneriaceae, Melastomataceae, and Urticaceae, there is a distinct lack in variety in Africa compared to South America or Asia (Aubréville 1955, 1975; Richards 1973; Brenan 1978; Blanc 1995). Few of these families are well represented in the pollen record, but palm pollen does survive well (Maley 1996). Maley documents three periods when diversity and abundance of pollen from palms in Africa were significantly reduced: the Cretaceous-Tertiary boundary, the end of the Eocene, and the end of the Miocene. At these times there was a corresponding rise in the abundance of grass pollen, suggesting that the climate dried out and forest gave way to more open habitats (see figure 1.1). While South America and Southeast Asia would also have been affected by these global climatic changes, it seems that they were more severe in Africa. Today there are only 117 species (17 genera) of palm in Africa, Europe, and Arabia combined, compared to 132 species (29 genera) in Madagascar, Mascarene Islands, and Seychelles; 837 species (64 genera) in South America; and 1,385 species (97 genera) in the eastern tropics (Moore 1973), although 65 million years ago all three continents probably had similar diversity.

Richards (1973) suggested that climatic change was partly responsible for Africa being the "odd man out" but also indicated that humans had played an important role in decreasing species diversity in Africa. Gentry (1988) emphasized this point, demonstrating that some African forests are of comparable diversity to South American forests with similar rainfall, but that others seem impoverished. Gentry found particularly high diversity in a plot in Madagascar, which has been separated from Africa since the end of the Cretaceous. Thus the surrounding ocean would have reduced the impact of climate change, which so influenced mainland vegetation (and, furthermore, humans arrived in Madagascar only at about 1500 B.P. [A.D. 500]). Indeed, about 8,500 plant species are known to occur on Madagascar (Brenan 1978; White 1983a), and there may well be over 10,000, of which about 80% are endemic. Pannell and White (1988) suggest that physiographic diversity, which produces rapid climatic gradients, may be one of the keys to the diversity of floras, driving speciation (see also Gentry 1988, 1989) and facilitating altitudinal shifts during periods of climatic change. Compared to other landmasses, Africa has been relatively stable geologically since the breakup of Gondwanaland, while regions with higher plant species diversity have been relatively unstable. For example, New Guinea, which is particularly species rich, possesses great physiographic diversity, with rapid climatic gradients (as is the case in Madagascar), and because of its geological instability, new habitats have been forming constantly throughout its geological history, creating new niches and driving speciation.

White (1979) divided the Guineo-Congolian region into three subregions, in line with previous classifications of, for example, Lebrun (1961), Aubréville (1962), and Léonard (1965). Following White's terminology, these were the Upper Guinea, Lower Guinea, and Congolia subregions, the first two separated by the Dahomey interval and the second two by the Sangha River interval (see figure 1.2). The nature of these two intervals remains unclear, and there is some debate about the importance of the Dahomey interval.

The Dahomey interval consists of wooded grasslands, with isolated forest patches of comparable species composition to some of the drier forests in Ghana (Ern 1988). It seems likely that vegetation in this area is severely degraded (Engler 1910; White 1979). Rainfall is between 733 mm and 1,300 mm per year, which makes forest vegetation particularly susceptible to alteration by human activity and fire

(White 1979), and today all surviving forest fragments are highly threatened by human deforestation (J. F. Oates, pers. comm.). It seems likely that this region would have been particularly susceptible to changes in climate and human demography in the past (see also T. B. Hart, this volume). The presence of remnant forest patches suggests that the opening of this grassland interval may be a fairly recent development. This is supported by Brenan's (1978) analysis of distribution patterns of plants, which puts the main transition between Guinean and Congolian domains at the Cross River in Nigeria, about 650 km to the east (see figure 1.2; see also Hall 1981).

The Sangha River interval is an area about 400 km across where mean annual rainfall varies between 1,600 and 1,800 mm. This area is currently forested, and rainfall data suggest that it should be less sensitive to climatic variation than the forests of the Dahomey interval. However, the interval is biogeographically well defined (e.g., Hamilton et al., this volume), and it is possible that during a drier climatic phase a grassland corridor between the Sudanian and Zambezian regions may have opened up in this area (White 1979).

Sosef (1994) has studied distribution patterns of *Begonia* species of the sections *Scutobegonia* and *Loasibegonia* in Africa. These two sections share the feature of having fruits, which remain attached to the parent plant. Fruits gradually disintegrate, thereby releasing their seeds close to the parent plants. Although secondary dispersal by water (streams or heavy rains), or in mud sticking to the legs or fur of passing animals, cannot be ruled out, this is likely to be minor. Furthermore, most species are self-incompatible, so chance dispersal events have to involve at least two individuals if a new population is to establish. Most of the 40 species in these sections are found in moist, lowland forests in shaded conditions and are strongly hygrophyllous, so they are unlikely to survive outside moist forests in times of environmental change. Sosef mapped the ranges of all species in these two sections and compared their distributions to areas of endemism previously mapped using other groups of plants and animals.

There is a remarkable correlation between the two maps. Overall distribution of these two sections of *Begonia* corresponds to the Guineo-Congolian region of White (1983a), but individual species can be used to divide the region beyond his three subdivisions. These subdivisions correspond closely to those proposed by Grubb and Maley in this book, as well as those proposed elsewhere (see the many references in Sosef 1994). Hence the *Scutobegonia* and *Loasibegonia* provide a good example of how the Guineo-Congolian region can be divided into separate biogeographical units, although, being essentially restricted to a single microhabitat (shady undergrowth and streams in closed-canopy rain forest), they are of limited use for defining vegetation types.

### THE AFROMONTANE REGION

The Afromontane archipelago-like regional center of endemism occurs in a series of isolated patches from the Loma Mountains and Tingi Hills in Sierra Leone (11° W) to the Ahl Mescat Mountains in Somalia (49° E) and from the Red Sea Hills in the Sudan Republic (17° N) to the Cape Peninsula in South Africa (34° S). In the tropics, Afromontane communities occur above 2,000 m altitude, but in areas where the climate is more oceanic, such as in the West Usambara Mountains of Tanzania, they can descend to 1,200 m. To the south, where latitude compensates for altitude, they descend progressively, occurring at only a few hundred meters above sea level in the Cape Region. In the mountains of western Cameroon and the Angolan highlands, many lowland species dilute the montane communities, complicating classification.

In Afromontane rain forest the mean annual rainfall generally exceeds 1,000 mm, varying from about 1,250 mm to more than 2,500 mm. This is supplemented to a lesser or greater extent by condensation from clouds. There is generally a dry season lasting between one and five months, but mists during the dry season are frequent, and most plant species are evergreen. Frosts may occur occasionally but are not severe. Climate is dependent on the size and configuration of the mountains and on the distance from the sea or other moisture sources. Vegetation varies locally as a function of aspect, exposure, incidence of frost, and soil depth.

Estimates suggest at least 4,000 species of plants, of which about 3,000 are endemic or almost so. There are two endemic families (Barbeyaceae and Olinaceae), and about one-fifth of tree genera are restricted to the region (e.g., *Hagenia*). The most characteristic tree species are *Aningeria adolfi-friedericii*, *Chrysophyllum gorungosanum*, *Cola greenwayi*, *Cylicomorpha parviflora*, *Diospyros abyssinica*, *Drypetes gerrardii*, *Entandophragma excelsum*, *Ficalhoa laurifolia*, *Hallea (Mitragyna) rubrostipulata*, *Myrianthus holstii*, *Ochna holstii*, *Ocotea usambarensis*, *Olea capensis*, *Parinari excelsa* (which extends to lowland forests), *Podocarpus latifolius*, *Prunus africana*, *Strombosia schleffleri*, *Syzygium guineense* subsp. *afromontanum*, *Tabernaemontana johnstonii*, and *Xymalos monospora*.

The physiognomy is similar to parts of the Guineo-Congolian lowland rain forest: the tallest trees are 25–45 m high. The middle stratum of narrow crowned trees is 14–30 m tall, and the lower stratum is 6–15 m. Below, there is a poorly differentiated shrub layer and sparse herb layer, consisting mostly of forest grasses. Lianas and strangling epiphytes are abundant, ferns are numerous, and such herbs as *Begonia*, *Impatiens*, *Peperomia*, and *Streptocarpus* are widespread. The presence of tree ferns (*Cyathea*) and conifers (*Podocarpus*) distinguishes Afromontane rain forests from those of the Guineo-

Congolian region. Subdivision of Afromontane rain forest into distinctive forest types and communities is possible to some extent (White 1983a; Letouzey 1985), although, as for Guineo-Congolian rain forest, no synthetic classification is available.

During dry climatic phases linked to glacial maxima, temperatures fell and the montane plant community descended in altitude. It seems that the climate during these periods would have been one similar to that currently experienced in coastal central Africa during the long dry season, when constant cover of stratiform clouds blocks out the sun but provides little rainfall, resulting in cool, dry but relatively humid conditions (see Maley 1996:48). During these periods the montane flora became more widespread than it is currently, although remnants of previous extensions of the montane flora persist at low altitude in certain places (e.g., Maley et al. 1990). White (e.g., 1981, 1990) demonstrated the existence of a "Southern Migratory Track," a series of mountains connecting east and west Africa by means of the Zambezi-Zaire watershed and then northward through upland (c. 1,000 m altitude) parts of Congo, Gabon, and southern Cameroon. Here, Afromontane nomads occur as numerous isolated satellite populations of individuals or small associations mixed with non-Afromontane species. These satellite populations can be up to 1,500 km from the nearest other known locality, although many areas remain underexplored and many new populations may well be discovered.

Some of these disjunctions may be explained by long-distance seed dispersal. Evidence for this hypothesis comes from the presence of populations of mainland African plant species on such outlying islands as Madagascar, which separated from the continent before modern species evolved (see figure 1.1), or Sao Tomé, which was never connected (Pannel and White 1988). Other disjunctions probably occurred because in the past, montane communities expanded in cool dry periods and then retracted as the climate became warmer and wetter. As a result, isolated patches emerged on high mountains and in climatic anomalies where local humidity was sufficient to maintain parts of the montane flora. The relative importance of these two processes remains unresolved (White 1993b).

## Vegetation Classification Within the Guineo-Congolian Region

In 1887 Mary Kinglsey wrote: "Nor do I recommend African forest life to anyone. Unless you are interested in it and fall under its charm...it is like being shut up in a library whose books you cannot read, all the while tormented, terrified and bored. But if you do fall under its spell it takes all the color out of other kinds of living" (adapted from Kingsley 1897). In 1971 Frank White wrote: "Most taxonomy (and also much chorology and ecology) would be both prohibitively time-consuming and deadly dull if it were not possible to detect patterns, at least tentatively, by using perception. In this sense, perception is based on the capacity of the human eye and mind to detect pattern using large amounts of visual and factual data before they have been consciously analyzed... but without subsequent rigorous analysis, perception is of limited value" (adapted from White 1971). And in 1983, White (1983a) wrote: "The classification of Guineo-Congolian forest is difficult."

Chorology can be used to define areas that share unique species assemblages, but there is great variability in physiognomy within homogeneous chorological units, as a function of climate, soils, topography, drainage, and disturbance. Such variability will affect plant distribution at a finer scale than that considered above, but is of great interest to plant and animal ecologists alike, as it will determine patterns of abundance of plants and have an impact on animal densities and seasonal movements (e.g., Oates et al. 1990; White 1994b). In the 1983 AETFAT memoir, White restricted his discussion to a limited number of easily identified physiognomic units. He recognized four main variants of Guineo-Congolian rain forest:

1. Hygrophilous coastal evergreen rain forest, which occurs in areas with annual rainfall typically between 2,000 and 3,000 mm (or more) and where most tree species are evergreen, shedding their leaves intermittently.

2. Mixed moist semi-evergreen rain forest, where annual rainfall is typically between 1,600 and 2,000 mm. This is either spread out fairly evenly through the year (as in DR Congo) or the annual dry season corresponds to a period when moist air from the sea maintains high humidity levels (as in west Africa). Some tree species are evergreen, but many are briefly deciduous.

3. Single dominant moist evergreen and semi-evergreen rain forest, which tend to be limited in extent, where one (sometimes two) caesalpiniaceous trees dominate (see, e.g., T. B. Hart, this volume).

4. Drier peripheral semi-evergreen rain forest, where rainfall is between 1,200 and 1,600 mm but dry-season humidity remains high. Most individuals of the commoner large trees are deciduous during a well-defined dry season, but any individual is deciduous for a period of only a few weeks, and there is little synchrony in leaf-fall within the tree community as a whole. Many individuals acquire new foliage on some branches before others have shed their old leaves, so they are never completely bare. This forest type has often been referred to elsewhere as "semi-deciduous" or even "deciduous," although it is, in fact, essentially evergreen.

On the AETFAT vegetation map (White 1983b) it was possible to map only five vegetation formations; the wetter and the drier Guineo-Congolian rain forests; transitional forests (toward the eastern limit of the region, on the lower slopes of the western rim of the Great Rift Valley); swamp forest; and a mosaic of rain forest and secondary grasslands (see below and White, this volume, chapter 11). Other vegetation classes were recognized, such as monodominant forests, secondary forests of different ages, and various edaphic formations, but these were not included on the map.

Letouzey's classic study of vegetation in Cameroon (Letouzey 1968, 1985) and his accompanying vegetation map, including 267 categories, with 182 subdivisions of the forest zone and 35 of the Afromontane and submontane zone, document the variability of the vegetation within the Guineo-Congolian region. Letouzey's classification was based on the perception of an experienced botanist, who had taken the time to read almost all those books in Mary Kingsley's library, and was backed up with carefully recorded observations and quantitative botanical inventories. Experienced botanists can recognize in Gabon and Congo most of the vegetation types he describes, but in no other country in the African forest zone, with the possible exception of Ghana (see Hall and Swaine 1981; Hawthorne 1996), has the vegetation been studied and described to the same extent.

In Ghana, Hall and Swaine (1981; see also Ntiamoa-Baidu, this volume) used the ordination technique of reciprocal averaging (Hill 1973) to evaluate the relations between vegetation plots from different parts of the forest zone and to aid forest classification. They divided the Ghanaian forests into several categories (wet evergreen, upland evergreen, moist evergreen, moist semi-deciduous, and dry semi-deciduous, with a number of additional subdivisions) on the basis of tree species distribution and abundance. More recently, van Rompaey (1993) employed a similar method, detrended correspondence analysis (Hill 1979), to classify forests in Liberia and southwest Ivory Coast into a similar framework. Both these studies illustrate how one can use modern analytical techniques to make objective and ecologically meaningful classifications of African rain forests. However, to date these are the only examples of such countrywide studies (although van Rompaey, pers. comm., now plans to apply his methods to the whole of west Africa). So, until regional analyses become available, we have to make do with patchy detail and general simplification of a complex system. By comparing White's (1983b) map with those of Letouzey (1985), Hall and Swaine (1981), and van Rompaey (1993), we can at least think about priorities for future work in the African forest zone. Figure 1.4 presents a synthesis of current knowledge about distribution of vegetation types across the African tropical forest zone.

## The Impact of Humans
## on the African Rain Forest System

When the space shuttle passes over the forest zone of Ghana on a clear day, the passengers within look down on a mosaic of isolated forest patches (the protected forest reserves) surrounded by extensive secondary vegetation. The forest reserves have unnatural straight boundaries, where humans have cleared previous forests for agriculture up to their very limits (see satellite image in Hawthorne and Abu Juam 1995). For most of us this is not at all surprising, for we are "bulldozer man," not "stone age man," and the combined technologies of chainsaw and shotgun are well known as the conservationists' bane. West Africa is currently undergoing rapid deforestation, and predictions for the future are extremely gloomy (Barnes 1990). Humans today are creating

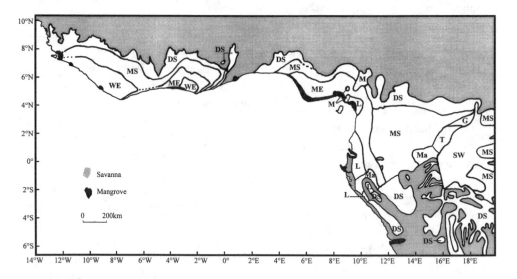

Figure 1.4. Synthetic vegetation map of the African rain forest (adapted from Maley 1990; Swaine and Hall 1986; van Rompaey 1993). WE = wet evergreen; ME = moist evergreen (these two vegetation types are characterized by an abundance of trees in the Caesalpiniacaeae); MS = moist semi-deciduous; DS = dry semi-deciduous; L = degraded forests and savanna colonization characterized by a predominance of *Lophira alata, Sacoglottis gabonensis,* and / or *Aucoumea klaineana*; Ma = Marantaceaec forests characterized by dense herbaceous understory dominated by species of Marantaceae and Zingiberaceae; G = forests dominated by Gilbertiodendron *dewevrei*; T = mosaic of Marantaceae and Gilbertiodendron forest types; M = montane forest; SW = swamp forest.

small forest islands surrounded by young secondary regrowth and artificial grasslands. Grass pollen is likely to be as abundant in the sediments of the Niger River as it has ever been. Other chapters in this book will address these modern problems, but here we consider the impact of humans on forest systems in the past and how it continues to affect vegetation today.

According to the evidence we have to date, the genus *Homo* probably appeared about 2.4 million years ago in east Africa. The australopithecines are thought to have evolved about 5 million years ago, perhaps forced from a forested habitat into more open areas, where they diverged from the other African great apes as the climate became dry and the Sahara formed (Wood 1992; see also figure 1.1). It seems likely that the climatic changes, which shaped the vegetation of Africa, also played a major role in our own evolution, yet the question arises, at what point would humans have begun to have an influence upon the vegetation?

The first evidence of tool making by hominids dates back at least 2 million years, and perhaps as much as 2.6 million years at Hadar in modern Ethiopia (Potts 1992), although in presently forested central Africa the earliest record is dated at about 400,000 years ago (Oslisly and Peyrot 1992; Oslisly, this volume). Early tools were handheld devices designed to help hominids obtain and process food. Indeed, axes that could have been used to cut wood were not invented until the recent stone age, that is, in the past 12,000 years or so, and became common only in Neolithic remains, as of 4,000–5,000 B.P. (see, e.g., Oslisly, this volume).

The first evidence of the use of fire by humans is dated to about 500,000 years ago, in China, while hearths first appeared in France about 300,000 years ago (Gowlett 1992). Hominids' adoption of fire is likely to have had far more important implications for vegetation than their use of stone tools. If hominids living in Africa 500,000 years ago were, for example,

burning vegetation in order to improve their ability to hunt, they might well have contributed to and increased the rate of vegetation modification initiated by climatic change.

The influence of prehistoric man on African vegetation, particularly rain forests, is difficult to evaluate. One would suspect that early hominids, before they mastered the use of fire, were unlikely to have had any greater direct effect on vegetation than do their modern relatives, the gorillas and chimpanzees. Thereafter, it was probably not until the past 10,000 years or so that humans began to cut into the forest with axes, and this is unlikely to have had a great impact until the advent of iron and agriculture. We can gauge the possible extent of this impact from more recent archaeological discoveries. In central Gabon the "strange Booué savannas" (Aubréville 1967), enclosed within the Guineo-Congolian rain forest, have long been an enigma, said by some to be natural (see, e.g., Aubréville 1967, despite Aubréville's belief to the contrary in previous papers) and by others to be anthropic in origin (see, e.g., White 1983a).

In similar savannas in Congo, Schwartz et al. (1996) used $\delta^{13}C$ analyses to show that savannas opened up from about 3,000 B.P., suggesting that a combination of a dry Holocene climatic phase, poor soils, and human fires resulted in forest degradation and the opening of grasslands since maintained by fire. Recent studies using $\delta^{13}C$ analyses in soil organic matter, calibrated using carbon dates, date the Booué (or Lopé) savannas to about 8,000–9,000 B.P. (Oslisly et al. 1996, 1997, but see the critique in Schwartz 1997). Their emergence may have been related to changes in regional climatic patterns that occurred at this time (Maley 1996), resulting in the establishment of a rain shadow effect in the Lopé area. However, there are numerous Stone Age village sites in the area, which date to this period (see, e.g., Oslisly, this volume) and it may be that rising human populations corresponded to the end of the last glacial maximum and that their fires contributed to vegetation changes related to climate. It seems likely that savanna fires lit by Stone Age, Neolithic, and Iron Age peoples in the region maintained the savannas (Oslisly and White 1996; Oslisly, this volume; White, this volume, chapter 11). Iron Age remains are particularly abundant in this region, and it is rare that soil profiles on hilltops in savanna or forest (preferred sites for villages) are found without both charcoal and remains of iron smelting (figure 1.5; see, e.g., Oslisly and Dechamps 1994; Tutin et al. 1996). Indeed, many of the forests in this region of central Gabon seem to date to about 1,400 B.P., when human populations crashed and, in the absence of (anthropic) fire, forest colonized grasslands.

Elsewhere in Africa there is evidence of high human populations crashing, allowing forests to regenerate. In Okomu, in southwestern Nigeria, Jones (1955, 1956) documented extensive remains of pottery and charcoal (mostly oil palm kernels) in a layer starting at 25 cm below the modern soil surface, in what were then considered some of the finest mahogany forests in the British Empire (Richards 1952). Jones suggested that the forest had grown back after a human population crash some 300–400 years previously, but charcoal from his study site was recently dated at 750–700 B.P. (White and Oates 1999). In northern Congo, Fay (1997) reported a similar phenomenon in forests also rich in mahoganies. He and his colleagues found large numbers of oil palm nuts in forests that were thought to be undisturbed and where *Elaeis guineensis* is currently restricted to human habitations. Carbon dates of 100 palm nuts were between 2,400 and 900 B.P., peaking between 1,800 and 1,700 B.P. and then declining sharply, suggesting a crash in the human populations cultivating oil palms. Given the tens of thousands of palm nuts and the extent of their distribution, Fay concludes that much of northern Congo, southeast Cameroon, and southwest Central African Republic was cultivated up to about 1,600 B.P. It seems that a population crash similar to that observed at Lopé

Figure 1.5. A *Syzygium staudtii* of 1 m diameter growing in a furnace used to smelt iron ore. The furnace, dated to 710 B.P., was one of about fifty found in an area covering about 10 ha in seemingly pristine forest in the center of the Lopé Reserve, Gabon (Oslisly and White 1996). Note the clay pipes used to introduce air from a bellows into the base of the furnace. Drawing by Kate Abernethy.

(and at about the same time) resulted in widespread forest regeneration, although further archaeological exploration is needed to confirm this hypothesis. Similar results have been obtained in southwest Cameroon, where *Elaeis guineensis* pollen in lake sediments rose from 2,730 B.P. to a peak around 1,600 B.P. and then declined (see, e.g., Reynaud-Farrera et al. 1996), and in de Maret's (1996) analysis of the clustering of carbon dates from archaeological sites in the central African region.

Evidence of forest fires, at least some of which are thought to have been related to humans, has been reported in Ituri in eastern DR Congo (Hart et al. 1996). Here almost 70% of soil pits contained charcoal at depths up to 80 cm and had radiocarbon ages between 130 and 4,190 B.P., mostly in the past 2,000 years. Identification of charcoal to species level and comparison with the modern flora suggest that significant shifts in forest composition had occurred as a result of these fires. Mbida's discovery of banana (*Musa*) phytoliths on pottery from an archaeological dig just outside Douala, dated to about 2200 B.P., shows that cultivation of bananas began in the region more than two millennia ago (Mbida 1996). Cultivation of bananas could have resulted in great increases in human population densities and would have involved significant modification of forest habitats (Vansina 1990).

Hence, a picture of widespread and significant human impact on quaternary African forests is emerging wherever palynologists, ecologists, and archaeologists team up (see also Maley 1992; Schwartz 1992). Human presence is likely to increase the frequency of fires, but lightning strikes do occasionally ignite forest trees (Tutin et al. 1996), and in some grassland areas natural savanna fires are frequent (see, e.g., Stander et al. 1993). It is possible, therefore, that during drier periods in the past, particularly since hominids began to use fire about 500,000 years ago, fire played a key role in determining the pattern and pace of vegetation change in Africa (see Swaine 1992).

No ecologist working in the African rain forest today can ignore the possibility that significant change may have affected a study area (or species) in the relatively recent past (particularly if this is measured relative to the potential life span of an emergent tree). The Sangha River interval, for example, may well prove to have been under extensive agriculture about 1,700 years ago, as was the case in Nouabale-Ndoki (Fay 1997). Indeed, Africa's perceived low plant species diversity may be related to the extent of human activity over the past few thousand years, or may even prove in part to be an artifact of sampling in areas that are actually relatively young secondary forests, with extensive archaeological remains just below the surface of the soil (an idea that has its origins in Jones's [1955, 1956] classic study in Okomu and Richards' [1952] text on the rain forest). Some of the most intriguing puzzles in plant ecology in African rain forests may prove to be related to human activities and population dynamics through the Iron Age. These include such enigmas as the origins of monodominant forests like the *Gilbertiodendron dewevrei* forests of Ituri and elsewhere in central Africa (Hart et al. 1996); the lack of regeneration of certain common species of emergent tree in Cameroon (Letouzey 1968; Newbery and Gartlan 1996); and the interdigitation of semi-evergreen, single-dominant moist evergreen and drier-peripheral semi-evergreen forests, the latter containing many faster-growing species that colonize secondary vegetation (Maley 1990; Wilks, pers. comm. [for northeast Gabon]). Multi-disciplinary research teams may be needed to tease apart the answers.

## On the Nature of Forest Refugia

The survival of forest refugia in equatorial Africa during periods of climatic stress has been well documented, but our understanding of the precise nature and distribution of these refugia remains vague and at times contentious (see, e.g., White 1993b:395, warning "those who see things in black and white to treat refuge theory with care"). When thinking about the influence of past vegetation change on modern distribution patterns, one has to take into account a number of factors: the tens of millions of years that have shaped the African rain forest system; the number of dry and humid climatic phases that are known to have affected Africa during the course of the Pleistocene; the severity of the most recent glacial maximum, which might be expected to have the strongest influence on contemporary distribution patterns; and the ability of each plant (or animal) species or group to expand out of a refuge into favorable habitats.

Over the past twenty years or so, numerous studies of plant and animal distribution have contributed to our concept of forest refugia (see, e.g., Sosef 1994; Maley 1996). The pattern that is emerging is somewhat more complex than early models, which proposed a limited number of discrete refugia, centered on Sierra Leone/Liberia, Ivory Coast/Ghana, Cameroon/Gabon, and eastern DR Congo (Hamilton 1976), although these areas continue to feature in more recent scenarios. As far as plants are concerned, the approach of Sosef (1994) and Rietkerk et al. (1995) of mapping current distribution of plants with limited dispersal capabilities seems very promising. Sosef's findings have provided support for Maley's (1987) proposed refugia, with minor modifications (see Maley 1996), although Rietkerk et al. question Maley's map. Reitkert et al. (1995) used herbarium records of collections of caesalpinioidous trees, whose seeds are dispersed by explosive dehiscence. Dispersal is therefore restricted to a number of tens of meters (maximum observed to date is 60 m for *Tetraberlinia moreliana;* van der Burgt 1997), and trees of this subfamily can be considered as indicators of refuges. Rietkerk et al. found that distributions of Caesalpin-

ioideae suggested that refugia were larger than those proposed by Maley. A recent study in Lopé (White et al. 1998) used some of the same caesalpinioidous tree species to map older forests in a forest-savanna mosaic. Caesalpinioideae with dehiscent pods were restricted to gallery forests close to permanent rivers and streams and were thought to represent a network of forest refugia that had survived a previous dry climatic phase. This study may prove to unite the ideas of Maley, Sosef, Rietkerk, and Hamilton into a more robust refuge theory that would retain the larger refugia but add a more diffuse peripheral region of gallery forests and small forest patches in favorable sites. Hygrophilous *Begonia* species would be unlikely to survive in narrow galleries and small forest patches, which would be susceptible to seasonal drying, while caesalpinaceous trees could survive, given their less fragile physiology and the drought resistance conveyed by ectomycorrhizae (see McKey 1994). Such ideas fit well with the fluvial refuge proposed by Colyn et al. (1991) in the Zaire basin and with comments made by Grubb (this volume) based on animal distributions. However, if future herbarium-based analyses are to test these ideas, plant collectors will have to take more detailed ecological data on collection locations than they currently tend to do.

## Forest Ecology and Natural History

In 1796 the great Scottish explorer and plant collector Mungo Park was stripped of most of his clothes and belongings by robbers. He wrote of his thoughts at the time:

I sat for some time, looking around me with amazement and terror. Whichever way I turned nothing appeared but danger and difficulty. I saw myself in the midst of a vast wilderness, in the depth of the rainy season; naked and alone; surrounded by savage animals, and men still more savage. I was 500 miles from the nearest European settlement. All these circumstances crowded at once on my recollection; and I confess that my spirits began to fail me. I considered my fate as certain, and that I had no alternative, but to lie down and perish.... At this moment... the extraordinary beauty of a small moss, in fructification, irresistibly caught my eye... for though the whole plant was not larger than the top of my fingers, I could not contemplate the delicate conformation of its roots, leaves and capsula without admiration. Can that Being (thought I), who planted, watered and brought to perfection, in this obscure part of the world, a thing which appears of so small importance, look with unconcern upon the situation and the sufferings of creatures formed after his own image?—surely not!

He collected the moss, "pulled himself together" (perhaps not in that order), and continued his exploration of Nigeria. The moss, a species of *Fissidens,* can still be seen today in the herbarium at Kew (Keay et al. 1964:9).

When one reads the works of Darwin (e.g., 1859, on geographical distribution, which is particularly relevant to this discussion) and other early biologists, their attention to detail and broad knowledge of and interest in natural history are immediately apparent. Modern scientific endeavor seems to have little time for natural history, yet Darwin ably demonstrated its value. His colleague J. D. Hooker (1862, 1864) described the occurrence of closely related plants on the mountains of Fernando Po, Cameroon, and Abyssinia (Ethiopia), and in temperate Europe. As White (1990) acknowledges, Darwin and Hooker had laid the foundation of his Southern Migratory Track long before botanical collection caught up with their farsighted ideas.

One of the main handicaps hampering modern ecological research in the African rain forest is the lack of knowledge of the natural history of the plants that occur within it and their interactions with animals (see, e.g., Hawthorne 1995a). Prance and White (1988:61) noted that "detailed studies of the taxonomy and reproductive biology of the bat-pollinated genus *Parkia,* have

shown that in the architecture and structure of the inflorescence there is a wealth of morphological detail closely related to the precise mechanism of pollen transfer, and that it is readily translatable into taxonomically useful characters. Indeed, it is only when function is understood in *Parkia* that a convincing taxonomy becomes possible." Going beyond this recognition of the need to understand function in taxonomy, F. White stressed the need for a better knowledge of the ecology of seed dispersal and habitat preference of plants if the dual influences of history and biology in modern plant distributions are to be teased apart (e.g., White 1983a; Pannell and White 1988).

We currently know little about the habitat preferences of most plant species found in the forest zone, although Hawthorne's (1995b) pioneering work, based on inventories of 330,000 trees in Ghana, will perhaps inspire further much-needed observation. Extending this theme, we know little about the rate of growth of forest plants, the age at which different species begin to reproduce, what stimulates their flowering, their pollination agents (most tropical rain forest plants are pollinated by insects; Bawa et al. 1985), their seed dispersal agents (75% of plants in African rain forests rely on animals for seed dispersal; see, e.g., Alexandre 1980; Hall and Swaine 1981; White 1994b) and the potential distance of dispersal, and their germination requirements. If this information is to be obtained, painstaking observations will have to be carried out on the 8,000 or so plant species of the Guineo-Congolian region—quite a task. Such information is fundamental for our understanding of forest dynamics, for interpretation of the significance of species assemblages, for our ability to model past and future trends (see, e.g., Puig 1995), and for forest management.

It is not possible to treat all these topics in an introduction. Some will be developed in other chapters of this book. I will, however, present one thought-provoking example. *Cola lizae* (Sterculiaceae) is a medium-large tree described in 1987 (Hallé 1987). It is the most common tree in a study area of 50 km² in Lopé, Gabon. *Cola lizae* constitutes just over 25% of all stems ≥10 cm diameter at breast height (dbh) in this study area, occurring at a density of almost 10,000 km⁻² (Hallé 1987; Williamson 1988). It is a remarkable species, with leaves sometimes exceeding 100 cm across, and has a distribution restricted to within about 75 km of Lopé (see White, this volume, chapter 11). The only seed disperser of note is the western lowland gorilla, *Gorilla g. gorilla,* and each individual is estimated to disperse between 11,160 and 18,600 seeds per fruiting season (Tutin et al. 1991), transporting seeds up to 5 km from the parent tree in their guts. Most of these seeds are deposited, undamaged, in dung within nest sites. Nests are generally constructed in light gaps (Tutin et al. 1995), and gorillas flatten dense herbaceous vegetation during nest construction, creating optimal conditions for growth of seedlings of *Cola lizae* (Tutin et al. 1991; Voysey 1995). *Cola lizae* is one of a group of species in Lopé whose flowering is stimulated by minimum temperatures in the cool dry season (Tutin and Fernandez 1993b; Tutin and White 1998). During ten years of data collection, there were two years when dry-season temperatures did not fall to the required level and flowering (and fruiting) failed, and two years when aseasonal temperature lows occurred, followed by unexpected (aseasonal) flowering (and fruiting). It seems that flowering is sensitive to relatively small variation in temperature. The "magic" temperature for *Cola lizae* and other species showing the same pattern in Lopé is 19°C (measured in the forest at 1.5 m above the ground). In the current warming climate it is not unlikely that global temperatures will increase by several degrees in the coming century (see, e.g., Wigley et al. 1996). Had Tutin et al. started their study a century later, would they have found a few rare, apparently senescing *Cola lizae* trees, which failed to fruit or flower and could not therefore be described as a new species? If gorillas had been eliminated by hunters, making use of the roads built by commercial loggers, would *Cola lizae* have ceased to regenerate, leaving the J-shaped distribution of diameter classes that is characteristic of some dominant trees in other African rain forests (see, e.g., Letouzey 1968; Newbery and Gartlan 1996)? Similar studies of the natural history of African rain forest plants and their relationships with animals should be considered a priority if conservation actions are to be founded on a solid, or at least acceptable, knowledge base. Such studies will require detailed, long-term observations of both plants and animals, as well as close collaboration between taxonomists and animal and plant ecologists (Pannell and White 1988; White 1993a).

## Conservation and Management of the African Rain Forest System

Between 1953 and 1973, exactly 7,478 new plant species from Africa and Madagascar were described—nearly one per day (Léonard 1975). In White's Upper Guinea subregion, forest cover has been reduced from 313,000 km² to about 80,000 km² since the end of the nineteenth century (Parren and de Graaf 1995)—a loss of 75%. In Ivory Coast, forest cover fell from 86,000 km² to 13,000 km² between 1966 and 1990—a loss of about 85% in 24 years (Sayer et al. 1992). Predictions for the more intact central African forests are for the most part gloomy (Barnes 1990). What hope is there for conservation of the African forest system? How many species will become extinct before they are described?

Ghana is perhaps an example of the way forward (Ntiamoa-Baidu, this volume)—a dashing knight to inspire other nations, although not without its problems, particularly with regard to wildlife conservation (J. F. Oates, pers. comm.). Ghana has lost about 80% of its forest cover over the past 150 years or so (Hawthorne and Abu Juam 1995), and its forests are today mostly restricted to government-controlled forest reserves. A collaborative project between the British Overseas Development Administration and the Ghana Forestry Department has systematically sampled vegetation composition and status in 127 forest reserves, inventorying a sample of c. 330,000 trees. Most tree species have been classified by ecological profile and conservation status, and detailed management recommendations have been made, incorporating both timber production needs and conservation priorities (Hawthorne 1993, 1994, 1995a, this volume; Hawthorne and Abu Juam 1995). Ghana can now manage its remaining forests from a position of knowledge, which far surpasses that of any other African country. In those central African countries that contain the majority of surviving intact forests (see Wilkie and Laporte, this volume), no such nationwide inventories have been undertaken, and management decisions (if they can be so called) are made blindly.

In spite of the good database available, in many respects forest managers in Ghana have to work as much in the dark as managers elsewhere: growth rates and life expectancy of tropical trees are poorly known (see, e.g., Hawthorne 1995b); pollination, dispersal, and germination requirements are poorly known; plant-animal interactions, which would affect management decisions, have generally not been documented; and long-term community dynamics and responses to logging, the major form of forest management, are poorly understood. These gaps in our knowledge challenge the managers and conservationists alike. Hawthorne (1995b) makes a perceptive analysis of IUCN categories of conservation priority. These criteria have been developed by a conservation community that is infatuated with large (cuddly) mammals, but we

lack the specific knowledge to be able to apply the criteria to tropical forest plants. In Hawthorne's words: "It is not difficult to gain a first approximation to a statement of broad facts for a species: *Tieghemella heckelii* lives in Upper Guinean rain forests; malaria is not common in temperate regions; flu is more common in winter. Attempts at greater precision, however, start to impinge on a biological uncertainty principle" (Hawthorne 1995b:10). More knowledge is needed if we are to extend our precision beyond best guesses and generalities. Until our understanding and knowledge improve, our ability to manage or conserve will be limited. In this respect, managers and conservationists share goals and needs and should work closely together.

In Gabon, *Aucoumea klaineana* (Burseraceae) is the dominant tree in two-thirds of the country (see, e.g., Saint-Aubin 1963). It had long been called the ideal timber species, being fast growing and light demanding (see Brunck et al. 1990). Yet recent ecological studies have demonstrated that there is almost no regeneration of *Aucoumea* within the forest and that much of its current distribution seems to be related to human demographic changes in the late Iron Age (it establishes in old farms following shifting cultivation) and savanna colonization (White et al. 1998). Logging gaps are not usually sufficiently large to permit establishment of *Aucoumea* seedlings, so regeneration is currently restricted mostly to the edges of major logging roads, the length of the railway line built in the 1980s across the country, and a few limited areas around villages and in savannas protected from fire. If the Gabonese forest remains intact and one could return in two hundred years and map *Aucoumea* distribution, it would be principally in long belts only about 50m wide (the roads—which if no longer detectable would pose quite a puzzle to ecologists). Unless the future logging industry in Gabon is based on intensive silvicultural management rather than selective felling in natural forests, *Aucoumea* is unlikely to feature in future harvests. The "ideal timber tree" is actually an ephemeral vestige of past vegetation change. If such a conceptual about-face is possible in the characterization of *Aucoumea*, one of the best-studied trees in Africa (see Brunck et al. 1990), what of our knowledge and ability to manage other species?

### REFERENCES

Alexandre, D. Y. 1980. Caractère saisonnier de la fructification dans une forêt hygrophile de Côte d'Ivoire. *Revue Ecologie* 34:335–350.

Aubréville, A. 1950. *Flore forestière soudano-guinéenne: A.O.F.–Cameroun–A.E.F.* Soc. Ed. Géographiques, Paris.

———. 1955. La disjonction africaine dans la flore forestière tropicale. *C. R. Soc. Biogéographie* 278:42–49.

———. 1962. Savanisation tropicale et glaciations quaternaires. *Adansonia* 2:16–84.

———. 1967. Les étranges mosaïques forêt-savane du sommet de la boucle de l'Ogooué au Gabon. *Adansonia Série 2* 7:13–22.

———. 1975. Essais sur l'origine et l'histoire des flores tropicales africaines: Application de la théorie des origines polytopiques des Angiospermes tropicales. *Adansonia* 15:31–56.

Barker, P. F., and J. P Kennett. 1988. Résultats préliminaires de la campagne 113 du Joides-Resolution (Ocean Drilling Program) en mer de Weddell: Histoire de la glaciation Antarctique. *C. R. Acad. Sc. Paris, série II* 306:73–78.

Barnes, R. F. W. 1990. Deforestation trends in tropical Africa. *African Journal of Ecology* 28:161–173.

Bawa, K. S., S. H. Bullock, D. R. Perry, R. E. Colville, and M. H. Grayum. 1985. Reproductive biology of tropical lowland rain forest trees II: Pollination systems. *Amer. J. Bot.* 73:346–356.

Blanc, P. 1989. *Biologie des plantes de sous-bois tropicaux.* Université de Paris VI.

———. 1995. Disjonctions et singularités dans les flores hygrophiles de sous-bois en Afrique. Pages 26–38 in *Actes Coll. Intern. Phytogéographie Tropicale.* Publ. ORSTOM, Paris.

Boltenhagen, E., J. Dejax, and M. Salard-Cheboldaeff. 1985. Evolution de la végétation tropicale africaine du Crétacé à l'Actuel d'après les données de la palynologie. *Bull. Sect. Sc., Paris* 8:165–194.

Bonnefille, R. 1983. Evidence for a cooler and drier climate in the Ethiopian uplands towards 2.5 Myr ago. *Nature* 303:487–491.

Brenan, J. P. M. 1978. Some aspects of the phytogeography of tropical Africa. *Ann. Missouri Bot. Gard.* 65:437–478.

Broecker, W. S., and G. H. Denton. 1989. The role of ocean-atmosphere reorganization in glacial cycles. *Geochimica et Cosmochimica Acta* 53:2465–2501.

Brunck, F., F. Grison, and H. F. Maitre. 1990. *L'Okoumé: Monographie.* Centre Technique Forestier Tropical, Nogent-sur-Marne, France.

Colyn, M., A. Gautier-Hion, and W. Verheyen. 1991. A re-appraisal of palaeoenvironmental history in Central Africa: Evidence for a major fluvial refuge in the Zaire Basin. *Journal of Biogeography* 18:403–407.

Darwin, C. 1859. *On the origin of species by means of natural selection.* John Murray, London.

de Maret, P. 1996. Pits, pots and the far-west streams. Pages 318–323 in J. E. G. Sutton, ed. *The growth of farming communities in Africa from the equator southwards.* British Institute in Eastern Africa, London.

Doyle, J. A. 1978. Fossil evidence on the evolutionary origin of tropical trees and forests. Pages 3–30 in P. B. Tomlinson and M. H. Zimmermann, eds. *Tropical trees as living systems.* Cambridge University Press, Cambridge.

Doyle, J. A., C. L. Hotton, and J. V. Ward. 1990a. Early Cretaceous tetrads, zonasulculate pollen, and Winteraceae. I. Taxonomy, morphology and ultrastructure. *Amer. J. Bot.* 77:1544–1557.

———. 1990b. Early Cretaceous tetrads, zonasulculate pollen, and Winteraceae. II. Cladistic analysis and implications. *Amer. J. Bot.* 77:1558–1568.

Doyle, J. A., S. Jardiné, and A. Doerenkamp. 1982. Afropollis, a new genus of early Angiosperm pollen, with notes on the Cretaceous palynostratigraphy and paleoenvironments of northern Gondwana. *Bull. Centres Rech. Explor. Prod. Elf-Aquitaine* 6:39–117.

Engler, A. 1910. Die Pflanzenwelt Afrikas, inbesondere seiner tropischen Gebiete: Bd 1, Allgemeiner Überblick über die Pflanzenwelt Afrikas und ihre Existenzbedingungen. Pages 1–1029 in A. Engler and O. Drude, eds. *Die Vegetation der Erde, 9.* Engelmann, Leipzig.

Ern, H. 1988. Flora and vegetation of the Dahomey Gap—a contribution to the plant geography of west tropical Africa. *Monogr. Syst. Bot. Missouri Bot. Gard.* 25:517–520.

Fay, J. M. 1997. The ecology, social organization, populations, habitat and history of the western lowland gorilla (*Gorilla gorilla gorilla* Savage and Wyman 1847). Page 416 in *Department of Anthropology*, Washington University, St. Louis.

Gentry, A. H. 1988. Changes in plant community diversity and floristic composition on environmental and geographic gradients. *Ann. Missouri Bot. Gard.* 75:1–34.

———. 1989. Speciation in tropical forests. Pages 113–134 in L. B. Holm-Nielsen, I. C. Nielsen, and H. Balslev, eds. *Tropical forests: Botanical dynamics, speciation and diversity*. Academic Press, London.

———. 1993. Diversity and floristic composition of lowland tropical forest in Africa and South America. In L. Goldblatt, ed. *Biological relationships between Africa and South America*. Yale University Press, New Haven.

Gowlett, J. A. J. 1992. Tools—the Paleolithic record. Pages 350–360 in S. Jones, R. Martin, and D. Pilbeam, eds. *The Cambridge encyclopedia of human evolution*. Cambridge University Press, Cambridge.

Guinet, P., N. El Sabrouty, H. A. Soliman, and A. M. Omran. 1987. Etude des caractères du pollen des Légumineuses—Mimosoideae des sédiments Tertiaires du nord-ouest de l'Egypte. *Mém. Trav. Ecole Pratiq. Htes. Et., Inst. Montpellier* 17:159–171.

Guinet, P., and M. Salard-Cheboldaeff. 1975. Grains de pollen du Tertiaire du Cameroun pouvant être rapportés aux Mimosacées. *Boissiera* 24:21–28.

Guiraud, R., and J. C. Maurin. 1991. Le rifting en Afrique au Crétacé inférieur: Synthèse structurale, mise en évidence de deux étapes dans la genèse des bassins, relations avec les ouvertures océaniques péri-africaines. *Bull. Soc. Géol. France* 162:811–823.

Hall, J. B., and M. D. Swaine. 1981. *Distribution and ecology of vascular plants in a tropical rain forest: Forest vegetation in Ghana*. W. Junk Publ., The Hague.

Hallé, N. 1987. *Cola lizae* N. Hallé (*Sterculiaceae*), nouvelle espèce du Moyen Ogooué (Gabon). *Adansonia, série 4* 9:229–237.

Hamilton, A. C. 1976. The significance of patterns of distribution shown by forest plants and animals in tropical Africa for the reconstruction of upper Pleistocene palaeoenvironments: A review. *Palaeoecology of Africa* 9:63–97.

Hart, T. B., J. A. Hart, R. Dechamps, M. Fournier, and M. Ataholo. 1996. Changes in forest composition over the last 4000 years in the Ituri basin, Zaire. Pages 545–563 in

L. J. G. van der Maesen, X. M. van der Burgt, and J. M. van Medanbach de Rooy, eds. *The Biodiversity of African Plants*. Kluwer Academic Publishers, Dordrecht.

Hawthorne, W. D. 1993. *Regeneration after logging in Bia South GPR, Ghana*. Natural Resources Institute, Chatham, England.

———. 1994. *Fire damage and forest regeneration in Ghana*. ODA, London.

———. 1995a. *Categories of conservation priority and Ghanaian tree species*. World Conservation Monitoring Centre, Cambridge.

———. 1995b. *Ecological profiles of Ghanaian forest trees*. Tropical Forestry Papers. Oxford Forestry Institute, Oxford.

———. 1996. Holes and the sums of parts in Ghanaian forest: Regeneration, scale and sustainable use. *Proceedings of the Royal Society of Edinburgh* 104B:75–176.

Hawthorne, W. D., and M. Abu Juam. 1995. *Forest protection in Ghana*. IUCN, Gland, Switzerland.

Herngreen, G. F. W., and A. F. Chlonova. 1981. Cretaceous microfloral provinces. *Pollen and Spores* 23:441–555.

Hill, M. O. 1973. Reciprocal averaging: An eigenvector method of ordination. *Journal of Ecology* 54:427–432.

———. 1979. DECORANA—A FORTRAN program for detrended correspondence analysis and reciprocal averaging. Cornell University, Ithaca.

Hladik, A. 1986. Données comparatives sur la richesse spécifique et les structures des peuplements des forêts tropicales d'Afrique et d'Amérique. In J. P. Gasc, ed. *Vertébrés et forêts tropicales humides d'Afrique et d'Amérique*. Muséum National d'Histoire Naturelle, Paris.

Hooker, J. D. 1862. On the vegetation of Clarence Peak, Fernando Po. *J. Linn. Soc. Bot.* 6:1–23.

———. 1864. On the plants of the temperate regions of Cameroons Mountains and islands in the Bight of Benin. *J. Linn. Soc. Bot.* 7:171–240.

Jones, E. W. 1955. Ecological studies of the rain forest of southern Nigeria. IV. The plateau forest of the Okomu Forest Reserve, part 1. The environment, the vegetation types of the forest, and the horizontal distribution of species. *Journal of Ecology* 43:564–594.

———. 1956. Ecological studies of the rain forest of southern Nigeria. IV. The plateau forest of the Okomu Forest Reserve, part 2. The reproduction and history of the forest. *Journal of Ecology* 44:83–117.

Keay, R. W. J. 1959. *Vegetation map of Africa south of the tropic of Cancer: Explanatory notes with French translation by A. Aubréville and a coloured map 1:10 000 000.* Oxford University Press, Oxford.

Keay, R. W. J., C. F. A Onochie, and D. P. Stanfield. 1964. *Nigerian trees, volume 1.* Federal Department of Forest Research, Ibadan, Nigeria.

Kingsley, M. 1897. *Travels in West Africa.* Everyman, London.

Knaap, W. A. 1971. A montane pollen species from the upper Tertiary of the Niger delta. *J. Mining and Geol.* 6:23–29.

Lebrun, J. 1947. La végétation de la plaine alluviale au sud du lac Edouard: Exploration de Parc National Albert. Mission J. Lebrun (1937–1938). Report, Institut des Parcs Nationaux du Congo Belge, Brussels.

———. 1960. Sur la richesse de la flore de divers territoires africains. *Bull. Séanc. Acad. r. Sci. Outre-Mer* 6:669–690.

———. 1961. Les deux flores d'Afrique tropicale. *Mém. Acad. r. Sci. Outre-Mer* 32:1–81.

Léonard, J. 1965. Contribution à la subdivision phytogéographique de la région guinéo-congolaise d'après la répartition géographique d'Euphorbiacées d'Afrique tropicale. *Webbia* 19:627–649.

———. 1975. Statistiques des progrès accomplis en 21 ans de la connaissance de la flore phanérogamique afrique et malagache (1953–1973). *Boissiera* 24:15–19.

Letouzey, R. 1968. *Etude phytogéographique du Cameroun.* Encylopédie Biologique. Editions Paul Lechevalier, Paris.

———. 1985. *Notice de la carte phytogéographique du Cameroun au 1/500.000.:* Inst. Carte Intern. Végétation, Toulouse, and Inst. Rech. Agron., Yaoundé, Cameroon.

Maley, J. 1980. Les changements climatiques de la fin du Tertiaire en Afrique: Leur conséquence sur l'apparition du Sahara et de sa végétation. Pages 63–86 in M. A. J. Williams and H. Faure, eds. *The Sahara and the Nile.* A. A. Balkema Publ., Rotterdam.

———. 1987. Fragmentation de la forêt dense humide africaine et extension des biotopes montagnards au Quaternaire récent: Nouvelles données polliniques et chronologiques. Implications paléoclimatiques et biogéographiques. *Palaeoecology of Africa* 18:307–334.

———. 1990. Synthèse sur le domaine forestier africain au Quaternaire récent. Pages 383–389 in R. Lanfranchi and D. Schwartz, eds. *Paysage quaternaires de l'Afrique Centrale Atlantique.* ORSTOM, Paris.

———. 1992. Mise en évidence d'une péjoration climatique entre ca. 2500 et 2000 ans BP en Afrique tropicale humide. *Bull. Soc. Géol. France* 163:363–365.

———. 1996. The African rain forest—main characteristics of changes in vegetation and climate from the Upper Cretaceous to the Quaternary. *Proceedings of the Royal Society of Edinburgh* 104B:31–74.

Maley, J., G. Caballe, and P. Sita. 1990. Etude d'un peuplement résiduel à basse altitude de *Podocarpus latifolia* sur le flanc Congolais de Massif du Chaillu: Implications paléoclimatiques et biogéographiques. Etude de la pluie pollenique actuelle. Pages 336–352 in R. Lanfranchi and D. Schwartz, eds. *Payasages quaternaires de l'Afrique Centrale Atlantique.* ORSTOM, Paris.

Mbida, M. C. 1996. L'émergence de communautés villageoises au Cameroun meridional: Etude archéologique des sites de Nkang et de Ndindan. Doctoral thesis, Université Libre de Bruxelles.

McKey, D. 1994. Legumes and nitrogen: The evolutionary ecology of a nitrogen-demanding lifestyle. Pages 211–228 in J. I. Srent and D. McKey, eds. *Advances in legume systematics.* Royal Botanic Gardens, Kew.

Mercer, J. H. 1983. Cenozoic glaciation in the Southern Hemisphere. *Ann. Rev. Earth Planet. Sc.* 11:99–132.

Monod, T. 1957. *Les grands divisions chorologiques de l'Afrique, with a chorological map 1:34 000 000.* CCTA/CSA Publication No. 24.

Moore, H. E., Jr. 1973. Palms in the tropical forest ecosystems of Africa and South America. Pages 63–88 in B. J. Meggers, E. S. Ayensu, and W. D. Duckworth, eds. *Tropical forest ecosystems in Africa and South America: A comparative review.* Smithsonian Institution Press, Washington, D.C.

Morley, R. J., and K. Richards. 1993. Gramineae cuticle: A key indicator of Late Cenozoic climatic change in the Niger delta. *Rev. Palaeobot. Palyno.* 77:119–127.

Newbery, D. M., and J. S. Gartlan. 1996. A structural analysis of rain forest at Korup and Douala-Edea, Cameroon. *Proceedings of the Royal Society of Edinburgh* 104:177–224.

Oates, J. F., G. H. Whitesides, A. G. Davies, P. G. Waterman, S. M. Green, G. L. Dasilva, and S. Mole. 1990. Determinants of variation in tropical forest primate biomass: New evidence from West Africa. *Ecology* 71:328–343.

Oslisly, R., and R. Deschamps. 1994. Découverte d'une zone d'incendie dans la forêt ombrophile du Gabon *ca* 1500 BP: Essai d'explication anthropique et implications paléoclimatiques. *Comptes Rendus de l'Académie des Sciences de Paris, série II* 318:555–560.

Oslisly, R., and B. Peyrot. 1992. Un gisement du paléolithique inférieur: La haute terrasse d'Elarmekora (Moyenne vallée de l'Ogooué) Gabon. Problèmes chronologiques et paléogéographique. *Comptes Rendus de l'Académie des Sciences de Paris, série II* 314:309–312.

Oslisly, R., B. Peyrot, S. Abdessadok, and L. J. T. White. 1996. Le site de Lopé 2: Un indicateur de transition écosystémique ca 10 000 BP dans la moyenne vallée de l'Ogooué (Gabon). *C. R. Acad. Sci. Paris, série IIa* 323:933–939.

———. 1997. Réponse au commentaire de Dominique Schwartz sur la note: Le site de Lopé 2: Un indicateur de transition écosystémique ca 10 000 BP dans la moyenne vallée de l'Ogooué (Gabon). *C. R. Acad. Sci. Paris, série IIa* 325:393–395.

Oslisly, R., and L. J. T. White. 1996. La rélation homme/milieu dans la Réserve de la Lopé (Gabon) au cours de l'Holocene: Les implications sur l'environnement. In *Dynamique à long terme des écosystèmes forestiers intertropicaux* (ed. S. ECOFIT). ORSTOM, Paris.

Pannell, C. M., and F. White. 1988. Patterns of speciation in Africa, Madagascar, and the tropical Far East: Regional faunas and cryptic evolution in vertebrate-dispersed plants. *Monogr. Syst. Bot. Missouri Bot. Gard.* 25:639–659.

Parren, M. P. E., and N. R. de Graaf. 1995. *The quest for natural forest management in Ghana, Cote d'Ivoire and Liberia.* Tropenbos, Leiden.

Phillips, O. L., and A. Gentry. 1994. Increasing turnover through time in tropical forests. *Science* 263:954–958.

Potts, R. 1992. The hominid way of life. Pages 325–329 in S. Jones, R. Martin, and D. Pilbeam, eds. *The Cambridge encyclopedia of human evolution.* Cambridge University Press, Cambridge.

Poumot, C. 1989. Palynological evidence for eustatic events in the tropical Neogene. *Bull. Centres Rech. Explor. Prod. Elf-Aquitaine* 13:437–453.

Prance, G. T., and F. White. 1988. The genera of Chrysobalanaceae: A study in practical and theoretical taxonomy and its relevance to evolutionary biology. *Phil. Trans. R. Soc. Lond. B* 320:1–184.

Puig, H. 1995. Eléments pour un réflexion sur la modélisation de la forêt tropicale humide: A-t-on les connaissances requises? *Revue d'Ecologie* 50:199–208.

Reitsma, J. M. 1988. *Végétation forestière du Gabon: Forest Vegetation of Gabon.* Tropenbos Technical Series. Ede, The Netherlands.

Reynaud-Farrera, I., J. Maley, and D. Wirrmann. 1996. Végétation et climat dans les forêts du sud-ouest Cameroun depuis 4770 ans BP: Analyse pollinique des sédiments du lac Ossa. *C. R. Acad. Sc. Série 2a,* 322:749–755.

Richards, P. W. 1952. *The tropical rain forest.* Cambridge University Press, Cambridge.

———. 1973. Africa, the "odd man out." Pages 21–26 in B. J. Meggers, E. S. Ayensu, and W. D. Duckworth, eds. *Tropical forest ecosystems of Africa and South America: A comparative review.* Smithsonian Institution Press, Washington, D.C.

Rietkerk, M., P. Ketner, and J. J. F. E. de Wilde. 1995. Caesalpinioideae and the study of forest refuges in Gabon: Preliminary results. *Adansonia* 1–2:95–105.

Ruddiman, W. F., M. Sarnthein, J. Backman, J. G. Baldauf, W. Curry, L. M. Dupont, T. Janececk, E. M. Pokras, M. E. Raymo, B. Stabell, R. Stein, and R. Tiedemann. 1989. Late Miocene to Pleistocene evolution of climate in Africa and the low-latitude Atlantic: Overview of Leg 108 results. *Proceed. Ocean Drill. Prog., Sc. Res.* 108:463–484.

Saint-Aubin, G. D. 1963. *La Forêt du Gabon.* CTFT, Nogent-sur-Marne, France.

Salard-Cheboldaeff, M. 1981. Palynologie Maestrichtienne et Tertiaire du Cameroun: Résultats botaniques. *Rev. Palaeobot. Palyno.* 32.

Salard-Cheboldaeff, M., and E. Boltenhagen. 1992. La palynologie des évaporites d'Afrique équatoriale et ses rapports avec le paléoenvironnement. *J. Afric. Earth Sc.* 14:191–195.

Salard-Cheboldaeff, M., and J. Dejax. 1991. Evidence of Cretaceous to recent West African intertropical vegetation from continental sediment spore-pollen analysis. *J. Afric. Earth Sc.* 12:353–361.

Sayer, J. A., C. S. Harcourt, and N. M. Collins, eds. 1992. *The conservation atlas of tropical forests: Africa.* IUCN/WCMC, Gland, Switzerland, and Cambridge, England.

Schwartz, D. 1992. Assèchement climatique vers 3000 B.P. et expansion Bantu en Afrique Centrale Atlantique: Quelques réflexions. *Bull. Soc. Géol. France* 163:353–361.

———. 1997. Commentaire à la note de Richard Oslisly, Bernard Peyrot, Salah Abdessadok et Lee White: Le site de Lopé 2: Un indicateur de transition écosystémique ca 10 000 BP dans la moyenne vallée de l'Ogooué (Gabon). *C. R. Acad. Sci. Paris, série IIa* 325:389–391.

Schwartz, D., R. Dechamps, H. Elenga, A. Mariotti, and A. Vincens. 1996. Les savanes d'Afrique centrale: Des écosystèmes à l'origine complexe, spécifiques de l'Holocene supérieur. In ECOFIT, ed. *Dynamique à long terme des écosystèmes forestiers intertropicaux.* CNRS/ORSTOM, Paris.

Sosef, M. S. M. 1994. *Refuge Begonias: Taxonomy, phylogeny and historical biogeography of Begonia sect. Loasibegonia and sect. Scutobegonia in relation to glacial rain forest refuges in Africa.* Studies in Begoniaceae, Wageningen Agric. Univ. Papers, The Netherlands.

Stander, P. E., T. B. Nott, and M. T. Mentis. 1993. Proposed burning strategy for semi-arid African savanna. *African Journal of Ecology* 31:282–289.

Swaine, M. D. 1992. Characteristics of dry forest in West Africa and the influence of fire. *Journal of Vegetation Science* 3:365–374.

Swaine, M. D., and J. B. Hall. 1996. Forest structure and dynamics. Pages 47–93 in G. W. Lawson, ed. *Plant ecology in west Africa.* J. Wiley, Chichester, England.

Tutin, C. E. G., and M. Fernandez. 1993a. Composition of the diet of chimpanzees and comparisons with that of sympatric lowland gorillas in the Lopé Reserve, Gabon. *American Journal of Primatology* 30:195–211.

———. 1993b. Relationships between minimum temperature and fruit production in some tropical forest trees in Gabon. *Journal of Tropical Ecology* 9:241–248.

Tutin, C. E. G., R. J. Parnell, L. J. T. White, and M. Fernandez. 1995. Nest building by lowland gorillas in the Lopé Reserve, Gabon: Environmental influences and implications for censusing. *International Journal of Primatology* 16:53–76.

Tutin, C. E. G., and L. J. T. White. 1998. Primates, phenology and frugivory: Present, past and future patterns in the Lopé Reserve, Gabon. Pages 309–338 in D. M. Newbery, H. H. T. Prins, and N. Brown, eds. *Dynamics of populations and communities in the tropics,* vol. 37. Blackwell Science, Oxford.

Tutin, C. E. G., L. J. T. White, and A. Mackanga-Missandzou. 1996. Lightning strike burns large forest tree in the Lopé Reserve, Gabon. *Global Ecology and Biogeography Letters* 5:36–41.

Tutin, C. E. G., E. A. Williamson, M. E. Rogers, and M. Fernandez. 1991. A case study of a plant-animal relationship: *Cola lizae* and lowland gorillas in the Lopé Reserve, Gabon. *Journal of Tropical Ecology* 7:181–199.

Upchurch, G. R., and J. A. Wolfe. 1987. Mid-Cretaceous to Early Tertiary vegetation and climate: Evidence from fossil leaves and woods. Pages 75–105 in E. M. Friis, W.

G. Chaloner, and P. R. Crane, eds. *The origins of Angiosperms and their biological consequences.* Cambridge University Press, Cambridge.

Van der Burght, X. M. 1997. Explosive seed dispersal of the rainforest tree *Tetraberlinia moreliana* (Leguminosae-Caesalpinioideae) in Gabon. *Journal of Tropical Ecology* 13:145–151.

Van Rompaey, R. S. A. R. 1993. Forest gradients in West Africa: A spatial gradient analysis. Wageningen Agric. Univ., The Netherlands.

Vansina, J. 1990. *Paths in the rainforests: Towards a history of political tradition in Equatorial Africa.* James Curry, London.

Voysey, B. C. 1995. Seed dispersal by gorillas in the Lopé Reserve, Gabon. Ph.D. diss., University of Edinburgh, Scotland.

White, F. 1979. The Guineo-Congolian Region and its relationships to other phytocoria. *Bull. Jard. Bot. Nat. Belg.* 49:11–55.

———. 1981. The history of the Afromontane archipelago and the scientific need for its conservation. *African Journal of Ecology* 19:33–54.

———. 1983a. *The vegetation of Africa, a descriptive memoir to accompany the UNESCO/AETFAT/UNSO vegetation map of Africa.* Natural Resources Research. UNESCO, Paris.

———. 1983b. *UNESCO/AETFAT/UNSO vegetation map of Africa: Scale 1:5,000,000 (in color).* UNESCO, Paris.

———. 1983c. Long distance dispersal and the origins of the Afromontane flora. *Sonderbd. Naturwiss. Ver. Hamburg* 7:87–116.

———. 1990. *Ptaeroxylon obliquum (Ptaeroxylaceae),* some other disjuncts and the Quaternary history of African vegetation. *Bull. Mus. natn. Hist. nat., Paris, 4e ser. section B, Adansonia no. 2* 12:139–185.

———. 1993a. The AETFAT chorological classification of Africa: History, methods and applications. *Bull. Jard. Bot. Nat. Belg.* 62:225–281.

———. 1993b. Refuge theory, ice-age aridity and the history of tropical biotas: An essay in plant geography. *Fragm. Flor. Geobot. Suppl.* 2:385–409.

White, L. J. T. 1994a. Biomass of rain forest mammals in the Lopé Reserve, Gabon. *Journal of Animal Ecology* 63:499–512.

———. 1994b. Patterns of fruit-fall phenology in the Lopé Reserve, Gabon. *Journal of Tropical Ecology* 10:289–312.

White, L. J. T., and J. F. Oates. 1999. New data on the history of the plateau forest of Okomu, southern Nigeria: An insight into how human disturbance has shaped the African rain forest. *Global Ecology and Biodiversity Letters* 8:355–361.

White, L. J. T., R. Oslisly, K. Abernethy, and J. Maley. 1998. *Aucoumea klaineana:* A Holocene success story now in decline? Editions ORSTOM. UNESCO, Paris.

Wigley, T. M. L., R. Richels, and J. A. Edmonds. 1996. Economic and environmental choices in the stabilization of atmospheric $CO_2$ concentrations. *Nature* 379:240–243.

Williamson, E. A. 1988. Behavioural ecology of western lowland gorillas in Gabon. Ph.D. diss., University of Stirling, Scotland.

Wood, B. A. 1992. Evolution of the Australopithecenes. Pages 231–240 in S. Jones, R. Martin, and D. Pilbeam, eds. *The Cambridge encyclopedia of human evolution.* Cambridge University Press, Cambridge.

CHAPTER 2

# Human Dimensions of the African Rain Forest

Lisa Naughton-Treves and William Weber

Conservation of the African rain forest—the intelligent use and protection of this vast and diverse resource—demands that we expand our vision to see beyond the trees and animals to the human cultural, economic, and political factors integral to the African forest ecosystem. These factors have helped to shape the forest as we know it today, and they are intricately tied to its future fate. It is this larger landscape, with its hundreds of ethnic groups and twenty-three nation-states (table 2.1); with its farms, roads, towns, and cities; with its urgent economic needs, inadequate institutions, uncertain politics, and a history of outside exploitation, that provides the key to understanding and conserving the African rain forest.

## Forest People

To know the forest we must know its people, but who are the "forest people"? The most common image is that of the Mbuti, a population of hunter-gatherers studied and popularized by Colin Turnbull forty years ago (Turnbull 1961). Living within the Ituri forest of the Congo Basin, the Mbuti are among Africa's most ancient residents (Templeton 1991) and have retained a high degree of dependence on the forest resource base, and the skills to exploit it efficiently. Related to the Mbuti are dozens of other ethnic groups (such as the Efe, Aka, and

Asua) inhabiting the forests of nine African countries across the Congo Basin (Bailey et al. 1992). Although outsiders commonly refer to these groups as pygmies, this generic term is rejected by the Mbuti, Efe, and other peoples, and is inadequate to cover the wide diversity found in their languages, cultures, and ecological relationships with the forest (Bailey et al. 1992). Similarly, the term *hunter-gatherer* has never fully captured the complexity of these relationships, and is limited in its applicability to groups that not only practice various forms of agriculture but are increasingly connected with distant commercial markets as both traders and consumers (Bailey et al. 1992; see also Peterson, this volume; Curran and Tshombe, this volume.) Finally, it is likely that hunter-gatherers shared the rain forest environment with other groups, particularly fisherfolk, who were likely the first to clash with Bantu immigrants (Vansina 1990). Our knowledge concerning these ancient forest inhabitants remains limited, pending further archaeological research (Vansina 1990).

Hunter-gatherers no doubt traded with farming populations along the forest fringe for millennia (Vansina 1990; Bailey et al. 1992). As Vansina (1990) recounts, this trade relationship took on added complexity when the first Iron Age farmers entered the rain forest during the great Bantu migration around 3,000 B.P. Among

Table 2.1

# Rain Forest Conservation and Socioeconomic Data
# for Twenty-Three African Countries

| COUNTRY | ORIGINAL CLOSED TROPICAL MOIST FOREST AREA (KM²)[a] | TROPICAL MOIST FOREST AREA, 1992 (KM²)[b] | TROPICAL MOIST FOREST REMAINING (% ORIGINAL)[c] | ANNUAL LOSS OF NATURAL FOREST, 1981–1990 (%)[d] | NUMBER OF NATIONALLY PROTECTED AREAS, 1994[e] | LAND UNDER PROTECTED AREAS, 1994 (% AREA)[e] |
|---|---|---|---|---|---|---|
| **West Africa** | | | | | | |
| Benin | 16,800 | 424 | 2.5 | 1.2 | 2 | 7.0 |
| Côte d'Ivoire | 229,400 | 27,464 | 12.0 | 1.0 | 12 | 6.3 |
| Gambia, The | 4,100 | 497 | 12.1 | 0.8 | 5 | 2.3 |
| Ghana | 145,000 | 15,842 | 10.9 | 1.3 | 9 | 4.9 |
| Guinea | 185,800 | 7,655 | 4.1 | 1.2 | 3 | 0.7 |
| Guinea-Bissau | 36,100 | 6,660[q] | 18.4 | 0.7 | n.a. | n.a. |
| Liberia | 96,000[p] | 41,238 | 43.0 | 0.5 | n.a. | n.a. |
| Nigeria | 421,000 | 38,620 | 9.2 | 0.7 | 19 | 3.3 |
| Senegal | 27,700 | 2,045 | 7.4 | 0.6 | 9 | 11.3 |
| Sierra Leone | 71,700 | 5,064 | 7.1 | 0.6 | 2 | 1.1 |
| Togo | 18,000 | 1,360 | 7.6 | 1.4 | 11 | 11.9 |
| Total/Average | 1,251,600 | 146,869 | 11.7 | | 72 | |
| **Central Africa** | | | | | | |
| Burundi | 10,600 | 413 | 3.9 | 0.6 | 3 | 3.5 |
| Cameroon | 376,900 | 155,330 | 41.2 | 0.6 | 14 | 4.4 |
| CAR | 324,500 | 52,236 | 16.1 | 0.4 | 13 | 9.8 |
| Congo | 342,000 | 213,400[q] | 62.4 | 0.2 | 10 | 3.4 |
| Equatorial Guinea | 26,000 | 17,004[r] | 65.4 | 0.4 | n.a. | n.a. |
| Gabon | 258,000[p] | 227,500[s] | 88.2 | 0.6 | 6 | 4.1 |
| Rwanda | 9,400 | 1,554 | 16.5 | 0.2 | 2 | 13.3 |
| DR Congo | 1,784,000 | 1,190,737[r] | 66.7 | 0.6 | n.a. | n.a. |
| Total/Average | 3,131,400 | 1,858,174 | 59.3 | | 48 | |
| **Eastern Africa** | | | | | | |
| Kenya | 81,200 | 6,900[q] | 8.5 | 0.6 | 36 | 6.2 |
| Tanzania | 176,200 | 14,400[q] | 8.2 | 1.2 | 31 | 15.8 |
| Uganda | 103,400 | 7,400[t] | 7.2 | 0.9 | 31 | 9.6 |
| Total/Average | 360,800 | 28,700 | 8.0 | | 98 | |
| **Madagascar** | 275,086 | 41,715 | 15.2 | 0.8 | 36 | 1.9 |

Table 2.1 *(continued)*

| COUNTRY | EXISTING TROPICAL MOIST FOREST IN PROTECTED AREAS, 1992 (% OF REMAINING MOIST FOREST)[b] | PROTECTED AREA BUDGET, 1993 (U.S.$ PER KM²)[f] | LOGGING OF CLOSED, BROAD-LEAF FORESTS, 1981–1990 (ANNUAL % TOTAL)[g] | GNP PER CAPITA, 1995 (U.S.$)[h] | POPULATION DENSITY, 1995 (PER 1,000 HA)[i] | POPULATION GROWTH RATE, 1990–1995 (AVERAGE ANNUAL %)[i] | % POPULATION IN URBAN SETTING, 1995[k] | URBAN POPULATION GROWTH RATE, 1990–1995 (AVERAGE ANNUAL %)[k] |
|---|---|---|---|---|---|---|---|---|
| **West Africa** | | | | | | | | |
| Benin | n.a. | n.a. | 0.3 | 370 | 489 | 3.1 | 31 | 4.6 |
| Côte d'Ivoire | 25.8 | n.a. | 7.6 | 660 | 448 | 3.5 | 44 | 5.0 |
| Gambia, The | 20.0 | n.a. | 1.1 | 320 | 1,118 | 3.8 | 26 | 6.2 |
| Ghana | 6.0 | 97.6 | 0.7 | 390 | 767 | 3.0 | 36 | 4.3 |
| Guinea | 1.8 | n.a. | 0.5 | 550 | 273 | 3.0 | 30 | 5.8 |
| Guinea-Bissau | n.a. | n.a. | 0.6 | 250 | 382 | 2.1 | 22 | 4.4 |
| Liberia | 3.2 | n.a. | 1.7 | n.a. | 314 | 3.3 | 45 | 4.6 |
| Nigeria | 5.6 | 57.8 | 2.3 | 260 | 1,227 | 3.0 | 39 | 5.2 |
| Senegal | 41.4 | 28.5 | 0.0 | 600 | 432 | 2.5 | 42 | 3.7 |
| Sierra Leone | 0.2 | n.a. | 0.2 | 180 | 630 | 2.4 | 36 | 4.8 |
| Togo | n.a. | n.a. | 0.3 | 310 | 761 | 3.2 | 31 | 4.8 |
| Total/Average | | | | | | | | |
| **Central Africa** | | | | | | | | |
| Burundi | 91.8 | n.a. | 0.0 | 160 | 2,489 | 3.0 | 8 | 6.6 |
| Cameroon | 7.3 | 20.2 | 4.5 | 650 | 284 | 2.8 | 45 | 4.9 |
| CAR | 8.3 | 8.2 | 0.0 | 340 | 53 | 2.5 | 39 | 3.4 |
| Congo | 5.7 | 3.0 | 0.4 | 680 | 76 | 3.0 | 59 | 4.9 |
| Equatorial Guinea | 18.5 | n.a. | 0.3 | 380 | 143 | 2.6 | 42 | 5.9 |
| Gabon | 7.9 | 40.2 | 0.7 | 3,490 | 51 | 2.8 | 50 | 4.7 |
| Rwanda | 9.7 | n.a. | 0.4 | 180 | 3,223 | 2.6 | 6 | 4.2 |
| DR Congo | 5.3 | 11.7 | 0.0 | 120 | 194 | 3.2 | 29 | 3.9 |
| Total/Average | | | | | | | | |
| **Eastern Africa** | | | | | | | | |
| Kenya | n.a. | 524.1 | 0.6 | 280 | 497 | 3.6 | 28 | 6.8 |
| Tanzania | n.a. | 26.6 | 0.2 | 120[v] | 336 | 3.0 | 24 | 6.1 |
| Uganda | 60.6 | n.a. | 0.1 | 240 | 1,067 | 3.4 | 13 | 5.8 |
| Total/Average | | | | | | | | |
| **Madagascar** | 13.9 | n.a. | 0.3 | 230 | 254 | 3.2 | 27 | 5.8 |

Table 2.1 *(continued)*

| COUNTRY | NUMBER OF REFUGEES, 1997[l] | NUMBER OF INTERNALLY DISPLACED PEOPLE, 1997[l] | NUMBER OF RETURNEES 1997[l] | ADULT FEMALE LITERACY, 1990 (%)[m] | ADULT MALE LITERACY, 1990 (%)[m] | FEMALE POPULATION ENROLLED IN PRIMARY SCHOOL, 1993 (% OF AGE GROUP)[n] | MALE POPULATION ENROLLED IN PRIMARY SCHOOL, 1993 (% OF AGE GROUP)[n] | % HIV POSITIVE, 1997[o] |
|---|---|---|---|---|---|---|---|---|
| **West Africa** | | | | | | | | |
| Benin | 5,960 | n.a. | n.a. | 19 | 42 | 44[w] | 88[w] | 0.9 |
| Cote d'Ivoire | n.a. | n.a. | n.a. | 24 | 44 | 58 | 80 | 4.9 |
| Gambia, The | 6,924 | n.a. | n.a. | 20 | 48 | 61 | 84 | 1.1 |
| Ghana | 35,617 | n.a. | n.a. | 46 | 71 | 70[w] | 83[w] | 1.1 |
| Guinea | 663,854 | n.a. | n.a. | 18 | 45 | 30 | 61 | 1 |
| Guinea-Bissau | 15,401 | n.a. | n.a. | 36 | 63 | n.a. | n.a. | 1.1 |
| Liberia | 120,061 | 320,000 | 11,084 | 18 | 49 | n.a. | n.a. | 1.8 |
| Nigeria | 8,486 | n.a. | n.a. | 39 | 61 | 82[w] | 105[w] | 1.9 |
| Senegal | 65,044 | n.a. | n.a. | 19 | 39 | 50[w] | 67[w] | 0.9 |
| Sierra Leone | 13,532 | 654,600 | n.a. | 14 | 40 | n.a. | n.a. | 1.5 |
| Togo | 12,589 | n.a. | 73,283 | 30 | 61 | 81 | 122 | 3.9 |
| Total/Average | 947,468 | 974,600 | 84,367 | | | | | |
| **Central Africa** | | | | | | | | |
| Burundi | n.a. | 882,900 | 71,031 | 19 | 45 | 63 | 76 | 4.1 |
| Cameroon | 46,407 | n.a. | n.a. | 44 | 70 | n.a. | n.a. | 2.3 |
| CAR | 36,564 | n.a. | n.a. | 41 | 60 | n.a. | n.a. | 5.3 |
| Congo | 20,451 | n.a. | n.a. | 59 | 78 | n.a. | n.a. | 3.6 |
| Equatorial Guinea | n.a. | n.a. | n.a. | 61 | 86 | n.a. | n.a. | 0.6 |
| Gabon | n.a. | n.a. | n.a. | 45 | 68 | 136[w] | 132[w] | 2 |
| Rwanda | 25,257 | n.a. | 1,300,582 | 44 | 65 | 76[w] | 78[w] | 6.3 |
| DR Congo | 675,973 | n.a. | n.a. | 61 | 83 | n.a. | n.a. | 2 |
| Total/Average | 804,652 | 882,900 | 1,371,613 | | | | | |
| **Eastern Africa** | | | | | | | | |
| Kenya | 223,640 | n.a. | n.a. | 62 | 82 | 91 | 92 | 5.6 |
| Tanzania | 498,732 | n.a. | n.a. | 49 | 75 | 69 | 71 | 4.4[x] |
| Uganda | 264,294 | n.a. | n.a. | 44 | 70 | 83 | 99 | 4.5 |
| Total/Average | 96,666 | | | | | | | |
| **Madagascar** | n.a. | n.a. | n.a. | 73 | 88 | 72 | 75 | 0.1 |

Table 2.1 *(continued)*

a. Data from, White 1983 in MacKinnon and MacKinnon 1986 in Sayer et al. 1992. Estimates of "original" cover are problematic because of ambiguous definitions of "forest" and "original." "Original" cover in this case includes woodland mosaics.

b. Unless otherwise noted, data are from country maps in chapters 11–32 of Sayer et al. 1992.

c. Data were calculated by dividing tropical moist forest area (column 4) by original closed tropical moist forest area (column 3). Average values for regions were calculated using aggregate regional values from columns 3 and 4.

d. Data from FAO 1993, in WRI 1996. Note that "annual change figures reflect net deforestation, which is defined as the clearing of forest lands for all forms of agricultural uses (shifting cultivation, permanent agriculture and ranching) and for other land uses such as settlements, other infrastructure and mining. In tropical countries, this entails clearing that reduces tree crown cover to alterations, such as selective logging (unless the forest cover in permanently reduced to less than 10%), that can substantially affect forests, forest soil, wildlife and its habitat, and the global carbon cycle" (WRI 1996:223).

e. Data from World Bank 1997. Data, which may refer to years earlier than 1994, are the most recent reported by the World Conservation Monitoring Centre. "Nationally protected areas" are defined as "areas of at least 1,000 ha that fall into one of 5 management categories: scientific reserves and strict nature reserves; national parks of national significance (not materially affected by human activity); natural monuments and natural landscapes with some unique aspects; managed nature reserves and wildlife sanctuaries; and protected landscapes nand seascapes (which may include cultural landscapes)....[This definition] does not include sites protected under local or provincial law or areas where consumptive uses of wildlife are allowed. The data are subject to variations in definition and in reporting to the organizations, such as the World Conservation Monitoring Centre, that compile and disseminate them" (World Bank 1997:257).

f. Data from R. B. Martin, "Should Wildlife Pay Its Way?" (Keith Roby address, Department of National Parks and Wildlife, Perth, Australia, December 8, 1993), cited in Edwards 1995.

g. Data from FAO 1993 in WRI 1996.

h. Data from World Bank 1997.

i. Data calculated from FAO land area data and population figures provided by United Nations Population Division, *Interpolated national populations (the 1994 revision)*, on diskette (U.N., New York, 1994), in WRI 1996.

j. Data from United Nations Population Division, *Annual populations (the 1994 revision)*, on diskette (U.N., New York, 1993), in WRI 1996.

k. Data from United Nations Population Division, *Urban and rural areas, 1950–2025 (the 1994 revision)*, on diskette (U.N., New York, 1995), in WRI 1996. Percentages of the population in an urban setting refer to the "midyear population of areas defined as urban in each of the countries of the world. These definitions vary slightly from country to country" (WRI 1996:155). Urban growth rates include the effects of urban-rural migration.

l. Data from UNHCR 1997. Numbers are available only for countries with more that 5,000 people of concern to UNHCR. Refugees are defined as "persons recognized as refugees under the 1951 United Nations Convention and/or its 1967 Protocol, the 1969 Organization of African Unity (OAU) Convention, in accordance with the UNHCTR Statute, persons granted a humanitarian status and/or those granted temporary protection. Refugees may be recognized either on a group or *prima facie* basis or following individual determination" *(www.unhcr.ch/refworld/refbib/refstat/1998/98intro.htm#1.Introduction)*. Internally displaced people are defined as "persons displaced inside their country for reasons similar to that of refugees, and who have become of concern to UNHCR as a result of a request from the Secretary-General or the competent principal organs of the United Nations. In addition, IDPs may become of concern to UNHCR because of their proximity to UNHGCR-assisted refugee and/or returnee populations" *(www.unhcr.ch/refworld/refbib/refstat/1998/98-intro.htm#VI.Glossary)*.

m. Data from United Nations Educational Scientific and Cultural Organization (UNESCO), *Statistics on adult illiteracy: Preliminary results of the 1994 estimations and projections* (UNESCO Division of Statistics, *Statistical Issues (STE-16)* (October 1994, Paris), and personal communication, in WRI 1996.

n. Data from World Bank 1997. Primary school enrollment data are "estimates of the ratio of children of all ages enrolled in primary school to the country's population of primary school-age children. Although many countries consider primary school age to be 6 to 11 years, others use different age groups. Gross enrollment ratios may exceed 100% because some pupils are younger or older than the country's standard primary school age" (World Bank 1997:255).

o. Data from World Health Organization (1998), Global HIV/AIDS and STD surveillance: Report on the global HIV/AIDS epidemic—June 1998, *www.who.int/emc-hivglobalreport/index.html.* Percentages were calculated by dividing "estimated number of people living with HIV/AIDS, end 1997 (Adults and children)" by "Population 1997 (Total, thousands)" and rounding to the nearest tenth of one hundred.

p. Figures for Gabon and Liberia, which were originally heavily forested, are country areas rather than land areas.

q. Data from FAO 1988.

r. Figures include 7,945 km² of degraded lowland rain

forest in Equatorial Guinea and 86,547 km² in DR Congo (Sayer et al. 1992: table 9.6).

s. Data from IUCN 1990.

t. Data from Howard 1991.

u. "This figure has been calculated by adding Green, G.M. and Sussman, R.W. (1990) figure for eastern rain forest to that calculated for mangroves from Map 26.1 plus an estimated 400 km² for forest remaining in the Sambirano region" (Sayer et al. 1992: table 9.6).

v. Figure covers only mainland Tanzania (World Bank 1997:215).

w. Figures are for years other than those specified (World Bank 1997:226).

x. Figure refers to United Republic of Tanzania.

the various factors favoring Bantu expansion into forests was the introduction of the banana (*Musa* spp.) from Southeast Asia more than 2,000 years ago. Bananas are an ideal crop for tropical evergreen forests. Compared to the autochthonous yams (*Dioscorea cayenensis*) or oil palms (*Elias guineensis*), bananas have a much higher yield and a longer planting season, and require less labor, less extensive forest clearing, and less food preparation time. The banana so favored Bantu penetration of central African forests that Vansina (1990:61) has declared that the "impact of metallurgy was on the whole perhaps smaller than that of the humble banana." Much later, several American crops were cultivated in African forests, including manioc, groundnuts, beans, and maize, after the sixteenth century. The Bantu penetration of central African forests gradually but inexorably drew the Mbuti and related peoples into a dependent relationship with agriculturalists. Although some groups resisted and others fled deeper into the forest, most intensified their trade of wild game (or bushmeat) and fish for crops with Bantu farmers, or themselves adopted agriculture as a way of life (Vansina 1990; see also Almquist, this volume; Peterson, this volume; Curran and Tshombe, this volume). So powerful was the cultural and

economic impact of Bantu agriculture in central African forests that today most hunter-gatherers speak a Bantu-derived language; only faint traces of their former speech remain (Vansina 1990:65).

Today, roughly 12 million descendants of Bantu as well as Sudanic immigrants dwell in African rain forests, where they are organized into more than 400 different ethnic groups (Vansina 1990). They remain primarily forest farmers, and most still practice some form of swidden agriculture, clearing small family plots in a systematic, rotational manner across a given area of forest over which they and their people have established traditional use rights. Despite its derisive description as "slash and burn," this form of agriculture has sustained human cultures across Africa and most of the world's rain forest regions for centuries. All too often these forest farmers are ignored or excluded from considerations as "forest people," despite their long-term influence on African rain forest ecology and composition (Delehanty 1993). Since their expansion into rain forest habitat, African shifting agriculturalists have created a mosaic of primary and secondary vegetation. This anthropogenic habitat is favored by many terrestrial browsers and grazers, among them favored game species, including bush pigs, cane rats, and forest antelope (Wilkie and Finn 1990; Lahm 1995). Even forest elephants and some monkeys are drawn to fallows and secondary forest (Thomas 1991; Barnes 1996; Naughton-Treves et al. 1998). Forest farmers often focus trapping activities in swiddens or fallows (Koch 1968; Vansina 1990; Bahuchet and Garine 1990). Thus in African forests, hunting and trapping were often interlinked so that "a yam field also produced bushpigs and...antelopes" (Vansina 1990:90). Forest farmers throughout much of Africa continue to depend on hunting, fishing, and foraging to add animal protein and variety to their diet, as well as to compensate for seasonal shortages and inevitable crop failures (Mubalama 1996).

During the twentieth century, the twin forces of population growth and market penetration have altered human impacts on African forests. The extended fallow periods at the heart of the system gradually shortened, and in many zones, fallows were eventually abandoned in favor of more permanent, intensive farming. Abbreviating fallows and clearing more forest have led to an increase in secondary vegetation at the expense of old-growth forest. In many areas of sub-Saharan Africa, forest clearing has accelerated as soil fertility declines and degraded farmland is abandoned (FAO 1993). Most of the future forests in Africa, as elsewhere in the humid tropics, will be secondary (Corlett 1995).

Beyond the population directly inhabiting African forests, more than 1,000 distinct ethnic groups occupy the zone surrounding the central African rain forest block. Many of these are forest peoples only in a historical sense, the forest sensu stricto having been completely cleared from their lands. In the highlands of Rwanda, Burundi, and neighboring areas of Congo and Uganda, large tracts of forest were converted to permanent farmland more than a century ago (Hamilton 1984; Chapman and Chapman 1996). In many other areas, the forest has disappeared within the lifetimes of vil-

lage elders (Weber 1981). To the north and west of the central African forest block, various groups have long exploited the rich resources of a shifting mosaic of natural savanna and forest (see, e.g., Fairhead and Leach 1996). Yet virtually all these groups retain some connection with the forest and its people through the consumption and trade of forest products, as well as through cultural beliefs and traditions that originated in their former environment (see Naughton-Treves 1999 for an example of savanna-based cultural beliefs among recent forest settlers).

Finally, more than 200 million people live in the twenty-three African countries that contain the continent's remaining rain forest (see table 2.1). Many live in subhumid and semiarid environments with no historic or current connection to rain forests. Urbanization in sub-Saharan Africa has outpaced general population growth, so that today 40% of the population lives in cities of various sizes (WRI 1996; see table 2.1). Many of these urban dwellers are among the world's poorest citizens, and their numbers and needs are increasing much faster than those of remote forest-dwelling populations. For example, the fuelwood demands of rapidly growing urban populations are placing enormous pressure on Africa's forests (Kammen 1995). Although forested Africa does not suffer from the crisis conditions of fuelwood scarcity and severe environmental degradation found in the Sahel (Leach and Mearns 1988), accelerating demands by both commercial and domestic consumers are nonetheless depleting forest resources. For example, more than 90% of the total wood cut in Uganda during the early 1980s was used for fuel (Hamilton 1984). The urban demand for charcoal and the expanding road networks threaten even remote forests (Ribot 1993; Osei 1997). Urban markets for bushmeat also penetrate distant forests, particularly in central and west Africa (Falconer 1995; Wilkie and Morelli

1998; Lahm, this volume). Indeed, regional trade in bushmeat has intensified recently with violent conflict and political unrest in the Democratic Republic of Congo, Liberia, and Sierra Leone (Richards 1996; Wilkie and Morelli 1998).

We can debate which groups truly qualify as "forest people" and which do not. Yet all are citizens of nations where central authorities as well as global markets increasingly control the fate of the forest and its people.

## Forest Politics

Viewed from space, the African forest appears as a vast expanse—a green block as wide as the continent, fragmented along its eastern and western edges. Largely invisible are the political lines imposed on the forest and its people: national borders that may follow certain natural features but which have little or nothing to do with traditional territories. The colonial imposition and subsequent contestation of these territorial claims have powerfully changed relationships between people and their forests (Richards 1996; Osei 1997).

In precolonial times, three interlocking units of political organization lay at the heart of the African rain forest: (1) the district, an informal cluster of villages or households sharing common security concerns; (2) the village, typically averaging 100 inhabitants; and (3) the home or hearth, usually comprising 10 to 40 people (Vansina 1990:73). The relationships among these units displayed great flexibility and dynamism across the Congo Basin (Vansina 1990). Many districts were linked to larger networks based on trade connections. These associations generally fell short of any formal confederation, although political change and expanding trade opportunities on the eve of colonial conquest marked the birth of several forest-based chiefdoms, principalities, and kingdoms (Vansina 1990). Highly organized and hierarchical kingdoms arose along the forest-savanna fringe, such as the Tutsi in Rwanda,

the Nyoro-Kitara in the Albertine rift, and the Ashanti in Ghana (Taylor 1962; Paterson 1991). Of course, even larger kingdoms had long existed in outlying savanna regions (Vansina 1966). But within the rain forest itself, decentralized political entities developed in apparent response to the needs and resource use patterns of the people.

In their rush to carve up Africa, the European powers were guided far more by greed and fancy than by any understanding of the land and people over which they laid claim. One result was to reduce the diverse pool of traditional political entities in the rain forest realm to fewer than twenty colonies. In the process, they divided the forest, its wildlife, and its people along artificial and disruptive lines (Richards 1996; Osei 1997). Imposed national boundaries often enclosed hundreds of ethnic groups, each with its own distinct language—more than four hundred in Nigeria alone—and few with any shared history prior to the colonial period (Osei 1997). Today the variability in sociopolitical systems and land-use practices within nation-states often exceeds that between nation-states. These internal differences undermine national-level political processes and often lead to violent civil conflict. Internal divisions also challenge efforts to build national policies for forest use and conservation.

Each of the colonial powers imposed foreign administrative structures and institutions—political, economic, religious—on the lands and people under its rule. Across much of the forest zone, as throughout the continent, traditional land tenure and resource use systems were superseded by the process of alienation, or the transfer of use rights to the colonial government (Anderson and Grove 1987). Centralized Forest Departments were given vast, unchecked powers to assure efficient exploitation of the forest resource base. Wildlife was typically claimed as Crown property, and local hunting was prohibited (Naughton-Treves

1999). The British and Belgian colonies established parallel bureaucracies to create and manage national parks, but most of these reserves were situated in savanna and mountain environments, not in the rain forest. The French, meanwhile, combined protected areas and wildlife management under Water and Forestry, which was part of the Ministry of Agriculture. These colonial management institutions continue largely intact today.

Independence did not rectify colonial boundaries or put an end to inherited foreign land management institutions. From the 1960s through the 1980s, most African forest and game departments followed top-down, blueprint management strategies (Little and Brokensha 1987). Commercial logging was often a management priority, and there was minimal interest in local people's usufruct rights or ethnobotanical knowledge (Fairhead and Leach 1996). These state-centered policies established during the colonial period served to alienate rural populations from forest management. Corruption and civil unrest further undermined state-centered policies, and many forests turned into open-access realms, plundered for short-term gains (Little and Brokensha 1987; Howard 1991).

The politics of forested Africa are likely to remain turbulent for some time. However, promising experiments in democracy are under way in some countries (such as Uganda), often reinforced by still-fresh memories of civil war and bloodshed (Museveni 1997). Concepts of minority and individual rights are also emerging in some countries (e.g., Burkina Faso) as essential corollaries of majority-rule democracy. Even some authoritarian states are permitting greater participation, along more traditional lines, at the village and local levels.

Shifting power relations within forested African countries will have a great influence on the fate of the forest. Building local support for forest conservation and redirecting rights and responsibilities to local populations are vital strategies. If rain forest conservation in Africa remains a Eurocentric or bureaucratic endeavor, it will not succeed (see Sindima 1995). Yet conservationists must relinquish simplistic prescriptions to return to indigenous, precolonial forest management systems. Such prescriptions ignore the linkages between local use and macro political and economic forces (Naughton-Treves 1999), and also ignore the profound demographic and economic changes of the past century.

## Forest Economics

Economic independence and security remain elusive goals for most African nations (Sindima 1995). The continent as a whole lags far behind other global regions in gross national product and in virtually every positive economic indicator, whereas it leads in per capita debt. Literacy levels are low, particularly for women, although many African nations are investing in primary education with promising results (see table 2.1). Export earnings from the forest region are dominated by timber sales, some mining, and a limited number of cash crops, such as coffee and cocoa, that are highly susceptible to fluctuations in international commodity markets. Foreign governments, international agencies, and multinational corporations all have enormous influence over domestic business interests. Recent liberalization and structural adjustment programs heighten this influence further. No single African nation has thus far surmounted these constraints to rival the emergent economies of Malaysia, Thailand, Indonesia, Chile, or Brazil.

Conditions are hardly better for subsistence production. Although food productivity indices have been rising, they have not kept pace with population growth. Of twenty-five countries cited by the FAO (1995) for their "low or critical food security indexes," twenty-two are from

Africa, including many from the forest zone. Per capita food aid in Africa far surpasses levels in South America or Asia (World Bank 1997).

Dating back to the days of trade in ivory, gold, and human beings, Africa has a long history of outside economic exploitation. This exploitation reached its organizational peak during the colonial period. Following independence, the Cold War changed the equation somewhat, but foreign assistance was all too often linked to preferential access to African raw materials. Today, with the end of East-West rivalries, competition is more narrowly focused on the economic sector, where Africa remains dependent on selling its primary resources within unfavorable world markets.

Many tropical rain forests are now targeted for mining, oil extraction, and large-scale logging (Bowles et al. 1998; see table 2.1). These activities have a serious impact on the forest and its wildlife, and perhaps an even larger impact on its people and their way of life. Mining and logging bring secondary and tertiary impacts of development through road construction, market demand for outside products, and the resultant need to further exploit the land or forest to pay for rising levels of consumption. With tragic frequency, war opens forests and wildlife to commercial exploitation tied to military authorities or even mercenary outfits. Social disruption too often follows in the form of alcoholism, prostitution, disease, and a general decline in traditional organization and authority. And when boom is followed by bust, local citizens are left to clean up the various messes left behind. As the cycle moves on to the next site, local populations again have little say and less control: powerful interests from Asia, Europe, and the Americas pay the bills, and central authorities in almost equally remote capitals—Yaoundé, Brazzaville, Libreville, and Accra—make the decisions.

### Forest Conservation

Global economic and political forces threaten the rain forest ecosystem. Yet other international influences also have an impact on the African rain forest. Conservation is one of those influences. From the time of their introduction in the colonial period, formalized conservation endeavors have represented a largely foreign constituency: individual citizens, private associations, and governments from abroad have all invested in efforts to save Africa's rain forests and wildlife as part of earth's biological heritage. Through their efforts and contributions, thousands of parks and reserves now exist to protect some of the world's greatest wildlife spectacles and ecosystems, especially in Africa's savanna and semi-arid regions. In the process, however, many foreign conservationists have ignored or acted directly against local citizens' subsistence needs and development concerns, thus creating animosity in communities neighboring protected areas (Adams and McShane 1992; Bonner 1993). Central authorities have often resorted to brutal tactics to protect parks and wildlife (Peluso 1993).

By the 1980s, it was clear that neither production-based forestry nor park preservation could wholly address the challenges of conserving African rain forests. Conservationists began to call for new approaches to forest management. Community-based forest and wildlife management has become a popular model, bolstered by political and ethical arguments regarding local resource access and self-determination (Western and Wright 1994). Campaigns to return Africa's wildlife (or forests) to Africans are peppered with references to the age-old wisdom, flexibility, and equity of indigenous management (Adams and McShane 1992; Balakrishnan and Ndhlovu 1992; Western 1997).

At the heart of community-based strategies is the question of property: Who is the legitimate owner of Africa's wildlife and forest (IIED 1994; Child 1995; Naughton-Treves and Sanderson 1995; Western 1997)? Many conservationists hope that providing local people with secure ownership of forests and forest-related products will discourage them from overhunting and destroying habitat. Yet while the state's failure to fully protect forest and wildlife is clearly described, rarely do conservationists identify the social group within present-day African society that would be best suited for stewardship of forests and animals (IIED 1994; Western and Wright 1994). In the popular conservation literature, the identity of the local community is too often described simply in terms of what it is not: not the state and not the market. Others refer vaguely to a socially cohesive and ethnically homogeneous group of forest stewards. Several contributors in this volume are involved in the difficult task of promoting community participation amid diverse and often divided local populations, including immigrants, indigenous people, refugees, merchants, and even military personnel (see table 2.1; see also Peterson, this volume; Curran and Tshombe, this volume; Ntiamoa-Baidu, this volume).

Fostering a sound process for local stakeholders to negotiate forest access rights and management responsibilities is an extremely important but highly political action. After all, the rain forest is much more than simply a "resource"; it may be a core element in a people's sociopolitical identity (Richards 1996). Where the state has lost its legitimacy, as is the case in some west and central African countries, securing local rights and responsibilities for forests becomes especially difficult.

Conservation strategies and actors have changed over the past two decades. There is growing local and national interest in conservation among citizens, scientists, and leaders of African countries. Most African rain forest parks have been created following independence

(Balmford et al. 1992). A growing cadre of African scholars and activists is assuming leadership of biodiversity conservation on the continent. Tragically, in conservation as in other professional realms in Africa, the high incidence of HIV debilitates the vital contribution of many young experts and leaders (see table 2.1).

Foreign and local conservationists alike increasingly recognize that biodiversity concerns cannot be addressed in isolation from local human populations (Weber 1994; Sibanda and Omwega 1996). This is expressed most clearly in the concept of integrated conservation and development, the dominant model in the field today (Western and Wright 1994). Across Africa's rain forests, critical experiments are under way in natural forest management, non-timber forest extraction, and tourism revenue sharing. Finally, today there is unprecedented international and national attention to rain forest ecosystems, and greater recognition of the value of their unequaled biodiversity. The combined effect of these changes is extremely positive and of critical importance for the future of the African forest.

The African rain forest is rapidly being transformed and cleared. Africa has already lost more than two-thirds of its original forest (see Wilkie and Laporte, this volume). Some of this loss is due to long-term climatic change, but much can be attributed to human agency. Commercial logging is directly responsible for the annual loss of more than 4% of the west African forest, and the industry has now moved aggressively into the central African region, backed by Asian and European investors (Sawyer 1992). Behind this front, many formerly logged areas have been converted to permanent settlements and cultivation. Others have suffered severe faunal depletion by hunters where logging roads have opened the forest to distant urban bushmeat markets (Lahm, this volume). In the east African highlands, population pressure that reaches 500 people/km² in some rural areas has driven intensive agricultural conversion, leaving scattered forest islands in a rising tide of banana, bean, and sorghum fields. Within the vast continental interior of the Congo Basin, an inadequate transportation network has limited intensive settlement and exploitation thus far, preserving both traditional cultures and the forest habitat on which they depend. Yet roads continue to spread like exposed veins and arteries across the heart of the forest, permitting outside interests and influences to penetrate to the core while providing outlets for newly valued forest products.

An aggressive conservation effort is required if these threats to the African rain forest are to be countered. Yet if understanding is an essential prelude to action, then we first face a monumental challenge in learning. Our understanding of the forest itself is in its infancy. Most species are unnamed and have never been studied. Only a few major groups of plants and animals have received systematic attention across the entire forest zone. Even among well-studied taxa, such as primates, our understanding of their ecological role and response to human disturbance remains incomplete (but see, e.g., Skorupa 1988; Struhsaker 1997). Complicating matters is the fact that there are relatively few rain forest biologists and ecologists working in Africa—and many fewer who are African citizens. Despite the greater need, international support for training and conservation research in African forests is woefully inadequate. According to a sample of fourteen African countries, protected area budgets average U.S.$71 per square kilometer, an order of magnitude less than that for the seven Asian countries sampled (Edwards 1995; see also table 2.1).

Solving conservation problems requires understanding that goes far beyond the forest and its wild species. The rain forest ecosystem includes millions of people who themselves are agents of change within the forests where they live, yet whose traditional cultures and economies will be the first casualties of continued forest destruction. It includes many millions more who live outside, whose needs and aspirations might be met only through greater exploitation of the forest resource base. Central authorities who make decisions about forest use are strongly influenced by these domestic development interests, as well as by opportunities—and pressures—from abroad. These, in turn, arise from global economic and political trends, which include growing pressures to both consume and conserve the remnant rain forests of Africa.

Understanding the many factors outlined above will take an unprecedented investment of time, people, and effort. Synthesizing and interpreting the results will take a degree of interdisciplinary genius and creativity that we have rarely imagined, let alone exhibited. Applying and adapting the results to effective conservation will require a long-term commitment and considerable flexibility. Recognizing the importance of the human dimensions of the rain forest is a critical first step.

ACKNOWLEDGMENTS

We thank R. Peterson, J. Delehanty, and A. Treves for their insightful comments and suggestions. E. Olson-Dedjoe compiled and organized the data for table 2.1.

REFERENCES

Adams, J. S., and T. O. McShane. 1992. *The myth of wild Africa*. W. W. Norton, New York.

Agrawal, A. 1997. *Community in conservation: Beyond enchantment and disenchantment*. Conservation and Development Forum, Gainesville, Fla.

Anderson, D., and R. Grove, eds. 1987. *Conservation in Africa*. Cambridge University Press, Cambridge.

Bahuchet, S., and I. Garine. 1990. The art of trapping in the rainforest. In C. M. Hladik, S. Bahuchet, and I. Garine, eds. *Food and nutrition in the African rainforest*. UNESCO, Paris.

Bailey, R., S. Bahuchet, B. Helwett, and D. Lewis. 1992. Forest people. In J. Sayer, C. Harcourt, and N. M. Collins, eds. *The conservation atlas of tropical forests—Africa*. Simon and Schuster, New York.

Balakrishnan, M., and D. E. Ndhlovu. 1992. Wildlife utilization and local people: A case-study in Upper Lupande Game Management Area, Zambia. *Environmental Conservation* 19(2):135–144.

Balmford, A., N. Leader-Williams, and M. Green. 1992. The protected areas system. In J. Harcourt, J. Sayer, and N. M. Collins, eds. *The conservation atlas of tropical forests—Africa*. Simon and Schuster, New York.

Barnes, R. F. W. 1996. The conflict between humans and elephants in the central African forests. *Mammal Review* 26(2):67–80.

Bonner, R. 1993. *At the hand of man—peril and hope for Africa's wildlife*. Knopf, New York.

Bowles, I., A. Rosefeld, C. Sugal, and R. Mittermeier. 1998. *Natural resource extraction in the Latin American tropics*. Conservation International, Washington, D.C.

Chapman, C. A., and L. J. Chapman. 1996. Mid-elevation forests: A history of disturbance and regeneration. Pages 385–400 in T. R. McClanahan and T. P. Young, eds. *East African ecosystems and their conservation*. Oxford University Press, New York.

Child, G. 1995. Managing wildlife successfully in Zimbabwe. *Oryx* 29:171–177.

Corlett, R. T. 1995. Tropical secondary forests. *Progress in Physical Geography* 19(2):159–172.

Delehanty, J. 1993. Review: The conservation atlas of tropical forests: Africa. *African Studies Review* 34:183–185.

Edwards, S. R. 1995. Conserving biodiversity: Resources for our future. In R. Bailey, ed. *The true state of the planet*. Free Press, New York.

Fairhead, J., and M. Leach. 1996. *Misreading the African landscape*. Cambridge University Press, Cambridge.

Falconer, J. 1995. Non-timber forest products in Southern Ghana. Report to World Wildlife Fund, U.S., Washington, D.C.

FAO. 1988. An interim report on the state of forest resources in the developing countries. Food and Agricultural Organization of the United Nations, Rome.

———. 1993. Forest resource assessment 1990: Tropical countries. Food and Agricultural Organization of the United Nations, Rome.

———. 1995. Assessment of the current world food security situation and medium term review. Item II of the Provisional Agenda, 20th Session, Committee on World Food Security. Food and Agricultural Organization of the United Nations, Rome.

Hamilton, C. A. 1984. *Deforestation in Uganda*. Oxford University Press, Nairobi.

Houghton, R. A. 1994. The worldwide extent of land-use change. *Bioscience* 44:305–313.

Howard, P. C. 1991. *Nature conservation in Uganda's tropical forest reserves*. IUCN, Gland, Switzerland.

IIED (International Institute for the Environment and Development). 1994. *Whose Eden?* Russell Press, Nottingham, England.

IUCN. 1990. *La conservation des écosystèmes forestiers du Gabon, basé le travail de C. Wilks*. IUCN, Gland, Switzerland.

Kammen, D. 1995. Cookstoves for the developing world. *Scientific American* 273:72–75.

Koch, H. 1968. *Magie et chasse dans la forêt camerounaise*. Berger-Lerrault, Paris.

Lahm, S. A. 1995. Survey of crop raiding animals. *Gnusletter* 14(2–3):22–23.

Leach, G., and R. Mearns. 1988. *Beyond the woodfuel crisis*. Earthscan Publications, London.

Little, P. D., and D. W. Brokensha. 1987. Local institutions, tenure and resource management in East Africa. In D. Anderson and R. Grove, eds. *Conservation in Africa*. Cambridge University Press, Cambridge.

Mubalama, L. 1996. An assessment of crop damage by large mammals in the Réserve de Faune à Okapis in the Ituri Forest, Zaire—with special emphasis on elephants. Report to the Wildlife Conservation Sociey, Bronx, N.Y.

Museveni, Y. 1997. *Sowing the mustard seed: The struggle for freedom and democracy in Uganda*. Macmillan, London.

Naughton-Treves, L. 1999. Whose animals? A history of property rights to wildlife in Toro, Western Uganda. *Land Degradation and Development* 10(2):311–328.

Naughton-Treves, L., and S. Sanderson. 1995. Property, politics and wildlife conservation. *World Development* 23(8):1265–1275.

Naughton-Treves, L., A. Treves, C. Chapman, and R. Wrangham. 1998. Temporal patterns of crop raiding by primates: Linking food availability in croplands and adjacent forest. *Journal of Applied Ecology* 36(4):596–606.

Osei, W. 1997. Human environmental impacts: Forest degradation and desertification. In S. Aryeetey-Attah, ed. *Geography of Subsaharan Africa*. Prentice Hall, Upper Saddle River, N.J.

Paterson, J. D. 1991. The ecology and history of Uganda's Budongo Forest. *Forest and Conservation History* 35:179–187.

Peluso, N. 1993. Coercive conservation? *Global Environmental Change* (June):119–217.

Ribot, J. 1993. Forest policy and charcoal production in Senegal. *Energy Policy* 21:559–585.

Richards, P. 1996. *Fighting for the rainforest*. Curry and Heinemann, London.

Sawyer, J. 1992. The timber trade. In J. Sayer, C. Harcourt, and N. M. Collins, eds. *The conservation atlas of tropical forests—Africa*. Simon and Schuster, New York.

Sayer, J., C. Harcourt, and N. M. Collins, eds. 1992. *The conservation atlas of tropical forests—Africa*. Simon and Schuster, New York.

Sibanda, B., and A. Omwega. 1996. Some reflections on conservation, sustainable development and equitable sharing of benefits from wildlife in Africa: The case of Kenya and Zimbabwe. *Southern Africa Journal of Wildlife Research* 26(4):175–181.

Sindima, H. 1995. Africa's agenda: The legacy of liberalism and colonialism in the crisis of African values. Greenwood Press, Westport, Conn.

Skorupa, J. P. 1988. *The effects of selective timber harvesting on rainforest primates in Kibale Forest Uganda*. Ph.D. dissertation, University of California, Davis.

Struhsaker, T. T. 1997. *Ecology of an African rain forest*. University Press of Florida, Gainesville.

Taylor, B. K. 1962. *The Western Lacustrine Bantu*. Sidney Press, London.

Templeton, A. R. 1991. Human origins and analysis of mitochondrial DNA sequences. *Science* 255: 737.

Thomas, S. C. 1991. Population densities and patterns of habitat use among anthropoid primates of the Ituri Forest, Zaire. *Biotropica* 23(1):68–83.

Turnbull, C. 1961. *The forest people*. Simon and Schuster, New York.

UNHCR (United Nations High Commission on Refugees). 1997. *The State of the World's Refugees,* *1997–1998: A Humanitarian Agenda*. Oxford University Press, Oxford.

Vansina, J. 1966. *Kingdoms of the savanna: A history of the central African states until European occupation*. University of Wisconsin Press, Madison.

———. 1990. *Paths in the rainforests*. University of Wisconsin Press, Madison.

WCMC (World Conservation Monitoring Centre). 1992. *Global biodiversity: Status of the earth's living resources*. Chapman and Hall, London.

Weber, W. 1981. Conservation of the Virunga gorillas: A socioeconomic perspective on habitat and wildlife preservation in Rwanda. Master's thesis, University of Wisconsin, Madison.

———. 1987. Socioecologic factors in the conservation of Afromontane forest reserves. Pages 205–229 in C. W. Marsh and R. A. Mittermeier, eds. *Primate conservation in the tropical rainforest*. Alan R. Liss, New York.

———. 1993. Ecotourism and primate conservation in African rain forests. In C. Potter, ed. *The conservation of genetic resources*. AAAS, Washington, D.C.

———. 1994. Biodiversity and rural development. In C. P. Green, ed. *Sustainable development: Population and the environment*. AED, Washington, D.C.

Western, D. 1997. *In the dust of Kilimanjaro*. Island Press, Washington, D.C.

Western, D., and M. Wright. 1994. *Natural connections*. Island Press, Washington, D.C.

Wilkie, D., and G. Morelli. 1998. The poor roads of the Ituri forest were bad for people but great for wildlife. *Natural History* (July–August):12, 14, 16, 18.

Wilkie, D. S., and J. T. Finn. 1990. Slash-burn cultivation and mammal abundance in the Ituri Forest, Zaire. *Biotropica* 22:90–99.

World Bank. 1997. *World development report 1997*. Oxford University Press, Oxford.

WRI (World Resources Institute). 1996. *World resources 1996–7: A guide to the global environment*. Oxford University Press, Oxford.

# Past and Present Trends in the African Rain Forest

# Overview of Part II

Lee J. T. White

T HE AFRICAN RAIN FOREST is a complex ecosystem, which has evolved over millions of years. Therefore, in order to understand its ecology and to be able to plan for conservation into the future, one needs to understand how the system has evolved through geological time. Furthermore, given that Africa is the cradle of humanity and that it is *our* actions that create the need for conservation, it is informative to assess humanity's historical and prehistorical impact on the African rain forest. The collection of chapters in this section is intended to set the scene for the remainder of the book, as well as to document the extent to which our knowledge of the past can help us predict and plan for changes in the future.

Chapter 3 provides a geological perspective on the conservation of African forests. It stresses that our knowledge of the past is fragmented and at times controversial, briefly compares what we know about Africa and South America, and speculates on the possible impact of greenhouse effects in the future. Chapter 4 asks the question "Does forest history matter for conservation?" Using a case study from Uganda, a country with good fossil pollen records, the authors suggest that conservation activities can be guided by a knowledge of forest history over geological time. The identification of forest refugia, which correspond to biodiversity hotspots in which forest cover has persisted in the past and is most likely to persist when the climate changes in the future, is considered to be particularly useful. Chapter 5 points out that climate change in the past has had a significant impact on the distribution of forest and savanna vegetation in Africa. A review of biogeographical literature is used to support the concept of forest refugia, and their importance for the evolution of the African rain forest as we see it today is discussed. The author then documents current knowledge about changes in forest cover through the Holocene and discusses the relative influence of climate, fire, and humanity. Chapter 6 provides a comprehensive review of the biogeography of mammals in the African rain forest and assesses to what extent mammal distributions correspond with current theories about the location of forest refugia. Chapter 7 emphasizes our lack of knowledge about the archaeology of areas covered by rain forest, but goes on to demonstrate that this is not because such remains do not exist. In part of central Gabon, stone tools dated to about 400,000 B.P. have been found, and a succession of Stone Age, Neolithic, and

47

Iron Age cultures have inhabited areas currently covered by rain forest and forest-savanna mosaic. This chapter suggests that humans have probably had a significant effect on at least part of the African rain forest ecosystem. Comparable studies should perhaps be undertaken throughout the African rain forest system, given the growing body of evidence that many forests have been significantly altered by humans in the past (see chapter 1). Chapter 8 reviews the current knowledge about the distribution and rates of loss of the African rain forest, pointing out that frequently cited works on the subject are flawed but that new technology is rapidly improving our ability to monitor change. The final chapter in the section predicts that African (and other) rain forests are likely to experience rapid climatic shifts and underscores the need for action to minimize our own contribution to these fluctuations.

Overall there is a fair degree of harmony in the views of the different authors of this section. All stress the need for further multidisciplinary research into the geological history of Africa's rain forests, in order to assess the impact of past climatic change and human (and hominid) activities. The majority feel that improving our knowledge of the past will help us to better plan for the future by locating protected areas in the places that are rich in biodiversity and that are likely to be least affected by future climatic shifts. We need to refine our knowledge of where refugia were located, and to better understand how species responded to forest contraction and expansion in the past. Daniel Livingstone, however, suggests in chapter 3 that all postulated maps of refuges should be viewed with extreme caution given the rudimentary state of our knowledge, echoing the sentiments of White (1993), who maintains that researchers have tended to oversimplify the refuge question and to develop hypotheses based on scant evidence. In chapter 6, Peter Grubb hypothesizes that mammal distribution patterns would be better explained by a diffuse network of refugia rather than the classic view of discrete forest blocks that survived during periods of climatic stress (see also chapter 1), and he emphasizes the importance that gallery forests and mountains have for conservation. The importance of montane habitats is also noted by Alan Hamilton, David Taylor, and Peter Howard in chapter 4. In chapter 3, Livingstone predicts that rainfall in tropical Africa may well increase as a result of global warming and that xeric communities (such as inselbergs, which act as refugia for savanna plants; see Reitsma et al. 1992) isolated within and around the African rain forest are likely to be particularly threatened. Indeed, several of the authors in this section voice concern about the possible effects of greenhouse gas emissions on climate. The consensus is that these effects are difficult to predict, but that we should be planning for the worst by choosing protected areas with high "ecological resilience" and by acting to reduce pollution.

Several of the authors consider genetic research to be a priority. Livingstone suggests that we need to know what sort of genetic bottlenecks different species experienced in the past in order to guide current exploitation, and it is clear that further research needs to be undertaken on the importance of isolation in refugia and of forest-savanna dynamics for speciation. Hamilton et al. refer to Fjeldsa's 1994 study, which demonstrated that bird species from recent radiations tend to be found in savannas rather than forests. An important publication by Smith et al. (1977) emphasizes the role of the forest-savanna ecotone in speciation. These studies suggest that more attention needs to be paid to ecotone areas, which, although not particularly rich in biodiversity, may play an important role in the evolution of rain forest species.

Given the rate at which the forest is being modified and destroyed by human activities, there is no time to spare if we are to act on the recommendations of these authors. Yet, to do so will require a significant increase in funding for research and conservation and a concerted effort to train scientists from the region to undertake these studies.

REFERENCES

Fjeldsa, J. 1994. Geographical patterns for relict and young species of birds in Africa and South America and implications for conservation priorities. *Biodiversity and Conservation* 3:207–226.

Reitsma, J. M., A. M. Louis, and J.-J. Floret. 1992. Flore et végétation des inselbergs et dalles rocheuses premier étude au Gabon. *Bull. Mus. Natl. Hist. Nat., Paris,* 4e sér. 14:73–97.

Smith, T. B., R. K. Wayne, D. J. Girman, and M. W. Bruford. 1977. A role for ecotones in generating rainforest biodiversity. *Science* 276:1855–1857.

White, F. 1993. Refuge theory, ice-age aridity and the history of tropical biotas: An essay in plant geography. *Fragm. Flor. Geobot. Suppl.* 2:385–409.

CHAPTER 3

# A Geological Perspective
# on the Conservation of African Forests

Daniel A. Livingstone

The direct impact of humanity on African forests is obvious and large. One is tempted to assume that other impacts, such as climatic change, are small by comparison. The geological record, however, shows that climatic change has affected African forests profoundly and rapidly many times during the period that modern tree species have existed, and is likely to do so again (Bradley 1985; Colinvaux et al. 1996).

We now add to this natural variability by changing the composition of the atmosphere. Increases in such gases as methane and carbon dioxide will certainly change the climate of the world in ways that are not yet predictable in detail but that are likely to have serious biological consequences. Greenhouse effects will not be quickly reversible, and because they will be superimposed on the natural variability of climate, they are likely to be quite pronounced before we can be sure that they have occurred (Kerr 1986). This chapter will summarize the geologic evidence of past changes in climate and consider the possible future.

## Global Evidence of Climatic Change

Large parts of the temperate zone, especially Western Europe, Canada, and adjacent parts of the United States, were overwhelmed by thick continental ice at least five times during the Quaternary (the past 1.7 million years) (Flint 1971). Each glacial advance, however, tends to destroy the evidence of earlier events, and a better geological record of ice ages comes from cores of marine sediment. These show that during Quaternary time there have been twenty-two glacial episodes, and that glacial conditions have been the Quaternary norm. Interglacial episodes, such as the one in which we live, are relatively short (Shackleton and Opdyke 1976).

Of the many indicators of past climate in marine sediment, none have been more useful than calcareous skeletons of planktonic foraminifera. These protozoa register ice ages by changes in their distribution, with cold-water species advancing toward the equator during each ice age, and by changes in their isotopic composition. Ice-age foraminifera contain relatively more of the heavy oxygen isotope of mass 18, partly because of the way temperature affects biological partitioning of the isotopes, and partly because the light isotope of mass 16 is preferentially bound up in glacier ice. If the isotopic changes are attributed largely to temperature, they suggest that tropical sea surface temperature fell by 6°C during each ice age (Emiliani 1955), but changes in tropical species assemblages suggest a fall of only 1–2°. Estimates of glacier ice mass are commonly adjusted to bring the isotopic estimates of temperature change down to this level (CLIMAP 1981).

In spite of these problems of temperature estimation, the isotopic record of the deep sea is an unambiguous indicator of past ice ages. It shows glaciation to be distinctly periodic, matching closely 20,000-, 40,000- and 100,000-year periods in irregularity of the earth's motion around the sun. Ice-rafted debris shows that there have been shorter quasi-period changes in glaciers, perhaps due to internal oscillations of the ice-ocean-atmosphere system.

A shorter but more detailed record of climatic change comes from cores drilled through the ice of Antarctica and Greenland. This evidence supports the astronomical periodicity of marine cores and also the existence of higher-frequency change, but it also shows that a very large part of the temperature change at the beginning and end of each ice age—perhaps as much as half of the total temperature difference between full glacial and full interglacial conditions—occurs within a few decades (Alley et al. 1993).

There is a serious unresolved discrepancy between various estimates of ice-age cooling at low latitudes, which run from 1–2°C to 8–12°C (Broecker 1995). This discrepancy highlights how poorly we comprehend the climatic dynamics of the earth, and how poorly we understand the thermal history of tropical forest.

The 1–2° estimate is based on statistical correlation between modern sea surface temperature and the foraminiferal assemblages of the tops of marine cores, which yields a transfer function relating temperature to foraminifera. The transfer function is applied to fossil assemblages to give temperature estimates. The underlying statistical correlation has never

been subjected to experimental test, and may be due partly to lack of a strictly equatorial element in the planktonic fauna (Livingstone 1993) and partly to loss of the youngest sediment from the core tops during coring (Emiliani 1992).

This isotopic estimate of temperature change is very sensitive to assumptions about the volume of glacier ice. New data on sea-level change suggest that ice volume has been overestimated (Clark et al. 1995), at least provided that no great part of it consisted of floating shelf-ice. A limit is placed on such revisionism by the isotopic composition of fossil foraminifera from the deep sea, where the water is still so cold that it could not get more than three degrees colder without freezing. It is difficult to reconcile the evidence from the open ocean with cooling of as much as 4–5° during each ice age.

Near shore there is a good case for substantial cooling. Corals in the Barbados indicate 5° of cooling (Guilderson 1994), and so does the distribution of hydroid coelenterates around peninsular Florida (Deevey 1950). On land, the indications of substantial ice-age cooling are even stronger: it is not easy to reconcile most of the terrestrial evidence with a cooling of so little as 5°, and some of it seems to demand cooling close to 10°.

During the last ice age, glaciers on tropical high mountains descended to much lower altitudes than they do today. Change in altitude of the firn line on the three highest East African mountains suggests drops in temperature of 4–6°, assuming no difference in snowfall (Osmaston 1975). The ice-age climate was dry, not wet, and a plausible correction for reduced snowfall would demand cooling of 8–9° to account for the fall of firn lines (Livingstone 1993). Similar changes of firn line seem to have prevailed in the Andes and in highland New Guinea. Thompson et al. (1995) have estimated from ice cores the temperature at which snow formed over the Andes, and they conclude that the ice-age temperature was 8–12° cooler than it is today. No African glacier ice old enough to provide such an estimate of cooling has been found.

Pollen analyses of lake cores from highland tropical Africa are commonly interpreted by their authors to suggest temperature lowering of >5°. The analyses are certainly consistent with such cooling, but most changes in pollen could be attributed, at least partly, to drought. Olive pollen near sea level at Bosumtwi in Ghana may be a reliable temperature indicator, but it could be explained by cooling of >3–4° (Maley and Livingstone 1983).

Many montane species of grasses are found on Mount Cameroon and also on the mountains of Ethiopia and East Africa. The topmost mountain grasses, by contrast, tend to be unique to each highland area (Clayton 1976). A time cool enough for montane grasses to spread across Africa over the intervening plateau, but not cool enough to permit the mountaintop grasses to do so, would explain this pattern. Because of the peculiar shape of Africa, with very small, mostly volcanic mountains rising sharply above the surrounding plateau, a slight reduction in temperature would have almost no effect on the amount of warm country that grasses would have to cross to get from East Africa to Cameroon. Cooling of >10° would do the job. By the same token, endemism of mountaintop grasses suggests that the climate did not cool by so much as 15°.

Some 30% of ice-age fossil cuticles of grass leaves at Bosumtwi, near sea level in Ghana, come from the temperate pooid subfamily (Talbot et al. 1984). They resemble native grasses restricted to such high altitudes on Mount Cameroon that > 7° of cooling would be needed to distribute them across the intervening lowland.

Few oceanographers are prepared to accept tropical sea surface cooling of more than a few degrees during the last ice age. Few terrestrial geologists are prepared to accept less than 5°,

and some would prefer 10°. Global circulation models will not reconcile such temperature contrasts between the tropical ocean and the tropical land (Manabe and Broccoli 1990).

## Moisture in Tropical Africa During the Last Ice Age

Before 12,500 B.P. every part of tropical Africa for which we have evidence was drier than it has been since. Lake levels were lower, lake waters more concentrated, windblown sand more widespread, and vegetation more xeric. At several sites there are suggestions of a moisture oscillation around 10,000 B.P., coeval with the Younger Dryas late-glacial oscillation in Europe (Kendall 1969). Since then there have been substantial changes in available moisture, with the wettest time falling before 6,000 B.P. in equatorial and northern tropical Africa, and since 6,000 B.P. in Madagascar and southern tropical Africa (Livingstone 1975).

## Changes in Tropical African Forest

Fossil pollen preserved in lake sediments provides most of our information about the history of African forest. Problems of identification are not trivial, but the most serious problems arise in interpreting the pollen diagrams once the fossils have been identified and counted. The standard method of fossil pollen interpretation is to use many hundreds of surface samples of lake sediment to develop a statistical relation between climate, vegetation, and pollen rain. This does not work well in tropical Africa, where lakes are not widespread or where vegetation is a fine-grained topographically controlled mosaic (Laseski 1983).

Most good tropical African fossil pollen sites are situated in and around the East African Rift and the Cameroonian Rift. It is particularly difficult to find suitable basins in areas of lowland evergreen forest, where most standing waters are riverain lakes of no great age and ill-defined or variable pollen catchment area. This patchy distribution of basins prevents us from using pollen to map the past vegetation of tropical Africa on a continent-wide scale.

In length and resolution of our record, however, we have an enormous advantage over most of the world. Many tropical African lakes are permanently stratified and anoxic in their depths. Without animals or strong currents to rework the sediment, the stratigraphic record can be very detailed. In principle, it should be possible to distinguish the deposits of each season of each year, if there were a compelling need to do so. Many lakes are old enough to provide a record of a million years, and the ancient lakes of the East African Rift, with kilometers of sediment, should contain a record of many millions of years (Cohen et al. 1993).

The surest conclusion to emerge from African pollen studies so far is that there have been significant changes in tropical forest vegetation during the past 15,000–30,000 years. Trees were especially common 5,000–10,000 B.P. and especially scarce between 12,000 and about 18,000 B.P. in almost all equatorial localities. These changes were dramatic at sites moist enough to support forest today, and more muted in present grassland or woodland localities, but the vegetation changed wherever it has been studied. At sites close to an ecotone between forest and moorland or wooded grassland, changes were comparable to those in temperate or arctic vegetation during the same time. African forests were able to re-establish themselves 10,000–12,000 B.P. on soils from which they had been driven by the cool, dry late Pleistocene climate. It may be true over human timescales that the removal of tropical forest leads to soil degeneration that prevents its return, but it is certainly not true over geological time (Livingstone 1975).

African vegetational changes of late Quaternary time do not seem to have been accompanied by great changes in regional floras. The range extensions we have found, and the ones implied by the biogeographic evidence, mostly involve plants presently restricted to high altitudes, which grew much closer to sea level when the climate was cooler. To find a fossil pollen flora very different from a modern regional one, we must go all the way back to the Pliocene (Yemane et al. 1987).

## The Question of Refugia

With new information coming in on the actual vegetational history of tropical America, the idea of Pleistocene forest refugia seems less plausible than it once did in its Amazonian homeland (Colinvaux 1993). Yet this idea may still be pertinent to Africa, if only in the sense that the abundance of forest trees was formerly much lower in at least some places. The species themselves are much older than the last dry period of their eclipse, so obviously they persisted somewhere.

It is reasonable to suspect that the refuges were in the places that are the wettest and the richest in species today. We cannot, however, map such refuges in convincing detail from pollen evidence. Too few suitable sedimentary basins are known, much less investigated, in the presently forested parts of tropical Africa. We may never know from stratigraphic evidence whether refuges were the size of Rwanda or the size of Congo. Small populations of many forest tree species may have persisted locally in particularly favorable microhabitats, or even in extensive areas that register in the pollen record as grassland, wooded grassland, or woodland. Non-palynologists would be well advised to view with skepticism all published maps of forest refugia, including any in this volume, that purport to represent conclusions based on evidence. Such maps are no more than tentative hypotheses that are not yet disproved by fossil evidence.

To conserve African forest species in the face of human exploitation, we need to know what sort of genetic bottlenecks the populations have survived in the geologic past, and how rapidly they have been able to spread from their Pleistocene refuges. At present we cannot tell the population sizes or areas of the refuges, much less the speed at which taxa expanded out of them. Without data on how fast or how far African forest trees adjusted their ranges in the face of natural climatic perturbations, one can offer no useful speculation about their ability to cope with artificial ones.

One difficulty in predicting greenhouse effects from the history of natural climatic change is that the processes are fundamentally different. Ice ages seem to be primarily the result of astronomical changes in the distribution of solar radiation over the surface of the earth. One consequence of those changes is the alteration of atmospheric carbon dioxide, resulting in worldwide modulation of temperature. What we face now is an increase in greenhouse gases with almost no change in astronomical pattern. This is combined with unprecedented reductions in forest by harvesting and conversion to agriculture. Even if we knew the distribution and genetic structure of Pleistocene vegetation very well and had accurate measurements of the rates and kinds of adjustments that African forests have made during the Holocene, we would be able to predict the future course of the forest with only a low level of reliability.

### Antiquity of Human Interaction with Forest in Africa

Hominids, *Homo,* and *H. sapiens* all seem to have arisen in Africa, and certainly have lived there for a very long time. It is interesting to speculate about the antiquity of serious human effects on the forest and vice versa. The archaeological record of presently forested parts of Africa is short, and Clark (1967) believes that people were unable to exploit the forest environment before development of the Sangoan culture, perhaps 100,000 years ago. The Sangoan tool kit includes artifacts that are plausible digging and woodworking tools, and Sangoan is the earliest culture with sites that are presently forested. So far we have no pollen records close to forested Sangoan sites to tell if they were also forested during Sangoan time.

In Europe, prehistoric human influence on the vegetation was relatively easy to demonstrate because Neolithic agriculturalists introduced exotic crops and weeds with distinctive and abundant pollen (Iversen 1949). Most indigenous African crops have local wild relatives, and few produce pollen that is both abundant and distinctive, so human effects on vegetation are much harder to find. The strongest suggestion of anthropogenic change is a decline in forest pollen and concomitant rise in grass at many sites from about 3,000 B.P. This could result from human clearing or a drier climate or some combination of the two. In Uganda there has been both a decrease in the rate of deposition of forest pollen and an increase in the rate of deposition of grass pollen (Kendall 1969). In Madagascar, where the first people arrived only a thousand years ago, the change started at the same time as it did in continental Africa, so it must have had a climatic cause (Burney 1993). Apparently there is a climatic component in the shift from trees to grass, even though part of it must surely be due to human agriculture in some places.

### Comparison with South America

All of those who work with the flora of Africa are aware of its poverty in contrast with the flora of South America. To the many explanations of that poverty already at hand, another emerges from paleoclimatic study of the two continents. Africa was much drier during the last ice age, apparently because it is more dependent on monsoonal flows for its rain, and monsoonal flows were weaker when the seasonal temperature difference between land and sea was less than it is now. There is good evidence from Indian Ocean cores that the Asian monsoon was much reduced during that time

(Prell and Kutzbach 1992), and reduced monsoonal flows are the most likely explanation for the late Pleistocene African dry period. In addition to having a much smaller area than South America that is wet enough to support rain forest today, tropical Africa was very much drier still during much of Quaternary time.

In parts of the world where geological evidence from land and sea is in agreement, global circulation models can be used to predict the likely climatic consequences of current artificial changes in the atmosphere. In the tropics, where our comprehension of past temperature distribution and its dynamic consequences is so obviously flawed, prediction must be less certain. It seems likely, however, that tropical temperatures and effective precipitation will rise. I hesitate to predict beneficial results from any sharp perturbation of an ecological organized system, but perhaps the effects of global warming on African forest, which would benefit from a warmer, wetter climate, will not be all bad. It is the more xeric communities around the forest, and those of higher latitudes and higher altitudes, which would be threatened by wetter and warmer conditions, that seem most at risk from greenhouse global warming.

REFERENCES

Alley, R. B., D. A. Meese, C. A. Shuman, A. J. Gow, K. C. Taylor, P. M. Grootes, J. W. C. White, M. Ram, E. D. Waddington, P. A. Mayewski, and G. A. Zielinski. 1993. Abrupt increase in Greenland snow accumulation at the end of the Younger Dryas event. *Nature* 362:527–529.

Bradley, R. S. 1985. *Quaternary Paleoclimatology*. Allen and Unwin, Boston.

Broecker, W. S. 1995. *The Glacial World According to Wally*. Lamont-Doherty Earth Observatory of Columbia University, Palisades, N.Y.

Burney, D. A. 1993. Late Holocene environmental changes in arid Southwestern Madagascar. *Quaternary Research* 40:98–106.

Clark, J. D. 1967. *The Atlas of African Prehistory*. University of Chicago Press, Chicago.

Clark, P. U., D. R. MacAyeal, J. T. Andrews, and P. J. Bartlein. 1995. Ice sheets play important role in climate change. *Eos* 76:265–270.

Clayton, W. D. 1976. The chorology of African mountain grasses. *Kew Bulletin* 31:273–288.

CLIMAP Project Members. 1981. Seasonal reconstructions of the Earth's surface at the last glacial maximum. Geological Society of America, Map and Chart Series MC-36.

Cohen, A. S., M. J. Soreghan, and C. A. Scholz. 1993. Estimating the age of formation of lakes: An example from Lake Tanganyika, East African Rift system. *Geology* 21:511–514.

Colinvaux, P. A. 1993. Diversity and floristic composition of lowland tropical forest in Africa and South America. Pages 500–547 in P. Goldblatt, ed. *Biological Relationships Between Africa and South America*. Yale University Press, New Haven.

Colinvaux, P. A., P. E. De Oliveira, J. E. Moreno, M. C. Miller, and M. B. Bush. 1996. A long pollen record from lowland Amazonia: Forest and cooling in glacial times. *Science* 274:85–88.

Deevey, E. S. 1950. Hydroids from Louisiana and Texas, with remarks on the Pleistocene biogeography of the western Gulf of Mexico. *Ecology* 31:334–367.

Emiliani, C. 1955. Pleistocene temperatures. *Journal of Geology* 63:538–578.

———. 1992. Pleistocene paleotemperatures. *Science* 257:1462.

Flint, R. F. 1971. *Glacial and Quaternary Geology*. Wiley, New York.

Guilderson, T. P., R. G. Fairbanks, and J. L. Rubenstone. 1994. Tropical temperature variations since 20,000 years ago: Modulating interhemispheric climate change. *Science* 263:663–665.

Iversen, J. 1949. The influence of prehistoric man on vegetation. *Danmarks Geologiske Undersogelse*, ser. 4, 3(6):1–25.

Kendall, R. L. 1969. An ecological history of the Lake Victoria Basin. *Ecological Monographs* 39:121–176.

Kerr, R. A. 1986. Greenhouse warming still coming. *Science* 232:573–574.

Laseski, R. A. 1983. Modern pollen data and Holocene climate change in Eastern Africa. Ph.D. dissertation, Brown University.

Livingstone, D. A. 1975. Late Quaternary climatic change in Africa. *Annual Review of Ecological Systems* 6:249–280.

————. 1993. Evolution of African climate. Pp. 455–472 in P. Goldblatt, ed. *Biological Relationships Between Africa and South America*. Yale University Press, New Haven.

Maley, J., and D. A. Livingstone. 1983. Extension d'un élément montagnard dans le partie sud du Ghana (Afrique de l'Ouest) au Pleistocène supérieur et à l'Holocène inférieur: Premières données polliniques. *C.R. Acad. Sci., Paris,* t. 296, ser. 2:1287–1292.

Manabe, S., and A. J. Broccoli. 1990. Mountains and arid climates of middle latitudes. *Science* 247:192–195.

Osmaston, H. A. 1975. Models for the estimation of firn-lines of present and Pleistocene glaciers. Pp. 218–245 in R. Peel, M. Chisholm, and P. Hagget, eds. *Processes in Physical and Human Geography*. Heinemann, London.

Prell, W., and J. E. Kutzbach. 1992. Sensitivity of the Indian monsoon to forcing parameters and implications for its evolution. *Nature* 360:647–651.

Shackleton, N. J., and N. D. Opdyke. 1976. Oxygen isotope record for Pacific planktonic foraminifera for the last million years. *Geological Society of America Memoir* 145:449–464.

Talbot, M. R., D. A. Livingstone, P. G. Palmer, J. Maley, J. M. Melack, G. Delibrias, and S. Gulliksen. 1984. Preliminary results from sediment cores from Lake Bosumtwi, Ghana. *Palaeoecology of Africa* 16:173–192.

Thompson, L. G., E. Mosley-Thompson, M. E. Davis, P.-N. Lin, K .A. Henderson, J. Cole-Dai, J. F. Bolzan, and K. B. Liu. 1995. Late glacial stage and Holocene tropical ice core records from Huascaran, Peru. *Science* 269:46–50.

Yemane, K., C. Robert, and R. Bonnefille. 1987. Pollen and clay mineral assemblages of a Late Miocene lacustrine sequence from the northwestern Ethiopian Highlands. *Paleogeography, Paleoclimatology and Paleoecology* 60:123–141.

# CHAPTER 4

# Hotspots in African Forests as Quaternary Refugia

Alan Hamilton, David Taylor, and Peter Howard

It has been claimed that biological hotspots in African forests—places with relatively high concentrations of species and endemics—are places where forest survived during climatically arid periods in the past (Hamilton 1982; Rodgers et al. 1982). The two most notable hotspots in terms of numbers of species lie on either side of the Congo forest block, in Cameroon/Gabon and eastern Congo; there are other major hotspots in west Africa and near the east African coast. At least some of these hotspots are broadly defined and, at a more detailed level, break up into a number of smaller areas (Kingdon 1980). The last major arid episode in Africa occurred during the last world ice maximum, centered on 18,000 B.P. Geomorphological evidence from tropical Africa indicates that at that time there were lower lake levels, more extensive dune fields, and other factors suggesting great expansion of areas of more arid climate in comparison with the present (Hamilton 1982). Supporters of the refugium theory see great reduction in forest cover at this time. There is also evidence that many, though not all, earlier Quaternary ice ages were dry in Africa, although at a more detailed level it is also known that at least the last ice age and subsequent postglacial were not climatically uniform but contained episodes of relatively dry and wet periods (Hamilton and Taylor 1991).

It has been argued that determination of the sites of Quaternary refugia in tropical forests is extremely helpful for devising strategies for conservation of forest biodiversity (Hamilton 1981). The argument is that, in the face of a predicted surge in forest destruction and degradation, the scant conservation resources that are available are best concentrated on trying to preserve forests in hotspots, that is, Quaternary refugia. But the question arises, if we know where hotspots are, does it matter whether they are Quaternary refugia or not? In fact, we believe that it does matter and, furthermore, that there is an urgent need to determine more about forest history to help guide our actions. Here are some reasons why:

- The earth will continue to be subject to climatic forcing as a consequence of orbital variations. Setting aside possible interference with natural processes by humans, it seems certain that there will be more global ice ages and more dry periods in Africa. An advantage of treating modern biogeographical patterns not just as given phenomena but as products of dynamic environmental systems, is that it makes prediction of future states possible (given sufficient information), which could clearly be of immense value in planning for the long term.

- A historical, systems-based approach allows predictions to be made concerning patterns of diversity in groups of forest organisms that have been little studied, based on patterns shown by better-studied groups.
- The location of places that may be hotspots, but about which little is known, may sometimes be predicted.
- Strategies for conservation of germ plasm, for example of timber trees, will be more efficient if forest history is taken into account, depending, in part, on theoretical understanding of how genetic variations are influenced by cycles of forest expansion and contraction.

Fjeldså (1994) has extended the subject of forest history in relation to conservation of biodiversity, emphasizing a distinction between places where species originate and those where species accumulate once they have evolved. According to analysis of the phylogenic age of tropical and southern African birds, based on DNA hybridization, species of recent radiations are generally found in savannas rather than forests. Within the forest environments, montane forests such as those of the Rwenzori-Virunga area (part of which is in western Uganda) are more significant for species origination than lowland forests, which tend to be areas of species accumulation. This study strengthens arguments for conserving certain particular patches of montane forest, as in the Uganda–DR Congo border region, but suggests less urgency over conservation of lowland forest areas with the same degree of localized concentration, because lowland forests are relatively extensive and many lowland forest bird species, in Africa at least, are reported to be remarkably tolerant to habitat disturbance.

Existing models of Quaternary forest history in tropical Africa may be criticized on several grounds (see, e.g., Connor 1986). These models are based to a significant extent on analysis of modern patterns of biological distribution, with the inherent danger of circularity in reasoning ("a forest is rich in species and endemics, showing that the area was once a refugium"; "there was a refugium in the area, which is why the modern forest is rich in species and endemics"). Another weakness is inadequate knowledge of the taxonomy of many groups, including patterns of variation in relation to geographical distribution. Despite the problems, it is worthwhile advancing models of forest history on the basis of existing data. Such models serve to stimulate further scientific work and are an interim basis for conservation action, pending refinements. Additionally, taxonomy and the mapping of distributions are not isolated "basic sciences" from which analyses of historical biogeography may flow: these subjects will themselves advance quicker if they modify their research strategies in the light of advances in historical biogeography.

## Distribution of Forest Organisms in Uganda and Its Historical Significance

Uganda is taken as an example to describe current knowledge of the distribution of forest organisms, considered in relation to relevant palynological data. The purpose is to show that, for this case at least (which is relatively well documented), there is good fossil evidence for substantial changes in forest cover and that biogeographical patterns have some relation to forest history.

Forest is thought to have once covered about 20% of Uganda (figure 4.1) but is now reduced to about 4% through clearance by humans. There are two main naturally forested regions—a zone to the north of Lake Victoria and (perhaps once connected to it) a zone in, and near, parts of the Rift Valley in the west and southwest. Isolated mountains in the north and east are forested at higher altitudes. Phytogeographically, lowland forest in Uganda belongs to the

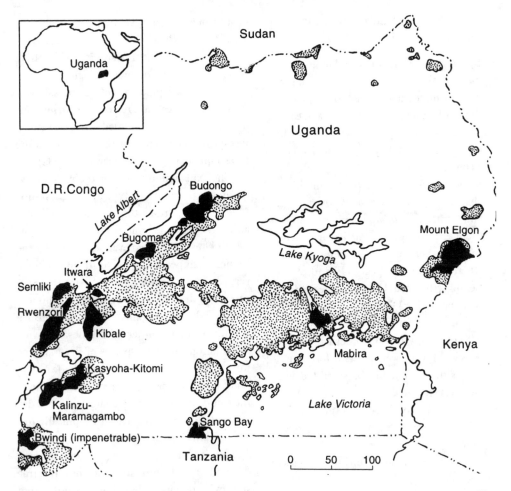

Figure 4.1. Uganda, showing the locations of principal forest reserves (black) and areas that presumably were once forested (stippled).

Guineo-Congolian phytochorion (White 1983); it can be considered to be an eastern (and floristically impoverished) extension of the main central African forest of Congo. The lowland forest passerine avifauna of Uganda is also an impoverished eastern extension of that in eastern Congo (Diamond and Hamilton 1980).

Hamilton (1974) analyzed patterns of distribution shown by forest trees in Uganda. The distribution of each tree was determined according to the four flora-areas used for the *Flora of Tropical East Africa* (1952 et seq.). Species were divided into lowland and mon-

tane, and patterns of distribution of the two groups were analyzed separately. This is a first step in trying to distinguish patterns of possible historical significance from those related to modern environmental factors. Patterns shown by montane species were further considered on a zone-by-zone basis. The assigning of species to lowland and montane groups was arbitrary to the extent that there are no altitudes in Uganda at which the floristic composition of the forests changes abruptly; instead, the forests form an altitudinal continuum (Hamilton 1975).

Forest trees in Uganda show the following patterns of distribution:

- Species of lowland forest and moist lower montane forest show a very clear-cut gradient of increasing species poverty from the southwest toward the north and east. Modern climatic gradients may contribute to this trend, but they appear inadequate as a complete explanation. For example, climatic factors alone cannot account for the absence of such species as *Parinari excelsa* and *Carapa grandiflora* in submontane forest in eastern Uganda. Particularly high concentrations of species not found elsewhere in Uganda appear in Bwindi and Semliki (Bwamba) Forests in the extreme west, bordering DR Congo, and, to a lesser extent, in the Sango Bay forests on the edge of Lake Victoria, bordering Tanzania.
- Species of dry lower-montane forest and high-altitude montane forest tend to be widely distributed.
- Disjunct distributions often appear to be relatable to modern environmental conditions.
- There are very few endemic tree species in Uganda.

To explain these patterns (the last two of which were not contained in Hamilton 1975), it has been proposed that many species of lowland forest and lower-altitude montane forest may have spread relatively recently from the southwest to other parts of the country, slow rates of spread of some species contributing to formation of the observed gradients of increasing species poverty. There is no evidence that Uganda includes the richest part of the total gradient in floristic diversity (which in fact seems to lie in eastern Kivu, DR Congo); there is no major hotspot or (interpreted historically) major refugium for lowland or moist lower-montane forest in Uganda during the postulated period of forest withdrawal. Yet minor

centers of forest survival may have been situated at or near Bwindi, Semliki, and Sango Bay. Most lowland-forest tree species, which have fruits or seeds that on morphological grounds seem poorly adapted to wide dispersal away from parent trees, have restricted distributions in Uganda, being confined to the western forests; this gives some further support for the forest-spread hypothesis (Hamilton 1982).

The explanation given for the wide distributions of many species typical of highest-altitude montane forest is that the climate at higher altitudes, being moderated by orographic rainfall and lower potential evaporation under the reduced temperatures, is relatively uniform regardless of whether a mountain is situated in a wetter or drier climatic region. Thus, this forest type is buffered against climatic change to a greater extent than other forest types. The wide distributions of many species of dry montane forest are seen as partly due to the wide occurrence of dry climates in the past and partly to the ability of many species characteristic of dry montane forest to survive in edaphically dry sites within climatically wet areas.

Howard (1991) has carried out a more detailed study of the distribution of forest tree species in Uganda, based on records from twelve individual forests. He has also compiled existing data and collected fresh data on the distributions of forest birds, diurnal primates, and butterflies (swallowtail family plus selected charaxids) (tables 4.1 and 4.2). His main conclusions can be summarized as follows:

- With one exception (butterflies in Bwindi Forest), no forest is known to support more than two-thirds of the country's species of any one of these four groups of organisms.
- The number of species in any particular forest is related partly to altitude (four altitude zones are recognized, based on species composition).
- Forests of the west, notably Bwindi and

Table 4.1

## Numbers of Species Belonging to Four Groups of Organisms in Twelve Forests in Uganda

| FOREST RESERVE | INDICATOR GROUP | | | |
| --- | --- | --- | --- | --- |
| | TREES | BIRDS | PRIMATES | BUTTERFLIES (SELECTED GROUPS) |
| Maximum possible number of species (i.e., total in entire country) | 427 | 329 | 12 | 71 |
| Kibale | 209 | 177 | 8 | 45 |
| Semliki | 168 | 216 | 8 | 51 |
| Budongo | 240 | 159 | 5 | 42 |
| Kalinzu-Maramagambo | 242 | 181 | 6 | 40 |
| Bugoma | 158 | 144 | 6 | 42 |
| Bwindi (impenetrable) | 163 | 214 | 7 | 57 |
| Kasyoha-Kitomi | 204 | 104 | 6 | 21 |
| Itwara | 143 | 87 | 6 | 25 |
| Sango Bay | 170 | 119 | 6 | 45 |
| Mabira | 202 | 151 | 2 | 39 |
| Mount Elgon | 112 | 144 | 2 | 36 |
| Rwenzori | 75 | 89 | 4 | 1 |

*Source:* Data from Howard 1991, table 2.2.

*Note:* Locations of forests are shown on figure 4.1.

Semliki, have an outstanding number of species of relatively restricted distribution in the country (these two forests are given high "conservation importance" scores, followed by Budongo, Kalinzu-Maramagambo, and Kibale; see table 4.2). As far as trees are concerned, this result backs up results of the earlier study by Hamilton (1974). The pattern is even more sharply obvious for birds, butterflies, and primates.

In considering the results of Howard's survey in relation to that of Hamilton, it must be remembered that the basis for comparison is not exactly the same; the sample areas used by Howard are smaller, and within each area there is, on average, a narrower range of vegetation types and hence potential species complements.

With the detailed records given by Howard, it is possible to evaluate the influence of environmental, as opposed to historical, factors for determining the distributions of species. According to a classification of the forests by overall species composition, forests of "high intermediate altitudes" (Kalinzu, Kasyoha-Kitomi, Kibale, and Itwara) belong to one forest type, and those of "low intermediate altitudes" (Budongo, Bugoma, and Mabira) belong to another. This second type occurs both in the west (Budongo and Bugoma), relatively close to the hypothetical refugium in DR Congo, and in the east (Mabira), which is at a greater distance. It might be supposed that relatively few species would be recorded only from Mabira, compared with from either of the other two forests. The results do indeed sug-

Table 4.2

## Conservation Importance Scores of Various Forests for Four Groups of Organisms in Uganda

| FOREST RESERVE | BIRDS | PRIMATES | BUTTERFLIES (SELECTED GROUPS) | ANIMAL GROUPS AVERAGE SCORE | TREES | OVERALL SPECIES IMPORTANCE SCORE |
|---|---|---|---|---|---|---|
| Kibale | 45 | 85 | 43 | 58 | 71 | 64 |
| Semliki | 100 | 100 | 70 | 90 | 77 | 83 |
| Budongo | 41 | 23 | 40 | 35 | 100 | 67 |
| Kalinzu/Maramagambo | 43 | 31 | 41 | 38 | 94 | 66 |
| Bugoma | 33 | 32 | 40 | 35 | 53 | 44 |
| Bwindi (impenetrable) | 86 | 76 | 100 | 87 | 84 | 85 |
| Kasyoha-Kitomi | 20 | 31 | 20 | 24 | 71 | 47 |
| Itwara | 16 | 31 | 21 | 23 | 46 | 34 |
| Sango Bay | 25 | 68 | 48 | 47 | 76 | 61 |
| Mabira | 34 | 14 | 36 | 28 | 76 | 52 |
| Mount Elgon | 47 | 9 | 36 | 31 | 86 | 58 |
| Rwenzori | 38 | 39 | 27 | 35 | 42 | 38 |

*Source*: Data from Howard 1991, table 2.3.

*Note*: Locations of forests are shown in figure 4.1. The importance score is calculated as the number of species present, each record of a species being divided by the number of forests out of the twelve studied in which it occurs. The results are scaled so that the forest with the highest score equals 100.

gest this but are not very clear cut; they can be summarized as follows:

- Trees: The exception to expectation is Bugoma, perhaps because the flora is believed to be underrecorded in comparison with the other two forests.
- Birds: Mabira has an unexpected abundance of records of bird species, probably because it is a favorite site for bird watchers living in Kampala.
- Primates: Distributions are well known, and the pattern follows expectation.
- Butterflies: The pattern follows expectation but is not very clear cut because there are many widely distributed species.

## Fossil Evidence for Upper Quaternary Forest History in Uganda

The forest history of Uganda is better known from the fossil record for the period during and since the last world ice maximum (18,000 B.P.) than that of any other African country. Pollen evidence is available for the following places:

- Kigezi, in the southwest, with excellently dated pollen diagrams from two sites (Hamilton et al. 1986; Taylor 1990, 1992) and additional diagrams for three other localities (Morrison and Hamilton 1974).
- Rwenzori, on the DR Congo border in the midwest, with one long and two short diagrams (Livingstone 1967; Hamilton 1982).
- Lake Albert, in the west, to the north of Rwenzori (Sowunmi 1991).

- Lake Victoria, with one long diagram (Kendall 1969; Hamilton 1982).
- Mount Elgon, with three sites covering most of the period under consideration (Hamilton 1987).

Many of the pollen diagrams provide evidence of both past vegetation near the sample sites and vegetation at greater distances, the latter reconstructed from pollen thought to have been transported to the sites from relatively far away. Reconstructed vegetation for four of the localities is shown in table 4.3. There seems to have been little or no lowland forest in Uganda before 12,000 B.P., at about which date it started to spread. No difference in the date of spread has been identified between sites in the west (near the refuge area) and in the east of the country, as might perhaps be expected; this could be because forest spread did not occur through the advance of a single forest front but rather through the wide dissemination of some plants with very efficient dispersal mechanisms and then through the expansion and coalescence of forest nuclei. This finding clearly fits the story of lowland forest history inferred from the modern distributions of forest organisms. Regarding montane forest, sediments in the Kigezi Highlands dating to the time of the last world glaciation contain much greater quantities of montane forest pollen than do sediments at the same age on Mount Elgon, indicating a moister environment in the west than the east; this, too, fits with the reconstructed history of forests, as inferred from modern distributions of forest organisms. This pattern of greater quantities of montane forest pollen toward the west during the last ice age is also confirmed by pollen diagrams from just outside Uganda, from Rwanda in the west (Hamilton 1982) and the Cherangani Hills in the east (Coetzee 1967).

## Some Implications for Conservation

Determination of the modern patterns of distribution of organisms is an important element in the formation of conservation strategies. We do need to know what the species are and where they live, so that we can target conservation resources to certain places, based on such considerations as high numbers of species, high numbers of endemics, and complementarity. Introducing historical considerations into the analysis can add fresh elements to the equation, such as giving extra consideration to phylogenetically more distinct taxa or to sites of active evolutionary radiation. In the case of African forests, much remains to be done. There is an urgent need for basic taxonomic research and for further information on the distributions of species and genetic diversity within species. Moreover, there is clearly much to be gained from introduction of new technologies, such as DNA hybridization (Fjeldså 1994).

The significance of considering forest history during past ice-age climatic fluctuations is that it allows us to estimate where forest is most likely to survive during future glacial episodes, which are predicted to occur. We can hypothesize that forest will survive best during the next ice age in the same places as it did in the last. According to present understandings, the localities of past refugia are determined partly by fundamental features of terrestrial geography, such as the presence of rising ground interrupting air streams blowing ultimately from the sea, as well as by features of oceanic and atmospheric circulation, which are likely to be influenced in similar ways during future ice ages as they have been in the past.

Apart from the macro-events of glacial climatic changes, African forests are, or will be, subject to at least two other types of climatic change that should be taken into consideration in devising conservation strategies. Some of these are consequent on increase in greenhouse

Table 4.3.

## A Summary of Vegetation History from Pollen Analysis, West to East Across Uganda, 20,000–0 B.P.

| DATE (KYR B.P.) | KIGEZI HIGHLANDS–MONTANE ENVIRONMENT (1,800–2,300M) | KIGEZI HIGHLANDS–LOWLAND ENVIRONMENT | RWENZORI–MONTANE ENVIRONMENT | RWENZORI–LOWLAND ENVIRONMENT | LAKE VICTORIA–LOWLAND ENVIRONMENT | ELGON–MONTANE ENVIRONMENT | ELGON–LOWLAND ENVIRONMENT |
|---|---|---|---|---|---|---|---|
| Present vegetation | Agriculture/moist lower-montane forest | Forest/agriculture/savanna | Moist montane communities | Forest/agriculture/savanna | Forest/agriculture/savanna | Moist and dry montane communities | Savanna agriculture/forest |
| 0–2 | Agriculture | Lowland forest present | Somewhat drier moist montane communities | Agriculture | Reduction of moist forest (drier climate and/or agriculture) | Expansion of drier montane forest types | Expansion of drier types of lowland forest and woodland |
| 2–4 | Expansion of drier montane forest types | Lowland forest present | Somewhat drier moist montane communities | Expansion of drier lowland forest | Reduction of moist forest (drier climate and/or agriculture) | Expansion of drier montane forest types | Expansion of drier types of lowland forest and woodland |
| 4–6 | Moist lower-montane forest (after 11,000 B.P.) | Lowland forest present | Moist montane vegetation types | Lowland forest present | Lowland forest present | Moist montane forest extensive (from at least 11,000 B.P.) | Lowland forest and woodland present |
| 6–8 | Moist lower-montane forest (after 11,000 B.P.) | Lowland forest present | Moist montane vegetation types | Lowland forest present | Lowland forest present | Moist montane forest extensive (from at least 11,000 B.P.) | Lowland forest and woodland present |

| | | | | | | | |
|---|---|---|---|---|---|---|---|
| 8–10 | Moist lower-montane forest (after 11,000 B.P.) | Lowland forest present | Moist montane vegetation types | Lowland forest present | Lowland forest present | Moist montane forest extensive (from at least 11,000 B.P.) | Lowland forest and woodland present |
| 10–12 | Upper moist montane forest (to 11,000 B.P.) | Lowland forest present | Moist montane vegetation types | Lowland forest present | Lowland forest present | No direct evidence (14,000–11,000 B.P.) | No direct evidence (14,000–11,000 B.P.) |
| 12–14 | Ericaceous Belt; some upper montane forest | Lowland forest absent | Montane vegetation belts lowered, drier types | Lowland forest absent | Lowland forest absent | No direct evidence (14,000–11,000 B.P.) | No direct evidence (14,000–11,000 B.P.) |
| 14–16 | Ericaceous and dry montane scrub | Lowland forest absent | Montane vegetation belts lowered, drier types | Lowland forest absent | Lowland forest absent | Grassland; virtually no montane forest | Lowland forest absent |
| 16–18 | Ericaceous and dry montane scrub | Lowland forest absent | No direct evidence | No direct evidence | No direct evidence | Grassland; virtually no montane forest | Lowland forest absent |
| 18–20 | Ericaceous and dry montane scrub | Lowland forest absent | No direct evidence | No direct evidence | No direct evidence | Grassland; virtually no montane forest | Lowland forest absent |

*Sources*: Data from Hamilton 1982, 1987; Kendall 1966; Livingstone 1967; Taylor 1990, 1992.

*Note*: Evidence from pollen believed to be of local and long-distance origins, respectively, is used to reconstruct local and wider (lowland) environments.

gases. These changes can be predicted to result generally in a warming and moistening of the tropical African climate, the moistening because of the greater influence of oceanic evaporation, as indicated by long-term paleoclimatic evidence. The most significant conclusion for conservation of African forest biodiversity is to ensure, where possible, that protected areas contain wide ranges of altitudes, to permit the vertical movement of species to higher elevations.

The other major climatic change envisaged is an aridification of many forest areas as a result of changes to the energy and water balances due to vegetational clearance and changes in surface albedos, resulting from human activities. Such climatic changes as reduced rainfall but sometimes very wet years, reduced reliability in rainfall, and a general warming in the climate are widely predicted. One conservation implication is the desirability of concentrating resources on wetter areas, but the conservation community should also not ignore the wider and more demanding conclusion: that conservation of the tropical African environment depends on the successful application of very widespread measures aimed to retain or restore a dense vegetation cover, both in places naturally under forest and elsewhere.

For Uganda at least, the refugium theory of the origin of hotspots in African forests is well supported by the fossil evidence. Further confirmation from other parts of Africa would be desirable, although few, if any, countries are as well endowed with suitable sample sites for pollen analysis as is Uganda. Meanwhile, it seems valid to make conservation decisions on the basis that the refugium theory of hotspot origin is correct, pending evidence to the contrary.

REFERENCES

Coetzee, C. G. 1967. Pollen—analytical studies in east and southern Africa. *Palaeoecology of Africa* 3:1–146.

Connor, E. F. 1986. The role of Pleistocene forest refugia in the evolution and biogeography of tropical biotas. *Trends in Ecology and Evolution* 1:165–168.

Diamond, A. W., and A. C. Hamilton. 1980. The distribution of forest passerine birds and Quaternary climatic change in tropical Africa. *Journal of Zoology, London* 191:379–402.

Fjeldså, J. 1994. Geographical patterns for relict and young species of birds in Africa and South America and implications for conservation priorities. *Biodiversity and Conservation* 3:207–226.

Flora of Tropical East Africa. 1952 et seq. Crown Agents, London, and Balkema, Rotterdam, The Netherlands.

Hamilton, A. C. 1974. Distribution patterns of forest trees in Uganda and their historical importance. *Vegetatio* 29:31–35.

———. 1975. A quantitative analysis of altitudinal zonation in Uganda forests. *Vegetatio* 30:99–106.

———. 1981. The Quaternary history of African forests: Its relevance to conservation. *African Journal of Ecology* 19:1–6.

———. 1982. *Environmental History of East Africa*. Academic Press, London.

———. 1987. Vegetation and climate of Mt Elgon during the late Pleistocene and Holocene. *Palaeoecology of Africa* 18:283–304.

Hamilton, A. C., and D. Taylor. 1991. History of climate and forests in tropical Africa during the last 8 million years. *Climatic Change* 19:65–78.

Hamilton, A. C., D. Taylor, and J. C. Vogel. 1986. Early forest clearance and environmental degradation in southwest Uganda. *Nature* 320:164–167.

Howard, P. C. 1991. *Nature Conservation in Uganda's Tropical Forest Reserves*. IUCN, Gland, Switzerland.

Kendall, R. L. 1969. An ecological history of the Lake Victoria Basin. *Ecological Monographs* 39:121–176.

Kingdon, J. S. 1980. The role of visual signals and face patterns in African forest monkeys (guenons) of the genus *Cercopithecus*. *Transactions of the Zoological Society of London* 35:425–475.

Livingstone, D. A. 1967. Postglacial vegetation of the Rwenzori Mountains in Equatorial Africa. *Ecological Monographs* 37:25–52.

Morrison, M. E. S., and A. C. Hamilton. 1974. Vegetation and climate in the uplands of south-western Uganda during the later Pleistocene period, 2. Forest clearance and other vegetational changes in the Rukiga Highlands during the past 8000 years. *Journal of Ecology* 62:1–31.

Rodgers, W. A., C. F. Owen, and K. M. Homewood. 1982. Biogeography of East African forest mammals. *Journal of Biogeography* 9:41–54.

Sowunmi, M. A. 1991. Late Quaternary environments in equatorial Africa: Palynological evidence. *Palaeoecology of Africa and the Surrounding Islands* 22:213–238.

Taylor, D. 1990. Late Quaternary pollen records from two Ugandan mires: Evidence for environmental change in the Rukiga Highlands of south west Uganda. *Palaeogeography, Palaeoclimatology and Palaeoecology* 80:283–300.

———. 1992. Pollen evidence from Muchoya Swamp, Rukiga Highlands Uganda, for abrupt changes in vegetation during the last ca. 21,000 years. *Bull. Soc. Géol. France* 163:77–82.

White, F. 1983. The vegetation of Africa: A descriptive memoir to accompany the UNESCO/AETFAT/UNSO vegetation map of Africa. UNESCO, Paris.

# CHAPTER 5

# The Impact of Arid Phases on the African Rain Forest Through Geological History

Jean Maley

There were great variations in paleoenvironment and vegetation in the African tropics toward the end of the Neogene. These can be explained by global temperature changes, particularly the cooling phase that occurred at the same time as Antarctic and Arctic glacial advances (Maley 1980, 1996a, 1996b). Pollen in the sediments of the Niger delta reflects changes in the rain forest and adjacent savannas during this period (Poumot 1989; Morley and Richards 1993). Analysis of pollen in sediments suggests that several important extensions of savanna vegetation, associated with reductions of the extent of the rain forest, occurred at the end of the Miocene, between 8 and 5 million years ago. There is evidence of similar changes in east Africa (Harris 1993, based on the paleofauna). This period was characterized by large and rapid eustatic fluctuations directly associated with variations of polar ice extent (Haq et al. 1987). However, near the Gulf of Guinea (Poumot 1989; Morley and Richards 1993), as well as in east Africa (Bonnefille 1983), the main change occurred around 2.5 million years ago, at the same time as the first large Arctic glacial advance. In fact, all the large glacial advances in the Arctic—including the last one, which culminated around 18,000 B.P.—have

had major impacts on humid tropical Africa (Maley 1996a, 1996b).

During the Pliocene and Pleistocene, changes in marine $\delta^{18}O$ show a progressive increase in the amplitude of glacial variations. Two main stages occurred, the first around 2.5 million years ago and the second about 800,000 years ago (Ruddiman et al. 1989), when there was a strong increase of climatic variability, characterized by stronger cooling phases with a periodicity of 100,000 years (Start and Prell 1984). Pollen data based on marine records in the Gulf of Guinea (covering the past 250,000 years) and lacustrine records in forest areas (covering the past 25,000 years) demonstrate the impact of these variations on vegetation at the end of the Quaternary. There was a maximum reduction of the rain forest between 20,000 and 15,000 B.P. (Maley 1996a). Indirect information about certain aspects of the history and dynamics of the rain forest can also be obtained from the modern biogeography of various groups of animals and plants (Hamilton 1976; Maley 1987; Colyn 1987, 1991; Sosef 1994).

Climate is highly dependent on latitude, the distribution of continents, and especially regional geographical factors (mountains, depressions, and so on), all of which were relatively stable during the period of interest (the end of the Neogene). Therefore, climatic changes in the past probably affected particular regions repeatedly in similar ways. This means that during arid phases, fragmentation of the forest must have reoccurred numerous times, especially over the past 800,000 years. Various sedimentary and micropaleontological recordings in the Gulf of Guinea demonstrate that cyclic worldwide climatic events have repeatedly had a marked impact on the African equatorial zone (Prell et al. 1976; Jansen et al. 1984, 1986; Ruddiman et al. 1989; Bonifay and Giresse 1992). Therefore, the end of the most recent climatic cycle, from 25,000 B.P. to the present, for which abundant data exist, could provide a model for the main fluctuations that have affected the African rain forest over the past 800,000 years. The two extremes of this model are (1) a phase of maximum fragmentation, when forest cover was at a minimum, and (2) a phase of maximum extension, which peaked between approximately 9,000 and 4,000 B.P. Furthermore, the intermediary periods, before and after each of these phases, are characterized by a number of phases of regression and recolonization (Maley 1996a). This suggests that during 80–90% of the past 800,000 years the African rain forest was less extensive and more fragmented than at present.

## Fragmentation of the Rain Forest

The study of paleoenvironments and their variations has often indicated phases of intense aridity. Stone lines that are found at the base of numerous soil profiles indicate that during several periods of the upper Pleistocene, extensive areas that are currently forested lost a large part of their tree cover and severe soil erosion occurred, especially between 70,000 and 40,000 B.P., as well as at the very end of the Pleistocene (Lanfranchi and Schwartz 1991; Maley 1993, 1996b; Runge 1997).

The study of fossil pollen, preserved in lacustrine sediments accumulated regularly over millennia, can be used to reconstitute the history of vegetation around the lakes. This technique has yielded important insights into the fragmentation of the forest. Pollen analyses at Lake Bosumtwi (Ghana) have demonstrated the regional disappearance of the forest from about 19,000 to 15,000 B.P. (Maley 1987, 1989, 1991). In contrast, pollen analyses at Lake Barombi Mbo have demonstrated that forest persisted in this region of west Cameroon between 20,000 and 15,000 B.P. (figure 5.1; see also Maley 1987, 1991; Brenac 1988; Giresse et al. 1994; Maley and Brenac 1998a).

The fragmentation of the forest block during arid periods resulted in the extension of various plant assemblages from more or less xeric habitats into the zone currently occupied by rain forest. This is clearly demonstrated by pollen analyses for the last arid maximum (18,000 B.P.) in the two aforementioned locations: in Bosumtwi the pollen of taxa from open environments completely dominated at that time, whereas at Barombi Mbo these taxa increased considerably, although they almost never dominated. Today, relics of xeric vegetation persist in isolated refuges where unfavorable soil or local climatic conditions have prevented recovery of moist rain forest, as on Mount Nimba (Ivory Coast and Guinea) (Schnell 1952), in the Crystal Mountains (Gabon) (Reitsma et al. 1992), or even in the large savanna enclave of La Lopé, in central Gabon (Aubréville 1967; White 1992). On the large rocky hill of Nkoltsia, near Bipindi in southeastern Cameroon, Villiers (1981) described a prairie of cactiform Euphorbiaceae and a low, open forest with numerous *Julbernardia letouzeyi*, which both he and Letouzey (1983) associate with the southern zambezian dry forest type.

Pollen analyses undertaken at Bosumtwi and Barombi Mbo, both located in lowland parts of equatorial Africa below 500 m altitude,

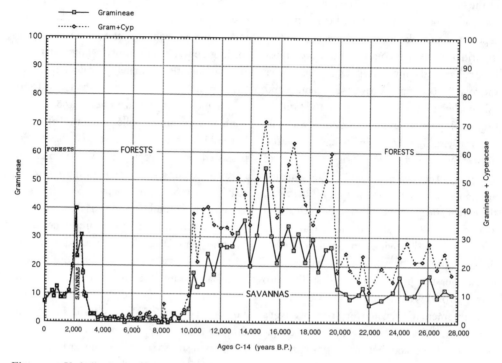

Figure 5.1. Variation in savannas and forests in western Cameroon during the recent Quaternary recon-
stituted from analysis of pollen in sediments of Lake Barombi Mbo, near Kumba (Maley and Brenac
1998a). A core about 23 m long was taken from this crater lake, located at about 300 m altitude in lowland rain forest about 40 km northeast of Mount Cameroon and 60 km from the sea. Pollen analyses
were undertaken on 82 samples, 12 of which were radiocarbon dated to establish the chronology of the
core (Maley and Brenac 1998a). The figure shows relative changes in tree pollen compared to pollen
from grasses (unbroken line). The dotted line represents the percentage of Cyperaceae pollen added to
that for Gramineae. The Gramineae pollen are almost exclusively from herbaceous plants found on well-
drained soils, characteristic of open savanna vegetation. The Cyperaceae are mostly aquatic. This figure
allows us to depict the different phases of forest and savanna expansion over the past 28,000 years. We
can see that from 28,000 to c. 20,000 B.P. forest cover was similar to that seen today (the percentages are
similar). From 20,000 to 10,000 B.P., during the last major global and regional cold period, the climate
was much drier, resulting in replacement of forest by savanna vegetation. However, in the Barombi Mbo
region grass pollen remained at about 30–40% and never dominated tree pollen. This suggests that the
landscape at the time was a forest-savanna mosaic in which forests were more widespread. This has been
confirmed by an analysis of carbon isotopes (Giresse et al. 1994). This landscape probably corresponds
to the forest refuges thought to have occurred in equatorial Africa during this period (Maley 1996a;
Sosef 1994). From c. 9,500 to 2,800 B.P. forest was at its most dominant; between 2,800 and 2,000 B.P.
there was a brief and intense period of forest retreat and savanna expansion. This climatic pejoration
was probably due to increased seasonality associated with a shorter rainy season (see text). Finally,
forests have expanded from about 2,000 B.P. up to the present.

have demonstrated that a montane forest taxon, *Olea capensis* (syn. *Olea hochstetteri*), extended into the lowlands before the Holocene. Furthermore, this species and two other montane forest taxa, *Podocarpus cf. latifolius* and *Ilex mitis*, extended into the lowlands on the Batéké Plateau north of Brazzaville before the Holocene and possibly before 20,000 B.P. (Elenga et al. 1991, 1992, 1994; Maley and Elenga 1993). As is the case for xeric vegetation, refuges of montane vegetation, with *Olea capensis* and *Podocarpus latifolius*, can still be found today at low altitudes (see White 1981). The largest stand located was at Kouyi, between 600 and 700 m altitude on an old ironcrust plateau in the Massif du Chaillu in Congo, which is very unfavorable for the growth of rain forest (Maley et al. 1990b). These montane extensions at low altitudes can be explained only as relics of periods in the past when decreased temperatures occurred during certain arid phases, especially the one that culminated around 18,000 B.P., when temperature was estimated to be 3–4°C below the present yearly mean (Maley and Livingstone 1983; Maley 1987, 1989, 1991, 1996a; Maley and Elenga 1993). Studies in the mountains of east Africa have demonstrated similar values (Bonnefille et al. 1992; Vincens et al. 1993). Thus, the entire zone currently occupied by rain forest has experienced extensive changes in the environment and vegetation in the past.

## Biogeography and History of the African Rain Forest: The Refuge Debate

Current climatic conditions in the equatorial region covered by rain forest are relatively homogeneous with respect to average temperature and humidity, although there is a gradient between a more humid zone near the Gulf of Guinea, with some extensions toward the central part of the Congo Basin, and a somewhat less humid peripheral zone in the north and south. The principal vegetation formation in the most humid zone is evergreen rain forest,

characterized by the predominance of Caesalpiniaceae, whereas the peripheral zone is mostly semi-deciduous forests dominated by Ulmaceae and Sterculiaceae (White 1983; Maley 1990).

This general pattern has numerous exceptions that can be analyzed from a historical viewpoint. In the most humid zone the distribution of species is not homogeneous, and it is particularly noteworthy that the main biogeographic gaps are not systematically located where important gaps occur at present, either for climatic or for geographic reasons (large rivers, mountain chains, and so one). The best example and the one that has been most studied is the Dahomey interval, where Sudanian savannas extend all the way to the sea and interrupt the western forest block for approximately 200 km in Togo and Benin. In fact, this gap seems to play only a minor role in the distribution of typical forest plants (Léonard 1965; Brenan 1978; White 1979), small mammals (Robbins 1978; Grubb 1982), birds (Moreau 1963, 1966; Crowe and Crowe 1982), batrachians (Amiet 1987), or insects (Carcasson 1964; Lachaise et al. 1988). These data show that the distribution limits of few forest species occur at the Dahomey interval. The most significant gap in the Guineo-Congolian forest block occurs in eastern Nigeria along the Cross River. As a result, Brenan (1978) distinguished the vegetation of the Guinean Domain, to the west of the Cross River, and the Congolian Domain, to the east (figure 5.2). Thus, there seems to be no direct relationship between the magnitude of these geographical gaps and their biogeographical impact. The geographical gap formed by the Cross River is minute compared to the scope of the biogeographical phenomenon, while the inverse holds for the Dahomey interval. This is why most of the authors who have discussed these questions have suggested that the biogeographical gap of the Cross River, like other less important gaps, might be

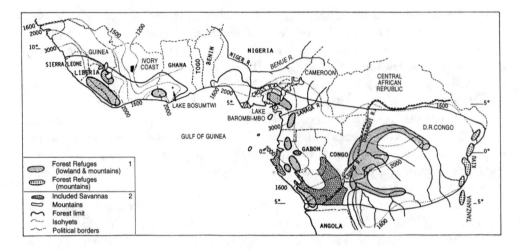

Figure 5.2. Schematic map of the African forest refuges during the last maximum arid phase (around 18,000 B.P.), according to various authors mentioned in the text. *Key to inset:* (1) c. 18,000 B.P.; (2) present-day environment. Figure adapted from Maley 1987, figure 1, and completed following Van Rompaey (1993) for the western forest block, Sosef (1994) for Gabon and west Congo, and Colyn (1987, 1991) for the Congo basin.

the result of previous fragmentations of the forest block as a result of past climatic changes (see Maley 1987, 1989, 1996a, 1996b).

Biogeographers who have studied the present distribution of wildlife in the African forest have concluded that four main sectors have high biodiversity, meaning that they are richer in species and endemic taxa than others.

- The richest sector is that which comprises the evergreen Biafran forest (close to the Bay of Biafra in the Gulf of Guinea). This extends from the Cross River to the Sanaga and, farther south, includes part of the forest of southwest Cameroon, the west of Gabon, and the Chaillu and Mayombe mountains in Gabon and western Congo (Aubréville 1968; Letouzey 1968, 1985; Brenan 1978; White 1983; Breteler 1990; Sosef 1994). In the east of Nigeria, Biafran extensions run all the way to the Niger delta (Hall 1981).
- A less rich sector is located at the western limit of the forest block. It consists of a

large part of the forests to the south of the Guinea Ridge, Liberia, and the far west of the Ivory Coast (Guillaumet 1967; Van Rompaey 1993). The southwestern forests of Ghana and adjacent parts of the Ivory Coast are also notable for the presence of various endemics (Hall and Swaine 1981; Swaine and Hall 1986).

- Two sectors in the Congo Basin have been identified on the basis of their relative richness, particularly for endemic mammals (especially simian primates; see Kingdon 1980; Grubb 1982; Colyn 1987, 1991; Colyn et al. 1991) and vascular plants (Ndjélé 1988). The first sector is located in the central part of the basin, mainly to the east of the large river, and the second, a mostly mountainous sector, is on the eastern side of the basin bordering the Kivu-Tanganyika Rift.

Most biogeographers conclude that this high biodiversity and endemism can be explained by the fact that these regions have been little affected, if at all, by arid phases and that they represent "refuges" for forest biotopes. Schematic maps of the supposed refuges have been published by various authors (see figure 5.2): Aubréville (1949, 1962), Moreau (1969), Hamilton (1976), Mayr and O'Hara (1986), Maley (1987), Sosef (1994), Rietkerk et al. (1995), and, for the Congo Basin, Grubb (1982) and Colyn (1987, 1991). The large central refuge was riparian, associated with the permanent fluviatile flux of the Congo River and its main tributaries (Giresse et al. 1982). This refuge is thought to have resulted not from regional climatic conditions but from conditions in its distant headwaters, particularly on the eastern ridge bordering the rift. Research by Colyn (1987, 1991) has shown that this region also contained refuges of mountain forests. This hypothesis seems to be confirmed by the botanical data of Ndjélé (1988) and by pollen analyses undertaken at Lake Mobutu in Uganda (Sowunmi 1991), and in Burundi (Bonnefille and Riollet 1988). These mountain forest refuges seem to have benefited from conditions typical of the cloud forest (see Maley and Elenga 1993).

As far as the locations of refuges are concerned, the study by Sosef (1991, 1994) of two sections of African begonias is particularly important and commendable. The forty-six species and subspecies of begonias included in the study are strictly limited to non-riparian rain forest environments, at low altitude of the Guineo-Congolian Domain. They are fragile plants that live only in very shady forest understory. The great majority are endemic to a few very small areas. A phylogenetic tree was established by modern cladistic methods. Then, by combining the cladogram with distribution records, Sosef obtained a historical hierarchy of the areas where these begonias occur at present. On the basis of this analysis Sosef (1994:116–121) conjectured: "During the last glacial era (18,000 years BP), they will

have survived almost exclusively in the refuges." Because the seeds are subject to very limited dispersal and these begonias sprout only at the base of the mother plants, Sosef (1994:117) concluded: "The location of the former refuges may be deduced from their present-day distribution." The areas of these refuges concur with those that had already been proposed by Maley (1987) for the Cameroon-Gabon sector, adding two new refuge areas, a very small one in the Doudou Mountains on the western edge of the Massif du Chaillu, and another larger one in Mayombe in western Congo (Sosef 1994). Because these plants do not live in riparian environments (gallery forests), it is logical that Sosef's map (1994) does not include refuges toward the center of the Congo Basin, identified by Colyn's (1987, 1991) study of forest monkeys, which do occur in gallery forests.

The work of the ethnologist Serge Bahuchet (1993, esp. fig. 6) is also relevant to discussions about refuges. He concluded that the pygmies of Central Africa were affected by the arid phase that peaked around 18,000 B.P., because their tribes fall into three main groups that correspond to the principal refuges of (from west to east) south Cameroon/Gabon, the central Congo Basin, and the eastern Congo Basin.

## Refuges and Speciation

Although various data are consistent with the fragmentation of the forest block and the maintenance of refuges during the most arid periods, there is much debate about the role played by these successive isolations (in refuges) in the appearance of new species or subspecies. Two principal lines of thought emerge. One, expressed by a minority, more or less refutes the idea that speciation is linked to fragmentation and isolation of forest biotopes. The other, on the contrary, suggests that these isolations are the origin of a great number of taxa. Many publications have discussed these issues; here I mention some that represent all or part of the issue (Endler 1982; Haffer 1982, 1993; Mayr and O'Hara 1986; White 1993).

Mayr and O'Hara (1986) argue that modern biogeographical gaps in the African rain forest are not spread spatially (when the sampling unit is small enough) but are limited to restricted areas, such as the Cross River gap. They believe that this evidence refutes Endler's (1982) hypothesis that these isolations resulted only in parapatric speciation (in areas that were partially linked) and that either the diversity and the great number of ecological niches in forest environments (see Richards 1969) or ecotones between forest and savanna (trees: Aubréville 1949; birds: Smith et al. 1997) or between lowland and upland formations (Gentry 1989) would generate sufficient isolation for the emergence of new species. However, it seems that the two explanations—speciation in environments isolated ecologically (niches) and speciation in geographically isolated environments (vicariance)—especially under the effect of arid phases, are not incompatible and indeed seem to complement each other (see Haffer 1982; Blandin 1987; Gentry 1989; Smith et al. 1997).

Given the problems of dating the appearance of different forest taxa, it is difficult to come to any conclusions in this debate about speciation. Yet it is interesting to note that many studies have resulted in reliable hypotheses about the appearance of the main genera of leguminous plants, especially the Caesalpiniaceae. The fact that numerous genera and species of leguminous plants are common throughout the humid African forests, including the relict rain forests bordering the Indian Ocean (in the Zanzibar-Inhambane area), clearly shows that they were once linked in a large area of continuous distribution (Aubréville 1968; White 1993; Van Rompaey and Oldeman 1996). Polhill et al. (1981) suggest that numerous genera of Mimosoideae and Caesalpiniaceae, which live in the African rain forest, date back to the beginning of the Tertiary (Paleogene, from 65 to 23 million years ago). In fact, palynological studies have revealed that various present-day leguminous forest plants can be associated with taxa from the Eocene, Oligocene, and Miocene (Guinet et al. 1987; Salard-Cheboldaeff and Dejax 1991). This indicates that they persisted through the arid phases of the end of the Neogene and the Quaternary (Maley 1996a), although numerous taxa must have disappeared during these arid phases.

The high species richness of the African rain forest can be explained by, on one hand, the conservation of a large stock of genera and species going back to the beginning of the Tertiary or even, for some, to the separation of the continent of Gondwana during the upper Cretaceous. Persistence of the same forest blocks during the most arid phases would favor the conservation of very ancient taxa. On the other hand, the species richness could be explained by the later fragmentation of a large forest block that had its origins in the Paleogene (Maley 1996a).

This phenomenon of repeated fragmentation, with refuges always in the same location, could explain the numerous endemic species or genera. If one compares Cowan and Polhill's (1981) catalogue for the four main forest sectors with high biodiversity of Caesalpiniaceae with Van Rompaey's (1993, app. 1) list of large trees found in the forests of southeast Liberia and southwest Ivory Coast (the presumed zone of the western refuge), it appears that from a total of forty-three to forty-five Caesalpiniaceae, twenty-three to twenty-five (53–55%) are endemic. Because arid to very arid periods have occurred in equatorial latitudes since the end of the Neogene, it seems that at least some of these endemic taxa go back to that era. The greater prevalence of climatic cycles associated with phases of high aridity over the past 800,000 years could also have resulted in further evolution of new taxa.

It seems that for genetic (or for other still unknown) reasons, some taxa or groups of taxa develop faster than others at certain moments of their geological history (see, for example, the difference between the ammonites, a dynamic group, and the nautilus, a closely related group that has been very stable). Studies suggest this is the case for begonias and monkeys in the African forests. Indeed, Sosef (1994), for the begonias, and Colyn (1987, 1991), for the monkeys of the Congo Basin, came to the conclusion that new species and subspecies may have appeared in the refuges resulting from the most recent arid phase. Genetic research using modern techniques to test these ideas should be considered a priority.

## Fragmentation of the Tropical Rain Forests During Warm Interglacial Periods

### COMPARISON BETWEEN EQUATORIAL AFRICA AND AMAZONIA

Palynological and sedimentological studies carried out over the past decade on sediments from various lakes in the forest zones of Amazonia and Africa provide important evidence about the evolution of vegetation and the paleoenvironments during the upper Quaternary. A comparison of the evolution of these two forest blocks shows that between 15,000 and 20,000 B.P. a phase of high aridity occurred along with a synchronous regression of the forest in Africa (see earlier discussion) and Amazonia (Absy et al. 1991; Servant et al. 1993; Van der Hammen and Absy 1994). From about 15,000 years ago phases of recolonization began in both the Amazonian and African forest blocks; the forests in both regions are presumed to have totally reestablished at the same time, around the beginning of the Holocene (c. 9,500 B.P.; Sifeddine et al. 1994). After that, while the forest in Africa was maintained or expanding (Maley 1991, 1996a), there appears to have been a new regression of part of the Amazonian forest from about 8,000 B.P. This was especially marked at Carajas, the region at the

southeast periphery of the forest block, which at present is less humid than the rest of the forest (Absy et al. 1991; Servant et al. 1993) and where the regression reached its peak around 5,000 B.P. In Ecuador, however, in the more humid piedmont region in the Andes, the forest existed from 7,000 B.P. onward (the base of the core was studied by Bush and Colinvaux [1988]). The Amazonian forest around Carajas recolonized to reach limits near the present ones about 3,700 B.P., although there were some fluctuations (Sifeddine et al. 1994). Evidence from sediments has shown that the regression of the forest in the Carajas region was characterized by the development of *Piper*, a pioneer forest taxon, and a great abundance of fine carbonized plant debris, demonstrating that fire played a role in the creation of openings in the forest (Absy et al. 1991; Martin et al. 1993; Sifeddine et al. 1994).

Remarkably, at the time the Amazonian forest had again reached an extension close to the present-day one, at c. 3,700 B.P., rainfall declined abruptly in the western part of the African forest block, as shown by a dramatic 120 m regression of Lake Bosumtwi in Ghana (Talbot et al. 1984). The forest around this lake was not directly affected, but the peripheral forests retreated, especially in present-day Benin and Togo, forming a new gap (the Dahomey interval, or Dahomey Gap) in the forest strip north of the Gulf of Guinea (Maley 1991). In the central part of the forest block, pollen data from lakes in Cameroon (for Barombi Mbo, see Maley 1992; Giresse et al. 1994; for Mboandong, see Richards 1986; for Ossa, see Reynaud-Farrera et al. 1996; for Njupi, see Zogning et al. 1997), western Congo (Elenga et al. 1992, 1994; Schwartz 1992), southeastern Uganda (Taylor 1990), and the region north of Lake Tanganyika (Vincens 1993) demonstrate that a pronounced decline of forest pollen types occurred relative to those from open environments (mainly Gramineae), starting around 3,000–2,800 B.P. and peaking around

2,500–2,000 B.P. (Maley 1992, 1997; Maley and Brenac 1998a; see also figure 5.1).

In the eastern part of the Guineo-Congolian Domain, similar data have been obtained in Ituri, a forest sector in the northeast Congo Basin (T. Hart and J. Hart, pers. comm.; Hart et al. 1996). Identification and dating of charcoal (R. Deschamps, pers. comm.) from the upper part of the soil in the region of Epulu, located more than 100 km from the present forest-savanna boundary, has demonstrated that over the past 2,500 years there have been frequent fires, which have burned forest trees. Further-more, the charcoal of two trees typical of savanna environments was found to occur there between c. 2,500 and 2,000 B.P. (Hart et al. 1996).

Runge (1995, 1997) studied soil profiles between Osokari and Walikale along the road from Kisangani to Bukavu to the south of Epulu. Fossil tree trunks were found in the saprolite below the main "stone line" which gave radiocarbon dates of 17,000–18,000 B.P. By comparing these data with those obtained in the deep-sea fan (delta) of the Congo river (Giresse et al. 1982), one can conclude that this stone line probably formed during another major arid phase at the end of the Pleistocene, around 11,000 B.P. (Maley 1996b; Maley and Brenac 1998a). Runge (1995, 1997) described another detrital and coarse layer containing charcoal dated to 2,000 B.P., similar to a layer observed in western Congo at Kakamoeka in the Mayombe (Maley and Giresse 1998).

Pollen and sedimentological data from other parts of tropical Africa confirm that the afore-mentioned facts are not isolated, and demon-strate that a phase of generalized aridification affected the majority of the African tropics, starting around or a little after 3,000 B.P. and culminating between 2,500 and 2,000 B.P. This arid phase affected the savannas of the Sahelian zone (Chad and Senegal: Maley 1981; Lézine 1989) and the more southern woody savannas of Cameroon and Nigeria, and is associated with intensive erosion. The term "climatic pejoration" is often used to describe this arid phase, because it seems to have resulted from a more marked seasonality of rainfall, probably due to an increase in squall lines (large cumuli-form clouds) to the detriment of other types of rain (that is, mainly "monsoon" rains) associ-ated with a hot, evaporating atmosphere (see Maley 1981, 1982, 1997; Maley and Elenga 1993; Maley et al., in press). As a result of this increase in squall lines there was probably a reduction in the length of the rainy season but probably not in the amount of annual precipita-tion (see Maley and Brenac 1998a; Maley et al., in press). Such phenomena have been observed in certain recent years (for instance, in 1983) and have been linked to large variations of sea-sur-face temperatures in the Gulf of Guinea (Suchel 1988; Mahé and Citeau 1993; Maley and Elenga 1993; Maley 1997).

The available sedimentological and pollen data for the end of this climatic pejoration, about 2,000 years ago, show that a humid cli-mate re-established in tropical Africa and that rain forest vegetation recovered. However, this recovery was very slow relative to the speed at which the forest retreated between 2,800 and 2,000 B.P., and the recovery was associated with large fluctuations and setbacks (Maley and Brenac 1998a).

The intense erosion that occurred during the climatic pejoration resulted in the deposi-tion of coarse sediments constituting the base of a "low terrace" in valley bottoms (Unit 1 below). This terrace has been observed in the woody savannas of the Chad Basin as well as in the forest zone of south Cameroon, Gabon, Congo, and DR Congo (Maley 1981, 1992; Maley and Brenac 1998b). During the past two millennia there has been further accumulation in the low terrace, which can be divided into four units (for more details, see Maley and Brenac 1998b):

1. Basal unit consisting of very coarse sediments (c. 2,800–2,000 B.P.)
2. Fine sediments frequently containing fossil wood and charcoal (c. 2,000–1,200 B.P.)
3. More sandy sediments frequently eroding Unit 2 (c. 1,200–800 B.P.)
4. Fine to sandy sediments (c. 800 B.P. to recent centuries)

A new phase of erosion intervened in recent centuries which became particularly strong at the beginning of the twentieth century (Maley 1981).

## THE ROLE OF FIRE

The role played by fire, especially in the forest, must have been important during the climatic pejoration discussed above. The situation in Africa was probably similar to the the situation in Carajas, Amazonia, described earlier. At present, fires started by humans sweep the peripheral savannas but do not penetrate the forest edge (Swaine 1992). They are stopped by the dense, low vegetation (lianas, ferns, and shrubs) that occurs along the forest edge and remains green during the dry season (Spichiger and Pamard 1973). However, in unusually arid years, when the dry season is extended, fires can penetrate the peripheral semi-deciduous forests. Hall and Swaine (1981) and Swaine (1992) describe a "fire zone" in Ghana, where fires penetrate forests adjacent to savannas approximately every fifteen years. In 1983 there were many reports of an exceptional propagation of forest fires in the Ivory Coast (Bertault 1990), Ghana (Hawthorne 1991; Swaine 1992), and the Nyong valley of Cameroon (Amougou 1986). Contemporary examples of spontaneous fires under similar conditions occurred in Borneo during the very dry El Niño year of 1982–1983, when fires raged through normally fire-resistant humid evergreen forest (Goldammer and Seibert 1990). Nevertheless, the propagation of fires within the African for-

est block is an uncommon phenomenon, as is the case for natural fires started in the forest by lightning, which generally remain confined (for example, Tutin et al. [1994] observed a *Pterocarpus soyauxii*, 40 m high, which burned in a forest in central Gabon).

In the rare event when a fire affects part of the forest, forest regeneration is rapid (assuming that no other fire occurs during the following months or years; see Swaine 1992), with a characteristic phase when large monocotyledons belonging to the Marantaceae and Zingiberaceae dominate, smothering pioneer trees, which are much reduced in density (De Foresta 1990; White, this volume, chapter 11). This stage corresponds to "clear forests" (forests with open canopy and few mature trees), which have been described by various authors and which sometimes cover considerable areas, as in Lopé (in Gabon), northern Congo, and the extreme east of Cameroon (see Letouzey 1968, 1985; Maley 1990). It has been suggested that the development of these "Marantaceae forests" reflects the passage of forest fires (Swaine 1992). However, these clear forests also develop naturally during the recolonization of savannas by the forest, as in Lopé, Gabon (White 1992, 1995, and this volume, chapter 11) or in the Mayombe, Congo (De Foresta 1990). The large areas of peripheral savannas that presently surround the forest block in the north and the south, almost without interruption, and which are characterized by very low density of savanna trees (Letouzey 1968, 1985; White 1983), might be the consequence of successive retreats of the forest block during the climatic pejoration that occurred in the late Holocene, between c. 3,700 and 2,000 B.P. (Maley et al., in press). The frequent occurrence of fires during this period, especially between 2,500 and 2,000 B.P. and also later in eastern DR Congo (see above) and in central Gabon (see Oslisly, this volume), may have destroyed the young regenerating forest dominated by the Marantaceae, Zingiberaceae, and pioneer trees, and thus could explain the occurrence of grass savannas observed today, where typical savanna trees seem to have had difficulty becoming established because of the frequency of fires.

Whereas fire destroys forests more or less instantaneously, their recovery over vast areas is a slow process. Wherever accurate observations have been made at the periphery of the African forest, recolonization of savannas has been shown to be under way, particularly since the beginning of the twentieth century (Maley 1990). This phenomenon has surprised observers in that it takes place despite the annual propagation of fires started by humans, and, even more astonishing, traditional agricultural practices and livestock passages further this phenomenon by removing vegetation cover and hence limiting the propagation and the intensity of fires. These processes have been well documented over the past forty years in the V-Baoulé, Ivory Coast (see Miège 1966; Spichiger and Pamard 1973; Blanc-Pamard and Peltre 1984; Gautier 1990), and more recently in the Lopé Reserve (White 1995) and in Odzala at the northern end of the Bateke Plateau (Dowsett-Lemaire 1996). In Cameroon south of the Adamaoua, the process was described in detail by the highly experienced botanist Letouzey (1968, 1985). The phenomenon of forest colonization, which he observed over more than forty years, occurs mainly in fallow areas after abandonment of the fields by the farmers, without an intermediary stage of savanna vegetation, as has often been suggested. In effect, temporary agricultural occupation promotes forest development in certain regions. This phenomenon of afforestation of savannas can be seen clearly by comparing several sets of aerial photographs and satellite images and occurs over a large area of up to one million hectares south of Adamaoua, in the basins of the Kadei, Lom-Pangar-Djerem, and Ndjim-Mbam-Noun-Kim (Letouzey 1985; see

also Youta-Happi 1998). A comparable large forest extension also occurred in Central African Republic around Bangassou (TREES Project 1998). In conclusion, it appears that in the twentieth century the tropical forests still had a considerable dynamism. Thus it is reasonable to suggest that the extensive peripheral band of semi-deciduous forest in southern Cameroon has developed over the past two millennia as a result of establishment and development of pioneer forest formations.

## Humans and the Changes in the Environment

The possible role of humans in forest retreat or fragmentation through the Holocene must be considered (here I shall discuss only Africa). During the upper Quaternary, groups of pygmies seem to have wandered throughout the African forest, but because of their lifestyle and their low numbers, they leave almost no trace (Bahuchet 1993). In contrast, Bantu farmers have cultivation practices that modify vegetation locally (the slash-and-burn method), and they may leave various archaeological traces (stone or iron tools, or pottery). Therefore, it is important to know how long the forest block of central Africa has been inhabited by farmers, following its recovery at the beginning of the Holocene. In a synthesis of the archaeology of the central Congo Basin, Eggert (1992, 1993) states that the oldest pottery level, the Imbonga Horizon, occurred between approximately 2,500 and 2,000 B.P. Further to the west, in coastal regions of Gabon, remains of settlements associated with archaeological remains of the Neolithic kind appeared as early as 4,000–5,000 B.P. (Clist 1990; Oslisly and Peyrot 1992; Oslisly and Fontugne 1993). However, the first period of relatively high density of settlements in Gabon occurred later and corresponds with the introduction of iron technology, probably by Bantu, starting approximately 2,500 B.P. (Oslisly and Peyrot 1992; Oslisly and Fontugne 1993; see also Oslisly, this volume).

Schwartz (1992) reviewed archaeological evidence for the region comprising south Cameroon, Gabon, and the forests of western Congo. He found evidence that metallurgy spread to all the areas between Yaoundé and Pointe-Noire (1,000 km) between approximately 2,250 and 2,100 B.P. Such a rapid spread of this new technology over the course of one or two centuries seems incompatible with the presence of dense, uninterrupted forest cover (Schwartz 1992). Schwartz hypothesized that the rapid spread of Bantu peoples carrying iron would have been possible only if there had been multiple openings in the forest block as a result of paleoclimatic changes during the late Holocene. The climatic pejoration described above, which affected all of central Africa and which culminated between 2,500 and 2,000 B.P., corresponds well with the period of iron expansion and also with the first pottery level (the Imbonga Horizon) in the central Congo Basin (Eggert 1993). Thus, it can be assumed that the numerous openings and regressions of the forest caused by this unfavorable climatic phase, which was also associated with the frequent occurrence of forest fires, might have favored and even caused this migration of the Bantus. However, from approximately 2,000 B.P. onward, when the climate became humid again, the available data show that the forest recovered even though the Bantu continued to live in the African forest and to practice their agricultural techniques.

Several pollen analyses have demonstrated that the period spanning the last two to three millennia was also characterized by increased abundance of the oil palm (*Elaeis guineensis*) (Elenga et al. 1992; Vincens 1993; Reynaud-Farrera et al. 1996; Bengo 1996; White et al. 1996; Zogning et al 1997; Maley and Brenac 1998a, 1998b; Maley and Giresse 1998; Maley, in press; J. M. Fay, pers. comm.). The oil palm is an indigenous tree whose typical pollen has been identified in deposits dating to the Eocene in

Guinea/Konakry and then to the Miocene from Senegal to Nigeria and Uganda (Maley and Brenac 1998a, 1998b). During the Holocene a first large extension phase occurred between approximately 2,800 and 2,200 B.P., with a second culminating between approximately 1,200 and 700 B.P. These two main extension phases were associated with large disturbances within the rain forest initiated by climatic changes that were linked to erosive phenomena. The two extension phases were a result of the pioneer nature of the oil palm, which becomes abundant in the forest-savanna mosaic and in regenerating forests (Letouzey 1968; Swaine and Hall 1986). Although the oil palm is extensively used by African peoples, there is little or no tradition of cultivation of this species (Zeven 1967; Eggert 1993). Instead, bunches of palm nuts are collected from naturally occurring palm stands, which may be managed to some extent by clearing of the understory (see Haxaire 1996). While collecting bunches of palm fruits, humans accidentally drop seeds, hence contributing to the regeneration of the oil palm, which is always prolific, and to the welfare of various species of animals (chimpanzees, tauracos, parrots, rats, and so on; see Schnell 1952). The true domestication of the oil palm is a recent phenomenon (Zeven 1972). Hence, it seems safe to conclude that the traditional exploitation of the oil palm corresponds to a "proto-culture" closely resembling the hunter-gatherer society of modern pygmies. It seems safe to suggest that the natural periods of expansion of the oil palm over the past three millennia were particularly favorable for the human populations who later undertook the Bantu migration in west central Africa.

The aridity of the climate during a period of global cooling was the main reason for the large forest fragmentations between 15,000 and 20,000 B.P. The forest retreats and fragmentations that occurred in Amazonia and in Africa during different periods of the Holocene, a relatively warm period, are also of climatic origin, but in several areas the propagation of large fires must have speeded up this process. In the African forest block, the last extensive regression seems to have occurred between 2,500 and 2,000 B.P., when deep openings appeared particularly in areas with the highest climatic variability. It is possible that the fragmentation of the forest block in the Cameroon-Gabon-Congo region provoked some of the Bantu migrations. The mosaic aspect of the present-day forests of Congolian type (characterized by a mixture of evergreen and semi-deciduous forests) and even of some evergreen Biafran formations, like those in Cameroon (see Letouzey 1985; Maley 1990), probably date back to the late Holocene and may be linked to these phenomena. Climatic trends that were out of phase in Africa and Amazonia during the Holocene were linked to fluctuations of sea-surface temperatures (Maley 1989, 1991, 1995, 1997; Fontaine and Bigot 1993), to the zonal atmospheric circulation of Walker type (see Flohn 1984, 1986), and especially to phenomena of the El Niño type in the Pacific and tropical Atlantic Oceans (Martin et al. 1993; Maley 1997; Maley et al., in press).

Climatological models of global warming during the twenty-first century (see Stager, this volume) predict that a 4°C rise in global air temperature would result in a 30% increase in evaporation with only 12% more rain in inland parts of tropical Africa (Rind 1995). This could lead to the catastrophic retreat and destruction of the African rain forest, as occurred between 2,800 and 2,000 B.P., during a warm climatic phase that is thought to have resulted in a large increase in convective rainfall, particularly in equatorial Africa (see the earlier discussion and Maley 1997; Maley et al., in press).

The devastation of vast areas of dense forest in Indonesia and Amazonia during the summer of 1998 are thought largely to be the result of a very strong El Niño climatic event result-ing from unusually warm sea-surface temperatures. A similar strong El Niño event occurred in 1983. Unusual climatic conditions associated with these two El Niño events resulted in reduced rainfall and, in particular, an extended dry season, which allowed savanna fires to penetrate deeply into the forest, causing considerable damage. Such savanna fires are generally blocked at the forest edge by a community of pioneer and lianescent plant species. In unusually dry years these plants perish, and in some places fires penetrate the forest edge and burn into the dry leaf litter in the forest understory. Observations in Indonesia and in Ghana (see Hawthorne 1991; Swaine 1992) have shown that areas that have been subject to commercial logging are particularly susceptible to fire at these times. It is therefore a priority for conservationists and forest managers alike to undertake further studies of the fire risk in logged forests and to evaluate the possibility of leaving unlogged areas to act as natural firebreaks within blocks of logged forest (see also the recommendations in Stager, this volume).

ACKNOWLEDGMENTS

Several data sets and conclusions used in this chapter were obtained within the scope of various programs conducted with ORSTOM/IRD (Institut Français de Recherche pour le Développement) and since 1992 with both the ECOFIT program funded by CNRS and ORSTOM and the CAMPUS/Cameroon program funded by the Ministry of Cooperation, Paris. The author has also benefited from discussions with Wildlife Conservation Society scientists: Terese and John Hart (Ituri, DR Congo) and Lee White (La Lopé, Gabon). This is Institut des Sciences de l'Evolution de Montpellier (ISEM/CNRS-UMR-5554) contribution 95-043.

REFERENCES

Absy, M. L., A. L. Cleef, M. Fournier, L. Martin, M. Servant, A. Sifeddine, M. F. Da Silva, F. Soubies, K. Suguio, B. Turcq, and T. Van der Hammen. 1991. Mise en évidence de quatre phases d'ouverture de la forêt dense dans le sud-est de l'Amazonie au cours des 60.000

dernières années. *Comptes Rendu Académie Sciences*, Paris, série 2, 312:673–678.

Amiet, J. L. 1987. Aires disjointes et taxons vicariants chez les Anoures du Cameroun: Implications paléoclimatiques. *Alytes* 6:99–115.

Amougou, A. 1986. Etude botanique et écologique de la vallée inondable du Haut-Nyong et ses affluents. Ph.D. thesis, University of Yaoundé, Cameroon.

Aubréville, A. 1949. *Contribution à la paléohistoire des forêts de l'Afrique tropicale*. Société Edition Géographique, Maritime et Coloniale, Paris.

———. 1962. Savanisation tropicale et glaciations quaternaires. *Adansonia* 2:16–84.

———. 1967. Les étranges mosaiques forêt-savane du sommet de la boucle de l'Ogooué au Gabon. *Adansonia* 7:13–22.

———. 1968. Les Césalpinioidées de la flore Cameroun-Congolaise: Considérations taxinomiques, chorologiques, écologiques, historiques et évolutives. *Adansonia* 8:147–175.

Bahuchet, S. 1993. History of the inhabitants of the central African rain forest: Perspectives from comparative linguistics. Pages 37–54 in C. Hladik, A. Hladik, O. Linares, H. Pagezy, A. Semple, and M. Hadley, eds. *L'alimentation en forêt tropicale: Interactions bioculturelles et applications au développement*. Parthenon and UNESCO, Paris.

Bengo, M. D. 1996. La sédimentation pollinique dans le sud Cameroun et sur la plateforme marine, à l'époque actuelle et au Quaternaire récent: Etudes des paléoenvironnements. Ph.D. thesis, Univ. Montpellier-2.

Bertault, J. G. 1990. Comparaison d'écosystèmes forestiers naturels et modifiés après incendie en Côte d'Ivoire. Page 6 in H. Puig, ed. *Atelier sur l'aménagement et la conservation de l'écosystème forestier tropical humide*. Séminaire de Cayenne, UNESCO, Paris.

Blanc-Pamard, C., and P. Peltre. 1984. Dynamique des paysages préforestiers et pratiques culturales en Afrique de l'Ouest (Côte d'Ivoire Centrale). *Mémoire ORSTOM, Le développement rural en question* 106:55–74.

Blandin, P. 1987. Evolution des écosystèmes et spéciation: Le rôle des cycles climatiques. *Bulletin d'Ecologie, Paris* 18:59–61.

Bonifay, D., and P. Giresse. 1992. Middle to late Quaternary sediment flux and post-depositional processes between the continental slope off Gabon and the mid-Guinean margin. *Marine Geology* 106:107–129.

Bonnefille, R. 1983. Evidence for a cooler and drier climate in the Ethiopian uplands towards 2.5 Myr ago. *Nature* 303:487–491.

Bonnefille, R., F. Chalie, J. Guiot, and A. Vincens. 1992. Quantitative estimates of full glacial temperatures in equatorial Africa from palynological data. *Climate Dynamics* 6:251–257.

Bonnefille, R., and G. Riollet. 1988. The Kashiru pollen sequence (Burundi): Palaeoclimatic implications for the last 40,000 yr B.P. in tropical Africa. *Quaternary Research* 30:19–35.

Brenac, P. 1988. Evolution de la végétation et du climat dans l'Ouest Cameroun entre 25.000 et 11.000 ans BP. Actes Xème Symposium Association des Palynologues Langue Française, *Travaux Section Sciences and Tech. Institut Français Pondichéry* 25:91–103.

Brenan, J. P. M. 1978. Some aspects of the phytogeography of tropical Africa. *Annals of the Missouri Botanical Garden* 65:437–478.

Breteler, F. J. 1990. Gabon's evergreen forest: The present status and its future. *Mitteilungen Institut Allg. Botanik Hamburg* 23:219–224.

Bush, M. B., and P. A. Colinvaux. 1988. A 7000-year pollen record from the Amazon lowlands, Ecuador. *Vegetatio* 76:141–154.

Carcasson, R. H. 1964. A preliminary survey of the zoogeography of African butterflies. *East African Wild Life J.* 2:122–157.

Clist, B. 1990. Des derniers chasseurs aux premiers métallurgistes: Sédentarisation et débuts de la métallurgie du fer (Cameroun, Gabon, Guinée Equatoriale). Pages 458–478 in R. Lanfranchi and D. Schwartz, eds. *Paysages Quaternaires de l'Afrique Centrale Atlantique*. Didactiques, ORSTOM, Paris.

———. 1995. *Gabon: 100 000 ans d'histoire*. Centre Culturel Français de Libreville and Sepia Publ., Paris.

Colinvaux, P. A., P. E. De Oliveira, J. E. Moreno, M. C. Miller, and M. B. Bush. 1996. A long pollen record from lowland Amazonia: Forest and cooling in Glacial times. *Science* 274:85–88.

Colyn, M. M. 1987. Les primates des forêts ombrophiles de la cuvette du Zaïre: Interprétations zoogéographiques des modèles de distribution. *Revue Zoologie Africaine* 101:183–196.

———. 1991. L'importance zoogéographique du bassin du fleuve Zaïre pour la spéciation: Le cas des Primates Simiens. *Annales Musée Royal Afrique Centrale, Sciences Zoologiques, Tervuren*, 264:1–250.

Colyn, M. M., A. Gautier-Hion, and W. Verheyen. 1991. A re-appraisal of palaeoenvironmental history in Central Africa: Evidence for a major fluvial refuge in the Zaire basin. *Journal of Biogeography* 18:403–407.

Cowan, R. S., and R. M. Polhill. 1981. Taxonomic part. Detarieae and Amherstieae. Pages 117–142 in R. M. Polhill and P. H. Raven, eds. *Advances in Legume Systematics.* Royal Botanic Gardens, Kew.

Crowe, T. M., and A. A. Crowe. 1982. Patterns of distribution, diversity and endemism in Afrotropical birds. *Journal of Zoology, London* 198:417–442.

De Foresta, H. 1990. Origine et évolution des savanes intramayombiennes (Congo), II: Apports de la botanique forestière. Pages 326–335 in R. Lanfranchi and D. Schwartz, eds. *Paysages Quaternaires de l'Afrique Centrale Atlantique.* Didactiques, ORSTOM, Paris.

Dowsett-Lemaire, F. 1996. Composition et évolution de la végétation forestière au Parc National d'Odzala, Congo. *Bulletin Jardin Botanique National de Belgique* 65:253–292.

Eggert, M. K. 1992. The Central African rain forest: Historical speculation and archaeological facts. *World Archaeology* (special issue on the humid tropics) 24:1–24.

———. 1993. Central Africa and the archaeology of the equatorial rain forest: Reflections on some major topics. Pages 289–329 in T. Shaw, P. Sinclair, B. Andah, and A. Okpoko, eds. *The Archaeology of Africa: Food, Metals and Towns.* Routledge, London.

Elenga, H., D. Schwartz, and A. Vincens. 1992. Changements climatiques et action anthropique sur le littoral congolais au cours de l'Holocène. *Bulletin Société Géologique de France* 163:83–90.

———. 1994. Pollen evidence of late Pleistocene-Holocene vegetation and climate changes in West Equatorial Africa: Analysis of continental sequences of Bateke Plateaux (Congo). *Palaeogeography, Palaeoclimatology and Palaeoecology* 109:345–356.

Elenga, H., A. Vincens, and D. Schwartz. 1991. Présence d'éléments forestiers montagnards sur les Plateaux Batéké (Congo) au Pléistocène supérieur: Nouvelles données palynologiques. *Palaeoecology of Africa* 22:239–252.

Endler, J. A. 1982. Pleistocene forest refuges: Fact or fancy. Pages 641–657 in G. T. Prance, ed. *Biological Diversification in the Tropics.* Columbia University Press, New York.

Flohn, H. 1984. Tropical rainfall anomalies and climatic change. *Bonner Meteo. Abhandlungen* 31:107.

———. 1986. Singular events and catastrophes now and in climatic history. *Naturwissenschaften* 73:136–149.

Fontaine, B., and S. Bigot. 1993. West African rainfall deficit and sea surface temperatures. *International Journal of Climatology* 13:271–285.

Gautier, L. 1990. Contact forêt-savane en Côte d'Ivoire centrale: Evolution du recouvrement ligneux des savanes de la Réserve de Lamto (sud du V-Baoulé). *Candollea* 45:627–641.

Gentry, A. H. 1989. Speciation in tropical forests. Pages 113–134 in L. B. Holm-Nielsen, I. C. Nielsen, and H. Balslev, eds. *Tropical Forests, Botanical Dynamics, Speciation and Diversity.* Academic Press, London.

Giresse, P., G. Bongo-Passi, G. Delibrias, and J. C. Duplessis. 1982. La lithostratigraphie des sédiments hémipélagiques du delta du fleuve Congo et ses indications sur les paléoclimats de la fin du Quaternaire. *Bulletin Société Géologique de France* 24:803–815.

Giresse, P., J. Maley, and P. Brenac. 1994. Late Quaternary palaeoenvironments in the lake Barombi Mbo (Cameroon) deduced from pollen and carbon isotopes of organic matter. *Palaeogeography, Palaeoclimatology and Palaeoecology* 107:65–78.

Giresse, P., J. Maley, and K. Kelts. 1991. Sedimentation and palaeoenvironment in crater Lake Barombi Mbo, Cameroon, during the last 25,000 years. *Sedimentary Geology* 71:151–175.

Goldammer, J. G., and B. Seibert. 1990. The impact of droughts and forest fires on tropical lowland rain forest of East Kalimatan. In J. G. Goldammer, ed. *Fire in the Tropical Biota.* Ecological Studies 84. Springer-Verlag, Berlin.

Grubb, P. 1982. Refuges and dispersal in the speciation of African mammals. Pages 537–553 in G. T. Prance, ed. *Biological Diversification in the Tropics.* Columbia University Press, New York.

Guillaumet, J. L. 1967. Recherches sur la végétation et la flore de la région du Bas-Cavally (Côte d'Ivoire). *Mémoire ORSTOM* 20

Guinet, P., N. El Sabrouty, H. A. Soliman, and A. M. Omran. 1987. Etude des caractères du pollen des Légumineuses-Mimosoideae des sédiments Tertiaires du nord-ouest de l'Egypte. *Mémoire Travaux Ecole Pratique Hautes Etudes, Institut de Montpellier* 17:159–171.

Haffer, J. 1982. General aspects of the Refuge theory. Pages 641–657 in G. T. Prance, ed. *Biological Diversification in the Tropics.* Columbia University Press, New York.

———. 1993. Time's cycle and time's arrow in the history of Amazonia. *Biogeographica* 69:15–45.

Hall, J. B. 1981. Ecological islands in south-eastern Nigeria. *African Journal of Ecology* 19:55–72.

Hall, J. B., and M. D. Swaine. 1981. *Distribution and Ecology of Vascular Plants in a Tropical Rain Forest: Forest Vegetation in Ghana*. W. Junk, The Hague.

Hamilton, A. C. 1976. The significance of patterns of distribution shown by forest plants and animals in tropical Africa for the reconstruction of upper Pleistocene palaeoenvironments: A review. *Palaeoecology of Africa* 9:63–97.

Haq, B. U., J. Hardenbol, and R. P. Vail. 1987. Chronology of fluctuating sea levels since the Triassic. *Science* 235:1156–1167.

Harris, J. 1993. Ecosystem structure and growth of the African savanna. *Global and Planetary Change* 8:231–248.

Hart, T. B., J. A. Hart, R. Dechamps, M. Fournier, and M. Ataholo. 1996. Changes in forest composition over the last 4000 years in the Ituri basin, Zaire. Pages 545–563 in L. J. G. Van der Maesen, X. M. Van der Burght, and J. M. Van Medenbach de Rooy, eds. *The Biodiversity of African Plants*. Kluwer, Dordrecht.

Hawthorne, W. D. 1991. Fire damage and forest regeneration in Ghana. ODA, London.

Haxaire, C. 1996. Le vin de palme et la noix de kola: Nourritures paradoxales, médiateurs de la communication avec les dieux. Pages 923–938 in M. Hladik, A. Hladik, O. Linares, H. Pagezy, A. Semple, and M. Hadley, eds. *L'alimentation en forêt tropicale: Interactions bioculturelles et applications au développement*. Parthenon and UNESCO, Paris.

Hooghiemstra, H., and T. Van Der Hammen. 1998. Neogene and Quaternary development of the Neotropical rain forest: The forest refugia hypothesis, and a literature overview. *Earth-Science Rev.* 44:147–183.

Jansen, J. H. F., A. Kuijpers, and S. R. Troelstra. 1986. A mid-Brunhes climatic event: Long-term changes in global atmosphere and ocean circulation. *Science* 232:619–622.

Jansen, J. H. F., T. C. Van Weering, R. Gieles, and J. Van Iperen. 1984. Middle and late Quaternary oceanography and climatology of the Zaire-Congo fan and the adjacent eastern Angola basin. *Netherlands Journal of Sea Research* 17:201–248.

Kingdon, J. S. 1980. The role of visual signals and face patterns in African forest monkeys of the genus Cercopithecus. *Transactions of the Zoological Society of London* 35:425–475.

Lachaise, D., M. L. Cariou, J. R. David, F. Lemeunier, L. Tsacas, and M. Ashburner. 1988. Historical biogeography of the *Drosophila melanogaster* species subgroup. *Evolutionary Biology* 22:159–225.

Lanfranchi, R., and D. Schwartz. 1991. Les remaniements de sols pendant le Quaternaire supérieur au Congo: Evolution des paysages dans la région de la Sangha. *Cahier ORSTOM, série Pédologie* 26:11–24.

Léonard, J. 1965. Contribution à la subdivision phytogéographique de la région guinéo-congolaise d'après la répartition géographique d'Euphorbiacées d'Afrique tropicale. *Webbia* 19:627–649.

Letouzey, R. 1968. Etude phytogéographique du Cameroun. *Encyclopédie Biologique*, Paris.

———. 1983. Quelques exemples camerounais de liaison possible entre phénomènes géologiques et végétation. *Bothalia* 14:739–744.

———. 1985. *Notice de la Carte Phytogéographique du Cameroun au 1/500.000*. Inst. Carte Internationale Végétation, Toulouse, and Inst. Recherche Agronomique, Yaoundé.

Lézine, A. M. 1989. Late Quaternary vegetation and climate of the Sahel. *Quaternary Research* 32:317–334.

Mahé, G., and J. Citeau. 1993. Interactions between the ocean, atmosphere and continent in Africa, related to the Atlantic monsoon flow: General pattern and the 1984 case study. *Veille Climatique Satellitaire* 44:34–54.

Maley, J. 1980. Les changements climatiques de la fin du Tertiaire en Afrique: Leur conséquence sur l'apparition du Sahara et de sa végétation. Pages 63–86 in M. A. J. Williams and H. Faure, eds. *The Sahara and the Nile*. Balkema, Rotterdam.

———. 1981. Etudes palynologiques dans le bassin du Tchad et paléoclimatologie de l'Afrique nord-tropicale de 30.000 ans à l'époque actuelle. *Travaux and Documents ORSTOM* 129.

———. 1982. Dust, clouds, rain types and climatic variations in tropical North Africa. *Quaternary Research* 18:1–16.

———. 1987. Fragmentation de la forêt dense humide africaine et extension des biotopes montagnards au Quaternaire récent: Nouvelles données polliniques et chronologiques: Implications paléoclimatiques et biogéographiques. *Palaeoecology of Africa* 18:307–334.

———. 1989. Late Quaternary climatic changes in the African rain forest: The question of forest refuges and the major role of sea surface temperature variations. Pages 585–616 in M. Leinen and M. Sarnthein, eds. *Paleoclimatology and Paleometeorology: Modern and Past Patterns of Global Atmospheric Transport*. NATO Adv. Sc. Inst. Ser. C, Math. and Phys. Sc., Kluwer, Dordrecht.

———. 1990. Histoire récente de la forêt dense humide Africaine et dynamisme actuel de quelques formations forestières. Pages 367–382 in R. Lanfranchi et D. Schwartz, eds. *Paysages Quaternaires de l'Afrique Centrale Atlantique*. Didactiques, ORSTOM, Paris.

———. 1991. The African rain forest vegetation and palaeoenvironments during late Quaternary. *Climatic Change* 19:79–98.

———. 1992. Mise en évidence d'une péjoration climatique entre ca. 2500 et 2000 ans BP en Afrique tropicale humide. *Bulletin Société Géologique France* 163:363–365.

———. 1993. The climatic and vegetational history of the equatorial regions of Africa during the Upper Quaternary. Pages 43–52 in T. Shaw, P. Sinclair, B. Andah, and A. Okpoko, eds. *The Archaeology of Africa: Food, Metals and Towns*. Routledge, London.

———. 1995. Holocene changes in the African rain forest: Paleomonsoon and sea surface temperature variations. In *14th Intern. Quaternary Congress*. Berlin.

———. 1996a. The African rain forest: Main characteristics of changes in vegetation and climate from the upper Cretaceous to the Quaternary. Pages 31–73 in I. J. Alexander, M. D. Swaine, and R. Watling, eds. *Essays on the Ecology of the Guineo-Congo Rain Forest*. Proceedings Royal Society of Edinburgh B 104.

———. 1996b. Fluctuations majeures de la forêt dense humide Africaine au cours des vingt derniers millénaires. Pages 55–76 in C. M. Hladik, A. Hladik, O. Linares, H. Pagezy, A. Semple, and M. Hadley, eds. *L'alimentation en forêt tropicale: Interactions Bioculturelles et Perspectives de Développement*. Parthenon and UNESCO, Paris.

———. 1997. Middle to Late Holocene changes in tropical Africa and other continents: Paleomonsoon and sea surface temperature variations. Pages 611–640 in H. N. Dalfes, G. Kukla, and H. Weiss, eds. *Third millennium BC Climate Change and Old World Collapse*. NATO Adv. Sc. Inst. Series., vol. 40, Springer-Verlag, Berlin.

———. In press. L'expansion du Palmier à huile (*Elaeis guineensis*) en Afrique centrale au cours des 3 derniers millenaires: Nouvelles données et interpretations. In H. Pagezy, ed. *L'homme et la forêt tropicale*. Travaux Société Ecologie Humaine, Bergier, Paris.

Maley, J., and P. Brenac. 1998a. Vegetation dynamics, palaeoenvironments and climatic changes in the forests of West Cameroon during the last 28,000 years BP. *Review of Palaeobotany and Palynology*, 99:157–187.

———. 1998b. Les variations de la végétation et des paléoenvironnements du Sud Cameroun au cours des derniers millénaires: Etude de l'expansion du palmier à huile. Pages 85–97 in J. P. Vicat and P. Bilong, eds. *Géo-*

*sciences au Cameroun*. GEOCAM, 1, Presses Univ. Yaoundé.

Maley, J., P. Brenac, S. Bigot, and V. Moron. In press. Variations de la végétation et des paléoenvironnements en forêt dense africaine au cours de l'Holocène: Impact de la variation des températures marines. In *Dynamique à long terme des écosystemes forestiers intertropicaux*. Symposium ECOFIT, Publ. UNESCO, Paris.

Maley, J., G. Caballé, and P. Sita. 1990. Etude d'un peuplement résiduel à basse altitude de *Podocarpus latifolius* sur le flanc congolais du Massif du Chaillu: Implications paléoclimatiques et biogéographiques. Etude de la pluie pollinique actuelle. Pages 336–352 in R. Lanfranchi and D. Schwartz, eds. *Paysages Quaternaires de l'Afrique Centrale Atlantique*. Didactiques, ORSTOM, Paris.

Maley, J., and H. Elenga. 1993. The role of clouds in the evolution of tropical African mountain paleoenvironments. *Veille Climatique Satellitaire* 46:51–63.

Maley, J., and P. Giresse. 1998. Etude d'un niveau argileux organique du Mayombe (Congo occidental) riche en pollens d'*Elaeis guineensis* et daté d'environ 2800 ans BP: Implications pour les paléoenvironnements de l'Afrique Centrale. Pages 77–84 in J. P. Vicat and P. Bilong, eds. *Géosciences au Cameroun*. GEOCAM, 1, Presses Univ. Yaoundé.

Maley, J., and D. A. Livingstone. 1983. Extension d'un élément montagnard dans le sud du Ghana (Afrique de l'Ouest) au Pléistocene supérieur et à l'Holocène inférieur: Premières données polliniques. *Compte Rendu Académie des Sciences, Paris*, série 2, 296:1287–1292.

Martin, L., M. Fournier, P. Mourguiart, A. Sifeddine, B. Turcq, M. L. Absy, and J. M. Flexor. 1993. Southern oscillation signal in South American palaeoclimatic data of the last 7000 years. *Quaternary Research* 39:338–346.

Mayr, E., and R. J. O'Hara. 1986. The biogeographic evidence supporting the Pleistocene forest refuge hypothesis. *Evolution* 40:55–67.

Miège, J. 1966. Observations sur les fluctuations des limites savanes-forêts en basse Côte d'Ivoire. *Annales Faculté des Sciences*, Dakar 19:149–166.

Moreau, R. E. 1963. The distribution of tropical African birds as an indicator of past climatic changes. Pages 28–42 in F. C. Howel and F. Bourlière, eds. *African Ecology and Human Evolution*. Aldine Press, Chicago.

———. 1966. *The Bird Faunas of Africa and Its Islands*. Academic Press, New York.

———. 1969. Climatic changes and the distribution of forest vertebrates in West Africa. *Journal of Zoology*, London 158:39–61.

Morley, R. J., and K. Richards. 1993. Gramineae cuticle: A key indicator of Late Cenozoic climatic change in the Niger delta. *Review Palaeobotany and Palynology* 77:119–127.

Ndjélé, M. 1988. Principales distributions obtenues par l'analyse factorielle des éléments phytogéographiques présumés endémiques dans la flore du Zaire. *Monography of Systematic Botany, Missouri Botanical Garden* 25:631–638.

Nepstad, D. C., A. Verissimo, A. Alencar, C. Nobre, and L. Eirivelthon. 1999. Large-scale impoverishment of Amazonian forests by logging and fire. *Nature* 398:505–508.

Oslisly, R. 1993. Préhistoire de la moyenne vallée de l'Ogooué (Gabon). *Travaux et Documents ORSTOM* 96.

———. 1994. The middle Ogooué valley: Cultural changes and palaeoclimatic implications of the last four millennia. *Azania* 29:324–331.

Oslisly, R., and M. Fontugne. 1993. La fin du stade néolithique et le début de l'âge du fer dans la moyenne vallée de l'Ogooué au Gabon: Problèmes chronologiques et changements culturels. *Compte Rendu Académie des Sciences, Paris*, série 2, 316:997–1003.

Oslisly, R., and B. Peyrot. 1992. L'arrivée des premiers métallurgistes sur l'Ogooué, Gabon. *African Archaeological Review* 10:129–138.

Polhill, R. M., P. H. Raven, and C. H. Stirton. 1981. Evolution and systematics of the Leguminosae. Pages 1–34 in R. M. Polhill and P. H. Raven, eds. *Advances in Legume Systematics*, pt. 1. Royal Botanical Gardens, Kew.

Poumot, C. 1989. Palynological evidence for eustatic events in the tropical Neogene. *Bulletin Centre Recherche Exploration and Production Elf-Aquitaine* 13:437–453.

Prell, W. L., J. V. Gardner, A. W. H. Be, and J. D. Hays. 1976. Equatorial Atlantic and Caribbean foraminiferal assemblages, temperatures and circulation: Interglacial and glacial comparisons. *Geological Society of America Memoir* 145:247–266.

Reitsma, J. M., A. M. Louis, and J. J. Floret. 1992. Flore et végétation des inselbergs et dalles rocheuses: Première étude au Gabon. Bulletin Museum National d'Histoire Naturelle. *Adansonia* 14:73–97.

Reynaud, I., and J. Maley. 1994. Histoire récente d'une formation forestière du sud-ouest Cameroun à partir de l'analyse pollinique. *Compte Rendu Académie des Sciences, Paris*, série 2, 317:575–580.

Reynaud-Farrera, I., J. Maley, and D. Wirrmann. 1996. Végétation et climat dans les forêts du sud-ouest Cameroun depuis 4770 ans BP: Analyse pollinique des sédiments du lac Ossa. *Compte Rendu Académie des Sciences, Paris*, série 2, 322:749–755.

Richards, K. 1986. Preliminary results of pollen analysis of a 6,000 year core from Mboandong, a crater lake in Cameroun. *Hull University Geography Department Miscellaneous Series* 32:14–28.

Richards, P. W. 1969. Speciation in the tropical rain forest and the concept of the niche. *Biological Journal of the Linnean Society* 1:149–153.

Rietkerk, M., P. Ketner, and J. J. F. E. De Wilde. 1995. Caesalpinioideae and the study of forest refuges in Gabon: Preliminary results. Bulletin du Museum National d'Histoire Naturelle, Paris, *Adansonia* 17:95–105.

Rind, D. 1995. Drying out the tropics. *New Scientist* 5:36–40.

Ritchie, J. C. 1986. Climate change and vegetation response. *Vegetatio* 67:65–74.

Robbins, C. B. 1978. The Dahomey Gap: A reevaluation of its significance as a faunal barrier to West African high forest mammals. *Bulletin of the Carnegie Museum of Natural History* 6:168–174.

Ruddiman, W. F., M. Sarnthein, J. Backman, J. G. Baldauf, W. Curry, L. M. Dupont, T. Janececk, E. M. Pokras, M. E. Raymo, B. Stabell, R. Stein, and R. Tiedemann. 1989. Late Miocene to Pleistocene evolution of climate in Africa and the low-latitude Atlantic: Overview of Leg 108 results. *Proceedings Ocean Drilling Program, Scientific Results* 108:463–484.

Runge, J. 1995. New results on Late Quaternary landscape and vegetation dynamics in eastern Zaire (Central Africa). *Zeitschrift Geomorphologie, Suppl. Bd.* 99:65–74.

———. 1997. Altersstellung und paläoklimatische Interpretation von Decksedimenten, Steinlagen (Stone Lines) und Verwitterungs-bildungen in Ost-Zaire. *Geo-Öko-Dynamik* 18:91–108.

Salard-Cheboldaeff, M., and J. Dejax. 1991. Evidence of Cretaceous to recent West African intertropical vegetation from continental sediment spore-pollen analysis. *Journal of African Earth Science* 12:353–361.

Schnell, R. 1946. Note sur le palmier à huile, sa répartition et sa dissémination dans la région forestière. *Notes Africaines, IFAN,* Dakar 31:30–31.

———. 1952. Végétation et flore de la région montagneuse du Nimba (A.O.F.). *Mémoire Institut Français d'Afrique Noire* 22.

————. 1976. *Introduction à la phytogéographie des pays tropicaux, 3: La flore et la végétation de l'Afrique tropicale.* Gauthier-Villars, Paris.

Schwartz, D. 1992. Assèchement climatique vers 3000 B.P. et expansion Bantu en Afrique Centrale Atlantique: Quelques réflexions. *Bulletin Société Géologique de France* 163:353–361.

Semmanda, I., and A. Vincens. 1993. Végétation et climat dans le bassin du lac Albert (Ouganda, Zaïre) depuis 13.000 ans B.P: Apport de la palynologie. *Compte Rendu Académie des Sciences,* Paris, série 2, 316:561–567.

Servant, M., J. Maley, B. Turcq, M. L. Absy, P. Brenac, M. Fournier, and M. P. Ledru. 1993. Tropical forest changes during the late Quaternary in African and South American lowlands. *Global and Planetary Change* 7:25–40.

Sifeddine, A., F. Frohlich, M. Fournier, L. Martin, M. Servant, F. Soubies, B. Turcq, K. Suguio, and C. Volkmer-Ribeiro. 1994. La sédimentation lacustre indicateur de changements des paléoenvironnements au cours des 30.000 dernières années (Carajas, Amazonie, Brésil). *Compte Rendu Académie des Sciences,* Paris, série 2, 318:1645–1652.

Smith, T. B, R. K. Wayne, D. J. Girman, and M. W. Bruford. 1997. A role for ecotones in generating rainforest biodiversity. *Science* 276:1855–1857.

Sosef, M. S. M. 1991. New species of begonia in Africa and their relevance to the study of glacial rain forest refuges. *Wageningen Agriculture University Papers, Studies in Begoniaceae* 4:120–151.

————. 1994. Refuge begonias: Taxonomy, phylogeny and historical biogeography of *Begonia* sect. *Loasibegonia* and sect. *Scutobegonia* in relation to glacial rain forest refuges in Africa. *Wageningen Agriculture University Papers, Studies in Begoniaceae* 5.

Sowunmi, M. A. 1991. Late Quaternary environments in equatorial Africa: Palynological evidence (Lake Mobutu Sese Seko). Pages 213–238 in A. Ballouche and J. Maley, eds. *Proceed. First Symp. African Palynology: Palaeoecology of Africa* 22.

Spichiger, R., and C. Pamard. 1973. Recherches sur le contact forêt-savane en Côte d'Ivoire: Etude du recru forestier sur des parcelles cultivées en lisière d'un îlot forestier dans le sud du pays Baoulé. *Candollea* 28:21–37.

Start, G. G., and W. L. Prell. 1984. Evidence for two Pleistocene climatic modes: Data from DSDP site 502. New perspectives in climate modelling. Pages 3–22 in A. L. Berger and C. Nicolis, eds. *Developments in Atmospheric Science.* Elsevier, London.

Suchel, J. B. 1988. Les anomalies climatiques des années 1983 et 1984 au Cameroun. *Publication Association Internationale de Climatologie, Université d'Aix en Provence* 1:131–142.

Swaine, M. D. 1992. Characteristics of dry forest in West Africa and the influence of fire. *Journal of Vegetation Science* 3:365–374.

Swaine, M. D., and J. B. Hall. 1986. Forest structure and dynamics. Pages 47–93 in G. W. Lawson, ed. *Plant Ecology in West Africa.* J. Wiley, Chichester.

Talbot, M. R., D. A. Livingstone, P. G. Palmer, J. Maley, J. M. Melack, G. Delibrias, and S. Gulliksen. 1984. Preliminary results from sediment cores from Lake Bosumtwi, Ghana. *Palaeoecology of Africa* 16:173–192.

Taylor, D. M. 1990. Late Quaternary pollen records from two Ugandan mires: Evidence for environmental change in the Rukiga Highlands of southwest Uganda. *Palaeogeography, Palaeoclimatology and Palaeoecology* 80:283–300.

TREES Project. 1998. Vegetation map of central Africa. European Joint Research Center, Ispra, Italy.

Tutin, C. E., L. J. White, A. Mackanga-Missandzou, and M. Fernandez-Puente. 1994. Coup de foudre fatal pour un seigneur de la forêt: Observation d'un phénomène naturel au coeur de la Réserve de la Lopé, au Gabon. *Canopée, Bulletin ECOFAC, Programme Européen, Libreville* 1:9.

Van der Hammen, T., and M. L. Absy. 1994. Amazonia during the last glacial. *Palaeogeography, Palaeoclimatology and Palaeoecology* 109:247–261.

Van Rompaey, R. S. 1993. Forest gradients in West Africa: A spatial gradient analysis. Ph.D. thesis, Wageningen Agriculture University.

Van Rompaey, R. S., and R. A. Oldeman. 1996. Analyse spatiale du gradient floristique arborescent dans les forêts de plaine du SE Libéria et SW Côte d'Ivoire. Pages 352–364 in J. L. Guillaumet, M. Belin, and H. Puig, eds. *Phytogéographie tropicale: Réalités et perspectives.* ORSTOM, Paris.

Vernet, J. L., L. Wengler, M. E. Solari, G. Ceddantini, M. Fournier, M. P. Ledru, and F. Soubies. 1994. Feux, climats et végétations au Brésil central durant l'Holocène: Les données d'un profil de sol à charbons de bois (Salitre, Minas Gerais). *Compte Rendu Académie des Sciences, Paris,* série 2, 319:1391–1397.

Villiers, J. F. 1981. Formations climatiques et rélictuelles d'un inselberg inclus dans la forêt dense camerounaise. Doctoral thesis, Université de Paris VI.

Vincens, A. 1993. Nouvelle séquence pollinique du Lac Tanganyika: 30,000 ans d'histoire botanique et climatique du bassin nord. *Review of Palaeobotany and Palynology* 78:381–394.

Vincens, A., F. Chalie, R. Bonnefille, J. Guiot, and J. J. Tiercelin. 1993. Pollen-derived rainfall and temperature estimates from Lake Tanganyika and their implication for Late Pleistocene water levels. *Quaternary Research* 40:343–350 and 41:253–254.

White, F. 1979. The Guineo-Congolian region and its relationships to other phytocoria. *Bulletin du Jardin Botanique National de Belgique* 49:11–55.

———. 1981. The history of the Afromontane archipelago and the scientific need for its conservation. *African Journal of Ecology* 19:33–54.

———. 1983. *The vegetation of Africa.* UNESCO, Paris.

———. 1993. Refuge theory, ice-age aridity and the history of tropical biotas: An essay in plant geography. *Fragmenta Floristica and Geobotanica, Suppl.* 2 (2):385–409.

White, L. J. 1992. Vegetation history and logging disturbance: Effects on rain forest mammals in the Lopé Reserve. Ph.D. Thesis, University of Edinburgh.

———. 1995. Réserve de la Lopé: Etude de la végétation. Report, ECOFAC, Gabon.

White, L., K. Abernethy, R. Oslisly, and J. Maley. 1996. L'Okoumé (*Aucoumea klaineana*): Expansion et déclin d'un arbre pionnier en Afrique Centrale Atlantique au cours de l'Holocène. Pages 195–198 in M. Servant, ed. *Dynamique à long terme des écosystèmes forestiers intertropicaux.* Symposium ECOFIT, ORSTOM, and CNRS.

Youta-Happi, J. 1998. Arbres contre Graminées: La lente invasion de la savane par la forêt au centre-Cameroun. Thesis, Université de Paris-Sorbonne.

Zeven, A. C. 1964. On the origin of the oil palm (*Elaeis guineensis* Jacq.). *Grana Palynologica* 5:121–123.

———. 1967. The semi-wild oil palm and its industry in Africa. Agricultural Research Report 689, Wageningen University.

———. 1972. The partial and complete domestication of the oil palm (*Elaeis guineensis*). *Economic Botany* 26:274–279.

Zogning, A., P. Giresse, J. Maley, and F. Gadel. 1997. The Late Holocene palaeoenvironment in the Lake Njupi area, west Cameroon: Implications regarding the history of Lake Nyos. *Journal of African Earth Sciences* 24:285–300.

# CHAPTER 6

# Endemism in African Rain Forest Mammals

Peter Grubb

The regional diversity of African rain forest mammals has been interpreted historically in terms of changing climatic conditions experienced during the Pleistocene (for example, Booth 1958a). From studies of their present-day patterns of endemism, it has been inferred that forest taxa survived cool, dry climatic conditions in small refuges. Referring to existing areas of endemism as "refuges," though, prejudges their status; these areas may reflect the occurrence of former refuges, but it does not follow that there is a one-to-one correspondence between them.

## Biogeography of Forest Zones

In discussing forest mammals, it is desirable to refer to geographically defined sectors of the biome. Division of forest in Africa by the Dahomey Gap into Western (Upper Guinea or Guinean) and Central (Lower Guinea or Congolean) Zones (White 1983) results in two very unequal entities. The larger Central Zone, though, can be further divided into three main sectors delimited by the course of the Congo River: West-Central (Cameroon-Gabon, continuing northwest into Nigeria, petering out in southeast Benin); East-Central (north and eastern DR Congo); and South-Central (southern DR Congo, including the bulk of the Cuvette Centrale). Between the Albertine and Gregory Rifts, forests scattered through the savanna belong with the East-Central Zone on faunal grounds, but elsewhere—from Somalia to Cape Province—they form a diffuse Eastern Zone. It is possible therefore to recognize a fivefold division of the Afrotropical rain forest (figure 6.1): Western, West-Central, East-Central, South-Central, and Eastern Zones (Grubb 1978, 1982).

## The Mammal Fauna

There are about 270 species of mammals belonging to 120 genera in the lowland rain forest and another 100 species confined to montane rain forest (Wilson and Reeder 1993). These mammal genera can be identified by major taxonomic group:

Pangolins (Pholidota): *Phataginus*
Otter-shrews (Tenrecidae): *Micropotamogale, Potamogale*
Golden moles (Chrysochloridae): *Chlorotalpa, Chrysochloris, Chrysospalax*
Shrews (Soricidae): *Congosorex, Crocidura, Myosorex, Paracrocidura, Ruwenzisorex, Scutisorex, Suncus, Surdisorex, Sylvisorex*
Fruit bats (Pteropodidae): *Casinycteris, Eidolon, Epomops, Hypsignathus, Lissonyc-*

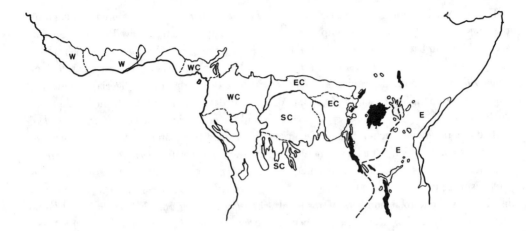

Figure 6.1. Outline of the African rain forest biome, with principal forest zones: W = Western Zone; WC = West-Central Zone; EC = East-Central Zone; SC = South-Central Zone; E = Eastern Zone. The thick dotted line in eastern Africa is "Kingdon's Line." Thin dotted lines elsewhere mark important within-zone river barriers (see figure 6.2).

teris, *Megaloglossus, Myonycteris, Nanonyc-teris, Pteropus, Rousettus, Scotonycteris*

Insectivorous bats (Microchiroptera): *Chaerephon, Glauconycteris, Hipposideros, Kerivoula, Mimetillus, Mops, Myopterus, Myotis, Nycteris, Pipistrellus, Rhinolophus, Saccolaimus, Scotophilus*

Pottos and galagos (Lorisiformes): *Arctocebus, Euoticus, Galago, Galagoides, Otolemur, Perodicticus, Pseudopotto*

Cattarhine primates (Cattarhini): *Allenopithe-cus, Cercocebus, Cercopithecus, Colobus, Gorilla, Lophocebus, Mandrillus, Miopithe-cus, Pan, Procolobus*

Squirrels (Sciuridae): *Allosciurus, Epixerus, Funisciurus, Heliosciurus, Myosciurus, Paraxerus, Protoxerus*

Dormice (Gliridae): *Graphiurus*

Mice and rats (Muridae): *Beamys, Colomys, Cricetomys, Delanymys, Dendromus, Den-droprionomys, Deomys, Dephomys, Gram-momys, Heimyscus, Hybomys, Hylomyscus, Lamottemys, Leimacomys, Lemniscomys, Lophuromys, Malacomys, Mus, Oenomys, Praomys, Prionomys, Stochomys, Thamnomys*

Scaly-tailed squirrels (Anomaluridae): *Anom-alurops, Anomalurus, Idiurus, Zenkerella*

Porcupines (Hystricidae): *Atherurus*

Elephant-shrews (Macroscelidea): *Petrodro-mus, Rhynchocyon*

Otters and ratel (Mustelidae): *Aonyx, Lutra, Mellivora*

Mongooses (Herpestidae): *Atilax, Bdeogale, Crossarchus, Liberiictis, Xenogale*

Civets and genets (Viverridae): *Civettictis, Genetta, Osbornictis, Nandinia, Poiana*

Cats (Felidae): *Panthera, Profelis*

Aardvark (Tubulidentata): *Orycteropus*

Hyraxes (Hyracoidea): *Dendrohyrax*

Elephants (Proboscidea): *Loxodonta*

Hippopotamuses and pigs (Suiformes): *Hexa-protodon, Hylochoerus, Potamochoerus*

Ruminants (Ruminantia): *Boocercus, Cephalo-phus, Hyemoschus, Neotragus, Okapia*

There is considerable regional diversity in these fauna, and many species have restricted distributions. Some are widespread, though, and 89 are present both west of the Volta River, in Upper Guinea, and in Cameroon or further east, and hence are found in both West and Central Zones. Not all of these 89 range right across the major forest blocks. For example, some species (the fruit bat *Scotonycteris ophiodon*, the mangabey *Cercocebus torquatus*, squirrels of the genus *Epixerus*, and the dormouse *Graphiurus nagtglasii*) reach the West-Central Zone but no farther east.

Fauna of these zones include many endemics, but the extent to which zone limits should be regarded as the boundaries for major centers of endemism requires further study. In the following account, I cite total numbers of mammal species present in each zone, numbers of species thought to be confined to the zone, endemic genera, and some widespread endemic zoogeographic species and allospecies.

In the Western (Guinean) Zone 134 species of mammals occur, including eleven near-endemics (ranging into Nigeria but not further east) such as *Hexaprotodon* (pigmy hippo), *Procolobus verus* (olive colobus), and *Cephalophus niger* (black duiker), and 34 strict endemics, such as *Allosciurus, Liberiictis, Dephomys, Cercopithecus diana* (diana monkey), *Anomalurus pelii*, and *Genetta johnstoni*. The Sassandra River separates richer western and somewhat poorer eastern faunas, while the Volta River, cutting off the end of the Upper Guinea forest, is an important boundary for squirrels, though less so for other mammals. It is also the westernmost limit of some Central Zone species, which thereby marginally enter the Western Zone.

There are 270 species in the Central (Congolean) Zone, of which about 170 are confined to it. Endemics largely or entirely restricted to lowland forest and found in at least two of the zone's three divisions include *Potamogale, Paracrocidura, Casinycteris, Lophocebus* (mangabeys), *Miopithecus* (talapoin), *Gorilla* (gorilla), *Deomys, Stochomys, Colomys, Okapia* (okapi), *Chlorotalpa leucorhina, Glauconycteris argentatus, Cercopithecus neglectus* (de Brazza's monkey), *Idiurus zenkeri, Graphiurus christyi, G. surdus, Herpestes naso* (long-nosed mongoose),

*Bdeogale nigripes* (black-footed mongoose), *Aonyx congicus* (small-clawed otter), *Cephalophus nigrifrons*, and *C. leucogaster* (duikers).

The 207 species of the West-Central Zone include the 11 that reach Nigeria, 19 endemics confined to the Bamenda Highlands, and 40 lowland endemics, including *Euoticus, Arctocebus, Mandrillus* (drill and mandrill), *Myosciurus, Zenkerella, Prionomys, Dendroptrionomys, Heimyscus, Suncus remyi, Myotis* cf. *megalopus, Galagoides alleni, Epixerus wilsoni, Funisciurus isabella, F. lemniscatus, F. leucogenys*, and *Lophuromys nudicaudus*. Like the Volta, the Niger River separates some predominantly Western from Central Zone taxa. The Sanaga River is a boundary between such well-defined allotaxa as drill and mandrill *Mandrillus leucophaeus* and *M. sphinx*, or duikers *Cephalophus ogilbyi* and *C. callipygus*, and is the southern or northern boundary for others. There are more lowland species south of the Sanaga than north of it. The Ogooué is also a faunal boundary, but the fauna occurring south of this river are poorly known. In the West-Central Zone as a whole, some species (for example, the mandrill and the squirrel *F. leucogenys*) are more or less confined to the valleys of westward-draining rivers. Other species occur predominantly in the basins of eastward-draining tributaries of the Congo River (such as the mangabey *Cercocebus agilis*). Others (red colobus, *Procolobus pennantii; Cercopithecus neglectus;* cusimanse, *Crossarchus platycephalus;* golden cat, *Profelis aurata;* forest hog, *Hylochoerus meinertzhageni;* dwarf antelope, *Neotragus batesi;* bongo, *Boocercus eurycerus*) have distribution gaps. These patterns are consistent with a former division between coastal refuges and an inland fluvial refuge during "interpluvial" times (as discussed below).

In the East-Central Zone, 192 species occur. This includes as many as 32 highland endemics but only 20 lowland endemics, including *Scutisorex* (armored shrew), *Osbornictis, Crocidura caliginea, Galago matschiei, Rhynchocyon (cirnei)*

*stuhlmanni, Paraxerus boehmi, P. alexandri, Loph-uromys luteogaster, Malacomys verschureni,* and *Genetta victoriae* (giant genet). Lowland species occur in highland forest with or without systematic differentiation, or do not range into the highlands at all (Rahm 1972; Colyn 1991). Unlike the situation in the West Central Zone, many lowland endemic species are widely distributed, even such rare ones as the water civet (*Osbornictis piscivora*). The Aruwimi-Ituri, Lindi, and Maiko Rivers form important faunal boundaries for East-Central primates, and there is also regional endemism south of these rivers among mice (*Hylomyscus*) and colobus monkeys (Colyn 1987, 1988, 1991). The colobus monkeys have a hiatus in their distribution curiously dovetailing with the limited lowland range of the eastern gorilla, *Gorilla gorilla graueri* (Emlen and Schaller 1960). Elements of the East-Central fauna extend beyond the forest block into East Africa (Kingdon 1971; Rodgers et al. 1982; Grubb 1983), with some subspeciation.

There are 99 species in the South-Central Zone, with 13 endemics, including *Congosorex, Allenopithecus* (Allen's swamp monkey), *Lophocebus (albigena) aterrimus* (black mangabey), *Pan paniscus* (bonobo), *Petrodromus (tetradactylus) tordayi, Funisciurus congicus, F. poolii, Lophuromys huttereri, Anomalurops (beecrofti) schoutedeni,* and *Crossarchus ansorgei.* The low total number of species is almost certainly misleading, in view of the very limited collecting of small mammals in the area, although a number of widespread forest species are absent south of the Congo River (e.g., *Cephalophus leucogaster* and *Neotragus batesi*).

Only 107 species are found in the Eastern Zone, with 22 lowland and 34 upland endemics—a large proportion of endemics (52%) in view of the total area of forest involved. Many endemics are vicarious or extremely localized. Included are *Surdisorex, Otolemur* (greater galagos), *Beamys, Myonycteris relicta, Pteropus voeltzkowi,* a number of

species of *Galagoides* (dwarf galagos), seven species of *Paraxerus* (tree squirrels), *Praomys delectorum, Dendrohyrax validus,* and four species of *Cephalophus* (duikers). Lowland taxa occupy the Zanzibar-Inhambane and Tongaland-Pondoland regional mosaics (White 1983). For them, the Rufiji River appears to be a significant faunal barrier (Kingdon 1974). Some are eurytopic, ranging into habitats other than rain forest (such as the squirrel *Paraxerus ochraceus*). Scattered upland forests provide discontinuous links with the East-Central Zone. Counterintuitively, the stenotopic nonvolant upland specialists (such as shrews, *Myosorex, Sylvisorex,* and *Crocidura* species) are phylogenetically closely linked with congeners in Albertine and even Bamenda or Angolan highlands, as are the potentially much more vagile avifauna (Moreau 1966).

Most of the East African endemics are separated from the Central Zone (Grubb 1983) by what may be termed Kingdon's Line (see figure 6.1; Grubb et al. 1999), following Kingdon's zoogeographical analysis (Kingdon 1971). Species that range right across the African forest blocks and reach the lowland coastal forest of Kenya and Tanzania are few. Indeed, apart from some bats (such as *Lissonycteris angolensis, Nycteris grandis, Hipposideros cyclops,* and *Mops brachyptera*) and taxa that extend widely into the savannas (for instance, the blotched genet, *Genetta pardina*), there is really only one example, the scaly tailed squirrel *Anomalurus derbianus.* Otherwise, widespread and even eurytopic species, including *Phataginus tricuspis* (tree pangolin), *Hypsignathus monstrosus* (hammerheaded bat), and *Atherurus africanus* (brushtailed porcupine), though present throughout both Western and Central Zones, are absent east of Kingdon's Line, perhaps because they were unable to pass through an upland-forest bottleneck to reach the coastal forests. Of 20 non-volant species found both in the Congolean lowland forest and east of Kingdon's

Line, and which are not widespread outside sylvan habitats, 13 occur also in upland forest, perhaps proving the point.

### Vicariance Within Forest Zones

From the examples given, it is evident that rivers are often barriers to ranges or mark vicariance of allotaxa within forest zones. The Central Zones have indeed been defined by the Congo and its affluents (figure 6.2). Vicariance within forest zones is evidently a feature of primate geography (Grubb 1990), which offers more than thirty examples. There is a statistically significant tendency for related allopatric primate taxa to differ at the subspecies level within forest zones but at the conventional species level between forest zones. There appear to be twelve cases of within-zone subspeciation among squirrels (Sciuridae, Anomaluridae), but in only five instances are the taxa well

marked. There are only three examples where carnivores or ungulates have speciated within single forest zones. Examples of within-zone vicariance among lowland-forest bats, dormice, murids, or shrews may exist, but at the species level none are apparent (Wilson and Reeder 1993), and at the subspecies level they are poorly known (although there are many instances of vicariance among small mammals of highland forests). Generally, non-primates do not have the complexity of allospecies or subspecies shown by primates (a "taxon effect"). They may be replaced by vicarious congeners in different forest zones, but they have not speciated so much within these zones. Consequently, vicariance of many primate taxa and some nonprimates can occur within the ranges of more widespread zonal endemics, or of species occurring in several zones. For example, the monotypic black duiker (*Cephalophus*

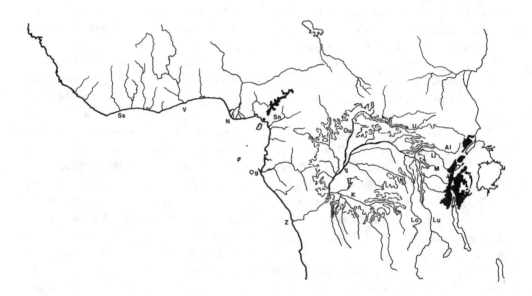

Figure 6.2. Map of western and central Africa, showing locations of some postulated major Pleistocene forest refuges: Bamenda Highlands and Albertine Highlands, indicated by black areas (showing land above 1,500 m elevation) in the west and east of the map, respectively; and Cuvette Centrale (indicated by dotted line enclosing land below 500 m). The following rivers are shown: AI = Aruwimi-Ituri; K = Kasai; Li = Lindi; Lo = Lomami; Lu = Lualaba; M = Maiko; N = Niger; Og = Ogooué; Ou = Oubangui; Sn = Sanaga; Ss = Sassandra; U = Uele; V = Volta; and Z = Zaire (Congo).

*niger*) occurs throughout west Africa, even as far east as the Niger River, but found within its range are allopatric spot-nosed monkeys *Cercopithecus petaurista buettikofferi*, *C. p. petaurista*, and *C. erythrogaster*. Such cases could provide a basis for a hierarchical classification of mammal distributions.

## Centers of Endemism

When account is taken of within-zone vicariance, zones can be partitioned into smaller units. These may be termed "centers of endemism." These (minor) centers have been recognized largely from the study of primate geography, in particular the work of Booth (1958a, 1958b) and Colyn (1987, 1988, 1991). Using higher primates as indicator species, eleven lowland and two highland centers can now be differentiated in the Western and Central Zones (figure 6.3), while additional centers can be distinguished in the Eastern Zone. Non-primate endemics can be found for nearly all

centers, even though they are unlikely to have allopatric representatives within the zone. Centers are, by definition, characterized by endemic species or subspecies, but they are also characterized by geographical isolates of non-endemic taxa and absences of taxa that are nonetheless present in adjacent centers. Members of a particular center share distributions, share status as vicarious representatives of their respective zoogeographical species, and share river barriers as limits to their distributions.

Centers and some of the endemic taxa that define them are identified below, updated from Grubb 1990, where longer lists of endemic subspecies are provided. Inadequate knowledge of distributions and infraspecific systematics makes it impossible to provide reliable numbers of endemic species and subspecies.

- The Liberian Center is an area lying west of the Sassandra River that possesses at least twelve species absent from the next center, including such endemics as *Microp-*

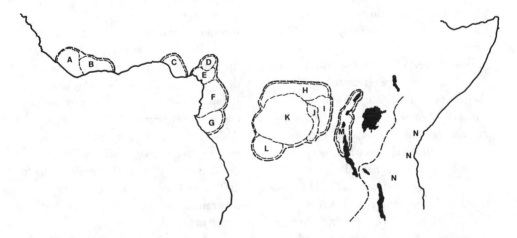

Figure 6.3. Schematic diagram of minor centers of mammalian endemism (A to N), grouped within putative former major centers of endemism ("refuges") in the African rain forest biome, as follows: Guinea Refuge, with Liberian (A) and Gold Coast (B) Centers; outlying Niger River Center (C); Cameroon-Gabon Refuge, with Bamenda Highland (D), West Cameroon (E), Rio Muni (F), and Ogooué (G) Centers; Fluvial Super-Refuge, with Ubangi-Uele (H), Kivu (I), Lomami-Lualaba (J), Salongo (K), and Kasai (L) Centers; Albertine Refuge, with the single Albertine Center (M); and Eastern Refuges with minor centers (N), not shown in detail.

*otamogale lammotei, Cephalophus zebra,* and
*C. jentinki.*

- The Gold Coast Center is relatively weak,
with only one endemic species, the white-
thighed colobus, *Colobus vellerosus* (and that
only an allospecies), and five monkey and
three squirrel subspecies. It lacks some of
the species present in the Liberian Center.
It would go unrecognized if taxa at the conventional species level alone were under
consideration (Grubb 1982, 1990). Nevertheless, it is still being treated as if it were a
significant refuge (see, e.g., Hamilton 1988).

- The Niger River Center is emerging as a
distinct unit, particularly from work by C.
B. Powell, who discovered a new subspecies of red colobus (*Procolobus
pennantii*) in the Niger Delta (Grubb and
Powell 1999). The fauna includes endemic
species (for example, the monkeys *Cercopithecus erythrogaster* and *C. sclateri*) and
subspecies or isolates of both western and
eastern affinity, but the contribution of the
western is greater (such as an isolate of the
olive colobus *Procolobus verus* and a subspecies of the pygmy hippopotamus *Hexa-
protodon liberiensis*). So although it is geographically part of the West-Central Zone,
faunistically the center is related to the
Western (Guinean) Zone (Grubb 1990).
This lack of correspondence between geographically defined forest zones and
regional faunas is also seen among birds,
where west African avifaunal regions
extend into Nigeria (Diamond and Hamilton 1980; Crowe and Crowe 1982).

- The West-Cameroon Center (including
Bioko) has endemic allospecies (angwan-
tibo, *Arctocebus calabarensis;* needle-clawed
bushbaby, *Euoticus pallidus;* red colobus,
*Procolobus preussi;* red-eared guenons, *Cercopithecus erythrotis;* Preuss's guenon, *C.
preussi;* drill, *Mandrillus leucophaeus*) and
several endemic subspecies.

- The Bamenda Highland Center covers dis-
continuously distributed, isolated highland
forests. There is evidence of provincialism
within the fauna (Hutterer et al. 1992).
Endemic genera and zoogeographic
species include *Paraxerus (Montisciurus)
cooperi, Lamottemys okuensis,* and *Crocidura
eisentrauti.* Other endemics include
*Chrysochloris (stuhlmanni) balsaci,*
allospecies of shrews (*Myosorex, Sylvisorex, Crocidura*) and mice (*Dendromys,
Hybomys, Lemniscomys, Praomys,
Hylomyscus, Lophuromys*), and one primate
(isolate of *Galagoides thomasi*).

- In the Rio Muni Center, endemics include
*Prionomys batesi, Euoticus elegantulus, Arc-
tocebus aureus,* and *Cephalophus callipygus.*

- Ogooué Center endemics include the little-
known squirrel *Funisciurus duchaillui,* the
recently discovered monkey *Cercopithecus
solatus,* and several subspecies, including
the recently named white-legged duiker
*Cephalophus ogilbyi crusalbum.*

- Assigned to the Ubangi-Uele Center are
five widespread primate taxa that occur in
the watershed between the Oubangui and
Congo Rivers and are not found south of
the Aruwimi-Ituri or Lindi Rivers. Like
the next center, this one is weakly defined.
It remains to be seen whether any non-primates can also be included.

- Rather few primate subspecies are strictly
confined to the Kivu Center, but distinctive East-Central endemics are shared with
the previous center. The relationship
between the two is therefore unlike the situation in the Western Zone.

- Taxa endemic to the massifs of the Albertine Highland Center include genera
(*Ruwenzisorex, Delanymys, Thamnomys*),
zoogeographic species (*Micropotamogale
ruwenzorii, Funisciurus carruthersi, Lophuromys rahmi, L. medicaudatus, L. woosnami,
Mus bufo*), and allospecies (*Paracrocidura*

*maxima, Heliosciurus ruwenzorii*). Provincialism within the center is evident from the scattered unreviewed literature. The faunas of several massifs are poorly known.

- The Salongo Center is the left-bank sector of the Cuvette Centrale, limited to the south by the Kasai and to the east by the Lomami, with *Procolobus (pennantii) tholloni, Cercopithecus dryas, Praomys minor,* and *Malacomys lukolelae* among endemics.
- Endemic to the Kasai Center are *Congosorex polli, Funisciurus bayoni,* and *Miopithecus talapoin.* The *Miopithecus talapoin* and the squirrel *Funisciurus pyrropus pembertoni,* while absent from the Salongo Center, are represented by related taxa in the West-Central Zone.
- The Lomami-Lualaba Center is a cul-de-sac into which some endemic primate taxa have percolated from the south (Colyn 1988, 1991). The fauna is not well known and may also include the squirrel *Paraxerus antoniae,* mice (*Hylomyscus* and *Lophuromys* species; Colyn 1991), and a bat (*Chaerephon gallagheri*) known from a single specimen. The *Lophuromys* species has recently been described as *L. huttereri,* with a single record from just outside this center (Verheyen et al. 1996).
- East African Centers have been recognized as subdivisions of the Eastern Zone (Grubb 1983, 1990), but further work is needed to define them.

Naming zones and centers is a preliminary exercise. Zones and centers are objective entities, but they do not account for all the diversity of mammalian distributions. Many taxa are not confined to a single center of endemism or forest zone but are found in some combination of centers or zones. Of monkeys in the Western and Central Zones, for example, 45 taxa are restricted to a single center, 15 occur in two centers, and 7 in more than two. Of montane forest

species, 85% are confined to a single zone, but for lowland forest species the figure is 52%. In the light of cases where allotaxa occur within the range of a single more widespread taxon, it is possible to rank distributions hierarchically, but this classification is also imperfect. A more comprehensive interpretation of mammal distributions will eventually emerge, probably with the assistance of vicariance analysis (Stephen J. Gooder, pers. comm.). Already, however, distributional data have provided a basis for historical analysis.

Numbers of highland forest species disproportionately inflate the figures for mammalian endemicity in Central and Eastern Zones, and a similar pattern of diversity is recorded for forest birds. Analyses of the African avifauna (Diamond and Hamilton 1980; Crowe and Crowe 1982) agree in recognizing an area between 10° E and 250° E within the Central Forest Zone both north and south of the Congo River, which is faunistically deficient and lacks any endemic bird species. The area is flanked to the northwest (Cameroon-Gabon) and east (eastern DR Congo) by regions with more species and many endemic species. Because of the large quadrats (with sides of 5° or 4°) employed in the studies, the pattern partly reflects boosting of diversity by the inclusion of montane forest species in mainly lowland forest quadrats. But it also results from discontinuities in distributions of lowland species across central Africa. Some of these cases may be spurious, given that occurrences in the intervening area may not yet have been revealed, but many are accepted as genuine (Mayr and O'Hara 1986). Similar distributional discontinuities occur in mammals (e.g., *Anomalurus pusillus* or *Cephalophus leucogaster*). Also contributing to a local deficiency in the mammalian fauna of the Central Zone is the low diversity of the South-Central sector, but, as already suggested, this may be partly an artifact.

Prigogine (1988) identified a group of lowland birds, apparently overlooked by other authors, whose distributions are confined within the Congo Basin. This area therefore includes endemics after all. There are Central Zone mammals with similar distributions (such as the fruit bat *Casinycteris argynnis*), although no avian parallel has yet been noticed for the more restricted South-Central mammalian endemicity. With better distributional data, both endemism and reduced diversity among central African birds and mammals could be more critically compared, but already the historical biogeography of this zone has been the subject of varying interpretations.

Speculation concerning the Pleistocene history of African forest mammals has addressed the nature and location of refuges during adverse paleoclimatic conditions (Booth 1958a; Kingdon 1971, 1980; Hamilton 1976, 1981, 1988; Grubb 1978, 1982; Colyn 1988, 1991; Colyn et al. 1991). In what has become a widely quoted interpretation, present-day centers of diversity (core areas) are supposed to correspond to past refuges. In west Africa, putative refuges have an essentially lowland facies and relate to minor centers of endemism, but for central Africa they correspond with major centers of endemism in West-Central and East-Central Zones. These refuges have been sited well away from the central Congo drainage basin and are associated with montane areas (see, e.g., Hamilton 1988).

For birds, the faunal deficiency in the central Congo basin is believed to indicate that many species have still not completed dispersal into this area—which is assumed once to have been deforested—from their Pleistocene refuges (Diamond and Hamilton 1980; Crowe and Crowe 1982). With particular reference to mammals of the West-Central and South-Central Zones, a different interpretation has more recently been presented (Colyn 1991; Colyn et al. 1991). Colyn has shown that lowland and montane mammal faunas have been confused, inflating the apparent diversity and endemism of the eastern DR Congo fauna and suggesting the

former existence of the East-Central Refuge, which would have had to harbor both lowland and montane species. But the endemic lowland species and subspecies of the Congo Basin—including at least 37 taxa of higher primates—could not have survived the last Quaternary arid climatic phase crowded on the Albertine Highlands and their foothills, together with the endemic highland forest fauna. It would be unreasonable to suggest that pan-Congolian patterns of vicariance could have been transferred wholesale across many river valleys and from distant places like the Kasai River, to be compressed into such small refuge areas. Nor could all the lowland taxa have dispersed into their present ranges and differentiated within the limited time available between the amelioration of environmental conditions and the present day. Colyn (1991) proposed instead that apart from montane refuges, there was also a major fluvial refuge (or super-refuge) in the Central Zone. Some authors had already recognized a lowland refuge, confined either to the swamp forest of the Cuvette Centrale or to a more extensive tract in the South-Central Zone (Laurent 1973; Kingdon 1980; Grubb 1982), where indeed Prigogine (1988) thought that his Central Forest birds must have survived "interpluvial" conditions. Colyn's super-refuge was not restricted to the South-Central Zone but extended through the whole Congo Basin. It would have been partitioned by lower reaches of larger tributaries, which continued to delimit centers of endemism.

In light of Colyn's work and doubts expressed elsewhere (Grubb 1982), we may question whether a "hard" refuge theory is entirely satisfactory anyway. Need all refuges have been totally isolated from each other, completely separated by savanna? We would expect savanna mammals to disperse widely between forest refuges of this kind, yet there is little indication of such a phenomenon. A refuge theory could account for present-day vicariance, but so could river barriers. Faunal discontinuities still occur even within continuous blocks of forest. Some of these discontinuities surely have an ecological basis and cannot all represent a failure to disperse out of former refuges. Conversely, populations that apparently had become isolated in refuges could still have been connected by gene flow, as suggested by extensive present-day dispersion of certain mammals beyond the forest, along riverine groves (see, e.g., Fay 1985). There are many widespread forest taxa whose weak geographical variation suggests that they were never split into refuge populations (see, e.g., Colyn and Van Rompaey 1994). So, while some "hard" refuges clearly existed and still exist, a comprehensive refuge theory may be neither necessary nor sufficient to account for many present-day patterns of diversity.

It is accepted that during the last cool, dry climatic phase, the number of discrete minor refuges among peripheral forests (Elgon, Mulanje, Jos Plateau, Kilimanjaro, and many others) increased and forest disappeared from some of these sites (Diamond and Hamilton 1980; Rodgers et al. 1982). Developing ideas expressed elsewhere (Grubb 1982; Colyn 1991), I suggest that both continuity and discontinuity could have existed between faunas of the main rain forest block, but many taxa were more restricted in distribution and more taxa had discontinuous populations than at present. The effect could be for major centers of endemism to become more distinct and more clearly separated by areas of lower diversity. Whereas these major centers amounted to refuges and still included minor centers of endemism, the blocks of lowland forest did not split up completely. They were more restricted in area and more dissected, especially where forest persisted only along the larger rivers. The Guinean and Central (Congolean) Forests could have become even more separated than they are at present. With lowered tempera-

tures, there were changes in the disposition of montane vegetation and perhaps whole montane forest ecosystems.

To the west, montane vegetation extended in range (White 1981, 1993; Maley 1987; Livingstone 1993) not only in the Bamenda Highlands but also in the Monts de Cristal and the Massif du Chaillu in Gabon. This affected the dispersion of lowland forest mammals of the present-day West-Cameroon, Rio Muni, and Ogooué Centers, which probably survived somewhat in isolation from the super-refuge. In the east, the Albertine montane forests may have expanded but became detached from the surviving lowland forest (Colyn et al. 1991), allowing upland races of lowland species to differentiate. The following are the refuges that possibly developed (see figure 6.3): Guinea Refuge (with Liberian and Gold Coast Centers) and an outlying Niger River Refuge; Cameroon-Gabon Refuge (with West Cameroon, Bamenda, Rio Muni, and Ogooué Centers); Fluvial Super-Refuge (with Ubangi-Uele, Kivu, Lomami-Lualaba, Salongo, and Kasai Centers) and some minor marginal refuges; Albertine Refuge; and a number of East African Refuges. Differences in topography account for the contrasts between presumed refuges peripheral to the Fluvial Super-Refuge: the Cameroon-Gabon Refuge would have included lowland and highland communities, but the Albertine Refuge, only montane habitats.

The historical model presented above accounts for more phenomena than a "hard" refuge theory, but it is based entirely on mammal distributions and must be tempered by hypotheses relying on other sources if it is to have a more general application.

## Conservation Recommendations

The pattern of forest zones and mammalian centers of diversity has a bearing on conservation strategy. Both the conservationist and the pure scientist require consolidation of research in mammalian biogeography—an action plan, no less. Of course data collection, systematic revision, exploration, and mapping are still under way, but more coordination is needed. Collecting of small mammals continues and is necessary to provide faunal inventories, but with the shortage of systematists, there is a lag in the description of material. Some valuable systematic studies on African forest mammals have been completed, but many more are needed, particularly at the subspecies level so that we can confidently use the subspecies as a well-established unit in biogeographical studies.

Exploration is still warranted. Even though important surveys are being carried out, knowledge of African forest mammals is imperfect: many small species have not been seen since first collected. New taxa are still being discovered: fifteen new species and subspecies of primates alone have been named in recent decades or are awaiting description. After the Ogooué Center was recognized (Grubb 1978, 1982), a new species of monkey (*Cercopithecus solatus*) was located there in 1984 (Harrison 1988). This and other discoveries indicate the potential for a proactive approach. We should search for new taxa in poorly known centers of diversity and in other areas where we may expect to find them, such as cul-de-sacs between major rivers (Colyn 1991), forests on isolated hills or in the uplands of Ethiopia, and any area where endemics have already been located yet the fauna is not fully known. It is worthwhile to look out for vicarious taxa: new discoveries often fill gaps in the known distribution of species groups. Even new centers of endemism have been recently identified. Colyn (1987, 1988) discovered new primate taxa that define the Lomami-Lualaba Center. While this chapter was being revised, C. B. Powell (pers. comm.) reported further discoveries in the Niger River Center. A new red colobus monkey is being described (Grubb and Powell 1999), and we now know that endemism east and west

of the lower Niger River defines subcenters within the Niger River Center: the geography of diversity is more complex than we thought. There may be more subcenters or even centers to be revealed elsewhere, as Colyn (1991) has suggested.

Accurate mapping of mammal distributions is needed to arrive at better descriptions of regional diversity. The relationship between faunal richness and endemism still requires more study. Statistical examination of the center concept needs better data, so that smaller quadrats can be used in analysis. Mapping programs are being undertaken, for instance by the Danish Center for Tropical Biodiversity, University of Copenhagen (Neil Burgess, pers. comm.), but fuller use should be made of the great resources provided by museum specimens. Information on mammal distribution is replete with misspellings and incorrect coordinates of place names, so mapping schemes must be able to handle frequent additions and corrections to the database. Exploratory research is also needed to advance distributional knowledge: relatively well known regions (southern Cameroon or the Aruwimi-Ituri basin in DR Congo, for example) are separated by huge poorly known tracts, such as the forests south of the Ogooué in Gabon and Congo (DR Congo), scattered forests in northwestern Angola, or the whole Cuvette Centrale. The need for recording quality data must be made clear. New and interesting distribution records are announced in the unpublished data, yet their significance is not always recognized and they evade the abstracting journals. Field identifications are still being published without adequate evidence for their reliability, and an improvement in the standard of these reports is desirable.

There has been little monitoring or updating of advances in systematics and biogeography of African forest mammals, so it would be fruitful if an appropriate umbrella organization could oversee or guide pure research in these areas. A regular column in an appropriate journal could adopt this role and speed up communication with conservationists. It could alert them to developments but also assess their significance.

It is hoped that conservation strategy will be influenced by our emerging understanding of the structure underlying regional diversity. The effectiveness of protected areas should be evaluated by the extent to which they represent centers of mammalian endemism. A new protected area should be planned for any center that lacks one. Conservation measures will continue to develop, but awareness of fundamental diversity patterns is always needed.

REFERENCES

Booth, A. H. 1958a. The Niger, the Volta and the Dahomey Gap as geographical barriers. *Evolution* 12:48–62.

———. 1958b. The zoogeography of West African primates: A review. *Bulletin de l'Institut Français d'Afrique Noire (A)* 20:587–622.

Colyn, M. M. 1987. Les primates des forêts ombrophiles de la Cuvette du Zaire: Interprétations zoogéographiques des modèles de distribution. *Revue de Zoologie Africaine* 101:183–196.

———. 1988. Distribution of guenons in the Zaire-Lualaba-Lomami river system. Pages 104–124 in A. Gautier-Hion, F. Bourlière, J.-P. Gautier, and J. Kingdon, eds. *A Primate Radiation: Evolutionary Biology of the African Guenons.* Cambridge University Press, Cambridge.

———. 1991. L'importance zoogéographique du Bassin du Fleuve Zaire pour la spéciation: Le cas des primates simiens. *Annalen. Zoologische Wetenschappen. Koninklijk Museum voor Midden-Afrika Tervuren, Belgie* 264.

Colyn, M., A. Gautier-Hion, and W. Verheyen. 1991. A reappraisal of palaeoenvironmental history in Central Africa: Evidence for a major fluvial refuge in the Zaire Basin. *Journal of Biogeography* 18:403–407.

Colyn, M., and H. Van Rompaey. 1994. Morphometric evidence of the monotypic status of the African long-nosed mongoose *Xenogale naso* (Carnivora, Herpestidae). *Belgian Journal of Zoology* 124:175–192.

Crowe, T. M., and A. A. Crowe. 1982. Patterns of distribution, diversity and endemism in Afrotropical birds. *Journal of Zoology* 198:417–442.

Diamond, A. W., and A. C. Hamilton. 1980. The distribution of forest passerine birds and Quaternary climatic change in tropical Africa. *Journal of Zoology* 191:379–402.

Emlen, J. T., and G. B. Schaller. 1960. Distribution and status of the mountain gorilla (*Gorilla gorilla beringei*)— 1959. *Zoologica* 45:41–52.

Fay, J. M. 1985. Range extensions for four *Cercopithecus* species in the Central African Republic. *Primate Conservation* 6:63–68.

Grubb, P. 1978. Patterns of speciation in African mammals. *Bulletin of Carnegie Museum of Natural History* 6:152–167.

———. 1982. Refuges and dispersal in the speciation of African forest mammals. Pages 537–553 in G. T. Prance, ed. *Biological Diversification in the Tropics.* Columbia University Press, New York.

———. 1983. The biogeographic significance of forest mammals in Eastern Africa. *Annalen. Zoologische Wetenschappen. Koninklijk Museum voor Midden-Afrika Tervuren, Belgie* 237:75–86.

———. 1990. Primate geography in the Afrotropical forest biome. Pages 187–214 in G. Peters and R. Hutterer, eds. *Vertebrates in the Tropics.* Alexander Koenig Zoological Research Institute and Zoological Museum, Bonn.

Grubb, P., and C. B. Powell. 1999. Discovery of red colobus monkeys (*Procolobus badius*) in the Niger Delta with the description of a new and geographically isolated subspecies. *Journal of Zoology* 248:67–73.

Grubb, P., O. Sandrock, O. Kullmer, T. M. Kaiser, and F. Schrenk. 1999. Relationships between eastern and southern African mammal faunas. Pages 253–267 in T. G. Bromage and F. Schrenk, eds. *African Biogeography, Climate Change, and Human Evolution.* Oxford University Press, New York.

Hamilton, A. C. 1976. The significance of patterns of distribution shown by forest plants and animals in tropical Africa for the reconstruction of upper Pleistocene palaeoenvironments; a review. *Palaeoecology of Africa* 9:63–97.

———. 1981. The Quaternary history of African forests: Its relevance to conservation. *African Journal of Ecology* 19:1–6.

———. 1988. Guenon evolution and forest history. Pages 13–34 in A. Gautier-Hion, F. Bourlière, J.-P. Gautier, and J. Kingdon, eds. *A Primate Radiation: Evolutionary Biol-*

*ogy of the African Guenons.* Cambridge University Press, Cambridge.

Harrison, M. J. S. 1988. A new species of guenon (genus *Cercopithecus*) from Gabon. *Journal of Zoology* 215:561–575.

Hutterer, R., F. Dieterlen, and G. Niklaus. 1992. Small mammals from forest islands of eastern Nigeria and adjacent Cameroon, with systematical and biogeographical notes. *Bonner Zoologische Beitrage* 43:393–414.

Kingdon, J. 1971. *East African Mammals: An Atlas of Evolution in Africa.* Volume 1. Academic Press, London.

———. 1974. *East African Mammals: An Atlas of Evolution in Africa.* Volume 2, part A ("Insectivores and Bats"). Academic Press, London.

———. 1980. The role of visual signals and face patterns in African forest monkeys (guenons) of the genus *Cercopithecus. Transactions of the Zoological Society, London* 35:425–475.

Laurent, R. F. 1973. A parallel survey of equatorial amphibians and reptiles in Africa and South America. Pages 259–266 in B. J. Meggers, E. S. Ayensu, and W. D. Duckworth, eds. *Tropical Forest Ecosystems in Africa and South America.* Smithsonian Institution Press, Washington, D.C.

Livingstone, D. A. 1993. Evolution of African climate. Pages 455–472 in P. Goldblatt, ed. *Biological Relationships Between Africa and South America.* Yale University Press, New Haven.

Maley, J. 1987. Fragmentation de la forêt dense humide africaine et extensions des biotopes montagnards au Quaternaire récent: Nouvelle données polliniques et chronologiques: Implications paleoclimatiques et biogéographiques. *Palaeoecology of Africa* 18:307–334.

Mayr, E., and R. J. O'Hara. 1986. The biogeographic evidence supporting the Pleistocene forest refuge hypothesis. *Evolution* 40:55–67.

Moreau, R. E. 1966. *The Bird Faunas of Africa and Its Islands.* Academic Press, New York.

Prigogine, Alexandre. 1988. Speciation pattern of birds in the central African forest refugia and their relationship with other refugia. *Proc. Int. Ornithol. Congr.* 19:2537–2546.

Rahm, U. 1972. Note sur la répartition, l'écologie et la régime alimentaire des sciuridés au Kivu (Zaire). *Revue de Zoologie et de Botanique Africaines* 85:321–339.

Rodgers, W. A., C. F. Owen, and J. M. Homewood. 1982. Biogeography of East African forest mammals. *Journal of Biogeography* 9:41–54.

Verheyen, W. N., M. Colyn, and J. Hulselmans. 1996. Reevaluation of the *Lophuromys nudicaudus* Heller, 1911 species-complex with a description of a new species from Zaire (Muridae—Rodentia). *Bulletin van het Koninklijk Belgisch Instituut voor Natuurwetenschappen, Biologie* 66:241–273.

White, F. 1981. The history of the Afromontane archipelago and the scientific need for its conservation. *African Journal of Ecology* 19:33–54.

———. 1983. The vegetation of Africa: A descriptive memoire to accompany the UNESCO/AETFAT/UNSO vegetation map of Africa. UNESCO, Paris.

———. 1993. Refuge theory, ice-age aridity and the history of tropical biotas: An essay in plant geography. *Fragmenta Floristica et Geobotanica, Supplementum* 2:385–409.

Wilson, D. E., and D. M. Reeder. 1993. *Mammal Species of the World: A Taxonomic and Geographic Reference.* 2d ed. Smithsonian Institution Press, Washington, D.C.

# CHAPTER 7

# The History of Human Settlement in the Middle Ogooué Valley (Gabon)

## Implications for the Environment

Richard Oslisly

Africa is extremely rich in prehistoric archaeological remains. These include the extraordinary geological deposits of eastern Africa, where many important human fossils have been discovered, as well as the extensive areas of rock art in the Sahara and in southern Africa. However, there is a strong bias in the literature toward regions with open countryside, which are more easily explored. Little information exists for the tropical forested regions of central Africa, long considered a hostile environment that had resisted human settlement.

Early archaeological research in central Africa was sporadic. The forest vegetation soon covers any remains of settlements, and early discoveries were mostly by informed geologists working on large engineering projects. More systematic archaeological research began in the 1960s, particularly in the context of the "Bantu phenomenon" of the past five thousand years, when agriculturalists from just south of the Sahara expanded into eastern and southern Africa. As this research developed, it became apparent that late Quaternary climate changes (see Maley 1991; Maley, this volume) of the Würm glaciation (70,000–12,000 B.P.—the recent Pleistocene period) and the following

interglacial of the Holocene (12,000 B.P. to the present) had played an important role in determining where and how people lived (Mortelmans and Monteyne 1962; De Ploey 1963).

In this chapter I shall present the results of sixteen years of work on the paleoenvironment and prehistory of the middle valley of the Ogooué, to build the first chronological framework of the history of people living in this region of central Gabon and to discuss their cultural traditions and relations with the forest. Findings from central Gabon will then be compared to research elsewhere to discuss the implications of human settlement and use on the African forest zone in general.

From a map of central Africa it is apparent that areas of southern Cameroon—bordered by the Atlantic to the west and the great swamps of the Congo Basin to the east—represent a likely channel for human population movements south out of the Bantu homeland in southeastern Nigeria and western Cameroon (figure 7.1). This is especially so because during periods in the past, much of the present forest cover of the region was replaced by a more open landscape (see, e.g., Maley, this volume).

The natural funnel of ridge lines running

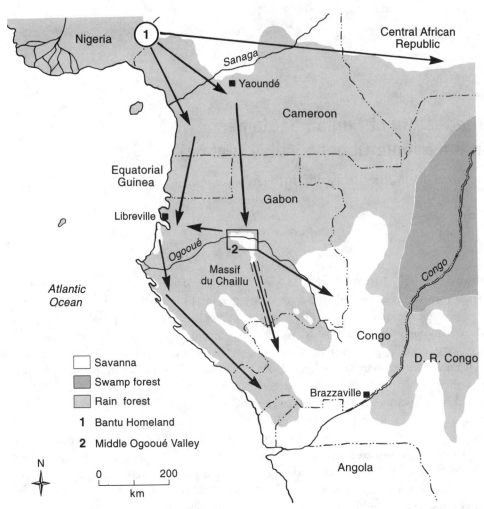

Figure 7.1. Migratory routes of the Bantu in west central Africa. Arrow surrounded by dotted lines indicates a savanna passage colonized by forest during the Holocene and the population hiatus. Adapted from Phillipson 1985 and Oslisly 1995 (corrected version). Reprinted with the permission of Cambridge University Press.

north–south from the north of Gabon is interrupted in the center of the country by the middle valley of the river Ogooué (0°–0° 30' S; 11°–12° E), which runs parallel to the equator in a series of rapids and falls. The region is delimited to the north by the equator and to the south by the first forested foothills of the Massif du Chaillu. It is an island of open savannas and gallery forests covering about 1,000 km² (see White, this volume, chapter 11).

As it is thought that the rapid spread of Bantu peoples may have occurred during a dry climatic phase between 3,500 and 2,000 B.P., when savanna vegetation was more extensive in central Africa (Maley 1992; Schwartz 1992), it seems likely that the middle valley of the Ogooué would have been a key staging area, because of its central location and the presence of a large river.

The history presented here has taken many

years to tease apart, and we continue to make new and exciting discoveries on a yearly, if not monthly basis. For prehistoric humans, the evergreen forest represented a challenging environment, the height and density of the canopy making orientation difficult. It was, however, an important source of food, and people probably traveled significant distances to gather fruits and nuts or to hunt and trap animals, as their closest relatives, gorillas and chimpanzees, do today (Tutin et al. 1991; Tutin and Oslisly 1995). It is extremely difficult to follow riverbanks because the vegetation is dense, but there are generally numerous elephant trails along the ridges between rivers, which even today are often adopted by humans. Following such trails, one catches occasional glimpses, through gaps in the trees, of distant peaks or valleys, which improve one's perception of topography and distance. Elephant trails were thus likely key elements for early human migrations through the forest zone, particularly considering the general north-south orientation of the major ridge lines in Gabon. Today's roads, be they national highways or foresters' trails, also tend to follow ridge lines, often revealing numerous remains of past human habitation. These ridge lines tend also to be rich in iron-bearing geological formations and could also have attracted Iron Age peoples for this reason.

### The Early Stone Age (400,000–120,000 B.P.)

There are numerous Stone Age sites in the middle valley of the Ogooué, including the oldest known stone tools in central Africa, discovered in the upper terrace of the Ogooué at Elarmékora (Oslisly and Peyrot 1992a) and estimated to date to around 400,000 B.P., in correlation with the mid-Bruhnes climatic change (Jansen et al. 1986). Unfortunately, as is the case elsewhere in central Africa, most sites are in river terraces or other stone lines, and therefore their stratigraphy is typically highly disturbed. Because of the strongly acidic soils (with pH generally 4–5), there are no bones or other organic remains to enable radio-chronological dating. Only for sites dating from the past 5,000 years are detailed radio-chronological, archaeological, and paleoecological data available.

### The Middle Stone Age (120,000–12,000 B.P.)

During the long dry period, the Maluekian (figure 7.2), which lasted from 70,000 to 40,000 B.P., many of the central African forests gave way to savannas (Roche 1979). During this period, brutal but infrequent rainstorms caused severe erosion, which deposited large quantities of stones, pebbles, and gravel in riverbeds. Peoples of the middle Stone Age lived close to these rivers, using the stones to fashion rough tools, including the choppers, picks, and scrapers of the Sangoan industry, which are common in riverbanks and around ancient lakes in central Africa. These industries tend to occur in flood-plain deposits, which are by nature very disturbed and remain poorly understood.

The following period, the Ndjilian, from 40,000 to 30,000 B.P., was humid and favorable for forest growth (Caratini and Giresse 1979), but very few human remains are known from this period.

From 30,000 to 12,000 B.P., equatorial Africa suffered a severe drought, the Leopoldvillian climatic phase, when forests were restricted to river galleries and moist mountain ranges (Maley 1991). Sea level was more than 120 m below its current position, revealing vast beaches, which were molded by the wind into dunes along the coasts of Equatorial Guinea, Gabon, and Congo. Sporadic rains again caused severe erosion, washing away clay and sand layers and revealing the base rock. Rising rivers carried stones and pebbles, which accumulated in the middle terraces, about 10 m higher than they are today. Hunter-gatherer peoples of this Lupembian period lived on hill-

| Chronological references (years) | Industry | Palaeoclimate periods | Major events |
|---|---|---|---|
| Present | Late Iron Age (800 BP to present) | | |
| 500 | | | Forest expansion |
| 1,000 | Hiatus (1,400-800 BP) | | Humid phase (from 2,000 BP) |
| 1,500 | | Kibangian B (3,500 BP to present) | |
| 2,000 | Early Iron Age (2,500-1,500 BP) | | |
| 2,500 | | | Dry phase (3,000-2,000 BP) Savanna expansion |
| 3,500 | Neolithic stage (4,500-2,500 BP) | | Anthropic influence |
| 4,500 | | | |
| | Tshitolian complex (12,000-4,500 BP) | Kibangian A (12,000-3,500 BP) | Humid phase Forest expansion |
| 10,000 | Late Stone Age | | |
| | Lupembian complex | Leopoldvillian (30,000-12,000 BP) | Dry phase |
| 30,000 | Middle Stone Age | | Savanna expansion |
| | ? | Ndjilian (40,000-30,000 BP) | Forest expansion Humid phase |
| 40,000 | | | |
| | Sangoan complex Stone-line Industry | Maluekian (70,000-40,000 BP) | Dry phase Savanna expansion |
| 70,000 | Middle Stone Age | | |
| | ? | Pre-Maluekian (Eemien stage) | Humid phase |
| 120,000 | | | |

Figure 7.2. Chronology of major events in the middle Ogooué valley from 120,000 B.P.

tops close to water and made more sophisticated stone tools. Their throwing weapons, made out of stone collected from numerous rock deposits, suggest that they were living in an open savanna environment in which visibility was good. One finds these industries at the top of alluvial deposits, but to date there is no site in central Africa of known stratigraphy to provide a chronological reference for the Maluekian, Ndjilian, or Leopoldvillian period (Lanfranchi 1990).

## Holocene Cultures

### THE LATE STONE AGE (12,000–4,500 B.P.)

Around 12,000 B.P., in the Kibangian period, the earth became warmer and the climate moister, and once again forest colonized the savanna, sea level rose, and large areas of mangroves developed. Humans of this period made small tools, or microliths, from stone and demonstrated their improved technology with the advent of the bow and arrow. Remains from the late Stone Age are much more abundant than those of the middle Stone Age and include the well-known Tshitolian industry. These comprise many sites along the valley of the Ogooué (Oslisly et al. 1994), as well as on either side of the lower course of the Congo River on the Téké Plateau and the Kinshasa plain in the Democratic Republic of Congo (Cahen and Mortelmans 1973), in the Niari Valley, Congo (Bayle des Hermens and Lanfranchi 1978), and the north of Angola (Ervedosa 1980). These hunter-gatherers lived mostly on hilltops in savanna areas but also inhabited caves and rock shelters.

Remains from the late Stone Age are occasionally found within the sandy-clayey layer above the stone line (figure 7.3), which tends to be 1–2 m deep in the middle Ogooué region when it is present at all (in many places this layer has been eroded away). Within this layer, stone industries have been dated to between 9,000 and 6,000 B.P., placing this soil formation in the

Kibangian (see figure 7.2). All late Stone Age sites in the middle valley of the Ogooué are located on hilltops, principally in savanna but sometimes on the forest crest lines, with the Lélédi 10 site dated at 5,700 B.P. (Oslisly and White, in press). They often appear as large eroded patches in the savanna, visible on aerial photographs, and in many areas erosion has washed away all remains. It seems that this erosion process, which is ongoing, began during a climate change in the Kibangian B (3,500–2,000 B.P.; Maley 1992).

These late Stone Age industries evolved from the Lupembian industry, which was characterized by small hand axes, picks, large arrowheads, core axes, and scrapers, but which also developed smaller, lighter tools called microliths. Their improved technology, which belongs to the Tshitolian complex, would have allowed the hunter-gatherers of the time to become more mobile than their ancestors.

During the late Stone Age, large areas of stone fragments accumulated where tools were fashioned, and these are often visible today as patches of erosion. Neolithic and Iron Age peoples dug refuse pits, and Iron Age peoples also built furnaces to extract iron from ore (figure 7.4). What landscape did these peoples live in? This question is difficult to answer, but a site that we have named Lopé 2 (Oslisly 1993a) gives some useful pointers. This site has unusually well preserved stratigraphy, with three layers of charcoal, the first two of which are associated with numerous flakes and stone tools made from milky quartz and jasper (Oslisly et al. 1996):

- Layer 1, between 30 and 40 cm, is dated to 6,760 B.P.
- Layer 2, between 60 and 70 cm, is dated to 9,170 B.P.
- Layer 3, between 100 and 110 cm, is dated to 10,320 B.P.

Charcoal in these layers has been identified

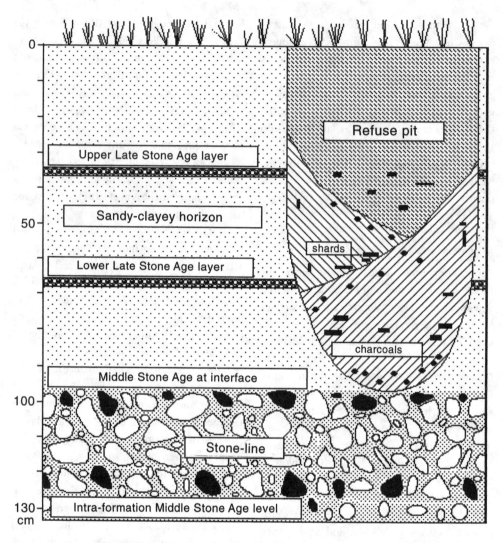

Figure 7.3. A classic example of an archaeological stratigraphic profile on a savanna hilltop in the middle Ogooué area. Early and middle Stone Age deposits are found in stone lines or in alluvial deposits. Holocene remains, including stone tools from the late Stone Age and refuse pits of Neolithic and Iron Age populations, are in the adjacent sandy–clayey horizon. In the middle valley of the Ogooué, hunter-gatherers of the late Stone Age, Neolithic peoples, and Iron Age peoples all tended to live on hilltops in the savanna, chosen for their dominant location with good visibility but always close to a source of water and to the forest. Hence most sites have several layers of habitation. Excavations must be meticulous, because remains from different eras are often in close proximity, particularly those in Neolithic and Iron Age refuse pits, which were sometimes dug into one another or which sometimes pass through remains from the late Stone Age. Adapted from Oslisly 1998.

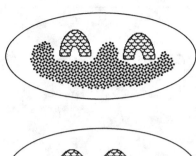

**Late Stone Age settlement**

Large concentration of stone
flakes around vegetal
settlement structures

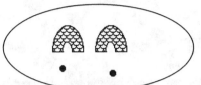

**Neolithic settlement**

A small settlement with few
people and one or two refuse
pits ● located on the hill top

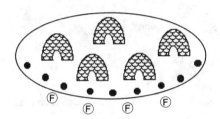

**Iron Age settlement**

There were numerous
settlement structures on the
hill top, with a belt of refuse
pits and iron furnaces Ⓕ
at the start of the slope

Figure 7.4. Schematic
representation of suc-
cessive human occu-
pations of a hilltop
during the three cul-
tural phases of the
Holocene.

as belonging to *Diogoa zenkeri* and *Strombo-siopsis tetrandra* (both Olacaceae), two species characteristic of mature rain forest in the region (Tutin et al. 1994). Analyses of levels of isotopic $\delta^{13}C$ indicate that savanna vegetation dominated down to 50 cm depth (c. 8,000 B.P.), but samples from 60–120 cm (c. 9,000 B.P. and older) gave values that were intermediate between forest and savanna (see Schwartz et al. 1986 for an explanation of this method). Hence these hunter-gatherer peoples lived in an open landscape of forest-savanna mosaic and chose to live on a hilltop. They are likely to have fed on plants collected from within the forest, cer-tainly hunted meat with bows and stone arrow-heads, and used the wood of forest trees for their fires.

TRANSITION TO THE NEOLITHIC

For the long Stone Age era, which ended around 5,000–4,500 B.P., there is nothing to suggest anything more than a gradual evolution

of technologies among a sedentary population. In contrast, the variety of cultures that appeared in the Neolithic and Iron Ages prob-ably reflects a series of waves of migration. The upper Holocene (5,000 B.P. to the present) saw a series of migrations of Bantu peoples from the grassy highlands of the Nigeria-Cameroon border (see figure 7.1) into central Africa fol-lowing two major routes (Phillipson 1985). The first, which appears to have been the most important, followed the area of forest-savanna contact to the north of the main forest block, leading first to the Great Lakes region of east Africa and then toward southern Africa. The second route was to the southeast, penetrating the equatorial forest in two very different direc-tions. One wave, probably the older, followed the coast, making use of a band of small savan-nas that run down from Equatorial Guinea through Gabon to Pointe Noire in Congo. Analyses of archaeological remains suggest that the first Neolithic groups traveled this way

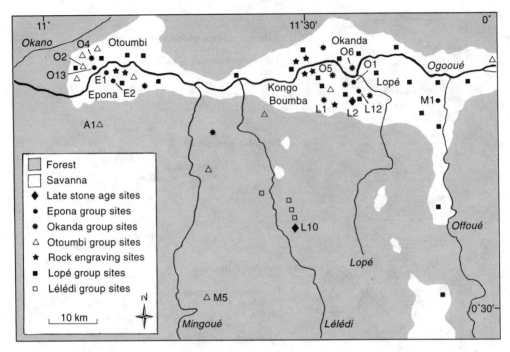

Figure 7.5. Map of late Stone Age, Neolithic, and Iron Age sites defined by pottery styles. Listing of sites: A1 = Anzem 1; E1 = Epona 1; E2 = Epona 2; L1 = Lindili 1; L2 = Lopé 2; L10 = Lélédi 10; L12 = Lopé 12; M1 = Makogué 1; M5 = Mingoué 5; O1 = Okanda 1; O2 = Otoumbi 2; O4 = Otoumbi 4; O5 = Okanda 5; O6 = Okanda 6; O13 = Otoumbi 13. Adapted from Oslisly 1998.

around 5,000–4,500 B.P. The second wave followed the many major ridge lines running north-south into the interior (Oslisly 1995).

### Neolithic Groups

It is possible that the Neolithic traits of polishing stone tools and making pottery evolved in situ, but their sudden, widespread appearance in the region about 4,500 B.P. suggests that the first Neolithic peoples arrived at that time from elsewhere. The earliest known sites are in the Okanda area located on ridges in the Massossou mountains, notably Okanda 5, dated to 4,500 B.P. (A. Assoko, pers. comm.) and Okanda 1, dated to 3,560 BP (figure 7.5). Excavations of refuse pits at Okanda 1 revealed a blackish substrate rich in charcoal and ashes from cooking fires, in which there were large pieces of pottery, a small pestle, the nucleus of a stone tool, and two flakes. The pottery remains are from spheroid receptacles characterized by uneven curvatures, some with annular bases and outward-bevelled, thicker top lips. The decorative, grooved patterns, made using a comb (figure 7.6), distinguish this pottery style from that of later Neolithic cultures.

Around 2,900 B.P. a new, more numerous Neolithic group arrived in the middle valley of the Ogooué. Eight sites (Ndjolé Cs, Ndjolé Pk5, Otoumbi 13, Epona 1 and 2, Okanda 6, Lopé 12, and Makogué 1) containing pottery of the same form and decoration, dated between 2,900 and 2,400 B.P. (the recent Neolithic or Epona group, Oslisly 1993a), have been discovered to date on hilltops close to the Ogooué, over a distance of about 100 km along its length, covering a total area of about 1,000 km². 

The pottery of this recent Neolithic stage is

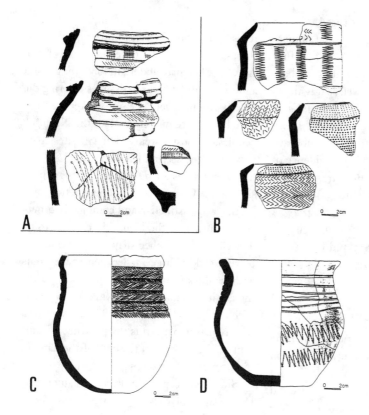

Figure 7.6. Neolithic pottery from the middle Ogooué valley: rims from Okanda 1 in the early Neolithic tradition (A); items from Otumbi 13 in the recent Neolithic Epona tradition, including keel pots (B), bilobed pot (C), and pot with out-curving lips (D). Adapted from Oslisly 1998.

stereotyped, with extreme constancy of form and decorative motifs over a long time period. This provides the first direct evidence of migration, since pottery of the same style in different places can be dated. Three types of vase can be defined (see figure 7.6): (1) keel pots, of which one fragment possesses decorated handles on its side; (2) bilobed pots, some of which have holes in their sides to enable them to be suspended; and (3) simple pots with out-curving lips. Bilobed vases with suspension holes and pots with out-curving lips similar in style to the pattern of the Epona group have been found on the coast at Lalala (Farine 1963) and at Okala, close to Libreville (Clist 1988), dated to a more recent period between 2,400 and 2,100 B.P. This suggests that there was cultural movement to the west of the middle Ogooué valley (Oslisly and Peyrot 1992b).

This hypothesis is further supported by the presence in the Lalala and Okala sites of polished stone axes made of amphibolite, a raw material available only in the middle valley of the Ogooué. In fact, tools made from this material are found throughout Gabon, demonstrating the extent to which these Neolithic groups traveled from their epicenter in the middle Ogooué valley.

Once again, these Neolithic groups inhabited hilltop sites, establishing a central platform around which one or two rubbish pits were dug (see figure 7.4). The settlements were invariably small, indicating low population density. Although rubbish pits were small in number, they have proved to be rich in information and archaeological remains. For example, one remarkable pit at the Otoumbi 13 site contained 13 kg of pottery, three grooved stones, one small pestle, two pitted stones (used to crack nuts), eleven polished stone axes, and numerous chisels and polished stone flakes. Refuse pits typically contain large quantities of charcoal from household fires, as well as fragments of the nuts of *Elaeis guineensis* (oil palm), *Coula edulis*, and *Antrocaryon klaineanum*. These small groups of people probably lived off forest resources, and to date there is no evidence of their practicing agriculture. Bone fragments in a pit at Otoumbi 13 (Oslisly 1993a) demonstrate that they hunted bushbuck (*Tragelaphus scriptus*) as well as some smaller mammals, but such remains are rarely preserved because of the acidic soils of the region.

## The Early Iron Age

### IRONWORKERS OF THE OKANDA TRADITION

Ironworking in west central Africa began in the Mandara mountains in northern Cameroon (MacEachern 1996) and in the area around Yaoundé in the south (Essomba 1989) around 2,600 B.P. The first evidence from the middle valley of the Ogooué is around 2,600–2,500 B.P. at Otoumbi 2 and Lopé 5, but it was not until around 2,300–2,100 B.P. that the Okanda tradition expanded, to the Otoumbi 4, Okanda 2 and 5, and Lindili 1 sites (see figure 7.5). The arrival of the new Okanda pottery style coincides with the abrupt disappearance of the Neolithic peoples. From a strictly chronological viewpoint, radiocarbon dates suggest that the first ironworkers may have coexisted with the last Neolithic peoples, but the range of dates for each group suggests that if they did cohabit, it was not for long (Oslisly and Fontugne 1993). There is no evidence of coexistence from cultural remains, suggesting that the superior weapons of the ironworkers enabled them to supplant the previous inhabitants.

These ironworking peoples invariably lived on hilltops, but their villages were much larger than those of the Neolithic cultures (see figure 7.4); the flat hillcrest was surrounded by belts of refuse pits, with furnaces nearby on the first slopes. These furnaces were built from clay to an average height of about 1 m above the ground and were ventilated through pipes entering the base at ground level. The ironworkers seem to have been more numerous than the Neolithic peoples; almost twenty known sites date between 2,300 B.P. and 1,800 B.P.

Pottery styles can be used to map the expansion of the Okanda group. Their pottery was particularly distinctive and completely different from that of the Neolithic Epona group. The closed and bilobed shapes disappeared, to be replaced by larger, taller, bell-shaped pots with handles (including one that was 50 cm high, with a capacity of 30 liters), decorated with characteristic patterns. The most typical feature of the pattern is the presence of concentric circles on or at the base of the handles (figure 7.7).

Similar concentric circles appear among the iconography of rock engravings at the Doda, Ibombi, Kongo Boumba, and Lindili sites (Kongo Boumba area) and at the Epona and Elarmekora sites (Epona area). These engravings were made by hammering pointed iron tools to form cuplike depressions in the rock. They cannot be dated directly, but the patterns correspond to those found on Okanda pottery dated between 2,300 and 1,700 B.P. (Oslisly and Peyrot 1993; Oslisly 1996). Of just over 1,600 engravings found to date, 67% are simple or concentric circles similar to those found on the handles of pottery (Oslisly 1993a, 1993b). In all, geometric shapes account for 75% of engravings, emphasizing the importance of abstract, symbolic characters. More realistic and vivid are animal representations, which account for a further 8%, portraying small quadrupeds or reptiles (figure 7.8); but there are no depictions of such large mammals as elephants, buffalo, or antelopes, as are found in engravings in the Sahara and southern Africa.

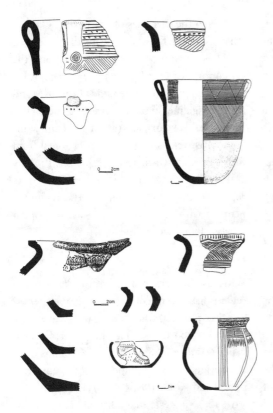

Figure 7.7. Iron Age pottery of the middle Ogooué valley: (*top*) pottery of the Okanda tradition (2,300–1,600 B.P.); (*bottom*) pottery of the Otoumbi tradition (1,900–1,500 B.P.). Adapted from Oslisly 1998.

Figure 7.8. Rock engravings of the Ibombi site: an animal figure represented as tanned hides with the body at the center, four lateral limbs, and a dorsal appendage forming the tail. Photo by R. Oslisly.

A third, poorly represented group includes such weapons and tools as throwing knives (the classic weapon of central Africa), spears, axes, and hunting nets.

Looking at the types of images portrayed, one can identify two styles, one abstract and symbolic, the other figurative. The dominance of symbolic figures suggests that the engravings served a magic or mystic purpose, illustrating events in an abstract form. Circular forms perhaps depict a culture inhabiting small open savannas enclosed by the forest.

IRONWORKERS OF THE OTOUMBI TRADITION

A new wave of ironworkers arrived in the middle Ogooué valley around 1,900–1,800 B.P. and took up residence on the hilltops vacated by the Okanda peoples. Apparently, after inhabiting the area for four or five centuries, they moved away to the south, following ridges through the forest to the savannas that descend toward the center of the Massif du Chaillu to the west of the Offoué river, and then on to the savannas of the upper Ogooué and northwest Congo. Such a scenario corresponds well with the more recent radiocarbon dates of iron smelting at Moanda in the upper Ogooué valley around 2,300–1,800 B.P. (Digombe et al. 1988) and on the Batéké plateau around 1,600–1,500 B.P. (Pinçon et al. 1995).

The Otoumbi ironworkers used the same smelting methods as the Okanda, but their pottery styles were new, with medium-sized pots that had out-curving lips and flat bases, decorated with indented lines, as well as small bowls with incurving lips (see figure 7.7). They were without handles and the decoration was more complex. A new style applied with a notched wheel appeared, and concentric circles were no longer featured. Similar pottery has been found 200 km further north in Gabon, at the Oyem 2 site in refuse pits dated to 2,300–2,200 B.P. (Clist 1989), and although data are few, the indications are that the Otoumbi ironworkers

arrived in the savannas of the middle Ogooué valley from the forests to the north.

These peoples expanded out from the Otoumbi area (see figure 7.5) between 1,900 and 1,600 B.P. along either side of the river and along ridgelines into the forest to the south, leaving tell-tale furnaces (at Anzem 1 and Mingoué 5) as well as large areas of charcoal from forest fires, dated to 1,500–1,400 B.P. (Oslisly and Dechamps 1994). These charcoal deposits represent the first evidence in the region of people with an itinerant slash-and-burn lifestyle living within the forest.

**The Hiatus**

Whereas the presence of ironworkers in the region has been confirmed for the period 2,500–1,500 B.P., radiocarbon dates suggest that between 1,400 and 800 B.P. the middle Ogooué valley was devoid of humans. No remains from this period have been found among seventy-five sites dated so far (Oslisly 1993a, 1995; figure 7.9). This suggests that a brutal phenomenon affected Iron Age peoples such that the region was empty until the arrival of the peoples of the recent Iron Age, who were the ancestors of populations present in the area today. The effect is not due to an anomaly in the calibration curve of $^{14}$C values with time, because it is essentially linear during the period in question (Stuiver and Becker 1993), and indeed the phenomenon is not restricted to the middle Ogooué valley; human remains dated during this period are rare throughout Gabon. The carbon dates that do fall between 1,400 and 800 B.P. are spread across the three coastal provinces, Nyanga, and the Haut Ogooué, but there is no evidence of human populations during the same period in the four other provinces of Gabon, which account for 46% of all carbon dates available for the time period (90 of a total of 194).

Was this hiatus caused by a severe epidemic? The tropics are certainly recognized as a region of crippling endemic diseases where sudden

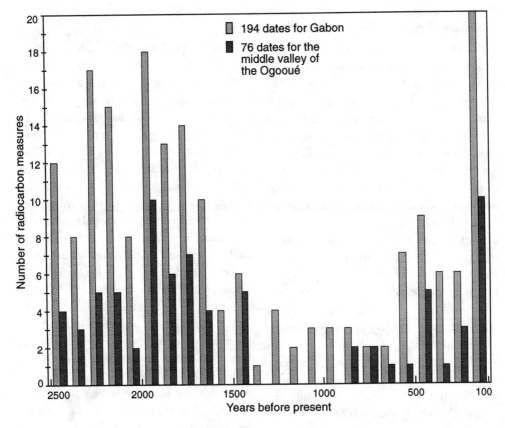

Figure 7.9. Distribution of uncalibrated radiocarbon dates between 2,500 and 100 B.P. Adapted from Oslisly 1998.

outbreaks can devastate human populations. For instance, early in the twentieth century bubonic plague devastated populations in many parts of Gabon (Deschamps 1962). At the start of my work, I was surprised to find sites that seemed almost intact, with artifacts lying on the surface of the ground as if suddenly abandoned. In an area where conditions were already difficult and settlements small, a serious illness could have decimated Iron Age communities in a very short time. This seems to be the most likely explanation for the population void and may even be the reason for the current low human population density. It would not be surprising if such an epidemic led

to a taboo, preventing repopulation of the area by surrounding peoples.

The six-hundred-year "human silence" most likely led to significant changes in the forest-savanna mosaic of the middle Ogooué valley. Anthropogenic fires, which currently maintain the savanna, would have been much less frequent, and the forest would have expanded dramatically, given that humid conditions favorable for forest growth have been prevalent in the region since 2,000 B.P. (Oslisly and White, in press; White, this volume, chapter 11).

## The Late Iron Age:
### Ironworkers of the Lopé and Lélédi Traditions

Ironworking populations reappeared in the middle Ogooué valley from 800 B.P. and again occupied hilltop dwelling sites. Ceramics of this period include large and small pots of flattened spherical shape with out-curved apexes and pots of uneven curvature, as well as clay pipes. The decoration of these pots is unique, with small circular motifs made with knotted strips of plant material arranged in herringbone patterns. The designs on pots form a band of variable width high on the sphere, and similar patterns are found on the clay bowls of pipes. This distinctive pattern has also been found on a pot from a cave at Lastoursville, 150 km upstream (Oslisly 1993a, 1995). Seeds of *Sesamum* cf. *calycinium* (Pedaliaceae—a species of sesame), which, when pounded, yield edible oil, were found on the inside of this pot. Carbon dates from sites containing this style of pottery confirm historical and linguistic studies of the Okandé peoples currently resident in the area (Ambouroué-Avaro 1981), which suggest that their ancestors arrived in the region around the fourteenth and fifteenth centuries.

This same style of pottery, named Lopé, has been found in an area greater than 1,500 km² in size and indicates that a unified cultural group occupied at least 250 km along the length of the Ogooué valley. This would have been possible because of these people's renowned skill at navigating the rapids of the Ogooué by dugout canoe, described by Savorgnan de Brazza in 1876 (Brunschwig 1966).

New and ongoing research has revealed numerous furnaces and extensive remains of slag along ridgelines opened up by logging roads in the middle valley of the Lélédi to the south of the Ogooué (between 0°15'–0°24' S and 11°28'–11°33' E; see figure 7.5) in a forested area at about 500 m altitude. Carbon dates fall between 800 and 200 B.P. Associated pottery found to date has been badly weathered, but the form, pattern, and composition are completely different from those of the Lopé tradition. Iron-smelting furnaces of these peoples consisted of holes dug to about a meter down into the sandy-clayey layer, into which alternate layers of iron ore and charcoal were loaded, with large clay pipes angled downward, through which air was pumped by means of a bellows into the base. Previous smelting technology used in the middle Ogooué valley had involved the erection of a clay structure above ground, which had to be broken open and destroyed to obtain the iron. This new method would have facilitated removal of metal and waste and subsequent reloading and reuse (Collomb 1977), distinguishing between these recent Iron Age peoples and those of the early period.

Hence, from the fourteenth through the nineteenth centuries, ironworkers in the region lived both in savanna (Lopé peoples) and within the forest (Lélédi tradition), extracting iron using the same technology but making pottery with different form and decoration. We will have to undertake further research to ascertain whether there was any commercial or cultural exchange between these two groups.

It is a sad reflection on "progress" that pottery traditions which had lasted about 4,500 years rapidly became extinct with the massive recent importation of containers from Europe. Pottery is no longer made or used in the middle valley of the Ogooué.

### Climatic Savannas Isolated Within the Forest

Following the long dry climatic phase of the Leopoldvillian (see figure 7.2), when savannas dominated the central African landscape, forests recovered in the humid Kibangian, reaching their present distribution around 6,000 B.P. (Maley 1987). Aubréville (1967) has already proposed that the "strange" middle Ogooué savannas (he termed them Booué) are a paleoclimatic relict. They have now been dated to around 9,000–8,000 B.P. using $\delta^{13}C$ and $\delta^{14}C$ analyses (Oslisly and White, in press; Oslisly et al. 1996), suggesting that they are a relict of the dry Leopoldvillian climatic phase, during which they would have been much more widespread. Further evidence that these are ancient savannas comes from the local presence of specialist savanna animals and plants: for example, such bird species such as the long-legged pipit, *Anthus pallidiventris;* the pectoral-patch cisticola, *Cisticola brunnescens;* and the black-faced canary, *Serinus capistratus*, which are thought unlikely to cross large expanses of forest (Christy and Clarke 1994; P. Christy, pers. comm.). Savanna frogs of the genus *Ptychadena*, considered good biological indicators of ancient grassland, are also present (Blanc 1998; C. Blank, pers. comm.), as is the bushbuck, *Tragelaphus scriptus*, a mammal that lives in savanna and woodland vegetation but which does not occur deep in the forest. The presence of these savanna specialists is good evidence that a savanna corridor once connected the middle Ogooué valley to Congo (see figure 7.1).

Identification of plant species from charcoal found in village sites known to have been located in the savanna demonstrates the presence of indicators of mature forest vegetation within the forest-savanna mosaic during the late Stone Age (White 1992, 1995). Furthermore, charcoal from the forest tree *Pterocarpus soyauxii* (Padouk) was favored by all Iron Age peoples for smelting ore and must have been obtained from within the forests of the period. This paints a picture of successive groups of peoples living in open savanna but making extensive use of nearby forests. It would have been their fires, lit to maintain their open habitat and, perhaps more important, to facilitate hunting of large savanna grazers, that prevented recolonization of the savannas by forest vegetation. Equally, during the hiatus when fires would, presumably, have been far less frequent, savanna areas were colonized rapidly by

forest species, illustrating the important ecological role played by humans (Oslisly and White, in press).

Research from the middle Ogooué valley does not provide evidence of true forest dwelling until the quite recent past. Even sites discovered within the forest today were actually located within savanna vegetation when inhabited (White and Oslisly, unpublished data). This emphasizes the need to take into account vegetation history when considering the dynamics of central African peoples, as well as when attempting to interpret the significance of archaeological remains (see also Eggert 1993).

Because of the lack of sites in forested central Africa with well-maintained stratigraphy dating to early or middle Stone Ages, we are currently unable to say much about the way these earlier peoples lived. From about 10,000 B.P. onward we possess a relatively detailed picture of population dynamics in the middle Ogooué valley, which demonstrates the arrival of a long sequence of civilizations, particularly from the Neolithic (c. 4,500 B.P.) onward. It seems that the major migrations of Bantu ironworkers were linked to a dry climatic phase in the Kibangian B (3,500–2,000 B.P.), which probably resulted in decreased forest cover and may have enabled these savanna-dwelling peoples to avoid the prospect of a daunting trip into extensive forest vegetation (Maley 1992; Schwartz 1992; see also Maley, this volume). These migrations were undertaken through Gabon following ridgelines (Oslisly 1995), although elsewhere river navigation was used (see, e.g., Eggert 1993). The migrating peoples seem to have favored areas with at least some savanna vegetation, reflecting their origins outside the forest ecosystem, and it is not surprising that the middle Ogooué valley was appealing to them. It seems that they systematically supplanted resident cultures, although it is possible that they also assimilated some local knowledge (see, e.g., Vansina 1990).

Elsewhere in forested Africa, as in the middle valley of the Ogooué, the recent Holocene was a time of successive waves of migrating Bantu peoples replacing resident populations. Hence, the overriding theme of human settlement in the middle Ogooué valley, as well as in other parts of Africa, is one of savanna-dwelling peoples who made use of the forest environment but who preferred to avoid dwelling within it as true "forest peoples." Therefore, this is not dissimilar to present patterns of migration within the forest zone, which tends to result in forest conversion.

This study demonstrates the rich and dynamic prehistorical past of the middle valley of the Ogooué, particularly during the Holocene, a period for which there are numerous archaeological remains. It also raises a number of questions, perhaps the most fundamental of which is: How important are these isolated savanna regions to the history of human populations in the middle valley of the Ogooué, and how has the long history of human habitation affected the savannas and the surrounding forest? Ongoing multidisciplinary research on archaeology, vegetation history, palynology, and plant ecology should provide answers and also reveal to what extent the history of human populations in the central African region as a whole mirrors that of the middle Ogooué valley.

ACKNOWLEDGMENTS

I thank the ECOFAC program, funded by the European Union and managed by AGRECO-GEIE, for supporting this research. I wish to express my gratitude to Lee White (Wildlife Conservation Society), Caroline Tutin (Center International de Recherches Médicales de Franceville-Gabon), and Bernard Peyrot for translation and constructive comments, and to Kate Abernethy, Michel Fernandez, and Alphonse Mackanga-Missandzou for help and cooperation at Lopé.

REFERENCES

Ambouroue-Avaro, J. 1981. *Un peuple gabonais à l'aube de la colonisation: Le Bas Ogowé au XIXème siècle.* Karthala, Paris.

Aubréville, A. 1967. Les étranges mosaïques forêt-savane du sommet de la boucle de l'Ogooué au Gabon. *Adansonia* 2(7):13–22.

Bayle des Hermens, R. de, and R. Lanfranchi. 1978. L'abri Tshitolien de Ntadi-Yomba (République du Congo). *L'anthropologie* 82(4):539–564.

Blanc, C. 1988. Reptiles et amphibiens de la réserve de faune de la Lopé. Rapport ECOFAC. Report.

Brunschwig, H. 1966. *Brazza explorateur de l'Ogooué (1875–1879).* Mouton, Paris.

Cahen, D., and G. Mortelmans. 1973. Un site Tshitolien sur le plateau des Batéké (République du Zaïre). *Annales du Musée Royal de l'Afrique Centrale, Sciences Humaines* 81.

Caratini, C., and P. Giresse. 1979. Contribution palynologique à la connaissance des environnements continentaux et marins du Congo à la fin du Quaternaire. *Comptes Rendus de l'Académie des Sciences de Paris* 288:379–382.

Christy, P., and W. Clark. 1994. *Guide des oiseaux de la Réserve de la Lopé.* ECOFAC, Libreville, Gabon.

Clist, B. 1988. Un nouvel ensemble néolithique en Afrique centrale: Le groupe Okala au Gabon. *Nsi* 3:59–65.

———. 1989. Vestiges archéologiques de fontes de fer dans la province du Woleu-Ntem au Gabon. *Nsi* 6:79–95.

Collomb, G. 1977. *Fondeurs et forgerons dans le bassin de l'Ogooué.* Ministère de la Culture et des Arts du Gabon, Libreville.

De Ploey, J. 1963. Quelques indices d'évolution morphologiques et paléoclimatiques des environs du Stanley-Pool (Congo). Studia Universatis Lovanium, Leopoldville, DR Congo.

Deschamps, H. 1962. *Traditions orales et archives au Gabon: Contribution à l'éthno-histoire.* Berger-Levrault, Paris.

Digombe, L., P. R. Schmidt, V. Mouleingui-Boukosso, J.-B. Mombo, and M. Locko. 1988. The development of an early Iron Age prehistory in Gabon. *Current Anthropology* 29(1):179–184.

Eggert, M. K. H. 1993. Central Africa and the archaeology of the equatorial rainforest: Reflections on some major topics. In T. Shaw, P. Sinclair, B. Andah, and A.

Okpoko, eds. *The Archaeology of Africa: Food, Metals and Towns.* Routledge, London.

Ervedosa, C. 1980. *Arqueologia Angolana.* Ediçoes 70, Lisbon.

Essomba, J. M. 1989. Dix ans de recherches archéologiques au Cameroun méridional (1979–1989). *Nsi* 6:33–57.

Farine, B. 1963. *Sites préhistoriques Gabonais.* Ministère de l'Information au Gabon, Libreville.

Jansen, J., A. Kuijpers, and S. Troelstra. 1986. A mid-Brunhes climatic event: Long term changes in global atmosphere and ocean circulation. *Science* 332:619–622.

Lanfranchi, R. 1990. Les industries préhistoriques en R.P. du Congo et leur contexte paléogéographique. In R. Lanfranchi and D. Schwartz, eds. *Paysages quaternaires de l'Afrique centrale atlantique.* ORSTOM, Paris.

MacEachern, S. 1996. Iron Age beginnings north of the Mandara mountains, Cameroon and Nigeria. Pages 489–495 in G. Pwiti and R. Soper, eds. *Aspects of African Archaeology* (Papers from the Tenth Congress of the Pan African Association for Prehistory and Related Studies—1995, University of Harare, Zimbabwe). University of Zimbabwe publications, Harare.

Maley, J. 1987. Fragmentation de la forêt dense humide africaine et extension des biotopes montagnards au Quaternaire récent: Nouvelles données polliniques et chronologiques. Implications paléoclimatiques et biogéographiques. *Palaeoecology of Africa* 18:307–334.

———. 1991. The African rain forest vegetation and palaeoenvironments during late Quaternary. *Climatic Change* 19:79–98.

———. 1992. Mise en évidence d'une péjoration climatique entre 2500 et 2000 ans BP en Afrique tropicale humide. *Bull. Soc. Géol. France* 163:363–365.

Mortelmans, G., and R. Monteyne. 1962. Le Quaternaire du Congo occidental et sa chronologie: Actes du IVème Congrès Panafricain de Préhistoire et de l'étude du Quaternaire. *Annales Sciences Humaines du Musée Royal de l'Afrique Centrale, Tervuren* 40:90–132.

Oslisly, R. 1993a. *Préhistoire de la moyenne vallée de l'Ogooué (Gabon).* TDM 96, ORSTOM, Paris.

———. 1993b. Rock art in Gabon: Petroglyphs in the Ogooué river. *Rock Art Research* 10(1):18–23.

———. 1995. The middle Ogooué valley, Gabon: Cultural changes and palaeoclimatic implications of the last four millennia. In J. Sutton, ed. *The Growth of Farming Communities in Africa from the Equator Southwards. Azania* 29–30:324–331.

———. 1996. The rock art of Gabon: Techniques, themes and estimation of its age by cultural association. Pages 361–370 in G. Pwiti and R. Soper, eds. *Aspects of African Archaeology* (Papers from the Tenth Congress of the Pan African Association for Prehistory and Related Studies—1995, University of Harare, Zimbabwe). University of Zimbabwe publications, Harare.

———. 1998. Hommes et milieux à l'Holocène dans la moyenne vallée de l'Ogooué (Gabon). *Bulletin de la Société Préhistorique Française* 95(1):93–105.

Oslisly, R., and R. Dechamps. 1994. Découverte d'une zone d'incendie dans la forêt ombrophile du Gabon *ca* 1500 BP: Essai d'explication anthropique et implications paléoclimatiques. *Comptes Rendus de l'Académie des Sciences de Paris* 318(2):555–560.

Oslisly, R., and M. Fontugne. 1993. La fin du stade néolithique et le début de l'âge du fer dans la moyenne vallée de l'Ogooué: Problèmes chronologiques et changements culturels. *Comptes Rendus de l'Académie des Sciences de Paris* 316(2):997–1003.

Oslisly, R., and B. Peyrot. 1992a. Un gisement du paléolithique inférieur: La haute terrasse d'Elarmekora (Moyenne vallée de l'Ogooué) Gabon: Problèmes chronologiques et paléogéographiques. *Comptes Rendus de l'Académie des Sciences de Paris* 314(2):309–312.

———. 1992b. L'arrivée des premiers métallurgistes sur l'Ogooué (Gabon). *African Archaeological Review* 10:129–138.

———. 1993. *Les gravures rupestres de la vallée de l'Ogooué (Gabon).* Sépia, Paris.

Oslisly, R., B. Peyrot, S. Abdessadok, and L. White. 1996. Le site de Lopé 2: Un indicateur de transition écosystémique *ca* 10.000 BP dans la moyenne vallée de l'Ogooué (Gabon). *Comptes Rendus de l'Académie des Sciences de Paris* 323(2):933–939.

Oslisly, R., M. Pickford, R. Dechamps, M. Fontugne, and J. Maley. 1994. Sur une présence humaine mi-Holocène à caractère rituel en grottes au Gabon. *Comptes Rendus de l'Académie des Sciences de Paris* 319(2):1423–1428.

Oslisly, R., and L. White. In press. La relation homme/milieu dans la réserve de la Lopé (Gabon) au cours de l'Holocène: Les implications sur l'environnement. In *Symposium dynamique à long terme des écosystèmes forestiers intertropicaux, ORSTOM, March 1996, Bondy.* UNESCO, Paris.

Phillipson, D. W. 1985. *African Archaeology.* Cambridge University Press, Cambridge.

Pinçon, B., R. Lanfranchi, and M. Fontugne. 1995. L'âge du fer dans les régions téké (Congo, Gabon): Datations radiométriques et interprétations historiques. *Comptes Rendus de l'Académie des Sciences de Paris* 320(2a):1241–1248.

Roche, E. 1979. Végétation ancienne et actuelle de l'Afrique centrale. *African Economy and History* 7:30–37.

Schwartz, D. 1992. Assèchement climatique vers 3000 BP et expansion Bantu en Afrique centrale atlantique: Quelques réflexions. *Bull. Soc. Géol. France* 163(3):353–361.

Schwartz, D., A. Mariotti, R. Lanfranchi, and B. Guillet. 1986. 13C/12C ratios of soil organic matter as indicators of vegetation changes in the Congo. *Geoderma* 39:97–103.

Stuivert, M., and B. Becker. 1993. High precision decadal calibration of the radiocarbon time scale AD 1950–6000 BC. *Radiocarbon* 35(1):35–65.

Tutin, C. E. G., M. Fernandez, M. E. Rogers, E. A. Williamson, and W. C. McGrew. 1991. Foraging profiles of sympatric lowland gorillas and chimpanzees in the Lopé Reserve, Gabon. *Philosophical Transactions of the Royal Society of London Series B* 334:179–186.

Tutin, C. E. G., and R. Oslisly. 1995. *Homo, Pan* and *Gorilla:* Co-existence over 60,000 years at Lopé in Central Gabon. *Journal of Human Evolution* 28:597–602.

Tutin, C. E. G., L. J. T. White, E. A. Williamson, M. Fernandez, and G. McPherson. 1994. List of plant species identified in the northern part of the Lopé Reserve, Gabon. *Tropics* 3:249–276.

Vansina, J. 1990. *Paths in the Rainforests.* James Curry, London.

White, L. J. T. 1992. Vegetation history and logging disturbance: Effects on rain forest mammals in the Lopé reserve, Gabon. Ph.D. thesis, University of Edinburgh, Scotland.

———. 1995. Etude de la végétation de la Réserve de la Lopé. Rapport final ECOFAC Gabon. Report.

# CHAPTER 8

# Forest Area and Deforestation in Central Africa

## Current Knowledge and Future Directions

David S. Wilkie and Nadine Laporte

Suitability of the central African climate for year-round plant growth and generally favorable geologic and topographic features have resulted in the formation of the second largest contiguous area of tropical moist forest (TMF) in the world (1.8 million km²). This immense biome constitutes about 15% of the world's remaining TMF (UNESCO 1978) and encompasses the entire countries of Gabon and Equatorial Guinea, much of Congo, Cameroon, and Democratic Republic of Congo (DR Congo), and the southwestern corner of the Central African Republic (figure 8.1). Although climatic conditions over much of central and west Africa indicate that the landscape is capable of supporting TMF, only in central Africa has much of the forest escaped the logger's chainsaw and the farmer's axe (table 8.1).

From a regional perspective, we can pose three basic questions about the state of tropical forests in central Africa: What do we know about the present extent and state of the dense humid forest? What are the major factors that result in a change in state and extent? and What tools do we have available to us to detect and monitor changes in the forest? For most of the nations that constitute central Africa, statistics on forest extent, clearing, and reforestation are woefully inadequate. Deforestation statistics, therefore, are presented in this chapter to highlight their variability as much as to provide basic information. Of greater importance is a discussion of the factors that contribute to a change in forest status, and a critique of available methods for evaluating the impact of these factors on a regional scale. Information for this chapter draws heavily from the results of a Biodiversity Support Program (BSP) study, funded by the United States Agency for International Development (USAID), of central Africa's role in global climate change. We also draw on information from the first phase of the Tropical Ecosystem Environmental Observation by Satellite (TREES) project of the Joint Research Center of the European Community (Italy) and the USAID Central Africa Regional Program for the Environment (CARPE).

In this chapter we attempt to report on the most recent studies and the capabilities and limitations of the most recent technologies, particularly with respect to satellite remote sensing. The pace of change, however, often means that by the time such survey chapters as this are published, the tools available to researchers and practitioners have advanced from those described. Fortunately, the advent and explosion of the Internet

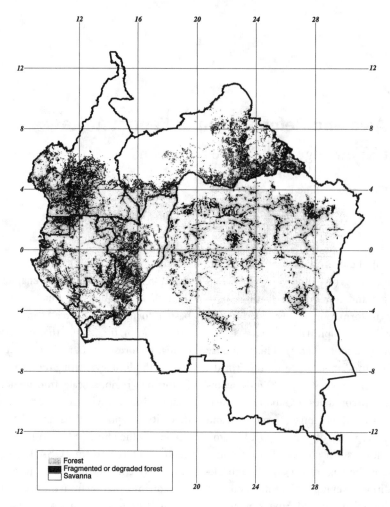

Figure 8.1. Vegetation map of central Africa derived from 1 km AVHRR imagery (1990s). From Laporte et al. 1998.

and the World Wide Web have dramatically increased our access to information on remote sensing and Geographical Information System (GIS) technologies, allowing us browse sources of remote sensing imagery, to locate and view the results of regional and global surveys of land cover, and to keep abreast of the latest advances in the field.

### Estimates of Tropical Moist Forest Area

Unlike the Amazon, where over the past ten years a time series of high-resolution satellite imagery has been gathered to assess forest extent and the rate and distribution of deforestation (Skole and Tucker 1993), for central Africa there has been a concerted mapping effort only since the mid-1990s (Malingreau et al. 1995; Justice et al. 1995; Laporte et al. 1998; Mayaux et al. 1998). Quantitative data on the distribution and extent of vegetation types within each country of central Africa are often incomplete, and the data that do exist vary in quality.

Table 8.1

## Forest Extent, Deforestation, and Reforestation Within Central Africa (1980s)

| COUNTRY | ALL FOREST AREA | % | DRY FOREST AREA | % | MOIST FOREST AREA | % | DEFORESTATION PER YEAR (KM²) | REFORESTED AREA (KM²) |
|---|---|---|---|---|---|---|---|---|
| Cameroon | 185 | 59 | 30 | 69 | 155 | 56 | 2,000–5,000 | 280 |
| CAR | 279 | 55 | 147 | 51 | 133 | 59 | 500–600 | 0 |
| Congo | 174 | 49 | 0 | 0 | 174 | 49 | 200–700 | 300–500 |
| Equatorial Guinea | 13 | 50 | 0 | 0 | 13 | 50 | 30 | — |
| Gabon | 173 | 35 | 0 | 0 | 173 | 35 | 150–600 | 300–500 |
| DRC | 832 | 57 | 91 | 54 | 741 | 57 | 2,000–4,000 | 400–600 |

*Sources:* Data from IIED 1988; Myers 1989; WRI 1991.

*Note:* % = % deforested; CAR = Central African Republic; DRC = DR Congo. Unless otherwise noted, areas are given in 10³ km².

THE FOOD AND AGRICULTURAL ORGANIZATION

The most frequently cited source of forest information for central Africa dates back to 1981, when the Food and Agricultural Organization (FAO) published, in conjunction with the United Nations Environment Program (UNEP), the results of the Tropical Forest Resources Assessment Project: Forest Resources of Tropical Africa (table 8.2; see Lanly 1981). From this study, FAO/UNEP hoped to determine the extent and state of forest resources within Africa and to assess rates of deforestation and afforestation. Central Africa data sources and data veracity were, however, as varied as the geography, socioeconomic situation, and institutional capacity of the constituent nations. For example, measures of forest extent in Cameroon were based on aerial photogrammetry and visual interpretation of a sample of Landsat MSS (80 m resolution at 1:1,000,000 scale) satellite images obtained over a two-year period (1973–1975). In contrast, forest estimates for DR Congo were based on a questionnaire sent to the national forestry agency and relied heavily on Devred's 1958 map, which was itself a compilation of maps and vegetation surveys developed in the 1930s. Lanly (1983) attempted to rank the quality of

data available on forest extent and deforestation of six nations within central Africa. Only data for Cameroon were ranked good.

Since the FAO report was published, it has become the major (or sole) source of forest area and deforestation data used by planners, conservation ecologists, and climate modelers. In fact, the FAO data are often the basis for unsourced tables and figures in secondary and tertiary articles that falsely give the impression of multiple data sources for central African forest statistics. More recently, the FAO has developed an inventory method that uses Landsat imagery to improve the estimates of forest area and rates of change (FAO 1996). This approach is discussed in more detail later.

OAK RIDGE NATIONAL LABORATORY

In 1990 the Oak Ridge National Laboratory (ORNL) completed a study for USAID to estimate carbon inventories and emissions for sub-Saharan Africa, as part of the primary project goal of estimating the extent to which sub-Saharan land-use change might be capable of mitigating global warming through additional carbon uptake (Graham et al. 1990). To achieve this goal, vegetation cover (forest, mixed forest-savanna, and savanna) was to be "deter-

Table 8.2
## Forest Extent in Central Africa (1980s)

| COUNTRY | CLOSED FOREST AREA | OPEN FOREST AREA | FALLOW AREA | ALL AREA |
|---|---|---|---|---|
| Cameroon | 179,200 | 77,000 | 16,900 | 317,200 |
| Central African Republic | 35,900 | 323,000 | 41,000 | 399,900 |
| Congo | 213,400 | — | 11,000 | 224,400 |
| Equatorial Guinea | 12,950 | — | 11,650 | 24,600 |
| Gabon | 205,000 | 750 | 15,000 | 220,750 |
| DR Congo | 1,056,500 | 718,400 | 184,000 | 1,958,900 |
| Total | 1,702,950 | 1,119,150 | 323,650 | 3,145,750 |

| COUNTRY | FOREST | WOODLAND | AGRIC-FALLOW | ALL |
|---|---|---|---|---|
| Cameroon | 205,871 | 122,727 | 102,440 | 431,038 |
| Central African Republic | 102,018 | 321,440 | 198,609 | 622,067 |
| Congo | 181,798 | 69,294 | 56,543 | 307,635 |
| Equatorial Guinea | 19,287 | 633 | 1,581 | 21,501 |
| Gabon | 144,908 | 21,763 | 77,094 | 243,765 |
| DR Congo | 897,002 | 1,084,415 | 208,831 | 2,190,248 |
| Total | 1,550,884 | 1,620,272 | 645,098 | 3,816,254 |

*Sources:* Data from Lanly 1981 (upper table) and Millington et al. 1991 (lower table).

*Note:* All figures are given in km².

mined" for each country using contemporary sources of information. After evaluating what recent timber inventory, forest resources, and exploitation maps were available, Graham and colleagues realized that methodological and land-cover classification differences precluded meaningful comparisons, and they decided to rely on data from the FAO/UNEP 1981 forest resources report as the basis for their deforestation projections. Graham and colleagues also decided to use the draft digitized FAO vegetation map (Lavenu 1987) and the White (1983) vegetation map for forest area coverage.

ADVANCED VERY HIGH RESOLUTION RADIOMETER NORMALIZED DIFFERENCE VEGETATION INDEX

Concerns about biomass fuel availability in sub-Saharan Africa resulted in a joint UNDP/World Bank Bilateral Energy Sector Management Program assessment of woody biomass standing stock and sustainable yield in sub-Saharan Africa (Millington et al. 1991). This study was one of the first attempts to map land-cover types in sub-Saharan Africa by interpretation of digital AVHRR (Advanced Very High Resolution Radiometer) Normalized Difference Vegetation Index (NDVI) data (Millington et al. 1991) at a ground resolution of 8 km. Although the forty-three biomass/land-cover classes generated in this study are not synonymous with FAO (Lanly 1981) and White (1983)

Table 8.3

## Percentage of Closed Forest Cover by Country

| COUNTRY | FAO = UNEP 1981 COUNTRY | FAO = UNEP 1981 REGION | MILLINGTON ET AL. 1981 COUNTRY | MILLINGTON ET AL. 1981 REGION |
|---|---|---|---|---|
| Cameroon | 39 | 11 | 44 | 13 |
| Central African Republic | 6 | 2 | 16 | 7 |
| Congo | 62 | 13 | 53 | 12 |
| Equatorial Guinea | 46 | 1 | 69 | 1 |
| Gabon | 80 | 12 | 56 | 9 |
| DR Congo | 47 | 62 | 40 | 58 |

*Note:* All figures are percentages. Country figures indicate the percentage of the area of each country that is covered by closed forest, whereas regional figures indicate the percentage of the region's total closed forest cover found in each country.

vegetation cover types (a perennial problem when compiling country-based information), they allow a very rough comparison when clumped into Forest and Woodland categories (see table 8.2).

Although the FAO and AVHRR-NDVI country-level estimates of forest cover differ from 12% to 54% (table 8.3), both show that DR Congo contains the largest area of closed forest in the region (table 8.4) and that closed forest covers from about 40% to 80% of each country (with the exception of the Central African Republic). The total area of all closed forests within the region was estimated at approximately 3,000,000 km² (approximately 30% of the conterminous United States). To further demonstrate the range of uncertainty associated with estimates of forest extent and deforestation within central Africa, data for each country are presented (see table 8.4).

### AVHRR-LAC/HRPT

Most recent studies of forest cover in central Africa have relied on AVHRR images for regional (national and multinational) forest surveys, owing to the frequent data acquisition (which increases the probability of acquiring cloud-free coverage), large area view, and relatively low cost. It is difficult, however, to acquire AVHRR data over areas with perennial cloud cover, and spatial resolution is often too coarse to detect finer-scale changes, particularly those characteristic of land-cover transformation by slash-and-burn settlers throughout much of central Africa. Nonetheless, degraded forest mosaics, including fields, fallow, and secondary forest, can be mapped at the regional scale. For example, estimates of the extent of degraded forest have been reported by country for the entire Congo Basin (Laporte et al. 1998).

As an important contribution to the International Geosphere Biosphere Program (IGBP 1992), a 1 km resolution vegetation map for the entire tropical belt was recently generated from AVHRR imagery (Malingreau et al. 1995; Mayaux et al. 1998). These vegetation maps provide a 1990s baseline for long-term monitoring at a regional scale. Reduced-resolution copies of the TREES project vegetation maps are available from the European Commission Joint Research web site. A similar effort has been undertaken for the central Africa regional maps by the USAID-funded

Table 8.4

## Difference in Estimates of Forest Cover for Central Africa

|  | FAO CLOSED/OPEN AREA (KM²) | MILLINGTON FOREST/WOODLAND AREA (KM²) | DIFFERENCE | % |
|---|---|---|---|---|
| Cameroon | 256,200 | 328,598 | -72,398 | 28 |
| Central African Republic | 358,900 | 423,458 | -64,558 | 18 |
| Congo | 213,400 | 251,092 | -37,692 | 18 |
| Equatorial Guinea | 12,950 | 19,920 | -6,970 | 54 |
| Gabon | 205,750 | 166,671 | 39,079 | 19 |
| DR Congo | 1,774,900 | 1,981,417 | -206,517 | 12 |

*Sources:* FAO data are from FAO 1981; Millington data are from Millington et al. 1981.

Central African Regional Program for the Environment (CARPE).

The TREES project and other studies (Justice et al. 1993; Laporte et al. 1995, 1998; Mayaux et al. 1998) used high-resolution imagery (Landsat TM and MSS) to verify AVHRR land-cover classifications as part of the BSP Central Africa Global Climate Change and Development project (BSP 1993). This effort has resulted in vegetation maps with four land-cover classes (forest, degraded forest, mixed forest-savanna, and savanna) for Cameroon and DR Congo. A comparison of cover-type areal estimates between the different approaches (table 8.5) has increased our knowledge of the extent and state of the forest within central Africa. The level of error in forest-cover estimates, based on comparison of the AVHRR data with a sample of higher-resolution Landsat MSS imagery and limited field surveys, ranges from 8% (overestimation in dense forest areas) to 21% (underestimation in forest-savanna transition areas). To improve forest-cover estimates using AVHRR, the TREES project developed a methodology to calibrate and correct the resulting classified images. After correction, the residual errors computed on an independent sample of high-resolution scenes varied

from 1% to 1.5% for central Africa (Mayaux and Lambin 1995).

AVHRR image analysis allows for a useful first approximation of the extent and state of forest resources in central Africa (table 8.6), but it is not ideal for detecting and characterizing forest change over time (Justice et al. 1993). Moreover, like all optical sensors, it is unable to image areas with perennial cloud cover. These difficulties have prompted the use of alternative data sources, which are described next.

### ERS-1 AND JERS-1 SAR

As the TREES project progressed toward completion of a "wall-to-wall" AVHRR image mosaic of central Africa, gaps in coverage of areas with perennial cloud cover became a critical concern. In late 1993 the TREES/ERS-1 1994 project was initiated to assess the usefulness of high-resolution (30 m) Synthetic Aperture Radar (SAR) imagery for regional forest monitoring, with an emphasis on discriminating forest from non-forest land cover (Malingreau and Duchossois 1996). With a mobile receiving station located in Libreville, Gabon, 477 scenes were acquired during the period from July 15 to August 28, 1994. The images were resampled to 100 m nominal spatial resolution to improve the signal-to-noise ratio,

Table 8.5

## Estimates of Areal Vegetation Extent from Cameroon and DR Congo

| | CAMEROON | | DEMOCRATIC REPUBLIC OF CONGO | |
| | FAO ('80s) | LAPORTE ('90s) | FAO ('80s) | LAPORTE ('90s) |
|---|---|---|---|---|
| Forest | 191,600 | 168,087 | 1,056,500 | 1,117,963 |
| Degraded forest | 45,350 | 64,218 | 78,000 | 102,821 |
| Mixed forest-savanna | 68,550 | 63,691 | 824,400 | 437,565 |
| Savanna | 22,550 | 24,709 | 30,700 | 34,299 |
| Total | 328,130 | 320,795 | 1,989,680 | 1,692,738 |

*Sources:* FAO data are from Lanly 1981; Laporte data are from Laporte et al. 1995.

Table 8.6

## Forest Extent Derived from AVHRR 1 km Resolution Imagery

| | LAPORTE ET AL. 1998 (1989–1990) | MAYAUX ET AL. 1998 (1990–1992) |
|---|---|---|
| Cameroon | 173,850 | 173,780 |
| Central African Republic | 60,897 | 60,370 |
| Congo | 224,615 | 239,160 |
| Equatorial Guinea | 16,207 | 18,110 |
| Gabon | 210,701 | 206,770 |
| DR Congo | 1,127,211 | 1,141,470 |

*Note:* All figures are in km².

while retaining sufficient detail for regional-scale analysis. The resulting mosaic covers the entire tropical forest domain of central Africa (>2500 km²) as well as the northern and southern forest-savanna transition zones. Acquisition of wall-to-wall imagery for the whole region within a two-month period demonstrates the utility of non-optical active imaging systems for tropical rain forest monitoring. Preliminary visual examination of the regional mosaic showed that the satellite-based SAR data were able to distinguish the boundaries between evergreen or semi-deciduous forest and the mixed seasonal forest-savanna in southern Congo Basin, and were able to detect savanna islands within the forest zone. The radar data were also able to discriminate some old-growth from postagricultural secondary-growth vegetation. However, rainfall immediately prior to data acquisition can affect the SAR signal, making it difficult to interpret the data.

Interpretation of SAR imagery within tropical forested regions of Africa is in its infancy, and an operational use for forest mapping and monitoring is still under development. Work by Dobson et al. (1995) and others suggests that land-cover classification using the experimental SAR satellites ERS-1 and JERS-1 is more accurate and at a higher spatial resolution than that generated by classification of NDVI data from multitemporal AVHRR imagery. ERS-1 data allowed differentiation of forest from nonforest distribution in central Africa but failed to distinguish different woody vegetation or degraded forest classes (de Gauwer and de Wulf 1997).

A new vegetation map of Africa, based on the JERS-1 SAR data, is under development through a collaboration between the Jet Propulsion Laboratory (JPL) and the TREES project. Validation of this map will attempt to merge optical and radar data and will be undertaken in collaboration with the University of Maryland and in-country partners identified through CARPE.

## Estimates of Forest Change and Deforestation

Slash-and-burn agriculture, logging, infrastructure development, mining, and fuelwood extraction all contribute, in different ways, to changes in forest extent and composition. The scale of the impact of these factors is a function of population density and growth rate, and regional and national economic conditions and policies. For example, forests close to major ports, such as Pointe-Noire in Congo, are often cut, whereas more isolated forests are high graded, with less than one or two trees extracted per hectare. Similarly, fuelwood gathering in rural areas with low population densities has no visible effect on the forest, whereas densely populated urban areas cause the development of "halos" of deforestation. In the case of Kinshasa, DR Congo, the halo extends over 100 km from the city center. In northeastern DR Congo, fluctuations in the world price of coffee, national tax policies, and a degrading infrastructure have resulted in the periodic expansion, abandonment, and rehabilitation of large commercial coffee plantations, and accompanying changes in the subsistence agricultural areas cleared by plantation laborers.

Land transformation in central Africa clearly varies in spatial and temporal scale. Choice of remote sensing imagery and the methodology for detecting and characterizing landscape change must, therefore, be appropriate to the scale of land transformation and the level of analysis (i.e., local, regional, or national). Given present extent and rate of deforestation in central Africa, the resolution of AVHRR imagery is unlikely to be sufficient for monitoring land transformation over much of the area. Attempts to measure landscape change in the region have thus employed high-resolution imagery when the temporal resolution of these data are sufficient (Stancioff and Pessutti 1981; Wilmet and Vennetier 1986; Castiaux et al. 1991).

### FOOD AND AGRICULTURAL ORGANIZATION

Two primary goals of the UN/FAO project were assessment of: (1) the state of the forest cover for the reference year 1990; and (2) the rate of change in forest cover between 1981 and 1990 (Singh 1990; FAO 1996). The assessment was conducted in two phases. The first phase was based on collecting and organizing forest-cover data made available by countries. Statistical (tabular) data were compiled within a database (FLORIS). Spatial data were compiled within a GIS. Forest-cover data contained within FLORIS were based on country assessments prepared on different dates. The FAO developed a deforestation model that combines forest-cover estimates with such ancillary data as human population growth, ecological zones, precipitation, and socioeconomic variables to correct data gathered on differing dates to the standard reference years of 1980 and 1990.

Country forest-cover estimates based on the model were made under the following scenarios: (1) when reliable multi-date inventories were available to calibrate the model and subsequently compute the standardized results; (2) when a reliable single-date inventory was available; and (3) when no reliable inventory data were available. In the third case the standardized results were computed using a general (uncalibrated) model. Estimates of baseline forest-cover area were extracted from existing vegetation maps, which were entered into scenario 2 above. Clearly, scenario 1 was the optimal approach, when possible.

The second phase of the project, performed to improve "traditional" FAO forest-cover estimates, was designed around a sampling approach using remote sensing imagery and a common system of classification and interpretation to estimate global deforestation rates over the previous decade (Singh 1990). For this phase, the project used a random sampling of forty Landsat MSS 1:250,000 color print pairs (one from 1980 and the other from 1990) covering central and southern Africa. The approach uses a geographical stratification and a second-stage forest-cover vegetation map. The Landsat images are visually interpreted, and change is determined using transparent overlays and dot-grid counts. Although this approach is simple, its cost effectiveness varies depending on the availability of imagery, the representativeness of the sampling scheme, and, most important, the ability of photo-interpreters (foresters from the region) to characterize land-cover types within the imagery and to detect land transformations between the image pairs. The skill and experience of photo-interpreters is vital to assessing

land transformation accurately. This method is more fully described elsewhere (FAO 1996).

The first phase of the study was completed in late 1994, and data in the form of national language "Country Briefs" are available on the FAO web site. Forest-cover estimates for 1980 and 1990 in central Africa (table 8.7) show the total change in cover over that time period, and an estimate of annual forest-cover change. All information for central African nations was generated using either scenario 2 or scenario 3 (i.e., the accuracy of the information is in doubt). Although the FAO has generated new estimates of forest cover for Africa, we are still left without confidence limits on the estimates for central Africa and are thus unsure whether they are an improvement over the 1981 FAO assessment. Comparison of table 8.4 and table 8.7 shows that FAO has not only generated new estimates of forest cover for 1990 but has revised (downward for all countries other than Equatorial Guinea) its original estimates for the extent of forest in 1980. A recent comparison (Mayaux et al. 1998) of FAO forest estimates with the 1990s estimates from TREES and the International Union for the Conservation of Nature (IUCN) supported this result.

Table 8.7

## Forest-Cover Change Estimates for Central Africa Between 1980 and 1990

|  | 1980 AREA (KM²) | 1990 AREA (KM²) | DIFFERENCE AREA (KM²) | % CHANGE |
|---|---|---|---|---|
| Cameroon | 215,690 | 203,500 | -1,220 | -0.6 |
| Central African Republic | 318,540 | 305,620 | -1,290 | -0.4 |
| Congo | 201,880 | 198,650 | -320 | -0.2 |
| Equatorial Guinea | 18,960 | 18,260 | -70 | -0.4 |
| Gabon | 193,980 | 182,350 | -1,160 | -0.6 |
| DR Congo | 1,205,970 | 1,132,750 | -7,320 | -0.6 |

*Sources:* FAO 1996 (FAO 1990 assessment). Reprinted by permission of the Food and Agriculture Organization of the United Nations.

BIODIVERSITY SUPPORT PROGRAM

To illustrate some of the methodological problems that can be expected in quantifying deforestation within central Africa, Justice et al. (1993) conducted a change detection study on four locations within the region chosen to provide a range of land-cover types (figure 8.2: Kanaga, Landsat Path/Row 177/64; Nyunzu 173/64; Ituri 174/58; Kutu 180/62). For each study site a search of the Landsat MSS archive was undertaken and near-anniversary images were selected at least nine years apart and with minimal cloud cover. Using unsupervised classification, visual labeling of classes, and pixel-by-pixel change detection, all four areas showed low rates of deforestation (< 0.1% per year). More important, the study showed how the low dynamic range and few spectral bands of Landsat MSS imagery resulted in considerable spectral overlap of land-cover types, and therefore an inability to discriminate among them. In addition, the relatively coarse pixel size (~80 m) produced an abundance of mixed pixels that were particularly prevalent in highly heterogeneous areas or areas where there were such linear features as gallery forests. The greatest limitation of this study was acknowledged to be lack of adequate ground information. Neither of the analysts conducting the classifications had firsthand knowledge of the areas, nor were such high-resolution products as aerial photography or videography available.

NASA LANDSAT PATHFINDER PROJECT

In 1990 the National Air and Space Administration (NASA), in conjunction with the United States Environmental Protection Agency and the United States Geological Survey EROS Data Center, began developing a process for using large amounts of high-resolution satellite imagery to map the rate of tropical deforestation (Asrar and Dokken 1993). The project focused initially on the Amazon but has been expanded as part of NASA's Earth Observing System activities to cover the tropical rain forests of central Africa and Southeast Asia. The Landsat Pathfinder Project is acquiring several thousand Landsat scenes at three points in time: mid-1970s, mid-1980s, and mid-1990s. Once the three-epoch data set is available, much more accurate estimates of forest extent and condition can be prepared and deforestation rates determined. Standardized land-cover classes and methods of analyses developed by the Pathfinder project will enhance the utility of the resulting vegetation-cover maps and will greatly facilitate the ability to conduct cross-country comparisons. Forest-extent coverages are now available for DR Congo and the Central African Republic for the 1980s and 1990s. Because of the persistence of clouds for most of Gabon, southern Cameroon, and Congo, and because of the absence of a Landsat receiving station in the region, a complete wall-to-wall map is still not available for these three countries. Products from the Pathfinder project can be downloaded from the NASA Pathfinder web site.

REGIONAL ENVIRONMENTAL INFORMATION PROJECT

The World Bank Regional Environment Information Management Project (REIMP) was implemented in 1997. The overall goal of the REIMP is to enhance the capacity of the six nations of central Africa to collaboratively monitor natural resource use and land-cover change, and with this information plan appropriate actions for natural resource management. Specific objectives are to establish a "demand-driven and action-oriented information system and to build capacity at local, national and regional levels to improve monitoring, land use planning, priority setting, and decision making for natural resource management, particularly for forest biodiversity conservation and management in the Basin" (Rantrua 1996). The project was financed

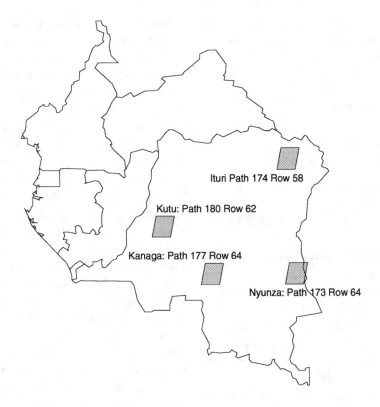

Figure 8.2. Map showing the locations of Landsat MSS scenes in the BSP study (Justice et al. 1993).

largely by the Global Environment Facility, although other multilateral and bilateral agencies (EU, UNDP, USAID, GTZ, FAC, ACDI, and AGCD) have expressed interest in participating, and others are already collaborating (FAO-Africover, NASA-Pathfinder, USAID-CARPE, and EU-ECOFAC). Links are also in place with non-governmental organizations (NGOs) active in the region (IUCN, WWF, and WCS).

The REIMP will acquire high-resolution satellite coverage for the basin, high-resolution aerial photography for urban areas, and regional 1:200,000 scale and local (urban) 1:50,000 scale topographic information. Data will be compiled or generated to create thematic maps of biodiversity, climate, soils, land use, and demography at 1:200,000 scale. Systems for managing, sharing, and using geographic data will be put in place, and personnel will be trained in the use of geographic databases for monitoring and decision making. The project will also work on increasing Internet connectivity to ease the exchange of data within and among nations. Each nation is expected to create a National Environmental Information Network (Réseaux Nationaux d'Information Environnementale—RNIE) to compile and exchange monitoring and planning information among governmental and non-governmental agencies within the country. Representatives from the national-level RNIEs would participate in a regional environmental information council (Conseil Régional de l'Information Environnementale—CRIE) that would attempt to develop collaborative solutions to transboundary environmental challenges. More information on the REIMP project can be found at its web site.

### Assessing the Reliability of Deforestation Estimates

Grainger (1993) offers a checklist of questions for assessing the reliability of deforestation-rate estimates. These include:

- Who made the estimate, and is it a primary or a secondary source?
- What type of forest and what kind of changes are included?
- Is the estimate based on measurements or subjective judgment?
- If measurement, what kind of remote sensing was used and what resolution?
- What were the dates of measurements?
- Was the whole country or region surveyed, and if not, was a statistical sampling methodology used?
- If a remotely sensed survey was compared with a map, on what measurements was the map based?
- Is the figure an estimate of an actual "historical" change or a projection of a possible future change?

Any study related to land-use and land-cover change should deal explicitly with these eight points in order to avoid confusion between estimates and to better understand the limitations of the data sets used.

Availability of and access to high-resolution satellite imagery for the Congo Basin are likely to improve in the future. However, the accuracy of estimates concerning forest cover and conditions will improve only if systems are put into place to integrate field survey information effectively into the process of classifying remote sensing imagery (Wilkie and Finn 1996). Although software from remote sensing image analysis can quickly categorize the landscape within an image into classes of similar spectral reflectance, only a knowledgeable individual using detailed ground survey information can label each category in complex landscapes. Thus although automatic digital classification analysis of remote sensing information is possible, without field-based information the human interpreter would often be unable to assign classes to a particular land-cover type. GPS-assisted low altitude aerial videography (Sidle and Ziewitz 1990; Marsh et al. 1994) will help greatly with image classification and validation, by providing rapid and accurate field survey information over relatively large areas. Because land transformation in central Africa is primarily the result of small-scale agricultural activities, however, accurate image interpretation will still require field demographic and land-use surveys conducted by trained individuals (Wilkie 1994). In-country personnel in natural resources and forestry departments and ministries have considerable field experience, and with fluency in local languages are usually best equipped to gather interview information. Yet in-country personnel may lack appropriate technical skills and usually lack the resources to conduct systematic surveys.

The key to enhancing the accuracy of, and confidence in, remote sensing image analysis estimates of forest cover in the Congo Basin is to develop strategies to team up the work of field-based conservation biologists with remote sensing experts and vegetation modelers. This will require that donors provide suitable training and logistical support to national departments and ministries to enable them to obtain the much-needed field data, and that NGOs, government agencies, and universities provide opportunities and incentives for scientists working at local and global levels to communicate and share information with one another.

Development of the International Geosphere Biosphere Program (IGBP) Data and Information System (IGBP-DIS) and the System for Analysis Research and Training

(START) is a major undertaking to establish networks of research scientists and to create global data sets essential to modeling climate change and monitoring land transformation at regional and global scales (Justice et al. 1995). The IGBP-DIS development and implementation process may provide the forum for improving collaboration among field scientists and researchers of global change. The United States Geological Survey, NASA, and IGBP-DIS should increase efforts to include field personnel of national and international conservation NGOs active within central Africa in the creation and validation of regional and global satellite-based land-cover data sets.

By more effectively combining the knowledge and experience of field-based researchers with those of remote sensing specialists, our ability to generate accurate maps of natural resource bases and land-cover change for the region will be enhanced greatly.

It is clear from the above program and data summaries that substantial variation exists in our estimates of forest extent, condition, and rate of deforestation. This should not be surprising given the size of the area involved, the differing categories used to classify forests, the range of factors that adversely impact central Africa's forests, and the limited infrastructure at a national level available to monitor forest resources. Despite the lack of accurate statistics, past studies agree that the forests of central Africa are subject to relatively low rates of deforestation compared to the rest of tropical forested Africa or the Amazon. Approximately

Table 8.8

## Programs and Their Associated Web Sites

| ABBREVIATION | NAME | WEB SITE |
|---|---|---|
| CARPE | Central Africa Regional Program for the Environment | http://carpe.umd.edu/ |
| EDC | EROS Data Center Landsat Data Distribution | http://edcdaac.usgs.gov/ |
| ERS-1 | European Remote Sensing Satellite | http://earth.esa.int/ERS/ |
| JERS-1 | Japanese Earth Resources Satellite | http://yyy.tksc.nasda.go.jp/Home/ Earth_Obs/e/jers_e.html |
| | Mapping activities, central Africa | http:// www-radar.jpl.nasa.gov |
| NASA/UMD Landsat Pathfinder | National Aeronautics and SpaceAdministration/ University of Maryland | http://www.geog.umd.edu/tropical/ |
| REIMP | Regional Environment Information Management Program | http://www.esd.worldbank.org/reimp/ |
| Space Imaging | Satellite imagery distributor (formerly EOSAT) | http://www.spaceimaging.com/ index.htm |
| SPOT | Système Pour l'Observation de la Terre | http://www.spotimage.fr/spot=us.htm |
| START | System for Analysis Research and Training | http://start.org |
| TREES | Tropical Ecosystem Environment oberservation by Satellite | http://ewse2.jrc.it/anonymous/ construct/build.pl/98692 |

0.2–0.6% of forest in central Africa is cleared annually, in contrast to 1% for the Ivory Coast and more than 0.9% on average for the rest of west Africa. However, with present rates of population growth, the rate and extent of deforestation in central Africa are likely to rise rapidly (Barnes 1990). Given the uncertainty of present estimates of forest cover for the region, the likelihood of increasing human impact on forests, and the unknown consequences of global warming on regional weather patterns, what strategies should be adopted to improve our assessment and monitoring of forest extent, composition, and rate of change? The first steps in establishing such a strategy include:

- coordination of national and international efforts to build and maintain spatial databases (satellite imagery, GIS forest maps, and so on);
- exchange of information and expertise between countries (through workshops, networks, the Internet, and so on);
- investigation of the operational use of such new tools as radar imagery or such new optical sensors as MODIS (moderate-resolution imaging spectroradiometer) for land-cover and change-detection studies;
- multi-sensor data fusion;
- multi-scale assessment of deforestation; and
- development of a spatially explicit model of deforestation combining sociocultural and economic factors.

Given the size of the forested regions in central Africa and the need for relatively quick but reliable estimates of areal extent and conversion rates, what is the most cost-effective source of forest information? Field surveys are extremely expensive and, though detailed, provide information at only the local level. Remote sensing, with its synoptic view and repeated coverage, is the most obvious choice for basin-wide forest surveys and has, most recently, been the key to improving estimates of forest cover and condition. Remotely sensed information can be drawn from a variety of sources: multi-spectral and radar imagery, aerial photography, and aerial videography. For a regional survey, how does one choose the best source of imagery? Appendixes 8.1 and 8.2 provide a brief outline of the features, advantages, and disadvantages of each source of remote sensing information. A more comprehensive treatment of the types and uses of remote sensing imagery for natural resources assessment and monitoring can be found at the web sites listed in table 8.8 and Wilkie and Finn 1996.

ACKNOWLEDGMENTS

Information for this paper was compiled during completion of a report titled *Central Africa Global Climate Change and Development* published by the Biodiversity Support Program, a USAID-funded consortium of World Wildlife Fund, the Nature Conservancy, and World Resources Institute. Thanks to Chris Justice and Ned Horning of the NASA Goddard Space Flight Center for collating the past and present uses of remote sensing in central Africa and for generating the Landsat TM image of northern Congo. Lee White and Scott Goetz provided many useful comments on earlier drafts of the manuscript.

REFERENCES

Asrar, G., and D. J. Dokken. 1993. *Earth Observation System reference handbook*. NASA, Washington, D.C.

Barnes, R. F. W. 1990. Deforestation trends in tropical Africa. *African Journal of Ecology* 28:161–173.

BSP (Biodiversity Support Program). 1993. *Central Africa: Global climate change and development—technical report*. Biodiversity Support Program, Washington, D.C.

Castiaux, N., M. Massart, and J. Wilmet. 1991. Environmental study of tropical African urban areas by multi-spectral satellite imageries (Lubumbashi in Zaire, central Africa). Twenty-fourth International Symposium on Remote Sensing, Rio de Janeiro, Brazil.

de Gauwer, V., and R. de Wulf. 1997. *Potential use of ERS-SAR data to discriminate natural and degraded woody vegetation types in central Africa*. European Space Agency, Florence.

de Grandi, G., M. Leysen, E. Nerzy, A. Bresati, F. Amar, and A. Fung. 1995. Wave scattering modeling of natural target under time evolution: A test case towards the physical interpretation of multi-temporal ERS-1 images over tropical forest. Pages 688–690 in *Proceedings: IGARSS conference*. International Geoscience and Remote Sensing Symposium, Florence.

Devred, R. 1958. La végétation du Congo Belge et du Ruanda-Urundi. *Bulletin de la Société Royale Botanique de Belgique* 65:409–468.

Dobson, M. C., F. T. Ulaby, and L. E. Pierce. 1995. Land-cover classification and estimation of terrain attributes using synthetic aperture radar. *Remote Sensing of Environment* 51:199–214.

FAO (Food and Agricultural Organization). 1994. *Tropical forest assessment 1990: Country briefs*.

———. 1996. *Forest resources assessment 1990*. FAO Forestry Paper 112.

FAO-UNEP. 1981. *Tropical forest resources assessment project: Forest resources of tropical Africa. Part 1: Regional synthesis*. FAO, Rome.

Graham, R. L., R. D. Perlack, A. M. G. Prasad, J. W. Ranney, and D. B. Waddle. 1990. *Greenhouse gas emissions in sub-Saharan Africa*. Oak Ridge National Laboratory, Oak Ridge, Tenn.

Grainger, A. 1993. Rates of deforestation in the humid tropics: Estimates and measurements. *Geographical Journal* 159:33–44.

IGBP. 1992. *Improved global data for land applications—a proposal for a new high resolution data set*. IGBP Report No. 20. International Geosphere Biosphere Program, Paris.

IIED. 1988. *Natural forest management for sustainable timber production*. IIED/ITTO, London.

Justice, C. O., G. B. Bailey, M. E. Maiden, S. I. Rasool, D. E. Strebel, and J. D. Tarpley. 1995. Recent data and information system initiatives for remotely sensed measurements of the land surface. *Remote Sensing of Environment* 51:235–244.

Justice, C. O., N. Horning, and N. Laporte. 1993. Remote sensing and GIS contributions to a climate change program in central Africa. In *Central Africa: Global climate change and development—technical Report*. Biodiversity Support Program, Washington, D.C.

Lanly, J. P. 1981. *Tropical forest resources assessment project (GEMS): Tropical Africa, Tropical Asia, Tropical America*. 4 vols. FAO/UNEP, Rome.

———. 1983. Assessment of forest resources of the tropics: A review article. *Forestry Abstracts* 44:6.

Laporte, N. T., S. J. Goetz, C. O. Justice, and M. Heinicke. 1998. A new land cover map of central Africa derived from multi-resolution, multi-temporal AVHRR data. *International Journal of Remote Sensing* 19:3537–3550.

Laporte, N., C. Justice, and J. Kendall. 1995. Mapping the dense humid forest of Cameroon and Zaire using AVHRR satellite data. *International Journal of Remote Sensing* 16:1127–1145.

Lavenu, F. 1987. Digitized vegetation map of Africa—descriptive memoir and map prepared for the department of forestry resources. FAO, Rome.

Malingreau, J. P., F. Achard, G. D'Souza, H. J. Stibig, J. D'Souza, C. Estreguil, and H. Eva. 1995. AVHRR for global tropical forest monitoring: The lessons of the TREES project. *Remote Sensing Reviews* 12:29–40.

Malingreau, J. P., and G. Duchossois. 1996. *The TREES/ERS-1 SAR '94 project*. European Commission Joint Research Centre, Ispra, Italy.

Marsh, S. E., J. L. Walsh, and C. Sobrevila. 1994. Evaluation of airborne video data for land-cover classification accuracy assessment in an isolated Brazilian forest. *Remote Sensing of Environment* 48:61–69.

Mayaux, P., F. Achard, and J. P. Malingreau. 1998. Global tropical forest area measurements derived from coarse resolution satellite imagery: A comparison with other approaches. *Environmental Conservation* 25:37–52.

Mayaux, P., and E. Lambin. 1995. Estimation of tropical forest area from coarse spatial resolution data: A two-step correction function for proportional errors due to spatial aggregation. *Remote Sensing of Environment* 53:1–16.

Millington, A. C., R. W. Critchley, T. D. Douglas, and P. Ryan. 1991. *Estimating woody biomass in sub-Saharan Africa*. World Bank, Washington, D.C.

Myers, N. 1989. *Deforestation rates in tropical forests and their climatic implications*. Friends of the Earth, London.

Rantrua, F. 1996. *Central African region: Regional Environmental Information Management Project (REIMP-CA)—project information document*. GEF/World Bank, Washington, D.C.

Sidle, J. G., and J. W. Ziewitz. 1990. Use of aerial videography in wildlife habitat studies. *Wildlife Society Bulletin* 18:56–62.

Singh, K. D. 1990. Design of a global tropical forest resources assessment. *Photogrammetric Engineering and Remote Sensing* 56(10):1353–1354.

Skole, D., and C. J. Tucker. 1993. Tropical deforestation and habitat fragmentation in the Amazon: Satellite data from 1978 to 1988. *Science* 260:1905–1910.

Stancioff, A., and J. Pessutti. 1981. Evaluation of the deforestation of the Nyunzu area by means of Landsat MSS imagery. Report on the accomplishments of ERTS Zaire in north Shaba, 1980–1981: Volume 2. Contract #AID-AFR-C-1483. USAID, Washington, D.C.

UNESCO. 1978. *Tropical forest ecosystems: A state-of-knowledge report*. UNESCO, Paris.

White, F. 1983. *The vegetation of Africa: A descriptive memoir to accompany the UNESCO/AETFAT/UNSO vegetation map of Africa*. UNESCO, Paris.

Wilkie, D. S. 1994. Importance of detailed field data to assess land use and land cover transformation in central Africa using Landsat MSS imagery. *Human Ecology* 22(3):379–403.

Wilkie, D. S., and J. T. Finn. 1996. *Remote sensing imagery for natural resources monitoring: A guide for first-time users*. Columbia University Press, New York.

Wilmet, J., and P. Vennetier. 1986. Croissance urbaine et évolution d l'environement à Brazzaville (Congo) et Kinshasa (Zaïre). *Photo-Interpretation* 86(5):37–43.

WRI. 1991. Forests and rangelands. Pages 101–120 in *World Resources 1990–91*. World Resources Institute, Washington, D.C.

Yamaguchi, Y., A. B. Kahle, H. Tsu, T. Kawakami, and M. Pniel. 1998. Overview and Advanced Spaceborne Thermal Emission and Reflection Radiometer (ASTER). *IEEE Transactions on Geoscience and Remote Sensing* 36:1062–1071.

APPENDIX 8.1

## Trade-Offs Associated with Various Forms of Image Data for Land-Cover Mapping

### Commercial Aerial Photography

Large-format aerial photography (i.e., at least 70 mm film stock) is used in most countries throughout the world to generate topographic and land-use maps. Thus, archival panchromatic, color, and IR aerial photography often can be purchased from national or state cartographic agencies. If photographs are not available for your area, commercial companies can be contracted to obtain photographs at the most appropriate time and scale. Aerial photography is most appropriate for applications that: have a small area or local perspective, < 1,000 km²; constitute once-off or base-line mapping; and are interested in detecting and identify-

ing small features, <10 m in diameter. The advantages of aerial photography include high spatial resolution, simple operation, and low cost of analysis equipment. The disadvantages of aerial photography include the limited spectral range of photographic film; limited digital analysis (which can be conducted only by scanning photographic prints or transparencies; difficulty in interpreting large volumes of data (large areas); high data acquisition and analysis costs per km² for areas greater than 1,000 km². Soon, new sensors launched by such private companies as Space Imaging will sell very high resolution satellite imagery (5 m).

### Non-Commercial Aerial Photography

Aerial 35 mm photography is most often obtained on an ad hoc basis by the researcher. Using your own 35 mm camera, oblique photographs are shot through the open windows of a rented small plane flying over the target area. Aerial 35 mm photography is most appropriate for applications that: involve one or a few small areas, <5 km²; require frequent, repeated coverage; and are interested in detecting and identifying very small features, <5 m in diameter. Advantages of 35 mm aerial photography include: inexpensive microscale surveys (sampling) and monitoring; superior spatial resolution (<1:500 scale possible); simple operation; and low cost of analysis equipment. Disadvantages are: geographic rectification is required for area estimation of features; digital analysis is possible only by scanning photographic prints or transparencies; color infrared film may require users to process and print their own negatives; photographic film has a limited spectral range; and it is difficult (in time and labor) to interpret large volumes of data (large areas).

Advent of digital cameras that merge traditional 35 mm photography with CCD digital data collection adds ease of digital processing to the advantages of aerial 35 mm photography. The Kodak DC 260 digital camera features a 1,536 × 1,024 pixel CCD sensor with 24-bit color. File size is 1.5 MB for a single black-and-white image and 4.5 MB for a 24-bit color image. The Minolta RD-175 offers a 1,528 × 1,146 three-CCD digital camera attached to a standard 35 mm camera lens system.

Though more expensive than photographic imaging and with somewhat lower pixel resolution, digital cameras provide for almost immediate access to images that can be transmitted electronically across telephone lines and computer networks, can be digitally processed and enhanced, and can be integrated easily into documents using desktop publishing software. The resolution of digital cameras is likely to match that of 35 mm photography in the near future. It should be noted, however, that as digital camera resolution increases, so do image storage requirements. For example, DC 260 requires 1.5 MB to store a single black-

and-white image, whereas a 1,732 × 1,732 resolution camera, matching the resolution of 35 mm film, would require 3.0 MB to store a single black-and-white image and more than 12 MB for a true color image. Relatively inexpensive digital cameras with lower pixel resolution are now also available.

### Aerial Videography

Aerial videography can be obtained much like 35 mm photography, by pointing a videocamera or camcorder through the window of a rented airplane. More often, however, the videocamera is mounted outside the window, pointing directly down, and is controlled remotely by the researcher within the plane (Sidle and Ziewitz 1990). Aerial videography is most appropriate for applications that: involve a few relatively small areas or samples along a transect or linear feature, <50 km²; require frequent, repeated coverage; and are interested in detecting and identifying relatively large features, >5 m in diameter. As with aerial photography, the spatial resolution of aerial videography can be increased by increasing the focal length of the camera lens or by reducing the altitude of the aircraft. Thus, although it is possible to use aerial videography to count poached elephant carcasses, this high spatial resolution comes at a cost of a very narrow field or view (i.e., narrow survey strip width). Advantages of aerial videography are: low cost—it is the least expensive system for small- to mid-scale surveys (sampling) and monitoring; ability to view imagery during image acquisition; the possibility of visual and digital analysis; simple operation; visible to near-IR spectral range of video cameras; and low cost of acquisition and analysis equipment. Disadvantages of aerial videography are: low pixel resolution; the requirement of relatively high light levels for image acquisition; and the requirement of geographic rectification for area estimation of features.

### Optical Satellite Imagery

Satellite imagery can be purchased from commercial companies (e.g., EOSAT and SPOT Image) or from government agencies (e.g., United States Geological Survey, Eurimage). Images can be obtained in digital form (i.e., the image is stored on tape or disk and must be transferred to, and viewed on, a computer) or as black-and-white or color prints and transparencies. Although at present, satellite imagery provides the most comprehensive regional-scale land-cover and land-use information globally, these data are not available for the years before 1972, when the first Landsat satellite was launched. Satellite imagery is most appropriate for applications that: have a large area, regional, or global perspective, >20,000 km²; require several spatially separate areas to be surveyed, monitored, or compared;

require frequent, repetitive coverage; require or can take advantage of multispectral data; and are interested in features larger than the spatial resolution of the imagery. Advantages of satellite imagery are: wide spectral range (UV–thermal); quantitative biophysical measurements from radiometric information obtained from calibrated sensors; wall-to-wall coverage; historical data; digital and visual analysis; digital and photographic (analog) output; ease of comparing different scales and wavelengths; semi-automated processing that makes use of full dynamic range of the data; and low cost. Disadvantages of satellite imagery are: low spatial resolution relative to airborne and ground-based sensors; high startup equipment costs.

### Side Looking Airborne Radar (SLAR) and Synthetic Aperture Radar (SAR) Imagery

Radar imagery can be obtained either by plane or by satellite. The preceding remote sensing systems all depend on the sun to illuminate landscape features. In contrast, radar systems provide their own source of illumination by transmitting an EMR signal in the microwave region. These microwaves are reflected back from landscape features and are detected by the sensor. The timing and intensity of the return signal are used to generate the final radar image. To overcome the cloud-cover problems associated with passive optical systems (e.g., Landsat MSS, TM, SPOT, and AVHRR), microwave remote sensing systems should improve spatially extensive data coverage. Advantages of radar are: the ability to obtain data regardless of weather conditions. Disadvantages of radar are: airborne equipment is expensive; satellite systems are primarily experimental; spatial resolution is moderate; image analysis and interpretation methods for natural resource management are not well developed; and radar provides information on terrain and vegetation texture and water content only.

SAR data are available through the European Space Agency in Frascati (Italy). As part of the JRC-ESA TREES project, wall-to-wall ERS-1 SAR mosaic for the entire central African region was built on the data acquired in 1994 using a mobile receiving station located in Libreville, Gabon (de Grandi et al. 1995). Furthermore, the Canadian satellite RadarSat, launched in late 1995, is providing C band (5.3 GHz) SAR satellite data with a ground resolution of 10–100 m ($1,600 per scene). And the Global Rainforest Mapping Project (GRFM) now distributes freely resampled data (100 m) from the Japanese Earth Resources Satellite (JERS-1).

SAR data are available for much of central and west Africa. JERS-1 data were acquired in January–March 1996 for area between 9° N and 9° S. The Congo River Basin was also covered during October–November 1996, when seasonally wet forests were inundated. Madagascar was acquired in January 1997. The JERS-1 SAR mission terminated on October 11, 1998.

### The Earth Observing System (EOS)

Earth Observing System (EOS) satellites are providing new imagery of the earth. Among the new sensors, the MODIS instrument supplies high spectral (thirty-six channels between 0.4 mm and 15 mm) and temporal (two-day cycle) resolution imagery at moderate spatial resolutions (250–1,000 m). These data provide a unique opportunity for regional and global land-cover change studies. On the same platform, the ASTER instrument will collect high spatial resolution (15–90 m) multi-spectral (visible through thermal IR) observations, providing substantial information for subpixel scaling analyses with other instruments (Yamaguchi et al. 1998).

### Other Sensors

The Russian Almaz satellite provides 15 m resolution radar imagery on a by-request basis (distributed by Hughes STX in Lanham, Maryland, and SPOT Image Corporation in Reston, Virginia), but the high cost of data ($0.9 per km²) is likely to preclude its use for regional surveys.

APPENDIX 8.2

## Choosing the Most Appropriate Imagery

Choice of an appropriate source of remote sensing information depends on the size of the area to be surveyed and the level of detail required. All imagery available for central Africa exhibits a trade-off between spatial detail (the smallest object that can be identified in the imagery) and spatial and temporal coverage. Figures 8.3 and 8.4 show this trade-off graphically. Notice how a single NOAA-AVHRR image has almost basin-wide coverage but its spatial resolution does not capture fine detail within the landscape. In contrast, aerial videography covers only an extremely narrow swath of the landscape but does so with great detail. Table 8.9 describes in very general terms the type of landscape features that can be identified at difference map scales, and table 8.10 summarizes the trade-offs among map scale, spatial resolution, and spatial coverage for the most commonly available sources of land-cover and land-use data for central Africa.

### Availability and Relative Costs of Different Remote Sensing Imagery for Regional Mapping

Wall-to-wall mapping of forest cover in the Congo Basin would require more than 40,000 aerial photographs (assuming 50% overlap) at a scale of 1:60,000; more than 700 SPOT images; 120 Landsat TM images; or only 2 AVHRR scenes. These figures do not include the duplicate images that need to be acquired to ensure cloud-free coverage or series of images necessary for vegetation seasonality or land-cover change analysis. For example, the TREES project had to acquire 150 AVHRR scenes to generate a relatively cloud-free forest map of the region.

Central African nations vary greatly in their archives of aerial photography. For example, Cameroon has a relatively complete (nationwide) archive of historical aerial photography, whereas Congo has only very limited photography. Contemporary aerial photographic coverage (within the past ten years) does not exist on a national or regional basis, and would be prohibitively expensive to obtain and analyze. Satellite remote sensing is thus the only feasible approach to monitoring forest cover over the whole basin. Several remote sensing systems are suitable for assessing central Africa's forest resources at different temporal and spatial resolutions (see table 8.10). The AVHRR sensor provides daily coverage at 4 km resolution, and 1 km resolution at only $90 per scene. The high temporal resolution offers the greatest opportunity for obtaining cloud-free imagery quickly and allows for very generalized vegetation mapping at a regional scale. The coarse spatial and spectral resolution limits its usefulness for change detection studies, however.

Although a series of high spatial resolution Landsat and SPOT satellites have been in orbit since 1973 and 1986, respectively, complete coverage for central Africa still does not exist because of dense cloud cover, the absence of a concerted effort to obtain data, and the high cost of obtaining imagery since Landsat was privatized in 1984 (prices rose from $200 per scene to $3,500). The future may not be so bleak. U.S. legislation (the Land Remote Sensing Policy Act of October 28, 1992, Public Law 10255) recognizes the failure of commercialization of the Landsat program by amending the Remote Sensing Commercialization Act of 1984. The amendment moved the Landsat program (acquisition, archiving, pricing, and distribution of imagery, and development of new satellite systems) to the National Aeronautics and Space Administration (NASA) and the Department of Defense (DOD) for joint management. The law provides access to Landsat imagery at marginal cost for U.S. agencies and researchers in the U.S. Global Climate Change Research Program and its international counterpart programs, researchers financially supported by U.S. government agencies, and international noncommercial organizations cooperating with the U.S. government on projects. By June 1999, Landsat 7 imagery was available to the public at the cost of duplication and handling (approximately $200). But the major constraint to Landsat data acquisitions for central Africa is still the absence of a permanent ground receiving station.

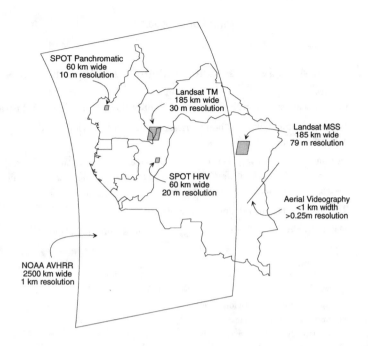

Figure 8.3. Spatial coverage of various remote sensing systems relative to the area of the Congo Basin.

Figure 8.4 Four images of the Mona Lisa, representing the detail visible within SPOT panchromatic 10 m resolution imagery (*top left*), Landsat TM 30 m resolution imagery (*top right*), Landsat MSS 79 m resolution imagery (*bottom left*), and NOAA AVHRR 1 km resolution imagery (*bottom right*).

Table 8.9

## A General Guide to What Features Can Be Identified at Various Scales

| SCALE | FEATURES |
|---|---|
| 1:500 | Plant species, size of individual trees, uses of buildings, function of industries |
| 1:5,000 | Volume of timber, wetland boundaries, outline of minor tributaries, transportation networks, property boundaries |
| 1:50,000 | Outline of areas of evergreen and deciduous trees, outline of areas of forest associations, direction of flow of water, outline of shorelines, major transportation routes, measurement of agricultural land |
| 1:500,000 | Regional vegetation and land-use classification |
| 1:5,000,000 | Major river systems, continental vegetation zones, continental cloud cover |

The SPOT system is fully operational, with four satellites in orbit. However, the price of SPOT imagery (from $1,400 to $3,000) and the relatively small area covered by each scene (60×60 km) would make a SPOT acquisition for the region almost five times the cost of Landsat TM. High-resolution coverage of the region may not therefore be feasible until Landsat 7 is fully operational.

Table 8.10

## Trade-Offs Between Spatial Resolution and Spatial Coverage for Various Remote Sensing Systems

| SENSOR SYSTEM | PLATFORM | MAP SCALE | SPATIAL RESOLUTION | TEMPORAL RESOLUTION | FIELD OF VIEW | IMAGE FORMAT |
|---|---|---|---|---|---|---|
| NOAA AVHRR | Satellite | >1:1,000,000 | 1.1 to 4 km | 12 hours | 2,700 km | Digital |
| Landsat MSS | Satellite | >1:500,000 | 79 m | 16 to 18 days | 185 km | Digital |
| Landsat TM | Satellite | >1:150,000 | 30 m | 16 days | 185 km | Digital |
| SPOT HRV | Satellite | >1:100,000 | 20 m | 5 to 26 days | 60 km | Digital |
| SPOT | Satellite | >1:50,000 | 10 m | 5 to 26 days | 60 km | Digital |
| Panchromatic radarsat | Satellite | >1:50,000 | 10 to 100 m | 3 to 24 days | 45–500 km | Digital |
| Photography | Aircraft | >1:500 | >0.10 m | Archive or on demand | <20 km | Hard copy |
| Videography | Aircraft | >1:500 | >0.25 m | Archive or on demand | <10 km | Digital |
| Digital photography | Aircraft | >1:500 | >0.25 m | Archive or on demand | <20 km | Digital |

# CHAPTER 9

# Climatic Change and African Rain Forests in the Twenty-First Century

J. Curt Stager

What lies in store for African rain forests in the twenty-first century as a result of climatic change? Much speculation is currently being focused on the issue of "global warming" and its potential effects on ecosystems around the world. Widespread assumptions that modern climatic changes are too rapid to be of natural origin, and that future climatic change can be controlled or prevented, are based largely on evidence that is extremely limited on spatial or temporal scales. Increasingly detailed paleo-ecological records challenge these assumptions (Taylor et al. 1993; Keigwin 1996; Laird et al. 1996a; Overpeck 1996) and provide a general idea of what the future climatic context of African forests will be like. In this admittedly incomplete review of current thinking on the subject, I shall show that natural climatic change can be far more rapid than is commonly assumed, and that much of it is both cyclic and inevitable.

## The Climatic Setting for African Rain Forests

Most of us think on timescales of days to months. For instance, economic models related to climate projections of 100–300 years are considered to be "very long-term" (Cline 1992). But one must look to millennial time-scales to fully grasp the natural range of climatic variability in the tropics. In considering the extremes over the last glacial-interglacial cycle, one notes that the tropics have tended to experience greater aridity during cool glacial periods and increased precipitation during warm interglacials (Nicholson and Flohn 1980; Clemens et al. 1996; Emeis et al. 1995). These climatic swings are driven primarily by the earth's orbital variations, such as the 23,000-year precession or "axis wobble" cycle (Kutzbach and Street-Perrot 1985; Prell and Kutzbach 1987; Clemens et al. 1996). Interestingly, if one surveys the past 250,000 years (Dansgaard et al. 1993), it appears that late Quaternary equatorial forests have spent more time adapting to cool (and presumably dry) glacial conditions than to warm (moist) interglacials or to intermediate conditions like those of today.

One of the most startling paleoclimatological findings in recent years is that natural climatic shifts can be extremely abrupt. This realization has matured with the advent of high-resolution sampling of ice and sediment records. For example, polar ice core records show that the last glacial period was not a smooth, monotonic cold excursion but was repeatedly punctuated by extremely abrupt warming spikes (Dansgaard et al. 1993; Taylor et al. 1993). During interglacial warmings of the past 200,000 years, temperatures over Greenland changed by as much as 10°C within a decade or so (White 1993). The Younger Dryas cooling event at the end of the Pleistocene plunged the North Atlantic region into (and out of) full glacial atmospheric conditions within twenty years or less (Mayewski et al. 1993). Even the late Holocene was disrupted by climatic fluctuations of a magnitude similar to those experienced during the twentieth century. For instance, sea-surface temperatures in the Sargasso Sea were approximately 1°C warmer than today during the Medieval Warm Period, about 1,000 years ago, and approximately 1°C cooler than today during the subsequent Little Ice Age, which ended during the nineteenth century (Keigwin 1996). The rates and magnitudes of such changes stand in stark contrast to earlier assumptions, which are still widely cited (Soulé 1995; IPCC 1996; Malcolm and Markham 1996; Stone 1996), that large climatic departures are naturally slow.

Widespread claims that the past 10,000 years (the Holocene) have been climatically stable are also potentially misleading. They are based largely on localized paleotemperature records that contrast enormous Pleistocene fluctuations against smaller Holocene changes, and in that context the Holocene has indeed been *relatively* more stable than the Pleistocene was (Taylor et al. 1993), especially if the word *climate* is intended to represent only temperature (IPCC 1996). However, local, regional, and global temperatures, precipitation, and wind regimes have in fact varied widely and rapidly over the past 10,000 years (Nicholson and Flohn 1980; Maley 1982; Heusser 1989; Overpeck 1989; Street-Perrott and Perrott 1990; Sukumar et al. 1993; van Campo and Gasse 1993; O'Brien et al. 1995; Emeis et al. 1995; Laird et al. 1996a, 1996b), and even after

considering the Holocene in comparison to the preceding >200,000 years of more dramatic changes, we have little evidence that such anomalously "stable" conditions should continue for much longer, even in the absence of humans.

## The Inevitability of Abrupt Climate Change in Africa

High-resolution sampling of sediment cores is becoming the norm among African paleoecological studies, and as a result we are discovering more and more evidence of abrupt environmental changes. Newly obtained Holocene records from East African lakes (~30–50-year resolution) display remarkably rapid, climatically driven variations in limnological conditions (Halfman et al. 1994; Stager et al. 1997; Stager 1998). Phytoplankton communities in Lake Victoria, for example, responded sensitively to pronounced decade-scale fluctuations in water-column mixing effectiveness (most likely related to wind or storm activity) over the past 7,000 years or so (figure 9.1; Stager et al. 1997). In relation to forests in and around the Victoria Basin, this would suggest that lake-generated rainfall patterns may also have been quite erratic through most of the Holocene. Historical records (Nicholson 1981) further indicate that brief, severe drought and flood episodes have occurred frequently throughout Africa over the past millennium.

Spectral analysis, a method of detecting and measuring cyclic components in long-term environmental records, shows that much of this temporal climatic variability can be described by the interactions of various periodicities that occur in marine and terrestrial records from within and around Africa (figure 9.2; Halfman et al. 1994; Sirocko et al. 1996; Stager et al. 1997). Several of these periodicities appear to be harmonics or combination tones of the Earth's orbital cycles (Sirocko et al. 1996; Mayewski et al. 1997). Some may represent

Calendar years before 2000 A.D.

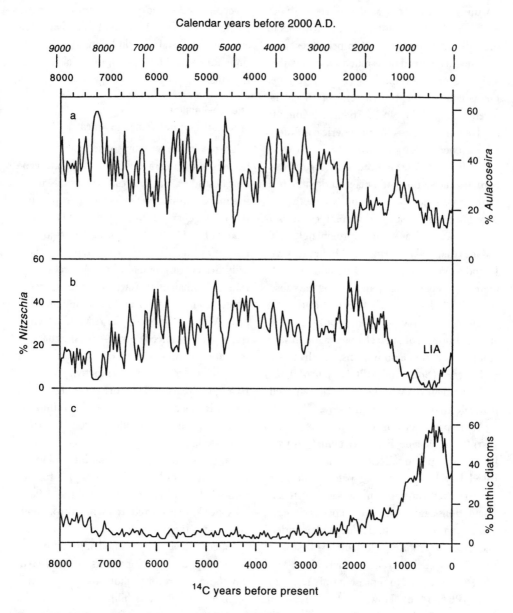

Figure 9.1. High-resolution (30–50-year interval) diatom series from a sediment core from Lake Victoria, east Africa, with both radiocarbon and calendar-year scales. Remarkably abrupt alternations between (a) *Aulacoseira*-dominated and (b) *Nitzschia*-dominated diatom communities apparently reflect rapid increases and declines of wind-driven or storm-driven mixing effectiveness in the lake. Benthic diatom percentages (c) increased with lake shallowing under arid conditions during the past millennium.

internal features of global or regional air-ocean circulation systems, while others, such as the globally distributed cycles of ~2,300–2,500, ~1,400–1,600, and ~11 years (Halfman et al. 1994; Bond and Lotti 1995; Mayewski et al. 1997; Stager et al. 1997), might represent variations in solar output, although this remains a point of debate (Damon and Jirikowic 1992).

How the interactions of relatively gradual cycles could cause extreme decadal-scale climatic shifts can be approximately represented by a wave analogy. Imagine surface waves of different frequencies traveling together across a lake. At certain points, the wave crests and troughs cancel each other out, forming relatively calm zones. At other points, crests (or troughs) are additive, forming sharp irregularities. The climatic equivalent of such "choppiness" may be responsible for much of the abruptness that is emerging from the latest high-resolution records (Mayewski et al. 1997).

Periodic and semi-periodic climatic oscillations operating today have been active for (at least) hundreds of thousands of years and will probably continue far into the future. The inescapable conclusion is that more or less cyclic, occasionally abrupt variability is the normal state of world climates (Bond and Lotti 1995; Mayewski et al. 1997). There are even indications that much of the well-documented global temperature increase of the past hundred years can be accounted for largely by natural climatic oscillations that appear to be bringing us out of the recent Little Ice Age cooling event (Heinrich 1988; Damon and Jirikowic 1992). The Little Ice Age itself closely resembles cyclic pulses of cooling and glacial activity that occurred repeatedly during the Pleistocene at about 7,000-year intervals, and which supposedly ended about 14,000 (2 × 7,000) years ago (Heinrich 1988).

If such speculations are accurate, then one might expect warming to continue into the twenty-first century even without forcing by anthropogenic greenhouse gas emissions. Far from providing an excuse to avoid measures to reduce human impacts on ecosystems, such information strongly confirms that immediate planning for future climatic instability is imperative.

## Future Projections

Today, human activity represents an additional driving mechanism behind climatic change, although its effects relative to natural variability remain uncertain. It is reasonable to assume that human inputs of greenhouse gases to the atmosphere could accelerate natural warming trends (IPCC 1996), but we still know too little about exactly how anthropogenic and natural forces interact in the world climatic system, and about the nature and timing of early human impacts on equatorial African ecosystems (Taylor 1993; Elenga et al. 1994). Climatic instability is only one of a constellation of disruptive factors, including invasions of species from other areas and global economic forces, that are at work in today's tropical forests, a fact that makes it increasingly difficult to find historical analogs for present conditions. As is often said, we are in the midst of an unprecedented planetary ecology experiment.

In spite of recent advances in climate modeling, it is extremely difficult to predict how climates will change in a particular region in the twenty-first century (Laird et al. 1996a), even if a precise amount of expected global average warming could be agreed upon. Global average temperature projections have little predictive value when applied to specific locales, in part because they are based on incomplete paleo-temperature records, but especially because they represent coarse-scale averages from which it is difficult to infer site-specific thermal conditions. Furthermore, temperature variations probably affect most terrestrial ecosystems less strongly than do the quantity and timing of precipitation. It is drought more than

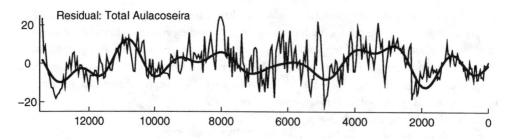

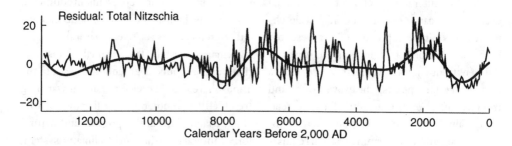

Figure 9.2. Summary of time series analyses of detrended diatom records from Lake Victoria, using calendar-year chronology (by D. Meeker; in Stager et al. 1997). Smooth oscillations superimposed on jagged diatom series are the sums of significant (99.9%) periodicities in those series, including cycles of 3,810–3,950, 2,370–2,550, and 1,590–1,770 years, which explain roughly one-third of the variability in the diatom records.

temperature change that brings episodic misery to today's Sahel. A massive shift from moist forest to *Celtis*-dominated woodlands in the Lake Victoria Basin around 7,500–7,000 years ago (Kendall 1969) was largely driven by a shift in rainfall seasonality. Unfortunately, in comparison to temperature, localized precipitation patterns respond to a frustratingly complex array of factors and have understandably received much less attention from modelers. Wind patterns, which are highly variable and chaotic and which interact with each other in both vertical and horizontal planes, are also extremely difficult to predict and model, although they can determine the nature and locations of entire weather systems (Moran 1992).

In spite of these severe limitations, some generalized speculation about future African climates is possible. Globally averaged warming in the twenty-first century, whatever its ultimate cause or amount, is likely to have abrupt climatic fluctuations associated with *cooling*. Equatorial African aridity and Sahelian flooding were associated with the Little Ice Age between the fourteenth and nineteenth centuries (Nicholson 1978; Bonnefille and Mohammed 1994); the onset of that historical climatic shift might even have been a trigger for the migrations of certain ironworking cultures in west Africa (see Oslisly this volume). If past temperature-precipitation relationships are still valid in a world more influenced by humans, then there might, conversely, be overall *increases* in equatorial African precipitation and Sahelian aridity associated with future warming trends.

Inevitable, potentially rapid climatic variations in the coming century will elicit site-specific responses from rain forest ecosystems. For instance, shifting climatic regimes might cause vertical migration of montane vegetation belts (see Maley 1989), and even slight variations in local temperatures might affect the composition of insect communities, with significant implications for the plants they interact with (Greenwood 1987; Parsons 1989a, 1989b). Climate-driven changes will most likely affect different regions of Africa differently, in a mosaic as complex as today's temporal and geographic rainfall distributions (Nicholson 1994). Although we are left woefully short of specifics, this information alone is sufficient to warrant making every effort to develop long-term, flexible forest management plans that allow for potentially extreme environmental variability (Bridgewater 1996).

## Recommendations

An important first step in developing far-sighted conservation plans for forest sites is to obtain as much information as possible about the natural range of variation in local vegetational communities over long time periods. Most rain forest species probably respond rather individualistically to climatic changes (Davis 1989), and both climatic variability and ecosystem responses to it tend to display great temporal and spatial non-uniformity (Moran 1992; Nicholson 1994; Holling 1996). Thus, in addressing future concerns, it is highly advisable to determine the *actual* past and present species dynamics of a particular forest region in addition to consulting theoretical models and analogs developed elsewhere. This amounts to asking, "How has the forest on *this* particular site responded to known climatic changes in the past?"

Generating such information must involve some aspects of paleoecology, usually through the coring of local lakes, bogs, or other wetlands. From pollen, charcoal, and plant macrofossils one can infer much about the nature and degree of stability of local vegetation. The remains of diatoms and other aquatic organisms can reveal much about lake responses to both long- and short-term climatic changes, from which one can begin to infer how local forests have responded to climatic variability in the past. Oral and written records of historical environmental conditions can also be useful in this regard (Nicholson 1981).

Because modern forests have reached their present states during the Holocene, roughly 10,000-year sediment records should be adequate to cover the natural range of climate-related variations in current versions of African rain forest ecosystems. It is important in studies of this sort to use as high a sampling resolution as possible in order to permit the detection of short-term variability. New sediment records should be calibrated with accelerator mass spectrometry (AMS) radiocarbon dating rather than more traditional "bulk" sediment dating, as AMS requires much smaller samples and thus reduces date uncertainty. Radiocarbon chronologies should also be converted to "real-time" calendar-year records through the use of standard, widely available calibration models (Bard et al. 1993; Stuiver and Reimer 1993). This conversion can make comparisons to other records simpler, especially those (such as historical, tree ring, and ice core chronologies) that use only calendar-year scales.

Above all, the paleoecological perspective shows that African climates are more dynamic than stable. The near-certain prospect of long-term climatic instability, both natural and anthropogenic, poses a challenge to conservationists to incorporate this reality into future models of tropical forest management, and provides an important context for efforts to incorporate ecological resilience into the design of parks and protected areas.

ACKNOWLEDGMENTS

Special thanks to J. Gould, S. Grimm, D. Livingstone, P. Mayewski, D. Meeker, Paul Smith's College, and J. Smol for editorial comments and technical assistance.

REFERENCES

Bard, E., M. Arnold, R. G. Fairbanks, and B. Hamelin. 1993. $^{230}$Th-$^{234}$U and $^{14}$C ages obtained by mass spectrometry on corals. *Radiocarbon* 35:191–199.

Bond, G. C., and R. Lotti. 1995. Iceberg discharges into the North Atlantic on millennial time scales during the last glaciation. *Science* 267:1005–1010.

Bonnefille, R., and U. Mohammed. 1994. Pollen-inferred climatic fluctuations in Ethiopia during the last 3000 years. *Palaeogeography, Palaeoclimatology, Palaeoecology* 109:331–343.

Bonnefille, R., J. C. Roeland, and J. Guiot. 1990. Temperature and rainfall estimates for the past 40,000 years in equatorial Africa. *Nature* 346:347–349.

Bridgewater, P. B. 1996. Protected area management in the face of climate change. *Parks* 6:4–13.

Clemens, S. C., D. W. Murray, and W. L. Prell. 1996. Nonstationary phase of the Plio-Pleistocene Asian monsoon. *Science* 274:943–948.

Cline, W. R. 1992. *The Economics of Global Warming*. Institute for International Economics, Washington, D.C.

Cohen, A. L., and P. D. Tyson. 1995. Sea temperature fluctuations during the Holocene on the south coast of Africa: Implications for terrestrial climate and rainfall. *Holocene* 5:306–312.

Damon, P. E., and J. L. Jirikowic. 1992. Solar forcing of global climate change? In R. E. Taylor, A. Long, and R. S. Kra, eds. *Radiocarbon after Four Decades: An Interdisciplinary Perspective*. Springer-Verlag, New York.

Dansgaard, W., S. J. Johnsen, H. B. Clausen, D. Dahl-Jensen, N. S. Gundestrup, C. U. Hammer, C. S. Hvidberg, J. P. Steffensen, A. E. Sveinbjörnsdottir, J. Jouzel, and G. Bond. 1993. Evidence for general instability of past climate from a 250-kyr ice-core record. *Nature* 364:218–220.

Davis, M. B. 1989. Insights from paleoecology on global change. *Ecological Society of America Bulletin* 70:222–228.

Elenga, H., D. Schwartz, and A. Vincens. 1994. Pollen evidence of late Quaternary vegetation and inferred climate changes in Congo. *Paleogeography, Palaeoclimatology, Palaeoecology* 109:345–356.

Emeis, K.-C., D. M. Anderson, H. Doose, D. Kroon, and D. Schulz-Bull. 1995. Sea-surface temperatures and the history of monsoon upwelling in the northwest Arabian Sea during the last 500,000 years. *Quaternary Research* 43:355–361.

Greenwood, S. R. 1987. The role of insects in tropical forest food webs. *Ambio* 5:267–271.

Halfman, J. D., T. C. Johnson, and B. P. Finney. 1994. New AMS dates, stratigraphic correlations and decadal climatic cycles for the past 4 ka at Lake Turkana, Kenya. *Palaeogeography, Palaeoclimatology, Palaeoecology* 111:83–98.

Heinrich, H. 1988. Origin and consequences of cyclic ice rafting in the northeast Atlantic Ocean during the past 130,000 years. *Quaternary Research* 29:142–152.

Heusser, C. J. 1989. Polar perspective of late-Quaternary climates in the Southern Hemisphere. *Quaternary Research* 32:60–71.

Holling, C. S. 1996. Engineering resilience vs. ecological resilience. In P. Schultze, ed. *Engineering Within Ecological Constraints*. National Academy Press, Washington, D.C.

IPCC. 1996. *Climate Change 1995. Impacts, Adaptations, and Mitigation of Climate Change: Scientific-Technical Analyses*. R. T. Watson, M. C. Zinyowera, R. H. Moss, and D. J. Dokken, eds. Cambridge University Press, Cambridge.

Keigwin, L. D. 1996. The Little Ice Age and Medieval Warm Period in the Sargasso Sea. *Science* 274:1504–1508.

Kendall, R. L. 1969. An ecological history of the Lake Victoria Basin. *Ecological Monographs* 39:121–176.

Kutzbach, J. E., and F. A. Street-Perrot. 1985. Milankovitch forcing of fluctuations in the level of tropical lakes from 18 to 0 kyr BP. *Nature* 317:130–134.

Laird, K. R., S. C. Fritz, E. C. Grimm, and P. G. Mueller. 1996a. Century-scale paleoclimatic reconstruction from Moon Lake, a closed-basin lake in the northern Great Plains. *Limnology and Oceanography* 41:890–902.

Laird, K. R., S. C. Fritz, K. A. Maasch, and B. F. Cumming. 1996b. Greater drought intensity and frequency before AD 1200 in the Northern Great Plains, USA. *Nature* 384:552–554.

Lamb, H. F., F. Gasse, A. Benkaddour, N. El Hamouti, S. van der Kaars, W. T. Perkins, N. J. Pearce, and C. N. Roberts. 1995. Relation between century-scale Holocene arid intervals in tropical and temperate zones. *Nature* 373:134–137.

Malcolm, J. R., and A. Markham. 1996. Ecosystem resilience, biodiversity and climate change: Setting limits. *Parks* 6:38–49.

Maley, J. 1982. Dust, clouds, rain types, and climatic variations in tropical North Africa. *Quaternary Research* 18:1–16.

―――. 1989. Late Quaternary climatic changes in the African rain forest: Forest refugia and the major role of sea surface temperature variations. Pages 585–616 in M. Leinen and M. Sarntheim, eds. *Paleoclimatology and Paleometeorology: Modern and Past Patterns of Global Atmospheric Transport*. Kluwer, Dordrecht.

Mayewski, P. A., L. D. Meeker, M. S. Twickler, S. Whitlow, Q. Yang, and M. Prentice. 1997. Major features and forcing of high-latitude northern hemisphere atmospheric circulation using a 110,000-year-long glaciochemical series. *Journal of Geophysical Research* 102:26345–26366.

Mayewski, P. A., L. D. Meeker, S. Whitlow, M. S. Twickler, M. C. Morrison, R. B. Alley, P. Bloomfield, and K. Taylor. 1993. The atmosphere during the Younger Dryas. *Science* 261:195–197.

Moran, J. M. 1992. The climatic future and lessons of the climatic past. In D. A. Dunnette and R. J. O'Brien, eds. *The Science of Global Change: The Impact of Human Activities on the Environment*. American Chemical Society, Washington, D.C.

Nicholson, S. E. 1978. Climatic variations in the Sahel and other African regions during the past five centuries. *Journal of Arid Environments* 1:3–24.

―――. 1981. The historical climatology of Africa. In T. M. L. Wigley, M. J. Ingram, and G. Farmer, eds. *Climate and History*. Cambridge University Press, Cambridge.

―――. 1994. Recent rainfall fluctuations in Africa and their relationship to past conditions over the continent. *Holocene* 4:121–131.

Nicholson, S. E., and H. Flohn. 1980. African environmental and climatic changes and the general atmospheric circulation in late Pleistocene and Holocene. *Climatic Change* 2:313–348.

O'Brien, S. R., P. A. Mayewski, L. D. Meeker, D. A. Meese, M. S. Twickler, and S. I. Whitlow. 1995. Complexity of Holocene climate as reconstructed from a Greenland ice core. *Science* 270:1962–1964.

Overpeck, J. T. 1989. Century- to millennium-scale climatic variability during the Late Quaternary. Pages 139–173 in R. S. Bradley, ed. *Global Changes of the Past*. UCAR/Office for Interdisciplinary Earth Studies, Boulder.

―――. 1996. Warm climate surprises. *Science* 271:1820–1821.

Parsons, P. A. 1989a. Century- to millennium-scale climatic variability during the Late Quaternary. Pages 139–173 in R. S. Bradley, ed. *Global Changes of the Past*. UCAR/Office for Interdisciplinary Earth Studies, Boulder.

―――. 1989b. Conservation and global warming: A problem in biological adaptation to stress. *Ambio* 18:322–325.

Prell, W. L., and J. E. Kutzbach. 1987. Monsoon variability over the past 150,000 years. *Journal of Geophysical Research* 92:8411–8425.

Sirocko, F., M. Sarntheim, H. Erlenkeuser, H. Lange, M. Arnold, and J. C. Duplessy. 1993. Century-scale events in monsoonal climate over the past 24,000 years. *Nature* 364:322–324.

Sirocko, F., D. Garbe-Schönberg, A. McIntyre, and B. Molfino. 1996. Teleconnections between the subtropical monsoons and high-latitude climates during the last deglaciation. *Science* 272:526–529.

Soulé, M. 1995. Health implications of global warming and the onslaught of alien species. *Wild Earth* 5:56–61.

Stager, J. C. 1998. Ancient analogs for recent environmental changes at Lake Victoria, East Africa. Pages 37–46 in *Environmental Change and Response in East African Lakes*. Kluwer, Dordrecht.

Stager, J. C., B. Cumming, and L. Meeker. 1997. A high-resolution, 11,400 year diatom record from Lake Victoria, East Africa. *Quaternary Research* 47:81–89

Stone, D. 1996. Impacts of climate change on selected ecosystems in Europe. *Parks* 6:25–37.

Street-Perrot, F. A., and R. A. Perrott. 1990. Abrupt climate fluctuations in the tropics: The influence of Atlantic Ocean circulation. *Nature* 343:607–612.

Stuiver, M., and P. J. Reimer. 1993. Extended $^{14}$C data base and revised CALIB 3.0 $^{14}$C age calibration program. *Radiocarbon* 35:215–230.

Sukumar, R., R. Ramesh, R. K. Pant, and G. Rajagopalan. 1993. A delta-C13 record of late Quaternary climate change from tropical peats in southern India. *Nature* 364:703–706.

Taylor, D. M. 1993. Environmental change in montane southwest Uganda: A pollen record for the Holocene from Ahakagyezi Swamp. *Holocene* 3:324–332.

Taylor, K. C., G. W. Lamorey, G. A. Doyle, R. B. Alley, P. M. Grootes, P. A. Mayewski, J. W. C. White, and L. K. Barlow. 1993. The "flickering switch" of late Pleistocene climate change. *Nature* 361:432–436.

van Campo, E., and F. Gasse. 1993. Pollen- and diatom-inferred climatic and hydrological changes in Sumxi-Co Basin (Western Tibet) since 13,000 yr B.P. *Quaternary Research* 39:300–313.

White, J. W. C. 1993. Don't touch that dial. *Nature* 364:186.

# Overview of Part III

Amy Vedder

ESSENTIAL TO SOUND CONSERVATION of any ecosystem is an understanding of its basic composition, distribution, dynamics, functions, and interrelationships with other neighboring systems. Without this knowledge, efforts to protect or conserve an ecosystem or its component species will be fraught with danger. Unfortunately, our collective knowledge of African rain forests lags well behind knowledge of any other forested region on earth. The chapters grouped in part III of this volume therefore represent a significant body of information on this lesser-known realm and should serve to inform us, guide us in conservation, and provide a basis for comparison with ecosystems elsewhere on earth.

In the following chapters, the biogeography, ecology, and dynamics of Africa's rain forest flora and fauna are presented. Although not all taxa are fully represented, this collection of chapters clearly reveals the great diversity of life found here. These chapters challenge earlier notions of African rain forest homogeneity and lesser species richness. The authors present some of the first studies of African forest dynamics, ecological adaptations of species, and interspecies relationships. They also point out the gaps in our knowledge.

The section begins with two chapters on forest vegetation. Although general patterns of vegetation associations have been well documented (White 1983 and chapter 1 in this volume), these chapters reveal new levels of heterogeneity in forest structure, composition, and function. In chapter 10, Terese B. Hart describes the dynamic coexistence between monodominant *Gilbertiodendron dewevrei* forest and mixed-dominant Caesalpiniaceae forest, the matrix within which *G. dewevrei* is located: a patchwork found throughout the vast Guineo-Congolian vegetation zone. In chapter 11, Lee J. T. White assesses the open-understory, herbaceous-dominated Marantaceae forest and concludes that it is one of several relatively long-lived successional stages of savanna recolonization. In each case, the pattern of vegetation is thought to be derived from human influence (see also chapters 5 and 7 in this volume). Recognition of such patterns, and of their origin and dynamics, is important to conservation planning. It guides us toward taking a landscape-level approach in our work and toward conserving medium-scale heterogeneity. It also points us toward several different kinds of conservation man-

agement options. Hart points out the vulnerable nature of a vegetation association such as *G. dewevrei*, a vulnerability that is further emphasized when
one considers the socioeconomic context of the same region (see chapters 21
and 30). Careful, full protection is proposed as the most effective means to
conserve this forest. In contrast, White suggests that conservationists should
consider active management of Marantaceae forest by assuring appropriate
disturbance regimes. Although caveats remain (see chapters 16 and 26), such
management would likely maintain the structure and the consequent benefits
that Marantaceae forest provides to the large mammals that thrive on the
abundance of its terrestrial foods.

The remaining chapters of this section focus on African forest fauna.
Organized by taxonomic groupings, the chapters cover biogeographic patterns, species and community ecology, and social behavior. Though data for
the taxa are presented separately, strikingly similar principles emerge. All
authors highlight the lack of basic information. They indicate the need for
straightforward studies, including species lists (herpetofauna: chapter 17),
species distributions (carnivores: chapter 14), basic ecology (fish: chapter 16),
and behavior (elephants: chapter 13). Yet each also provides a fascinating first
look at these topics. Chapters 12 and 14 present ecological studies of mammal
communities, examining ecological separation and interspecific interactions.
John A. Hart describes a diverse community of ungulates, which differ significantly in their habitat preferences, diet, movements, population densities,
and responses to disturbance. Despite the difficulties inherent in studying
elusive carnivores, Justina C. Ray outlines their importance in maintaining
community balance and notes their vulnerability to hunting (both of the carnivores themselves and of their prey). In Chapter 13, Andrea K. Turkalo and
J. Michael Fay allow us to peer into the social world of elephant behavior,
thanks to the "window" offered by an unusual vegetational feature of the forest: natural clearings. The authors present results of the first observational
study of the forest elephant subspecies, which differ from their savanna conspecifics in size, social structure, behavior, and ecology. Turkalo and Fay
emphasize the importance of large landscapes from which these elephants
come, the protection of such key sites as bais, the connectivity of these sites,
and the potential for monitoring populations in a new fashion.

The final three chapters in this section take the leap to a larger scale,
assessing the regional biogeographic patterns of several more taxa: avifauna
(chapter 15) fishes (chapter 16), and herpetofauna (chapter 17). The authors
highlight evidence of complex dynamics of forest contraction and expansion
of Pleistocene refugia, and ancient connections between the western and
eastern forests of central Africa (see also chapters 3–6). In each of these taxonomic assessments, areas of high species richness are identified, including a
previously undocumented, unprecedented variety of African forest herpetofauna (chapter 17). Species of freshwater fishes are found in numbers equivalent to the Amazon, but at much higher levels of endemism (chapter 16),

and high degrees of endemism are reported in particular for montane forest birds (chapter 15). The authors of these chapters all call for further basic work to be done: inventories, taxonomic definition, and first-round ecological studies of these lesser-known taxa. They note the potential value of these taxonomic groups as indicators of ecosystem change, habitat loss, and impacts of human use.

It is clear from these chapters that field biologists are disturbed by the loss of forest habitat. They also soundly and repeatedly raise concerns about faunal depletion (see also chapter 20) from bushmeat hunting (of duikers, herpetofauna, or larger birds), hunting for non-food commercial products (elephant ivory; ungulate, felid, and reptile skins), and the loss of prey, which leaves carnivore populations with diminished food sources. These issues underscore the importance of safeguarding key core habitats and species under protected status: thus the values of parks and reserves. These chapters also raise the possibility of new, complementary approaches to conservation. With an improved understanding of the dynamics of the open, transitional nature of some forest types, we may be able to design additional, more broadly applicable conservation systems that incorporate zones of careful resource use, including tightly managed selective logging. If hunting in these zones can be strictly controlled and for some species prohibited, it is possible that a significant portion of African forest fauna (including elephants, gorillas, and selected ungulates) can be simultaneously conserved on site. Success in these endeavors will depend on ensuring good management—of both protected areas and those areas carefully managed for use. Such success will also depend on continued research and careful monitoring, particularly in light of the limitations of our current understanding.

REFERENCE

White, F. 1983. *The vegetation of Africa: A descriptive memoir to accompany the Unesco/AET-FAT/UNSO vegetation map of Africa.* Unesco, Paris.

CHAPTER 10

# Forest Dynamics in the Ituri Basin (DR Congo)

Dominance, Diversity, and Conservation

Terese B. Hart

The growing understanding of the importance of history in determining forest composition in the tropics (Whitmore 1989; Hubbell and Foster 1990; Bush et al. 1992; Ritchie 1995) poses a new challenge to those interested in conserving the diversity of forest associations. Previously, the change from one type of "mature" forest to another in a single landscape would have been thought to indicate a change in the physical environment (Richards 1952; Ashton 1976). If history is more important than environment, however, some of the transitions in forest structure and composition that are characteristic of and significant to a region's overall biodiversity may occur under apparently homogeneous environmental conditions. The long-term maintenance of this diversity of forest types may require a rethinking of conservation strategies. A prerequisite for conservation planning is an understanding of the origin and dynamics of the transitions between forest types. If history is important, then one must ask, how ancient is the discontinuity of forest type? Is it stable in time?

A special challenge is presented in Africa, where a large part of the upland forest is a relatively uniform, rolling plateau. The topographic homogeneity may have reduced opportunities

for speciation (White 1978), but it has not precluded a diversification of forest associations and locally high species diversity (Gentry 1988; Makana et al. 1998).

In central Africa, near the northern limits of the forest, localized diversification of forest types has been attributed to differing forest histories. This is the case where adjacent forests represent different stages in a long-term succession as forest invades grassland (Aubréville 1966; White 1992; White, this volume, chapter 11). The younger forests are dominated by heliophile or shade-intolerant species, which do not regenerate well in their own shade (White, this volume, chapter 11). In Africa, the more mature forests of these successional clines are often characterized by dominance of Caesalpiniaceae, a family less represented in younger forest (Letouzey 1968; White, this volume, chapter 11). In contrast to the principal canopy species of the younger forests, these Caesalpiniaceous species are shade tolerant and maintain abundant stocking of juveniles under their canopies.

In other areas of the central African forest block, particularly in the northeastern sector, there are frequent transitions between forest types with as yet no historical explanation, even though they elude explanation by variation in the physical environment. These transitions are often far from the forest border. The case investigated in this chapter concerns the upland formations of the central Ituri Forest. Two adjacent and interdigitating forest types are dominated by different Caesalpiniaceous species. Both these forest types have been considered mature, based on forest inventories that reveal abundant stocking of the principal canopy species in the seedling and sapling size classes (Gérard 1960; Hart 1985, 1995), thus suggesting shade tolerance. The same species also have heavy seeds that are explosively dispersed. These life–history characteristics are typically associated with mature-phase species (Budowski 1970; Ewel 1980; Swaine and Whitmore 1988).

One of the Ituri forest types, the mbau forest, is monodominant, with *Gilbertiodendron dewevrei* making up more than 75% of the canopy stems. The second forest type, herein called mixed forest, has strong representation of the species *Julbernardia seretii* in the canopy. Monodominance by gregarious Caesalpiniaceae is not uncommon in African forests (Devred 1958; Letouzey 1968), and the forests dominated by *G. dewevrei* are particularly widespread and characteristically of high canopy dominance (Léonard 1952; White 1983). Mbau forest is most developed in the Democratic Republic of Congo (DR Congo), on the uplands (from 200 to 800 m altitude) forming an arc around the northern and eastern edges of the central basin. Its distribution extends, however, as small patches or as strips along watercourses, from southeastern Nigeria through Cameroon, Gabon, Congo, Central African Republic, and into the lowlands of the DR Congo Basin, thus spanning a wide range of climate and soil types. Dominance by *G. dewevrei* is found on soils derived from African shield rock, sedimentary rocks, and decomposing laterite with highly variable percentages of sand and clay.

Although monodominance has been reported in tropical forests worldwide (Richards 1952; Whitmore 1984; Hart et al. 1989), the condition has been considered exceptional in most of these species-rich ecosystems. In central Africa, where monodominant forests cover larger areas, the high dominance has been associated with forest maturity and the shade tolerance of the principal species (Louis 1947; Germain and Evrard 1956; Gérard 1960). A transition from forest dominated by one suite of Caesalpiniaceous species to forest dominated by another suite of Caesalpiniaceous species, as in the Ituri, presents an enigma. Historical explanations would not be expected, a priori, as both forests, being dominated by shade-tolerant species, do not fulfill predictions for a successional cline.

Evidence for past major changes in the composition of the Ituri Forest has been revealed by analyses of buried charcoal (Hart et al. 1996). Charcoal is widespread in the soils under forests of the Ituri. A selection of charcoal samples from different sites and soil depths was radiocarbon dated. The oldest of twenty-eight dates was just over 4,000 B.P., with most dates less than 2,000 B.P. Since the fires that created the charcoal, the composition of the forest has changed radically; some of the species best represented in the charcoal are no longer present today, although where they presently occur they are part of forest formations existing under climate similar to that of the Ituri Forest today. Other charcoal species—including *Cynometra alexandri,* in the Caeslapiniaceae—are still common today. The charcoal from more than four hundred sample pits did not include any formed from *G. dewevrei* wood. These pits were mostly in mixed forest, although thirty-six were from areas that are currently transitions between the two forest types. The overall conclusion of the study was that there had been significant species replacement that could not be explained by directional change in the regional climate. Is species replacement continuing today? If so, what pertinence does it have to the coexistence of the two forest types?

Using a different line of evidence—results from a study of short-term forest dynamics—I further examine, in this chapter, the possibility of a historical explanation for the co-occurrence of the Ituri Forest formations. If historical events have mediated this coexistence, it is likely that the species compositions of these formations are not stable (Hubbell and Foster 1990; Bush et al. 1992) and that ongoing species replacement would be determined by life-history characteristics of the coexisting species, as in typical successions (Prentice 1986; Pickett and McDonnell 1989). Although the life-history characteristics, as roughly portrayed by inventory data, indicate that the dominants of both forest types are mature-phase species, demographic data collected over time should give better resolution of species differences and could indicate that one forest is younger, in a successional sense, than the other. The dominant canopy species of the younger forest should exhibit characteristics that give it relatively greater resilience to disturbance and an advantage in invading earlier successional forests. It might grow faster to maturity, have a shorter life cycle, and/or produce more abundant, more widely dispersed propagules (Huston and Smith 1987). If dominance is a characteristic that increases with forest age or maturity, I would expect that the older formation would either have greater dominance on small plots, if dominance is cumulative over time, or larger monodominant patches, if the dominant species is spreading over time.

In the next section, I examine these predictions using ten-year growth and mortality data from permanent plots in the Ituri Forest, and I also consider evidence from previous studies concerning the Ituri Forest or its dominant species (Gérard 1960; Hart 1985, 1995; Hart et al. 1989, 1996; Conway 1992; Makana et al. 1998).

## The DR Congo Study Site

The study sites were all within 15 km of the village of Epulu, the administrative center of the 13,700 km² Okapi Wildlife Reserve. Epulu is in the center of the Ituri Forest; the nearest savanna border is at a distance of more than 150 km.

Previous work suggests that edaphic factors do not determine the transitions between the two major upland forest types on the study area, mbau forest dominated by *G. dewevrei* and mixed forest where the most abundant species is *J. seretii* (Gérard 1960; Hart 1985; Conway 1992). The topography is similar under both forest types, with soils derived from Precambrian shield rock (Sys 1972; Lavreau 1982). Mbau forest is an important forest type throughout the central and southern sectors of the Okapi Wildlife Reserve and in bordering forests to the south, west, and northwest. It generally occurs in a matrix of more mixed forest where dominance is variable. In the Okapi Wildlife Reserve, one of two Caesalpiniaceous species, *J. seretii or C. alexandri*, is generally most abundant in the canopy, although another Caesalpiniaceous species, *Erythrophleum suaveolens*, or species from other families, *Cleistanthus michelsonii* and *Uapaca guineensis*, may be locally gregarious.

Mean annual rainfall between 1987 and 1993 was 1,700 mm, with a minimum of 1,307 mm and a maximum of 2,084 mm. A single principal dry season, of variable severity, occurs between January and March (Hart 1995).

Permanent plots were set up in 1981–1982 at six separate sites, three in mixed forest and three in mbau forest. The closest two sites were more than 3 km apart (Hart 1995), each was an expanse of relatively uniform forest type (> 1 km²), and none had a known history of disturbance.

At each of the six forest sites, eight 25 × 25 m plots were laid along a transect with 75 m between plots. Where inappropriate habitat was encountered (stream crossings, major treefalls, and so on), an additional 35 m were added to the distance between plots. All trees greater than 10 cm diameter were identified, and the diameter measured at a height of 1.3 m (dbh) or just above buttresses. Each forest type was sampled in all twenty-four plots, for a total of 1.5 ha. Stems were re-measured in 1992–1993.

Sample sizes permitted only the most abundant species to be analyzed independently for growth. Within their respective forests, the dominant species were, therefore, compared to all other species combined. Where assumptions were met, parametric statistics were used to test hypotheses.

In addition to overall growth rate of the two species, growth at different points in their life cycle was pertinent. The life cycle can be divided somewhat arbitrarily into size classes. I considered that the categories 10–30, 30–50, and >50 cm dbh had biological significance in that the smallest size class is still likely to be overtopped by other trees, whereas the middle size class is likely to have some direct canopy sun and the largest is likely to be chiefly in direct sunlight. Comparisons were further complicated by the many zero values and even negative values (which were computed as zero values). For this reason, rather than direct comparison of growth increment, the values used were the numbers of individuals for which positive growth or no growth was recorded. Chi-square tests were used to test null hypotheses. Chi-square tests were also used to compare mortality rates. Further statistical information is found in the notes to this chapter.

Various attributes of the most abundant species support the thesis that both the mbau and the mixed communities are mature forests. First, in both forests, the species best represented in the canopy is also well represented as advance regeneration in the understory (table 10.1). In addition, these same species have relatively heavy seeds without special adaptations for transport by wind or animal and no seed dormancy (Hart 1985).

Table 10.1

**Stem Density of the Dominant Canopy Species in Monodominant Mbau Forest and Mixed Forest in the Ituri River Basin, DR Congo**

| | *GILBERTIODENDRON* IN MBAU FOREST | *JULBERNARDIA* IN MIXED FOREST |
|---|---|---|
| % of all stems | | |
| ≥10 <30 cm dbh | 73 | 21 |
| >30 cm dbh | 95 | 41 |
| >50 cm dbh | 93 | 28 |
| Stems ha$^{-1}$ | | |
| ≥10 <30 cm dbh | 207 | 119 |
| >30 cm dbh | 129 | 68 |
| >50 cm dbh | 75 | 16 |

There are, however, some interesting community-level differences between the forests. Mbau forest's basal area (approximately 36 m$^2$ ha$^{-1}$, 1993 values) is larger than that of mixed forest (approximately 32 m$^2$ ha$^{-1}$, 1993 values).[1] Nevertheless, both forests' basal area appeared larger than that of a nearby forty-year-old secondary forest (approximately 29 m$^2$ ha$^{-1}$; Hart 1985), although, because there was only one secondary site, no statistical comparison was done.

Another community-level difference was the degree of dominance. The most abundant species in mixed forest, *J. seretii*, made up 41% of the trees > 30 cm dbh and less than 30% of the largest trees (>50 cm dbh), whereas *G. dewevrei* in mbau forest accounted for more than 90% of both size classes (see table 10.1).

In mbau forest, the mean diameter increment for all surviving individuals of all species >10 cm dbh in 1982 (N = 376) was 3.9% over ten years. This percentage was heavily weighted by *G. dewevrei*, which accounted for 82% of all surviving stems and grew more slowly than most. All other species combined had an increment of 6.8%, whereas *G. dewevrei*, considered alone, had a mean diameter increment of 3.3% of its 1982 girth. On the mbau forest plots, the difference in growth between *G. dewevrei* and other species combined was significant.[2]

In mixed forest, mean diameter increment for surviving individuals of all species combined (N = 625) was 8.4%. The most abundant species, *J. seretii*, had relatively greater growth, with a mean growth of 12.5%. On mixed forest plots, the difference in growth between *J. seretii* and other species combined was significant.[3]

The growth of *J. seretii* in mixed forest plots was compared with that of *G. dewevrei* in mbau forest plots. The mean growth increments were found to be significantly different.[4]

As for growth at different points in the life cycle, two null hypotheses were tested with chi-square tests. The first $H_0$ is that different size classes of the focus species have similar proportions of individuals with positive growth. This $H_0$ was rejected for *J. seretii* ($\chi^2$ observed = 9.258, $df$ = 2; $\chi^2$ tab = 5.991, = 0.05) but could not be rejected for *G. dewevrei* ($\chi^2$ = 2.657). The second $H_0$ is that the two species, *G. dewevrei* and *J. seretii*, have similar proportions of individuals in specific size classes with positive growth. The two dominant species appear to differ in the two smaller size classes but not necessarily in the largest size class (for 10–30 cm dbh $\chi^2$ = 8.909, for 30–50 cm dbh $\chi^2$ = 4.119, and for >50 cm dbh $\chi^2$ = 0.362, $df$ = 1; $\chi^2$ tab = 3.841, $\alpha$ = 0.05). It appeared, therefore, that the two species differed with respect to the evenness of growth across different size classes (figures 10.1 and 10.2). Whereas the vast majority of *J. seretii* individuals under 50 cm dbh had positive growth (with only 5% showing no growth), 33% of the individuals over 50 cm dbh had no growth. *Gilbertiodendron dewevrei* showed less variation, with a relatively high proportion of no-growth individuals through all the size classes (24%, 11%, and 21%, respectively).

*Gilbertiodendron dewevrei* sustained low

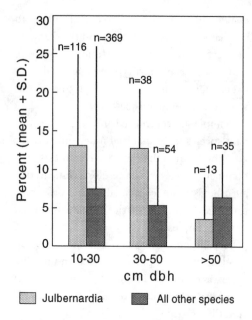

Figure 10.1. Tree growth, measured as the percentage increase in diameter between 1982 and 1992, of different size classes of the dominant species, *J. seretii*, and of all other species (combined) in the mixed forest of the Ituri Basin.

mortality over ten years as compared to other species in mbau forest (figure 10.3). *Gilbertiodendron* lost 3.4% of its 1982 population over 10 cm dbh, whereas all other species pooled lost 10.7%. By contrast, *J. seretii* had high mortality relative to associated species in mixed forest (figure 10.3). *Julbernardia* lost 16.9% of its 1982 population, while other species pooled lost 11.2%.

When mortality of the dominant species of the two forest types are compared, with all size classes combined, *J. seretii* suffered significantly higher mortality ($\chi^2$ corrected for continuity = 19.047, $df$ = 1, $P$ < 0.001). This difference was also evident when only the smaller size classes (>10 cm dbh, <30 cm dbh) were compared; *J. seretii*'s mortality was significantly greater than that of *G. dewevrei* ($\chi^2$ = 29.1, $df$ = 1, $P$ < 0.001). *Julbernardia* had 20% mortality compared to *G. dewevrei*'s 2% mor-

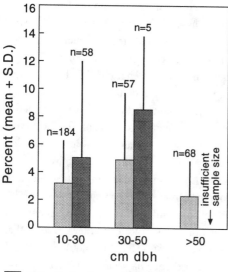

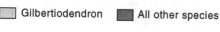

Figure 10.2. Tree growth, measured as the percentage increase in diameter between 1982 and 1992, of different size classes of the dominant species, *G. dewevrei*, and of all other species (combined) in the mbau forest of the Ituri Basin.

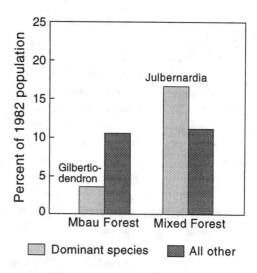

Figure 10.3. Mortality of stems >10 cm dbh between 1982 and 1992 in two forest types of the Ituri Basin. In both mbau and mixed forest types, the dominant canopy species is compared to all other species combined.

tality. In the larger size classes the difference was no longer significant.

Previous studies of dispersal (Hart 1985) and recruitment (Hart 1995) suggest other pertinent life-history differences between *G. dewevrei* and *J. seretii*. Although both species are gravity-dispersed, the greater mass of *G. dewevrei* seeds (mean = 30 g) limits the radius of its seedling shadow (between 6 m and 10 m) as compared to *J. seretii* (mean seed weight = 4 g), whose seedlings extended between 30 m and 40 m from the crown edge. Seedlings of both species are shade tolerant, but ten-year observations of tagged seedlings in the understory revealed a trend toward greater persistence in *G. dewevrei*, with nearly 50% of seedlings persisting ten years, as compared to *J. seretii*, with less than 35% surviving (Hart 1995).

The ten-year demographic data did show differences in the dynamic properties of the two dominant species. *Gilbertiodendron dewevrei* has a slower growth rate than *J. seretii* and more growth suppression in surviving understory individuals. *Julbernardia*'s change in growth pattern with increasing size class is likely related to its higher mortality rate as compared to *G. dewevrei*. The surviving smaller individuals of *J. seretii* had continuous growth but, once large, senesced rapidly. A shorter life cycle for the most abundant species in mixed forest is also consistent with the higher treefall frequency reported for that forest (Hart et al. 1989) and accounts for the more open canopy of mixed forest.

Particularly interesting are the differences that the dominant Caesalpiniaceous species show relative to other trees within their own forests. The two forests have considerable species overlap as regards subdominant elements (Makana et al. 1998). *Julbernardia seretii* grew rapidly relative to these species, whereas *G. dewevrei* grew slowly. According to J. P. Grime's scheme (1979), *J. seretii* has a competitive strategy within its community, as suggested not only by its rapid growth but also by its extensive canopy and root system and its shorter life expectancy. With its narrow crown, taproot, slow growth, and apparently greater longevity, *G. dewevrei* fits Grime's stress-tolerant strategy. According to Grime's scenario, in a landscape recovering from disturbance, competitive dominants are eventually replaced by "stress-tolerant" dominants.

Other differences in community characteristics and life-history traits are consistent with the potentially earlier recovery status of mixed forest. Basal area increases along successional gradients (Budowski 1970); mbau forest has the greater basal area. Although *J. seretii*'s reproductive strategies are not typical of pioneer species, relative to *G. dewevrei* its dispersal shows greater invasive ability and its recruitment less persistence at low light levels. The greater proportion of zero-growth *G. dewevrei* individuals in the subcanopy of mbau forest is indicative of greater shade suppression. It is particularly interesting that *G. dewevrei*, the species with greater zero growth, should also have the overall lower mortality, because the general case for tropical tree species is that zero growth is associated with senescence (Swaine et al. 1987). This may not be the case with the most stress tolerant of late-recovery species.

A less distant disturbance history for the mixed forest is consistent with observations made outside the Ituri Forest as well (Léonard 1952). Stands of forest dominated by *G. dewevrei* occur in a matrix of mixed forest more than 550 km west of the Ituri study sites. In this case the mixed forest is dominated by *Scorodophloeus zenkeri* (Caesalpiniaceae). Léonard (1952) observed that the mixed forest was more likely to have soil charcoal and pottery remains than the monodominant forest. These observations suggest that the pattern seen in the Ituri may be of wider relevance.

The charcoal record from the Ituri suggests numerous small forest fires, all of which burned forest that was structurally similar to the contemporary forests (Hart et al. 1996). The fire regime may have been intensified by increasing human occupation of the forest as well as climate fluctuations. If fire was the historical disturbance ultimately responsible for the mixed-mbau transitions, then we would expect less fire, or evidence for more ancient fire, in mbau forest. A systematic sampling of the substrate from the center of large mbau forest stands has not yet been undertaken, but the mbau forest in transition areas produced older charcoal than the mixed forest (Hart et al. 1996).

The presence of seeds of *G. dewevrei* is a prerequisite for the replacement of mixed forest by mbau forest. Abundant stocking of *J. seretii* in the shaded understory of mixed forest, along with the persistence and growth of this regeneration class, indicates that mixed forest is not successional in a classical sense. Whereas, in the absence of disturbance, mbau forest may encroach on mixed forest at transition areas, the invasive ability of *G. dewevrei* is very limited. Replacement of *J. seretii* is not a foregone conclusion, and a dynamic coexistence of the two forest types in regional landscapes may be permanent if the rate of seed invasion by *G. dewevrei* is counterbalanced by the frequency of major disturbances (Huston and Smith 1987).

The life-history differences between *G. dewevrei* and *J. seretii* indicate that, in the absence of major disturbance, *G. dewevrei* will encroach on mixed forest. Forest change in the Ituri Forest, however, is probably even slower than that in many other tropical forests (table 10.2). The dynamics of the mixed forest in the Ituri appears similar to that of other forests studied in the tropics. It has a calculated stand half-life of 53 years, as compared to stand half-lives of from 34 to 68 years for other forest sites across the tropics. Mbau forest, however, appears exceptional, with a stand half-life of 148 years.

Table 10.2

## Mortality Rates at the Ituri Forest Sites and Other Tropical Forest Sites

| SITE | PLOT AREA (HA) | INTERVAL (YRS) | MORTALITY (%) | ANNUAL MORTALITY[a] ($\lambda$) | STAND HALF-LIFE[b] ($T_{0.5}$) | ORIGINAL SOURCE |
|---|---|---|---|---|---|---|
| La Selva, Costa Rica | 12.4 | 13 | 23.2 | 0.0203 | 34 | Lieberman et al. 1985 |
| Kade, Ghana | 2.0 | 12 | 19.8 | 0.0177 | 38 | Swaine and Hall 1986 |
| BCI, Panama | | | | | | |
|   old forest | 2.0 | 5 | 5.2 | 0.0106 | 65 | Putz and Milton 1982 |
|   young forest | 5.0 | 5 | 8.7 | 0.0183 | 38 | Putz and Milton 1982 |
| Bukit Lagong, Malaysia | 1.6 | 10 | 12.3 | 0.0131 | 53 | Wyatt-Smith 1966 |
| Sepilok, Malaysia | 1.8 | 6 | 5.9 | 0.0102 | 68 | Nicholson 1965 |
| Central Amazon, Brazil | 5.0 | 5 | 5.6 | 0.0116 | 60 | Rankin-de-Merona et al. 1990 |
| Amazonia, Venezuela | 1.0 | 5 | 4.9 | 0.0102 | 68 | Uhl 1982 |
| Ituri, DR Congo | | | | | | |
|   mixed forest | 1.5 | 10.4 | 12.8 | 0.0132 | 53 | This chapter |
|   mbau forest | 1.5 | 10.6 | 4.8 | 0.0047 | 148 | This chapter |

*Source:* Adapted from Lieberman et al. 1985, whose bibliography includes all other sources mentioned in this table. Additional information from Rankin-de-Merona et al. 1990.

a The annual mortality rate, $\lambda$, is calculated as the slope of $\log_e$ survivorship $v$ time, $\lambda = \ln N_0 - \ln N_t / t$ (Lieberman et al. 1985).

b Stand half-life $= \ln(0.5)/\lambda$.

The mbau forest's slower turnover could conceivably result in slower-than-usual landscape changes in the Ituri.

### Implications for Conservation

The timescales for forest change elucidated in this chapter appear to limit our ability to predict the long-term outcomes of contemporary conservation decisions. Several pertinent results from this and previous vegetation studies in the Ituri Forest should, however, help us define some conservation guidelines:

- The defined transitions between mature forest types may not be stable. Progressive changes are not necessarily dependent on climatic variability or human activity. This

result suggests that in forest conservation it may be more appropriate to take a landscape approach than to focus on single species or particular forest types of interest. As change is to be expected, small islands are risky for long-term conservation; areas should be large enough to accommodate long-term changes. Just as individual species are more likely to be lost from small islands of habitat than larger islands of habitat (Miller and Harris 1977), so species associations, such as forest types with their characteristic dynamic properties and mix of species, are more likely to be lost in small islands of protected landscape than in larger islands of protected landscape.

- Growth rates, mortality rates, and other life-history characteristics vary between different mature-forest dominant tree species. Forests dominated by one or two Caesalpiniaceous species are relatively common in Africa's central forest zone and have been given a single designation, "single-dominant moist evergreen and semi-evergreen Guineo-Congolian rain forest" (White 1983); however, it is likely that they are not ecological equivalents and should be managed differently. Where two or more of these forests co–occur, the pattern of their interdigitation may be a pattern of continuing vegetation response following past disturbance. More information about the life-history characteristics of important species would help us to unravel the history of vegetation change in Africa's forest zone and help set priorities for future conservation.

- The life-history characteristics that contribute to *G. dewevrei*'s dominance in unexploited forests may make it more vulnerable to disappearance in exploited forests. *Gilbertiodendron*'s low resilience in a disturbed landscape is due to its very limited seed dispersal and particularly slow growth. In contrast, the related traits of large seeds or seedlings and low mortality under shaded conditions promote its increasing dominance in low-disturbance situations. To assure the long-term persistence of examples of this now widespread and characteristic African forest type, mbau forest must be included in some protected areas where cutting is not permitted.

### NOTES

1. Student's t-test, observed $t = 2.951$, $df = 4$, $P < .05$. The values compared were site means (three for each forest type). The test for heteroscedasticity did not reject $H_0$ of equal variances.

2. Paired comparisons, 2-way ANOVA, $Fs = 8.314$, $df = 1, 11$, $P < .05$. The values compared were the arc sine transformed mean percentage growth increments from each plot. Each plot considered provided a pair of values, the growth increment of the dominant species and that of all other species combined. A plot was not considered where fewer than three individuals either of the focus species or of other species combined were present. In all, twelve plots were included for the mbau forest comparison and twenty-one for the mixed forest comparison. In each case $H_0$ was "no difference" in the mean percentage between the focus species and all other species combined.

3. Paired comparisons, 2-way ANOVA, $Fs = 8.45$, $df = 1, 20$, $P < .01$. See note 2 for an explanation.

4. Approximate t-test, $ts = 7.207$, $df = 11$, $P < .001$. Values are the same as for the previous comparisons: arc sine transformed mean percentage growth increments from each plot; however, in this case there are no paired comparisons. Because of heteroscedasticity, an "approximate t-test" was used to test the equality of the means.

### REFERENCES

Ashton, P. S. 1976. Mixed dipterocarp forest and its variation with habitat in the Malayan lowlands: A re-evaluation at Pasoh. *Malaysian Forester* 39:56–72.

Aubréville, A. 1966. Les lisières forêt-savanne des régions tropicales. *Adansonia* 6:175–187.

Budowski, G. 1970. The distinction between old secondary and climax species in tropical central American lowland forests. *Tropical Ecology* 11:44–48.

Bush, M. B., D. R. Piperno, P. A. Colinvaux, P. E. DeOliveira, L. A. Krissek, M. C. Miller, and W. E. Rowe. 1992. A 14,300 year paleoecological profile of a lowland tropical lake in Panama. *Ecological Monographs* 62:251–275.

Conway, D. J. 1992. A comparison of soil parameters in monodominant and mixed forest in the Ituri Forest Reserve, Zaire. Honors Project, University of Aberdeen, Scotland.

Devred, R. 1958. La végétation forestière du Congo Belge et du Ruanda-Urundi. *Bull. de la Société Royale Forestière de Belgique* 6:408–468.

Ewel, J. 1980. Tropical succession and manifold routes to maturity. *Biotropica* 12:2–7.

Gentry, A: H. 1988. Changes in plant community diversity and floristic composition on environmental and geographical gradients. *Annals of the Missouri Botanical Gardens* 75:1–34.

Gérard, P. 1960. Etude écologique de la forêt dense à *Gilbertiodendron dewevrei* dans la region de l'Uele. Publ. INEAC (Inst. Natl. Etude Agron. Congo), sér. sci. no. 87, Brussels.

Germain, R., and C. Evrard. 1956. Etude écologique et phytosociologique de la forêt à *Brachystegia laurentii*. Publ. INEAC (Inst. Natl. Etude Agron. Congo), sér sci. no. 67, Brussels.

Grime, J. P. 1979. *Plant Strategies and Vegetation Processes.* Wiley, Chichester, England.

Hart, T. B. 1985. The ecology of a single-species-dominant forest and of a mixed forest in Zaire, Africa. Ph.D. dissertation, Michigan State University, East Lansing.

———. 1995. Seeds, seedlings, and sub-canopy survival in monodominant and mixed forests of the Ituri Forest, Africa. *Journal of Tropical Ecology* 11: 443–459.

Hart, T. B., J. A. Hart, R. Dechamps, M. Fournier, and M. Ataholo. 1996. Changes in forest composition over the last 4000 years in the Ituri Basin, Zaire. Pages 545–563 in L. J. G. van der Maesen, X. M. van der Burgt, and J. M. van Medenbach de Rooy, eds. *The Biodiversity of African Plants: Proceedings of the XIVth AETFAT Congress, 22–27 August 1994, Wageningen, The Netherlands.* Kluwer Academic Press, Dordrecht, The Netherlands.

Hart, T. B., J. A. Hart, and P. G. Murphy. 1989. Monodominant and species-rich forests of the humid tropics: Causes for their co-occurrence. *American Naturalist* 133(5):613–633.

Hubbell, S. P., and R. B. Foster. 1990. Structure, dynamics and equilibrium status of old growth forests on Barro Colorado Island. Pages 522–541 in A. H. Gentry, ed. *Four Neotropical Forests.* Yale University Press, New Haven.

Huston, M., and T. Smith. 1987. Plant succession: Life history and competition. *American Naturalist* 130:168–198.

Lavreau, J. 1982. Etude géologique du Haut-Zaire: Genèse et évolution d'un segment lithosphérique archéen. *Annales—Musée Royale de l'Afrique Centrale* 8, *Sciences Géologiques* 88. Koninklijk Museum voor Midden-Africa, Tervuren, Belgium.

Léonard, J. 1952. Les divers types de forêts du Congo Belge. *Lejeunia* 16:81–93.

Letouzey, R. 1968. *Etude Phytogéographique du Cameroun: Encyclopédie Biologique LXIX.* Paris, Editions Paul Lechevalier.

Lieberman, D., M. Lieberman, R. Peralta, and G. S. Hartshorn. 1985. Mortality patterns and stand turnover rates in a wet tropical forest in Costa Rica. *Journal of Ecology* 73:915–924.

Louis, J. 1947. Contribution à l'étude des forêts équatoriales congolaises: Comptes rendus de la Semaine Agricole de Yangambi. Publ INEAC (Inst. Natl. Etude Agron. Congo), hors série, Brussels.

Makana, J.-R., T. B. Hart, and J. A. Hart. 1998. Forest structure and diversity of lianas and understory treelets in monodominant and mixed stands in the Ituri Forest, Democratic Republic of Congo. Pages 429–446 in F. Dallmeier and J. A. Comiskey, eds. *Forest Biodiversity Research, Monitoring and Modeling—Conceptual Background and Old World Case Studies,* vol. 20. Man and the Biosphere Series. Parthenon Publishing Group, Pearl River, N.Y.

Miller, L. D., and L. D. Harris. 1977. Isolation and extirpations in wildlife reserves. *Biological Conservation* 12:311–315.

Pickett, S. T. A., and M. J. McDonnell. 1989. Changing perspectives in community dynamics: A theory of successional forces. *Trends in Ecology and Evolution* 4:241–245.

Prentice, C. 1986. Vegetation response to past climatic variation. *Vegetatio* 67:131–141.

Putz, F. E., and K. Milton. 1982. Tree mortality rates on Barro Colorado Island. Pages 95–100 in E. G. Leigh, Jr., A. S. Rand, and D. M. Windsor, eds. *The Ecology of a Tropical Forest: Seasonal Rhythms and Long-Term Changes.* Smithsonian Institution Press, Washington, D.C.

Rankin-de-Merona, T., R. W. Hutchings, and T. E. Lovejoy. 1990. Tree mortality and recruitment over a five-year period in undisturbed upland rainforest of the central Amazon. Pages 573–584 in A. H. Gentry, ed. *Four Neotropical Rainforests.* Yale University Press, New Haven.

Richards, P. W. 1952. *The Tropical Rain Forest: An Ecological Study.* Cambridge University Press, Cambridge.

Ritchie, J. C. 1995. Current trends in studies of long-term plant community dynamics. *New Phytol.* 130:469–494.

Swaine, M. D., D. Lieberman, and F. E. Putz. 1987. The dynamics of tree populations in tropical forest: A review. *Journal of Tropical Ecology* 3:359–366.

Swaine, M. D., and T. C. Whitmore. 1988. On the definition of ecological species groups in tropical rain forests. *Vegetatio* 75:81–86.

Sys, C. 1972. Caractérisation morphologique et physico-
chimique des profits types de l'Afrique Centrale. L'Institut National pour l'Etude Agronomique du Congo
(I.N.E.A.C.), hors série, Brussels.

Uhl, C. 1982. Tree dynamics in a species rich tierra firme
forest in Amazonia, Venezuela. *Act. Cien. Venez.*
33:72–77.

White, F. 1978. The taxonomy, ecology, and chorology of
African Ebenaceae. I. The Guineo-Congolian species.
*Bull. Jard. Bot. Nat. Belg.* 49:11–55.

———. 1983. The vegetation of Africa: A descriptive
memoir to accompany the UNESCO/AETFAT/UNSO
vegetation map of Africa. UNESCO, Paris.

White, L. J. T. 1992. Vegetation history and logging damage: Effects on rain forest mammals in the Lopé Reserve,
Gabon. Ph.D. dissertation, University of Edinburgh.

Whitmore, T. C. 1984. *Tropical Rain Forests of the Far
East.* 2d ed. Clarendon, Oxford.

———. 1989. Changes over twenty-one years in the
Kolombangara rain forests. *Journal of Ecology*
77:469–483.

# CHAPTER 11

# Forest-Savanna Dynamics and the Origins of Marantaceae Forest in Central Gabon

Lee J. T. White

Plants belonging to the families Marantaceae (arrowroot) and Zingiberaceae (gingers) occur throughout the African tropical rain forest region. In some areas they are a major component of the understory, to the extent that Marantaceae Forest is recognized as an important forest type in west and central Africa (see, e.g., Koechlin 1964; Letouzey 1968; Maley 1987, 1990; de Foresta 1990; Swaine 1992; Hawthorne 1994; White et al. 1995; Fay 1997). In addition to the dense ground cover dominated by species of Marantaceae and Zingiberaceae (henceforth referred to collectively as "herbs"), Marantaceae Forest is characterized by a generally open appearance above the understory owing to low densities of medium-sized trees.

Several species of herbs (Fay [1997] extends the definition of herbs, or terrestrial herbaceous vegetation [THV], to include all ground-rooted species in the subclass Monocotyledonae) provide important food items for apes (*Gorilla gorilla*, *Pan troglodytes*, and *Pan paniscus*), mandrills (*Mandrillus sphinx*), and elephants (*Loxodonta africana*) in the forests of central Africa (Jones and Sabater-Pi 1971; Kano 1983; Badrian and Malenky 1984; Kano and Mulavwa 1984; Hoshino 1986; Carroll 1988; Rogers et al. 1988; Williamson et al. 1990; Malenky and Stiles 1991; Tutin et al. 1991; Wrangham et al. 1991, 1993; Tutin and Fernandez 1993a; White et al. 1993). These plants become particularly important at times of the year when fruit is scarce, when they are a "keystone" food (*sensu* Terborgh 1986) for all these species. Marantaceae Forest in central Gabon supports a mammalian biomass up to almost 6,000 kg/km$^2$, about four times greater than that in forests with open understory (Prins and Reitsma 1989; White 1994). There is evidence for high densities of elephants and gorillas in other Marantaceae Forests (Carroll 1988; Fay 1989), and it has been suggested that the availability of abundant herbaceous foods may have influenced the evolution of social organization and terrestriality in African great apes (Wrangham 1986).

The occurrence of stands of herbs in African rain forests has been attributed variously to: the passage of fire (Hall and Swaine 1981; Swaine 1992; Hawthorne 1994); savanna recolonization (Letouzey 1968; de Foresta 1990; White et al. 1995); association with permanent water (e.g., Rogers and Williamson 1987; Wrangham et al. 1993); such forest disturbance as logging, cultivation, or natural gaps (Letouzey 1968; Calvert 1985; Carroll 1988; de Foresta 1990; White et al. 1995); and elephants (Guillaumet 1967; Calvert 1985). In some central African forests, however, there is no obvious explanation for the occurrence of these stands (D. Harris, Central African Republic, pers. comm.). The stands vary greatly in size. Where herbs grow as a result of disturbance or close to water, they form narrow belts along roads or rivers, or patches up to several hectares in extent. In contrast, in parts of central and western Africa there are forest formations in which herbs dominate understory vegetation over areas of hundreds or thousands of square kilometers (see, e.g., Letouzey 1968; Swaine 1992; see Maley 1990 for a review). It is these more extensive formations that I refer to here as Marantaceae Forest.

In Ghana, Marantaceae Forests develop subsequent to the passage of fires lit by humans through semideciduous forests. Fires have altered about 30% (1,400 km²) of this forest formation (Swaine 1992; Hawthorne 1994). The fires kill most of the smaller plants but leave many upper-canopy trees unharmed. Some herbs colonize soon after the passage of the fire and are able to grow rapidly into a dense understory, which may interfere with further regeneration. Swaine (1992) has maintained that all central African Marantaceae Forests are "clearly" caused by fire. Other authors, however, have suggested that Marantaceae Forests represent a stage in the natural succession from savanna to forest, which occurs when colonization of savannas by forest species is not blocked by regular fires (Letouzey 1968; de Foresta 1990; White et al. 1995).

Given the large extent of Marantaceae Forests and their value to large mammals, they should be considered a conservation priority. Their superficial resemblance to degraded forest has meant that they have been largely overlooked in the past (White et al. 1995). In order to understand the role that Marantaceae Forest

has played in the evolution of African rain forest taxa, and in order to design sensible management measures for the future, we have to understand its origins and dynamics. In the Lopé Reserve, Gabon, Marantaceae Forest is one of the dominant vegetation types and some savannas are currently being colonized by forest vegetation (Tutin et al. 1994; White et al. 1995). A study was undertaken to document the development of Marantaceae Forest and to evaluate whether it is formed after savanna colonization or if alternative hypotheses, such as one involving fires penetrating the forest from the savanna, could better explain observed vegetation patterns.

Research was carried out in the Lopé Reserve, central Gabon, at the Station d'Etudes des Gorilles et Chimpanzés (SEGC) at 0°12' S, 11°36' E. Lopé has long been known for its "strange forest-savanna mosaic" (Aubréville 1967). Most of the reserve is covered by semievergreen lowland tropical rain forest, but there are 800 km² of savanna and forest-savanna mosaic along its northern and eastern limits (figure 11.1). These savannas, completely enclosed within the Congolian forest block, have recently been dated to about 9,000 years B.P. using $\delta^{13}C$ analyses and $^{14}C$ dating (Oslisly et al. 1996, 1997; Oslisly, this volume; note that Oslisly et al.'s conclusions have been challenged by Schwartz [1997]). Today, as in the past, they are maintained by annual fires lit by humans, but where fires are not lit, the forest is currently advancing into the savanna (White 1992; White et al. 1998). As in other parts of central Africa (Letouzey 1968; de Foresta 1990), vegetation around these savannas is predominantly Marantaceae Forest. In Lopé the Marantaceae Forest extends up to about 20 km from the savanna edge and is then replaced by forest types with open understories (see figure 11.1).

Lopé lies in an area of low rainfall compared to much of Gabon (EDICEF 1983). Mean annual rainfall at SEGC is 1,503 mm (1984–1997), although there is considerable interannual variation in the amount and distribution of rainfall (Tutin and Fernandez 1993b). An important feature of the annual cycle is a major dry season from mid-June to mid-September, but the timing of its onset and duration varies from year to year. December–February are generally relatively dry, but a second dry season is not well defined. Mean relative humidity varies little through the year and does not drop below 70%. Temperatures also vary little over the year, but mean monthly minima and maxima are lowest during the major dry season, when cloud cover is almost constant (Saint-Vil 1977), with a range of 20–23°C and 26–33°C, respectively, over nine years (Tutin and Fernandez 1993b).

Vegetation in Lopé is a complex mosaic of plant associations (White 1992). Within and adjacent to the forest-savanna mosaic, five major vegetation types can be recognized by differences in structure and species composition:

- Savanna vegetation, maintained by annual fires, is dominated by grasses (to date, seventy species have been identified; dominant species include *Anadelphia afzeliana [=Pobeguinea arrecta]*, *Hyparrhenia diplandra*, *Andropogon pseudapricus*, and *Ctenium newtonii*), with such shrubs as *Crossopteryx ferruginea*, *Nauclea latifolia*, *Psidium guineensis*, and *Bridelia febrifuga* patchily distributed in some areas (Descoings 1974; White et al. 1998; Alers and Blom, in White 1995). Large continuous areas of Savanna are restricted to low altitudes (below 250 m), while small isolated patches occur within the forest either around these zones or on hilltops with altitudes of about 250–450 m (IGN 1985).
- Colonizing Forest occurs adjacent to savannas in areas protected from fire. Such

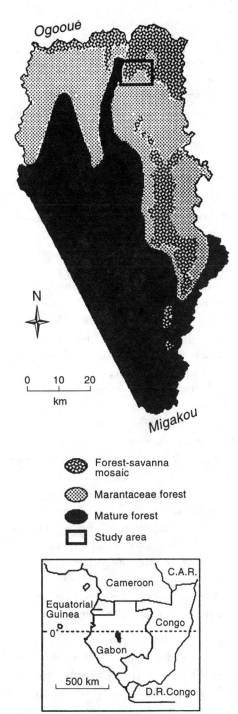

Figure 11.1. Distribution of main vegetation types in the Lopé Reserve (vegetation types were mapped using radar images produced in 1992).

shrubs as *Psidium guineensis, Psychotria vogeliana,* and *Antidesma vogelianum* become common, as do the trees *Aucoumea klaineana, Lophira alata, and Sacoglottis gabonensis,* all of which are able to invade savanna and thrive where protected from fire (Tutin et al. 1994; White et al. 1998). Colonizing individuals are generally small (compared to those of the same species in the forest), twisted, and often branch low. Ground vegetation is predominantly grasses, but some herbs occur.

- Monodominant Forest is dominated (> 75% of basal area) by one or a combination of *Aucoumea klaineana, Lophira alata,* or *Sacoglottis gabonensis,* but some other species occur (e.g., *Klainedoxa gabonensis*). Many individuals are crooked or branch low, and most are smaller than is usual for their species. Ground vegetation is sparse, but herbs become more common.

- Marantaceae Forest is still dominated by *Aucoumea klaineana* and *Lophira alata,* but by a smaller number of larger individuals. Herbs are abundant, especially *Haumania liebrechtsiana,* which covers the ground to a height of up to about 2 m and forms dense tangles and vine towers on some trees, which reach a height of 10 m or more.

- In Mixed Marantaceae Forest, greater numbers of tree species are present, adding to the structural complexity and species diversity. *Aucoumea klaineana* and *Lophira alata* are no longer dominant. Herbaceous vegetation remains abundant.

Figure 11.2 shows representative profile diagrams (see Richards 1952) for these vegetation types. White et al. (1995) hypothesized that these five vegetation types represent a succession from savanna to mature forest.

The five vegetation types were mapped in a 50 km² section of the north of the Lopé Reserve. Aerial photographs from 1982 and radar images taken in 1992 were available for the entire study area, and there were also some aerial photographs from 1957 (all three at 1:50,000 scale). It was possible to distinguish on aerial photographs and radar images between Savanna, Colonizing Forest, Monodominant Forest, and Marantaceae Forest, but the two types of Marantaceae Forest could not be distinguished from one another. A map of 1:10,000 scale was produced by tracing visible vegetation boundaries from aerial photographs and radar images of 1:50,000 scale and enlarging. Locations of watercourses, old logging roads, and major elephant trails were marked onto these maps using unpublished records from SEGC, aerial photographs, and an existing topographic map and by direct mapping using a sighting compass and a "topofil" measure. The entire study area was then walked, and each vegetation type was mapped with the aid of a compass, topofil, and GPS unit, by walking its perimeter.

Five patches of Colonizing Forest were selected in different parts of the study area. This is an important vegetation type for gorillas and chimpanzees at certain times of the year (Tutin et al. 1995), and workers at SEGC had visited all these areas regularly during a period of ten years before this study. Their background knowledge, data on growth rates of tree species found in these areas (White, Williamson, Harrison, and Reitsma, in White 1995), and aerial photographs taken in 1957 were used to estimate the time since each area had last been burned. A 40 × 20 m plot was located randomly within each of the five patches.

The plots were located in the field, demarcated, and divided into 10 × 10 m subplots. All trees and lianas with diameter ≥5 cm at 1.3 m above the ground (diameter at breast height, or dbh) were labeled with a numbered aluminum tag and identified (see Dallmeir 1992 for details of how problematic individuals were treated). For each individual, the dbh was measured

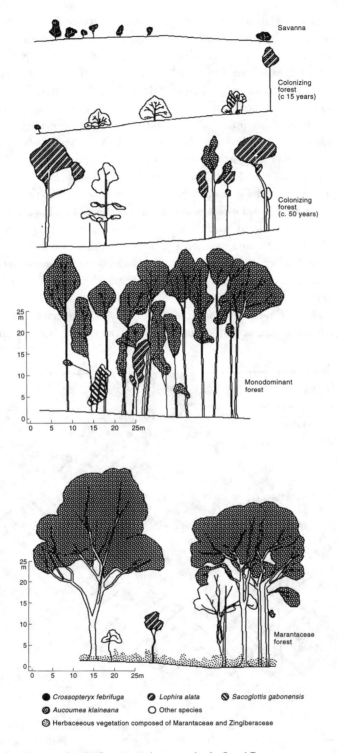

Savanna

Colonizing
forest
(c 15 years)

Colonizing
forest
(c. 50 years)

Monodominant
forest

Marantaceae
forest

● *Crossopteryx febrifuga*      ◐ *Lophira alata*      ◒ *Sacoglottis gabonensis*

◉ *Aucoumea klaineana*      ○ Other species

◉ Herbaceeous vegetation composed of Marantaceae and Zingiberaceae

Figure 11.2. Profile diagrams for the five vegetation types in the Lopé Reserve.

using a diameter tape. A scale drawing of the plot was made and the position of each individual marked. A vertical projection of the canopy of each individual was drawn by measuring the distance from the trunk to the edge of the canopy on the four compass points and then sketching in the shape of the canopy.

In the first (southeast or southwest corner, depending on orientation of the plot) 10 × 10 m subplot in each plot, all individuals with diameter ≥1 cm at 10 cm above the ground were measured, identified, and labeled with a numbered aluminum tag (attached with wire). The position of each individual was indicated on a scale drawing. The number of stems of all species of Marantaceae and Zingiberaceae in each square meter within the entire plot was noted. For each plot in Colonizing Forest, a control in the nearest Savanna, located at about

the same distance from the forest edge, was established and enumerated in the same way.

Five relatively large patches of each of Monodominant, Marantaceae, and Mixed Marantaceae Forest were selected such that the distance between patches of the same forest type was equal to or greater than that between different forest types. A 40 × 20 m plot was located randomly in each patch, and data were taken for plants ≥5 cm dbh. In addition, ten 2 × 2 m subplots were chosen randomly within each plot, and stems of each species of Marantaceae and Zingiberaceae present were counted.

The seedlings of *Aucoumea klaineana* and *Lophira alata* in a 5 m strip along a series of transects established across vegetation transitions (White et al. 1998) were counted in each meter in order to assess the effect of herbs on regeneration of these two species. Non-parametric statistical

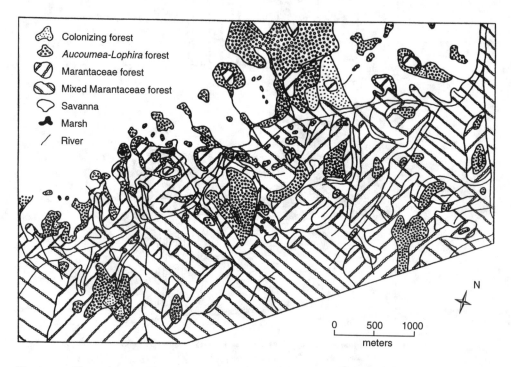

Figure 11.3. Vegetation map illustrating the various vegetation types in the forest-savanna mosaic. Symbols for vegetation types are based on Iron Age rock engravings found in the Lopé region; see Oslisly and Peyrot 1993.

tests (Siegel and Castellan 1988) were used to test for statistically significant differences in species composition between forest types.

Figure 11.3 shows the distribution of the five vegetation types in a representative section comprising about half the study area. Colonizing Forest was present adjacent to savanna in some places (but separated by a natural barrier such as a marsh or river) or in small isolated patches within the forest. Each patch of Colonizing Forest was enclosed by Monodominant Forest (where it was not in contact with the savanna), and in each case there were concentric rings of Marantaceae Forest and then Mixed Marantaceae Forest around this (in many areas Monodominant Forest did not enclose Colonizing Forest but was surrounded by concentric rings of Marantaceae and Mixed Marantaceae Forest).

The five plots in Colonizing Forest were in dense vegetation containing many shrubs and small trees, with decreased grass cover compared to Savanna vegetation. These plots could be arranged in order of the time elapsed since they were last burned, on the basis of a judgment of observed changes in the structure of vegetation since 1983 (C. E. G. Tutin, pers. comm.) and differences in two sets of aerial photographs (1957 and 1982). For convenience, they were numbered successively from one to five (youngest to oldest).

Table 11.1 shows the number of individuals of all species ≥1 cm diameter that occurred at least ten times in either Savanna or Colonizing Forest 10 × 10 m subplots. In Colonizing Forest these were a mixture of twenty-six trees, five shrubs, and four lianas, while in the Savanna the species represented were one tree and five

Table 11.1

## Abundance of Species That Occurred at Least Ten Times in Savanna or Colonizing Forest Plots

| SPECIES | FAMILY | TOTAL COUNTED COLONIZING | SAVANNA | LIFEFORM |
|---|---|---|---|---|
| *Psidium guineense* | Myrtaceae | 311 | 25 | Shrub |
| *Aucoumea klaineana* | Burseraceae | 144 | 2 | Tree |
| *Uapaca guineensis* | Euphorbiaceae | 73 | — | Tree |
| *Psychotria vogeliana* | Rubiaceae | 59 | — | Shrub |
| *Barteria fistulosa* | Passifloraceae | 39 | — | Tree |
| *Crossopteryx ferruginea* | Rubiaceae | 24 | 281 | Shrub |
| *Duboscia macroceras* | Tiliaceae | 22 | — | Tree |
| *Klainedoxa gabonensis* | Irvingiaceae | 22 | — | Tree |
| *Triumfetta cordifolia* | Tiliaceae | 20 | — | Shrub |
| *Swartzia fistulosa* | Caesalpiniaceae | 17 | — | Tree |
| *Lophira alata* | Ochnaceae | 16 | — | Tree |
| *Antidesma vogelianum* | Euphorbiaceae | 14 | 1 | Shrub |
| Other species | | 117 | 9 | |
| Total number of individuals | | 891 | 318 | |
| Number of species | | 63 | 6 | |

shrubs. The mean number of individuals in Colonizing Forest was 178, with a mean of 24 species per plot, while in Savanna there were fewer (Mann-Whitney U-test, two-tailed, $P =$ 0.02–0.01), with means of 64 individuals and 3 species per plot (Mann-Whitney U-test, two-tailed, $P =$ 0.01–0.002). It is immediately evident that Colonizing Forest has many more shrubs and seedlings or saplings than Savanna, except for the fire-resistant shrub *Crossopteryx ferruginea*, which regenerates well in burned savanna but occurs at lower densities in Colonizing Forest. Several tree species that are common in the forest at Lopé (White 1992; Tutin et al. 1994), particularly *Aucoumea klaineana* and *Uapaca guineensis*, were well represented in the small plots.

Plots in Savanna contained only two species of shrub that grew to ≥5 cm dbh and were essentially open grassland vegetation (see figure 11.2). Table 11.2 lists the first ten species ≥5 cm dbh ranked in terms of basal area in burned and unburned savanna plots (all five plots in each vegetation type combined). Only the fire-resistant shrubs *Crossopteryx ferruginea* and *Nauclea latifolia* occurred in both Savanna and Colonizing Forest, but *Nauclea* was relatively uncommon in both samples. Densities of *Crossopteryx ferruginea* were lower in Colonizing Forest, but the mean diameter was higher, approaching statistical significance (Mann-Whitney U-test, one-tailed, $P =$ 0.06) and individuals were significantly taller in unburned areas (Mann-Whitney U-test, one-tailed, $P =$ 0.000005). In some instances, shading from taller colonizing tree species was apparently killing both *Nauclea latifolia* and *Crossopteryx ferruginea*.

The number and size of individuals varied between plots, as did the number of species. Sample sizes are not large—the five plots represent 0.4 ha in each habitat type. White (1992) found that in samples of 2.5 ha, species area curves continued to rise. However, these small samples do reveal some interesting trends. Some species occurred in only one or two vegetation types. Some of these were rare, and this may be because of the small sample sizes, but others were species typical of Savanna (e.g., *Crossopteryx ferruginea*) or Colonizing Forest that did not occur in Marantaceae Forest, or were species found only in the Marantaceae Forest types (e.g., *Dacryodes buettneri*). Species present in three or more vegetation types were most abundant in either Colonizing Forest, Monodominant Forest, or one of the Marantaceae Forest types and showed patterns consistent with successional change between these vegetation types. For example, the first four species in table 11.2 show peak densities in either Colonizing or Monodominant Forest and then become progressively less abundant.

Figure 11.4 shows the distribution of girth classes in each vegetation type. There was a successive increase in number and proportion of individuals in higher diameter classes from Savanna to Mixed Marantaceae Forest. Individuals of *Lophira alata* and *Aucoumea klaineana* in each vegetation type were divided into diameter classes (figures 11.5a and 11.5b). In Colonizing Forest young *Lophira* trees with dbh ≤30 cm were common, but there were few larger individuals. In Monodominant Forest there are comparable numbers of young trees but some larger (older) individuals. In Marantaceae and particularly Mixed Marantaceae Forest there were fewer young individuals (that is, less regeneration under the developing canopy and dense Marantaceae) but some quite large (old) trees occurred. For each of Colonizing Forest, Monodominant Forest, and Marantaceae Forest, distribution by size class did not differ significantly from a Poisson distribution, indicating random distribution between size classes, but for Mixed Marantaceae Forest there is a significant difference ($P =$ 0.0002), owing to the occurrence of several large individuals. The distribution of *Lophira* diameters supports the theory that the

Table 11.2

## Species Ranked in the First Ten by Basal Area in at Least One of the Vegetation Types

| SPECIES | FAMILY | SAVANNA | | | COLONIZING | | | MONODOMINANT | | | MARANTACEAE | | | MIXED | | |
|---|---|---|---|---|---|---|---|---|---|---|---|---|---|---|---|---|
| | | BA | NO | RANK | BA | NO | RANK | BA | NO | RANK | BA | NO | RANK | BA | NO | RANK |
| Crossopteryx ferruginea | Rubiaceae | 1.45 | 420[a] | 1 | 0.89 | 153[a] | 5 | | | | | | | | | |
| Nauclea latifolia | Rubiaceae | 0.01 | 3 | 2 | 0.21 | 8 | 9 | | | | | | | | | |
| Aucoumea klaineana | Burseraceae | | | | 2.05 | 5 | 1 | 26.74 | 210 | 1 | 38.88 | 113 | 1 | 2.2 | 8 | 7 |
| Lophira alata | Ochnaceae | | | | 1.58 | 128 | 2 | 4.28 | 155 | 2 | 1.41 | 115 | 4 | 3.21 | 25 | 4 |
| Barteria fistulosa | Passifloraceae | | | | 1.39 | 215 | 3 | 0.79 | 95 | 4 | 0.13 | 10 | 16 | 0.03 | 5 | 38 |
| Xylopia aethiopica | Annonaceae | | | | 0.91 | 53 | 4 | 0.35 | 18 | 7 | 0.15 | 18 | 14 | 0.01 | 3 | 48 |
| Antidesma vogelianum | Euphorbiaceae | | | | 0.56 | 60 | 6 | 0.05 | 8 | 26 | | | | | | |
| Pauridiantha efferata | Rubiaceae | | | | 0.35 | 58 | 7 | 0.10 | 20 | 17 | 0.02 | 5 | 32 | | | |
| Maprounea membranacea | Euphorbiaceae | | | | 0.34 | 30 | 8 | 0.22 | 5 | 11 | 0.03 | 5 | 28 | | | |
| Klainedoxa gabonensis | Irvingiaceae | | | | 0.20 | 15 | 10 | 0.29 | 5 | 9 | | | | | | |
| Cola lizae | Sterculiaceae | | | | | | | 1.48 | 73 | 3 | 3.05 | 170 | 2 | 4.12 | 245 | 3 |
| Erythroxylum mannii | Erythroxylaceae | | | | 0.12 | 15 | 13 | 0.54 | 8 | 5 | | | | | | |
| Sacoglottis gabonensis | Humiriaceae | | | | 0.09 | 18 | 15 | 0.46 | 20 | 6 | | | | | | |
| Xylopia quintasii | Annonaceae | | | | | | | 0.32 | 60 | 8 | 0.20 | 20 | 13 | 0.12 | 10 | 28 |
| Diospyros dendo | Ebenaceae | | | | | | | 0.28 | 48 | 10 | 0.89 | 123 | 7 | 0.60 | 123 | 16 |
| Pentaclethra eetveldeana | Mimosaceae | | | | | | | | | | 1.84 | 10 | 3 | | | |

Table 11.2 (continued)

| SPECIES | FAMILY | SAVANNA | | | COLONIZING | | | MONODOMINANT | | | MARANTACEAE | | | MIXED | | |
|---|---|---|---|---|---|---|---|---|---|---|---|---|---|---|---|---|
| | | BA | NO | RANK | BA | NO | RANK | BA | NO | RANK | BA | NO | RANK | BA | NO | RANK |
| *Hylodendron gabunense* | Caesalpiniaceae | | | | | | | | | | 1.32 | 5 | 5 | 0.21 | 3 | 24 |
| *Pentaclethra macrophylla* | Mimosaceae | | | | | | | | | | 1.30 | 5 | 6 | 1.92 | 3 | 9 |
| *Canarium schweinfurthii* | Burseraceae | | | | | | | | | | 0.84 | 3 | 8 | | | |
| *Polyalthia suaveolens* | Annonaceae | | | | | | | | | | 0.40 | 40 | 9 | 0.63 | 38 | 15 |
| *Diospyros polystemon* | Ebenaceae | | | | | | | | | | 0.37 | 20 | 10 | 1.36 | 40 | 12 |
| *Dacryodes buettneri* | Burseraceae | | | | | | | | | | | | | 6.42 | 25 | 1 |
| *Pterocarpus soyauxii* | Papilionaceae | | | | | | | | | | | | | 5.03 | 5 | 2 |
| *Pycnanthus angolensis* | Myristicaceae | | | | | | | | | | | | | 2.80 | 3 | 5 |
| *Celtis tessmannii* | Ulmaceae | | | | | | | | | | | | | 2.53 | 8 | 6 |
| *Testulea gabonensis* | Luxembourghiaceae | | | | | | | | | | | | | 1.96 | 3 | 8 |
| *Scottellia coriacea* | Flacourtiaceae | | | | | | | 0.03 | 3 | 31 | 0.10 | 8 | 20 | 1.59 | 3 | 10 |
| Totals | | 1.46 | 423 | 2 | 9.56 | 860 | 35 | 37.75 | 990 | 49 | 52.63 | 853 | 40 | 41.38 | 748 | 57 |

*Note:* BA = basal area (m² ha⁻¹); No = density (individuals ha⁻¹); Rank = rank by basal area in that vegetation type.

a   One individual that had recently died was present.

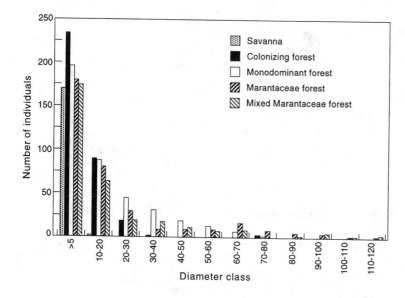

Figure 11.4. Distribution of diameter classes for trees in each vegetation type.

four vegetation types represent a succession of increasing age, with a gradual buildup of older individuals along the succession.

For *Aucoumea* the relationship is less clear. In the Colonizing Forest sample there were only two individuals, both of which were quite large (50–60 cm dbh), although there were a great many seedlings and saplings. In Monodominant Forest, diameters do not differ significantly from a Poisson distribution, showing random distribution of trees of different sizes, but there were no trees ≤10 cm dbh, suggesting that there was little regeneration occurring. In Maranta-ceae Forest the distribution was strongly skewed toward larger (older) individuals, and in Mixed Marantaceae Forest only a few old individuals survived, with no regeneration.

Figure 11.6 shows scale drawings of vertical projections of the canopies of all trees and lianas ≥5 cm dbh in selected plots. The top row shows a typical Savanna plot and plots in Colonizing Forest that have not been burned for 12–15 and about 25–30 years, respectively. The bottom row shows a plot from each of Monodominant, Marantaceae, and Mixed Marantaceae Forests. The plot in Monodominant For-

est could be dated to about 60 years since burning, as it was an enclosed savanna with obvious colonization under way in the 1957 aerial photographs. As for the diameter data, these diagrams illustrate graphically the apparently ordered change from Savanna to Mixed Marantaceae Forest.

Table 11.3 shows that Marantaceae and Zingiberaceae were slow to establish, and table 11.4 gives mean densities of all species of Marantaceae and Zingiberaceae in sample plots in each vegetation type. Animals disperse the seeds of *Megaphrynium* spp., *Aframomum* spp., and *Renealmia* spp., although these species also reproduce vegetatively. The fruits of *Haumania liebrechtsiana* and *Hypselodelphis violacea* are dehiscent capsules, which project dry seeds a few meters. Their seeds are eaten by gorillas and chimpanzees but are crunched up and are therefore rarely or never viable in dung. These two species seem to rely predominantly on vegetative reproduction, advancing as a front. Hence, the animal-dispersed species are the first to appear, because they are regularly dispersed over large distances and are specifically carried into Colonizing Forest during periods

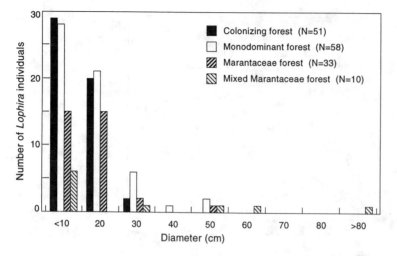

Figure 11.5a.
Diameter classes
for *Lophira alata* in
each vegetation
type.

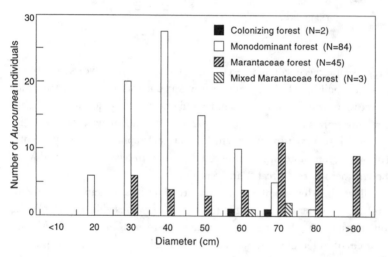

Figure 11.5b.
Diameter classes
for *Aucoumea
klaineana* in each
vegetation type.

when animals are attracted by the fruits of such species as *Psychotria vogeliana* and *Antidesma vogelianum* (see Tutin et al. 1995). They are later joined, and to some extent replaced, by the vegetative herb species.

In Marantaceae and Mixed Marantaceae Forests herbs form an impenetrable tangle of vegetation up to a height of about 2 m, climbing to more than 10 m in vine towers (particularly of *Haumania liebrechtsiana*) that often choke smaller trees. In these vegetation types 17% (N = 370) of trees <10 m tall were covered in dense vine towers, and there were often dead

trees of this size completely swamped and presumably killed by *Haumania liebrechtsiana*. Vine towers are often used by elephants as a means to pull trees over in order to feed on their foliage, and a further 28% of trees <10 m tall had been broken at some stage and resprouted.

For five profiles that crossed between Monodominant and Marantaceae Forest boundaries, the number of seedlings of *Aucoumea klaineana* and *Lophira alata* was significantly higher in Monodominant Forest. For *Aucoumea klaineana*, seedling density was 1,808 ha$^{-1}$ compared to 59 ha$^{-1}$ in Marantaceae Forest, and corre-

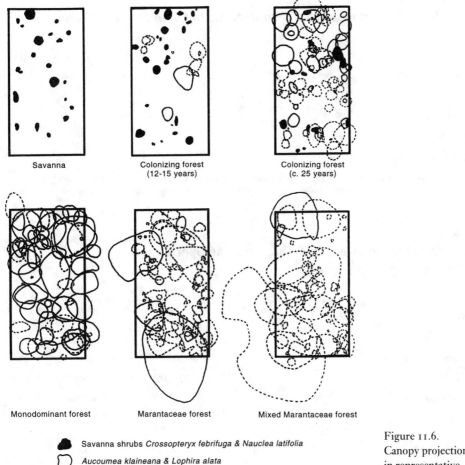

Savanna

Colonizing forest
(12-15 years)

Colonizing forest
(c. 25 years)

Monodominant forest

Marantaceae forest

Mixed Marantaceae forest

Savanna shrubs *Crossopteryx febrifuga* & *Nauclea latifolia*

*Aucoumea klaineana* & *Lophira alata*

Other species

Figure 11.6.
Canopy projections
in representative
vegetation plots.

sponding values for *Lophira alata* were 1,089 ha$^{-1}$ and 197 ha$^{-1}$, respectively. These figures demonstrate the strong inhibitive effect of high densities of Marantaceae and Zingiberaceae on the germination of these two species and, indeed, their general inhibition of regeneration.

The picture that emerges is one of a continuum between Savanna and Marantaceae Forest. Once savanna fires cease, many new species are able to colonize. As more individuals establish and grow, such forest species as *Aucoumea klaineana*, *Lophira alata*, *Barteria fistulosa*, and

*Xylopia aethiopica* replace the savanna shrubs *Crossopteryx febrifuga* and *Nauclea latifolia*. Later, *Aucoumea klaineana* and *Lophira alata*, the only two potentially large rain forest trees that establish and grow well in unburned savanna in the study area (elsewhere in Lopé, *Sacoglottis gabonensis* behaves similarly; see White 1992; White et al. 1998), come to dominate the vegetation and form the upper canopy of the forest. As these trees grow taller, they begin to shade out many of the colonizing shrubs and remaining grasses, and create conditions favorable for the establishment of other

Table 11.3

## Densities of Marantaceae and Zingiberaceae in Colonizing Vegetation

| | | NUMBER OF STEMS | | | | |
| SPECIES | FAMILY | PLOT 1 (12–15) | PLOT 2 (15–20) | PLOT 3 (20–25) | PLOT 4 (25–30) | PLOT 5 (c. 50) |
|---|---|---|---|---|---|---|
| *Aframomum sericeum* | Zingiberaceae | 17 | 92 | 191 | 866 | 884 |
| *Renealmia macrocolea* | Zingiberaceae | 0 | 0 | 0 | 0 | 25 |
| *Renealmia cincinnata* | Zingiberaceae | 0 | 0 | 0 | 0 | 14 |
| *Costus afer* | Zingiberaceae | 0 | 7 | 0 | 0 | 9 |
| *Megaphrynium* spp. | Marantaceae | 0 | 19 | 0 | 0 | 39 |

*Note:* Numbers in parentheses in column headings indicate the estimated years since each plot was last burned.

Table 11.4

## Densities of Marantaceae and Zingiberaceae Species in Various Vegetation Types

| | | | DENSITY (STEMS HA⁻¹) | | |
| SPECIES | SAVANNA | COLONIZING FOREST | MONODOMINANT FOREST | MARANTACEAE FOREST | MIXED MARANTACEAE FOREST |
|---|---|---|---|---|---|
| **Marantaceae** | | | | | |
| *Haumania liebrechtsiana* | 0 | 0 | 100 | 30,300 | 32,450 |
| *Megaphrynium* spp. | 0 | 145 | 18,250 | 10,650 | 23,900 |
| *Hypselodelphis violacea* | 0 | 0 | 600 | 2,350 | 1,800 |
| **Zingiberaceae** | | | | | |
| *Aframomum* spp. | 0 | 5,125 | 2,300 | 550 | 650 |
| *Renealmia* spp. | 0 | 95 | 1,650 | 50 | 0 |

forest species, including herbs. A continuous canopy forms, consisting exclusively of early colonizing trees that grow up in a tightly packed, even-aged stand. Within this vegetation, *Aucoumea klaineana* attains dominance comparable to *Gilbertiodendron dewevrei* in the monodominant forests of central Africa (see T. Hart, this volume).

Densities of *Megaphrynium* spp. become high in Monodominant Forest, but it is not until the next stage in the succession that *Haumania liebrechtsiana* becomes abundant. There is no evidence to suggest that this transition is due to forest fires in Lopé. In Ghana, scarring from the passage of fires through the forest is evident on many of the large trees in Marantaceae Forest (M. D. Swaine, pers. comm.), but this has not been observed in Lopé. When a large emergent tree in Lopé was struck by lightning and caught fire, the flames did not spread, although a hot fire burned for two days (Tutin et al. 1997). The remains of this fire were characteristic and would be easily identified if encountered elsewhere. In addition, one would predict that if fire were the cause of the transition, Marantaceae Forest would dominate most areas adjacent to savanna. Instead, these areas tend to be dominated by Colonizing Forest or Monodominant Forest (see figure 11.3).

There are marked differences in composition and particularly structure of the upper canopy in the first type of Marantaceae Forest compared to Monodominant Forest; notably, this Marantaceae Forest shows decreased numbers of *Lophira alata* and *Aucoumea klaineana* individuals and increased canopy size of these two species. A possible explanation for this change is high mortality, because trees that have grown in an even-aged stand will tend to die at about the same time, and the resulting gaps create ideal conditions for an explosion in herb biomass. By monitoring permanent plots established during and before this study in all vegetation types found in Lopé, we will gradually obtain a more detailed picture of the changes. However, there is good evidence that the Marantaceae Forests of central Gabon represent a stage in the development of mature forest following savanna colonization.

There has been a long history of climatic change in central Africa resulting in periodic movements of forest-savanna boundaries (see, e.g., Hamilton 1982; Maley 1990, 1991). In Lopé, Marantaceae Forests extend up to 20 km from the savanna edge in places (see figure 11.1) and are then replaced by Mature Forests with increased species diversity and structural complexity and decreased densities of herbs. Oslisly and Dechamps (1994) found evidence of fires about 20 km south of the current savanna-forest interface in Lopé, which dated from 1,400–1,500 B.P. However, these were closely associated with remains of iron-smelting furnaces dated to the same period and were probably caused by humans. Two tree species identified from the charcoal, *Sapium ellipticum* and *Erythroxylum* sp., are currently restricted to the forest-savanna mosaic (Tutin et al. 1994). This suggests that the savannas extended further south at that time.

Working on the assumption that all Marantaceae Forests in Lopé are the result of savanna colonization, there would have to have been a period in the fairly recent past when savanna fires were absent, or rare enough to allow the forest to advance up to 20 km into the savanna. This would be unlikely if human populations had continued to live in the area, because they would burn savannas regularly. However, Oslisly (1993, 1996, and this volume) reports evidence of a population crash at about 1,400 B.P., which lasted until about 800 B.P.: there are numerous archaeological remains from the Lopé area both before and after this period, but no evidence of human activity between these dates (see Oslisly, this volume). Such a population crash, coinciding as it did with a relatively humid period in Africa's climatic history (see Maley 1992), would have allowed the forest to advance rapidly into the savanna from its southern limit at the time. This scenario suggests that a great deal more of the vegetation in the Lopé Reserve was savanna up to 1,500 years ago. This area is delimited today by the distribution of Marantaceae Forests (see figure 11.1). Today these forests support high mammalian biomass (White 1994).

De Foresta (1990) described a similar situation to Lopé at Makaba, Congo—where Marantaceae Forests are found around enclosed savannas—as did Letouzey (1968) for Cameroon. Both these authors suggested that the Marantaceae Forest was a product of savanna colonization. It seems reasonable to hypothesize that at times when central African forests were expanding out of refugia into savanna, Marantaceae Forest was much more widespread than it is today. If this is so, this forest is likely to have played a major role in the evolution of African rain forest animals. Gorillas in particular rely upon species of Marantaceae and Zingiberaceae for food and nesting materials (White et al. 1995), and forest elephants (*L. a. cyclotis*) seem ideally suited to life in Marantaceae forest, with their small body size, downward-pointing tusks, and low shoulders. A study by Smith et al. (1997)

demonstrated that forest savanna ecotones might play a key role in speciation. It is perhaps no coincidence that one of the commonest trees in Marantaceae forest, *Cola lizae*, and the dominant ginger, *Aframomum sericeum*, are species new to science, described from collections made in Lopé. Dowsett-Lemaire and Pannell (1996) report a similar finding in Odzala, Congo, where a new species of tree, *Diospyros whitei*, is common in Marantaceae forests.

Savanna colonization does not explain the origins of all areas of Marantaceae Forest, as Swaine (1992) points out. The sequence of vegetation types seen when forest colonizes savanna after fires stop, and when fire penetrates the forest, will not necessarily be mirror images of one another, but are likely to share many of the same species. They certainly both involve development of a dense understory dominated by herbs (see Swaine 1992; Hawthorne 1994), suggesting that herbs have become an important component of forest structure both in periods when the African forest has been expanding out of refuges and when it has been contracting in the face of drought and fire (see Maley, this volume).

Perhaps such catastrophes as fire, shifting cultivation, mechanized logging, or high mortality in an even-aged stand of savanna colonizers, which results in increased light levels in the forest understory, are the key to the origins of Marantaceae Forest. The species composition of herbs in any forest, and their seed dispersal mechanisms, may give a clue as to the origins. Herbs dispersed by animals, particularly those that use disturbed areas, will be better colonizers than dehiscent species and therefore better able to establish in short-lived openings. There is a need for comparable studies in Marantaceae Forests throughout the African rain forest system if we are to understand its origins, dynamics, and current distribution. Given the importance of Marantaceae Forest as a habitat for several threatened large mammal species,

such studies should be considered a priority. In that extensive Marantaceae Forests in several parts of central Africa seem to represent a transitional stage between savanna and mature forest, they may gradually be replaced by vegetation types that support lower mammalian biomass (see White 1994). As such, artificial maintenance, or even creation, of Marantaceae Forests could become important tools for managers of isolated protected areas in the future. For this to become a viable management option, our understanding of the dynamics of this vegetation formation needs to improve.

ACKNOWLEDGMENTS

This study was undertaken with financial support from the Wildlife Conservation Society and the ECOFAC program funded by the European Community (EC: DGVIII) and managed by AGRECO-GEIE, while the author was resident at the Station d'Etudes des Gorilles et Chimpanzés of the Centre International de Recherches Médicales de Franceville. I thank Michel Fernandez, Caroline Tutin, Kate Abernethy, and Joachin Dibakou for help in the field and Gordon McPherson for help with botanical identifications. I am especially grateful to Jean-Hubert Eyi-Mbeng and Alphonse Mackanga-Misandzou and the Direction de la Faune for assistance and permission to work in the Lopé Reserve. Mike Fay, John and Terry Hart, Hilary Simons Morland, Mike Swaine, Caroline Tutin, and Liz Williamson commented on early versions of the manuscript.

REFERENCES

Aubréville, A. 1967. Les étranges mosaïques forêt-savane du sommet de la boucle de l'Ogooué au Gabon. *Adansonia Série 2*, 7(1):13–22.

Badrian, N. L., and R. K. Malenky. 1984. Feeding ecology of *Pan paniscus* in the Lomako Forest, Zaire. In R. L. Susman, ed. *The pygmy chimpanzee: Evolutionary biology and behaviour*. Plenum Press, New York.

Calvert, J. J. 1985. Food selection by western lowland gorillas (*G. g. gorilla*) in relation to food chemistry. *Oecologia* 65:236–246.

Carroll, R. W. 1988. Relative density, range extension, and conservation potential of the lowland gorilla (*Gorilla gorilla gorilla*) in the Dzangha-Sangha region of southwestern Central African Republic. *Mammalia* 52(3):309–323.

Dallmeir, F., ed. 1992. Long term monitoring of biological diversity in tropical areas: methods for establishment and inventory of permanent plots. MAB Digest 11. UNESCO, Paris.

de Foresta, H. 1990. Origine et évolution des savanes intramayombiennes (R.P. du Congo), II: Apports de la botanique forestière. Pages 326–335 in R. Lanfranchi and D. Schwartz, eds. *Paysages quaternaire de l'Afrique Centrale Atlantique*. ORSTOM, Paris.

Descoings, B. 1974. *Les savanes du Moyen-Ogooué, région de Booué (Gabon)*. CNRS, Montpellier.

Dowsett-Lemaire, F., and C. M. Pannell. 1996. A new *Diospyros (Ebenaceae)* from the Congo Republic. *Bull. Jard. Nat. Belg. Bull. Nat. Plantentuin Belg.* 65:399–403.

EDICEF. 1983. *Géographie et cartographie du Gabon*. J. Barret, ed. EDICEF, Paris.

Fay, J. M. 1989. Partial completion of a census of the western lowland gorilla (*Gorilla g. gorilla* Savage and Wyman) in southwestern Central African Republic. *Mammalia* 53:203–215.

———. 1997. The ecology, social organization, populations, habitat and history of the western lowland gorilla (*Gorilla gorilla gorilla* Savage and Wyman 1847). Ph.D. dissertation, Washington University, St. Louis.

Guillaumet, J.-L. 1967. *Recherches sur la végétation et la flore de la région du Bas-Cavally (Côte d'Ivoire)*. ORSTOM, Paris.

Hall, J. B., and M. D. Swaine. 1981. *Distribution and ecology of vascular plants in a tropical forest: Forest vegetation of Ghana*. W. Junk, The Hague, The Netherlands.

Hamilton, A. C. 1982. *Environmental history of East Africa*. Academic Press, London.

Hawthorne, W. 1994. *Fire damage and forest regeneration in Ghana*. ODA, London.

Hoshino, J. 1986. Feeding ecology of mandrills (*Mandrillus sphinx*) in Campo Animal Reserve, Cameroon. *Primates* 27:248–273.

IGN. 1985. Carte au 1:50,000. SA-32-VI Booué, République Gabonaise. France, IGN.

Jones, C., and J. Sabater-Pi. 1971. Comparative ecology of *Gorilla gorilla* (Savage and Wyman) and *Pan troglodytes* (Blumenbach) in Rio Muni, West Africa. *Bibliotheca Primatologia* 13:1–96.

Kano, T. 1983. An ecological study of the pygmy chimpanzees (*Pan paniscus*) of Yalosidi, Republic of Zaire. *International Journal of Primatology* 4:1–31.

Kano, T., and M. Mulavwa. 1984. Feeding ecology of the pygmy chimpanzees (*Pan paniscus*) of Wamba. In R. L.

Susman, ed. *The pygmy chimpanzee: Evolutionary biology and behaviour*. Plenum Press, New York.

Koechlin, J. 1964. Marantacées, Zingibéracées. In A. Aubréville, ed. *Flore du Gabon*, vol. 9. Muséum National d'Histoire Naturelle, Paris.

Letouzey, R. 1968. *Etude phytogéographique du Cameroun*. Encyclopédie Biologique 69. Editions Paul Lechevalier, Paris.

Malenky, R. K., and E. W. Stiles. 1991. Distribution of terrestrial herbaceous vegetation and its consumption by *Pan paniscus* in the Lomako Forest, Zaire. *American Journal of Primatology* 23:153–169.

Maley, J. 1987. Fragmentation de la forêt dense humide africaine et extension des biotopes montagnards au Quaternaire récent: Nouvelles données polliniques et chronologiques. Implications paléoclimatiques et biogéographiques. *Palaeoecology of Africa* 18:307–334.

———. L'histoire récent de la forêt dense humide africaine: Essai sur le dynamisme de quelques formations forestières. Pages 367–382 in R. Lanfranchi and D. Schwartz, eds. *Paysages quaternaires de l'Afrique Centrale Atlantique*. ORSTOM, Paris.

———. 1991. The African rain forest vegetation and paleoenvironments during the late quaternary. *Climatic Change* 19:79–98.

———. 1992. Mise en évidence d'une péjoration climatique entre ca. 2500 et 2000 ans B.P. en Afrique tropicale humide. *Bull. Soc. Géol. France* 163(3):363–365.

Oslisly, R. 1993. *Préhistoire de la moyenne vallée de l'Ogooué (Gabon)*. ORSTOM, Paris.

———. 1996. The Middle Ogooué valley, Gabon, from 3500 years BP: Cultural changes and palaeoclimatic implications. *AZANIA* 14–15:324–331.

Oslisly, R., and R. Dechamps. 1994. Découverte d'une zone d'incendie dans la forêt ombrophile du Gabon *ca* 1,500 BP: Essai d'explication anthropique et implications paléoclimatiques. *C. R. Acad. Sci. Paris, Sér. II* 318:555–560.

Oslisly, R., and B. Peyrot. 1993. *Les gravures rupestres de la vallée de l'Ogooué (Gabon)*. Editions Sépia, St.-Maur, France.

Oslisly, R., B. Peyrot, S. Abdessadok, and L. J. T. White. 1996. Le site de Lopé 2: Un indicateur de transition écosystémique *ca* 10.000 BP dans le moyenne vallée de l'Ogooué (Gabon). *C. R. Acad, Sc. Paris, Sér. IIa* 323:933–939.

———. 1997. Réponse au commentaire de Dominique Schwartz sur la note: Le site de Lopé 2: Un indicateur de transition écosystémique *ca* 10.000 BP dans la moyenne vallée de l'Ogooué (Gabon). *C. R. Acad. Sci. Paris, Sér. IIa* 325:393–395.

Prins, H. H. T., and J. M. Reitsma. 1989. Mammalian biomass in an African equatorial rain forest. *Journal of Animal Ecology* 58:851–861.

Richards, P. W. 1952. *The tropical rain forest: An ecological study.* Cambridge University Press, Cambridge.

Rogers, M. E., and E. A. Williamson. 1987. Density of herbaceous plants eaten by gorillas in Gabon: Some preliminary data. *Biotropica* 19(3):278–281.

Rogers, M. E., E. A. Williamson, C. E. G. Tutin, and M. Fernandez. 1988. Effects of the dry season on gorilla diet in Gabon. *Primate Report* 22:25–33.

Saint-Vil, J. 1977. Les climats du Gabon. *Annales de l'Université National du Gabon* 1:101–125.

Schwartz, D. 1997. Absence de preuve d'un comportement original des savanes de la Lopé (Gabon): Commentaires sur la Note de R. Oslisly et al. *C. R. Acad. Sci. Paris, Sér. II* 325:389–391.

Siegel, S., and N. J. Castellan. 1988. *Nonparametric statistics for the behavioural sciences.* McGraw-Hill, New York.

Smith, T. B., R. K. Wayne, D. J. Girman, and M. W. Bruford. 1997. A role for ecotones in generating rainforest biodiversity. *Science* 276:1855–1857.

Swaine, M. D. 1992. Characteristics of dry forest in West Africa and the influence of fire. *Journal of Vegetation Science* 3:365–374.

Terborgh, J. 1986. Keystone plant resources in the tropical forest. In M. E. Soulé, ed. *Conservation biology: The science of scarcity and diversity.* Sinauer Associates, Sunderland, Mass.

Tutin, C. E. G., and M. Fernandez. 1993a. Composition of the diet of chimpanzees and comparisons with that of sympatric lowland gorillas in the Lopé Reserve, Gabon. *American Journal of Primatology* 30:195–211.

———. 1993b. Relationships between minimum temperature and fruit production in some tropical forest trees in Gabon. *Journal of Tropical Ecology* 9:241–248.

Tutin, C. E. G., M. Fernandez, M. E. Rogers, E. A. Williamson, and W. C. McGrew. 1991. Foraging profiles of sympatric lowland gorillas and chimpanzees in the Lopé Reserve, Gabon. *Philosophical Transactions of the Royal Society of London B* 334:179–186.

Tutin, C. E. G., R. M. Ham, L. J. T. White, and M. J. S. Harrison. 1997. The primate community of the Lopé Reserve, Gabon: Diets, responses to fruit scarcity, and

effects on biomass. *American Journal of Primatology* 42:1–24.

Tutin, C. E. G., R. J. P. Parnell, L. J. T. White, and M. Fernandez. 1995. Nest building by lowland gorillas in the Lopé Reserve, Gabon: Environmental influences and implications for censusing. *International Journal of Primatology* 16:53–76.

Tutin, C. E. G., L. J. T. White, E. A. Williamson, M. Fernandez, and G. McPherson. 1994. List of plant species identified in the northern part of the Lopé Reserve, Gabon. *Tropics* 3:249–276.

White, L. J. T. 1992. Vegetation history and logging disturbance: Effects on rain forest mammals in the Lopé Reserve, Gabon. PhD dissertation, University of Edinburgh.

———. 1994. Biomass of rain forest mammals in the Lopé Reserve, Gabon. *Journal of Animal Ecology* 63(3):499–512.

White, L. J. T., R. Oslisly, K. Abernethy, and J. Maley. 1998. *Aucoumea klaineana:* A Holocene success story now in decline? Editions ORSTOM.

White, L. J. T., M. E. R. Rogers, C. E. G. Tutin, E. A. Williamson, and M. Fernandez. 1995. Herbaceous vegetation in different forest types in the Lopé Reserve, Gabon: Implications for keystone food availability. *African Journal of Ecology* 33:124–141.

White, L. J. T., C. E. G. Tutin, and M. Fernandez. 1993. Group composition and diet of forest elephants, *Loxodonta africana cyclotis* Matschie 1900, in the Lopé Reserve, Gabon. *African Journal of Ecology* 31(3):181–199.

Williamson, E. A., C. E. G. Tutin, M. E. Rogers, and M. Fernandez. 1990. Composition of the diet of lowland gorillas at Lopé in Gabon. *American Journal of Primatology* 21:265–277.

Wrangham, R. W. 1986. Ecology and social relationships in two species of chimpanzee. In D. I. Rubenstein and R. W. Wrangham, eds. *Ecological aspects of social evolution.* Princeton University Press, Princeton.

Wrangham, R. W., N. L. Conklin, C. A. Chapman, and K. D. Hunt. 1991. The significance of fibrous foods for the Kibale Forest chimpanzees. *Philosophical Transactions of the Royal Society of London B* 334:171–178.

Wrangham, R. W., M. E. Rogers, and G. I-Basuta. 1993. Ape food density in the ground layer in Kibale Forest, Uganda. *African Journal of Ecology* 21:49–57.

# CHAPTER 12

# Diversity and Abundance in an African Forest Ungulate Community and Implications for Conservation

John A. Hart

The diversity of African savanna ungulates (including members of the orders Artiodactyla and Perissodactyla) is well known, and their ecology is the subject of several reviews (Owen-Smith 1982; McNaughton and Georgiadis 1986). By contrast, African forest ungulate communities have received much less attention. Ecological studies of individual forest species—including several duikers, *Cephalophus* spp. (Dubost 1980; Feer 1988); chevrotain, *Hyemoschus aquaticus* (Dubost 1978); pygmy antelope, *Neotragus batesi* (Feer 1979); okapi, *Okapia johnstoni* (Hart and Hart 1989); and buffalo, *Syncerus caffer nanus* (Molloy 1997)—have been undertaken, but the basic biology of even some of the more widespread species, including pigs and a number of duikers, remains almost unknown.

On a continental scale, the African forest ungulate fauna is less diverse than that of the savannas. Some twenty-nine currently recognized species of ten genera occur in African forests, compared with forty-three species of twenty-eight genera in African savannas (Owen-Smith 1982). Nevertheless, on a local scale, forest ungulate communities can be very diverse, including up to sixteen co-occurring species in some west and central African forests, compared with about nineteen species (excluding elephants) in the most diverse savanna areas (McNaughton and Georgiadis 1986). Overall, more species of ungulates have been recorded in African forests than in Asian or Neotropical forests. The African duikers of the genus *Cephalophus*, totaling seventeen species, is the most species-rich group of forest ungulates globally.

Bourlière (1973) and Emmons et al. (1983) have provided a generalized description of African forest ungulate trophic relations and community structure. Estimates of forest ungulate population densities have been more difficult to achieve, in particular in habitat mosaics (Prins and Reitsma 1989). In the 1990s, White (1994a) and Tutin et al. (1997a) provided estimates of ungulate densities at one complex multihabitat site, Lopé, in Gabon. Few studies to date have attempted to evaluate factors limiting ungulate abundance and community structure.

Many of the forest ungulates are intensively exploited as a source of game meat over most of the African forest zone (Colyn et al. 1987; Feer 1993; Robinson and Bennett 2000). At the same time, deforestation and alteration of forest habitats through logging and agriculture are accelerating. The need for a more systematic understanding of the ecology of forest ungulates, and of factors that affect their abundance, is essential if these communities are to be managed and conserved.

## The Réserve de Faune à Okapis

In this chapter I provide information on the community structure and abundance of ungulates of the Réserve de Faune à Okapis (RFO) in Ituri Forest, northern Democratic Republic of Congo (DR Congo). Fieldwork for this review covers the period from 1980 to 1996 and was conducted in the central Ituri Forest, in the area now included in the RFO (figure 12.1). The RFO covers 13,700 km² and, with fifteen species, contains one of the more diverse assemblages of forest ungulates.

Based on an interpretation of satellite imagery, Wilkie (1989) determined that 85–90% of the RFO area is covered with closed-canopy forest. Active agriculture and recent fallow, mainly restricted to the vicinity of the two roads bordering and crossing the reserve, constitute between 5% and 10% of the area. Outcrops, a distinctive habitat mosaic with open, xerophilic vegetation, including small pockets of grassland, occur in a scattered band through the north central sectors of the reserve. Forested swamps covering from less than 1 to more than 50 ha in area occur throughout the reserve along large flood plains and in the impeded drainage of smaller streams. Microhabitats not recorded on the satellite image but noted on ground surveys include open meadows, wallows, and mineral licks of varying permanence created by actions of animals (referred to collectively hereafter as *edos*), riverine meadows, and multi-tree blowdowns.

Forests in Ituri are classified as shade-tolerant, evergreen to semi-deciduous moist tropical forests (White 1986). Mature forests are dominated by *Cynometra alexandri* over much of the central and northern RFO. Areas of monodominant *Gilbertiodendron dewevrei* (mbau) occur in the south and west. *Julbernardia seretti* is co-dominant with *C. alexandri* and occurs with *G. dewevrei* over most of the central sectors of the reserve (Hart and Bengana 1996). Mean annual rainfall at three sites in the central Epulu sector averages between 1,680 and 1,783 mm (annual range 1,304 to 2,085). Lowest rainfall occurs from December through March, with the wettest months on average being April, May, October, and November (Hart and Carrick 1996).

Human occupation of the RFO area dates back to more than 12,000 B.P. (Mercader 1997). Currently, human populations average about 2 per km² over the RFO area; however, most human activity is localized along the roads crossing the forest (Hart and Hall 1996).

During the study period, hunting in the RFO focused on duikers and varied in intensity. Hunters in some areas supplied meat for local and regional markets, but this trade remained limited owing to poor transportation conditions in the forest. Small ungulate hunting was mainly by drive hunts with nets in southern and central forest, and by archery in the north. Use of snares occurred at a number of sites. Ungulate species other than duikers and chevrotain were not targeted by hunters in most areas, and other than duikers, populations of all other ungulates were subject to almost no exploitation, in particular in sectors remote from the roads. Hart and Bengana (1996) provide a more detailed account of the RFO's large mammals, and of the distribution and impact of human activities. Further information on the distribution and intensity of hunting, and its impact on ungulates, can be found in Wilkie et al. 1997 and Hart 2000.

Data analyzed for this chapter were gathered from two sources: (1) large-scale, transect-based surveys of the distribution and abun-

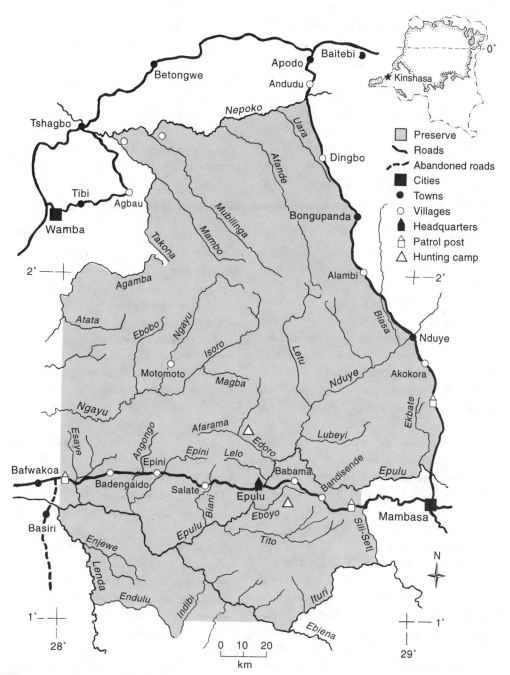

Figure 12.1. The Okapi Wildlife Reserve (Réserve de Faune à Okapis, or RFO), Ituri Forest, DR Congo.

dance of large mammals and habitats in the RFO (1993–1996), and (2) monitoring of ungulate populations and fruit availability in two study areas, in the central (Epulu) sector of the RFO (1991–1996).

With the exception of the okapi, a species found only in northeastern DR Congo, all the ungulate species of the Ituri Forest have wide distributions. The species composition of the ungulate community of the Ituri Forest is similar to that found in many other forest areas in central Africa.

## LARGE-SCALE SURVEYS

Line transect estimates of relative abundance and distribution of ungulate sign were conducted at fifty-four sites (figure 12.2). An average of 3.5 transects of 5 km in length (968 km total) was surveyed on a single visit at each site. Transects followed compass lines, were measured using hip chains, and were laid out with minimal cutting and disturbance. Observations were made by a four-person team. Data collected included feeding sign and habitat type, recorded at 100 m intervals along the transect, and counts of droppings (dung) and direct sightings of animals, recorded continuously. Perpendicular distances to the transect line were measured for each observation of dung and animal. Further details of methods can be found in Hart and Bengana 1996 and Hart and Hall 1996.

Habitat preferences were assessed for each species using a ratio developed in Neu et al. 1974 of percentage utilization of a given habitat (percentage of dung observations) divided by the percentage availability of the habitat, and calculated as: $(d_i/D)/(h_i/H)$ where $d_i$ is the number of dung piles recorded in habitat $i$ ($h_i$) and $D$ and $H$ are total recorded dung and habitat samples respectively. Ratios >2.0 are defined as strong positive selection for a given habitat $i$.

## POPULATION MONITORING

Populations were monitored over the period between 1991 and 1996 on the Lenda Study Area (LSA), 12 km southeast of Epulu, established in 1990 and covering 15 km², and the Edoro Study Area (ESA), 22 km northwest of Epulu, established in 1986 and covering about 45 km². Both study areas are managed by the Congolese National Parks Institute (ICCN) research and training center, CEFRECOF, based at Epulu. There is no hunting on the study areas, which are regularly patrolled.

The LSA is located in a large area of monodominant mbau forest, with contiguous stands of *Gilbertiodendron dewevrei* covering more than half the area. The ESA is located in an area of mixed and older secondary formations. Dominant species include *Cynometra alexandri, Julbernardia seretii, Cleistanthus michelsonii,* and *Erythrophleum suaveolens.* A total of 409 species of trees (>1 cm diameter at 1.5 m), 243 species of lianas (>2 cm diameter at 1.5 m), and 21 species of strangler figs (*Ficus* spp.) have been identified on four 10 ha inventories (20 ha per site) on the ESA and LSA. Further information on these inventories and the floristic composition of the two sites is available in Makana et al. 1998.

Ungulate populations were monitored on the ESA and LSA by line transects (all ungulate species recorded) and on 5 km² areas within both study areas by drive counts with nets that captured duiker, chevrotain, and pygmy antelope on forest blocks of known area.

Transect methods followed those of large mammal surveys above. Estimates of dung production and decay rates needed to convert dung counts to estimates of animal densities were made from observations of captive animals (duiker and okapi) or observations of dung piles in the forest (red river hogs, okapi, and buffalo). Estimates of density ($D$) were calculated for each species as $D = N/(AD)$, where

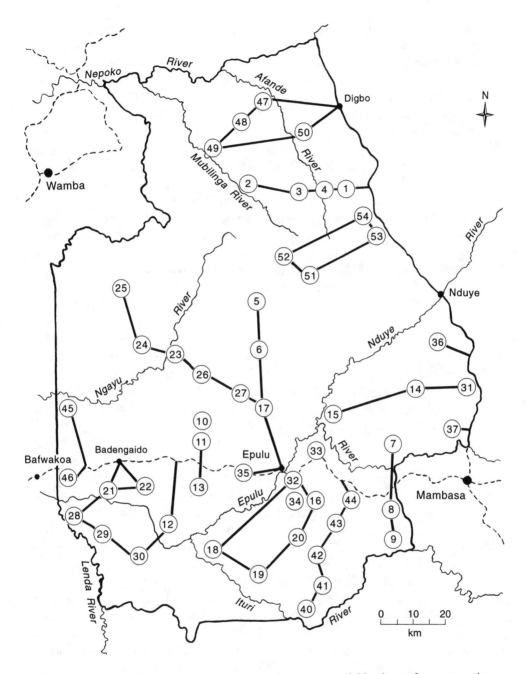

Figure 12.2. Locations of large–mammal inventory surveys: 1994–1996. Numbers refer to survey sites.

$N$ equals number of dung piles/km² for species $i$, $A$ equals the mean daily dung production rate for the species, and $D$ equals the mean disappearance rate of a dung pile in days averaged across wet and dry seasons.

Additional observations were made of selected ungulate species at edos to determine general occurrence on the study areas by different species at different seasons. Further details on census methods are provided in Hart 2000 and Hart and Hall 1996.

Fruit availability on the ESA and LSA was assessed each month from 1991 through 1995 on the same 5,000 m × 1 m transects (2 transects, 10 km total per site). Fallen fruits and seeds, as well as selected flowers observed on the forest floor and known to be eaten by duiker (all referred to collectively hereafter as fruit), were recorded. The area of each fruit patch falling on the transect was measured and the weight of available fruit estimated. Observations of single fruit items were assigned a patch area of 1 m². Fruit abundance (kg/ha) and diversity (Shannon-Weiner index, H') were developed for monthly and seasonal samples.

Other ungulate-related research conducted on the two study areas between 1986 and 1996 includes radiotelemetry studies of okapi (Hart and Hart 1989), duikers, chevrotain, and pygmy antelope (Hart 2000), leopard predation (Hart et al. 1996), and small ungulate health status (Karesh et al. 1995). Surveys of small ungulate population densities, hunting offtake, diet (rumen contents), and food availability were conducted in the Epulu area between 1980 and 1983 (Hart 1986, 2000; Koster and Hart 1988).

### The Ituri Forest Ungulate Community

Fifteen species of ungulates have been recorded in the RFO (table 12.1). These fall into three main trophic categories based on dominant food types in diet: species eating mainly fruit and seeds, termed frugivores; species eating mostly foliage, termed folivores; and mixed feeders having diets that include major components of more than one food type. In addition to fruits and foliage, diets of Ituri Forest ungulates also include fallen flowers, fungi, tubers, roots, resin, and gum. At least four species are known to feed on carrion, at least occasionally. Other food types, in particular insects (termites), may also be eaten, but their contribution cannot be evaluated at present.

The Ituri Forest ungulate community is evenly divided between folivores (seven species) and frugivores (seven species), with one species, the red river hog *Potamochoerus porcus*, being a mixed feeder. Folivores include four species of bovidae (the buffalo, two Tragelaphine antelope, and the pygmy antelope), a forest giraffe (okapi), one pig (giant hog *Hylochoerus meinertzhageni*), and the hippopotamus *Hippopotamus amphibius*. Most folivores are large (adult body size >80 kg), the only exception being the pygmy antelope (3 kg), the smallest of the forest ungulate community. Dicotyledonous browse dominates the diets of all folivores with the exception of the hippopotamus and buffalo. Frugivores are concentrated in the small body-weight range (5–60 kg) and further concentrated taxonomically in the genus *Cephalophus* (six duiker species), the chevrotain being the only frugivore not a duiker. The diet of the red river hog is dominated by fruits and seeds but also contains significant contribution of roots and tubers as well as fungi, carrion, and possibly other animal matter, including earthworms.

Many of the forest ungulates are solitary or with diffuse social groupings. Maximum observed group sizes of seven of the fifteen species were three or fewer animals. Only three folivores—the buffalo, the giant hog, and the bongo *Tragelaphus euryceros*—and the omnivorous red river hog regularly occurred in cohesive groups of more than three individuals. The

Table 12.1
# Ungulates of the Réserve de Faune à Okapis, Ituri Forest, DR Congo

| SPECIES[a] | COMMON NAME | WEIGHT (KG) | DIET[b] | | | DATA BASE[d] | MAXIMUM GROUP SIZE |
|---|---|---|---|---|---|---|---|
| | | | FOLIAGE | FRUIT/SEED | OTHER[c] | | |
| *Hippopotamus amphibius* | Hippopotamus | 1,000 | 2 | 0 | 0 | S | ? |
| *Hylochoerus meinertzhageni* | Giant hog | 175 | 2 | 1 | 1 (r) | S,D,O | 6 |
| *Potamochoerus porcus* | Red river hog | 70 | 1 | 2 | 2 (r,fg,c) | S,D | 50 |
| *Hyemoschus aquaticus* | Water chevrotain | 11 | 1 | 2 | 1 (fg,fl) | S,D,G | 2 |
| *Syncerus caffer* | Buffalo | 350 | 2 | 1 | ? | S,D,O | 18 |
| *Okapia johnstoni* | Okapi | 250 | 2 | 0 | 0 | S,G,O | 3 |
| *Tragelaphus euryceros* | Bongo | 150 | 2 | 1 | ? | S | 8 |
| *Tragelaphus spekei* | Sitatunga | 80 | 2 | ? | ? | S,D | ? |
| *Cephalophus monticola* | Blue duiker | 5 | 1 | 2 | 1 (fg,fl,g) | G,O | 4 |
| *Cephalophus nigrifrons* | Black-fronted duiker | 14 | 1 | 2 | 1 (fl) | G,O | 2 |
| *Cephalophus leucogaster* | White-bellied duiker | 17 | 1 | 2 | 1 (fg,fl) | G,O | 3 |
| *Cephalophus (callipygus) weynsi* | Weyn's duiker | 18 | 1 | 2 | 1 (fg,fl,c) | G,O | 5 |
| *Cephalophus dorsalis* | Bay duiker | 22 | 1 | 2 | 1 | G,O | 2 |
| *Cephalophus sylvicultor* | Yellow-backed duiker | 60 | 1 | 2 | 1 (fl,c) | G,O | 2 |
| *Neotragus batesi* | Pygmy antelope | 3 | 2 | 0 | 0 | S,G | 2 |

a   Species names were taken from Kingdon 1997.

b   Ocurrence in diet: o = not present in diet; 1 = minor or seasonal occurrence in diet; 2 = major dietary component; ? = inadequate basis for assessment.

c   Other food items: flower (fl), fungi (fg), tubers or roots (r), gum or resin (g), carrion (c).

d   Data base for diet classification: S = feeding sign; D = dung analysis; G = gut contents; O = observations of feeding.

grouping patterns of the hippopotamus and the sitatunga, *Tragelaphus spekei,* are poorly known in the RFO. The hippopotamus is rare and localized, although it probably lives in groups where it occurs. The sitatunga, though apparently widespread, does not form large groups.

**Large-Scale Surveys: Relative Abundance and Habitat Selection of Forest Ungulates**

Eight major habitat types were identified on surveys. These were classified on the basis of successional stage and distinguished by substrate type, canopy composition and closure, and understory composition and structure (table 12.2). The relative abundance of these habitat types in the large-scale survey sample is shown in figure 12.3.

Three mature forest types were distinguished: a mixed heterogeneous forest with *Cynometra alexandri,* the most frequent canopy dominant, achieving at some sample sites up to 30% dominance in the canopy; *mbau* forest, a monodominant forest in which *Gilbertiodendron dewevrei* achieves 60–85% of the canopy; and hill forest, a forest dominated by Meliaceae, Ulmaceae, and *Cynometra alexandri,* found on dry slopes with shallow soils. Together these forest types were recorded on 53% of survey sites sampled.

Two forest types with varying, often-broken canopy closure and dense understory were classed provisionally as secondary forests, although their successional status was uncertain at some sites. In one type, liana thickets predominated in the understory, and in the second, a herbaceous understory dominated by Zingiberaceae and Commelinaceae occurred. Together these forest types represented 32% of sample sites. Upland forests with Maranta-

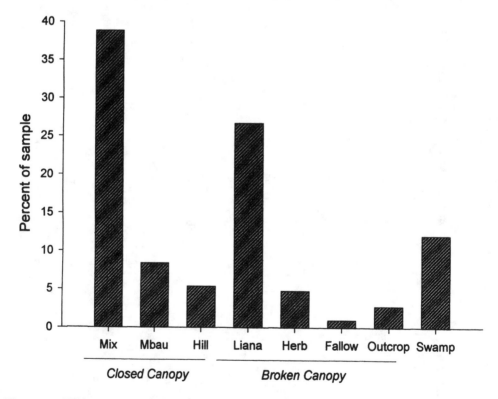

Figure 12.3. Habitat types sampled on large-mammal inventory surveys.

Table 12.2

## Ungulate Habitat Classification Used in Large Mammal Surveys in the Okapi Wildlife Reserve, Ituri Forest

| HABITAT | SUCCESSIONAL STAGE | SUBSTRATE | CANOPY | UNDERSTORY |
|---|---|---|---|---|
| mixed forest | mature | well drained | closed | open to closed |
| monodominant *Gilbertiodendron* forest | mature | well drained | closed | open |
| hill forest | mature | shallow soils, slopes | closed | open |
| forest with herbaceous understory | secondary | well drained | broken | closed |
| forest with liana understory | secondary | well drained | broken to closed | closed |
| fallow | regenerating | well drained | open to broken | closed |
| outcrop forest, thicket, and savannas | varied | rock | open to broken | open to closed |
| swamp forest | varied | impeded drainage | closed to open | open to closed |

ceae-dominated understory are uncommon in the RFO.

Areas of regenerating shifting cultivation were classified as fallow and comprised 2% of the sample. Two habitat types occurring on distinctive substrates were outcrop mosaics of forest, thicket, and grass, mainly associated with the granite inselbergs in the northern RFO and representing 3.5% of the sample; and swamp forests of varying floristic composition, occurring in areas of impeded drainage or in seasonally inundated flood plains. Climbing palms were a frequent component of swamp forests; however, raffia palms (*Raphia* sp.) occur only in scattered and restricted areas over most of the RFO. About 11% of sample sites were classified as swamp forest.

Relative frequency of ungulate dung recorded on transects is shown in table 12.3. Dung of duikers, pygmy antelope, and chevrotain was likely to have been confounded on transects, especially in older samples. In addition, dung decay rates were likely to have varied both between species and within the same species at different seasons, thus making it difficult to relate dung frequency with animal abundance directly. Despite these limitations in the data set, several striking patterns emerge: duiker dung was the most frequently encountered ungulate sign across the entire RFO, recorded in all habitat types and representing 77% of a total 6,931 dung piles observed. Despite their widespread occurrence, it was not possible to distinguish a strong habitat preference for duikers as a group, although one species, the black-fronted duiker *Cephalophus nigrifrons*, preferred swamp forests. After duiker dung, red river hog and okapi dung were the most frequently observed, and were recorded in all major habitat types. Okapi showed a strong preference for closed-canopy habitats. Excluding pygmy antelope and chevrotain, whose dung was likely confused with duikers, dung of all the remaining species was uncommon and in

some cases strongly localized. Buffalo dung was recorded on 22% of transects and in a wide range of habitats, but buffalo showed a strong selection for thickets and open grassy patches in outcrop habitats. Giant hog dung was even more infrequently encountered (8% of transects), but this species also showed a strong preference for outcrop mosaics. Sitatunga were very uncommon but showed a preference for fallow areas near small streams. Only two samples of bongo dung were recorded, and no dung of hippopotamus was found.

In the RFO, open environments, including fallows and outcrops, were selected by at least four species. The importance of open environments for the Ituri Forest ungulates can be further shown by data on dung records, feeding sign occurrence, and direct observations of animals across a range of clearing types from treefall gaps (ranging in area from 0.1 to 0.5 ha) to outcrop savannas, some of which were estimated at 50 ha in area (table 12.4). Although most of these clearings were too small and discontinuous to be classified as major habitat types, they were nevertheless recorded on transects. Infrequent use of clearings may have been overlooked for some species, and use may be underestimated for uncommon species, such as the bongo and sitatunga, but frequent track sign, dung, or direct observations of animals were unlikely to have been mistakenly identified or overlooked.

The origins of forest clearings varied from small ephemeral features such as treefall gaps, which are widely distributed and relatively frequent, to blowdowns, which include multiple treefalls over larger areas caused by high-force localized storms. Larger, more permanent features such as flood-plain meadows and outcrop savannas are infrequent microhabitats in the RFO and are localized along large rivers or restricted to dry inselbergs. Faunal and human dynamics are implicated in two types of clearings in the RFO: edos, which are often created

Table 12.3

## Transect Dung-Count Frequencies and Habitat Associations of Ungulates in the Okapi Wildlife Reserve

| SPECIES | DIET CATEGORY | TOTAL DUNG | OCCURRENCE (% TRANSECTS) | HABITAT ASSOCIATIONS[a] | | | | |
|---|---|---|---|---|---|---|---|---|
| | | | | CLOSED CANOPY | BROKEN CANOPY | FALLOWS | OUTCROP | SWAMP |
| duikers (6 spp.) | frugivore | 5,346 | 98 | + | + | + | + | + |
| red river hog | mixed | 584 | 71 | + | + | + | + | + |
| okapi | folivore | 518 | 87 | ++ | + | + | + | + |
| pygmy antelope | folivore | 243 | 34 | + | + | ++ | | |
| chevrotain | frugivore | 113 | 29 | + | + | | | |
| buffalo | folivore | 98 | 22 | + | + | + | ++ | + |
| giant forest hog | mixed | 18 | 8 | + | + | + | ++ | |
| sitatunga | folivore | 9 | 4 | | + | ++ | | |
| bongo | folivore | 2 | 1 | + | + | | | |
| hippopotamus | folivore | not recorded | — | | | | | |
| Total | | 6,931 | 100 | | | | | |

a   Habitat associations: + signifies "recorded as present", ++ signifies "selected habitat." Selected habitats are those for which the ratio of percent used to percent available is greater than 2.0 (see text).

Table 12.4
# Clearing Types and Utilization by Ungulates in the Okapi Wildlife Reserve, Ituri Forest

| TYPE | AREA (HA) | SUBSTRATE AND UNDERSTORY VEGETATION | OKAPI | GIANT HOG | RED RIVER HOG | DUIKER & CHEVROTAIN | PYGMY ANTELOPE | SITATUNGA | BONGO | BUFFALO | HIPPO |
|---|---|---|---|---|---|---|---|---|---|---|---|
| treefall gap | 0.1–0.5 | wet/dry, subcanopy treelet and shrub; canopy coppice and liana | ++ | | | + | ++ | | | | |
| gardens/fallow | 0.2–2.0 | dry, herbaceous, and woody thicket, crop plants | + | | ++ | + | ++ | ++ | | + | |
| blowdown | 0.5–2.5 | dry, herbaceous, and woody thicket | ++ | | + | | + | + | | + | |
| edo (wallow/ mineral dig) | 0.5–2.5 | wet, varied herbaceous and shrub | + | ++ | + | + | | | + | ++ | |
| flood-plain meadow | 0.5–3.0 | wet, varied herbaceous | | | + | + | | + | | ++ | ++ |
| outcrop savanna | 0.5–50.0 | dry, short grass and thicket | | ++ | + | + | | | | ++ | |

*Note:* Utilization categories of clearings by ungulate species *i*: (+) Limited use, ≤10% total survey dung, feeding sign, or observation of species *i*; (++) Concentrated use, > 10% total survey dung, feeding sign, or observation of species *i*.

or maintained by elephants, and human gardens opened by shifting cultivators.

Most clearing types are used by a range of species, and several—notably gardens, edos, and the inselberg savannas—are points of concentration for more than one species. Direct observation of animals feeding, or evidence of browsed vegetation and other feeding sign, was the most frequently recorded evidence of ungulate use of different clearing types. Wallowing and consumption of presumably mineral-rich earth was recorded in some edos. Although all ungulate species used clearings, only folivores and the mixed feeder red river hog showed a concentrated use of these sites. Species-specific variability in clearing use suggests niche differentiation among folivore species. Okapi and pygmy antelope selected treefalls and gardens or fallows, while buffalo and giant hogs selected larger, more open edos, flood-plain meadows, and savannas. Pygmy antelope and sitatunga showed a concentrated use of gardens, as did the red river hog, which was the only ungulate reported to raid food crops frequently.

## Population Monitoring: Ungulate Densities and Fruit Availability in Two Closed-Canopy Forest Types

Seasonal differences in the availability of fruits (including also seeds and selected flowers) on the ESA and LSA study areas are shown in figure 12.4 (top). Fruit abundance varied between seasons and between years in both mixed and mbau forest. The greatest variability in quantities of fruits on the forest floor was found in mbau forest. Very high mean fruit availability was recorded in the period September through November, during periods of mast *G. dewevrei* seed fall. Major mast occurred only in one year, 1992. During the preceding year and from 1993 to 1995, mbau mast was limited or completely failed. During the high-mast year, food availability was also heightened during the months

leading up to seed fall, as large quantities of *G. dewevrei* flower buds and unripe seeds were recorded on transects. During years when *G. dewevrei* mast was not present, available quantities of fallen fruits and seeds were reduced by as much as a factor of 50 or 100 (see also Blake and Fay 1997). Fruit and seed levels were consistently very low during the dry season (December through February), with periods when the standing fruit crop was almost nonexistent, extending into March during years with low levels of *G. dewevrei* flowering.

Compared with mbau forest, mean fruit availability in mixed forest was lower but also less variable between years. Periods of mast *C. alexandri* and *J. seretii* seed fall occurred, but standing crops of seed did not reach levels recorded for *G. dewevrei* in mbau forest. Fruit availability was lowest in the dry season, but mean fruit abundance was higher in the mixed forest than in mbau forest at that time of year.

The effect of monodominance on fruit availability is further demonstrated by a comparison of available fruit diversity in the two forest types (figure 12.4 [bottom]). Fruit diversity was consistently lower in mbau forest than in mixed forest over the entire year, and was very low during the dry season, when available fruit biomass was also lowest.

In summary, the availability of fruit, seed, and flower resources to ungulates in both mixed and mbau forest is strongly driven by both the seasonality and the interannual variability in flowering, fruit, and seed production of trees and lianas on the study areas. The mass flowering and seed production of a single species, *G. dewevrei*, dominates trends of food availability in the mbau forests. *G. dewevrei* seeds completely rot, sprout, or have been consumed within three months of seed fall (Hart 1995). Periods of mast seed abundance thus are brief and are separated by longer and irregular periods of very low food availability.

Estimates of ungulate populations on the

*J. A. Hart*

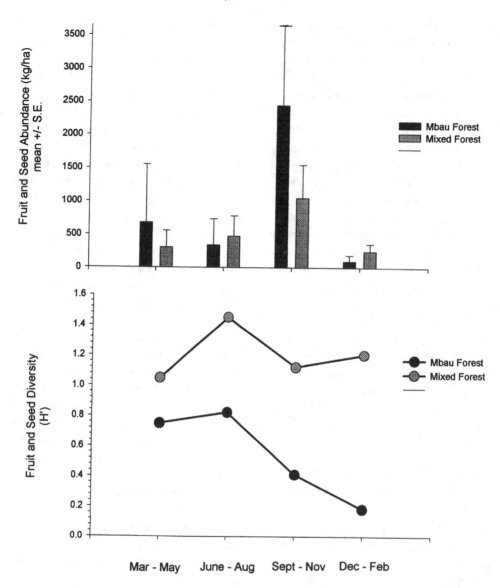

Figure 12.4. Fruit availability (*top*) and diversity (*bottom*) in mixed forest (Edoro Study Area) and monodominant *Gilbertiodendron* (mbau) forest (Lenda Study Area) in the Ituri Forest.

ESA and LSA between 1991 and 1996 are provided in table 12.5. Thirteen ungulate species were recorded on the ESA and eleven on the LSA over the study period. Hippopotamus were not present at either site, and neither the bongo nor the sitatunga were recorded on either study area. Mean total ungulate population density in the mixed forest of the ESA was estimated at about 51 animals per km² (984 kg/km²). Mean density in the monodominant mbau forest of the LSA was estimated at about half this level, 25 animals per km² (491 kg/km²). Frugivores dominate the biomass at both the mixed and mbau forest sites. Lower mean biomass of both frugivores and red river pigs in mbau forest compared to mixed forest contributed to the differences in ungulate biomass between the ESA and LSA. In contrast, the contribution of folivores to total biomass was relatively similar at both sites (figure 12.5).

Differences in ungulate species composition between the two study areas reveal differences in the dynamics of the mixed and monodominant forest communities. Itinerant species—those moving in and out of the study areas or those whose home ranges were too large to be included in the area surveyed—occurred on both ESA and LSA. However, these wide-ranging species showed higher fluctuations in abundance in the mbau forest on LSA than in the mixed forest on ESA. Densities of red river hogs varied by a factor of 80 (mean densities 0.1/km²–8.0/km²) and buffalo densities varied by a factor of 50 (mean densities 0.0/km²–5.0/km²) over the study period on LSA, with highest densities of both species occurring during periods when *G. dewevrei* mast was available. On ESA, mean densities of red river hogs (1.0/km²–5.8/km²) and buffalo 50 (0.0/km²–0.5/km²) both varied by a factor of only about 5 over the study period. Okapi densities were less variable and were also similar on ESA (mean 0.5/km²) and LSA (mean 0.3/km²).

Two species, the chevrotain and giant hog, were recorded on ESA but not on LSA. The absence of the chevrotain on LSA was due to lack of available microhabitats (larger streams), which occurred on ESA but were absent from LSA. The giant hog may have occurred on LSA in the edos but was not recorded during surveys, and it would be uncommon and likely itinerant in any case.

Although the Ituri Forest's fifteen species of ungulates are divided almost evenly between folivores and frugivores, the ungulate numbers and biomass are dominated by frugivores. Dung of duikers and chevrotain—as well as of

the red pig, for whom seeds and fruit are a major dietary component—composed the majority of dung records at most sites in the RFO, in particular where hunting pressure was not high. The only common and widespread folivores were the okapi and pygmy antelope. The remaining folivores were uncommon and either nomadic or restricted to limited habitat types. At no site were folivores the most abundant species. Frugivore biomass accounted for 62% and 58% of the total ungulate biomass on ESA and LSA, respectively.

## Habitat Limitations of Folivores

At two recently surveyed sites, Lopé, Gabon (White 1994a; Molloy 1997; Tutin et al. 1997b) and Odzala, Congo (Vanleeuwe et al. 1998), a number of the folivorous ungulates that are rare or uncommon in Ituri are abundant and regularly aggregate in large concentrations. At Odzala, herds of up to forty buffalo, as well as groups of giant forest hog, bongo, and sitatunga, are frequently encountered in *bais*, a complex of animal-maintained clearings varying in size, vegetation cover, and proximity to water, and linked by well-marked animal trails and corridors (Vanleeuwe and Gautier-Hion 1998; Vanleeuwe et al. 1998). Similar concentrations of these species are also recorded in clearings at neighboring sites in Congo (Nouabale-Ndoki) and the Central African Republic (Dzanga), although the geomorphologic structure and the origin of some of the clearings at these sites may vary (Paul Elkan and Sarah Elkan, pers. comm.). At Lopé, bais are not a prominent landscape feature, but large areas of the Lopé contain a mosaic of forest islands colonizing Marantaceae forest and savanna patches that are used by buffalo, bushbuck (a species not present in the RFO), duikers, and sitatunga. The most common folivore in this mosaic is the buffalo. Herds of up to sixteen use patches of savanna from 0.5 to 3.0 km², and densities have been estimated at 4.3/km² (Mol-

Table 12.5

## Ungulate Abundance in Mixed Forest and Monodominant *Gilbertiodendron dewevrei* Forest Near Epulu (1991–1996)

| SPECIES | CENSUS METHOD[a] | MIXED FOREST (EDORO STUDY AREA) | | | MONODOMINANT *GILBERTIODENDRON* FOREST (LENDA STUDY AREA) | | |
| --- | --- | --- | --- | --- | --- | --- | --- |
| | | SAMPLES | MEAN DENSITY (NO / KM²) | RANGE (NO / KM²) | SAMPLES | MEAN DENSITY (NO / KM²) | RANGE (NO / KM²) |
| Hippopotamus | none | — | not present | — | — | not present | — |
| Giant forest hog | T,E | 4 | itinerant | 0.0–0.2 | 4 | not recorded | — |
| Red river hog | D, T, E | 7 | 2.5 | 1.0–5.8 | 7 | 1.0 | 0.1–8.0 |
| Buffalo | T, E | 4 | itinerant | 0.0–0.5 | 4 | itinerant | 0.0–5.0 |
| Okapi | T | 3 | 0.5 | 0.2–1.1 | 4 | 0.3 | 0.2–0.9 |
| Bongo | T | 3 | not recorded | — | 3 | not recorded | — |
| Sitatunga | T | 3 | not recorded | — | 3 | not recorded | — |
| Blue duiker | D | 4 | 20.6 | 16.0–25.0 | 6 | 10.2 | 6.3–15.6 |
| Water chevrotain | D | 4 | 1.4 | 0.3–2.0 | 6 | not recorded | — |
| Black-fronted duiker | D | 4 | 2.0 | 1.1–3.0 | 6 | 1.3 | 1.1–2.8 |
| White-bellied duiker | D | 4 | 5.3 | 3.5–6.0 | 6 | 6.9 | 3.0–9.2 |
| Weyn's duiker | D | 4 | 11.2 | 8.3–15.6 | 6 | 0.6 | 0.0–1.0 |
| Bay duiker | D | 4 | 2.7 | 1.8–4.1 | 6 | 1.9 | 0.3–3.1 |
| Total red duikers & chevrotain | | 4 | 22.6 | 17.5–25.9 | 6 | 10.7 | 5.6–14.9 |
| Yellow-backed duiker | D | 4 | 1.6 | 0.9–2.6 | 6 | 0.5 | 0.0–1.5 |
| Pygmy antelope | D | 4 | 2.5 | 1.0–3.0 | 6 | 1.0 | 0.0–1.8 |
| Totals | | | 50.5 | | | 24.7 | |

a   Census method: transect (T), drive (D), edo observations (E).

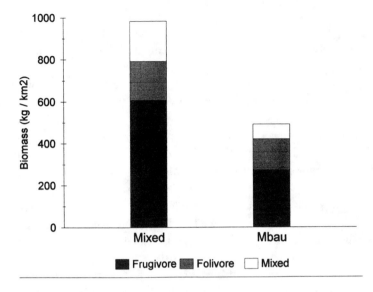

Figure 12.5. Mean ungulate biomass in mixed and monodominant *Gilbertiodendron* (mbau) forest types in the Ituri Forest.

loy 1997). At Lopé, giant hog and bongo are not recorded.

In the RFO, large bais comparable to those found in Odzala do not occur, and the edos, which are generally smaller clearings that are maintained by elephants in the RFO, are small, are less permanent, and may lack forage resources available in the richer bais further west. The lack of suitable clearings and the remoteness of most of the RFO from major savanna ecotones appear to be the most important factors contributing to the rarity of tragelaphine antelope, hippopotamus, and giant hog over most of the RFO. The only permanent savannas in the RFO are those restricted to the inselbergs and the small, isolated herbaceous thickets and meadows along seasonally inundated flood plains of larger streams and rivers.

In spite of the small areal extent of the inselberg mosaic of savanna and forest in the northern RFO, both buffalo and giant forest hog may achieve local densities comparable to those recorded at Odzala and Lopé. In the RFO, buffalo also form herds seasonally in areas of mbau mast, where they feed on germinating seeds and seedlings. Local densities at these times can

approach those reported for some savanna ecotone sites. Unlike their use of savannas, however, where the same small herd can occupy a site for extended periods of time (Molloy 1997), concentrations of buffalo in mast seed areas appear to be ephemeral and mobile, corresponding with the brief period of abundance of the mast resources. Giant hog and bongo also use mbau mast in monodominant *Gilbertiodendron* stands in northern Congo (Blake and Fay 1997). Seasonal concentrations of giant hog and bongo were not recorded in mast-producing *G. dewevrei* in the RFO, however.

The okapi is an exception to other forest folivores in the Ituri. This species does not frequently visit large clearings, inselberg savannas, or even gardens, and rarely appears in the open. Although okapi feed preferentially in small treefall gaps and on the edges of blowdowns, most of their food intake is acquired from the forest subcanopy, where the species feeds selectively on mature leaf of a wide range of evergreen understory treelets, shrubs, and seedlings. This food resource is widespread and seasonally constant, if generally sparse (Hart and Hart 1989). The okapi's feeding

niche appears to be unique among Ituri folivores and may explain why the okapi is widespread across the RFO and why local population densities, though low, are locally constant
over time.

The only other widespread folivore is the
pygmy antelope. Unlike the okapi, this species
appears to be truly dependent on clearings.
Pygmy antelope feed selectively on new leaves
in Gabon (Feer 1979) and have been observed
feeding on similar browse in the RFO (J. Hart,
unpublished observations). The small body
size and limited total intake needs of the pygmy
antelope allow this species to use small treefall
gaps, the edges of garden, and other small
clearings in the forest subcanopy, a niche not
available to larger-bodied species because of
their larger food intake needs.

In summary, use of clearings by folivorous
ungulates in the Ituri Forest is linked to differences in feeding ecology and body size. Ungulate folivory ranges from highly selective feeding on dicotyledonous leaf flush (pygmy antelope and possibly sitatunga), selective feeding
on mature dicotyledonous foliage (okapi),
mixed feeding (bongo and giant hog), and bulk
grazing with a high intake of monocotyledo-
nous foliage (buffalo and hippopotamus). Further clarification of the feeding ecology of
these species is needed to establish how clearings are used, and the importance of closed-
canopy areas in the ecology of each species. It
would appear that with the exception of a few
species, specifically the okapi and pygmy antelope, folivorous ungulates have had limited
success in occupying the closed forest.

## Dynamics of Frugivorous Ungulate Populations

The availability of fallen fruits, seeds, and flowers, unlike the availability of evergreen understory foliage, is highly variable seasonally. An
annual period of low fruit availability characterizes most moist tropical forests studied to
date, including those in Africa (van Schaik et al.

1993; White 1994b; Tutin et al. 1997b). In the
RFO, seasonal differences in fruit availability
were more accentuated in monodominant
mbau forests than in mixed forest. The diversity of fallen fruit recorded on transects was
lower, and variations in fruit abundance were
consistently greater, in mbau forest than in
mixed forest. In mbau forest during years with
no *G. dewevrei* mast, standing crops of fallen
fruit were very low, and during the dry season
almost totally absent. Periods of low food availability were of longer duration and more severe
in mbau forest than in mixed forest.

Limitations in fruit resources have had a
marked effect on the populations of frugivorous
ungulates. Red river hogs were all but absent
from the mbau forests of the LSA during periods of low fruit availability but appeared in
eruptive numbers during periods of mbau
mast, and seasonally when fruits and seeds of
the common latex lianas (*Landolphia* spp.) were
available.

Total duiker populations were about twice
as high on average in the mixed forests than in
the mbau forests. The major differences in
duiker numbers between the ESA and LSA can
be explained by the much lower densities in
mbau forest of blue duiker and Weyn's duiker,
*Cephalophus callipygus weynsi*, the two species
most common in mixed forest. Both these
species are more social than other duikers,
occurring in groups of up to four (blue duiker)
or five (Weyn's duiker) in mixed forest. Average
densities of blue duiker in mbau forest were
about half those in mixed forest. Weyn's
duiker, with mean densities of over 11/km² in
the mixed forest, was rare and perhaps only
marginally present in the mbau forest.

Population densities of frugivorous duikers
in the Ituri Forest fall within the ranges of those
recorded at several sites in west central Africa
but tend to be on the low side (table 12.6). It is
unclear whether the very high densities reported
for some locations, such as Makokou in Gabon,

refer to estimates obtained from a single, limited site or to estimates applicable over larger areas. Low densities of some species—such as blue duikers, *Cephalophus monticola*, at Lopé—may reflect inadequate sampling (Tutin et al. 1997b).

Not all duiker species responded to variability in food resources in the same way. Mean densities of black-fronted duiker and bay duiker, *Cephalophus dorsalis*, averaged respectively 35% and 30% lower in mbau forest than in mixed forest. The black-fronted duiker prefers swamp forest, which comprises approximately equal areas on both the ESA and the LSA. However, swamp forests are dominated by *G. dewevrei* on the LSA, whereas on the ESA, swamp forests contain more diverse stands of trees. Mean densities of white-bellied duiker, *Cephalophus leucogaster*, averaged 30% higher on the LSA in mbau forest than on the ESA in mixed forest. It is not clear why mbau forest might support higher densities of white-bellied duiker, but it is possible that this species is at a competitive disadvantage in mixed forests, with their higher overall duiker densities.

The linkages between duiker populations and food availability were mediated by differences in predation and dispersal on the two study areas (Hart 2000). Annual per capita probabilities of predation (mainly by leopards) on radio-collared red duikers averaged from 30% to more than 100% higher in mixed forest on the ESA than in mbau forest on the LSA (Hart 2000). In contrast, probabilities of long-distance dispersal (>1.0 km) varied in the opposite direction, with dispersal rates ranging from 0.12 to 0.43 (mean 0.22) individuals/km²/year for four duiker species in mbau forest on the LSA and from 0.0 to 0.27 (mean 0.13) individuals/km²/year for five duiker species in mixed forest on the ESA (Hart, in press).

Despite average lower predation on duikers on the LSA, high seasonal rates of predation occurred during periods of extreme food shortage in non-mast years. It remains to be determined whether these higher predation rates were due to increased exposure of duikers foraging to find limited food resources, or whether other factors, in particular disease, might also have played a role.

Karesh et al. (1995) report seropositive rates of up to 100% for a range of bacterial and viral diseases in blood samples from four species of duikers on both the ESA and LSA. Positive antibody titres against five *Leptospira interrogans* serotypes ranged from 17% to 70% in the two study areas. Leptospirosis is a disease caused by a spirochete that is associated with elevated levels of spontaneous abortions, acute death, or hemolytic crisis in a number of other ungulate species. None of the seropositive animals showed pathologies at the time of their capture, and the positive titres may be only indicators of exposure or cross-reactivity with other serovars or pathogens. Nevertheless, it is possible that infections may have caused increased morbidity and reduced fertility in animals that otherwise appeared healthy. Although disease has been implicated or hypothesized as a cause of population decline in primates (Tutin et al. 1997a) and savanna ungulates (Sinclair 1977), the impact of disease on forest ungulate populations remains insufficiently investigated to date.

## Conservation of Forest Ungulates

Two factors—hunting for the bushmeat trade and habitat loss through logging or shifting agriculture—have a direct impact on forest ungulate populations and represent major challenges to the conservation and management of almost all species across the entire African continent. At least three species of duiker—Ader's, Jentink's, and Zebra—are currently threatened by extinction globally (East 1990).

The effect of hunting on duikers has been the subject of a number of studies (Feer 1993; Lahm 1993; FitzGibbon et al. 1995; Wilkie et

Table 12.6

# Composition and Abundance of Duiker Faunas at Selected Lowland Forest Sites

| SITE | FOREST TYPE | HUNTING PRESSURE | NUMBER OF SPECIES | DENSITIES (NO./KM²) | | | SOURCE |
| --- | --- | --- | --- | --- | --- | --- | --- |
| | | | | BLUE DUIKERS[a] | RED DUIKERS[b] | TOTAL | |
| Korup, NP, Cameroon | mixed semi-evergreen | high | 2 | 15.5 | 19–30 | 35–45 | Payne 1992 |
| Makokou, Gabon | mixed semi-evergreen | none | 4 | 62–78 | >22 | >84 | Feer 1988; Dubost 1980 |
| Northeast Gabon | mixed semi-evergreen | none–moderate | 4 | 30–53 | 3–13 | 33–64 | Lahm 1993 |
| Lopé, Gabon | varied savanna ecotone | none | 4 | 0.3–6 | 13 | 5–20 | White 1994a; Tutin et al. 1997a |
| Ituri, DR Congo | mixed semi-evergreen | none–high | 4 | 11–25 | 8–25 | 19–50 | Koster and Hart 1988; Hart 2000 |
| Ituri, DR Congo | monodominant | none | 4 | 6–16 | 6–15 | 12–31 | Hart 2000 |
| Budongo, Uganda | mature Cynometra | low | 2 | 3–8 | low | >5 | Plumptre 1994 |
| Budongo, Uganda | secondary logged | low | 2 | 2–6 | low | >4 | Plumptre 1994 |

a  Blue duiker, Cephalophus monticola

b  Red duikers include from one to four species of Cephalophus, depending on the site. Korup, NP: C. dorsalis; Makokou, northeast Gabon, Lopé: C. dorsalis, C. callipygus. C. leucogaster; Ituri: C. dorsalis, C. weynsi, C. leucogaster, C. nigrifrons; Budongo: C. weynsi.

al. 1997; Hart 2000). Several studies have shown that the same hunting regime is likely to have a greater impact on larger duikers, in particular the medium-sized red duikers, than on the very small blue duiker (Lahm 1993; Jeanmart 1998; Hart 2000). Not only are hunted populations of duikers reduced, but communities are shifted toward a greater predominance of blue duikers. Some larger duikers, such as the white-bellied duiker, appear particularly vulnerable to hunting, and populations may be severely reduced under even moderate hunting pressure (Delvingt 1997; Hart 2000).

Species-specific differences in vulnerability to overhunting are likely linked to a variety of factors, including reproductive rates, rates of movement (hence exposure to snares and hunters), and risk of predation (Feer 1993; Hart 2000). Whereas some of these factors have been documented, the basic reproductive biology—including birth intervals, gestation periods, and growth rates—of many duiker species remain inadequately known, thus limiting the opportunity to directly model duiker productivity in relation to hunter offtake.

The effect of logging on forest ungulates has been investigated mainly indirectly, through its facilitation of hunting (Wilkie et al. 1992). There is less information on the direct impact that tree removal by selective logging or shifting cultivation has on ungulate populations. Plumptre's (1994) surveys in logged and unlogged Cynometra forests in the Budongo Reserve, western Uganda, revealed marginally lower blue duiker densities on logged as compared to unlogged sites. In Borneo, two species of mouse deer, *Tragulus javanicus* and *T. napu*, tragulids similar in their basic ecology to the African duikers and chevrotain, had lower densities in logged than in unlogged lowland forest (Heydon and Bulloh 1997). In contrast to these results, Wilkie and Finn (1990) suggest that secondary forests regenerating after shifting cultivation may support higher densities of

some ungulates, including duikers, than do primary forest stands in the northern Ituri Forest. Differences between sites in the areal extent of disturbance, numbers of trees removed, and the composition of secondary forests make it difficult to generalize from the available data about the effect of logging or shifting agriculture on ungulate populations in tropical forests.

Tutin et al. (1997b) have attempted to evaluate the possible effect of forest loss and fragmentation on forest mammals in Gabon by comparing population densities recorded in isolated forest fragments and galleries with population densities of the same species in adjacent contiguous forest. Although natural fragments supported high densities of a number of species, including ungulates, it is not clear that human-engendered forest fragmentation resulting from logging or agriculture would lead to comparable conditions. Forest islands in a largely deforested landscape would likely be heavily exploited and impoverished, and are rarely remnants of intact habitat. In addition, formerly forested areas cleared by farmers are unlikely to support the same potential for animal dispersal that may be important in maintaining population densities of ungulates in natural mosaics (Hart 2000).

Knowledge of the relatively undisturbed Ituri Forest ungulate fauna suggests that the conservation of many forest ungulate species will depend on the ability of animals to move within a landscape that has been variably modified by humans and contains patches of differential food resource availability and habitat suitability. The ability to manage—and mitigate—human impact, including zones of disturbance and exploitation, at more than one spatial scale is likely to be critical in assuring conservation of even sedentary species. Variability in predation pressure and dispersal rates creates spatial variations in population densities even in unhunted areas (Hart 2000). Irregularities in the seasonal availability of foods such as mbau mast, and the temporal dynamics of plant succession and natural disturbance, point to a dynamic relationship between forest ungulates and their environments that is also likely to vary over time at any one location. A landscape approach to forest ungulate conservation will become increasingly necessary as Africa's forest environments become more fragmented.

ACKNOWLEDGMENTS

Terese Hart, Andrew Plumptre, Lee White, John Robinson, Amy Vedder, and John Oates constructively criticized early versions of this chapter. My thanks to Andrew Plumptre and Paul and Sarah Elkan for their contributions of information on duiker densities and ungulates in clearings. Nicole Gottdenker provided helpful suggestions and information about leptospirosis. Tom Gambel generously assisted in preparing figures.

REFERENCES

Blake, S., and J. M. Fay. 1997. Seed production by *Gilbertiodendron dewevrei* in the Nouabale-Ndoki National Park, Congo, and its implications for large mammals. *Journal of Tropical Ecology* 14:885–891.

Bourlière, F. 1973. The comparative ecology of rain forest mammals in Africa and tropical America: Some introductory remarks. Pages 279–292 in B. Meggars, E. Ayensu, and D. Duckworth, eds. *Tropical Forest Ecosystems in Africa and South America: A Comparative Review*. Smithsonian Institution Press, Washington, D.C.

Colyn, M., A. Dudu, and M. Mankoto. 1987. Exploitation du petit et moyen gibier des forêts ombrophile du Zaire. *Nature et Faune* 3(4): 22–39.

Delvingt, W. 1997. *La Chasse Villageoise: Synthèse Régionale des Etudes Réalisées durant la Première Phase du Programme ECOFAC au Cameroun, au Congo et en RCA*. ECOFAC/AGRECO-CTFT, Brussels.

Dubost, G. 1978. Un apercu sur l'écologie du chevrotain aquatique, *Hyemoschus aquaticus*. *Mammalia* 42:1–62.

———. 1980. L'écologie et la vie sociale du céphalophe bleu (*Cephalophus monticola* Thunberg), petit ruminant forestier africain. *Zeitschrift für Tierpsychologie* 54:205–266.

East, R. 1990. *Antelopes: Global Survey and Regional Action Plans, Part 3. West and Central Africa*. IUCN/SSC, Gland, Switzerland.

Emmons, L., A. Gautier-Hion, and G. Dubost. 1983. Community structure of the frugivorous-folivorous mammals of Gabon. *Journal of Zoology, London* 199:209–222.

Feer, F. 1979. Observations écologiques sur le néotrague de Bates (*Neotragus batesi*) du nord-est du Gabon. *Terre et Vie* 33:159–239. UNESCO Man and the Biosphere Series 13: 691–708.

———. 1988. Stratégies écologiques de deux espèces de bovidés sympatriques de la forêt sempervirente africaine (*Cephalophus callipygus* et *C. dorsalis*): Influence du rythme d'activité. Ph.D. dissertation, Université Pierre et Marie Curie, Paris.

———. 1993. The potential for sustainable hunting and rearing of game in tropical forests. In C. M. Hladik, A. Hladik, O. Linares, H. Pagezy, A. Semple, and P. Hadley, eds. *Tropical Forests, People and Food: Biocultural Interactions and Applications to Development*. UNESCO Man and the Biosphere Series 13.

FitzGibbon, C., H. Mogaka, and J. Fanshawe. 1995. Subsistence hunting in Arabuko-Sokoke Forest, Kenya and its effects on mammal populations. *Conservation Biology* 9:1116–1126.

Hart, J. 1986. Comparative ecology of a community of frugivorous forest ungulates in Zaire. Ph.D. dissertation, Michigan State University, East Lansing, Mich.

———. 2000. Impact and sustainability of indigenous hunting in the Ituri Forest, Congo-Zaire: A comparison of unhunted and hunted duiker populations. Pages 106–153 in J. Robinson and E. Bennett, eds. *Hunting for Sustainability*. Columbia University Press, New York.

Hart, J., and F. Bengana. 1996. *Exploration de la Réserve de Faune à Okapis: Distribution des Grands Mammifères et la Présence Humaine dans la Réserve*. Centre de Formation et de Recherche en Conservation Forestière (CEFRECOF) Technical Report no. 2.

Hart, J., and P. Carrick. 1996. *Rainfall and Temperature in the Epulu Sector, Réserve de Faune à Okapis*. Centre de Formation et de Recherche en Conservation Forestière (CEFRECOF) Technical Report no. 1.

Hart, J., and J. Hall. 1996. Status of eastern Zaire's forest parks and reserves. *Conservation Biology* 10:316–327.

Hart, J., and T. Hart. 1989. Ranging and feeding behavior of okapi (*Okapia johnstoni*) in the Ituri Forest of Zaire: Food limitation in a rainforest herbivore? *Zoological Society of London Symposia* 61:31–50.

Hart, J., M. Katembo, and K. Punga. 1996. Diet, prey selection and ecological relations of leopard and golden cat in the Ituri Forest, Zaire. *African Journal of Ecology* 34:364–379.

Hart, T. 1995. Seed, seedling and sub-canopy survival in monodominant and mixed forests of the Ituri Forest, Africa. *Journal of Tropical Ecology* 11:443–459.

Heydon, M., and P. Bulloh. 1997. Mousedeer densities in a tropical rainforest: The impact of selective logging. *Journal of Applied Ecology* 34:484–496.

Jeanmart, P. 1998. *Tentative d'Elaboration d'un Plan de Gestion de la Chasse Villageoise dans la Réserve de Faune du Dja.* ECOFAC/AGRECO, Libreville, Gabon.

Karesh, W., J. Hart, T. Hart, T. House, A. Torres, E. Dierenfeld, W. Braselton, H. Puche, and R. Cook. 1995. Health evaluation of five sympatric duiker species (*Cephalophus* spp.). *Journal of Zoo and Wildlife Medicine* 26:485–502.

Kingdon, J. 1997. *The Kingdon Field Guide to African Mammals.* Academic Press, New York.

Koster, S., and J. Hart. 1988. Methods of estimating ungulate populations in tropical forests. *African Journal of Ecology* 26:117–126.

Lahm, S. 1993. Utilization of forest resources and local variation of wildlife populations in northeastern Gabon. Pages 213–226 in C.M. Hladik, A. Hladik, O. Linares, H. Pagezy, A. Semple, and P. Hadley, eds. *Tropical Forests, People and Food: Biocultural Interactions and Applications to Development.* UNESCO Man and the Biosphere Series 13.

Makana, J.-R., T. Hart, and J. Hart. 1998. Forest structure and diversity of lianas and understory treelets in monodominant and mixed stands in the Ituri Forest, Democratic Republic of Congo. Pages 429–446 in F. Dallmeier and J. Comiskey, eds. *Forest Biodiversity Research, Monitoring and Modeling.* UNESCO Man and the Biosphere Series 20.

McNaughton, S., and N. Georgiadis. 1986. Ecology of African grazing and browsing mammals. *Annual Review of Ecology and Systematics* 17:39–65.

Mercader, J. 1997. Bajo el Techo Forestal: La Evolución del Poblamiento en el Bosque Ecuatorial del Ituri, Zaire. Ph.D. dissertation, Universidad de Madrid.

Molloy, L. 1997. Forest buffalo, *Syncerus caffer nanus* and burning of savannas at Lopé Reserve, Gabon. Master's thesis, University of Florida, Gainesville.

Neu, E. W., C. R. Beyers, and J. M. Peek. 1974. A technique for analysis of utilization-availability data. *Journal of Wildlife Management* 38:541–545.

Owen-Smith, N. 1982. Factors influencing the consumption of plant products by large herbivores. Pages 359–404 in G. Huntley and B. Walker, eds. *Ecology of Tropical Savannas.* Springer-Verlag, New York.

Payne, J. C. 1992. A field study of techniques for estimating densities of duikers in Korup National Park, Cameroon. Master's thesis, University of Florida, Gainesville.

Plumptre, A. 1994. The effects of long term selective logging on blue duikers in the Budongo Forest Reserve, Uganda. *Gnusletter* 13:15–16.

Prins, H., and J. Reitsma. 1989. Mammalian biomass in an African equatorial rainforest. *Journal of Animal Ecology* 58:851–861.

Robinson, J., and E. Bennett, eds. 2000. *Hunting for Sustainability.* Columbia University, New York.

Sinclair, A. R. 1977. *The African Buffalo: A Study of Resource Limitation of Populations.* University of Chicago Press, Chicago.

Tutin, C. E. G., R. Ham, L. White, and M. Harrison. 1997a. The primate community of the Lopé Reserve, Gabon: Diets, responses to fruit scarcity, and effects on biomass. *American Journal of Primatology* 42:1–24.

Tutin, C. E. G., L. J. T. White, and A. Mackanga-Missandzou. 1997b. The use by rain forest mammals of natural forest fragments in a equatorial African savanna. *Conservation Biology* 11:1190–1203.

Vanleeuwe, H., S. Cajani, and A. Gautier-Hion. 1998. Large mammals at forest clearings in the Odzala National Park, Congo. *Revue d'Ecologie (Terre et Vie)* 53:185–194.

Vanleeuwe, H., and A. Gautier-Hion. 1998. Forest elephant paths and movements at the Odzala National Park, Congo: The role of clearings and Marantaceae forests. *African Journal of Ecology* 36:174–182.

van Schaik, C., J. Terborgh, and S. Wright. 1993. The phenology of tropical forests: Adaptive significance and consequences for primary consumers. *Annual Review of Ecology and Systematics* 24:353–377.

White, F. 1986. *La Végétation de l'Afrique.* ORSTOM-UNESCO, Paris.

White, L. 1994a. Biomass of rainforest mammals in the Lopé Reserve, Gabon. *Journal of Animal Ecology* 63:499–512.

———. 1994b. Patterns of fruit-fall phenology in the Lopé Reserve, Gabon. *Journal of Tropical Ecology* 10:289–312.

Wilkie, D. 1989. Human settlement and forest composition within the proposed Okapi rainforest reserve in northeastern Zaire. Unpublished report. WWF, Gland, Switzerland.

Wilkie, D., B. Curran, R. Tshombe, and G. Morelli. 1997. Modeling the sustainability of subsistence farming and hunting in the Ituri Forest of Zaire. *Conservation Biology* 12:137–147.

Wilkie, D., and Finn, J. 1990. Slash-burn cultivation and mammal abundance in the Ituri Forest, Zaire. *Biotropica* 22:90–99.

Wilkie, D., J. Sidle, and G. Boundzanga. 1992. Mechanized logging, market hunting and a bank loan in Congo. *Conservation Biology* 6:570–580.

# CHAPTER 13

# Forest Elephant Behavior and Ecology

Observations from the Dzanga Saline

Andrea K. Turkalo and J. Michael Fay

Previous long-term studies of the African elephant based on direct observation have been limited to the savanna subspecies of African elephant, *Loxodonta africana africana* (Douglas-Hamilton 1972; Moss 1988). Although high densities of forest elephants, *Loxodonta africana cyclotis,* Matschie 1900, are known to inhabit the central African forests and are estimated to constitute one-third to one-fourth of the entire African elephant population (Michelmore et al. 1989), they have been poorly studied because of the difficulty in observing them through thick vegetation for extended periods of time. Studies have therefore been based on the examination of secondary evidence or opportunistic encounters resulting in sparse scientific literature about social and population structure and behavior. Until now, most studies of the subspecies *cyclotis* have been limited to examination of spoor to collect data on population status, foraging habits, and crop depredation (Dudley et al. 1992), estimation of elephant densities based on dung counts (Short 1983; Mertz 1986; Barnes and Jenson 1987; Carroll 1988; Fay and Agnagna 1991), seasonal movements (Short 1983; White 1994), determination of diet by examination of feeding trails and dung piles (Short 1981; Tchamba and Seme

1993), seed dispersal (Alexandre 1978; Powell, in prep.), and genetics (Georgiadis et al., in prep.).

## Clearings: A Window on the Forest

In the forests of the southwestern Central African Republic and northern Congo, naturally occurring clearings have been observed to attract many species of forest mammals. These clearings, generally located on watercourses, are known to be abundant in mineral salts (Fay et al., unpublished data; Klaus, pers. comm.), attracting large numbers of mammals and birds on a regular basis. Because of the unimpeded viewing possibilities these salines offer, they provide ideal places for studying certain aspects of the biology of elephants and other species of mammals, such as bongo (*Tragelaphus euryceros*) and western lowland gorilla (*Gorilla gorilla gorilla*).

Traditionally these clearings have provided hunting areas for the local human populations, and although animal sign is evident, animals frequent many of these clearings only at night because of hunting pressure. Several protected clearings, situated within the newly created national parks of Nouabale-Ndoki and Dzanga-Ndoki in Congo and the Central African Republic, respectively, are serving as study areas of western lowland gorillas and forest elephants. With sound management and protection of these clearings, scientific studies based on direct observation can provide previously unattainable information about forest species. Similar clearings occur across the African forest system, from Nigeria (James 1954) to eastern DR Congo (Hall et al., in prep.), so with appropriate protection, comparative studies could be undertaken throughout the system.

The Dzanga clearing, offering unimpeded viewing of forest elephants, is providing the ideal milieu for the first long-term study of forest elephants based on direct observation of known individuals. The objective of the Dzanga study is to provide information on several aspects of the forest elephant's biology: population structure, demographics, morphometrics, genetics, social structure, and social behavior.

According to the local Ba-benjellé Pygmies, the Dzanga saline has always attracted large numbers of forest mammals, particularly the forest elephant, *Loxodonta africana cyclotis*. Elephant density, based on dung counts proximate to the saline, is estimated to be 0.016 to 2.63 elephants/km (Carroll 1988), placing the area among the most densely populated elephant zones in the dense forest of central Africa.

The clearing, measuring 500 × 200 m, consists of a sandy pan through which a small river, the Dzanga, drains. The pan is wet the entire year, but during the months of December through March it is relatively dry, enabling elephants to dig sizable holes in search of mineral-rich water. Vegetation is absent in the pan except for species of Cyperaceae and such aquatic herbs as *Ludwigia* sp. (Onagraceae) found along its wet edges. On its drier edges, grasses predominate, leading into tree species of the forest edge: *Berlinia grandiflora*, *Erythrophoelum suaveolens*, and *Lophira lanceolata*. Precipitation averages 1,500 mm per annum, with a long wet season occurring between the months of April and November and a dry season occurring during the months of December through March. A small dry season occurs in July or August. Numerous clearings exist in the vicinity of this pan, but the Dzanga pan attracts more elephants on a consistent basis than any other saline in the region.

The study of the Dzanga elephant population was initiated in 1990, when it was ascertained that individual elephants could be identified and subsequently reidentified in the clearing. Observations begin daily at 1400 hours, when elephants begin to enter the saline after feeding in the forest. Observers in the saline have noted a variety of elephant behaviors, including drinking, fighting, bathing, mating, grazing, sleeping, and excavating holes in the saline to search for the mineral-rich substrate. This movement into the saline from the forest continues throughout the night, and numbers can surpass one hundred individuals before dawn. In the early daylight hours, the elephants begin to leave the saline in search of food in the forest, thus completing their daily cycle. Daily observations are concluded at 700 hours.

Observations are made from a fixed, elevated platform located on the western edge of the clearing. Sightings are made with a spotting scope. The same system used in identifying savanna elephants (Douglas-Hamilton 1972; Moss 1988) is being employed in the Dzanga study: individual identity cards are established, which contain drawings recording such salient characteristics as ear holes and tears, tusk length, and tail shape, and note such other features as sex, age, size, and body scarring. Each time a previously identified individual is encountered in the saline, the date is recorded on its card. In the case of a family group, which consists of an adult female and her offspring, each individual is assigned a card on which its characteristics are drawn and observation dates recorded. When the observer arrives at the saline at 1400, numbers of elephants and other species are noted and elephants present are identified. As other elephants enter the saline, the observer identifies them and notes the time, entrance location, and, in the case of groups, individuals present. Exact counts of elephants and any other mammal species present in the saline are recorded every thirty minutes.

## Population Structure

Between April 1990 and August 1998, the number of elephants identified at the Dzanga clearing continued to increase (figure 13.1). As of 1998, new individuals were appearing less frequently in the saline, and the accrual curves for both the males and females were slowly leveling off. Most elephants—85% of the males and 89% of the females—had been resighted at least a second time. Females, however, were sighted more frequently than males. Sixty-seven of the adult females, as opposed to only eighteen males, had been observed in the saline more than one

hundred times. The maximum number of sightings was 309 for females and 169 for males. Because males were sighted less frequently than females, they are thought to range farther. During the entire study period—from January 1991 through September 1998—the average number of sightings per male was 20.1 times, whereas the average per female was 32.3 times.

Several limitations exist in such a study, where one waits for animals to appear in a certain location. Births are recorded when a previously identified female appears with a new infant, and the age of the infant is approximated from when the female was last sighted without the newborn. Thus the age of the infant at first viewing can vary from between several days to several years. In addition, mortality rate is not possible to determine, since the absence of an individual from the clearing does not assure that the animal is dead. In several instances, both

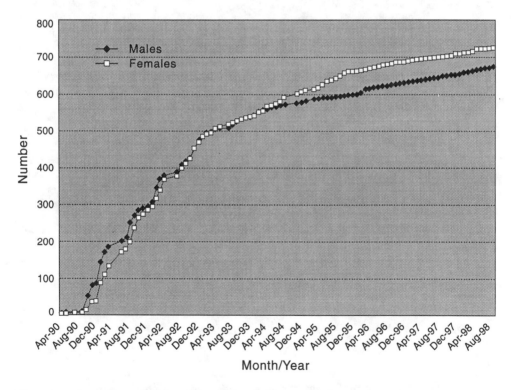

Figure 13.1. Accrual rate of elephants identified at the Dzanga clearing.

adult males and adult females (which are believed to be more sedentary) appeared in the clearing after a three-year absence.

## Social Organization

Previous observations have emphasized that forest elephants—unlike savanna elephants, which travel together in large groups—form small groups of two to five individuals (Merz 1981; Short 1981; Dudley et al. 1992; White et al. 1993). At Dzanga this is generally the case, but several large groups do exist, with one group of eleven regularly observed. The average group size observed in the Dzanga clearing, excluding females who have been observed to

be solitary, is 3.2 (N = 521, range 2–18). This figure corresponds closely to data collected on forest elephants in the Lopé Reserve in Gabon, where the mean group size was found to be 2.8 (White et al. 1993). Figure 13.2 provides a breakdown of the family units or groups (a family unit is defined as one or more adult females and their offspring).

The most common family unit observed in the Dzanga clearing is a female accompanied by one or two offspring. The biggest groups have been accurately determined by repeated observations, but although they are permanent, they are not stable. The larger groups, composed of smaller family units, periodically separate into

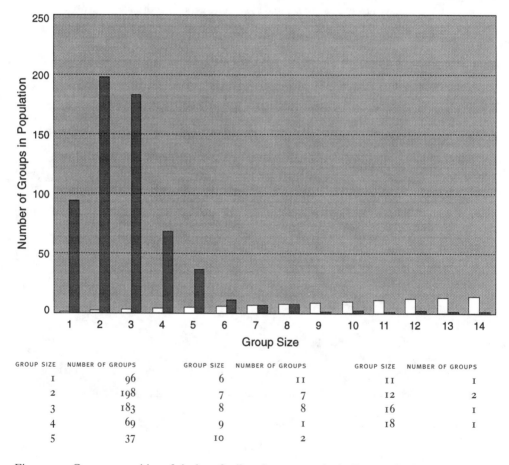

| GROUP SIZE | NUMBER OF GROUPS | GROUP SIZE | NUMBER OF GROUPS | GROUP SIZE | NUMBER OF GROUPS |
|---|---|---|---|---|---|
| 1 | 96 | 6 | 11 | 11 | 1 |
| 2 | 198 | 7 | 7 | 12 | 2 |
| 3 | 183 | 8 | 8 | 16 | 1 |
| 4 | 69 | 9 | 1 | 18 | 1 |
| 5 | 37 | 10 | 2 | | |

Figure 13.2. Group composition of elephant family units or groups in the Dzanga clearing.

their individual families, only to regroup at a later date. Whether the separation and reuniting of these bigger groups is seasonal is yet to be determined.

The composition of family units observed in the saline fluctuates, particularly when males in the family unit reach the juvenile age. Unlike savanna elephants, which form tightly knit groups whose males become solitary upon attaining adulthood (Moss 1988), forest juvenile males and females start disassociating from their family groups earlier, particularly if a younger sibling is born. Two well-known subadult females have been observed to join unrelated family groups where a newborn is present. In these groups, these young females allo-mother newborns, although there is no evidence that any familial relationship exists.

Forest adult males have never been observed in bachelor herds, as seen in the savanna subspecies (Moss 1988). In the saline, adult males are always observed alone, never associating with other males for longer than a couple of hours. Subadult and juvenile males also do not form permanent associations with other males. Unlike older adult males, however, subadult and juvenile males socialize with males of their age in the saline, where they form small groups that feed on mineral salts, keeping in proximity to each other, periodically "trunking" and "jousting." These temporary groups last only a few hours.

The older adult bulls continually test and establish their hierarchy in the saline, particularly in the dry season when they are able to excavate huge holes in search of mineral-rich water. The bulls who excavate the holes hold exclusive rights to their use when present. The higher a bull is in the hierarchy, the more dominant he will be in maintaining his place in a particular hole, where he will spend hours ingesting mineral-rich soil and water. Bulls lower in the hierarchy have been observed to wait more than three hours at the edge of a hole before a dominant bull gives up his spot. When he does

so, the waiting bulls scramble to establish their position in the hole. This hierarchy is clearly understood and respected by bulls present in the saline, with little fighting observed unless two evenly matched bulls confront each other.

The perception at present is that much of the jockeying between individual males at the various holes in the saline is not a result of competition for the resource but rather a testing ground for social hierarchy. Only further study at the saline will reveal the real importance of these water holes in determining dominance hierarchy. Each encounter by bulls is noted, along with its result. This information will be used to establish the order of hierarchy in the Dzanga bull population.

### Birth Rate

To determine the birth rate in the Dzanga population, one must take into account several factors: identification and reidentification of individual females, dates on which new offspring are first identified, and subsequent births to females that have been previously identified (in order to determine an approximate birth interval). Since the inception of the study in January 1991, 301 newborns have been recorded to 270 previously identified adult females. Of the 270 mothers, 30 of the females gave birth twice during the period from January 1991 to September 1998, with 1 female having given birth three times. The birthing interval is now determined to be between 3.5 and 4 years. As additional data are collected, a more accurate birth interval and demographics of the Dzanga elephant population will be determined.

### A Discussion of Forest Elephant Behavior and Ecology

POPULATION STRUCTURE

Although the accrual rate for both males and females in the Dzanga population is approaching an asymptote, new individuals are still being identified in the saline. Based on a figure of 80%

individuals identified, a minimum population that frequents the saline can be calculated at about 3,200. With new identification becoming more infrequent, the total will continue to increase, which may increase the minimum population to more than 3,200. If the elephant population in the Dzanga area is more than one elephant per square kilometer (Carroll 1988), which is considerably higher than all the previous dung count surveys would indicate, this would mean that every elephant from an area larger than the entire Dzanga-Ndoki Park, comprising 1,220 km², visits the Dzanga clearing.

Populations of savanna elephants previously studied, as in Amboseli (Moss 1988), show a much higher percentage of adult females than males, owing to selective hunting for ivory. The relatively close ratio of adult females to adult males (733:665) observed in the Dzanga study may indicate either that there is little hunting pressure on adult males or that females are equally hunted. Elephant poaching is a continuing activity in the area, but information is scarce on the sexes killed. Of the several poached carcasses located in the park, all have been males. In the forest environment, decomposition is rapid and few carcasses are found. Hunting in the forest is also less selective than in the savanna because of poor visibility.

SOCIAL ORGANIZATION

Compared to the savanna elephant subspecies, forest elephants form smaller social units, averaging 3.2 individuals. Both male and female juveniles and subadults have been identified and observed to circulate without familial ties. At the same time, known offspring (subadults or juveniles of both sexes of a previously identified female) have been observed to leave and then return to their natal groups after an extended absence. This phenomenon of subadult males leaving their familial groups has led to the belief in the existence of a "pygmy" race of elephants. During five and a half years

of direct observation, many of these elephants have been sighted. Because of their solitary nature and precocial tusk development, their presence seems to have confirmed the existence of a smaller race of elephant. This "myth" of the pygmy elephant has been perpetuated by white hunters, who have had the most contact with the two subspecies of African elephant. Genetic sampling of the Dzanga population in January 1992 has shown that a "pygmy" genotype does not exist: analysis of mDNA provides no evidence of genetic distinctiveness (Georgiadis et al., unpublished data).

Group size is determined by several factors in an animal's environment: predation, food availability, and competition (Davies and Krebs 1981). In the forests of central Africa the leopard is the only predator of elephants besides humans, and is unable to kill any elephant beyond the infant size. Unlike the savanna, which offers the mega-herbivores large areas of such quality food as the Poaceae, the forest offers patchy food sources such as fruit and certain species of monocotyledons. Because the forest group is not faced with the problem of predation, as in the savanna, it would be in the interest of the group to minimize its numbers in order to secure adequate food quantities, particularly during the fruiting season when food resources are abundant but localized (White et al. 1993; White 1994).

## Conservation Implications

Biological research in tropical forests has always been impeded by inaccessibility and limited possibilities for viewing resident species. Therefore, information based on direct, systematic observation is scarce for many forest mammals, and most information is based instead on opportunistic observations and secondary evidence such as tracks, dung analysis, and inspection of feeding sites. Yet the forests of central Africa contain clearings that attract large forest mammals, which congregate to socialize and to exploit concentrations of mineral salts and monocotyledons. Because of the concentrating effect these forest clearings have on many species of forest animals, including forest buffalo (*Syncerus caffer nanus*), western lowland gorilla (*Gorilla gorilla gorilla*), bongo (*Tragelaphus euryceros*), and sitatunga (*Tragelaphus spekei*), they present unique opportunities to study forest mammals using direct sustained observation.

What role can these clearings play in conserving forest species? In an environment such as the forest, where visibility is poor if not fleeting, clearings such as Dzanga can be, and are, the window of the forest—a place where observations, sustained systematically over time, can provide valuable management information on the demography and behavior of forest species.

In the first stage of managing a natural area, surveillance of forest clearings would be the quickest way to evaluate the numbers, population structure, and species present in the region. Long-term surveillance would serve as a tool for evaluating protection and management techniques employed by park managers over longer periods of time.

The Dzanga study, eight years after its inception, is not only providing data on demographics, behavior, and seasonal use of an area by a significant forest elephant population based on direct observation of known individuals, but is also monitoring numbers and behavior of other species that visit the saline. Combining similar kinds of information from such species as sitatunga, red river hogs, and gray parrots should provide a means of indirectly monitoring the status of multiple species as well as the forest as a whole.

### REFERENCES

Alexandre, D.-Y. Le rôle disseminateur des elephants en Forêt de Tai, Côte d'Ivoire. *Terre et vie* 32:47–72.

Barnes, R. F. W., and K. L. Jensen. 1987. How to count elephants in forests. *IUCN African and Elephant Specialist Group Technical Bulletin* 1:1–6.

Carroll, R. W. 1988. Elephants of the Dzanga-Sangha dense forest of southwestern Central African Republic. *Pachyderm* 10:12–15.

Douglas-Hamilton, I. 1972. On the ecology and behavior of the African Elephant. Ph.D. dissertation, Oxford University.

Dudley, J., A. Memsah-Ntiamoah, and D. Kpelle. 1992. Forest elephants in a rainforest fragment: Preliminary findings from a wildlife conservation project in southern Ghana. *African Journal of Ecology* 30:116–124.

Fay, J. M. 1991. An elephant (*Loxodonta africana cyclotis*) survey using dung counts in the forests of the Central African Republic. *Journal of Tropical Ecology* 7:25–36.

Fay, J. M., and M. Agnagna. 1991. A population survey of forest elephants (*Loxodonta africana cyclotis*) in northern Congo. *African Journal of Ecology* 29:177–187.

Krebs, J., and J. Davies. 1988. *Introduction to Behavioral Ecology*. Blackwell, Cambridge, Mass.

Merz, G. 1981. Recherches sur la biologie de nutrition et les habitats préférés de l'éléphant de forêt, *Loxodonta africana cyclotis*, Matschie 1900. *Mammalia* 45:299–312.

———. 1986. The status of the forest elephant, *Loxodonta africana cyclotis*, Matschie 1900, in the Gola Forest Reserves, Sierra Leone. *Biological Conservation* 36:83–94.

Michelmore, F., K. Beardsley, R. F. W. Barnes, and I. Douglas-Hamilton. 1989. Elephant population estimates for the central African forests. In S. Cobb, ed. *The Ivory Trade and the Future of the African Elephant*. International Development Centre, Oxford.

Moss, C. 1988. *Elephant Memories*. Ballantine, New York.

Short, J. 1981. Diet and feeding behavior of the forest elephant. *Mammalia* 45:177–185.

———. 1983. Density and seasonal movements of forest elephant (*Loxodonta africana cyclotis*, Matschie) in Bia National Park, Ghana. *African Journal of Ecology* 21:175–184.

Tchamba, M., and P. Seme. 1993. Diet and feeding behaviour of the forest elephant in the Santchou Reserve, Cameroon. *African Journal of Ecology* 31:165–171.

White, L. J. T. 1994. *Sacoglottis gabonensis* fruiting and the seasonal movements of elephants in the Lopé Reserve, Gabon. *Journal of Tropical Ecology* 10:121–125.

White, L. J. T., C. E. G. Tutin, and M. Fernandez. 1993. Group composition and diet of forest elephants, *Loxodonta africana cyclotis*, Matchie 1900, in the Lopé Reserve, Gabon. *African Journal of Ecology* 31:181–189.

# CHAPTER 14

# Carnivore Biogeography and Conservation in the African Forest

## A Community Perspective

Justina C. Ray

The geographical distributions of animal species and the resultant patterns of local diversity are a reflection of evolutionary, historical, and geographical elements (Ricklefs and Schluter 1993). A present-day snapshot of an ecological community is the result of a suite of factors, such as climatic and habitat associations of individual species, past and present ecological interactions, and regional geographical history. The distribution of mammals within the African forest zone provides an excellent illustration of this complex of factors. Although the rain forest biome stretches continuously over most of the central part of the continent, relatively few species range across its entire extent (Happold 1996). As a result, the assemblage of species in one locale may differ markedly from that in another. A subject of great interest is how species in these communities have interacted over time, and what ecological factors have facilitated their coexistence. This chapter examines these questions for the case of the mammalian carnivores.

Unfortunately, ecological information on African forest carnivores is rudimentary, and knowledge about many species is derived from a few museum specimens. In contrast to their savanna counterparts, carnivores in rain forest habitats can be easily overlooked. With the exception of a few isolated tracks or scats, one can pass many months without even a glimpse of a mongoose, genet, civet, or leopard. They are difficult to lure into traps or to census using traditional methods that target primates and ungulates (see, e.g., Prins and Reitsma 1989; White 1994a). As a result, biogeographic and ecological studies of the African forest Carnivora are sparse and have been limited to investigations of a few species (Charles-Dominique 1978; Colyn and Van Rompaey 1994a, 1994b; Ray 1997).

At the same time, the forests of Africa are under increasing pressure from such human activities as clearance for agriculture, hunting, and logging (Happold 1996). The resulting transformation of the landscape and attendant habitat changes can have serious implications for carnivores and the communities of which they are part. For example, the invasion of new predators into an area following modification of the habitat, or the preferential removal of others through hunting, may result in a new competitive balance and other changes that ripple through the community. Our capacity to understand or even document these impacts is severely limited by a lack of understanding of the nature of intact communities.

The aims of this chapter are threefold. First, existing information on the biogeography of African Carnivora inhabiting the forest biome is summarized. Second, the general results from a study of ecological separation among eight carnivores in the Dzanga-Sangha Reserve in southwestern Central African Republic are presented, and their implications for other communities throughout the African forest biome are considered. Finally, the major issues threatening the conservation of African forest carnivores are outlined and discussed. The focus of this chapter is the carnivores of mainland Africa; the unique assemblage of herpestids and viverrids inhabiting the island of Madagascar is not discussed. Unless otherwise noted, all taxonomic names follow Wozencraft 1993.

## Biogeography

The carnivores inhabiting the forest regions of Africa display a wide range of distributions: some are broad and cover the entire forest belt,

whereas others are more limited. Of the twenty-four species known to occur in African forests, eight are widespread throughout the Congo Basin, west Africa, and east African forest remnants (table 14.1; figure 14.1). Six of these species are also found in the savanna regions of the continent, given suitable ecological conditions (that is, streams or adequate cover). Only two of the eight (golden cat, *Profelis aurata*, and palm civet, *Nandinia binotata*) occur exclusively in forested habitats (Rosevear 1974; Kingdon 1997). The distributions of the remaining carnivore species are more restricted.

Distributional patterns of many African mammals appear to have been shaped by historical factors relating to alternating phases of cool-dry and hot-wet climate and the resulting shifts in the distribution of rain forest habitat (Prance 1982; Mayr and O'Hara 1986; Happold 1996). It appears that "refuges" of forest habitat with relatively constant conditions survived during periods of forest retraction. The forest biota were periodically restricted to these isolated habitats, from which dispersal is presumed to have occurred upon the return to pre-

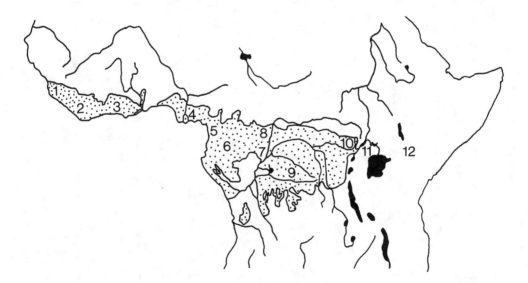

Figure 14.1. Localities for which carnivore community species composition is listed in table 14.1. Stippling indicates the distribution of moist forest. Redrawn by Jay Malcolm from UNESCO 1978.

Table 14.1

# Carnivore Communities in Twelve Selected Areas in African Moist Forests

| 1. SIERRA LEONE | 2. LIBERIA | 3. E. CÔTE D'IVOIRE | 4. S.E. NIGERIA | 5. W. CAMEROON | 6. N. GABON |
|---|---|---|---|---|---|
| *Aonyx capensis* | *Aonyx capensis* | *Aonyx capensis* | *Aonyx congica* | *Aonyx congica* | *Aonyx congica* |
| *Crossarchus obscurus* | *Crossarchus obscurus* | *Crossarchus obscurus* | *Crossarchus platycephalus*[1] | *Crossarchus platycephalus* | |
| *Poiana richardsoni* | *Poiana richardsoni* | *Poiana richardsoni* | *Poiana richardsoni* | *Poiana richardsoni* | *Poiana richardsoni* |
| | | | *Herpestes naso* | *Herpestes naso* | *Herpestes naso* |
| | *Genetta johnstoni* | *Genetta cristata*[2] | *Genetta servalina* | *Genetta servalina* | |
| | *Liberiictis kuhni* | | *Bdeogale nigripes* | *Bdeogale nigripes* | *Bdeogale nigripes* |

| 7. CONGO | 8. S.W. C.A.R | 9. S.C. DR CONGO | 10. N.E. DR CONGO | 11. W. UGANDA | 12. E. AFRICA |
|---|---|---|---|---|---|
| *Aonyx congica* | *Aonyx congica* | *Aonyx congica* | *Aonyx congica* | *Aonyx congica* | *Aonyx capensis* |
| *Crossarchus platycephalus*(?) | *Crossarchus platycephalus* | *Crossarchus ansorgei* | *Crossarchus alexandri* | *Crossarchus alexandri* | |
| *Poiana richardsoni* | *Poiana richardsoni* | *Poiana richardsoni* | *Poiana richardsoni* | | |
| *Herpestes naso* | *Herpestes naso* | *Herpestes naso* | *Herpestes naso* | *Herpestes naso* | |
| *Genetta servalina* | *Genetta servalina* | *Genetta servalina* | *Genetta servalina* | *Genetta servalina* | |
| *Bdeogale nigripes* | *Bdeogale nigripes* | *Bdeogale nigripes* | *Bdeogale nigripes* | *Bdeogale nigripes* | *Bdeogale jacksoni/crassicauda* |
| | | *Crossarchus alexandri* | *Osbornictis piscivora* | *Aonyx capensis* | |
| | | | *Genetta victoriae* | | |

*Sources:* Data from Allen 1924; Bates 1905; Booth 1960; Coetzee 1971; Colyn and Van Rompaey 1994a, 1994b; de Beaufort 1965; Dowsett and Granjon 1991; Ewer 1973; Haltenorth and Diller 1977; Happold 1987; Hayman (in Sanderson 1940); Kingdon 1977; Malbrant and Maclatchy 1949; Michaelis 1972; Perret and Aellen 1956; Rahm 1966; Rosevear 1974; Schouteden 1948; Schreiber et al. 1989; Taylor 1989; Wozencraft 1993.

*Note:* The twelve selected areas are shown on figure 14.1. Eight species are found in all areas, and are not listed in this table: *Atilax paludinosus, Civettictis civetta, Genetta maculata, Lutra maculicollis, Mellivora capensis, Nandinia binotata, Profelis aurata,* and *Panthera pardus.* Of these eight, all but *Nandinia binotata* and *Profelis aurata* are also found in savanna habitats.

1  *Crossarchus platycephalus* is not recognized as a different species from *C. obscurus* by Wozencraft (1993), but see Colyn et al. 1995 and Goldman 1984.

2  *Genetta cristata* is considered a subspecies of *G. servalina* by Wozencraft (1993) but is widely treated as a distinct species (Heard and Van Rompaey 1990; IUCN 1996; Powell and Van Rompaey 1998)

vious environmental conditions (Grubb 1978; Colyn et al. 1991). The ultimate extent of the range of a given species would then depend on its dispersal abilities, speed of re-colonization, and ecological and physical barriers. The little work that has been done examining subspeciation among Carnivora downplays the importance of rivers as barriers to distribution (*Crossarchus* spp., Colyn and Van Rompaey 1994a; *Herpestes (Xenogale) naso*, Colyn and Van Rompaey 1994b). Many carnivores are tolerant of a wide range of ecological conditions. This, combined with the fact that they often use large home ranges relative to their body size in combination with rigid spacing patterns and territorial defense, may account for a relatively rapid speed of re-colonization (Grubb 1978).

Among the carnivores that do not range across the entire forest biome (table 14.2), a variety of distributional patterns are apparent. Some species, such as the servaline genet (*Genetta servalina*) and the long-nosed mongoose (*Herpestes naso*), are found throughout the Congo Basin but are absent from the Upper Guinea forests and east Africa (beyond western Uganda). The small-clawed otter (*Aonyx congica*) is replaced by its congener *Aonyx capensis* in western and eastern Africa; the two species are apparently sympatric in western Uganda (Butynski 1984). Two closely related forms of the black-footed mongoose (genus *Bdeogale*) are adapted to forest (*B. nigripes* and *jacksoni*) but are allopatric. A species typically found in savanna (*B. crassicauda*) is present in eastern African coastal forests (Kingdon 1977; Taylor 1986). The African linsang has a disjunct distribution, with distinct forms in west and central Africa. *Poiana richardsoni liberiensis* occurs in suitable habitat in Upper Guinea forests (Taylor 1989; Hoppe-Dominik 1990), but this taxon (considered by Crawford-Cabral [1993] to be a distinct species) is absent east of the Niger River and in northern Cameroon and southwestern Central African Republic. The genus reappears with *P. richardsoni*

Table 14.2

## Carnivores Occurring in Three Broadly Defined Habitat Types Within the African Forest Biome

| FOREST INTERIOR | FOREST EDGE | AQUATIC HABITATS |
|---|---|---|
| *Bdeogale nigripes* | *Civettictis civetta* | *Aonyx* spp. |
| *Crossarchus* spp. | *Genetta maculata* | *Atilax paludinosus* |
| *Genetta johnstoni* | *Mellivora capensis* | *Lutra maculicollis* |
| *Genetta servalina* | | *Osbornictis piscivora* |
| *Genetta victoriae* | | |
| *Herpestes naso* | | |
| *Liberiictis kuhni* | | |
| *Nandinia binotata* | | |
| *Panthera pardus* | | |
| *Poiana richardsoni* | | |
| *Profelis aurata* | | |

*Sources:* Data from various sources listed in table 14.1.

*Note:* Although forest-edge species are seldom if ever found in closed forest, several true forest species appear to be tolerant of secondary-forest and edge conditions.

*leightoni* in southern Cameroon and eastward (Sanderson 1940; Coetzee 1971).

For some species, further range extensions may just be a question of time. The cusimanse (genus *Crossarchus*) is represented by four species, the distributions of which correspond with proposed "principal faunal regions" in the Upper Guinea forest and the two ends of the Congo Basin (see Colyn and Van Rompaey 1994a). It has been proposed that the large gap in distribution between allopatric members of the genus reflects "incomplete" colonization rather than the presence of restrictive barriers (Colyn and Van Rompaey 1994a). More recently, the presence of *C. platycephalus* was discovered east of its proposed distributional limit in eastern Cameroon, indicating perhaps colonization from the Atlantic coastal basins (Colyn et al. 1995). This species, which is often numerically dominant and hence visible where it occurs, may be less so on the edge of its range. In the southwestern Central African Republic, local BaAka use a distinct name (*Ekaila*) to refer to an uncommon group-living mongoose in the area. This, as well as a sighting by J. Malcolm (pers. comm.), shed doubt on the assumption of Colyn et al. (1995) that *Crossarchus* does not occur to the west of the Ubangui River. Likewise, the range of *Herpestes naso* has been extended to include territory west of the Cross River, formerly thought to be the western limit of its range (Colyn and Van Rompaey 1994b). It is unclear whether these examples illustrate range expansion through increased colonization or improved knowledge of species distributions. The latter would not be surprising given that these are often elusive creatures.

A few Carnivora illustrate unusual patterns of endemism. The giant genet (*Genetta victoriae*) and the aquatic genet (*Osbornictis piscivora*) are species endemic to eastern DR Congo in the "Ituri-Maniema forest refuge" (Allen 1924; Rahm 1966; Kingdon 1989). The Liberian mongoose (*Liberiictis kuhni*) and Johnston's

genets (*Genetta johnstoni*) are known from no more than ten specimens each from southern Guinea, eastern Liberia, and western Ivory Coast (Schreiber et al. 1989; Hoppe-Dominik 1990; Colyn et al. 1998). Not one of the four is known to be a dominant member of any community. The presence of each may attest to relative climatic stability of the regions in which it is found, whereas the absence in neighboring areas may be due to the presence of ecological competitors (Kingdon 1989).

## Community Structure and Organization

A subject of general interest is the factors that facilitate the ecological coexistence of members of a carnivore community. One possibility is differing patterns of resource use, along such axes as diet, habitat, and activity (Schoener 1974; Jaksic et al. 1981). In complex assemblages, ecological separation along more than one resource dimension can be expected. Because so little is known of the ecology of many forest-dwelling carnivores, the role that any one species plays in a community is often difficult to determine. With the exception of Charles-Dominique's (1978) classic study on the palm civet (*Nandinia binotata*), no forest carnivore has been the subject of detailed long-term research. Most ecological information has been gleaned from scattered analyses of scats and stomach contents and from assumptions based on morphology and anecdotal observations. These assumptions are often erroneous, as highlighted by two recent studies of little-known members of the order. Whereas the foot morphology and diurnal foraging habits of the cusimanse (*Crossarchus obscurus*) suggest a terrestrial existence, radio-collared individuals in Tiwai Island, Sierra Leone, rested in tree holes 10–20 m from the ground (Olson, in prep.). Presumably because of its forest-dwelling and solitary lifestyle (thereby fitting the ancestral herpestid prototype), the long-nosed mongoose (*Herpestes naso*) had always been assumed to be nocturnal (Rosevear 1974; Haltenorth and Diller 1977). A recent study revealed it to be wholly diurnal in the Dzanga-Sangha Forest of the southwestern Central African Republic (Ray 1997). Both findings have important implications for the ecological niche each species fills within its community.

Questions of sympatry obviously depend on the scale of inquiry. Not all carnivores found in the forested zones of Africa are actually found deep in closed forest, and many species are specialized to habitats within the forest, such as wetlands (see table 14.2). For example, when superimposed, the distribution maps of *Civettictis civetta*, *Atilax paludinosus*, and *Nandinia binotata* look very similar—all three species range widely throughout African forests. However, at a finer scale of resolution, it is clear that habitat divisions play a role (Kingdon 1997). *Civettictis* is common along forest edges close to human habitation and seldom ventures into closed forest. *Nandinia* is a true forest species, and *Atilax*, although present in both forest and savanna habitats, is dependent on water. This does not mean, however, that all closed forest species will be loathe to venture near forest edges; *Nandinia binotata*, for example, is said to visit fields near forest edges (Rahm 1966; Naughton-Treves 1998; L. White, pers. comm.).

## A Case Study:
### The Dzanga-Sangha Carnivore Community

A large collection of scats from the Dzanga-Sangha Reserve (Central African Republic), representing at least eight carnivore species, was assembled over a two-year period. Sixty-two percent (666 of 1,066) of the scats could be positively identified to species (table 14.3) through a combination of techniques, including scat diameter, associated field sign, microscopic examination of predator hair, and thin-layer chromatography of bile acids (Ray 1996). These results allowed for the first examination of food resource utilization patterns among members of an African forest carnivore community. Analyses of scat contents, in conjunction with data gathered from radio-tracking,

Table 14.3

## Carnivore Species Whose Food Habits Were Studied from 1992 to 1994 in the Dzanga-Sangha Reserve of Southwestern Central African Republic

| SCIENTIFIC NAME | COMMON NAME | BODY WEIGHT RANGE (KG) | NUMBER OF SCATS COLLECTED[a] | SOURCES FOR BODY WEIGHT |
|---|---|---|---|---|
| *Herpestes naso* | long-nosed mongoose | 2–3.2 | 346 | a |
| *Bdeogale nigripes* | black-footed mongoose | 2–3 | 21 | b, c, d |
| *Atilax paludinosus* | marsh mongoose | 2.5–4 | 49 | a, c, d, e |
| *Genetta servalina* | servaline genet | 1–1.2 | 35 | b, d |
| *Civettictis civetta* | African civet | 7–15 | 30 | b, c, d, f |
| *Nandinia binotata* | African palm civet | 1.5–4.1 | 8 | c, g, h |
| *Profelis aurata* | African golden cat | 5–12 | 17 | c, i |
| *Panthera pardus* | leopard | 30–60 | 160 | c, j |

Sources: (a) Ray 1996, (b) Allen 1924, (c) Kingdon 1977, (d) Rosevear 1974, (e) Baker 1992, (f) Ray 1995, (g) Charles-Cominique 1978, (h) Ray, unpublished data, (i) Van Mensch and Van Bree 1969, and (j) Bailey 1993. For the 1992–1994 study, see Ray 1996.

a   Number of scats, out of 1,066 total, identified as belonging to each species.

field sign, and the literature, were used to examine the ecological factors that might account for the maintenance of this rich assemblage.

Even at a relatively gross level of identification, where the dominant (highest volume component) food class in each scat was tabulated for each species, obvious differences in the diets of Dzanga-Sangha carnivores could be discerned (figure 14.2). As expected, mammals were the dominant food items in felid scats. Arthropods and mammals were important components of all small carnivore diets; the genets were the most carnivorous and least insectivorous of the group. The marsh mongoose (*Atilax paludinosus*) fed on herpetofaunal and fish prey significantly more than other species, while fruit was most important for civets and palm civets.

Detailed analyses of hair and teeth of mam-

malian remains resulted in the identification of at least forty-eight prey species (Ray 1996). Golden cats and leopards differed from each other, and from other carnivores, with respect to the weights of mammalian prey taken (figure 14.3). More than 80% of prey items appearing in mongoose and genet scats were under 1 kg, whereas for golden cats and leopards, the figures were 52% and 10%, respectively. Nearly half of leopard prey weighed between 1 and 5 kg, and only two items found in scats were from animals weighing more than 20 kg. However, small viverrids and herpestids also consumed larger prey (between 5 and 20 kg) to some extent.

Body weights of Dzanga-Sangha small carnivores, particularly the three mongoose species and the palm civet, were broadly overlapping. This very likely is a reflection of a lack of selective pressure for prey size separation, as has been suggested to be the case in a number of carnivore guilds (Dayan and Simberloff 1996). The most probable explanation is that mam-

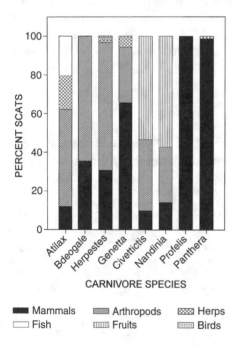

Figure 14.2. Proportion of scats from Dzanga-Sangha carnivores containing each of six possible broadly defined prey classes (mammals, arthropods, herpetofauna, fish, fruit, and birds) as the dominant (largest-volume) component.

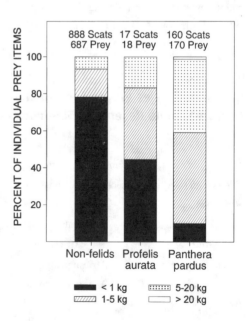

Figure 14.3. Weight distribution of mammalian prey identified in leopard (*Panthera pardus*), golden cat (*Profelis aurata*), and non-felid carnivore scats from the Dzanga-Sangha Reserve.

malian prey, which exhibit a wide range in size, were much less important than small-bodied arthropods in the diets of mongooses, civets, and palm civets. Indeed, size differences among the three most carnivorous species—genet, golden cat, and leopard—were large and correlated with the differing weight distributions of their prey. Similar size differences were obtained for sympatric felids in Peru (Emmons 1987), eastern DR Congo (Hart et al. 1996), and India (Karanth and Sunquist 1995).

Evidence gathered from radiotelemetry, field observations, and the literature demonstrated that ecological separation among non-felid carnivores was not restricted to the food resource dimension. Other aspects of component species' ecology, such as activity regimen, degree of arboreality, and habitat use, were also important (figure 14.4). *Herpestes naso* was the sole diurnal carnivore among the eight examined and apparently the numerically dominant member of the community (Ray 1997). Its closest potential food competitors were well separated in space and time: genets (*Genetta*) and black-footed mongooses (*Bdeogale*), though both nocturnal, exploited different microhabitats, the former

spending most of its time in the trees and the latter being largely ground dwelling (Rahm 1972; Rosevear 1974; Kingdon 1977).

The marsh mongoose (*Atilax*) was restricted to streamside and associated swamp habitat and had a specialized semi-aquatic diet (Ray 1997). Although the civet (*Civettictis*) and palm civet (*Nandinia*) were both frugivorous, they were well separated by their use of different forest strata and by the restriction of the former to forest edges. In fact, competition between *Nandinia* and non-carnivores, such as primates and rodents (arboreal, nocturnal insectivore-frugivores; Emmons et al. 1983), may be of greater importance and needs further investigation.

## Implications for Other African Forest Carnivore Communities

The little that is known about the food habits of other forest mongooses indicates that insectivory is important and, as shown in the Dzanga-Sangha study, may facilitate ecological coexistence among closely related members of a community. While small mammal and herpetological remains are usually present in the stomach contents of assorted herpestids, diets

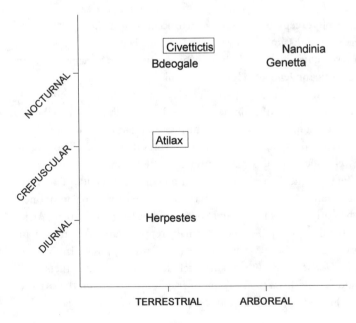

Figure 14.4. Separation among Dzanga-Sangha non-felid carnivore species along three ecological dimensions: activity regime, degree of arboreality, and habitat. Carnivores enclosed in boxes demonstrated restricted use of the study area: *Atilax paludinosus* was semi-aquatic, and *Civettictis civettictis* was found only in forest-edge habitats.

are generally dominated by hard and soft invertebrates (see Taylor 1986; Goldman 1987; Goldman and Taylor 1990; Van Rompaey and Colyn 1992). The results of Waser's (1980, 1981) work in the Serengeti has demonstrated that a prey base that renews itself rapidly following exploitation will be able to support a high number of predators. Highly insectivorous carnivores may be partitioning the arthropod prey base by foraging in different microhabitats, as suggested by Dickman (1988), or by foraging at different times of the day on this potentially highly renewable resource. Alternatively, one cannot exclude the possibility that in environments as rich as tropical forests, broad dietary overlap may be allowed by an abundant prey base that is not limiting predator populations.

The dentition of *Genetta* species points to a more carnivorous diet (Ewer 1973), and dietary analyses have confirmed this prediction, although arthropods are also usually present (see Ewer 1973; Delibes et al. 1989; Hamdine et al. 1993; Le Jacques and Lodé 1994; Stuart and Stuart 1998). Concerning differences among forms within the genus, Rosevear (1974:181) has stated that all genets can be "safely assumed …[to be]… broadly similar in habits except for such differences that may arise from dissimilarity in habitat." Although researchers often record the presence of several species of genets in their study areas (see, e.g., Allen 1924; Hoppe-Dominik 1990), taxonomic revisions of the genus have reduced greatly the number of recognized species (see Wozencraft 1989, 1993). Nevertheless, species classification of genets remains confused (Powell and Van Rompaey 1998). With the exception of northeastern DR Congo, no more than two species of genet are likely to occur sympatrically, and separation is probably achieved by divergent use of the forest habitat. For example, *Genetta maculata* (also called *G. pardina* and *G. rubiginosa;* Powell and Van Rompaey 1998) is widespread

throughout a variety of forest habitats in Africa (Rosevear 1974) and avoids open grassland (Fuller et al. 1990). However, it is less arboreal (Rosevear 1974) and less common within the high forest than *Genetta servalina* (Malbrant and Maclatchy 1949; Rahm 1966; Kingdon 1977; Davies 1990).

Next to nothing is known of the food habits of the African linsang (*Poiana richardsoni*), but its placement in a genus separate from *Genetta* has been called into question (Crawford-Cabral 1993). Its distribution is apparently patchy, and it is locally abundant in certain areas (Rosevear 1974). In forests where it does occur, *Poiana* is generally found in the highest branches of the canopy (Rahm 1966, 1972; Charles-Dominique 1978), which is permitted by its small body size (de Beaufort 1965).

Leopard (*Panthera pardus*) and golden cats (*Profelis aurata*) are sympatric throughout African forests and are the only felids within the forest interior. Golden cats are rarely sighted and have been assumed to be rare. However, the relative plenitude of museum specimens suggests that this species can be locally common (Malbrant and Maclatchy 1949; Mensch and van Bree 1969; Kingdon 1977). Both felids have been known to occur both on the forest edge and in relatively undisturbed areas in the interior (Rahm 1966; Rosevear 1974; Kingdon 1977; Hoppe-Dominik 1984; Bailey 1993; Jenny 1996). The Dzanga-Sangha study, and a study on the ecological relationship between the two felids in Ituri Forest (Hart et al. 1996), indicate that these carnivores depend heavily on mammalian prey separated primarily on the basis of prey size. Although the diversity of felids in African forests is low compared to South America and Southeast Asia (Hart et al. 1996), Africa is home to many representatives of other carnivore families (Herpestidae, Viverridae, Mustelidae) that include mammals in their diets.

A number of carnivores in the African forests frequent watercourses. Again, when examined carefully, the microdistributions of the various species overlap a good deal less than range maps might indicate. For example, *Lutra macullicolis* is known principally from the larger rivers in Africa, whereas the small-clawed otters (*Aonyx congica* and *A. capensis*), though larger in body size, inhabit the smaller tributaries (Rosevear 1974; Haltenorth and Diller 1977). There is evidence from the Dzanga-Sangha Reserve that the marsh mongoose (*Atilax paludinosus*) and *Aonyx congica* do not frequent the same small streams within the forest (Ray 1997). An analysis of scats belonging to *Aonyx capensis*, *Lutra maculicollis*, and *Atilax paludinosus*, found in a site in South Africa when the three species were forced to cohabit a relatively small range during drought conditions, has demonstrated clear dietary separation (Somers and Purves 1996). The effect on community structure in northeastern DR Congo by the addition of yet another aquatic carnivore, *Osbornictis piscivora*, is unknown; its diet is presumed to be dominated by aquatic prey (see Van Rompaey 1988). However, it must be noted that carnivores adapted to a semi-aquatic existence often live in microsympatry. For example, study of three otter species in west-central Thailand revealed broad overlap in habitat use, diet, and morphology; coexistence was attributed to a rich food resource base (Kruuk et al. 1994).

Food habits of *Civettictis civetta*, *Atilax paludinosus*, *Nandinia binotata*, and *Panthera pardus* in the Dzanga-Sangha study area bore close resemblance to diets in other parts of their ranges (see Charles-Dominique 1978; Hoppe-Dominik 1984; Baker 1992; Bailey 1993; Ray 1995; Hart et al. 1996). However, this does not necessarily mean that the role a given species plays within the community remains the same from one region to the next. For example, some evidence indicates that the abundances of some carnivore species vary markedly across their ranges. For instance, in most communities

where *Crossarchus* spp. occur, they are the numerically dominant members of the forest carnivore community. In eastern DR Congo, *Crossarchus alexandri* was described as the most common carnivore (Carpaneto and Germi 1989) and represented the great majority (67%) of the 113 carnivores hunted successfully by BaMbuti pygmies (Hart and Timm 1978). *Crossarchus ansorgei* constituted 83% of the carnivore meat surveyed in regional markets in southern DR Congo (Colyn et al. 1988). In western Africa, *Crossarchus obscurus* was said to be the most common carnivore (Booth 1960; Olson, pers. comm.). However, in the outer limits of their range *Crossarchus* species are relatively rare (Colyn et al. 1995). In these areas, such species as *Herpestes naso* may be numerically dominant, as is the case in southwestern Central African Republic, where *Crossarchus* is at least very rare (Ray 1997). Bates (1905:76) remarked that *Herpestes naso* was the most frequently killed "small animal" in southern Cameroon. In contrast, in areas where *Crossarchus* spp. are abundant, *Herpestes naso* does not seem to occur at high densities (Hart and Timm 1978; Carpaneto and Germi 1989). The success of *Herpestes naso* in the Dzanga-Sangha forest may be facilitated by its sole occupancy of the diurnal, terrestrial, predominantly insectivorous niche. It would be useful to investigate if and how its role within the community changes depending on the presence and abundance of *Crossarchus*.

The relative abundance of *Herpestes naso* in a community may also influence the ecology of *Atilax paludinosus*. The marsh mongoose is widespread throughout Africa, and although its presence is always associated with aquatic habitats, the degree of its restriction to this habitat varies from one locale to the next. In southern (Rosevear 1974; Baker 1989; Skinner and Smithers 1990) and western (L. White, pers. comm.) African savanna habitats, and in western forests (Olson, in prep.), individual

*Atilax* have been recorded moving relatively long distances between streambeds. In the Dzanga-Sangha forest, however, no sign of this species was found away from streambeds, and a male was never located outside a narrow band of swamp habitat during a year of radio-tracking (Ray 1997). The fact that *Herpestes naso* is absent in southern and western Africa and present at high densities in the Dzanga-Sangha forest may account for this pattern.

In summary, the role a species plays within an assemblage will depend both on historical factors that account for the presence or absence of potential competitors and on the ecological relations between sympatric members of its guild. Moreover, the little evidence available suggests that dominance relationships may vary from one region to the next. Many questions remain unanswered. For example, what factors account for the persistence of *Osbornictis piscivora, Genetta victoriae,* and *Liberiictis kuhni,* none of which are obvious "replacements" for other allopatric species?

## Carnivore Conservation

### MONITORING AFRICAN FOREST CARNIVORES

Traditionally, most attention has been focused on larger members of the order (Schaller 1996), in the hopes that conserving viable populations of the top predators in a community will take care of the needs of the smaller members (see, e.g., Myers 1986). As a result, small carnivores are often ignored in surveys and management plans.

Since the IUCN Small Carnivore Specialist Group issued its "action plan" (Schreiber et al. 1989), efforts to assess the ecology and status of the world's mustelids, viverrids, and herpestids have increased. Of ten African small carnivore taxa (excluding felids) listed as "known or likely to be threatened" (Schreiber et al. 1989), eight are from the forest biome (table 14.4). All were listed as a result of their relatively restricted distributions. A priority for future research is the development of reliable and cost-efficient tech-

niques for inventory and monitoring of forest carnivore species such as these. Currently no protocol exists in tropical forests even for the assessment of presence or absence of particular carnivore species, much less their relative abundances. Regions in the Congo Basin that have thus far escaped large-scale human-induced impacts can be used as a baseline to monitor the structure of carnivore communities in fragmented landscapes, such as the Upper Guinea rain forests (Schreiber et al. 1989).

Many useful techniques developed for censusing temperate forest carnivores (Raphael 1994; Zielinski and Kucera 1995) will be of little use in the tropics (particularly snow-tracking!). Frequent rain makes the use of track beds difficult, and even if one were to wait until the dry season, laterite soil can be difficult to keep moist for clear track impressions. Relying on trapping can be a costly and time-consuming endeavor, particularly because many species are notoriously trap-shy. Because sightings of carnivores are rare during traditional transect surveys, the principal source of information on carnivores has been traditionally derived from interviews with local people, expeditions with local hunters, and market surveys (see, e.g., Schlitter 1974; Hart and Timm 1978; Colyn and Gevaerts 1986; Colyn et al. 1988, 1995; Taylor 1989; Davies 1990; Hoppe-Dominik 1990; Ashraf et al. 1993).

Yet efforts to employ standardized survey sampling units for censusing and monitoring forest carnivores (Zielinski and Kucera 1995; J. Weaver, pers. comm.) will be of considerable future utility in tropical habitats. Some methods such as camera trapping (see, e.g., Kucera and Barrett 1993) hold special promise: flash-equipped autofocus cameras can be placed near trails and along transects and set to take photos of any animal that crosses a tripping device. Another indirect method for monitoring forest carnivore populations is to collect hair systematically through hair traps (J. Weaver, pers. comm.)

Table 14.4

## Species and Subspecies of African Forest Carnivores Listed as of "Conservation Concern"

| TAXON | DISTRIBUTION | IUCN THREAT CATEGORY |
|---|---|---|
| *Aonyx congica* | Congo Basin | Near threatened |
| *Bdeogale crassicauda omnivora* | Coastal rainforest of Kenya | Endangered |
| *Bdeogale jacksoni* | Central Kenya and southeastern Uganda | Vulnerable |
| *Crossarchus ansorgei* | Left bank of Congo River and southward; northern Angola | Data deficient |
| *Genetta johnstoni* | Liberia and southeastern Guinea | Data deficient |
| *Genetta cristata* | Southeastern Nigeria, western Cameroon | Endangered |
| *Genetta victoriae* | Northeastern DR Congo | Not listed |
| *Liberiictis kuhni* | Northeastern Liberia, western Côte d'Ivoire, southeastern Guinea | Endangered |
| *Osbornictis piscivora* | Northeastern DR Congo | Data deficient |
| *Poiana richardsoni liberiensis* | Liberia, Côte d'Ivoire, Sierra Leone | Data deficient |

*Sources:* Data from the IUCN/SSC Small Carnivore Specialist Group (Schreiber et al. 1989) and the Otter Specialist Group (Foster-Turley et al. 1990). IUCN Red List Threat Categories are from IUCN 1996.

or scats. A sizable scat collection is readily available once a search image has been developed. Standardized effort (that is, a fixed number of collectors, over a fixed route, over a fixed area) could yield valuable information on relative abundances of component carnivore species (Putnam 1984). However, the chances of locating scats on roads or well-worn trails are much better than on randomly developed transects, a situation that may lead to species biases (Ray 1996). Such methods depend on the continued development of reliable species identification techniques. Thin-layer chromatography, though relatively expensive and time consuming, was important in the Dzanga-Sangha study in differentiating between the scats of similarly sized carnivores (Ray 1996). Scat and hair identification using DNA techniques (Höss et al. 1992; Kohn and Wayne 1997; J. Weaver, pers. comm.) will likely prove extremely useful in the future.

Aside from the monitoring of carnivore populations, the collection and analysis of scats constitute a valuable technique for assessing and monitoring the composition and breadth of the prey base. This was particularly evident in the case of the Dzanga-Sangha shrew fauna. Only four species were captured by trap or by hand during a two-year period, but sixteen species were identified through the recovery of skull remains in feces (Ray and Hutterer 1996; Ray 1998). This extraordinary shrew diversity may not be unique to this area but may simply reflect the greater efficiency of the carnivore as a "moving trap" relative to traditional collecting methods. A series of immobile pitfall traps is presumably less able to sample the diverse array of microhabitats found in such complex habitat mosaics as tropical forests (Ray and Hutterer 1996).

### CONSERVATION ISSUES

Schreiber et al. (1989) listed habitat destruction and hunting as the main threats facing small carnivores. Habitat destruction occurs generally through the clearing of fields and logging.

Logging in the Congo Basin generally occurs at low levels (<25% of basal area removed), resulting in relatively low levels of incidental damage (Wilkie et al. 1992; White 1994b) and better chances for regeneration than in areas to the east (Struhsaker et al. 1996; Struhsaker 1997) and west (White 1994b). Tracks and scats of most Dzanga-Sangha carnivores were found on logging roads; leopards apparently traveled along these routes, and mongooses crossed them frequently. There was no evidence from radio-tracking that *Herpestes naso* avoided logged areas within their ranges.

Although this might indicate that at least some carnivore species lived comfortably within the context of human-induced disturbance, other evidence suggests that such a conclusion may be premature. For example, negative impacts of selective logging on such prey as duikers (Struhsaker 1997) and primates (see, e.g., Johns and Skorupa 1987; Struhsaker 1997) may have important implications for their predators. Negative direct effects on forest carnivore population have been demonstrated in areas subjected to higher intensities of forest exploitation. In a selectively logged forest in Borneo where approximately 7% of large trees were removed, densities of civets were significantly lower than in adjacent unlogged forest. The full species complement still existed in the logged forest, however (Heydon and Bulloh 1996). Commercial logging (especially clearcutting) has often been cited as the dominant factor in the decline of North American mustelids (Berg and Kuehn 1994). Small cuts that are dispersed throughout the forest apparently exert fewer adverse effects on martens and fishers (Thompson and Harestad 1994). Yet small carnivores, as diet generalists and opportunists, can find food in a variety of habitats, and the increased prey biomass that often arises in logged areas may actually serve to increase predator populations (Brainerd et al. 1994). However, habitat preferences are not dictated solely by foraging behavior; selection of resting sites may be dependent on particular features of mature forests.

Some carnivore species may be vulnerable because of their dependence on certain habitats (Johns 1988). For example, results from radio-tracking of mongooses have highlighted the importance of wetlands for both *Atilax paludinosus* and, to a lesser extent, *Herpestes naso* (Ray 1997). Other species, such as *Bdeogale nigripes* and perhaps *Genetta servalina*, may depend on mature forests for their survival; *Nandinia binotata* may be affected if large numbers of key fruit trees are removed.

The increasing advance of human settlement often leads to conflicts between humans and forest wildlife, particularly over raids of crops and livestock. Leopards, for example, are well known for their use of habitat edges and their ability to live close to humans. In South Asia, where protected areas are becoming more and more insular, leopards have been documented to be switching from wild to domestic prey of the same size classes (Seidensticker et al. 1990). Concomitant with the expanding human population in east Africa, golden cats have been more frequently invading chicken coops (Kingdon 1977). Palm civets have been cited as important crop-raiding species in the Kibale Forest, Uganda (Naughton-Treves 1998), and in eastern DR Congo (Rahm 1966).

The Upper Guinea forest region (Ivory Coast, Liberia, and Sierra Leone) has been listed as a "core area" for conservation of small carnivores. Forest clearing in the area is leading to encroachment by savanna-like habitats. This, in turn, enables potential "invaders" from neighboring savanna habitats (table 14.5) to expand their ranges into formerly forested areas (Happold 1987, 1996), resulting in a dramatic shift in carnivore communities. Indeed, *Herpestes ichneumon*, a savanna-dwelling mongoose that requires dense undergrowth (Rosevear 1974), has expanded into forests of west Africa

(Skinner and Smithers 1990). Rosevear (1974) also writes of a specimen of *Genetta thierryi* found in a locale that was nominally a closed-forest zone. Because so little forest remained (owing to human activity), he assumed that it was a recent immigrant from nearby open habitats. The relative proportion of forest to edge and invader species over time may signal the degree of degradation of the forest habitat. The fact that several true forest species (*Herpestes naso*, *Bdeogale nigripes*, and *Genetta servalina*) do not occur outside the continuous Congo Basin may indicate the extent of fragmentation of the eastern and western forests.

Hunting for subsistence and local markets is ubiquitous in the African forest biome. The extent to which carnivores are preferred game meat is variable throughout the continent and is not necessarily a function of decreased availability of more traditional species. Although carnivores were caught in cable snares and nets on occasion, they were not deliberately hunted in the Dzanga-Sangha Reserve (Noss 1997, 1998). Out of 17,571 carcasses censused from markets in Equatorial Guinea, only 121 (less than 1%) were carnivores (Fa et al. 1995). Similarly, small carnivores represented only 1.4% of all game in the southwest Ivory Coast (Hoppe-Dominik 1990). By contrast, African civet (*Civettictis civettictis*) carcasses were present in Sierra Leone markets on most mornings

during a one-year survey (Davies 1990) and are apparently among the preferred prey of villagers living in the northern Congo (M. Fay, pers. comm.) and in western DR Congo (L. White, pers. comm.). Carnivores were actively pursued with dogs in the Ituri Forest of DR Congo, and mongooses were "common game" for Mbuti pygmies (Hart and Timm 1978; Carpaneto and Germi 1989). Mongooses (*Crossarchus* spp.) and genets were also plentiful in markets near Kisangani (Colyn et al. 1988). Hunting poses the most serious threat to carnivore populations in the Upper Guinea rain forests, where they are already exposed to the adverse effects of forest fragmentation (Schreiber et al. 1989).

Even if some carnivores are not directly threatened by hunting activities, several of their most important prey species may suffer. Human impacts in apparently intact forest should not be underestimated (Redford 1992), particularly when human hunters are competing with animal predators (Jorgenson and Redford 1993). Roads built to accommodate selective logging activities inevitably serve to open up the forest to hunting pressures (Wilkie et al. 1992). For example, in the Dzanga-Sangha area, three of the four principal game species for indigenous and immigrant hunters (Noss 1997, 1998) were also the most frequent prey items found in leopard scats. Two of these were likewise the most important prey for golden

Table 14.5

## Potential Carnivore "Invaders" of the African Forest Biome from Savanna Habitats

| SCIENTIFIC NAME | COMMON NAME | DISTRIBUTION |
| --- | --- | --- |
| *Felis libyca* | African wild cat | widespread |
| *Felis serval* | serval | widespread |
| *Galerella sanguinea* | slender mongoose | widespread |
| *Genetta thierryi* | Thierry's genet | West Africa |
| *Herpestes ichneumon* | Egyptian mongoose | widespread |
| *Ichneumia albicauda* | white-tailed mongoose | widespread |

cats (Ray 1996). Given that hunting is occurring in the area at levels deemed unsustainable for these populations (Noss 1998), the result can only be negative for the region's larger carnivores, particularly in the face of human population increases.

In spite of the relatively continuous nature of the tropical moist forest belt that blankets the central region of Africa, only one-third of the carnivore species found in the biome range throughout it. This pattern, evident also in other forest biota, is shaped by a combination of historical factors and community interrelationships. As a result, the carnivore communities in different sites exhibit differences in species composition and abundances. Component species in a given community display habitat divisions not discernible at the scale of a range map. A study on ecological separation among eight sympatric carnivores in the Congo Basin demonstrated clear dietary differences between most members of the community, suggesting a broad potential resource base. Although many carnivores are known for their opportunistic and eurytopic natures, it is nevertheless clear that many species will experience adverse direct or indirect effects from the twin threats of habitat conversion and hunting, and that these may greatly affect the community of which they are part. Future research must focus on the basic ecology of forest carnivores and their relationships with sympatric members of the order. Additional efforts are required to develop effective methods to monitor the small carnivores of African forests. The important contributions that carnivores make to the biological diversity of an area cannot be overlooked simply because they are inconspicuous members of the biota.

ACKNOWLEDGMENTS

Fieldwork in the Central African Republic was supported by grants from the Wildlife Conservation Society and the Conservation, Food and Health Foundation, and a Fulbright Scholarship. Maka Sylvestre, Bokombi Franco, Mokonzo Etienne, Mokoko Marc, Etubu Gaston, Singale Jerome, and Balayona Joseph provided invaluable field assistance, particularly in the collection of carnivore scats. This work would not have been possible without the support of the government of the Central African Republic and the World Wildlife Fund. Jay Malcolm, Amy Vedder, and Lee White provided numerous helpful comments on the manuscript. Jay Malcolm drew figure 14.1.

REFERENCES

Allen, J. A. 1924. Carnivora collected by the American Museum Congo Expedition. *Bulletin of the American Museum of Natural History* 47:73–281.

Ashraf, N. V. K., A. Kumar, and A. J. T. Johnsingh. 1993. Two endemic viverrids of the western Ghats, India. *Oryx* 27:109–114.

Bailey, T. N. 1993. *The African Leopard: Ecology and Behavior of a Solitary Felid.* Columbia University Press, New York.

Baker, C. M. 1989. Feeding habits of the water mongoose *Atilax paludinosus. Zeitschrift für Säugetierkunde* 54:31–39.

———. 1992. *Atilax paludinosus. Mammalian Species* 408:1–6.

Bates, G. L. 1905. Notes on the mammals of southern Cameroons and the Benito. *Proceedings of the Zoological Society of London* 1:65–85.

Berg, W. R., and D. W. Kuehn. 1994. Demography and range of fishers and American martens in a changing Minnesota landscape. Pages 262–271 in S. W. Buskirk, A. S. Harestad, M. G. Raphael, and R. A. Powell, eds. *Martens, Sables, and Fishers: Biology and Conservation.* Cornell University Press, Ithaca.

Booth, A. H. 1960. *Small Mammals of West Africa.* Longman Group, London.

Brainerd, S. M., J. O. Helldin, E. Lindstrom, and J. Rolstad. 1994. Eurasian pine martens and old industrial forest in southern boreal Scandinavia. Pages 343–354 in S. W. Buskirk, A. S. Harestad, M. G. Raphael, and R. A. Powell, eds. *Martens, Sables, and Fishers: Biology and Conservation.* Cornell University Press, Ithaca.

Butynski, T. M. 1984. Ecological survey of the Impenetrable (Bwindi) Forest, Uganda: Recommendations for its conservation and management. Unpublished report. Wildlife Conservation Society, Bronx.

Carpaneto, G. M., and F. P. Germi. 1989. Mustelidae and viverridae from north-eastern Zaire: Ethnozoological research and conservation. *Mustelid and Viverrid Conservation* 1:2–4.

Charles-Dominique, P. 1978. Ecologie et vie sociale de *Nandinia binotata:* Comparison avec les prosimiens sympatriques du Gabon. *La Terre et Vie (Revue d'Ecologie)* 32:477–528.

Coetzee, C. G. 1971. Order Carnivora, part 8. In J. Meester and H. W. Setzer, eds. *The Mammals of Africa: An Identification Manual.* Smithsonian Institution Press, Washington, D.C.

Colyn, M., P. Barrière, P. Formenty, O. Perpète, and H. Van Rompaey. 1998. First confirmation of the presence of the Liberian mongoose, *Liberiictis kuhni,* in Côte d'Ivoire. *Small Carnivore Conservation* 18:12–14.

Colyn, M., M. Dethier, P. Ngegueu, O. Perpète, and H. Van Rompaey. 1995. First observations of *Crossarchus platycephalus* (Goldman, 1984) in the Zaire/Congo System (Dja River, southeastern Cameroon). *Small Carnivore Conservation* 12:10–11.

Colyn, M., A. Dudu, and M. Mbaelele. 1988. Données sur l'exploitation du petit et moyen gibier des forêts ombrophiles du Zaïre. *Nature et Faune* 3:22–39.

Colyn, M., A. Gautier-Hion, and W. Verheyen. 1991. A re-appraisal of paleo-environmental history in central Africa: Evidence for a major fluvial refuge in the Zaire basin. *Journal of Biogeography* 18:403–407.

Colyn, M., and H. Gevaerts. 1986. *Osbornictis piscivora* Allen, 1919, deux nouvelles stations de récolte dans la sous-région de la Tshopo (Haut Zaïre). *Bull. Inst. R. Sci. Nat. Belg.* 56:9–11.

Colyn, M., and H. Van Rompaey. 1994a. A biogeographic study of cusimanses (*Crossarchus*) (Carnivora, Herpestidae) in the Zaire Basin. *Journal of Biogeography* 21:479–489.

———. 1994b. Morphometric evidence of the monotypic status of the African long-nosed mongoose *Xenogale naso* (Carnivora, Herpestidae). *Belgian Journal of Zoology* 124:175–192.

Crawford-Cabral, J. 1993. A comment on the systematic position of *Poiana. Small Carnivore Conservation* 9:8.

Davies, G. 1990. The Gola project in Sierra Leone: Report on the viverrids. *Mustelid and Viverrid Conservation* 2:4–5.

Dayan, T., and D. Simberloff. 1996. Patterns of size separation in carnivore communities. Pages 243–266 in J. L. Gittleman, ed. *Carnivore Behavior, Ecology and Evolution.* Volume 2. Cornell University Press, Ithaca.

De Beaufort, F. 1965. Répartition et taxonomie de la poiane (Viverridae). *Mammalia* 29:275–280.

Delibes, M., A. Rodriguez, and F. F. Parreno. 1989. Food of the common genet (*Genetta genetta*) in northern Africa. *Journal of Zoology (London)* 218:321–326.

Dickman, C. R. 1988. Body size, prey size, and community structure in insectivorous mammals. *Ecology* 69:569–580.

Dowsett, R. J., and L. Granjon. 1991. Liste préliminaire des mammifères du Congo. *Tauraco Research Report* 4:297–310.

Emmons, L. H. 1987. Comparative feeding ecology of felids in a Neotropical rainforest. *Behavioral Ecology and Sociobiology* 20:271–283.

Emmons, L. H., A. Gautier-Hion, and G. Dubost. 1983. Community structure of the frugivorous-folivorous forest mammals of Gabon. *Journal of Zoology (London)* 199:209–222.

Ewer, R. F. 1973. *The Carnivores.* Cornell University Press, Ithaca.

Fa, J. A., J. Juste, J. Perez del Val, and J. Castroviejo. 1995. Impact of market hunting on mammal species in Equatorial Guinea. *Conservation Biology* 9:1107–1115.

Foster-Turley, P., S. Macdonald, and C. Mason. 1990. *Otters: An Action Plan for Their Conservation.* International Union for the Conservation of Nature and Natural Resources, Gland, Switzerland.

Fuller, T. K., A. R. Biknevicius, and P. W. Kat. 1990. Movements and behavior of large spotted genets (*Genetta maculata* Gray 1830) near Elmenteita, Kenya (Mammalia Viverridae). *Tropical Zoology* 3:13–19.

Goldman, C. A. 1984. Systematic revision of the African mongoose genus *Crossarchus* (Mammalia: Viverridae). *Canadian Journal of Zoology* 62:1618–1630.

———. 1987. *Crossarchus obscurus. Mammalian Species* 290:1–5.

Goldman, C.A., and M. E. Taylor. 1990. *Liberiictis kuhni. Mammalian Species* 348:1–3.

Grubb, P. 1978. Patterns of speciation in African mammals. *Bulletin of the Carnegie Museum of Natural History* 6:152–167.

Haltenorth, T., and H. Diller 1977. *A Field Guide to the Mammals of Africa.* Collins, London.

Hamdine, W., M. Thevenot, M. Sellami, and K. De Smet. 1993. Régime alimentaire de la genette (*Genetta genetta* Linne 1758) dans le Parc National du Djurdjura, Algérie. *Mammalia* 57:9–12.

Happold, D. C. D. 1987. *The Mammals of Nigeria.* Clarendon Press, Oxford, England.

———. 1996. Mammals of the Guinea-Congo rain forest. *Proceedings of the Royal Society of Edinburgh* 104B:243–284.

Hart, J. A., M. Katembo, and K. Punga. 1996. Diet, prey selection and ecological relations of leopard and golden cat in the Ituri Forest, Zaire. *African Journal of Ecology* 34:364–379.

Hart, J. A., and R. M. Timm. 1978. Observations on the aquatic genet in Zaire. *Carnivore* 1:130–132.

Heard, S., and H. Van Rompaey. 1990. Rediscovery of the crested genet. *Mustelid and Viverrid Conservation* 3:1–4.

Heydon, M. J., and P. Bulloh. 1996. The impact of selective logging on sympatric civet species in Borneo. *Oryx* 30:31–36.

Hoppe-Dominik, B. 1984. Etude du spectre des proies de la panthère, *Panthera pardus*, dans le Parc National de Tai en Côte d'Ivoire. *Mammalia* 48:477–487.

———. 1990. On the occurrence of the honey-badger (*Mellivora capensis*) and the viverrids in the Ivory Coast. *Mustelid and Viverrid Conservation* 3:9–13.

Höss, M., M. Kohn, S. Pääbo, and W. Schröder. 1992. Excrement analysis by PCR. *Nature* 359:199.

IUCN. 1996. *IUCN Red List of Threatened Animals.* International Union for the Conservation of Nature and Natural Resources, Gland, Switzerland.

Jaksic, F. M., H. W. Greene, and J. L.Yanez. 1981. The guild structure of a community of predatory vertebrates in central Chile. *Oecologia* 49:21–28.

Jenny, D. 1996. Spatial organization of leopards *Panthera pardus* in Taï National Park, Ivory Coast: Is rainforest habitat a "tropical haven"? *Journal of Zoology (London)* 240:427–440.

Johns, A. D. 1988. Effects of "selective" timber extraction on rainforest structure and composition and some consequences for frugivores and folivores. *Biotropica* 20:31–37.

Johns, A. D., and J. P. Skorupa. 1987. Responses of rainforest primates to habitat disturbance: A review. *International Journal of Primatology* 6:157–191.

Jorgenson, J. P., and K. H. Redford. 1993. Humans and big cats as predators in the Neotropics. *Symposia of the Zoological Society of London* 65:367–390.

Karanth, K. U., and M. E. Sunquist. 1995. Prey selection by tiger, leopard and dhole in tropical forests. *Journal of Animal Ecology* 64:439–450.

Kingdon, J. 1977. *East African Mammals: An Atlas of Evolution in Africa.* Volume IIIA, *Carnivores.* University of Chicago Press, Chicago.

———. 1989. *Island Africa: The Evolution of Africa's Rare Animals and Plants.* Princeton University Press, Princeton.

———. 1997. *The Kingdon Field Guide to African Mammals.* London: Academic Press.

Kohn, M. H., and R. K. Wayne. 1997. Facts from feces revisited. *Trends in Ecology and Evolution* 12:223–227.

Kruuk, H., B. Kanchanasaka, S. O'Sullivan, and S. Wanghongsa. 1994. Niche separation in three sympatric otters, *Lutra perspicillata, Lutra lutra* and *Aonyx cinerea* in Huai Kha Khaeng, Thailand. *Biological Conservation* 69:115–120.

Kucera, T. E., and R. H. Barrett. 1993. The Trailmaster camera system for detecting wildlife. *Wildlife Society Bulletin* 21:505–508.

Le Jacques, D., and T. Lodé. 1994. L'alimentation de la genette d'Europe, *Genetta genetta* L. 1758 dans un bocage de l'ouest de la France. *Mammalia* 58:383–389.

Livingstone, D. A. 1982. Quaternary geography of Africa and the refuge theory. Pages 523–536 in G. T. Prance, ed. *Biological Diversification in the Tropics.* Columbia University Press, New York.

Malbrant, R., and A. Maclatchy. 1949. *Faune de l'Equateur Africain Français, Tome II: Mammifères.* Paris: Paul Lechevalier.

Mayr, E., and R. J. O'Hara. 1986. The biogeographic evidence supporting the Pleistocene forest refuge hypothesis. *Evolution* 40:55–67.

Mensch, P. J. A. van, and P. J. H. Van Bree. 1969. On the African golden cat, *Profelis aurata* (Temminck, 1827). *Biologica Gabonica* 5:235–269.

Michaelis, B. 1972. Die Schleichkatzen (Viverriden) Afrikas. *Säugetierkundliche Mitteilung* 20:1–110.

Myers, N. 1986. Conservation of Africa's cats: Problems and opportunities. Pages 437–446 in S. D. Miller and D. D. Everett, eds. *Cats of the World: Biology, Conservation, and Management.* National Wildlife Federation, Washington, D.C.

Naughton-Treves, L. 1998. Predicting patterns of crop damage by wildlife around Kibale National Park, Uganda. *Conservation Biology* 12:156–168.

Noss, A. J. 1997. The economic importance of communal net hunting among the BaAka of the Central African Republic. *Human Ecology* 25:71–89.

———. 1998. The impacts of cable snare hunting on wildlife populations in the forests of the Central African Republic. *Conservation Biology* 12:390–398.

Perret, J. L., and V. Aellen. 1956. Mammifères du Cameroun de la collection J. L. Perret. *Revue Suisse de Zoologie* 63:395–450.

Powell, C. B., and H. Van Rompaey. 1998. Genets of the Niger Delta, Nigeria. *Small Carnivore Conservation* 19:1–7.

Prance, G. T., ed. 1982. *Biological Diversification in the Tropics.* Columbia University Press, New York.

Prins, H. H. T., and J. M. Reitsma. 1989. Mammalian biomass in an African equatorial rain forest. *Journal of Animal Ecology* 58:851–861.

Putnam, R. J. 1984. Facts from faeces. *Mammal Review* 14:79–97.

Rahm, U. 1966. Les mammifères de la forêt equatoriale de l'est du Congo. *Annales Musée Royal de l'Afrique Centrale ser. 8 Sciences Zoologiques* 149:39–121.

———. 1972. Zur verbreitung und ökologie der säugetiere des Afrikanischen Regenwaldes. *Acta Tropica* 29:452–473.

Raphael, M. G. 1994. Techniques for monitoring populations of fishers and American martens. Pages 224–240 in S. W. Buskirk, A. S. Harestad, M. G. Raphael, and R. A. Powell, eds. *Martens, Sables, and Fishers: Biology and Conservation.* Cornell University Press, Ithaca.

Ray, J. C. 1995. *Civettictis civetta. Mammalian Species* 488:1–7.

———. 1996. Resource use patterns among mongooses and other carnivores in a Central African rainforest. Ph.D. thesis, University of Florida, Gainesville.

———. 1997. Comparative ecology of two African forest mongooses, *Herpestes naso* and *Atilax paludinosus. African Journal of Ecology* 35:237–253.

———. 1998. Temporal variation of predation on rodents and shrews by small African forest carnivores. *Journal of Zoology (London)* 244:363–370.

Ray, J. C., and R. Hutterer. 1996. Structure of a shrew community in Central African Republic based on the analysis of carnivore scats, with the description of a new *Sylvisorex* (Mammalia: Soricidae). *Ecotropica* 1:85–97.

Redford, K. H. 1992. The empty forest. *BioScience* 42:412–422.

Ricklefs, R. E., and D. Schluter. 1993. Species diversity: Regional and historical influences. Pages 350–363 in R. E. Ricklefs and D. Schluter, eds. *Species Diversity in Ecological Communities.* University of Chicago Press, Chicago.

Rosevear, D. R. 1974. *The Carnivores of West Africa.* British Museum of Natural History, London.

Sanderson, I. T. 1940. The mammals of the north Cameroons forest area being the results of the Percy Sladen expedition to the Mamfe Division of the British Cameroons. *Transactions of the Zoological Society of London* 24:623–725.

Schaller, G. B. 1996. Introduction: Carnivores and conservation biology. Pages 1–10 in J. L. Gittleman, ed. *Carnivore Behavior, Ecology and Evolution.* Volume 2. Cornell University Press, Ithaca.

Schlitter, D. A. 1974. Notes on the Liberian mongoose, *Liberiictis kuhni* Hayman, 1958. *Journal of Mammalogy* 55:438–442.

Schoener, T. W. 1974. Resource partitioning in ecological communities. *Science* 185:27–39.

Schouteden, H. 1948. Faune du Congo Belge et du Ruanda-Urundi, 1: Mammifères. *Annales Musée Royal du Congo Belge ser. 8 Sciences Zoologiques* 1:331.

Schreiber, A., R. Wirth, M. Riffel, and H. Van Rompaey. 1989. *Weasels, Civets, Mongooses, and Their Relatives: An Action Plan for the Conservation of Mustelids and Viverrids.* International Union for the Conservation of Nature and Natural Resources, Gland, Switzerland.

Seidensticker, J., M. E. Sunquist, and C. McDougal. 1990. Leopards living at the edge of the Royal Chitwan National Park, Nepal. Pages 415–423 in J. C. Daniel and J. Serrao, eds. *Conservation in Developing Countries: Problems and Prospects.* Proceedings of the Centenary Seminar, Bombay Natural History Society, Bombay.

Skinner, J. D., and R. H. N. Smithers. 1990. *The Mammals of the Southern African Subregion.* University of Pretoria, Republic of South Africa.

Somers, M. J., and M. G. Purves. 1996. Trophic overlap between three syntopic semi-aquatic carnivores: Cape clawless otter, spotted-necked otter, and water mongoose. *African Journal of Ecology* 34:158–166.

Struhsaker, T. T. 1997. *Ecology of an African Rain Forest: Logging in Kibale and the Conflict Between Conservation and Exploitation.* University Press of Florida, Gainesville.

Struhsaker, T. T., J. S. Lwanga, and J. M. Kasenene. 1996. Elephants, selective logging and forest regeneration in the Kibale Forest, Uganda. *Journal of Tropical Ecology* 12:45–64.

Stuart, C., and T. Stuart. 1998. Notes on the diet of four species of viverrid in a limited area of southern Namaqualand, northern Cape, South Africa. *Small Carnivore Conservation* 19:9–10.

Taylor, M. E. 1986. Aspects on the biology of the four-toed mongoose, *Bdeogale crassicauda. Cimbebasia (A)* 8:187–193.

———. 1989. New records of two species of rare viverrids from Liberia. *Mammalia* 53:122–125.

Thompson, I. D., and A. S. Harestad. 1994. Effects of logging on American martens, and models for habitat management. Pages 355–367 in S. W. Buskirk, S. Harestad, M. G. Raphael, and R. A. Powell, eds. *Martens, Sables, and Fishers: Biology and Conservation.* Cornell University Press, Ithaca.

UNESCO. 1978. *Tropical Forest Ecosystems.* UNESCO-UNEP, Paris.

Van Rompaey, H. 1988. *Osbornictis piscivora. Mammalian Species* 309:1–4.

Van Rompaey, H., and M. Colyn. 1992. *Crossarchus ansorgei. Mammalian Species* 402:1–3.

Waser, P. M. 1980. Small nocturnal carnivores: Ecological studies in the Serengeti. *African Journal of Ecology* 18:167–185.

———. 1981. Sociality or territorial defense? The influence of resource renewal. *Behavioral Ecology and Sociobiology* 8:231–237.

White, L. J. T. 1994a. Biomass of rain forest mammals in the Lopé Reserve, Gabon. *Journal of Animal Ecology* 63:499–512.

———. 1994b. The effects of commercial mechanized selective logging on a transect in lowland rainforest in the Lopé Reserve, Gabon. *Journal of Tropical Ecology* 10:313–322.

Wilkie, D. S., J. G. Sidle, and G. C. Boundzanga. 1992. Mechanized logging, market hunting, and a bank loan in Congo. *Conservation Biology* 6:570–580.

Wozencraft, W. C. 1989. Classification of the recent Carnivora. Pages 569–593 in J. L. Gittleman, ed. *Carnivore Behavior, Ecology, and Evolution.* Cornell University Press, Ithaca.

———. 1993. Order Carnivora. Pages 279–348 in D. E. Wilson and D. M. Reeder, eds. *Mammal Species of the World: A Taxonomic and Geographic Reference.* Smithsonian Institution Press, Washington, D.C.

Zielinski, W. J., and T. E. Kucera. 1995. American marten, fisher, lynx, and wolverine: Survey methods for their detection. General Technical Report PSW-GTR-157. Pacific Southwest Research Station, Forest Service, U.S. Department of Agriculture, Albany, Calif.

# CHAPTER 15

# African Forest Birds

## Patterns of Endemism and Species Richness

F. Dowsett-Lemaire and R. J. Dowsett

Moreau (1966) discussed regional patterns of species richness in Africa, but without presenting a firm zoogeographical framework and without publishing the lists of species on which he based his arguments. A suitable and accurate description of vegetation types across the continent was also lacking until White's (1983a) original and detailed work that describes the vegetation types within a clearly defined floristic framework. White's memoir provides zoologists with an excellent environmental tool with which patterns of animal species distribution can be suitably re-examined. Indeed, an analysis of bird distribution patterns in the light of White's floristic subdivisions shows great similarities between phytochoria and avian zoochoria (Dowsett-Lemaire and Dowsett 1998).

Evergreen or semi-evergreen rain forest in Africa is the main, or one of the main, vegetation types of three phytogeographical regions (White 1983a): the Afromontane and Guineo-Congolian regions and the coastal belt of eastern Africa. All three are important centers of endemism for both plants and birds. This chapter describes patterns of species richness and distribution in these three forest zones, with particular emphasis on the Guineo-Congolian region, as several recent papers have already dealt at some length with the avifauna of the Afromontane and Eastern regions (Dowsett 1986; Dowsett-Lemaire 1989; Dowsett-Lemaire and Dowsett 1989, 1990, 1998). We also address briefly the taxonomic affinities of the species within and outside Africa and their conservation status.

In the maps in this chapter, localities are mainly plotted by the most accurate coordinates (latitude and longitude) available, but less exact scales (degree squares, 30' × 30' or 15' × 15' squares) are used for atlas data (published and unpublished) and for localities that cannot be traced precisely. The maps have been produced using the computer program DMAP.

### Biogeographical Concepts and Definitions

White (1983a) divided the African flora into eighteen geographically exclusive regional centers of endemism and transition zones (figure 15.1). As far as the forest avifauna is concerned, our treatment of the geographical coverage of the Guineo-Congolian region and Afromontane archipelago follows largely that defined by White. One exception is that the forests of the Angolan scarp and highlands are considered here as forming an integral, albeit impoverished, part of the Afromontane region, whereas White considered their inclusion as problematical and requiring further study. The lowland forests of eastern Africa were initially included in two "regional mosaics" (centers of low endemism): the Zanzibar-Inhambane region and, south of it, the Tongaland-Pondaland region (see figure 15.1). Further analysis of the forest flora of eastern Africa, including the lowland forests of Malawi, has prompted Dowsett-Lemaire and White (in Dowsett-Lemaire et al., in press) to combine and extend these two regional mosaics into an "Eastern" archipelago-like region that encompasses small forest enclaves along the eastern side of Africa from Ethiopia to the eastern Cape, extending westward into the Somalia-Masai region, the eastern half of the Zambezian region, and the eastern fringes of the Sudanian, Lake Victoria Basin, and Kalahari-Highveld regions. The cumulative distributions of Eastern forest birds correspond well with this newly defined Eastern region.

Each of these three forest zones—the Guineo-Congolian, the Afromontane, and the Eastern—is characterized by a number of bird species strictly confined to it, that is, true endemics. It is also necessary to recognize the concept of a near-endemic species, one that occurs predominantly in one floristic region but extends marginally into neighboring areas or more distant satellite populations. Thus forest birds that are predominantly Guineo-Congolian but extend into forest enclaves of the neighboring transition zones—east to the Guineo-Congolian enclaves of Uganda and western Kenya and south to similar forest islands in Angola, southern DR Congo, and northwestern Zambia (Mwinilunga)—are near-endemics. Similarly, Afromontane endemic bird species are strictly confined to the Afromontane massifs—with a lower altitudinal limit that varies with latitude and the proximity of the cooling effects of ocean currents. Afromontane near-

endemics designate those species that are predominantly montane but can also be found in small forest outliers at intermediate elevations. The additional category of sub-Afromontane is included here for those species that occur at medium elevations on some mountains or escarpments but are normally absent from lowland forests. Eastern near-endemics are those that extend slightly further west of the Eastern archipelago, for example, into western Zambia, Uganda, or the Cape.

### The Endemic Avifauna of the Three Forest Regions: Species Richness and Distribution Patterns

Appendixes 15.1–15.3 present the list of bird species confined (or nearly so) to each of the three forest regions. Nomenclature follows Dowsett and Forbes-Watson 1993, except that *Bleda eximia* and *notata* are now considered separate species (Chappuis and Erard 1993). A different taxonomic treatment and improved knowledge of the range of some local species would no doubt change the status of some species, but the overall trend is clear.

Almost all stenotopic forest birds in the Afrotropics belong to one or more of the three forest regions defined above. The few exceptions include mainly species of the transition zones around the Guineo-Congolian region: *Cossypha heinrichi* has a small range in the western Guineo-Congolian/Zambezian transition zone; *Ptyrticus turdinus* is more widespread, in both the Guineo-Congolian/Zambezian and Guineo-Congolian/Sudanian transition zones; and *Cossypha polioptera* is in both of these and also the lower Victoria Basin. A bulbul of coastal forest and thicket, *Phyllastrephus fulviventris*, has a somewhat anomalous distribution, centered on Angola, which crosses three phytochoria—Zambezian, Guineo-Congolian/Zambezian, and Guineo-Congolian—with a recent record from southern Gabon (Sargeant 1993).

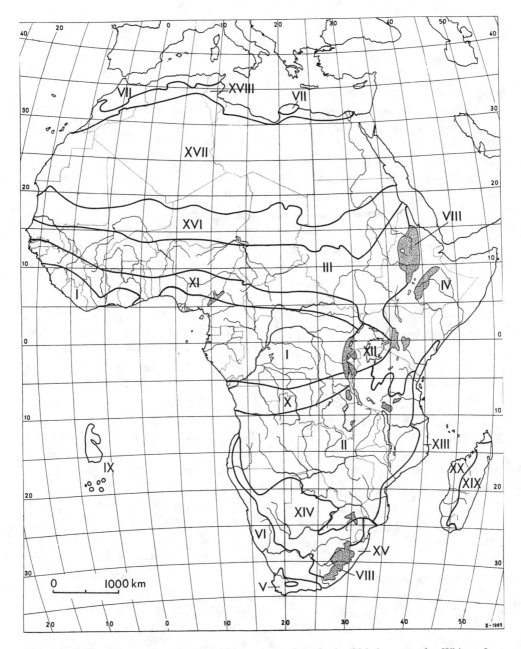

Figure 15.1. The eighteen phytochoria of Africa and two phytochoria of Madagascar, after White 1983a. Those relevant to this chapter are: (I) Guineo-Congolian center of endemism; (II) Zambezian center of endemism; (III) Sudanian center of endemism; (IV) Somalia-Masai center of endemism; (VIII) Afromontane center of endemism; (XI and X) Guineo-Congolian/Sudanian and Guineo-Congolian/Zambezian transition zones; (XII) Lake Victoria Basin regional mosaic; (XIII and XV) Zanzibar-Inhambane and Tongaland-Pondaland regional mosaics; (XIV) Kalahari-Highveld transition zone. For other zones, see the caption to figure 1.2.

## THE EASTERN FOREST AVIFAUNA

The Eastern forest avifauna (forty species) is very impoverished, in an area of extreme habitat fragmentation and relative isolation from the great block of Guineo-Congolian forests. Several species of narrow ecological niche in lowland forest are limited to one or a few localities only (for example, *Otus irenae* and *Anthus sokokensis* are both exclusively coastal in Kenya and northern Tanzania). Some ascend the mountains to medium altitudes but may be limited in range through specialized feeding habits (like the barbet *Stactolaema olivacea*, a fig specialist common in only some of the fig-dominated forests; see figure 3 in Dowsett-Lemaire and Dowsett 1998). By contrast, some species that occur also in scrub forest or thicket and under open canopy are reasonably widespread (for example, *Trochocercus cyanomelas* and *Nicator gularis*; see maps in Hall and Moreau 1970).

## THE AFROMONTANE FOREST AVIFAUNA

The Afromontane forest avifauna (166 species) is considerably more diverse than the Eastern and occupies a much wider range of forest islands, scattered from Ethiopia to the Cape, with the most isolated groups of mountains in Cameroon and Angola. The mountains of eastern Africa can also be subdivided into five regional systems based on patterns of local speciation (Dowsett 1986; Dowsett-Lemaire 1989). Not surprisingly, the most isolated mountain system of Cameroon and Nigeria has the highest percentage of local endemism, nearly 50% (table 15.1). However, several of the local endemics—*Phyllastrephus poliocephalus, Andropadus montanus, A. tephrolaemus, Cossypha isabellae, Kupeornis gilberti, Nesocharis shelleyi*—have their closest relatives among east African montane birds, and in fact some of them have in the past been treated as conspecific with their east African vicariants.

A striking feature of the Afromontane avifauna (and flora) is indeed its relative uniformity across the continent. Many species are present in two or more regional mountain systems; the distribution of a widespread passerine species (the babbler *Pseudoalcippe abyssinica*) is shown in figure 15.2. Twenty-seven species of the Cameroon group are found again across the Congo Basin in the mountains of the Albertine Rift some 2,000 km distant (see table 15.1).

Diamond and Hamilton's statement (1980: 379) that "there is no evidence of past connections between currently isolated montane forests" is not supported by the facts (see, e.g., Elenga and Vincens 1990). Several essentially Afromontane species of trees and birds survive today in small satellite populations that provide stepping stones between different mountain ranges. Of the twenty-seven montane birds found in both Cameroon and eastern DR Congo, nine are also found in relict populations away from the mountains (see table 15.1), mainly in mid-altitude forest along the DR Congo–Zambezi watershed and on some hills in Lower Guinea between Cameroon and Angola. These show a connecting route very similar to the "southern migratory track" of several Afromontane trees (White 1981, 1983b). One of these species, the warbler *Apalis jacksoni*, also survives in a few places along a northern route (figure 15.3). The small population centered on the plateau around the Cameroon-Gabon-Congo border is found very locally in tall-canopy forest with a cool microclimate (as indicated by the abundance of *Usnea* lichens). It is racially distinct from the birds inhabiting the high mountains of Cameroon and Nigeria (race *bambuluensis*) and apparently identical to the populations of the east African mountains. The isolated Angolan birds are slightly different: they are presently known from mid-altitude forest in the north (another connecting route?) and not from the higher mountains of the south. Connecting routes may have been used more than once in the past, and much remains to be discovered, as several important massifs of intermediate altitude in Gabon and Congo remain to be explored, especially the Chaillu and Mounts de Cristal (see also Dowsett-Lemaire and Dowsett 1996).

### THE GUINEO-CONGOLIAN FOREST AVIFAUNA

Finally, the Guineo-Congolian forest avifauna is the most species rich, with 250 endemic and near-endemic species (appendix 15.3). This total differs from the 216 given by Louette (1990)—which is perhaps a slip of the pen for 235 (the total of species listed in his four tables). On one hand, Louette includes some non-forest birds (e.g., *Cisticola anonymus, Estrilda atr-*

*icapilla*) and others that are not confined to the Guineo-Congolian region (e.g., *Bostrychia olivacea, Pitta angolensis*). On the other hand, he seems to have overlooked some species that ought to be included (e.g., *Francolinus ahantensis, Sarothrura pulchra*).

Today this vast forest region is naturally separated into two parts by the drier Dahomey Gap of savanna vegetation, but nevertheless the majority of Guineo-Congolian species (144, or 57%) occur throughout, from Upper Guinea to the Congo Basin, which suggests that the Dahomey Gap has not always been deforested.

The distribution of more localized species can conveniently be analyzed by subdividing

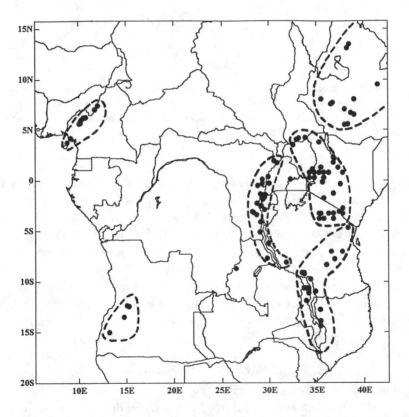

Figure 15.2. Distribution of the Afromontane endemic babbler *Pseudoalcippe abyssinica*, present in six of the seven regional mountain systems (Cameroon/Nigeria, Angola, Ethiopia, Sudan/Kenya, Albertine Rift, and Tanzania/Malawi).

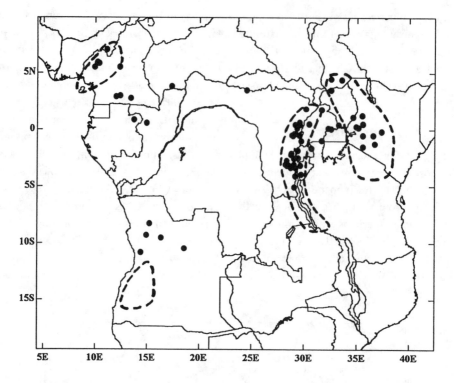

Figure 15.3. Distribution of the Afromontane near-endemic warbler *Apalis jacksoni*, a species abundant in the high montane forests of the Albertine Rift and Sudan-Kenya systems, more local in the Cameroon-Nigeria system, and with satellite populations at medium altitudes in northern Angola, in the lower Victoria Basin, and on the "northern connecting route."

the regions into three zones: Upper Guinea, Lower Guinea, and the Congo Basin (or Congolia) *sensu* White (1979) rather than Moreau (1966), whose "Lower Guinea" included the Congo Basin. However, the limit between Upper and Lower Guinea is not clear cut. Some Upper Guinea species go no further east than the Dahomey Gap (e.g., *Agelastes meleagrides, Scotopelia ussheri, Campethera maculosa, Bathmocercus cerviniventris, Prinia leontica, Apalis sharpii, Illadopsis rufescens, Picathartes gymnocephalus, Lamprotornis cupreocauda*). An example of this pattern of distribution is given by the two rockfowls *Picathartes* spp. (figure 15.4). However, a substantial number of Upper Guinea forest species extend to Nigeria in the surroundings of the Nigeria–western Cam-

eroon border (e.g., *Ceratogymna elata, Thripias pyrrhogaster, Bleda canicapilla, Macrosphenus kempi, Diaplorophyia blissetti, Nectarinia adelberti, Malimbus scutatus*). The example of two *Malimbus* spp. is shown in figure 15.5.

Similarly, the limits between Lower Guinea and Congolia are near the Gabon-Congo watershed for some species (*Batis minima, Dyaphorophyia chalybea, Picathartes oreas*, and *Malimbus racheliae* have still not been found east of this watershed during recent fieldwork in northern Congo and southern Central African Republic by P. Christy and ourselves) but are a little further east, near the Sangha or Ubangui Rivers, for other species (e.g., within the genera *Tauraco* and *Illadopsis*). Two green turacos of Upper and Lower Guinea drop out, respec-

Table 15.1

## Afromontane Bird Species Found in the Forests of the Cameroon and Nigerian Highlands

| LOCAL ENDEMICS | PRESENT IN WEST AND EAST AFRICAN MOUNTAINS |
|---|---|
| Francolinus camerunensis | Columba arquatrix* |
| Tauraco bannermani | C. albinucha |
| Psalidoprocne fuliginosa | Aplopelia larvata* |
| Andropadus montanus | Apaloderma vittatum |
| A. tephrolaemus | Pogoniulus coryphaeus |
| Phyllastrephus npoensis | Campethera tullbergi |
| P. poliocephalus | Coracina caesia |
| Cossypha isabellae | Zoothera crossleyi* |
| Bradypterus bangwaensis | Sheppardia bocagei* |
| Phylloscopus herberti | Cossyphicula roberti |
| Urolais epichlora | Bradypterus lopezi* |
| Apalis bamendae | Apalis jacksoni* |
| Poliolais lopezi | A. pulchra |
| Kupeornis gilberti | A. cinerea* |
| Nectarinia ursulae | Elminia albiventris |
| N. oritis | Kakamega poliothorax |
| Speirops lugubris | Pseudoalcippe abyssinica |
| Laniarius atroflavus | Nectarinia preussi |
| Malaconotus kupeensis | Dryoscopus angolensis* |
| M. gladiator | Laniarius fuelleborni |
| Ploceus bannermani | Onychognathus walleri |
| Nesocharis shelleyi | Ploceus melanogaster |
| | P. insignis* |
| | Cryptospiza reichenovii |
| | Serinus burtoni |
| | Linurgus olivaceus |

Note: An asterisk indicates Afromontane near-endemics with satellite populations (or migrants: Columba arquatrix) recorded away from the mountains. In addition, the swift Schoutedenapus myoptilus is widespread in all the mountain groups but in the Cameroon group is restricted to Bioko.

tively, just west of the Sangha River in Congo (Tauraco macrorhynchus, pers. obs.) and just east of the Ubangui in extreme northwest DR Congo (T. persa, Snow and Louette 1981), to be replaced by T. schuetti in the Congo Basin. The babbler Illadopsis cleaveri is replaced on the other side of the Ubangui and Congo Rivers by its close relative I. albipectus (figure 15.6).

Of this total of 250 Guineo-Congolian forest species, many are found in all or at least two of the three subregions defined above, and each subregion is about as species-rich as the others: Upper Guinea (west of the Dahomey Gap) has 183 forest species, Lower Guinea has 202, and Congolia has 206 (see appendix 15.3). But the number of local endemics is very uneven: Upper Guinea has twenty-nine endemics and near-endemics (that is, those that extend east to Nigeria), Lower Guinea has seven (two of which, Chlorocichla falkensteini and Dyaphorophyia chalybea, reach Angola and are more appropriately termed near-endemics), and Congolia has sixteen (of which eight extend into the Guineo-Congolian enclaves of the lower Victoria Basin: Francolinus nahani, Tauraco schuetti, Phyllastrephus hypochloris, Zoothera oberlaenderi, Eremomela turneri, Diaphorophyia jamesoni, Illadopsis albipectus, Ploceus weynsii). Lower Guinea and Congolia jointly have sixty-eight species that are missing from Upper Guinea. Far fewer species (seven) are shared between Upper Guinea and Lower Guinea while being absent from Congolia: Tauraco macrorhynchus, T. persa, Gymnobucco calvus, G. peli, Batis poensis, Illadopsis cleaveri, and Anthreptes gabonicus.

More surprisingly, two passerine species are common to Upper Guinea and the Congo Basin but missing from Lower Guinea: Erythropygia leucosticta and Parmoptila rubrifrons (Hall and Moreau 1970). Whereas the latter is replaced in Lower Guinea by a close relative, P. woodhousei, the Erythropygia's niche there apparently remains empty. Another pair of species, the forest coucals Centropus anselli and

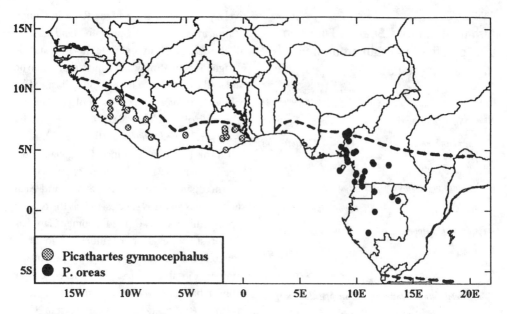

Figure 15.4. An example of two vicariant species of the Guineo–Congolian forest region that replace each other on opposite sides of the Dahomey Gap.

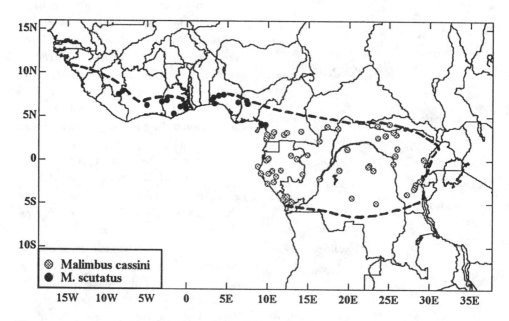

Figure 15.5. An example of two vicariant species of the Guineo–Congolian forest region that replace each other east of the Dahomey Gap, in western Cameroon.

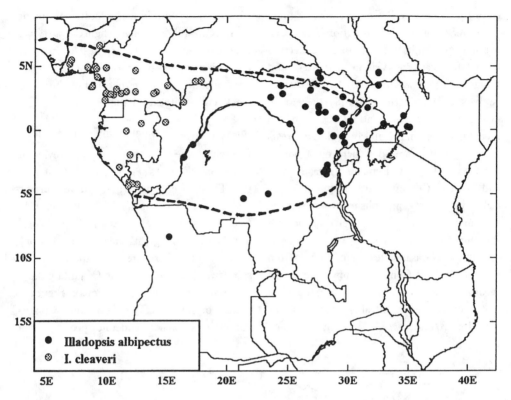

Figure 15.6. An example of two vicariant species of the Guineo–Congolian forest region that replace each other on opposite sides of the Ubangui and Zaire Rivers.

*C. leucogaster*, shows a distribution pattern (Snow 1978) not very dissimilar to that of the *Parmoptila. Centropus leucogaster* is the only species in Upper Guinea east to Nigeria, but further east it is confined to the northern and northeastern rims of the forest block, with *C. anselli* occupying the southern and central sections, south to northern Angola.

These somewhat loose boundaries between subcenters of endemism suggest that the Guineo-Congolian forest block has had a complex history of forest contraction and extension; moreover, some present limits may be determined more by ecological competition in areas of secondary contact between vicariant species than by natural barriers. The area of the Korup Forest, where some Upper Guinea

near-endemic species reach their eastern limits today, is not a natural barrier. Until recent deforestation by the human population, the main natural gaps further south were in the form of large rivers, and from our personal observations in Congo (on the Kouilou, Mambili, and Sangha Rivers, among others), these are readily crossed by forest birds of a large variety of species, even apparently weak-flying *Tauraco* spp.

The distribution maps published in the atlases of Hall and Moreau (1970) and Snow (1978) also show a discontinuous range between the western section of Lower Guinea and the eastern section of the Congo Basin for more than forty species. However, Louette (1984) rightly pointed out that many specimens from west-central DR Congo (most of them listed in Schouteden's publications) had been overlooked. Similarly, an important collection from northeastern Gabon (Brosset and Dragesco 1967) was ignored. In the case of twelve species (Louette 1984, 1987) there are in fact records from west-central DR Congo, and recent fieldwork in the poorly explored forests of southeastern Cameroon (pers. obs.), northern Congo (Dowsett-Lemaire 1997a, 1997b), and southern Central African Republic (especially by P. Christy, unpublished) has closed the gap further for these same twelve species. Of the thirty or so species for which Louette (1984:table 2) was unable to trace records from west-central DR Congo, we now have recent records for twenty-one of them from northern Congo or southern Central African Republic, including *Apaloderma aequatoriale*, for which

Louette (1987) did in fact relocate specimens from central DR Congo (figure 15.7). We believe that more fieldwork in west-central DR Congo is necessary before deciding that apparent distribution gaps in that part of the forest block reflect the real situation.

### Forest Avifauna of Wider Distribution in Africa

In each of the three forest regions reviewed above, one also finds species of wider distribution. These occur in two or more floristic regions and can be qualified as "pluriregional" (see Dowsett-Lemaire 1989). Their local importance varies greatly: in the Guineo-Congolian region there are only approximately sixty-two species of this kind (including those of secondary forest formations); they are greatly outnumbered by the Guineo-Congolian endemic or near-endemic species, which

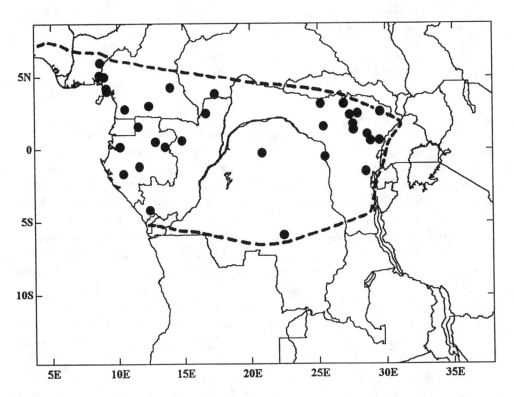

Figure 15.7. Continuous distribution of the trogon *Apaloderma aequatoriale* from Lower Guinea to the Congo Basin.

form about 80% of the local forest avifauna. Some of these pluriregional species are stenotopic forest birds not found outside the three regions of forest endemism (table 15.2). The Guineo-Congolian/Afromontane elements are those that extend beyond the Guineo-Congolian region into the montane forests of eastern Africa. The Guineo-Congolian/Eastern elements are rather fewer and are usually represented in east Africa by relatively isolated populations that are subspecifically distinct. The majority of pluriregional species, however, are less specialized ecologically and can also be found in rich woodland or thicket in several floristic regions: for example, some raptors (*Accipiter tachiro*), cuckoos (*Cuculus solitarius, C. clamosus, Chrysococcyx cupreus*), and owls (*Strix woodfordii*), among others, are equally common in primary Guineo-Congolian rain forest and in other habitats outside the region. It is of note that two of the three most numerous passerines of forest understory in the Guineo-Congolian region, *Andropadus virens* and *Nectarinia olivacea* (Brosset and Erard 1986; Dowsett-Lemaire and Dowsett 1991; Dowsett-Lemaire 1997a), extend not only to forest patches in the Eastern and Afromontane

regions but also to drier evergreen forest and thicket in the Zambezian region, and thus have very wide ranges (Hall and Moreau 1970).

In the Afromontane region, the local forest endemic and near-endemic species represent approximately 50% of the avifauna overall, but that proportion naturally increases with altitude on individual mountains (see Dowsett-Lemaire 1989). The Eastern forest region is much poorer in endemics, and the proportion of non-endemic species is greater: in lowland forests of Malawi and Tanzania, up to 80% or more of the species are pluriregional elements, many with a wide ecological range—exactly the reverse of what is observed in the Guineo-Congolian region. While several Eastern endemics ascend the mountains to moderate altitudes, two species are represented by lowland and montane populations in different parts of their range and are more appropriately qualified as Eastern/Afromontane elements (see table 15.2).

### Taxonomic Affinities of the Forest Avifauna

Dowsett (1986) has written in detail on the intra-African and extralimital relationships of the Afromontane avifauna. Most species have

Table 15.2

## Stenotopic Forest Birds Found in Two of the Three Forest Regions of Endemism

| GUINEO-CONGOLIAN/AFROMONTANE | GUINEO-CONGOLIAN/EASTERN | EASTERN/AFROMONTANE |
|---|---|---|
| *Bostrychia olivacea* | *Bubo poensis* | *Cercococcyx montanus* |
| *Francolinus squamatus* | *Neocossyphus rufus* | *Phylastrephus flavostriatus* |
| *Poicephalus gulielmi* | *Illadopsis rufipennis* | |
| *Phoeniculus bollei* | *Spermophaga ruficapilla* | |
| *Gymnobucco bonapartei* | | |
| *Andropadus gracilirostris* | | |
| *A. latirostris* | | |
| *Alethe poliocephala* | | |
| *Prinia bairdii* | | |
| *Nigrita canicapilla* | | |

their closest relatives in the African lowlands, while some have no obvious affinities. Outside Africa, affinities lie with the Palearctic for several genera (e.g., *Buteo, Turdus, Phylloscopus*) and with the Oriental region for others (e.g., the closest ally of the Itombwe Owl *Phodilus prigoginei* is *P. badius,* of southeast Asia). The broadbill *Pseudocalyptomena graueri,* also endemic to the Albertine Rift, is monotypic, but its nearest allies are also in the Oriental region, as is also probably the case for several genera of montane babblers (*Pseudoalcippe, Kupeornis, Lioptilus*) and some warblers (*Hemitesia, Orthotomus*). The recently discovered francolin of the Uzungwa Mountains in Tanzania, *Xenoperdix udzungwensis,* has its closest relatives in the hill-partridges of the Himalayas, *Arborophila* spp. (Dinesen et al. 1994).

The Eastern forest species are generally related to Afrotropical species elsewhere in the lowlands, in the Zambezian and especially Guineo-Congolian regions—genera *Tauraco, Otus, Campethera, Phyllastrephus, Sheppardia, Macrosphenus, Erythrocercus, Trochocercus, Batis,* and *Oriolus.* What is striking is the lack of endemic genera, whereas the Guineo-Congolian forest avifauna has thirty-two, covering fifty-eight species (table 15.3), and the Afromontane forest avifauna has twenty-two, covering thirty-four species.

There are few Guineo-Congolian birds that seem to have close relatives outside the Afrotropics, the most obvious exceptions being *Tigriornis* (apparently related to the Neotropical *Tigrisoma*), *Canirallus* (the one other member of this genus is in Madagascar), *Afropavo* (the peafowl *Pavo* spp. of the Oriental region), and *Sasia* (with Asian congeners).

Among the Guineo-Congolian genera with two or more species, the following are allopatric and are closely enough related to have been considered conspecific by some authorities: *Lobotos lobatus* and *L. oriolinus; Phyllastrephus baumanni* and *P. hypochloris; Bleda eximia* and *B.*

Table 15.3
### Genera Endemic to the Guineo-Congolian Forest Region

| FAMILY | GENERA (NUMBER OF SPECIES) |
|---|---|
| Ardeidae | *Tigriornis* (1) |
| Anatidae | *Pteronetta* (1) |
| Accipitridae | *Dryotriorchis* (1), *Urotriorchis* (1) |
| Phasianidae | *Afropavo* (1) |
| Numididae | *Agelastes* (2) |
| Rallidae | *Himantornis* (1) |
| Psittacidae | *Psittacus* (1) |
| Musophagidae | *Corythaeola* (1) |
| Strigidae | *Jubula* (1) |
| Bucerotidae | *Tropicranus* (1) |
| Lybiidae | *Buccanodon* (1) |
| Indicatoridae | *Melignomon* (2), *Melichneutes* (1) |
| Campephagidae | *Lobotos* (2) |
| Pycnonotidae | *Calyptocichla* (1), *Baeopogon* (2), *Ixonotus* (1), *Thescelocichla* (1), *Pyrrhurus* (1), *Bleda* (3), *Criniger* (4) |
| Turdidae | *Stizorhina* (1), *Stiphrornis* (1) |
| Muscicapidae | *Fraseria* (2) |
| Platysteiridae | *Megabyas* (1), *Dyaphorophyia* (6) |
| Timaliidae | *Phyllanthus* (1), *Picathartes* (2; incertae sedis) |
| Remizidae | *Pholidornis* (1; incertae sedis) |
| Ploceidae | *Malimbus* (all 10 species) |
| Estrildidae | *Parmoptila* (2) |

*binotata; Criniger barbatus* and *C. chloronotus; Prinia leucopogon* and *P. leontica; Dyaphorophyia blissetti, D. jamesoni,* and *D. chalybea; Picathartes gymnocephalus* and *P. oreas.* These can usefully be considered to form superspecies.

Variation at the infra-specific level is limited to a few fairly well marked subspecies (based on plumage colors rather than size, as a rule), and these are generally far removed geographically (e.g., between Upper and Lower Guinea). Louette (1991) examined some small samples of a number of species and found that what little recordable variation there was in bill- or wing-length varied independently of geography. Examination of rather larger samples of weights and wing-lengths of birds caught for ringing at a few scattered localities shows, too, that the few quantifiable differences seem to present no meaningful pattern. A few examples are given in table 15.4, based on our own data from southern and northern Congo (Kouilou and Odzala) and northern Zambia, those of Bowden (1986) from Cameroon, and of Mann (1985) from Kakamega in western Kenya. The only important difference shown between samples in table 15.4 concerns the remarkably small size of the *Alethe poliocephala* of Kakamega forest in western Kenya. Mann's (1985) data refer to the short-winged form *carruthersi*, the Cameroon and Congo birds to the nominate race. This variation in size is not shown by any other species in the region, and surprisingly, morphological variation provides little support for the existence of speciation processes.

## Conservation Status of the Afrotropical Forest Avifaunas

The Eastern lowland forests and several fragments of montane forest (especially in eastern Tanzania, southern Malawi, and parts of the Cameroon range) are by far the most threatened. Not very extensive to start with, they are now coming under increasing pressure for land in areas of high population density. As regards the great Guineo-Congolian forest blocks, the Upper Guinea zone has already lost most of its original forest, and all of it (without exception) is under very severe pressure. The eastern fringes of the Congo Basin are also suffering from much habitat destruction, and the recent tumultuous political upheavals around the Great Lakes can only make matters worse. It is true, therefore, that a great many of the endemic forest birds of these regions are in danger of at least local extinctions in the short to medium term. Table 15.5 lists, for the three regions considered here (Afromontane, Guineo-Congolian, and Eastern), those forest-associated species that have been listed as threatened or near-threatened (Collar and Andrew 1988; Collar et al. 1994), together with their classification in the Red Data Book (Collar and Stuart 1985).

We exclude from separate treatment here all forest birds that are confined to Indian Ocean or other islands, and also the following, which are treated by some authors as species but which we believe to be no more than subspecifically distinct:

- Threatened: *Bubo poensis vosseleri, Glaucidium capense albertinum, Stactolaema olivacea woodwardi, Turdus olivaceus helleri* and *T. o. ludoviciae, Zoothera camaronensis kibalensis, Sylvietta leucophrys chapini, Apalis rufogularis argentea, A. thoracica fuscigularis, A. t. lynesi, Apalis porphyrolaema kaboboensis, Platysteira peltata laticincta, Nectarinia afra prigoginei, Zosterops poliogaster kulalensis, Z. p. silvanus, Z. p. winifredae, Laniarius luehderi brauni,* and *L. l. amboimensis.*

- Near-threatened: *Zoothera piaggiae tanganjicae* and *Nectarinia loveridgei moreaui.*

The Red Data Book qualifications (see table 15.5) largely reflect our opinion that many species are more usefully classified as apparently rare or little known than as threatened. Much fieldwork is still needed in several forested regions not explored in recent years (such as the Angolan scarp and highlands, and the Itombwe massif of eastern DR Congo) to investigate the status of several little-known species. At the same time, conservation efforts must be concentrated on those relatively few species or regions that are clearly under threat at present.

By contrast, the forests of Lower Guinea and most of the Congo Basin and their avifau-

Table 15.4

## Some Comparative Morphological Measurements of Populations of Certain Forest Birds Found in the Guineo-Congolian Region (Not All Endemic)

| SPECIES | LOCATION[a] | WING-LENGTH | | WEIGHT | |
| | | N | MEAN (RANGE) (MM) | N[b] | MEAN (RANGE) (G) |
|---|---|---|---|---|---|
| *Alcedo leucogaster* | Cam. | 13 | 58.8 (56–61) | | 17.8 (15.5–20.3) |
| | Kou. | 20 | 56.4 (53–58) | | 14.2 |
| *Andropadus latirostris* | Cam. | 71 | 81.0 (75–89) | | 28.3 (22.3–34.2) |
| | Kou. | 107 | 82.1 (74.5–89) | | 27.6 (21.6–34.1) |
| | Odz. | 81 | 81.1 (75–90) | 15 | 27.1 (22.8–30) |
| | Kak. | 481 | 83.7 | | 28.8 |
| *Andropadus virens* | Cam. | 19 | 73.3 (68–82) | | 22.8 (19.1–27) |
| | Kou. | 132 | 73.9 (68–81) | | 22.3 (18–27) |
| | Odz. | 77 | 73.8 (68–79) | 16 | 23.9 (20.5–28.9) |
| | Kak. | 54 | 77.9 | | 24.4 |
| | Zam. | 169 | | | 24.6 |
| *Bleda eximia* | Kou. | 55 | 90.7 (81–97.5) | | 33.4 (28.6–40.7) |
| | Odz. | 20 | 93.0 (84–103) | 10 | 35.4 (31.2–39.5) |
| *Bleda syndactyla* | Odz. | 10 | 109.7 (103–121) | | |
| | Kak. | 26 | 105.4 | | 47.7 |
| | Zam. | 12 | | | 47.2 (38.5–55) |
| *Phyllastrephus icterinus* | Kou. | 30 | 73.5 (66.5–79) | | 18.1 (15.4–22) |
| | Odz. | 19 | 71.6 (67–77) | | |
| *Alethe diademata* | Cam. | 39 | 89.2 (82–98) | | 32.3 (28.4–37.6) |
| | Kou. | 48 | 90.1 (82–97) | | 31.4 (24.5–36.7) |
| | Odz. | 20 | 89.9 (76–97) | 12 | 33.7 (30–41) |
| *Alethe poliocephala* | Cam. | 105 | 92.4 | | 37.0 |
| | Kou. | 16 | 90.5 (86–93.5) | | 33.7 (28.5–36.8) |
| | Odz. | 17 | 90.2 (87–96) | | |
| | Kak. | 55 | 67.6 | | 15.8 |
| *Hylia prasina* | Kou. | 15 | 62.4 (54–69) | | 12.6 (10.8–14.5) |
| | Odz. | 14 | 63.1 (58–70) | | |
| *Terpsiphone rufiventer* | Cam. | 34 | 77.4 (73–83) | | 16.3 (14–19.6) |
| | Kou. | 13 | 76.5 (69.5–80.5) | | 16.5 (14.3–20.7) |
| *Illadopsis cleaveri* | Cam. | 19 | 74.8 (68–80) | | 32.6 (28.3–36) |
| | Kou. | 15 | 75.2 (71–79) | | 30.8 (28–34) |
| | Odz. | 15 | 76.2 (70.5–81) | | |
| *Illadopsis fulvescens* | Kou. | 15 | 72.4 (66–79) | | 27.0 (21.5–33.7) |
| | Odz. | 16 | 73.3 (66–78) | | |
| | Kak. | 13 | 73.2 | | 31.4 |
| *Illadopsis rufipennis* | Cam. | 13 | 73 (67–77) | | 27.1 (23–30.6) |
| | Odz. | 10 | 71.7 (65–75) | | |
| | Kak. | 24 | 68.0 | | 22.4 |

Table 15.4 *(continued)*

| SPECIES | LOCATION[a] | N | WING-LENGTH MEAN (RANGE) (MM) | N[b] | WEIGHT MEAN (RANGE) (G) |
|---------|-------------|---|------------------------------|------|-------------------------|
| *Nectarinia olivacea* | Cam. (M)[c] | 19 | 63.4 (61–66) | | 11.3 (10–12.8) |
| | Cam. (F)[c] | 10 | 56.7 (55–60) | | 9.6 (7.8–11.5) |
| | Kou. | 149 | 60.0 (52.5–69) | | 10.1 (8.3–13.2) |
| | Odz. | 65 | 59.6 (54–66) | | |
| | Kak. | 42 | 63.0 | | 10.8 |
| | Zam. | | | 24 | 10.7 (9.4–12.6) |
| *Spermophaga haematina* | Kou. | 18 | 68.7 (67–72.5) | | 23.0 (20–26) |
| | Odz. | 15 | 70.7 (65–87) | | |

a   Cam. = Cameroon (Bowden 1986); Kou. = Kouilou, Southern Congo (Dowsett-Lemaire and Dowsett 1991);
    Odz. = Odzala, Northern Congo (Dowsett-Lemaire and Dowsett, unpublished data); Kak. = Kakamega, Kenya
    (Mann 1985); Zam. = northern Zambia (Dowsett, unpublished data).

b   Data given only if sample is not the same as for wing-lengths.

c   M = male; F = female

Table 15.5

# Chorological Status of Threatened and Near-Threatened Afrotropical Forest Birds

| SPECIES | CHOROLOGY | STATUS | RED DATA BOOK[a] |
|---------|-----------|--------|------------------|
| *Francolinus nahani* | Guineo-Congolian | Threatened | Rare |
| *F. harwoodi* | Afromontane | Threatened | [Not classified] |
| *F. griseostriatus* | Sub-Afromontane | Threatened | [Not classified] |
| *F. camerunensis* | Afromontane | Threatened | Rare |
| *F. swierstrai* | Afromontane | Threatened | Indeterminate |
| *F. ochropectus* | Afromontane | Threatened | Endangered |
| *Xenoperdix udzungwensis* | Afromontane | Threatened | [Not classified] |
| *Afropavo congensis* | Guineo-Congolian | Threatened | Of special concern |
| *Agelastes meleagrides* | Guineo-Congolian | Threatened | Endangered |
| *Columba albinucha* | Sub-Afromontane | Near-threatened | |
| *Tauraco bannermani* | Afromontane | Threatened | Endangered |
| *T. ruspolii* | Afromontane | Threatened | Rare |
| *Phodilus prigoginei* | Afromontane | Threatened | Indeterminate |
| *Otus ireneae* | Eastern | Threatened | Endangered |
| *Scotopelia ussheri* | Guineo-Congolian | Threatened | Insufficiently known |
| *Caprimulgus prigoginei* | Sub-Afromontane | Threatened | [Not classified] |
| *Melignomon eisentrauti* | Guineo-Congolian | Threatened | Insufficiently known |
| *Indicator pumilio* | Afromontane | Near-threatened | |

Table 15.5 (continued)

| SPECIES | CHOROLOGY | STATUS | RED DATA BOOK[a] |
|---|---|---|---|
| Pseudocalyptomena graueri | Afromontane | Threatened | Rare |
| Psalidoprocne fuliginosa | Afromontane | Near-threatened | |
| Anthus sokokensis | Eastern | Threatened | Vulnerable |
| Lobotos lobatus | Guineo-Congolian | Threatened | Vulnerable |
| Coracina graueri | Afromontane | Threatened | [Not classified] |
| Andropadus montanus | Afromontane | Near-threatened | |
| Chlorocichla prigoginei | Sub-Afromontane | Threatened | Vulnerable |
| Phyllastrephus lorenzi | Guineo-Congolian | Near-threatened | |
| P. leucolepis | Guineo-Congolian | Threatened | [Not classified] |
| P. poliocephalus | Afromontane | Near-threatened | |
| Bleda eximia | Guineo-Congolian | Threatened | [Not classified] |
| Criniger olivaceus[b] | Guineo-Congolian | Threatened | Vulnerable |
| Zoothera guttata | Afromontane | Threatened | Rare |
| Z. oberlaenderi | Guineo-Congolian | Threatened | Rare |
| Alethe choloensis | Afromontane | Threatened | Endangered |
| Swynnertonia swynnertoni | Afromontane | Threatened | Rare |
| Sheppardia gunningi | Eastern | Threatened | Rare |
| S. gabela | Sub-Afromontane | Threatened | Indeterminate |
| S. montana | Afromontane | Threatened | Rare |
| S. lowei | Afromontane | Threatened | Rare |
| Xenocopsychus ansorgei | Afromontane | Near-threatened | |
| Bradypterus alfredi | Sub-Afromontane | Near-threatened | |
| Bathmocercus cerviniventris | Guineo-Congolian | Threatened | [Not classified] |
| B. winifredae | Afromontane | Threatened | Rare |
| Eremomela turneri | Guineo-Congolian | Threatened | Rare |
| Macrosphenus pulitzeri | Sub-Afromontane | Threatened | Indeterminate |
| Prinia leontica | Guineo-Congolian | Threatened | (newly added) |
| Apalis ruddi | Eastern | Near-threatened | |
| A. chariessa | Eastern | Threatened | [Not classified] |
| A. bamendae | Afromontane | Threatened | (newly added) |
| A. chirindensis | Afromontane | Near-threatened | |
| A. moreaui | Afromontane | Threatened | Rare |
| Melaenornis annamarulae | Guineo-Congolian | Threatened | Indeterminate |
| Muscicapa lendu[c] | Afromontane | Threatened | Rare |
| Batis fratrum | Eastern | Near-threatened | |
| Terpsiphone bedfordi | Guineo-Congolian | Near-threatened | |
| Modulatrix orostruthus | Afromontane | Threatened | Rare |
| Illadopsis rufescens | Guineo-Congolian | Near-threatened | |

Table 15.5 *(continued)*

| SPECIES | CHOROLOGY | STATUS | RED DATA BOOK[a] |
|---|---|---|---|
| *Kupeornis gilberti* | Afromontane | Threatened | Rare |
| *K. rufocinctus* | Afromontane | Near-threatened | |
| *K. chapini* | Sub-Afromontane | Near-threatened | |
| *Picathartes gymnocephalus* | Guineo-Congolian | Threatened | Vulnerable |
| *P. oreas* | Guineo-Congolian | Threatened | Rare |
| *Anthreptes reichenowi* | Eastern | Near-threatened | |
| *A. neglectus* | Eastern | Near-threatened | |
| *A. pallidigaster* | Eastern | Threatened | Rare |
| *A. rubritorques* | Afromontane | Threatened | Rare |
| *Nectarinia ursulae* | Afromontane | Near-threatened | |
| *N. rufipennis* | Afromontane | Threatened | Rare |
| *N. neergardi* | Eastern | Near-threatened | |
| *N. loveridgei* | Afromontane | Near-threatened | |
| *N. rockefelleri* | Afromontane | Threatened | Rare |
| *Laniarius turatii* | Guineo-Congolian | Near-threatened | |
| *Malaconotus kupeensis* | Afromontane | Threatened | Indeterminate |
| *M. gladiator* | Afromontane | Threatened | Rare |
| *M. monteiri* | Sub-Afromontane | Threatened | Indeterminate |
| *M. alius* | Afromontane | Threatened | Rare |
| *prionops alberti* | Afromontane | Threatened | [Not classified] |
| *Lamprotornus cupreocauda* | Guineo-Congolian | Near-threatened | |
| *Cinnyricinclus femoralis* | Afromontane | Threatened | [Not classified] |
| *Ploceus bannermani* | Afromontane | Threatened | Vulnerable |
| *P. batesi* | Guineo-Congolian | Threatened | Rare |
| *P. golandi* | Eastern | Threatened | Endangered |
| *P. aureonucha* | Guineo-Congolian | Threatened | Indeterminate |
| *P. flavipes* | Guineo-Congolian | Threatened | Vulnerable |
| *P. nicolli* | Afromontane | Threatened | Rare |
| *Malimbus ibadanensis* | Guineo-Congolian | Threatened | Endangered |
| *M. ballmanni* | Guineo-Congolian | Threatened | Indeterminate |
| *Cryptospiza shelleyi* | Afromontane | Threatened | [Not classified] |
| *Hypargos margaritatus* | Eastern | Near-threatened | |

*Sources:* Data from Collar and Andrew 1988; Collar et al. 1994. Red Data Book classifications are from Collar and Stuart 1985.

a  Species marked "Not classified" are not catalogued in the Red Data Book but have been added by Collar et al. (1994), some as a result of a taxonomic split.

b  *Criniger olivaceus* is now treated as conspecific with the widespread and common form *ndussumensis* of Lower Guinea (Dowsett and Dowsett-Lemaire 1993).

c  Including *itombwensis*.

nas do not present any immediate cause for concern. This is in spite of extensive (but selective) logging in places and the opening of roads into hitherto inaccessible areas. Selective logging creates clearings in the forest that are essentially well tolerated by the bird communities. The real danger here is of increased access to hunting, and large-scale elimination of such mammals as elephants, monkeys, and some antelopes. Some of the larger birds suffer at the same time. Large birds are being hunted for food, for example in the Man and Biosphere reserve of Dimonika, in the Congolese Mayombe, where local extinctions of some hornbills, guineafowls, and other species are evident (Dowsett-Lemaire and Dowsett 1991). It is highly desirable that further exploration be carried out in those parts of the Guineo-Congolian region that have scarcely been visited in recent years. There is a huge area between the Ubangui River and the Ituri Forest that has seen few ornithologists since the courageous pioneering efforts of Boyd Alexander (1907), the ill-fated J. S. Jameson (1890), and the German explorers F. Bohndorff and G. A. Schweinfurth.

REFERENCES

Alexander, B. 1907. *From the Niger to the Nile*. 2 vols. Edward Arnold, London.

Bowden, C. G. R. 1986. The use of mist-netting for studying forest birds in Cameroon. Pages 130–174 in S. N. Stuart, ed. *Conservation of Cameroon Montane Forests*. ICBP, Cambridge, England.

Brosset, A., and J. Dragesco. 1967. Oiseaux collectés et observés dans le Haut-Ivindo. *Biol. Gabonica* 3:59–88.

Brosset, A., and C. Erard. 1986. *Les Oiseaux des Régions Forestières du Nord-ouest du Gabon, Vol. 1: Ecologie et Comportement des Espèces*. Soc. Natl. Protection de la Nature, Paris.

Chappuis, C., and C. Erard. 1993. Species limits in the genus *Bleda* Bonaparte, 1857 (Aves, Pycnonotidae). *Z. zool. Syst. Evolut.-forsch.* 31:280–299.

Collar, N. J., and P. Andrew. 1988. *Birds to Watch: The*

*ICBP World Check-List of Threatened Birds*. ICBP, Cambridge, England.

Collar, N. J., M. J. Crosby, and A. J. Stattersfield. 1994. *Birds to Watch, 2: The World Check-List of Threatened Birds*. BirdLife International, Cambridge, England.

Collar, N. J., and S. N. Stuart. 1985. *Threatened Birds of Africa and Related Islands: The ICBP/IUCN Red Data Book, Part 1*. ICBP, Cambridge, England.

Diamond, A. W., and A. C. Hamilton. 1980. The distribution of forest passerine birds and Quaternary climatic change in tropical Africa. *Journal of Zoology, London* 191:379–402.

Dinesen, L., T. Lehmberg, J. O. Svendsen, L. A. Hansen, and J. Fjeldså. 1994. A new genus and species of perdicine bird (Phasianidae, Perdicini) from Tanzania: A relict form with Indo-Malayan affinities. *Ibis* 136:2–11.

Dowsett, R. J. 1986. Origins of the high-altitude avifaunas of tropical Africa. Pages 557–585 in F. Vuilleumier and M. Monasterio, eds. *High Altitude Tropical Biogeography*. Oxford University Press, New York.

Dowsett, R. J., and F. Dowsett-Lemaire. 1993. Comments on the taxonomy of some Afrotropical bird species. *Tauraco Research Report* 5:323–389.

Dowsett, R. J., and A. D. Forbes-Watson. 1993. *Checklist of Birds of the Afrotropical and Malagasy Regions, Vol. 1: Species Limits and Distribution*. Tauraco Press, Liège, Belgium.

Dowsett-Lemaire, F. 1989. Ecological and biogeographical aspects of forest bird communities in Malawi. *Scopus* 13:1–80.

———. 1997a. The avifauna of Odzala National Park, northern Congo. *Tauraco Research Report* 6:15–48.

———. 1997b. The avifauna of Nouabale-Ndoki National Park, northern Congo. *Tauraco Research Report* 6:111–124.

Dowsett-Lemaire, F., and R. J. Dowsett. 1989. Zoogeography and taxonomic relationships of the forest birds of the Cameroon Afromontane region. *Tauraco Research Report* 1:48–56.

———. 1990. Zoogeography and taxonomic relationships of the forest birds of the Albertine Rift Afromontane Region. *Tauraco Research Report* 3:87–109.

———. 1991. The avifauna of the Kouilou basin in Congo. *Tauraco Research Report* 4:189–239.

———. 1996. Découverte de *Phylloscopus budongoensis* et autres espèces à caractère montagnard dans les forêts d'Odzala (Cuvette congolaise). *Alauda* 64:364–367.

———. 1998. Parallels between F. White's phytochoria

and avian zoochoria in tropical Africa. *Proceedings of the F. White Memorial Symposium* (Oxford, 1996). Royal Botanic Gardens, Kew.

Dowsett-Lemaire, F., F. White, and J. D. Chapman. In press. *Evergreen Forest Flora of Malawi*. Royal Botanic Gardens, Kew.

Elenga, H., and A. Vincens. 1990. Paléoenvironnements quaternaires récents des Plateaux Bateke (Congo): Etude palynologique des depôts de la dépression du bois de Bilanko. Pages 271–282 in R. Lanfranchi and D. Schwartz, eds. *Paysages Quaternaires de l'Afrique Centrale Atlantique*. ORSTOM, Paris.

Hall, B. P., and R. E. Moreau. 1970. *An Atlas of Speciation in African Passerine Birds*. British Museum (Natural History), London.

Jameson, J. S. 1890. *The Story of the Rear Column of the Emin Pasha Relief Expedition*. Pater, London.

Louette, M. 1984. Apparent range gaps in African forest birds. *Proceedings of the Pan-African Ornithological Congress* 5:275–286.

———. 1987. Additions and corrections to the avifauna of Zaire (1). *Bulletin of the British Ornithological Club* 107:137–143.

———. 1990. Distribution patterns in African lowland forest birds. Pages 237–247 in G. Peters and R. Hutterer, eds. *Vertebrates in the Tropics*. Museum Alexander Koenig, Bonn.

———. 1991. Geographical morphometric variation in birds of the lowland equatorial forest of Africa. *Acta XX Congr. Intern. Ornithol.* 1:475–482.

Mann, C. F. 1985. An avifaunal study in Kakamega Forest, Kenya, with particular reference to species diversity, weight and moult. *Ostrich* 56:236–262.

Moreau, R. E. 1966. *The Bird Faunas of Africa and Its Islands*. Academic Press, London.

Sargeant, D. 1993. *Gabon: A Birders Guide to Gabon, West Africa*. Published privately.

Snow, D. W., ed. 1978. *An Atlas of Speciation in African Non-Passerine Birds*. British Museum (Natural History), London.

Snow, D. W., and M. Louette. 1981. Atlas of speciation in African non-passerine birds—Addenda and Corrigenda 2. *Bulletin of the British Ornithological Club* 101:336–339.

White, F. 1979. The Guineo-Congolian Region and its relationships to other phytochoria. *Bulletin du Jardin Botanique National de Belgique* 49:11–55.

———. 1981. The history of the Afromontane archipelago and the scientific need for its conservation. *African*

*Journal of Ecology* 19:33–54.

———. 1983a. *The Vegetation of Africa*. UNESCO, Paris.

———. 1983b. Long-distance dispersal and the origins of the Afromontane flora. Pages 87–116 in K. Kubitzki, ed. *Dispersal and Distribution: An International Symposium*. Sonderbände des Naturwissenschaftlichen Vereins in Hamburg 7. P. Parey, Hamburg.

APPENDIX 15.1

## Forest Bird Species Endemic to the Eastern Region

**ACCIPITRIDAE**
Southern banded snake eagle *Circaetus fasciolatus*

**MUSOPHAGIDAE**
Livingstone turaco *Tauraco corythaix**
Knysna lourie *Tauraco corythaix**
Fischer's turaco *Tauraco fischeri*
Purple-crested lourie *Tauraco porphyreolophus*

**STRIGIDAE**
Sokoke scops owl *Otus irenae*

**ALCEDINIDAE**
Mangrove kingfisher *Halcyon senegaloides*

**BUCERORIDAE**
Silvery-cheeked hornbill *Bycanistes brevis*

**LYBIIDAE**
White-eared barbet *Stactolaema leucotis*
Green barbet *Stactolaema olivacea*
Green tinkerbird *pogoniulus simplex*
Brown-breasted barbet *Lybius melanopterus*

**PICIDAE**
Knysna woodpecker *Campethera notata**
Mombassa woodpecker *Campethera mombassica*

**MOTACILLIDAE**
Sokoke pipit *Anthus sokokensis*

**PYCNONOTIDAE**
Sombre bulbul *Andropadus importunus**
Gray-olive bulbul *Phyllastrephus cerviniventris**
Fischer's bulbul *Phyllastrephus cerviniventris**
Tiny greenbul *Phyllastrephus debilis*

**TURDIDAE**
Gunning's akalat *Sheppardia gunningi*
Brown scrub robin *Erythropygia signata*
Eastern bearded scrub robin *Erythropygia quadrivirgata**

**SYLVIIDAE**
Kretschmer's longbill *Macrosphenus kretschmeri*
Rudd's apalis *Apalis ruddi*
White-winged apalis *Apalis chariessa*
Black-headed apalis *Apalis melanocephala*

**PLATYSTEIRIDAE**
Forest batis *Batis mixta*
Woodward's batis *Batis fratrum*

**MONARCHIDAE**
Livingstone's flycatcher *Erythrocercus livingstonei**
Little yellow flycatcher *Erythrocercus holochlorus*
Blue-mantled flycatcher *Trochocercus cyanomelas**

**NECTARINIIDAE**
Plain-backed sunbird *Anthreptes reichenowi*
Uluguru violet-backed sunbird *Anthreptes neglectus*
Amani sunbird *Anthreptes pallidigaster*
Grey sunbird *Nectarinia veroxii*
Neergard's sunbird *Nectarinia neergardii*

**ORIOLIDAE**
Green-headed oriole *Oriolus chlorocephalus*

**MALACONOTIDAE**
White-throated nicator *Nicator gularis*

**PRIONOPIDAE**
Chestnut-fronted helmet shrike *Prionops scopifrons*

**STURNIDAE**
Black-bellied glossy starling *Lamprotornis corruscus*

**PLOCEIDAE**
Clarke's weaver *Ploceus golandi*

**ESTRILDIDAE**
Pink-throated twinspot *Hypargos margaritatus*

*Note:* Asterisks denote near-endemics. Some birds are not in forest throughout their range, but also in thicket (mainly *Andropadus importunus*, *Apalis ruddi*, *Erythrocercus livingstonii*, *Nectarinia veroxii*), riparian woodland and forest edges (*Lybius melanopterus*) and one is mainly in mangrove (*Halcyon senegaloides*).

APPENDIX 15.2

## Forest Bird Species Endemic to the Afromontane Region

**ACCIPITRIDAE**
Forest buzzard *Buteo orephilus*

**PHASIANIDAE**
Harwood's francolin *Francolinus harwoodi*
Grey-striped francolin *Francolinus griseostriantus* (S)
Chestnut-naped francolin *Francolinus castaneicollis*
Jackson's francolin *Francolinus jacksoni*
Handsome francolin *Francolinus nobilis*
Mount Cameroon francolin *Francolinus camerunensis*
Swierstra's francolin *Francolinus swierstrai*
Djibouti francolin *Francolinus ochropectus*
Uzungwa francolin *Xenoperdis udzungwensis*

**COLUMBIDAE**
Rameron pigeon *Columba arquatrix**
White-naped pigeon *Columba albinucha* (S)
Cinnamon dove *Aplopelia larvata**

**PSITTACIDAE**
Yellow-fronted parrot *Poicephalus flavifrons*

**MUSOPHAGIDAE**
Bannerman's turaco *Tauraco bannermani*
Hartlaub's turaco *Tauraco hartlaubi*
White-cheeked turaco *Tauraco leucotis*
Prince Ruspoli's turaco *Tauraco ruspolii* (S)
Ruwenzori turaco *Ruwenzorornis johnstoni*

**TYTONIDAE**
Congo bay owl *Phodilus prigoginei*

**STRIGIDAE**
Long-eared owl *Asio otus*

**CAPRIMULGIDAE**
Prigogine's nightjar *Caprimulgus prigoginei* (S)

**APODIDAE**
Scarce swift *Schoutedenapus myoptilus*
Schouteden's swift *Schoutedenpau schoutedeni*

**TROGONIDAE**
Bar-tailed trogon *Apaloderma vittatum*

**MEROPIDAE**
Cinnamon-breasted bee-eater *Merops oreobates*

**LYBIIDAE**
Moustached green tinkerbird *Pogoniulus leucomystax*
Western green tinkerbird *Pogoniulus coryphaeus*

## INDICATORIDAE
Dwarf honeyguide *Indicator pumilio*

## PICIDAE
Fine-banded woodpecker *Campethera tullbergi*
Abyssinian woodpecker *Dendropicos abyssinicus*
Olive woodpecker *Mesopicos griseocephalus**

## EURYLAIMIDAE
African green broadbill *Pseudocalyptomena graueri*

## HIRUNDINIDAE
Cameroon Mountain saw-wing *Psalidoprocne fuliginosa*

## CAMPEPHAGIDAE
Grey cuckoo-shrike *Coracina caesia**
Grauer's cuckoo-shrike *Coracina graueri*

## PYCNONOTIDAE
Cameroon montane greenbul *Andropadus montanus*
Shelley's greenbul *Andropadus masukuensis*
Western mountain greenbul *Andropadus tephrolaemus*
Eastern mountain greenbul *Andropadus nigriceps*
Striped-cheeked greenbul *Andropadus milanjensis*
Joyful greenbul *Chlorocichla laetissima**
Prigogine's greenbul *Chlorocichla prigoginei* (S)
Cabanis's greenbul *Phyllastrephus cabanisi**
Cameroon olive greenbul *Phyllastrephus poensis*
Grey-headed greenbul *Phyllastrephus poliocephalus*

## TURDIDAE
Olive thrush *Turdus olivaceus**
Spotted ground thrush *Zoothera guttata* (S)*
Orange thrush *Zoothera gurneyi*
Crossley's ground thrush *Zoothera crossleyi**
Abyssinian ground thrush *Zoothera piaggiae*
Red-throated alethe *Alethe poliophrys*
White-chested alethe *Alethe fuelleborni*
Cholo alethe *Alethe choloensis*
Starred robin *Pogonocichla stellata*
Swynnerton's robin *Swynnertonia swynnertoni*
Bocage's robin *Sheppardia bocagei* (S)*
Equatorial akalat *Sheppardia aequatorialis*
Sharpe's akalat *Sheppardia sharpei*
Gabela akalat *Sheppardia gabela* (S)
Usambara ground robin *Sheppardia montana*
Iringa ground robin *Sheppardia lowei*
White-bellied robin-chat *Cossyphicula roberti*
Olive-flanked robin *Cossypha anomala*
Archer's ground robin *Crossypha archeri*
Mountain robin *Cossypha isabellae*
Cape robin *Cossypha caffra**
Rüppell's robin-chat *Crossypha semirufa*
Chorister robin *Cossypha dichroa**

Angola cave chat *Xenocopsychus ansorgei**

## SYLVIIDAE
Barratt's warbler *Bradypterus barratti*
Bangwa forest warbler *Bradypterus bangwaensis*
Evergreen forest warbler *Bradypterus lopezi**
Bamboo warbler *Bradypterus alfredi* (S)*
Mrs. Moreau's warbler *Bathmocercus winifredae*
Mountain yellow warbler *Chloropeta similis*
White-browed crombec *Sylvietta leucophrys*
Neumann's short-tailed warbler *Hemitesia neumanni*
Pulitzer's longbill *Macrosphenus pulitzeri* (S)
Brown woodland warbler *Phylloscopus umbrovirens*
Yellow-throated warbler *Phylloscopus ruficapilla*
Red-faced woodland warbler *Phylloscopus laetus*
Uganda woodland warbler *Phylloscopus budongoensis* (S)*
Black-capped woodland warbler *Phylloscopus herberti*
Brown parisoma *Sylvia lugens*
Green longtail *Urolais epichlora*
Mountain masked apalis *Apalis personata*
Black-throated apalis *Apalis jacksoni**
Black-collared apalis *Apalis pulchra*
Bar-throated apalis *Apalis thoracica**
Grey apalis *Apalis cinerea**
Chestnut-throated apalis *Apalis porphyrolaema*
Chestnut-headed apalis *Apalis chapini*
Bamenda apalis *Apalis bamendae* (S)
Chirinda apalis *Apalis chirindensis*
Long-billed apalis *Apalis moreaui*
Red-capped forest warbler *Orthotomus metopias*
White-tailed warbler *Poliolais lopezi*
Grauer's warbler *Graueria vittata*

## MUSCICAPIDAE
Yellow-eyed black flycatcher *Melaenornis ardesiacus*
Chapin's flycatcher *Muscicapa lendu*

## PLATYSTEIRIDAE
Cape batis *Batis capensis**
Ruwenzori batis *Batis diops*

## MONARCHIDAE
White-tailed crested flycatcher *Elminia albonotata*
White-bellied flycatcher *Elminia albiventris*

## TIMALIIDAE
Spot-throat *Modulatrix stictigula*
Dappled mountain robin *Modulatrix orostruthus*
Grey-chested illadopsis *Kakamega poliothorax*
Mountain illadopsis *Illadopsis pyrrhoptera*
African hill babbler *Pseudoalcippe abyssinica*
Bush blackcap *Lioptilus nigricapillus*
White-throated mountain babbler *Kupeornis gilberti*

Red-collared babbler *Kupeornis rufocinctus*
Chapin's babbler *Kupeornis chapini* (S)
Abyssinian catbird *Parophasma galinieri*

**PARIDAE**
Stripe-breasted tit *Parus fasciiventer*
White-backed black tit *Parus leuconotus*

**NECTARINIIDAE**
Banded green sunbird *Anthreptes rubritorques*
Ursula's mouse-colored sunbird *Nectarinia ursulae*
Cameroon blue-headed sunbird *Nectarinia oritis*
Ruwenzori blue-headed sunbird *Nectarinia alinae*
Rufous-winged sunbird *Nectarinia rufipennis*
Northern double-collared sunbird *Nectarinia preussi*
Eastern double-collared sunbird *Nectarinia mediocris*
Loveridge's sunbird *Nectarinia loveridgei*
Rockefeller's sunbird *Nectarinia rockefelleri*
Regal sunbird *Nectarinia regia*
Tacazze sunbird *Nectarinia tacazze*
Purple-breasted sunbird *Nectarinia purpureiventris*

**ZOSTEROPIDAE**
Black-capped speirops *Speirops lugubris*
Montane white-eye *Zosterops poliogaster*

**ORIOLIDAE**
Montane oriole *Oriolus percivali*
Abyssinian black-headed oriole *Oriolus monacha*

**MALACONOTIDAE**
Pink-footed puffback *Dryoscopus angolensis**
Yellow-breasted boubou *Laniarius atroflavus*
Fülleborn's black boubou *Lantiarius fuilleborni*
Olive bush shrike *Malaconotus olivaceus**
Mount Kupé bush shrike *Malaconotus kupeensis*
Doherty's bush shrike *Malaconotus dohertyi*
Lagden's bush shrike *Malaconotus lagdeni**
Green-breasted bush shrike *Malaconotus gladiator*

Monteiro's bush shrike *Malaconotus monteiri* (S)
Uluguru bush shrike *Malaconotus alius*

**PRIONOPIDAE**
Yellow-crested helmet shrike *Prionops alberti*
Gabela helmet shrike *Prionops gabela* (S)

**STURNIDAE**
Stuhlmann's starling *Poeoptera stuhlmanni*
Kenrick's starling *Poeoptera kenricki*
Waller's red-winged starling *Onychognathus walleri*
Abbott's starling *Cinnyricinclus femoralis*
Sharpe's starling *Cinnyricinclus sharpii*

**PLOCEIDAE**
Bannerman's weaver *Ploceus bannermani*
Bertram's weaver *Ploceus bertrandi*
Black-billed weaver *Ploceus melanogaster*
Strange weaver *Ploceus alienus*
Brown-capped weaver *Ploceus insignis**
Usambara weaver *Ploceus nicolli*

**ESTRILDIDAE**
White-collared olive-back *Nesocharis ansorgei* (S)
Little olive-back *Nesocharis shelleyi*
Red-faced crimsonwing *Cryptospiza reichenovii*
Abyssinian crimsonwing *Cryptospiza salvadorii*
Dusky crimsonwing *Cryptospiza jacksoni*
Shelley's crimsonwing *Cryptospiza shelleyi*

**FRINGILLIDAE**
Forest canary *Serinus scotops*
Brown-rumped seed-eater *Serinus tristriatus*
Thick-billed seed-eater *Serinus burtoni*
Oriole finch *Linurgus olivaceus*

*Note:* Asterisks denote near-endemics; (S) denotes sub-Afromontane species.

Forest Bird Species Endemic to the Guineo–Congolian Region

| | | UG | LG | C |
|---|---|:---:|:---:|:---:|
| **ARDEIDAE** | | | | |
| White-crested tiger heron | *Tigriornis leucolopus* | x | x | x |
| **THRESKIORNITHIDAE** | | | | |
| Spot-breasted ibis | *Bostrychia rara** | x | x | x |
| **ANATIDAE** | | | | |
| Hartlaub's duck | *Pteronetta hartlaubii** | x | x | x |
| **ACCIPITRIDAE** | | | | |
| Congo serpent eagle | *Dryotriorchis spectabilis** | x | x | x |
| Western little sparrowhawk | *Accipiter erythropus* | x | x | x |
| Chestnut-flanked sparrowhawk | *Accipiter castanilius** | x | x | x |
| Long-tailed hawk | *Urotriorchis macrourus* | x | x | x |
| Cassin's hawk eagle | *Spizaetus africanus** | x | x | x |
| **PHASIANIDAE** | | | | |
| Forest francolin | *Francolinus lathami** | x | x | x |
| Nahan's francolin | *Francolinus nahani** | — | — | x |
| Ahanta francolin | *Francolinus ahantensis** | x | — | — |
| Congo peacock | *Afropavo congensis* | — | x | — |
| **NUMIDIDAE** | | | | |
| White-breasted guineafowl | *Agelastes meleagrides* | x | — | — |
| Black guineafowl | *Agelastes niger* | — | x | x |
| Plumed guineafowl | *Guttera plumifera* | — | x | x |
| **RALLIDAE** | | | | |
| Nkulengu rail | *Himantornis haematopus* | x | x | x |
| White-spotted flufftail | *Sarothrura pulchra** | x | x | x |
| Grey-throated rail | *Canirallus oculeus* | *x* | *x* | *x* |
| **COLUMBIDAE** | | | | |
| Afep pigeon | *Columba unicincta** | x | x | x |
| Blue-headed wood dove | *Turtur brehmeri* | x | x | x |
| **PSITTACIDAE** | | | | |
| African gray parrot | *Psittacus erithacus** | x | x | x |
| Black-collared lovebird | *Agapornis swindernianus* | x | x | x |
| **MUSOPHAGIDAE** | | | | |
| Guinea turaco | *Tauraco persa** | x | x | — |
| Black-billed turaco | *Tauraco schuetti** | — | — | x |

| | | UG | LG | C |
|---|---|---|---|---|
| Verreaux's turaco | *Tauraco macrorhynchus* | x | x | — |
| Great blue turaco | *Corythaeola cristata** | x | x | x |
| **CUCULIDAE** | | | | |
| Dusky long-tailed cuckoo | *Cercococcyx mechowi** | x | x | x |
| Olive long-tailed cuckoo | *Cercococcyx olivinus** | x | x | x |
| Yellow-throated green cuckoo | *Chrysococcyx flavigularis* | x | x | x |
| Black-throated coucal | *Centropus leucogaster* | x | x | x |
| Gabon coucal | *Centropus anselli** | — | x | x |
| **STRIGIDAE** | | | | |
| Sandy scops owl | *Otus icterorhynchus* | x | x | x |
| Maned owl | *Jubula lettii* | x | x | x |
| Shelley's eagle owl | *Bubo shelleyi* | x | x | x |
| Akun eagle owl | *Bubo leucostictus* | x | x | x |
| Ussher's fishing owl | *Scotopelia ussheri* | x | — | — |
| Bouvier's fishing owl | *Scotopelia bouvieri* | — | x | x |
| Red-chested owlet | *Glaucidium tephronotum** | x | x | x |
| Sjöstedt's barred owlet | *Glaucidium sjostedti* | — | x | x |
| **CAPRIMULGIDAE** | | | | |
| Brown nightjar | *Caprimulgus binotatus* | x | x | x |
| Bates's nightjar | *Caprimulgus batesi* | — | x | x |
| **APODIDAE** | | | | |
| Chapin's spinetail | *Telacanthura melanopygia* | x | x | x |
| Sabine's spinetail | *Rhaphidura sabini** | x | x | x |
| Cassin's spinetail | *Neafrapus cassini* | x | x | x |
| Bates's swift | *Apus batesi* | — | x | x |
| **TROGONIDAE** | | | | |
| Bare-cheeked trogon | *Apaloderma aequatoriale* | — | x | x |
| **ALCEDINIDAE** | | | | |
| Shining-blue kingfisher | *Alcedo quadribrachys** | x | x | x |
| White-bellied kingfisher | *Alcedo leucogaster** | x | x | x |
| Dwarf kingfisher | *Ceyx lecontei** | x | x | x |
| Chocolate-backed kingfisher | *Halcyon badia* | x | x | x |
| **MEROPIDAE** | | | | |
| Black-headed bee-eater | *Merops breweri** | x | x | x |
| Blue-headed bee-eater | *Merops muelleri** | x | x | x |
| Black bee-eater | *Merops gularis** | x | x | x |
| **CORACIIDAE** | | | | |
| Blue-throated roller | *Eurystomus gularis** | x | x | x |

| | | UG | LG | C |
|---|---|:---:|:---:|:---:|
| **PHOENICULIDAE** | | | | |
| Forest wood hoopoe | *Phoeniculus castaneiceps** | x | x | x |
| **BUCEROTIDAE** | | | | |
| White-crested hornbill | *Tropicranus albocristatus** | x | x | x |
| Black dwarf hornbill | *Tockus hartlaubi* | x | x | x |
| Red-billed dwarf hornbill | *Tockus camurus** | x | x | x |
| Pied hornbill | *Tockus fasciatus** | x | x | x |
| White-tailed hornbill | *Bycanistes fistulator** | x | x | x |
| White-thighed hornbill | *Bycanistes cylindricus** | x | x | x |
| Black-and-white-casqued hornbill | *Bycanistes subcylindricus** | x | x | x |
| Black-wattled hornbill | *Ceratogymna atrata** | x | x | x |
| Yellow-casqued hornbill | *Ceratogymna elata** | x | — | — |
| **LYBIIDAE** | | | | |
| Naked-faced barbet | *Gymnobucco calvus** | x | x | — |
| Bristle-nosed barbet | *Gymnobucco peli* | x | x | — |
| Sladen's barbet | *Gymnobucco sladeni* | — | — | x |
| Yellow-spotted barbet | *Buccanodon duchaillui** | x | x | x |
| Speckled tinkerbird | *Pogoniulus scolopaceus** | x | x | x |
| Yellow-throated tinkerbird | *Pogoniulus subsulphureus** | x | x | x |
| Red-rumped tinkerbird | *Pogoniulus atroflavus** | x | x | x |
| Hairy-breasted barbet | *Tricholaema hirsuta** | x | x | x |
| Yellow-billed barbet | *Trachyphonus purpuratus** | x | x | x |
| **INDICATORIDAE** | | | | |
| Cassin's honeyguide | *Prodotiscus insignis** | x | x | x |
| Zenker's honeyguide | *Melignomon zenkeri* | — | x | x |
| Yellow-footed honeyguide | *Melignomon eisentrauti* | x | — | — |
| Spotted honeyguide | *Indicator maculatus** | x | x | x |
| Western least honeyguide | *Indicator exilis** | x | x | x |
| Willcocks's honeyguide | *Indicator willcocksi** | x | x | x |
| Lyre-tailed honeyguide | *Melichneutes robustus* | x | x | x |
| **PICIDAE** | | | | |
| Piculet | *Sasia africana** | x | x | x |
| Golden-backed woodpecker | *Campethera maculosa** | x | — | — |
| Buff-spotted woodpecker | *Campethera nivosa** | x | x | x |
| Brown-eared woodpecker | *Campethera caroli** | x | x | x |
| Gabon woodpecker | *Dendropicos gabonensis* | x | x | x |
| Fire-bellied woodpecker | *Thripias pyrrogaster** | x | — | — |
| Yellow-crested woodpecker | *Thripias xantholophus** | — | x | x |
| Elliot's woodpecker | *Mesopicos elliotii** | — | x | x |

|  |  | UG | LG | C |
|---|---|---|---|---|
| **EURYLAIMIDAE** |  |  |  |  |
| Rufous-sided broadbill | *Smithornis rufolateralis* | x | x | x |
| Grey-headed broadbill | *Smithornis sharpei* | — | x | x |
| **HIRUNDINIDAE** |  |  |  |  |
| Square-tailed saw-wing | *Psalidoprocne obscura** | x | x | x |
| Fanti saw-wing | *Psalidoprocne obscura** | x | — | — |
| Forest swallow | *Hirundo fuliginosa* | — | x | — |
| White-throated blue swallow | *Hirundo nigrita** | x | x | x |
| **CAMPEPHAGIDAE** |  |  |  |  |
| Petit's cuckoo-shrike | *Campephaga petiti** | — | x | x |
| Western wattled cuckoo-shrike | *Lobotos lobatus* | x | — | — |
| Eastern wattled cuckoo-shrike | *Lobotos oriolinus* | — | x | x |
| Blue cuckoo-shrike | *Coracina azurea* | x | x | x |
| **PYCNONOTIDAE** |  |  |  |  |
| Little gray greenbul | *Andropadus gracilis** | x | x | x |
| Ansorge's greenbul | *Andropadus ansorgei** | x | x | x |
| Cameroon somber greenbul | *Andropadus curvirostris** | x | x | x |
| Golden greenbul | *Calyptocichla serina* | x | x | x |
| Honeyguide greenbul | *Baeopogon indicator** | x | x | x |
| Sjöstedt's honeyguide greenbul | *Baeopogon clamans* | — | x | x |
| Spotted greenbul | *Ixonotus guttatus** | x | x | x |
| Falkenstein's greenbul | *Chlorocichla falkensteini** | — | x | — |
| Simple greenbul | *Chlorocichla simplex** | x | x | x |
| Swamp palm bulbul | *Thescelocichla leucopleura** | x | x | x |
| Leaflove | *Pyrrhurus scandens** | x | x | x |
| Baumann's greenbul | *Phyllastrephus baumanni* | x | — | — |
| Toro olive greenbul | *Phyllastrephus hypochloris** | — | — | x |
| Sassi's olive greenbul | *Phyllastrephus lorenzi* | — | — | x |
| Icterine greenbul | *Phyllastrephus icterinus** | x | x | x |
| White-winged greenbul | *Phyllastrephus leucolepis* | x | — | — |
| Xavier's greenbul | *Phyllastrephus xavieri** | — | x | x |
| White-throated greenbul | *Phyllastrephus albigularis** | x | x | x |
| Bristlebill | *Bleda syndactyla** | x | x | x |
| W. green-tailed bristlebill | *Bleda eximia* | x | — | — |
| E. green-tailed bristlebill | *Bleda notata* | — | x | x |
| Gray-headed bristlebill | *Bleda canicapilla* | x | — | — |
| Bearded greenbul | *Criniger barbatus* | x | — | — |
| Eastern bearded greenbul | *Criniger chloronotus* | — | x | x |
| Red-tailed greenbul | *Criniger calurus* | x | x | x |

| | UG | LG | C |
|---|---|---|---|
| White-bearded greenbul — *Criniger olivaceus* | x | x | x |

**TURDIDAE**

| | UG | LG | C |
|---|---|---|---|
| White-tailed ant thrush — *Neocossyphus poensis** | x | x | x |
| Rufous ant thrush — *Stizorhina fraseri** | x | x | x |
| Black-eared ground thrush — *Zoothera camaronensis* | — | x | x |
| Gray ground thrush — *Zoothera princei* | x | x | x |
| Oberlaender's ground thrush — *Zoothera oberlaenderi** | — | — | x |
| Fire-crested alethe — *Alethe diademata** | x | x | x |
| Forest robin — *Stiphrornis erythrothorax** | x | x | x |
| Akalat — *Sheppardia cyornithopsis** | x | x | x |
| Blue-shouldered robin — *Cossypha cyanocampter** | x | x | x |
| Forest scrub robin — *Erythropygia leucosticta** | x | — | x |

**SYLVIIDAE**

| | UG | LG | C |
|---|---|---|---|
| Black-headed rufous warbler — *Bathmocercus cerviniventris* | x | — | — |
| Black-faced rufous warbler — *Bathmocercus rufus** | — | x | x |
| Rufous-crowned eremomela — *Eremomela badiceps** | x | x | x |
| Turner's eremomela — *Eremomela turneri** | — | — | x |
| Green crombec — *Sylvietta virens** | x | x | x |
| Lemon-bellied crombec — *Sylvietta denti* | x | x | x |
| Grey longbill — *Macrosphenus concolor** | x | x | x |
| Yellow longbill — *Macrosphenus flavicans** | — | x | x |
| Kemp's longbill — *Macrosphenus kempi* | x | — | — |
| Violet-backed hyliota — *Hyliota violacea* | x | x | x |
| Green hylia — *Hylia prasina** | x | x | x |
| White-chinned prinia — *Prinia leucopogon** | — | x | x |
| White-eyed prinia — *Prinia leontica* | x | — | — |
| Masked apalis — *Apalis binotata* | — | x | x |
| Black-capped apalis — *Apalis nigriceps** | x | x | x |
| Buff-throated apalis — *Apalis rufogularis** | — | x | x |
| Sharpe's apalis — *Apalis sharpii* | x | — | — |
| Gosling's apalis — *Apalis gosling** | — | x | x |
| Yellow-browed camaroptera — *Camaroptera superciliaris** | x | x | x |
| Olive-green camaroptera — *Camaroptera chloronota** | x | x | x |

**MUSCICAPIDAE**

| | UG | LG | C |
|---|---|---|---|
| Nimba flycatcher — *Melaenornis annamarulae* | x | — | — |
| Forest flycatcher — *Fraseria ocreata* | x | x | x |
| White-browed forest flycatcher — *Fraseria cinerascens* | x | x | x |
| Olivaceous flycatcher — *Muscicapa olivascens* | x | x | x |
| Cassin's gray flycatcher — *Muscicapa cassini** | x | x | x |

| | | UG | LG | C |
|---|---|---|---|---|
| Little gray flycatcher | *Muscicapa epulata* | x | x | x |
| Yellow-footed flycatcher | *Muscicapa sethsmithi* | — | x | x |
| Dusky blue flycatcher | *Muscicapa comitata** | x | x | x |
| Tessmann's flycatcher | *Muscicapa tessmanni* | x | x | x |
| Sooty flycatcher | *Muscicapa infuscata** | — | x | x |
| Ussher's flycatcher | *Muscicapa ussheri* | x | — | — |
| Grey-throated flycatcher | *Myioparus griseigularis* | x | x | x |

**PLATYSTEIRIDAE**

| | | | | |
|---|---|---|---|---|
| Shrike-flycatcher | *Megabyas flammulatus** | x | x | x |
| Gabon batis | *Batis minima* | — | x | — |
| Ituri batis | *Batis ituriensis* | — | — | x |
| Fernando Po batis | *Batis poensis* | x | x | — |
| Yellow-bellied wattle-eye | *Dyaphorophyia concreta** | x | x | x |
| Red-cheeked wattle-eye | *Dyaphorophyia blissetti* | x | — | — |
| Jameson's wattle-eye | *Dyaphorophyia jamesoni** | — | — | x |
| Black-necked wattle-eye | *Dyaphorophyia chalybea** | — | x | — |
| Chestnut wattle-eye | *Dyaphorophyia castanea** | x | x | x |
| White-spotted wattle-eye | *Dyaphorophyia tonsa* | — | x | x |

**MONARCHIDAE**

| | | | | |
|---|---|---|---|---|
| Chestnut-capped flycatcher | *Erythrocercus mccalli* | x | x | x |
| Dusky crested flycatcher | *Elminia nigromitrata** | x | x | x |
| Blue-headed crested flycatcher | *Trochocercus nitens** | x | x | x |
| Rufous-vented paradise flycatcher | *Terpsiphone rufocinerea** | — | x | x |
| Bedford's paradise flycatcher | *Terpsiphone bedfordi* | — | — | x |
| Red-bellied paradise flycatcher | *Terpsiphone rufiventer** | x | x | x |

**TIMALIIDAE**

| | | | | |
|---|---|---|---|---|
| Brown illadopsis | *Illadopsis fulvescens** | x | x | x |
| Scaly-breasted illadopsis | *Illadopsis albipectus** | — | — | x |
| Blackcap illadopsis | *Illadopsis cleaveri* | x | x | — |
| Rufous-winged illadopsis | *Illadopsis rufescens* | x | — | — |
| Puvel's illadopsis | *Illadopsis puveli** | x | x | x |
| Capuchin babbler | *Phyllanthus atripennis** | x | x | x |
| White-necked picathartes | *Picathartes gymnocephalus* | x | — | — |
| Grey-necked picathartes | *Picathartes oreas* | — | x | — |

**PARIDAE**

| | | | | |
|---|---|---|---|---|
| Dusky tit | *Parus funereus** | x | x | x |

**REMIZIDAE**

| | | | | |
|---|---|---|---|---|
| Yellow-fronted penduline tit | *Anthoscopus flavifrons* | x | x | x |
| Tit-hylia | *Pholidornis rushiae** | x | x | x |

| | | UG | LG | C |
|---|---|---|---|---|
| **NECTARINIIDAE** | | | | |
| Fraser's sunbird | *Anthreptes fraseri** | x | x | x |
| Brown sunbird | *Anthreptes gabonicus* | x | x | — |
| Violet-tailed sunbird | *Anthreptes aurantium* | x | x | — |
| Yellow-chinned sunbird | *Anthreptes rectirostris** | x | x | x |
| Little green sunbird | *Nectarinia seimundi** | x | x | x |
| Bate's sunbird | *Nectarinia batesi** | x | x | x |
| Reichenback's sunbird | *Nectarinia reichenbachii** | x | x | x |
| Blue-throated brown sunbird | *Nectarinia cyanolaema** | x | x | x |
| Green-throated sunbird | *Nectarinia rubescens** | — | x | x |
| Buff-throated sunbird | *Nectarinia adelberti* | x | — | — |
| Tiny sunbird | *Nectarinia minulla* | x | x | x |
| Johanna's sunbird | *Nectarinia johannae* | x | x | x |
| Superb sunbird | *Nectarinia superba** | x | x | x |
| **ORIOLIDAE** | | | | |
| Western black-headed oriole | *Oriolus brachyrhynchus* | x | x | x |
| Black-winged oriole | *Oriolus nigripennis** | x | x | x |
| **MALACONOTIDAE** | | | | |
| Black-shouldered puffback | *Dryoscopus senegalensis* | — | x | x |
| Sabine's puffback | *Dryoscopus sabini* | x | x | x |
| Lühder's bush shrike | *Laniarius luehderi** | — | x | x |
| Turat's boubou | *Laniarius turatii* | x | — | — |
| Sooty boubou | *Laniarius leucorhynchus** | x | x | x |
| Gray-green bush shrike | *Malaconotus bocagei** | — | x | x |
| Fiery-breasted bush shrike | *Malaconotus cruentus* | x | x | x |
| Western nicator | *Nicator chloris** | x | x | x |
| Yellow-throated nicator | *Nicator vireo** | — | x | x |
| **PRIONOPIDAE** | | | | |
| Northern red-billed helmet shrike | *Prionops canicept* | x | x | x |
| **DICRURIDAE** | | | | |
| Shining drongo | *Dicrurus atripennis* | x | x | x |
| **STURNIDAE** | | | | |
| Narrow-tailed starling | *Poeoptera lugubris** | x | x | x |
| Forest Chestnut-winged starling | *Onychognathus fulgidus** | x | x | x |
| Copper-tailed glossy starling | *Lamprotornis cupreocauda* | x | — | — |
| Purple-headed glossy starling | *Lamprotornis purpureiceps** | — | x | x |
| **PLOCEIDAE** | | | | |
| Bates's weaver | *Ploceus batesi* | — | x | — |
| Orange weaver | *Ploceus aurantius** | x | x | x |

| | | UG | LG | C |
|---|---|---|---|---|
| Vieillot's black weaver | *Ploceus nigerrimus*\* | x | x | x |
| Weyns's weaver | *Ploceus weynsi*\* | — | — | x |
| Golden-naped weaver | *Ploceus aureonucha* | — | — | x |
| Yellow-mantled weaver | *Ploceus tricolor*\* | x | x | x |
| Maxwell's black weaver | *Ploceus albinucha* | x | x | x |
| Yellow-legged weaver | *Ploceus flavipes* | — | — | x |
| Preuss's golden-backed weaver | *Ploceus preussi*\* | x | x | x |
| Yellow-capped weaver | *Ploceus dorsomaculatus* | — | x | x |
| Blue-billed malimbe | *Malimbus nitens* | x | x | x |
| Crested malimbe | *Malimbus malimbicus*\* | — | x | x |
| Cassin's malimbe | *Malimbus cassini* | — | x | x |
| Ibadan malimbe | *Malimbus ibadanensis*ᵃ | — | — | — |
| Red-vented malimbe | *Malimbus scutatus* | x | — | — |
| Rachel's malimbe | *Malimbus racheliae* | — | x | — |
| Gola malimbe | *Malimbus ballmanni* | x | — | — |
| Red-headed malimbe | *Malimbus rubricollis*\* | x | x | x |
| Red-bellied malimbe | *Malimbus erythrogaster* | — | x | x |
| Red-crowned malimbe | *Malimbus coronatus* | — | x | x |
| **ESTRILDIDAE** | | | | |
| Red-headed flower-pecker | *Parmoptila woodhousei*\* | — | x | x |
| Red-faced flower-pecker | *Parmoptila rubrifrons* | x | — | x |
| Pale-fronted negrofinch | *Nigrita luteifrons*\* | — | x | x |
| Chestnut-breasted negrofinch | *Nigrita bicolor*\* | x | x | x |
| White-breasted negrofinch | *Nigrita fusconota*\* | x | x | x |
| Crimson seed-cracker | *Pyrenestes sanguineus* | x | — | — |
| Grant's bluebill | *Spermophaga poliogenys* | — | — | x |
| Bluebill | *Spermophaga haematina*\* | x | x | x |
| Totals (251 species) | | 184 | 203 | 207 |

*Note:* An *x* denotes the presence of the species in the subregions of Upper Guinea (UG), Lower Guinea (LG), and the Congo Basin (C). Asterisks denote near-endemics.

a    confined to Nigeria but taxonomic and chorological status uncertain.

# CHAPTER 16

# Fishes of African Rain Forests

## Diverse Adaptations to Environmental Challenges

Lauren J. Chapman

The extraordinary diversity of the terrestrial taxa in African rain forests that has captivated biologists for more than a century is paralleled by unique and highly diverse aquatic faunas. Africa has more than 2,900 known species of indigenous freshwater fishes, over 28% of which are endemic to the rain forest regions of central and west Africa (Roberts 1975; Daget et al. 1984, 1986, 1991; Lowe-McConnell 1988). Forest rivers harbor a rich assortment of species with extraordinary adaptations to the challenges imposed by the forest environment, in addition to a diverse suite of archaic and phylogenetically isolated groups (Roberts 1975; Lowe-McConnell 1988). Yet our knowledge of these faunas goes little beyond catalogues of species lists and a handful of ecological studies from very few sites. Deforestation, species introductions, and wetland degradation all pose serious threats to fishes of the African rain forests, but the nature and severity of their effects remain largely unknown. In this chapter, I review the striking characteristics of African forest fish faunas, with a focus on fish-forest interactions and adaptive responses to environmental challenges. I identify key areas for future study that would permit a more synthetic analysis of fish-forest interactions and end by briefly

discussing anthropogenic impacts that may precipitate declines in fish faunal diversity or shifts in community structure.

## Fishes of the Forest

### DEMOCRATIC REPUBLIC OF CONGO (DR CONGO) VERSUS AMAZON

Much of our knowledge and interest in fish-forest interactions is based on studies of the inundated forest (Igapo) of the Amazon floodplain, where huge areas of forest become inundated with seasonal flooding, and exploitation of the forest by fruit-eating fishes is integral to forest ecosystem function (Goulding 1980). In Africa, true forested floodplains similar to the Amazon are more or less confined to the Congo Basin and a few smaller river basins in Cameroon and Gabon. Other less heavily forested rivers occur in west Africa, Madagascar, and the small patches of forest in western Uganda (figure 16.1).

The two largest forest rivers in the world, the Congo and the Amazon, have more kinds of fishes than any other river basins, and both systems exhibit high endemism (Roberts 1973). They rank the highest in the world for catchment, length, and output (table 16.1), and both arise on very poor, leached podsolitic soils (Wel-

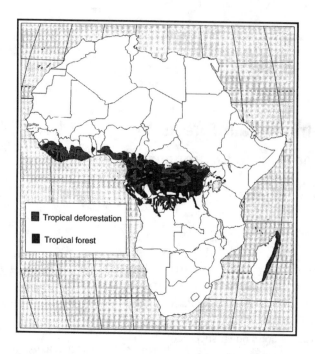

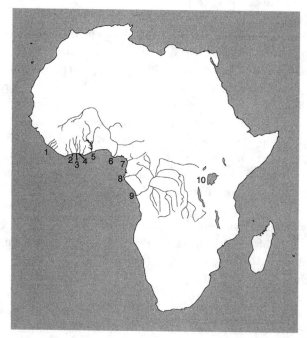

Figure 16.1. *Top:* Map of the African continent and Madagascar illustrating the major regions of tropical rain forest and the extent of deforestation in these areas. Adapted from *National Geographic Atlas of the World* (1992),NG MAPS/NGS Image Collection. *Bottom:* Location of selected rivers from the areas of tropical rain forest in Africa: (1) "Liberia"; (2) Sassandra; (3) Bandama; (4) Komoe; (5) Volta; (6) Niger; (7) Sanaga; (8) Ogooue; (9) Congo; and (10) "Uganda." Adapted from Skelton 1988.

comme 1985). However, beyond these characters the two systems are fundamentally different, and our broader knowledge of fish ecology and limnology in the Amazon cannot be translated to the Congo system. The Congo Basin covers an area of approximately 4 million km² and drains the largest expanse of tropical forest in Africa, yet only 34% of the basin is covered by forest, compared with 80% in the Amazon basin (Marlier 1973; Lowe-McConnell 1975; see also table 16.1). Its catchment has been affected by earth movements that have altered the course of the drainage and produced many abrupt changes of elevation (Beadle 1981). The Congo River is thus characterized by many more rapids and waterfalls than the Amazon (figure 16.2), limiting penetration of marine fishes into the basin (Marlier 1973; Lowe-McConnell 1975; Beadle 1981; see also table 16.1). In contrast, the Amazon has a very gentle slope at a low altitude and has consequently received many marine immigrants (Marlier 1973; Beadle 1981). The Congo exhibits only small fluctuations in water level (3 m) compared

with the Amazon (15m or more at Manaus; Lowe-McConnell 1975). This relative lack of fluctuation is due in part to the presence of lakes and swampy depressions in the upper course of the Congo River drainage that absorb the effects of heavy rains. This is not to say that the system is aseasonal; the water-level fluctuations that are bimodal in nature have marked effects on fish ecology in the region. Like the Amazon, the Congo River is characterized by seasonal flooding of extensive areas, particularly in the central basin, where the flooded river inundates huge areas of forest many kilometers from the main river (Lowe-McConnell 1975), creating an enormous area for seasonal exploitation by fishes and other aquatic organisms.

ENDEMISM

Endemism in the Congo region is impressive; exclusive of Lake Tanganyika, 669 species representing 25 families and 168 genera have been recorded in the entire basin, with 82% endemism at the specific level and 42 endemic genera (Poll 1959, 1973; Beadle 1981). In the

Table 16.1
## Characteristics of the Amazon and Congo River Basins

| CHARACTER | CONGO | AMAZON |
|---|---|---|
| Size | second largest in the world | largest in the world |
| Length | 4,650 km | 5,500 km |
| Output | 1,400,000 f³/sec | 3–4,000,000 f³/sec |
| Drainage discharge | 3,822,000 km² | 5,600,000 km² |
| Forest cover | 34% | 80% |
| Water level fluctuations | 3 m (central basin) | 15+m (Manaus) |
| Topography | highly variable slope (extensive rapids) | gentle slope |
| Species richness (fishes) | 560+ | 1,300+ |
| Marine immigrants | low | high |
| Human population density | 0–10 people/km² | 0–0.4 people/km² |
| Deforestation rate (% / yr) | 0.2 (Congo) | 0.4 (Brazil) |

*Sources:* Data from FAO 1981; Marlier 1973; Roberts 1973; World Resources Institute 1994; National Geographic Atlas of the World (1992).

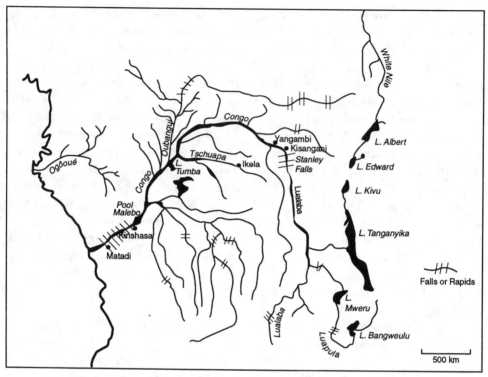

Figure 16.2. The Congo River Basin of Africa, indicating larger tributaries, major survey sites, and adjacent water systems. Adapted from Beadle 1981, courtesy of Pearson Education Limited. Tributaries east of Lakes Tanganyika, Kivu, Edward, and Albert are not included.

central basin of the Congo River (a region defined in Poll and Gosse 1963 and Poll 1973), which includes much of the drainage, Poll and Gosse (1963) reported 408 fish species from 24 families. New species emerge with new expeditions, and will be likely to do so for some time. Factors contributing to this high endemism may include a wide range of habitats and numerous barriers (rapids and falls) that have contributed to the prolonged isolation of fish communities (Beadle 1981).

This extraordinarily high level of endemism also occurs in several of the coastal rivers in the region, from Nigeria to DR Congo (Roberts 1975). Endemism is high at the generic level in Madagascar (14 of the 23 genera are endemic; Reinthal and Stiassny 1991) but lower at the specific level (35%, or 23 species). The total

richness of the fauna is also low (66 species; Roberts 1975). The rivers draining the small forests of western Uganda are characterized by widespread species and low endemism, although many of the environmental features (high relief, low water-level fluctuations) are shared with the Congo Basin (Greenwood 1966; Chapman, unpublished data).

FOREST VERSUS SAVANNA RIVERS

Most of what we know about the natural history and ecology of African freshwater fishes applies to savanna floodplain rivers; far less is known about forested rivers. This emphasis reflects the importance of savanna floodplain fisheries and the inaccessibility of some forested regions. For example, the forested floodplains of the Congo Basin below Kisangani, which culminate in the extensive Bangal swamps, tend to retain water to a greater extent than many savanna floodplains. However, fish ecology in this area is largely unknown; the region is almost completely unpopulated by people, and access is extremely difficult (Welcomme 1985).

The forested rivers differ in many respects from savanna rivers. Their floodplains tend to be restricted in area and in some areas internalized over the numerous islands that form between the braided channels in the larger rivers (Welcomme and de Merona 1988). Various factors may contribute to their more stable flow. Flood peaks may be softened by the long retention time of the water in the flooded forests; in some systems the water arrives at different times of the year from different tributaries; and in the equatorial rivers, there are two rainy seasons a year, producing a bimodal flood regime (Welcomme 1979, 1985; Welcomme and de Merona 1988). Forested river waters are frequently impoverished, with low conductivity and negligible silt load, and are often stained brown with humic acids. In combination with heavy shading of the river channel, this tends to produce waters with low levels of primary production (Marlier 1973; Welcomme 1985; Welcomme and de Merona 1988). The main food source is allochthonous plant and animal matter that falls into the water from the surrounding forest. These characteristics of forested rivers have a strong influence on the ecology of forest fish faunas.

Several families and genera of fishes are shared between forest and savanna river systems; however, many species of fishes are not shared, and some tend to have restricted ranges in the equatorial rain forest systems. In west Africa, for example, Daget and Iltis (1965) distinguished between widespread species found in several savanna rivers, and species much more restricted in range and found in the forested coastal rivers (Welcomme and de Merona 1988).

## DOMINANT TAXA AND HABITAT ASSOCIATIONS

The rain forest rivers of Africa are characterized by a diversity of habitats that differ in physico-chemical conditions and species associations. Major habitat types identified in the Congo Basin include the slow-flowing section of the main river, rapids of the main river, marginal waters of the shores, inundated forest, wetlands, and affluent rivers and streams (Poll 1959; Gosse 1963; Matthes 1964; Roberts and Stewart 1976). The main rivers host a greater diversity of microhabitats than smaller streams and wetlands, and are therefore characterized by higher species richness and diversity (figure 16.3). Different habitats pose different environmental challenges for fishes, and the diversity of habitats is reflected in a rich array of adaptations to permit exploitation of even the most extreme conditions.

In the Congo River Basin, detailed information on species richness and habitat associations is available for only a few sites, including Pool Malebo and nearby areas (Poll 1959); Yangambi near Kisangani (Gosse 1963); the Tschuapa River, a large tributary (Matthes 1964); Lake Tumba, a lake of the Congo River below its confluence with the Tschuapa River (Matthes 1964); and the rapids of the Lower Congo River between Kinshasa and Matadi (Roberts and Stewart 1976; see figure 16.2). The dominant families of fishes in these areas of the Congo Basin include the Mormyridae, Bagridae, Clariidae, Mochokidae, Cyprinidae, Characidae, Cichlidae, Citharinidae, and Anabantidae. These families are also represented in many of the other equatorial forest rivers in Africa (table 16.2).

The nocturnal mormyrids are extremely widespread in the rivers of Africa, with the exception of Madagascar (see table 16.2; figure 16.4). They are the dominant family in both richness (75 of the 408 species) and numbers of individuals in the slow-flowing waters of the central

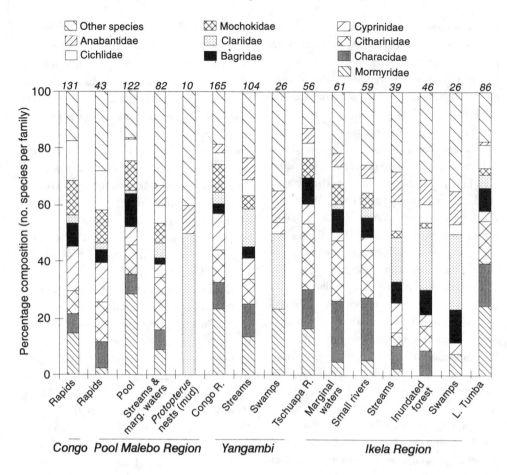

Figure 16.3. The percentage composition of fishes (number of species per family) in major habitats within four regions of the Congo River. The total number of species in each habitat type is indicated at the top of each bar. Data abridged from Poll 1959 (Pool Malebo), Gosse 1963 (Yangambi), and Matthes 1964 (Ikela region and Lake Tumba), compiled by Lowe-McConnell (1975); and from Roberts and Stewart 1976 (lower Congo rapids). The sites are indicated in figure 16.2.

Congo Basin (Poll and Gosse 1963; Beadle 1981). Mormyrids are well known for their exceptionally large cerebellum, related at least in part to their remarkable electrogenic and electrolocation capabilities (Bennett 1971; Hopkins 1981, 1986; Kramer 1990). Many mormyrids are characterized by tubular mouths with small openings that permit probing and extraction of benthic invertebrates (see figure 16.4). They occur in a variety of habitats, including streams and swamps (see figure 16.3), but they are principally

fluviatile, associating in shoals that are thought to be kept together by the signals from their electric organs (Beadle 1981; Hopkins 1986).

In the central Congo River Basin the catfishes represent 21% of the fish fauna (36 mochokid species, 27 bagrids, and 23 clariids), and many belong to endemic genera (Poll and Gosse 1963, 1973; Lowe-McConnell 1988; see also figure 16.4). The catfishes are mainly nocturnal and have specialized defenses, including a unique arrangement of the pectoral and dor-

Table 16.2

# Fish Families Represented in a Selection of African Rain Forest Rivers

| FAMILY | RIVER (% FOREST COVER) | | | | | | "LIBERIA"[a] | "UGANDA"[a] | AIR BREATHERS[b] |
| --- | --- | --- | --- | --- | --- | --- | --- | --- | --- |
| | CONGO (34%) | OGOOUÉ (100%) | SANAGA (100%) | KOMOE (44%) | BANDAMA (76%) | SASSANDRA (91%) | | | |
| Protopteridae | X | X | X | X | X | | | X | Yes |
| Polypteridae | X | X | X | X | X | X | X | X | Yes |
| Osteoglossidae | X | | | | **[c] | X | | | Yes |
| Pantodontidae | X | | | | | | | | Yes |
| Notopteridae | X | X | | X | X | X | X | | Yes |
| Mormyridae | X | X | X | X | X | X | X | X | |
| Clupeidae | X | | X | X | X | X | X | | |
| Kneriidae | X | X | | | | | X | | |
| Phractolaemidae | X | | | | | | | | Yes |
| Cyprinidae | X | X | X | X | X | X | X | X | |
| Citharinidae | X | X | X | X | X | X | X | X | |
| Hepsetidae | X | X | X | X | X | X | X | | |
| Characidae | X | X | X | X | X | X | X | X | |
| Bagridae | X | X | X | X | X | X | X | X | |
| Schilbeidae | X | X | X | X | X | X | X | | |
| Amphiliidae | X | X | X | X | X | X | X | X | |
| Clariidae | X | X | X | X | X | X | X | X | Yes |
| Malapteruridae | X | X | X | X | X | X | X | | |

Table 16.2. (continued)

| | RIVER | | | | | | "LIBERIA"[a] | "UGANDA"[a] | AIR BREATHERS[b] |
|---|---|---|---|---|---|---|---|---|---|
| | CONGO | OGOOUÉ | SANAGA | KOMOE | BANDAMA | SASSANDRA | | | |
| Mochokidae | X | X | X | X | X | X | | X | |
| Salmonidae | | | | | | | | **c | |
| Poecillidae | | | | | | | **c | **c | |
| Cyprinodontidae | X | X | X | X | X | X | X | X | |
| Mastacembelidae | X | X | X | X | X | X | X | | ? |
| Centropomidae | X | | | X | X | X | | | |
| Nandidae | | X | | | | | X | | |
| Cichlidae | X | X | X | X | X | X | X**d | X**d | |
| Eleotridae | X | X | | X | X | X | X | | |
| Gobiidae | | | | | X | X | | | |
| Anabantidae | X | X | X | X | X | | X | X | Yes |
| Channidae | X | X | X | X | X | X | X | X | Yes |
| Tetraodontidae | X | | | | | | | | |

*Sources:* Data from Greenwood 1966; Daget et al. 1973; Lowe–McConnell 1975; Roberts 1975; Skelton 1988; Hugueny 1989; Paugy et al. 1990; and Chapman and Chapman, unpubl. data.

*Note:* Locations of the rivers are indicated on figure 16.1. Note that many of the regions within "forest" rivers have been heavily deforested.

a  "Liberia" and "Uganda" include several small river systems within the forested regions of these countries.

b  "Air breathers" refers to families with air-breathing representatives.

c  Represented only by introduced species.

d  Represented by both native and introduced species.

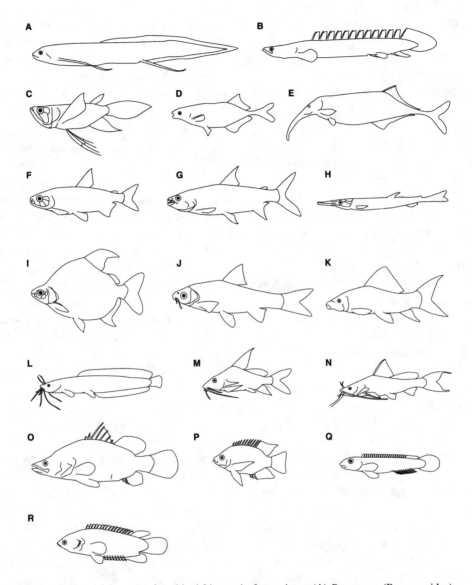

Figure 16.4. Selected fish genera found in African rain forest rivers: (A) *Protopterus* (Protopteridae); (B) *Polypterus* (Polypteridae); (C) *Pantodon* (Pantodontidae); (D) *Gnathonemus* (Mormyridae); (E) *Campylomormyrus* (Mormyridae); (F) *Alestes* (Characidae); (G) *Hydrocynus* (Characidae); (H) *Belonophago* (Citharinidae); (I) *Citharinus* (Citharinidae); (J) *Barbus* (Cyprinidae); (K) *Labeo* (Cyprinidae); (L) *Clarias* (Clariidae); (M) *Synodontis* (Mochokidae); (N) *Bagrus* (Bagridae); (O) *Lates* (Centropomidae); (P) *Oreochromis* (Cichlidae); (Q) *Gobiocichla* (Cichlidae); (R) *Ctenopoma* (Anabantidae). Drawings adapted from Greenwood 1966; Lowe-McConnell 1975; and Welcomme 1985.

sal fin spines (Alexander 1966; Roberts 1975). All have barbels and small eyes. The mochokids are primarily bottom dwellers, feeding from soft deposits with their ventral mouth. However, a few species habitually swim upside down, taking food or oxygen from the surface film (Roberts 1975; Chapman et al. 1994). The clariid catfishes are often found over muddy bottoms and are omnivorous feeders. Many have a respiratory organ in the branchial chamber, which permits them to use atmospheric air to meet their oxygen demands in oxygen-scarce waters (Liem 1987; Chapman 1995). Some members of the genus *Clarias* are capable of terrestrial locomotion during wetter periods (Bruton 1979; Liem 1987). The family Bagridae includes very large predatory catfishes and several small species that live in fast flowing waters of the Congo River (Poll 1957, 1959; Lowe-McConnell 1988).

The Cyprinidae is one of the most abundant and widespread freshwater fish families in Africa and is well represented in many of the rain forest rivers, with the exception of Madagascar. It is the dominant taxon in many of the small forest rivers of western Uganda (Chapman, unpubl. data) and represents 13% of the fish fauna in the central Congo River Basin (Poll and Gosse 1963; Lowe-McConnell 1988). The genus *Barbus* (figure 16.4) includes several species in the Congo Basin (Poll 1973), many of which are small species that feed primarily on benthic invertebrates and vegetation debris (Matthes 1964). Other cyprinids of the genus *Labeo* are mud and detritus feeders and are largely open water forms (Matthes 1964). In the Congo Basin, the cyprinids are most common in the rivers and streams and rare in wetlands (see figure 16.3). This differs from patterns observed in the small forest streams and rivers of Uganda, where some *Barbus* species are extremely tolerant of hypoxia and are abundant in wetland habitats (Chapman and Liem 1995).

The characiforms (characids, citharinids,

hepsetids) are widespread in the rain forest rivers of Africa, with the exception of Madagascar, and compose 18% of the fish fauna of the central Congo Basin (Poll and Gosse 1963; Lowe-McConnell 1988; see table 16.2 and figure 16.4). Most members of the family Characidae are silvery, iridescent, laterally compressed, open water fishes that are active in the daytime (Roberts 1973). In the Congo River, the group includes large piscivorous forms (e.g., *Hydrocynus*) in the open waters and smaller omnivorous species, plankton feeders, and surface insect feeders (Gosse 1963; Matthes 1964). The Citharinidae include an interesting group of deep-bodied, mud-sucking fishes that occupy the benthic zone of the open river (e.g., *Citharinus*, figure 16.4). Also included in this family is a remarkable group of long slender fishes, found mainly in the Congo Basin, which are specialized to feed on the fins of other fishes (e.g., *Phago* and *Belonophago*, Matthes 1961), and several species of macrophyte feeders (*Distichodus* spp.) that tend to occupy marginal waters (Matthes 1964). The family Hepsetidae includes only one species, the primitive characiform *Hepsetus odoe*, a large predatory fish.

Cichlids are found in most forest rivers in Africa (see table 16.2), representing 8% of the central Congo Basin fauna in terms of species richness (Poll and Gosse 1963; Lowe-McConnell 1988). However, with the exception of the Lower Congo rapids, where the endemism and abundance of cichlids are remarkably high, they seldom constitute a major portion of the individual fishes at a site, and species tend to be few at any one site (Roberts and Stewart 1976). A similar pattern occurs in the forested rivers of Uganda (Chapman, unpublished data). Endemicity of cichlids at the specific and generic levels is high in DR Congo, but the spectacular radiations that have produced large species flocks are limited to the Great Lakes of east Africa (Greenwood 1981, 1994; Kaufman et al. 1996). The cichlids include the tilapiines, which are important food fish (both when caught in the wild and when produced in aquaculture), as well as many other small species (see figure 16.4). In the Congo River Basin they are found in a variety of habitats, including the inundated forest zone, but seldom inhabit the hypoxic waters of swamps (see figure 16.3).

The Anabantidae is also a widespread family in equatorial forest rivers and is one of the few fish taxa to be well represented in swamps (see table 16.2 and figure 16.3). A labyrinthine respiratory organ in the suprabranchial cavity permits use of atmospheric air and enables survival in the deoxygenated water of dense swamps. Many anabantids feed on benthic invertebrates or surface insects of allochthonous origin (Matthes 1964; Greenwood 1966).

Many other families are represented in the rain forest rivers of Africa (see table 16.2 and figure 16.4). Some of the most widespread families tend to be species poor. This includes many air-breathing fishes that can be found in the dense interior of deoxygenated swamps, like the lungfishes (*Protopterus dolloi* and *P. annectens*) and polypterids (seven species; Poll 1973). The ability to disperse through swampy divides and their resilience to extreme physico-chemical conditions facilitated by air breathing may contribute to the widespread distribution and low endemicity of many air breathers.

Although specific associations of fishes characterize different habitats of the Congo Basin and other forested rivers in Africa (see figure 16.3), community composition changes over the year as fishes move to different habitats with the seasons (Lowe-McConnell 1975). In addition, community composition will vary among years owing to annual variation in the onset, magnitude, and duration of flooding. The dynamic nature of the community composition is not unique to forest rivers; in fact, it is probably more dramatic in savanna rivers, with their more extensive floodplains. Such seasonal shifts in habitat use and interannual variation in the relative abundance of species may be important in permitting high numbers of species to coexist (Lowe-McConnell 1975; Chapman and Chapman 1993).

## Responses to Environmental Challenges

### SEASONALITY

Water level is often the most seasonally varying environmental parameter in tropical rivers, the effects of which can match the importance of temperature in many temperate waters. Although water levels are less dynamic in most African forest rivers than in the Amazon region, the ecology and behavior of the fishes are still geared to changes in available habitat mediated by seasonal rains. The pronounced seasonality associated with biannual flooding in the equatorial rain forest rivers and annual flooding in rain forests of higher latitudes (such as Sierra Leone) is reflected in patterns of breeding, growth, and mortality.

The main effect of seasonal flooding is an increase in available habitat. The degree of expansion varies with the topography of the system, and the most pronounced effects can be observed in shallow forested wetlands and flat river basins bordered by low relief forest that becomes inundated. In high-gradient channels, like the rapids of the lower Congo, the effects of the floods are reflected in changes in current velocity, with less extensive habitat expansion. In intermittent forest streams, habitats may shift rapidly from fast-flowing waters during flooding to small isolated stagnant pools during the drier periods.

Seasonal flooding can also induce marked changes in the physico-chemical environment, but the nature of the shift varies among habitats. For example, dissolved oxygen is extremely low during the dry season in many wetlands and stagnant pools of intermittent forest streams. Studies in Kibale National Park of Uganda have demonstrated the strong interaction between seasonal flooding and habitat in contributing to variation in dissolved oxygen levels. Kibale Forest is drained by two major ever-flowing rivers, the Dura and Mpanga Rivers; both are tributaries of Lake George (figure 16.5). These rivers are choked by papyrus (*Cyperus papyrus*) swamps for several kilometers and are fed by numerous small forest streams. Average monthly oxygen levels are extremely low in the swamps but vary markedly with season. In the Rwembaita Swamp of central Kibale, oxygen averaged only 1.5 mg/L over a two-year period but showed a strong increase to more than 3.5 mg/L during peak flood conditions (figure 16.6). A small intermittent tributary showed

higher values of dissolved oxygen than the swamp (mean = 3.3 mg/L), with less dramatic seasonal variation than the swamp. In the ever-flowing Njuguta River, a tributary of the larger Mpanga River, dissolved oxygen was consistently high throughout the year (mean = 6.3 mg/L), with little seasonal variation (see figure 16.6). The dramatic changes in the oxygen environment of wetlands and other seasonally hypoxic (oxygen-scarce) waters may permit seasonal access to fishes otherwise unable to exploit these areas, or provide conditions suitable for seasonal spawning.

During the high-water period, many fishes move into the inundated forest and forest wet-

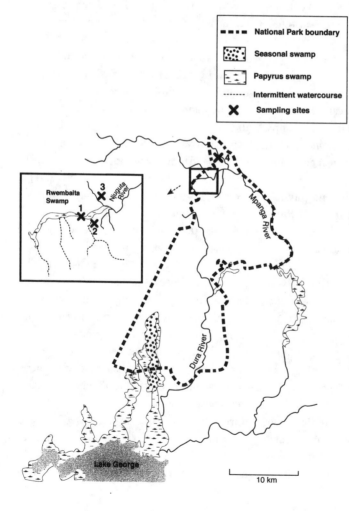

Figure 16.5. Map of the major river systems that drain Kibale Forest (Dura River and Mpanga River) and the sampling sites for dissolved oxygen concentration and the gill size of the cyprinid *Barbus neumayeri*. Site 1: Rwembaita Swamp, a dense papyrus swamp; site 2: Inlet Stream East, a small intermittent feeder stream of the Rwembaita Swamp; site 3: Njuguta River, an ever-flowing tributary of the Mpanga River; site 4: a forest site on an ever-flowing section of the Mpanga River. Dotted lines mark the boundaries of Kibale National Park.

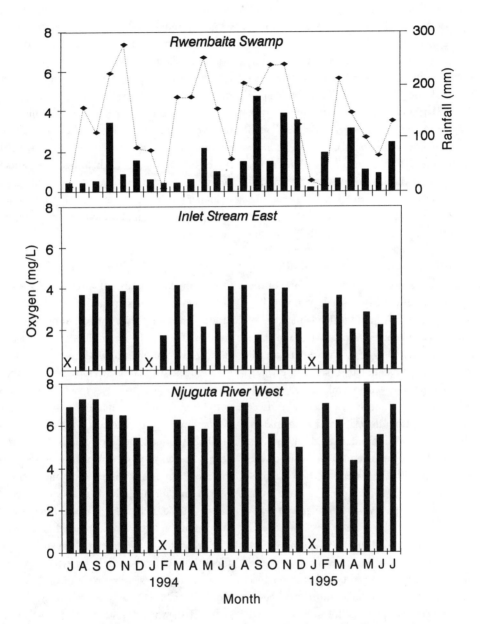

Figure 16.6. Mean dissolved oxygen levels (mg/L) for three sites from Kibale National Park: a papyrus swamp (Rwembaita Swamp; figure 16.5, site 1), a small intermittent stream (Inlet Stream East; figure 16.5, site 2), and an ever-flowing river (Njuguta River; figure 16.5, site 3). Each value represents the mean of duplicate samples at a series of sites in each system over a two-year period. Seasonal variation in oxygen levels in the Rwembaita Swamp are shown relative to the seasonal pattern of rainfall (mm). No data are available for the sampling periods indicated by $X$.

lands to feed and reproduce. Growth during the flood period is rapid because fishes are widely dispersed in new habitats where food is plentiful. In the Congo River, where there are two floods a year, fishes tend to have two breeding seasons, although it is unclear as to whether individual fish breed once per year or during both flood events (Matthes 1964; Lowe-McConnell 1975). Many fishes migrate upstream as the Congo River waters begin to rise and then move laterally into the flooded forest to release the young in the forest waters (Matthes 1964; Lowe-McConnell 1975). For example, in Lake Tumba, a large lake in the Congo River Basin (see figure 16.2), fishes penetrate tens of kilometers into the inundated forest to spawn and feed during high water. This seasonal strategy may contribute to the richness of the Lake Tumba fish fauna. The water of the lake is chemically impoverished, with very low primary productivity, but the floods allow fishes to find food and nursery areas far from the lake in the inundated forest (Lowe-McConnell 1975). In the Yangambi area (see figure 16.2), Gosse (1963) reported juveniles of many fish species using the inundation zones where insect larvae, rich bottom deposits, and zooplankters provide young with abundant food supplies. As water levels fall, the juveniles move out of the inundation zone to the floating meadows or marginal macrophytes, which provide cover from predators like *Hydrocynus* (Gosse 1963; Lowe-McConnell 1975).

Some fish are total spawners and have very discrete breeding seasons, while others are multiple spawners that tend to spawn from just prior to the onset of the floods until peak floods. Other species in equatorial forest rivers breed throughout the year but show strong peaks in spawning activity. For example, in the valley swamps and forest rivers of Kibale Forest, reproductively mature individuals of *Barbus neumayeri* and *Clarias liocephalus* are found throughout the year. However, there are definite peaks in reproductive activity coinciding with seasonal peaks of precipitation, when oxygen levels are relatively high (Chapman and Frankl, unpublished data). This seasonal pattern may permit the young to access seasonally available habitat and may minimize exposure of eggs to the extreme hypoxia that characterizes the dry season in the valley swamps.

The flood regimes are more predictable than other kinds of variability and have permitted adaptive responses to evolve. New habitat increases environmental heterogeneity, alters community composition, and may contribute to higher levels of species diversity than would be supported by a more stable regime (Roberts 1973).

### FISH-FOREST TROPHIC INTERACTIONS

One of the key features of forest waters is the dependence of fishes on food of allochthonous origin. The forest rivers are often black-water rivers with low pH, low conductivity, and a low silt load. In combination with the heavy shading induced by the forest canopy, the contribution of phytoplankton to primary production within forest rivers tends to be extremely low (Welcomme and de Merona 1988). Nutrients of allochthonous forest products (vegetation debris, insects, and so on) are therefore often the most important source of input into the system and form the basis of the food webs (Lowe-McConnell 1975; Welcomme 1985; Welcomme and de Merona 1988). Fishes must adapt themselves to the seasonal availability of allochthonous foods, and the main feeding time is the high-water season, when forest areas are more accessible (Lowe-McConnell 1975).

Much of what we understand about the use of flooded forest by fishes comes from studies of the inundated forest of the Amazon. Here many fishes move into the inundated forest zone to feed on forest products. For example, Goulding (1980) reported forty fruit and seed species (thirty-eight genera) that were exploited by fishes in the Rio Machado of the Amazon Basin.

Adaptations to fruit eating include the strongly developed dentition for seed crushing seen in some Amazonian species (e.g., *Colossoma* and *Myleus*), which act as seed predators. Other species ingest the seeds whole and serve as dispersal agents (Goulding 1980). Unfortunately, we understand far less about the inundated forest regions of African equatorial forest rivers. Specialized fruit eaters like those of the Amazon do not seem to be as apparent; the African fish faunas seem to lack taxa with strongly developed dentition that characterizes some of the Amazonian species (Goulding 1980). However, it is clear that several species rely on forest foods during the inundation period. Matthes (1964) reported 38% of the 127 fish species of the Ikela region and 26% of the 119 fishes in the Lake Tumba region of the Congo River as seasonal inhabitants of the inundated forest zone. Foods of terrestrial origin in the stomach contents of these fishes included vegetative material (leaves, twigs, flowers, fruits, seeds), terrestrial invertebrates (such as ants, termites, coleopterans, dipterans, and spiders), and detritus (Matthes 1964). Some fishes inhabiting the margin of the inundated forest also benefit from allochthonous forest foods that fall into the water or emerge floating from the interior.

RAPIDS

The 350 km of rapids in the Congo River between Stanley Pool (near Kinshasa) and Matadi (see figure 16.2) probably represents the most extensive area of rapids in the tropics (Beadle 1981). The rapids are characterized by an abundance of oxygen but very low light levels under the rocks, where some rapids fishes spend much of their time. Some remarkable endemic adaptations characterize the fish fauna of this region. The most common structural modification of the rapids fishes is reduction of the eyes. Of the sixty-six highly specialized rapids fishes that Roberts and Stewart (1976) found in their

study of the rapids of the lower Congo River, twenty-six species had small or minute eyes, five species had eyes reduced in size and partially or completely covered by skin and other tissues, and one species had no eyes. Other characteristic structural modifications include mouths modified as suckers to attach to rocks, dorso-ventral flattening of the body, horizontal positioning of paired fins (which are often enlarged), and loss of accessory air-breathing organs in some clariid catfishes (Welcomme 1985; Welcomme and de Merona 1988; figure 16.7). Members of the family Amphiliidae are well adapted to life in strong currents. They are elongated, streamlined fishes that possess sucker-like mouths or enlarged fins with which they can cling to the substrate (Welcomme 1985; Welcomme and de Merona 1988; see also figure 16.7). Elongated forms occur in other rapid-adapted fishes, including the polypterid *Calamoichthys calabaricus*, the mormyrid *Mormyrops longiceps* (see figure 16.7), and the cichlid *Gobiocichla wonderi* (Welcomme and de Merona 1988).

Most of the rapids fishes in the Congo Basin are endemic, and many of them are restricted to the extensive rapids between Pool Malebo and Matadi (Roberts and Stewart 1976). The families most represented in this radiation are the Mormyridae, Bagridae, Mochokidae, and Cichlidae. The radiation of endemic rapids species of cichlids is particularly surprising given that cichlids seldom comprise more than a few species in other riverine faunas.

Our estimates of the diversity and endemism of fish faunas of the rapids habitats is conservative, because these fishes are not well known, particularly those from the Ubangui, the Kasai, and the Lualaba regions of the Congo Basin (Roberts and Stewart 1976). Sampling is difficult, and surveys have been few. Additional sampling will no doubt yield new species in this region and novel adaptations to this extreme environment.

Figure 16.7. Adaptations to swift current in some African fishes that inhabit the rapids of the rain forest rivers: (A) *Gymnallabes typus* (elongated body); (B) *Mormyrops longiceps* (elongated body); (C) *Amphilius platychir* and (D) its sucker-like mouth; (E) *Chiloglanis micropogon* and (F) its elaborate sucker-like mouth. Drawings adapted from Welcomme and de Merona 1988, courtesy of IRD (Institut de Recherche pour le Développement).

## DEOXYGENATION

Oxygen scarcity (hypoxia) is widespread in tropical freshwaters, particularly in heavily vegetated swamps, marshes, inundated forest areas, and the dry-season pools of intermittent watercourses (Carter and Beadle 1930; Carter 1955; Kramer et al. 1978; Chapman and Kramer 1991; Chapman and Liem 1995; Chapman et al. 1998). In these habitats, the availability of oxygen is strongly influenced by the flood cycles, and these two factors create strong seasonal changes in habitat availability. Hypoxic habitats are well represented in the rain forests of Africa. Extensive wetlands, including enormous areas of swamp forest, lie in the central Congo Basin, north of the confluence of the Congo and Oubangui Rivers (see figure 16.2), and several shallow lakes (including Lake

Tumba) are associated with these swamps. Swamp forests are also prevalent in some lowland areas of west Africa, which grade into extensive mangrove in coastal areas. In the forests of east Africa, one finds river valleys choked with wetlands, which can extend for several kilometers. The permanent swamps in east Africa and parts of central Africa are dominated by papyrus (*Cyperus papyrus*) (Carter 1955). Along sections of the Congo River, *Vossia cuspidata* (hippo grass) replaces papyrus, and the permanent forested swamps of the central Congo Basin are dominated by *Cyrtosperma senegalense* and the palm *Raphia* (Howard-Williams and Gaudet 1985). Papyrus is relatively rare in west Africa, where it is replaced by *Cyrtosperma senegalense* and *Vossia*.

Heavily vegetated swamps where emergent

vegetation shades the water from light and wind are often characterized by very low oxygen conditions. For example, the dense canopy of *Cyperus papyrus* (which averages 4–5 m in height) intercepts more than 90% of incoming radiation (Thompson et al. 1979; Jones and Mithuri 1985). In combination with high rates of organic decomposition, these conditions contribute to extreme hypoxia in the dense interior of some papyrus swamps, which can average less than 1.0 mg/L for much of the year (Carter 1955; Chapman and Liem 1995; Chapman et al 1999; see also figure 16.6). Temporary peripheral swamps that form part of the inundation zone can be characterized by higher oxygen levels if the vegetation permits exposure to light and mixing. These less hypoxic wetlands are extremely important nursery and breeding areas for fishes on a seasonal basis.

One sees myriad adaptations (physiological, morphological, and behavioral) by fishes to the respiratory challenges imposed by oxygen-scarce waters. Permanent swamps tend to be inhabited by a specialized fauna adapted for life in deoxygenated waters. The development of air-breathing organs is more common in tropical freshwaters than anywhere else (Roberts 1975), and in Africa one finds air-breathing representatives in at least nine of the families of freshwater fishes (see table 16.2). All of the genera of air-breathing fishes are common in the swampy regions of equatorial forest (Roberts 1975; Beadle 1981). The largest air-breathing families of the equatorial forest are the Clariidae and the Anabantidae. Unlike gills, which seem to be evolutionarily homologous among species, air-breathing organs show remarkable diversity in their structural morphology and their origins (figure 16.8). These include such innovations as diverticula of the branchial chambers (as in *Clarias*, *Anabas*, and *Ctenopoma*) and modification of the air bladder (as in *Polypterus*, *Pantodon*, and *Phractolaemus*). Air-breathing fishes combine the use of

dissolved and atmospheric oxygen; however, there is great variation in the degree of dependence on atmospheric air and in the degree of the development of gills and air-breathing organs (Carter 1957; Johansen 1970). Some species, like the lungfishes (*Protopterus* spp.; see figure 16.4), a widespread taxa in equatorial forest swamps, are obligatory air breathers and will die without access to the surface. Other species, including the air-breathing *Clarias*, have well-developed gills and can meet their oxygen requirements using water breathing at higher oxygen levels.

Air breathing permits the lungfish to withstand extended periods of desiccation that can characterize dry-season conditions in some wetlands. Methods for surviving prolonged dry periods vary among lungfish species and include secreting subterranean cocoons (*Protopterus annectens*; see figure 16.8), lying in water-filled subsurface burrows (*Protopterus dolloi*; see figure 16.8), or burrowing into moist regions of the substrate (*Protopterus aethiopicus;* this species is also known to aestivate like *Protopterus annectens* [Greenwood 1987]).

Although air-breathing fishes tend to be widespread in swamps of equatorial Africa, some non-air breathers also survive in dense wetlands. For example, the small cyprinid, *Barbus neumayeri*, is found in the interior of papyrus swamps of Kibale Forest in Uganda, where it survives by virtue of a low critical oxygen tension, high hemoglobin, large gills, and extremely efficient use of aquatic surface respiration (Chapman and Liem 1995; Olowo and Chapman 1996). Even when a water column is devoid of oxygen, diffusion will maintain a microlayer of well-oxygenated water at the surface. Some water-breathing fishes, like *Barbus neumayeri*, skim the surface under extreme hypoxia to exploit the oxygen-rich surface film, a behavior referred to as aquatic surface respiration, or ASR (Kramer and Mehegan 1981; see figure 16.8). ASR is common among non-

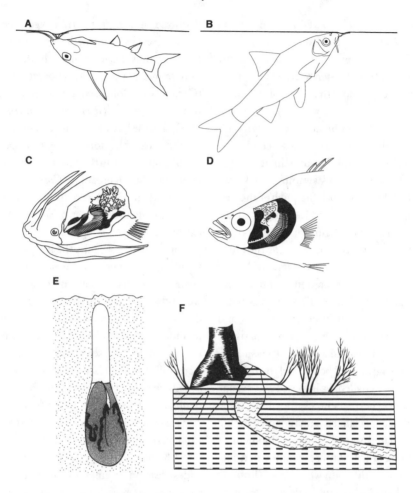

Figure 16.8. Adaptations to deoxygenation and desiccation in some African fishes that are found in rain forest rivers: (A) posture used by *Synodontis nigriventris* and (B) *Barbus neumayeri* during aquatic surface respiration; air-breathing organs derived from the branchial organs in (C) *Clarias* and (D) *Ctenopoma;* (E) position of an estivating and cocooned *Protopterus annectens;* and (F) dry-season burrow and wet-season breeding nest of *Protopterus dolloi* (dry soil is indicated by solid lines, wetter regions by interrupted horizontal lines, and water within the nest by sinuous lines). Drawings adapted from Chapman et al. 1994 (A); Marshall 1966 (C) ; Norris 1992 (D); Johnels and Svensson 1954 (E); and Brien et al. 1959 (F).

air-breathing fishes from hypoxic habitats (Gee et al. 1978; Kramer and McClure 1982; Kramer 1983; Winemiller 1989; Chapman et al. 1995); however, species of fishes with sub-terminal mouths (like many catfishes) face the problem of attaching their mouth to the surface film. The upside-down swimming habit of the central African mochokid catfish, *Synodontis*

*nigriventris*, is one solution to this problem (see figure 16.8). This catfish exhibits reverse countershading and can be found swimming inverted (Daget 1948), a habit often associated with exploitation of the water surface for feeding or respiration (Chapman et al. 1994).

Roberts (1973) noted an interesting link between oxygen-deficient habitats and parental care in fishes of the Amazon and Congo Basins. He observed that in these large tropical rivers, parental care occurs primarily in fishes in which the adults breed in swamps and other oxygen-deficient habitats. For example, in the African lungfish (*Protopterus* spp.), a nest is constructed and guarded by the male (Johnels and Svennson 1954; Bouillon 1961; Greenwood 1987). In the central Congo Basin *P. dolloi* excavates a burrow that receives air through a chimney-like structure, without which the eggs would be deprived of oxygen (see figure 16.8). The small anabantid fish *Ctenopoma damasi* and the characoid *Hepsetus odoe* construct floating nests of foam in which the eggs are supported (Berns and Peters 1969; Roberts 1973).

Although the dense interior of heavily vegetated wetlands in African forests is depauperate with respect to fish species richness, these habitats may still contribute to the maintenance of fish faunal structure and diversity for a number of reasons. First, wetlands provide important habitats for some fishes that are rare in other habitats (such as *Phractolamus*, *Ctenopoma*, and some clariid catfishes). Second, swamps may contribute to faunal diversification. For air-breathing fishes, papyrus swamps and other large swampy divides are not likely to be a barrier to dispersal. However, for hypoxia-intolerant water breathers, dense swamps may limit movement and serve as an isolating mechanism. For water breathers that can survive in the dense interior of papyrus, the use of, and dispersal through, papyrus are likely to be limited by oxygen availability and the efficiency of oxygen uptake in the species. This may result in geographical variation between swamp populations and populations from open-water sections of the drainage. For example, there are significant differences in the gill morphology, critical oxygen tension, respiratory behavior, and life-history parameters between *B. neumayeri* from a papyrus swamp in the Kibale

Forest of Uganda and *B. neumayeri* from other sites in the drainage with higher oxygen levels (Chapman and Liem 1995; Olowo and Chapman 1996). The relationship between the gill size of *Barbus neumayeri* and dissolved oxygen concentration at four sites in Kibale indicates that this variation can occur over very small geographical scales in response to local oxygen conditions (figure 16.9). Recent data have revealed very low dispersal rates between the swamp and river or stream populations, but one that would be theoretically sufficient to homogenize gene frequencies (Chapman et al. 1999). However, randomly amplified polymorphic DNA (RAPD) markers indicated significant genetic differentiation among sites, suggesting habitat-specific selection pressures on dispersers (Chapman et al. 1999).

Unfortunately, we understand very little about the ecology of fishes in the wetlands of African rain forests, particularly in the huge expanses of wetland in the central Congo Basin. Oxygen seems to play a major role in structuring wetland communities, but it is not the only factor of importance. The interaction between oxygen and other physicochemical parameters in affecting fish movement patterns and distribution has received little attention. Often wetlands and other hypoxic habitats, like areas of the inundated forests, serve as nursery areas. It is important that we begin to examine the strategies developed to cope with such environmental challenges during the early stages of life and how the conditions of these habitats affect growth and development. In addition, oxygen levels vary with seasonal flooding, permitting swamp access to many species that are unable to cope with dry-season conditions. To evaluate the importance of rain forest wetlands for fish ecology, evolution, and fisheries production, we need to examine, in much more detail and over a much broader geographical scale, the seasonal patterns of wetland use and physico-

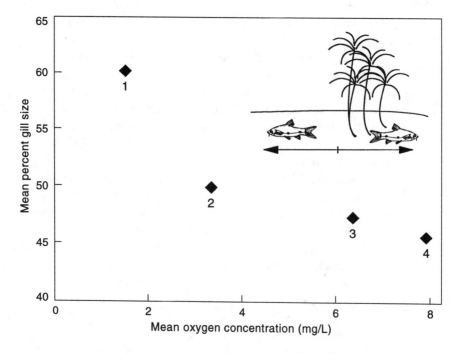

Figure 16.9. Scatter plot showing the relationship between mean percent gill size for four populations of *Barbus neumayeri* from Kibale National Park, Uganda, and mean dissolved oxygen concentration (mg/L). Mean percent gill size represents the relative position within the estimated range of total gill filament length for freshwater fishes, following Palzenberger and Pohla 1992. Values of dissolved oxygen represent the means of two years of monthly samples for the Rwembaita Swamp (1), the Inlet Stream East (2), and the Njuguta River (3), and the mean of three months of samples for the Mpanga River (4). Numbers refer to the locations indicated on figure 16.5.

chemical thresholds for movement in and out of wetland sites.

### Threats to Forest-Fish Interactions

Three of the most serious potential threats to the maintenance of fish faunal structure and diversity in African rain forests are deforestation, wetland degradation, and species introductions. However, the independent or interactive effects of these factors are largely unknown for many African waters. Fish biology in African rain forests suffers from a lack of baseline ecological studies, which would provide more solid grounds for predicting impacts of anthropogenic change. Nevertheless, from

what we understand about accelerating rates of land conversion in the region, the continued introductions of non-native species, and studies in other tropical regions, we can anticipate the nature and severity of such perturbations.

Deforestation in rain forest areas is particularly severe in west Africa, averaging 90% of the original forest cover from Sierra Leone to Nigeria. However, the central DR Congo region is also threatened from all directions, with loss of rain forest estimated at 57% (World Resources Institute 1994; see figure 16.1). Other areas of central Africa are experiencing similar loss. In Madagascar, the eastern rain forest region is largely deforested (84%

loss), a region including several small rivers that drain the forested slopes. In Uganda, remaining forests are primarily isolated fragments; deforestation has been severe, with an estimated 86% loss of tropical moist forest (World Resources Institute 1994; see Wilkie and Laporte, this volume).

Deforestation threatens forest fish faunas on several dimensions, indirectly through the effects of forest removal on water quality and flow regimes, and directly through loss of allochthonous input generated by the forest. Deforestation of the watershed leads to changes in the seasonal flood regime. In forested watersheds, vegetation and topsoil aid in retaining water. With removal of the forest and topsoil, flood peaks tend to become higher and shorter, because runoff is not delayed by the holding capacities of the forest. Dramatic changes in the flood regime can negatively impact fish populations that require a smoother seasonal transition (Welcomme 1985).

Increased turbidity is associated with increased sedimentation and siltation, and much of the increased silt load of tropical rivers in recent history has been associated with the deforestation of land in the upper regions of the watershed (Marlier 1973; Eckholm 1976). In the Ulu Segama rain forest of Malaysia, Douglas et al. (1992) found that logging resulted in a dramatic eighteenfold increase in sediment yield immediately after the bulk of the catchment had been logged. One year after logging, the system had not recovered, although sediment yields were much reduced, to 3.6 times those of the undisturbed forest. Increased sedimentation and higher turbidity can lead to the decline of plankton through a reduction in light penetration, and to a decline in fish production, as well as the disappearance of many benthic animals that are sensitive to mud on their integument and gills or lose their interstitial habitats to clogging by silt (Chutter 1968; Burns 1972; Marlier 1973). Clearing of the

forests also leads to increased sunlight and higher water temperatures. Marlier (1973) reported changes in chemical composition associated with increased solar insulation for small watercourses in the eastern Congo Basin. Waters from deforested areas had higher pH and a higher concentration of salts than protected waters.

We understand little about the potential effects of forest clearance on the productivity of fish populations in African rain forests. It is possible that productivity could increase, as was seen in the salmonid standing crops in deforested areas of the northwestern United States and in many benthic invertebrate communities that were invaded by tolerant taxa (Welcomme 1985). However, given that many rain forest rivers in Africa have relatively low primary productivity and rely largely on allochthonous input, destruction of the forest may result in a decline in aquatic productivity. In areas of the central Congo Basin where flooding allows fishes to find food, refuge, and breeding areas in the inundated forest, the consequences of deforestation may be particularly severe.

Although data are few on the impact of flooded forest conversion in African rain forests, the recent history of the Mekong system of Asia provides a basis on which to speculate. In Cambodia, the Mekong River floods into the Grand Lac, a seasonally fluctuating lake, and surrounding floodplain (formerly forest). Many fishes move into the Grand Lac as it fills and then move out into the inundated zone, returning to the lake and migrating downriver as waters fall in the dry season (Lowe-McConnell 1975). Chevey and Le Poulain (1940) found that growth of some cyprinids was faster in the flooded forest than in the open lake and the river, which they attributed to abundant fish food in the forest.

Subsequent to their study, the forest surrounding the Grand Lac area of the Mekong was cleared for agriculture and firewood. This clearance was accompanied by a decline of about 50% in the fishery in twenty-five years (Welcomme 1985). The decline was attributed to erosion and siltation in the Mekong Basin and the dramatic decline of allochthonous food arising from the deforestation. Siltation led to increased turbidity of waters, which in the case of the Grand Lac coincided with a drop in primary production reflected in decreased fish production. Welcomme (1985) reports that the process was reversed during the late 1970s when political troubles led to a collapse of the human population in the region, which permitted the forest around the Grand Lac to regenerate. Forest regeneration coincided with an increase in fish populations, which may have been related to the renewal of forest resources and reduced fishing pressure.

In Madagascar, forested freshwater regions have undergone some of the most severe habitat degradation in Africa (IUCN 1985; Reinthal and Stiassny 1991). In their survey of Madagascan fish communities in the heavily degraded forests of the central and eastern highlands, Reinthal and Stiassny (1991) reported severe attrition of the endemics. Despite intensive sampling in freshwater habitats, they recorded only six endemic species of the twenty-eight freshwater endemics known from Madagascar. However, deforestation is only one of two major factors that correlated with severe attrition of the freshwater fish fauna. Exotic species introductions was the other. Reinthal and Stiassny (1991) found that the fish communities were dominated by exotic species, and most sites were composed entirely of introduced species, the most common of which were the tilapiine *Oreochromis mossambicus* and two poeciliid fishes, *Xiphophorus helleri* and *Gambusia holbrooki*.

Within Africa, the transfer or introduction of non-native species has been widespread and continues (FAO/CIFA 1985). The primary purpose of such transfers has been to maintain or increase fish yield, although some introductions have been undertaken to expand sport fisheries or to exert biological control (Ogutu-Ohwayo and Hecky 1991). The most dramatic and catastrophic introduction involves the transfer of the Nile perch (*Lates niloticus*) into Lake Victoria. Lake Victoria is the largest tropical lake in the world and shares its waters with Uganda, Kenya, and Tanzania. It is not a rain forest environment, but it illustrates the basic cycles of events that can follow the introduction of non-native predators into aquatic systems. More than 50% of the endemic fishes disappeared from Lake Victoria between 1980 and 1986, and many are presumed extinct (Kaufman 1992; Witte et al. 1992a, 1992b; Kaufman et al. 1997). The introduced Nile perch is proposed to have been a major contributor to the mass extinction; the decline in endemic haplochromines is almost reciprocal with the increase in Nile perch (Ogutu-Ohwayo 1990a, 1990b; Kaufman 1992; Witte et al. 1992a, 1992b). As recently as 1978, the haplochromine fauna was intact; haplochromines contributed about 80% of the biomass and Nile perch less than 2%, with the remainder consisting of the introduced *Oreochromis niloticus* and native non-cichlids (Kaufman 1992). By 1983 in Kenya and by 1986 in Tanzania, the Nile perch constituted more than 80% of the catch (Kaufman 1992). The situation in Lake Victoria has been followed most closely by the international community, because of its economic importance and extremely high endemism. But similar changes have occurred with the introduction of the Nile perch into other lakes in the region (such as Lakes Kyoga and Nabugabo; see Ogutu-Ohwayo 1993; Chapman et al. 1996b) and with the introduction of other fish predators in other tropical freshwaters (such as the introduction of *Cichla ocellaris* into Gatun Lake, Panama; see Zaret and Paine 1973).

The distribution of introduced fishes in the

waters of African rain forests is not well known, and should be addressed. Sites of early surveys in the central Congo Basin (Pool Malebo, Lake Tumba, and the Tschuapa River) should be revisited to evaluate spread and impact of non-native species. In Kibale National Park of Uganda, two introduced fishes (*Poecilia reticulata* and *Oreochromis niloticus*) have been recorded in one of the major rivers draining the forest. At this point they are rare, with no apparent impact; however, spread is inevitable given the prevalence of introduced tilapia and other non-natives in Uganda and accelerating interest and donor support for aquaculture activities in the region. Fish introductions will continue to be made either for biological management or by accidental transfer, and it is therefore critical that we begin to understand the community dynamics associated with such transfers. Two issues are particularly relevant to the introduction of exotic fishes into aquatic systems. First, we must know what limits leaks of exotics from the intended site of introduction into other water bodies. Second, we must know the characteristics of refugia that can be exploited by indigenous species when confronted with an introduced predator or competitor (Chapman et al. 1996a, 1996b). The situation in Madagascar should be an alarm signal to the potential impacts of introductions into other African rain forest waters.

The continued degradation of wetlands in east and central Africa also poses problems for fish populations. In many areas of Africa, wetlands are harvested in a sustainable manner for a variety of purposes. However, an expanding and accelerating trend is large-scale drainage and conversion to agricultural land (Denny and Turyatunga 1992). Wetlands are also threatened by irrigation schemes, improved transport along waterways, industrial pollution, and mining extracts. Estimates of wetland loss range from 40% in Cameroon to 50% in DR Congo to 70% in Liberia (World Resources Institute 1994).

Continued degradation of wetlands will obviously threaten the functional values of wetlands in their water-holding capacity and effects on microclimates (Howard-Williams and Thompson 1985). But in addition, wetland conversion may contribute to a decline of fish diversity through loss of habitat, destruction of refugia, and faunal mixing. Permanent and seasonal swamps are important year-round habitats for some indigenous fishes and are seasonal feeding and breeding grounds for many species. Loss of wetlands means loss of habitat for these species, many of which are important to local fisheries. Of particular importance are the marginal wetlands or ecotones, where the emergent wetland meets the open waters of lakes and rivers. Here oxygen levels are higher because of interaction with the open waters of the river or lake, and fish do not encounter the respiratory challenges imposed by the dense swamp interior. The vast swampy regions that border the shores of Pool Malebo and marginal swamp pools in the central Congo Basin provide cover for many of the small fishes endemic to the region from larger predatory species that may not be able to feed effectively in the structural complexity of the swamp vegetation (Poll 1959). Loss of wetland may contribute to loss of refugia for prey species from swamp-intolerant predators. Finally, by limiting movement and demanding specialization for extreme hypoxia, wetlands may contribute to faunal diversification (Chapman and Liem 1995; Olowo and Chapman 1996; Chapman et al. 1999). Valley swamps in some forest rivers, like those of Uganda, create formidable barriers to the dispersal of some fishes. However, swamp channelization has been adopted as a means of creating agricultural land and improving transport through swamps. In the steeper valleys, the straight, cleared channels reduce hydraulic resistance and decrease the water-holding capacity of the wetland, which encourages fast removal and more dramatic seasonal flooding

(Denny and Turyatunga 1992). In addition, there is the question whether channelization will remove barriers among populations whose genetic integrity was maintained by swampy divides or facilitate the expansion of introduced species whose dispersal was limited by dense swamps.

Unfortunately, baseline biological data on the interactions of fish, forests, and wetlands are inadequate to make potent predictions with respect to the consequences of continued wetland loss. However, it is clear that wetland loss will precipitate change in fish communities and potentially lead to a decline in fish diversity and biomass.

In tropical rain forests, monitoring, regeneration, and restoration programs have focused primarily on terrestrial communities. However, in reviewing threats to fish-forest interactions in African rain forests, it is clear that aquatic systems are resources that, along with the forests, are subject to significant use and conversion. In addition, aquatic systems absorb the impacts of land-use change and thereby accumulate and consolidate information on terrestrial change over broad spatial scales in flowing river systems, and broad temporal scales in static lake environments (Hellawell 1986; Crisman 1988). Lakes and reservoirs have a history because of their tendency to accumulate materials. In contrast, the renewal of water in flowing systems reflects terrestrial input at any given time. The value of aquatic faunas and water quality as indicators of terrestrial change has long been recognized (Hellawell 1986; Crisman 1988); however, the vast majority of the work on aquatic bioindicators has focused on temperate systems. Future work in rain forest waters should include studies linking the degradation and restoration of tropical forest communities with aquatic ecosystem dynamics to empirically examine relationships between the degradation and restoration of a tropical forest and the structure and composition of aquatic communi-

ties. Predictive models of terrestrial change (land-cover and forest-community composition) based on characters of fish and benthic invertebrate communities would be extremely useful in anticipating and ameliorating the impacts of terrestrial change on aquatic communities.

ACKNOWLEDGMENTS

Funding for the research in Kibale Forest and other regions of Uganda has been provided by NSERC (Canada), USAID (Kampala), the Wildlife Conservation Society, the National Geographic Society, the National Science Foundation, and the University of Florida. Permission to conduct research in Uganda was granted from the Office of the President, Uganda, the National Research Council for Science and Technology (Uganda), the Ugandan Forest Department, and the Makerere University Biological Field Station, and field programs have greatly benefited from the assistance of numerous collaborators. I also express my gratitude to the Department of Zoology, Makerere University, and to the present and past directors of the Makerere University Biological Field Station. I thank Laurie Walz for her assistance with the figures. Lee White, Colin Chapman, and Amanda Rosenberger provided comments on an early version of this manuscript.

REFERENCES

Alexander, R. M. 1966. Structure and function in the catfish. *Journal of Zoology, London* 148:88–152.

Beadle, L. C. 1981. *The Inland Waters of Tropical Africa: An Introduction to Tropical Limnology*. Longman, London.

Bennett, M. V. L. 1971. Electric organs. Pages 347–491 in W. S. Hoar and D. J. Randall, eds. *Fish Physiology*, vol. 4. Academic Press, London.

Berns, S., and H. M. Peters. 1969. On the reproductive behaviour of *Ctenopoma muriei* and *Ctenopoma damasi* (Anabantidae). *Annual Report of the East African Freshwater Fisheries Research Organization* (1968):44–49.

Bouillon, J. 1961. The lungfish of Africa. *Natural History* 70:62–71.

Brien, P., M. Poll, and J. Bouillon. 1959. Ethologie de la reproduction de *Protopterus dolloi* Blgr. *Annales du Musée Royal du Congo Belge Tervuren, Série 8, Sciences Zoologiques* 71:3–21.

Bruton, M. N. 1979. The survival of habitat desiccation by air-breathing clariid catfishes. *Environmental Biology of Fishes* 4:273–280.

Burns, J. W. 1972. Some effects of logging and associated road construction on northern California streams. *Transactions of the American Fisheries Society* 101:1–16.

Carter, G. S. 1955. *The Papyrus Swamps of Uganda*. Heffer, Cambridge, England.

———. 1957. Air breathing. Pages 65–79 in M. E. Brown, ed. *The Physiology of Fishes*, vol. 1. Academic Press, London.

Carter, G. S., and L. C. Beadle. 1930. The fauna of the swamps of the Paraguayan Chaco in relation to its environment, 1: Physico-chemical nature of the environment. *Journal of the Linnean Society of Zoology* 37:205–258.

Chapman, L. J. 1995. Seasonal dynamics of habitat use by an air-breathing catfish (*Clarias liocephalus*) in a papyrus swamp. *Ecology of Freshwater Fish* 4:113–123.

Chapman, L. J., and C. A. Chapman. 1993. Fish populations in tropical floodplain pools: A re-evaluation of Holden's data on the River Sokoto. *Ecology of Freshwater Fish* 2:23–30.

Chapman, L. J., C. A. Chapman, D. A. Brazeau, B. McLaughlin, and M. Jordan. 1999. Papyrus swamps, hypoxia, and faunal diversification: Variation among populations of *Barbus neumayeri*. *Journal of Fish Biology* 54:310–327.

Chapman, L. J., C. A. Chapman, and M. Chandler. 1996a. Wetland ecotones as refugia for endangered fishes. *Biological Conservation* 78:263–270.

Chapman, L. J., C. A. Chapman, and T. L. Crisman. 1998. Limnological observations of a papyrus swamp in Uganda: Implications for fish faunal structure and diversity. *Verhandlungen Proceedings* 26:1821–1826.

Chapman, L. J., C. A. Chapman, L. S. Kaufman, and F. E. McKenzie. 1995. Hypoxia tolerance in twelve species of East African cichlids: Potential for low oxygen refugia in Lake Victoria. *Conservation Biology* 9:1274–1288.

Chapman, L. J., C. A. Chapman, R. Ogutu-Ohwayo, M. Chandler, L. Kaufman, and A. E. Keiter. 1996b. Refugia for endangered fishes from an introduced predator in Lake Nabugabo, Uganda. *Conservation Biology* 10:554–561.

Chapman, L. J., L. Kaufman, and C. A. Chapman. 1994. Why swim upside down? A comparative study of two mochokid catfishes. *Copeia* 1994:130–135.

Chapman, L. J., and D. L. Kramer. 1991. Limnological observations of an intermittent tropical dry forest stream. *Hydrobiologia* 226:153–166.

Chapman, L. J., and K. F. Liem. 1995. Papyrus swamps and the respiratory ecology of *Barbus neumayeri*. *Environmental Biology of Fishes* 44:183–197.

Chevey, P., and F. Le Poulain. 1940. La pêche dans les eaux douces du Cambodge. *Memoire l'Institut Oceanographique de l'Indochine* 5:193 pp.

Chutter, F. M. 1968. The effects of silt and sand on the invertebrate fauna of streams and rivers. *Hydrobiologia* 34:57–76.

Crisman, T. L. 1988. The use of subfossil benthic invertebrates in aquatic resource management. Pages 71–88 in W. J. Adams, G. A. Chapman, and W. G. Landis, eds. *Aquatic Toxicology and Hazard Assessment*. American Society for Testing and Materials, Philadelphia.

Daget, J. 1948. Les *Synodontis* (Siluridae) à polarité pigmentaire inversée. *Bulletin du Muséum National d'Histoire Naturelle* 20:239–243.

Daget, J., J.-P. Gosse, G. G. Teugels, and D. F. E. Thys van den Audenaerde, eds. 1991. *Check-List of the Freshwater Fishes of Africa*. Vol. 4. ISNB, Brussels; MRAC, Tervuren, Belgium; ORSTOM, Paris.

Daget, J., J.-P. Gosse, and D. F. E. Thys van den Audenaerde, eds. 1984. *Check-List of the Freshwater Fishes of Africa*. Vol. 1. MRAC, Tervuren, Belgium; ORSTOM, Paris.

———. 1986. *Check-List of the Freshwater Fishes of Africa*. Vol. 2. ISNB, Brussels; MRAC, Tervuren, Belgium; ORSTOM, Paris.

Daget, J., and A. Iltis. 1965. Poissons de Côte d'Ivoire (eaux douces et saumatres). *Mémoires de l'Institut Français d'Afrique Noire* 74:85 pp.

Daget, J., N. Planquette, and P. Planquette. 1973. Premieres données sur la dynamique des peuplements de poissons du Bandama (Côte d'Ivoire). *Bulletin du Muséum National d'Histoire Naturelle, Série 3, Ecologie générale* 151:129–143.

Denny, P., and F. Turyatunga. 1992. Ugandan wetlands and their management. Pages 77–84 in E. Maltby, P. J. Dugan, and J. C. Lefeuvre, eds. *Conservation and Development: The Sustainable Use of Wetland Resources: Proceedings of the Third International Wetlands Conference* (1988). IUCN, Rennes, France.

Douglas, I., T. Spencer, T. Greer, K. Bidin, W. Sinun, and W. W. Meng. 1992. The impact of selective commercial logging on stream hydrology, chemistry and sediment loads in the Ulu Segama rain forest, Sabah, Malaysia.

*Philosophical Transactions of the Royal Society of London B* 335:397–406.

Eckholm, E. P. 1976. *Losing Ground: Environmental Stress and World Food Prospects.* Norton, New York.

FAO. 1981. *Tropical Forest Resources Assessment Project (GEMS): Tropical Africa, Tropical Asia, Tropical America.* 4 vols. FAO/UNEP, Rome.

FAO/CIFA. 1985. *Introduction of Species and Conservation of Genetic Resources.* Committee for Inland Fisheries of Africa (CIFA), Lusaka, Zambia. CIFA 85/113. FAO, Rome.

Gee, J. H., R. F. Tallman, and H. J. Smart. 1978. Reactions of some great plains fishes to progressive hypoxia. *Canadian Journal of Zoology* 56:1962–1966.

Gosse, J.-P. 1963. Le milieu aquatique et l'écologie des poissons dans la région de Yangambi. *Annales du Musée Royal du Congo Belge Tervuren, Série 8, Sciences Zoologiques* 116:113–270.

Goulding, M. 1980. *The Fishes and the Forest.* University of California Press, Berkeley.

Greenwood, P. H. 1966. *The Fishes of Uganda.* Uganda Society, Kampala.

———. 1981. *The Haplochromine Fishes of the East African Lakes.* Kraus International Publications, Munchen.

———. 1987. The natural history of African lungfishes. *Journal of Morphology, Suppl.* 1:163–179.

———. 1994. The species flock of cichlid fishes in Lake Victoria and those of other African Great Lakes. *Archiv für Hydrobiologie, Beiheft, Ergenbnisse der Limnologie.* 44:347–354.

Hellawell, J. M. 1986. *Biological Indicators of Freshwater Pollution and Environmental Management.* Elsevier Applied Science Publishers, London.

Hopkins, C. D. 1981. On the diversity of electric signals in a community of mormyrid electric fish in West Africa. *American Zoology* 21:211–222.

———. 1986. Behavior of Mormyridae. Pages 527–576 in T. H. Bullock and W. Heiligenberg, eds. *Electroreception.* John Wiley and Sons, New York.

Howard-Williams, C., and J. J. Gaudet. 1985. The structure and functioning of African swamps. Pages 154–175 in P. Denny, ed. *The Ecology and Management of African Wetland Vegetation.* W. Junk Publishers, Dordrecht, The Netherlands.

Howard-Williams, C., and K. Thompson. 1985. The conservation and management of African wetlands. Pages 203–230 in P. Denny, ed. *The Ecology and Management of*

*African Wetland Vegetation.* W. Junk Publishers, Dordrecht, The Netherlands.

Hugueny, B. 1989. West African rivers as biogeographic islands: Species richness of fish communities. *Oecologia* 79:236–243.

IUCN. 1985. Madagascar: A conference for the future. *IUCN Bulletin Supplement* 3.

Johansen, K. 1970. Air breathing in fishes. Pages 361–413 in W. S. Hoar and D. J. Randall, eds. *Fish Physiology.* Vol. 4. Academic Press, New York.

Johnels, A. G., and G. S. O. Svensson. 1954. On the biology of *Protopterus annectens* (Owen). *Arkiv for Zoologi* 7:131–164.

Jones, M. B., and F. M. Muthuri. 1985. The canopy structure and microclimate of papyrus (*Cyperus papyrus*) swamps. *Journal of Ecology* 73:481–491.

Kaufman, L. S. 1992. Catastrophic change in species-rich freshwater ecosystems: The lessons of Lake Victoria. *Bioscience* 42:846–858.

Kaufman, L., L. J. Chapman, and C. A. Chapman. 1996. The great lakes. Pages 178–204 in R. McClanahan and T. P. Young, eds. *East African Ecosystems and Their Conservation.* Oxford University Press, New York.

———. 1997. Evolution in fast forward: Haplochromine fishes of the Lake Victoria region. *Endeavour* 21:23–30.

Kramer, B. 1990. *Electrocommunication in Teleost Fishes.* Springer-Verlag, Berlin.

Kramer, D. L. 1983. Aquatic surface respiration in the fishes of Panama: Distribution in relation to risk of hypoxia. *Environmental Biology of Fishes* 8:49–54.

Kramer, D. L., C. C. Lindsey, G. E. E. Moodie, and E. D. Stevens. 1978. The fishes and the aquatic environment of the central Amazon basin, with particular reference to respiratory patterns. *Canadian Journal of Zoology* 56:717–729.

Kramer, D. L., and M. McClure. 1982. Aquatic surface respiration, a widespread adaptation to hypoxia in tropical freshwater fishes. *Environmental Biology of Fishes* 7:47–55.

Kramer, D. L., and J. P. Mehegan. 1981. Aquatic surface respiration, an adaptive response to hypoxia in the guppy, *Poecilia reticulata* (Pisces, Poeciliidae). *Environmental Biology of Fishes* 6:299–313.

Liem, K. F. 1987. Functional design of the air ventilation apparatus and overland excursions by teleosts. *Fieldiana (Zool.)* 37:1–29.

Lowe-McConnell, R. H. 1975. *Fish Communities in Tropical Freshwaters.* Longman, London.

————. 1988. Broad characteristics of the ichthyofauna. Pages 93–110 in C. Leveque, M. N. Bruton, and G. W. Ssentongo, eds. *Biology and Ecology of African Freshwater Fishes*. ORSTOM, Paris.

Marlier, G. 1973. Limnology of the Congo and Amazon rivers. Pages 223–238 in B. J. Meggers, E. S. Ayensu, and W. D. Duckworth, eds. *Tropical Forest Ecosystems in Africa and South America: A Comparative Review*. Smithsonian Institution Press, Washington, D.C.

Marshall, N. B. 1966. *The Life of Fishes*. World Publishing, Cleveland.

Matthes, H. 1961. Feeding habits of some Central African freshwater fishes. *Nature*. 192:78–80.

————. 1964. Les poissons du Lac Tumba et de la région d'Ikela: Etude systématique et ecologique. *Annales du Musée Royal du Congo Belge Tervuren, Série 8, Sciences Zoologiques* 126: 204 pp.

Norris, S. M. 1992. Anabantidae. Pages 837–847 in C. Leveque, D. Paugy, and G. G. Teugels, eds. *Faune des Poissons d'Eaux Douces et saumatres d'Afrique de l'Ouest*, vol. 2. Musée Royal de l'Afrique Centrale, Tervuren, Belgium. Collection Faune Tropical no. 28. ORSTOM, Paris.

Ogutu-Ohwayo, R. 1990a. The decline of the native fishes of Lakes Victoria and Kyoga (East Africa) and the impact of introduced species, especially the Nile perch, *Lates niloticus*, and Nile tilapia, *Oreochromis niloticus*. *Environmental Biology of Fishes* 27:81–96.

————. 1990b. Changes in prey ingested and the variations in the Nile perch and other fish stocks in Lake Kyoga and the northern waters of Lake Victoria (Uganda). *Journal of Fish Biology* 37:55–63.

————. 1993. The effects of predation by Nile perch, *Lates niloticus*, on the fish of Lake Nabugabo, with suggestions for conservation of endangered endemic cichlids. *Conservation Biology* 7:701–711.

Ogutu-Ohwayo, R., and R. E. Hecky. 1991. Fish introductions in Africa and some of their implications. *Canadian Journal of Fisheries and Aquatic Science* 48 (suppl.):8–12.

Olowo, J. P., and L. J. Chapman. 1996. Papyrus swamps and variation in the respiratory behaviour of the African fish *Barbus neumayeri*. *African Journal of Ecology* 34:211–222.

Palzenberger, M., and H. Pohla. 1992. Gill surface area of water-breathing freshwater fish. *Reviews in Fish Biology Fisheries* 2:187–216.

Paugy, D., C. Leveque, G. G. Teugels, R. Bigorne, and R. Romand. 1990. Freshwater fishes of Sierra Leone and Liberia. *Revue d'Hydrobiologie Tropicale* 23:329–350.

Poll, M. 1957. Les genres des poissons d'eau douce de l'Afrique. *Annales du Musée Royal du Congo Belge Tervuren, Série 8, Sciences Zoologiques* 54:1–191.

————. 1959. Recherches sur la faune ichthyologique de la region du Stanley Pool. *Annales du Musée Royal du Congo Belge Tervuren, Série 8, Sciences Zoologiques* 71:75–174.

————. 1973. Nombre et distribution géographique des poissons d'eau douce africains. *Bulletin du Muséum National d'Histoire Naturelle, Série 3* 150:113–128.

Poll, M., and J. P. Gosse. 1963. Contribution à l'étude systématique de la faune ichthyologique du Congo central. *Annales du Musée Royal du Congo Belge Tervuren, Série 8, Sciences Zoologiques* 116:45–110.

Reinthal, P. N., and M. L. J. Stiassny. 1991. The freshwater fishes of Madagascar: A study of an endangered fauna with recommendations for a conservation strategy. *Conservation Biology* 5:231–243.

Roberts, T. R. 1973. Ecology of fishes in the Amazon and Congo Basins. Pages 239–254 in B. J. Meggers, E. S. Ayensu, and W. D. Duckworth, eds. *Tropical Forest Ecosystems in Africa and South America: A Comparative Review*. Smithsonian Institution Press, Washington, D.C.

————. 1975. Geographical distribution of African freshwater fishes. *Zoological Journal of the Linnean Society* 57:249–319.

Roberts, T. R., and D. J. Stewart. 1976. An ecological and systematic survey of fishes in the rapids of the Lower Zaire or Congo River. *Bulletin of the Museum of Comparative Zoology* 147:239–317.

Skelton, P. H. 1988. The distribution of African freshwater fishes. Pages 65–91 in C. Leveque, M. N. Bruton, and G. W. Ssentongo, eds. *Biology and Ecology of African Freshwater Fishes*. ORSTOM, Paris.

Thompson, K, P. R. Shewry, and H. W. Woolhouse. 1979. Papyrus swamp development in the Upemba Basin, Zaire: Studies of population structure in *Cyperus papyrus* stands. *Botanical Journal of the Linnean Society* 78:299–316.

Welcomme, R. L. 1979. *Fisheries Ecology of Floodplain Rivers*. Longman, London.

————. 1985. River fisheries. FAO Fisheries Technical Paper 262, Rome.

Welcomme, R. L., and B. de Merona. 1988. Fish communities of rivers. Pages 251–276 in C. Leveque, M. N. Bruton, and G. W. Ssentongo, eds. *Biology and Ecology of African Freshwater Fishes.* ORSTOM, Paris.

Winemiller, K. O. 1989. Development of dermal lip protuberances for aquatic surface respiration in South American characid fishes. *Copeia* 1989:382–390.

Witte, F., T. Goldschmidt, P. C. Gouswaard, W. Ligtoet, M. J. P. van Oijen, and J. H. Wanink. 1992a. Species extinction and the concomitant ecological changes in Lake Victoria. *Netherlands Journal of Zoology* 42:214–232.

Witte, F. T. Goldschmidt, J. Wanink, M. van Oijen, K. Goudswaard, E. Witte-Mass, and N. Bouton. 1992b. The destruction of an endemic species flock: Quantitative data on the decline of the haplochromine cichlids of Lake Victoria. *Environmental Biology of Fishes* 34:1–28.

World Resources Institute (in collaboration with UNEP and UNDP). 1994. *World Resources, 1994–1995.* Oxford University Press, New York.

Zaret, T. M., and R. Paine. 1973. Species introduction in a tropical lake. *Science* 82:449–455.

# CHAPTER 17

# Herpetofauna of the African Rain Forest

Overview and Recommendations for Conservation

Dwight P. Lawson and Michael W. Klemens

Within the rain forests of southwestern Cameroon lie some of the world's most herpetologically diverse sites. For example, amphibian biodiversity at Korup National Park and the neighboring Banyang-Mbo Wildlife Sanctuary is very high: around ninety species have already been recorded (Lawson 1993), and this list could be expected to grow with additional inventory and taxonomic revisions. Cameroon is remarkably rich in amphibians. Approximately 10% of the world's known species have been reported from this country, which is roughly the size of Texas.

In Tanzania, forested habitat is found on mountains and in patches along the coast. Over millennia, these patches have become increasingly isolated because of climatic shifts (Lovett 1993a) and, more recently, human activities. Many of these forests, especially those of the ancient Eastern Arc Mountains, which stretch in a broken crescent from the Taita Hills in southeastern Kenya to the Udzungwa Mountains in southern Tanzania, are characterized by high degrees of endemism among flora (Hawthorne 1993; Lovett 1993b), millipedes (Hoffman 1993), Linyphiid spiders (Scharff 1993), amphibians (Howell 1993), and reptiles (Howell 1993). Poynton (1998) reported that

Tanzania has the richest Bufonid fauna (twenty-eight species) in sub-Saharan Africa; 57% of this diversity is composed of species restricted to forested habitats.

What is remarkable, however, is not only the rich herpetological diversity that has already been reported from sites that we and others are studying in Cameroon (DPL) and Tanzania (MWK), but how much still remains to be learned both there and throughout the African rain forest block. Only a fraction of Africa has been studied, and large tracts of forest, savanna, and montane grasslands have not even been cursorily inventoried.

The herpetological inventories that we are conducting at Korup, Banyang-Mbo, and in the Udzungwa Mountains National Park of Tanzania highlight both the richness of the African rain forests and our general lack of understanding of the herpetological diversity of Africa. We are only beginning to understand the historical factors that have generated the patterns of diversity being defined by a handful of herpetologists working at widely separated sites within the world's second largest rain forest. Species lists, which are essential to understanding the fundamental patterns of distribution and diversity, exist for only a few sites in Africa.

For example, most amphibian and reptile collecting that has taken place in eastern Africa since Arthur Loveridge's back-country expeditions of the late 1920s and 1930s has been conducted at sites that are relatively accessible. Most of these surveys have not been conducted in a systematic, repeatable manner and therefore cannot be considered comprehensive. For example, Sanje Falls in the Udzungwa Mountains National Park has been repeatedly visited by zoologists and ecotourists, yet on the ridge just to the south of Sanje Falls, a recent field expedition (Klemens et al. 1997; Klemens 1998), employing an array of collecting techniques, discovered a frog and a lizard not known from the Udzungwa Mountains.

Sadly, even before the basic task of cataloguing the diversity of amphibians and reptiles in forested Africa is even partially completed, many species will become extinct. Gone will be our opportunity to document phenomena of potential practical and theoretical value to future generations of humans. And we will have lost the incalculable ecological and aesthetic values of these interesting and important animals. Much of the vast African rain forest—long overshadowed by a focus on herpetological exploration in the Neotropics and suffering from a history of political instability, lack of research investment, and geographical inaccessibility—remains unexplored. In this chapter we shall present a brief overview of the history of herpetological exploration in two key areas of forested Africa, provide a glimpse of the historical biogeography of the herpetofauna of the African rain forests, discuss the ecological and societal importance of the African forest herpetofauna, and conclude with lessons and recommendations based on our own field research.

### History of Exploration

Much of our current understanding of Africa's mesic forest herpetofauna devolves from field research conducted during the colonial period, primarily at the southern and eastern periphery of the forest. Harvard's Museum of Comparative Zoology mounted a series of expeditions to the montane forests of eastern and southern Tanzania, producing a large volume of scientific reports and species descriptions (e.g., Barbour and Loveridge 1928; Loveridge 1932, 1933a, 1933b, 1935, 1936a, 1936b, 1942a, 1942b, 1944) as well as a checklist of the amphibians and reptiles of east Africa (Loveridge 1957). Fuchs (1939) traversed the montane grasslands and forests of the Ufipa Plateau, between Lakes Tanganyika and Rukwa. Uthmoller (1942) reported on the snakes of the more geologically recent volcanic mountains associated with the Rift Valley, including Kilimanjaro, Meru, and Hanang. Laurent made significant herpetological contributions in Angola (1954, 1964), the Great Lakes Region (1956), and the Virungas (1972). Roux-Estève and De Witte (1975) reported on the snakes of Virunga National Park, and Schmidt and Inger (1959) produced a detailed report of the amphibians of Upemba National Park. In 1938, Pitman produced a landmark study on the snakes of Uganda that was reprinted in 1974. The central portions of the forest remain the most poorly known. The American Museum's Lang-Chapin Congo Expedition resulted in three important monographs on the amphibia (Noble 1924) and reptiles (Schmidt 1919, 1923) of that region. Historically, research in the western portions of the forest was mostly limited to the edge of the forest, notably the coastal regions of Cameroon and Nigeria (Sjöstedt 1897; Werner 1897, 1898, 1899; Andersson 1903, 1905, 1907, 1909; Nieden 1908, 1910; Sternfeld 1908, 1917; Parker 1936; Sanderson 1936; Mertens 1937, 1938).

Post-colonial efforts in west and west-central Africa are largely attributable to several individuals. Amiet (e.g., 1972, 1977, 1978, 1980), in particular, has published extensively on the anurans of Cameroon. Perret (1966, 1977) has also provided considerable information on tropical anurans, in general, and Cameroon anurans, in particular. Knoepffler (1966, 1974) made important contributions to the herpetology of Gabon. He also produced an undated compilation of the amphibians and reptiles of the L'Ivindo Basin, where between 1975 and 1977 he recorded forty-seven species of amphibians and fifty-eight species of reptiles. Schiøtz (1967, 1975) made significant contributions to our knowledge of the tree frogs of both east and west Africa, as well as the amphibians of Nigeria (Schiøtz 1963). Stewart (1967) produced a valuable account of the frogs of Malawi, and Broadley and Howell (1991) contributed a checklist of Tanzanian reptiles. Poynton and Broadley's *Amphibia Zambesiaca* (1985a, 1985b, 1987, 1988, 1991) included portions of southern Tanzania in their studies, and has relevance to many of the species that occur at the southern and eastern periphery of the African rain forest. Stucki-Stern (1979) and Hughes and Barry (1969) made important contributions to our understanding of the snake fauna of Cameroon and Ghana, respectively.

Other, more recent publications serve to highlight the incompleteness of our understanding of the African forest herpetofauna. Joger (1990) noted that only half a dozen publications devoted to herpetofaunal collections from the Central African Republic have appeared over the past century. He reported 184 reptile and amphibian species from that country, a total nearly equal to the 172 species that Lawson (1993) reported from Korup National Park. This is not because the Central African Republic, a topographically and ecologically diverse country, has a particularly depauperate herpetofauna, but rather is a reflection of the lack of herpetological field exploration. De La Riva (1994) described a collection of frogs made at Monte Alén National Park in Equatorial Guinea. Twenty-four of the forty-nine species (49%) that he collected were the first records of those species for the mainland of that country.

An expedition to the Udzungwa Mountains National Park, Tanzania (Klemens et al. 1997), discovered major range extensions for the microhylid frog (*Probreviceps rungwensis*) and the arboreal lizard (*Holaspis guentheri*). Before this survey, *Probreviceps rungwensis* was known only from the type series of four specimens collected more than half a century ago on the Rungwe Plateau of southwestern Tanzania (Loveridge 1932). Beyond species lists and alpha-level taxonomy, knowledge of even the most basic life-history and ecology of the African rain forest herpetofauna is minimal. Klemens (1998) reported brachial and tympanic hypertrophy and the presence of tympanic nipples in breeding male *Arthroleptides martiensseni* (a suite of characters previously considered to be unique to the west African frog genus *Petropedetes*). All these discoveries were made during a twenty-day expedition that intensively sampled two sites, employing a wide array of collecting techniques, and serve to illustrate how much still remains to be studied and described in the African rain forest biome.

## African Herpetological Diversity

Considerable speculation has centered around the apparent lack of diversity of many groups of tropical African plants and animals as compared to the Neotropics (Moreau 1969; Laurent 1973; Richards 1973). Proposed explanations for this disparity in species richness include the hypothesized extinctions of lowland faunas during periods of severe forest reduction (Moreau 1963; Carcasson 1964) or more general observations on Africa's relative geological stability, which limited opportunities for the rapid creation of climatic gradients, which, in turn, drive rates of floral speciation (Richards 1973; Gentry 1988; Pannell and White 1988). On a pan-tropical scale, African forests do contain fewer described amphibian and reptile species than the larger and more topographically heterogeneous South and

Central American forests. However, such find-ings as the amphibian richness of Korup make it arguable that some of this perceived gulf between the Old and New Worlds is the result of a lack of basic inventories and collections, as well as a poor understanding of zoogeographic patterns for most of the African region, and the rudimentary and incomplete state of the alpha-level taxonomy of many groups.

Figure 17.1 compares the number of amphib-ian species described from four of the best-stud-ied African and Neotropical countries, which lie at comparable latitudes, encompass a similar range of elevations and equivalent habitat types, and have a comparable area of land mass. The dif-ferences in amphibian diversity illustrated in fig-ure 17.1 could be explained as a direct reflection of the higher diversity of the Neotropical fauna compared to that of Africa.

We, however, favor an alternative explana-tion for the disparity shown in figure 17.1.

Beginning in the 1960s, a stream of research efforts in the Neotropics steadily increased. Although scientists from the United States ini-tially led these efforts, their research played an important role in identifying and supporting nationals, who subsequently pursued careers in systematics and ecology. This steady increase in the development of national scien-tific capacity resulted in increased scientific output, especially in the fields of exploration, species description, and zoogeographic analy-ses. In Colombia and Ecuador this "second wave" began in the 1970s and continued, unabated, into the 1990s. In contrast, Cam-eroon and Tanzania (not unlike the rest of sub-Saharan Africa, with the notable exception of South Africa) have never been the beneficiaries of such intensive study, by either expatriates or nationals. Although the species-description curves for all four countries were roughly com-parable through the 1940s, the amount of

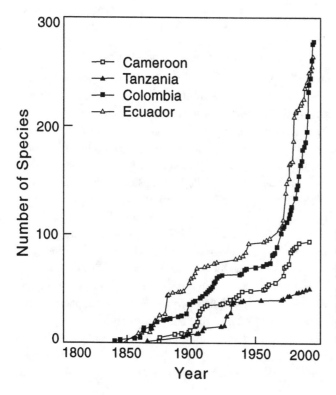

Figure 17.1. Cumulative number of recognized amphibian species described from Cameroon, Tanzania, Colombia, and Ecuador between 1800 and 1996. Synonyms were not included in this analysis. Data from Frost 1996.

research efforts in the Neotropics over the past forty years has resulted in a much better understanding of the speciation, distribution, and ecology of the Neotropical herpetofauna.

Because our alpha-level taxonomic understanding of many groups of African amphibians and reptiles is so rudimentary, available faunal lists usually underestimate biological diversity, as researchers tend to "pigeonhole" specimens from new localities into existing species descriptions. Several large, pan-African frog genera are in desperate need of taxonomic revision. The genera *Arthroleptis*, *Phrynobatrachus*, *Ptychadena*, and *Hyperolius*, for instance, are complex, species-rich groups lacking reliable diagnoses for many of their component species. Exacerbating these taxonomic problems is the lack of well-preserved comparative material from sites throughout the mesic forests, as well as series of material from single sites. Without such preserved material, it is impossible to assess the range of inter- and intra-population variation, develop taxonomic diagnoses, and conduct systematic revisions. Adequate collections of systematic reference material, including preserved specimens of various life stages, tape recordings, DNA samples, color photographs, and ecological notes are lacking for many parts of Africa. In addition, cross-indexing and retrieving data associated with natural history collections, whether for use by researchers, managers, or decision makers, is a vital, labor-intensive, and expensive precursor to taxonomic research as well as biodiversity conservation.

Rapid advances in the field of molecular systematics have made available new tools to assist in solving the huge backlog of alpha-level taxonomic problems that plague the African forest herpetofauna. However, because of the lack of comprehensive collections from a wide range of sites, these techniques have largely been applied to resolving theoretical higher-level taxonomic questions. Beyond species descriptions, preliminary checklists, and natural history notes scattered through the scientific and popular literature, little is known about the ecology and natural history of the amphibian and reptile species themselves. The need for more detailed information on the life-history and ecology of these species becomes critically apparent when considering how to effectively conserve and manage them.

## Biogeography

Africa's history of repeated and dramatic climatic shifts and their concomitant alteration of habitats has left a tangled legacy of interrelated faunal distributions. The distribution of the African mesic forest herpetofauna can be grossly divided into a series of highland centers showing varying degrees of historical interconnection and several distinctive lowland groupings. Decreased temperatures during the Pleistocene caused montane conditions to prevail over much of tropical Africa. Subsequent climatic changes caused a reduction of the montane habitat, stranding an ancestral highland fauna on isolated mountain ranges, allowing for independent radiations and the development of endemic species and the preservation of relict faunas.

The Eastern Arc Mountains of Tanzania and Kenya, the southwestern highlands of Ethiopia and the Bale Mountains, the Western Rift highlands of Rwanda, Uganda, and DR Congo, the Cameroon highlands, and the isolated west African peaks including Bioko Island, Mount Cameroon, and Mount Nimba are major montane forest centers of herpetological importance. Kingdon (1990) has summarized general highland relationships based on the distribution of various plant and animal groups. Among the amphibia, the distribution of the frog genus *Petropedetes* in Cameroon and the closely related (possibly conspecific) genus *Arthroleptides* in the Eastern Arc Mountains points to old biogeographical connections between east and west Africa (Klemens 1998). Similarly, the distribution of the four genera of "viviparous" toads (formerly all considered *Nectophrynoides*, partitioned by Dubois 1987) in the Eastern Arc Mountains of Tanzania (*Nectophrynoides*), Ethiopian highlands (*Altiphrynoides, Spinophrynoides*), and the Mount Nimba region of Liberia, Ivory Coast, and Guinea (*Nimbaphrynoides*) provides evidence of once widespread species-groups that now persist as relict endemics in montane forests across the continent. Obscuring the historical connections among highland areas is the presence of endemic radiations derived from ancestral lowland stock, such as exhibited by the rich diversity of chameleon species in the highlands of Cameroon and Tanzania.

Three major divisions of lowland herpetofauna are readily identifiable: a Congo Basin fauna stretching from eastern Cameroon and Gabon to the mountains of eastern DR Congo, a west African fauna from Senegal to western Nigeria, and a coastal fauna from the Cross River Basin of eastern Nigeria south along the coast through Cameroon and extending into Equatorial Guinea. These areas are not exclusive faunal zones, but have considerable species overlap. The differences among and between these zones has been ascribed to climatic shifts and prior forest refugia during times of severe forest contraction. The hypothesized refugia in eastern DR Congo, coastal west Africa (Cameroon, Gabon, and Equatorial Guinea), and possibly the Niger delta played a key role in maintaining forest species during periods of forest contraction, and the current fauna is undoubtedly heavily influenced by the descendants of these refugia. However, it is difficult to unravel the historical influences of these refugia from currently observed local ecological conditions. Agglomerations of once isolated forest blocks have allowed faunal interchange across much of forested central and west Africa,

and indeed, some now disjunct lowland areas show greater similarities to each other than they do to areas with which they are contiguous. Current lowland distributions, therefore, are a mixture of relict and refugial species and recent exchanges between and among formerly isolated faunas.

## Ecological Importance of the Herpetofauna

### DIVERSITY OF ECOLOGICAL ROLES

Reptiles and amphibians are a diverse and often understated component of the African rain forest environment. For example, Klemens (unpublished data) counted hundreds of litter-dwelling frogs (*Arthroleptis, Probreviceps*, and *Schoutedenella*) per day in pitfall traps in rain forest in Tanzania's Udzungwa Mountains. The high biomass attained by frogs in both primary and disturbed forest areas attests to their importance as a major invertebrate predator, and in turn, their importance as a food source for a wide variety of predators at a higher trophic level.

The African rock python, *Python sebae*, has an extensive range throughout sub-Saharan Africa, occurring in a wide variety of habitats from tropical forest to arid scrub, preying on small to medium-sized mammals such as duikers. The omnivorous forest tortoises, *Kinixys erosa* and *K. homeana*, serve as seed dispersal agents for trees and plants at Banyang-Mbo (Lawson, unpublished data). Janzen (1976) reported that the central African forests carried relatively low levels of reptile biomass when compared to the Neotropics. Yet the absence of any comprehensive ecological studies on the African forest herpetofauna renders such statements largely conjectural. Much more research will be required before any informed comparisons of either biodiversity or biomass can be made between the Neotropical and African forest herpetofaunas.

## PEST CONTROL AND PUBLIC HEALTH ISSUES

Snakes can be locally important predators on such injurious species as rats and mice, which occur in and around agricultural areas. For example, Lawson (1993) found that certain snakes at Korup and Banyang-Mbo, including vipers *Bitis nasicornis* and *Bitis gabonica*, attained higher densities in farms and plantations than in primary forest. At Banyang-Mbo, the African rock python, *Python sebae*, is the primary predator of adult cane rats, *Thryonomys swinderianus*, a highly destructive agricultural pest. The forest cobra, *Naja melanoleuca*, and Jameson's mamba, *Dendroaspis jamesonii*, are also common around human settlements and in agricultural areas. These elapids, as well as the vipers, pose a public health risk. Snake abundances in edge areas and farms may be a result of higher rodent densities and a concurrent depression of non-human predators (*sensu* Janzen 1976); however, increased microhabitat availability for these large heliothermic predators may also play an important role (Vitt and Colli 1994; Vitt et al. 1997).

### Human Food Resources

Many species of reptiles and amphibians are eaten by humans, and in some regions they serve as the primary source of protein (Butler and Shitu 1985; Lawson 1993; Klemens and Thorbjarnarson 1995). In southwestern Cameroon, villagers consume forest tortoises, *Kinixys erosa* and *K. homeana;* soft-shell turtles, *Trionyx triunguis;* mud turtles, *Pelusios niger* and *P. castaneus;* nile monitor lizards, *Varanus niloticus;* terrestrial vipers, *Bitis gabonica* and *B. nasicornis;* Jameson's mamba, *Dendroaspis jamesonii;* forest cobras, *Naja melanoleuca;* and rock pythons, *Python sebae*. The hairy frog, *Trichobatrachus robustus*, and the goliath frog, *Conraua goliath*, are considered delicacies, and even the small Pipid frogs (*Xenopus* and *Silurana*) are consumed (Lawson 1993).

Crocodiles are intensively hunted for food in southwest Cameroon and throughout much of central and west Africa (IUCN/SSC 1992). The dwarf crocodile, *Osteolaemus tetraspis*, is a favorite of hunters because of its relatively small size and the accessibility of the small streams and swamps that it inhabits (Klemens and Thorbjarnarson 1995). Monitor lizards (*Varanus*) are opportunistically hunted by the Mbuti in DR Congo (Klemens and Thorbjarnarson 1995).

### TRADITIONAL MEDICINES AND CULTURAL BELIEFS

Many of the conspicuous reptiles and amphibians figure prominently in local culture, especially the well-established system of beliefs and taboos. In Cameroon, chameleons are ascribed great powers and are greatly feared. In Tanzania, chameleons are regarded as especially evil and are often killed on sight (Klemens, pers. obs.). In the coastal forests of southeastern Tanzania, the Makonde frequently incorporate the chameleon in complex ebony and rosewood carvings that depict vampires, sorcerers, and piles of hideously deformed, twisted, and skeletonized people.

The rock python is also important in the local system of traditions and beliefs. Considered "community beef," a python kill is apportioned to villagers according to their social status and familial relationship (Lawson 1993). Python fat is used as a balm for a variety of ailments (Butler and Shitu 1985; Lawson 1993). In the Pare Mountains of Tanzania, when snakes are killed, the head is chopped off and buried separately, thereby preventing it from rejoining the body and coming back to life (Klemens, pers. obs). In eastern DR Congo, the Mbuti have strict protocols concerning the slaughtering of monitor lizards. These lizards are killed in running water, so as not to "disturb the honey season," as illustrated by Klemens and Thorbjarnarson (1995:291).

ECONOMIC RESOURCES

Increasingly, reptiles and amphibians collected in the African forests are appearing in the wildlife trade. Apart from the more traditional trades in crocodile leather, snake and lizard skins, tortoiseshell, bushmeat, and eggs, the trade in live animals for the expanding exotic pet market presents major challenges to the conservation and sustainable use of forest herpetofauna. Large numbers of hingeback tortoises (*Kinixys erosa, K. homeana*) and ball pythons (*Python regius*) are exported from west Africa. Chameleons are an extremely popular specialty item; countries that contain large numbers of endemic forest chameleons, including Tanzania and Madagascar, have been particularly hard hit by the wildlife trade. Rarely is even a cursory survey conducted to determine the status of targeted wild populations and whether they can sustain large-scale exploitation. Therefore, it is unclear what, if any, threat is posed by this growing trade in forest reptiles.

Nevertheless, a study (Klemens and Moll 1995) reporting the effects of ten years of uncontrolled exploitation of pancake tortoises, *Malacochersus tornieri*, is illuminating. Klemens and Moll found that populations of this crevice-dwelling savanna species had been severely depleted over a wide portion of northern Tanzania as a result of the wildlife trade. In addition, they reported minimal benefits to the local economy from this trade. Many of the commercially important reptiles (pythons, tortoises, and chameleons) are listed on appendix 2 of the Convention on International Trade in Endangered Species (CITES), which requires that exports can proceed only if the trade is conducted in a non-detrimental manner, although the performance standards for this requirement are rarely, if ever, addressed. As a result, an increasing number of African forest amphibians and reptiles have been flagged by the CITES Animals Committee as "Significant Trade Species," requiring in-depth studies to assess the sustainability of the current levels of trade.

## Lessons from the Field

Our ongoing amphibian and reptile surveys in Cameroon and Tanzania continue to identify new species, unexpected occurrences, and range extensions of already described species. As there are scant data on most life-history and ecological parameters of the African forest herpetofauna, almost all recent observations on reproduction, feeding ecology, and interspecific interactions are novel. Most, if not all, of this work has conservation value, and in some instances has directly influenced planning and management of protected areas.

That we continue to add additional and new species to our study areas, even after repeated and intensive sampling, indicates that there is still yet more to be found, and illustrates the difficulty of compiling a complete species list for any forested site in Africa. One important discovery is the occurrence of the diminutive toad, *Didynamipus sjostedti*, at Banyang-Mbo and Korup (Cameroon). Long considered the rarest toad in Africa, known from a few specimens from Fernando Po, southwestern Cameroon, and Mount Cameroon (Grandison 1981; Gartshore 1984), *Didynamipus* turns out to be exceedingly abundant in certain primary forest and selectively logged areas. Obviously, this species is more widely distributed than previously reported, although its distribution appears discontinuous and highly localized, often in small (< 1 ha) isolated patches. Appropriate forest management regimes will undoubtedly determine the future survival of these populations. Although they presently do not appear threatened, populations near Korup National Park and Banyang-Mbo lie within recently reactivated logging concessions, and their survival, therefore, is uncertain.

The occurrence of *Didynamipus* in southwestern Cameroon underscores the need for more comprehensive inventories of both protected and non-protected areas, to ensure that planning and management activities optimize opportunities to conserve a broad spectrum of biological diversity. This approach has good precedent at Tarangire, one of Tanzania's premier savanna parks. The critical importance of small, disjunct, ephemeral wetlands and the spring-fed portions of the Tarangire River for fish and amphibians, reported as part of a preliminary biodiversity survey of Tarangire National Park by Moehlman et al. (1995), resulted in diverting tourist use and relocating additional development away from these sensitive areas. These changes in park management regimes occurred because the management structure under which Tarangire operates had sufficient flexibility to respond to the challenges posed by the incorporation of new information. Because frogs respond dramatically to changes in vegetation around and within small wetlands, they may prove to be excellent tools to measure changes in species composition and richness, which often result from different forms and intensities of landscape management.

For many small species with limited home ranges, multiple-use reserves, considered too small and fragmented for most birds and larger mammals, may provide sufficient habitat to conserve such species as *Didynamipus*. However, the more prudent course of action in the long term would be to use as wide a range of species as possible when delineating the boundaries of protected areas and reserves, and to conduct research into what types of multiple-use practices are compatible with the maintenance of a wide spectrum of biodiversity. For example, preliminary data have demonstrated that certain types of extractive practices in temperate forests may be compatible with the survival of amphibian communities. While

Petranka et al. (1993) reported that clear-cutting adversely affects species richness and abundance of mesic forest salamanders, Mitchell et al. (1996) found that these impacts could be reduced through the implementation of various shelterwood cutting regimes—that is, by cutting only a portion of habitat at any point in time.

Direct exploitation of reptile and amphibian species, either commercially or for local consumption, poses additional problems and imperatives for research and conservation. Virtually nothing is known of the ecology and natural history of even the most readily recognized species. Without basic natural history information it is impossible to adequately address issues concerning exploited species and to make sound management decisions on levels of harvest and offtake.

Robinson and Redford (1994) have discussed and summarized approaches to measuring the sustainability of hunting in tropical forests. Five indexes are cited as measures of sustainability, some of which may be immediately applicable to exploited herpetofauna, while others point to our need for additional ecological and life-history information. So far, none have been applied specifically to amphibian or reptile species in forested Africa. It is worthwhile to discuss each briefly here and to examine how these indexes might apply to the design and implementation of future herpetological research.

• Population density comparisons. Relative indexes of a species' density are compared between hunted and unhunted sites, or between sites of differential management. Through these indexes, coupled with a knowledge of the species' life-history parameters, the "sustainability" of offtake can be evaluated. This type of study works well on species that are restricted to specific habitats (such as the crevice-dwelling pancake tortoise studied by Klemens and Moll 1995) or those that are found with sufficient frequency and repeatability, such as many species of frogs. However, for species that occur over a broad range of habitats in low densities and leave no identifiable sign (such as snakes and forest tortoises), it will be difficult to construct the indexes of abundance needed for these comparisons.

- Population density declines (at a single site where exploitation is occurring). These are possible long-term studies, particularly for amphibians. Again, there is a need here for refined survey and monitoring techniques. This genre of study is important for the evaluation of non-cyclical amphibian declines. Because of the concerns expressed about amphibian declines by many herpetologists, population density decline studies should include non-exploited amphibian species.

- Hunting yield comparisons and hunting yield changes. These assess population status over place and time by recording the number of kills made per hour of hunting effort. This may be an important technique to monitor exploited reptile populations. This technique would also prove fruitful for amphibians, provided that monitoring was also conducted of non-exploited populations as a control. For reptiles in particular, a hunting yield comparison would require a long-term study, as most reptile species are taken fortuitously and their effective abundance may vary drastically with season. Information of this nature has been gathered in several studies (e.g., Noss 1995); however, herpetofauna is usually analyzed by order (snakes, turtles, and so on), precluding more detailed comparisons of the effects of hunting practices on individual species.

- Age-structure comparisons. These exam-ine the distribution of age classes between exploited and non-exploited populations. The difficulty in applying this technique is that there are usually a variety of factors that contribute to declines, including hunting, wildlife trade, habitat fragmentation, and elevated levels of natural predation that often occur in disturbed habitats. This technique has proven extremely useful in demonstrating non-cyclical declines in populations of turtles and tortoises (Klemens 1989). In these species, low survivorship of eggs and possibly hatchlings is the norm, but is offset by the long reproductive period of the adults. Where the loss of reproductive adults is occurring in excess of recruitment of juveniles into the population, a skewed age-size-class ratio is exhibited. The resulting population is composed of aged adults, which produce a small number of hatchlings, and lacks the intermediate age-size classes of juveniles, subadults, and young adults.

Robinson and Redford (1991) discuss two models of sustainability of exploitation. Both the population analysis model and the population growth model require much more refined demographic and natural history data than are currently available for reptiles. Although amphibians violate the population analysis model's assumption of non-significant preharvest mortality, they would make interesting case studies of the population growth model. For example, it would be fairly straightforward to census and obtain the necessary reproductive information for the exploited hairy frog, *Trichobatrachus robustus*, and goliath frog, *Conraua goliath*, in Cameroon.

## Conservation and Future Research

Based on the above discussion, we make the following recommendations as high priorities for future research and conservation initiatives.

EXPLORATION OF POORLY KNOWN REGIONS AND
AREAS OF POTENTIAL ENDEMICITY

In the face of burgeoning human populations and increasing commercial forest exploitation, there is an immediate need to undertake exploratory inventories of little-known areas and regions likely to contain endemics, before our opportunity to study and conserve them is lost. This work may be most cost effective when conducted not just for herpetofauna but in conjunction with other taxa (birds, small mammals, fish, invertebrates, and so on). Apart from gathering much-needed baseline biological information, these activities, whenever possible, should contribute to building national capacity in conducting biological inventories and strengthening national institutions to both study and, in a sustainable manner, manage natural history collections and associated data.

Although we see great value in the expeditionary approach, which is often funded and in part staffed by expatriate researchers, as well as in short-term collecting expeditions, we caution that these collecting activities should be conducted in a manner that will ensure both participation and professional advancement of national scientists. The use of drift fences with pitfalls greatly increases both the number of species and the number of individuals captured at a site, and readily permits comparisons between contiguous habitat patches. Although labor intensive to install and monitor (a minimum stay of seven to ten days at each sampling site is required), this technique is the most reliable method to census the biodiversity and biomass of the small, often secretive, litter-dwelling and fossorial herpetofauna that is both diverse and abundant in the African rain forest. Longer-term inventories and surveys should be designed in such a way as to concurrently set up ecological monitoring programs for species that are rare, or that are heavily harvested, or have a demonstrated sensitivity to environmental change (such as tree frogs).

QUANTITATIVE SURVEYS
OF COMMUNITY COMPOSITION

In addition to exploration of new areas, quantitative surveys should be undertaken in such relatively well-known areas as field stations and protected areas to create a baseline of information to monitor ecological change over time. For the poorly studied mesic African forests, we cannot even speculate on the potential for the occurrence of non-cyclical amphibian decline, as reported from Australia, North America, and the Neotropics. It is impossible to discuss this phenomenon of worldwide concern as it might relate to tropical Africa because we lack an adequate understanding of the number of species, species distributions, and basic life-history parameters of the resident amphibians. Refuting or corroborating the phenomenon of amphibian decline on the African continent would provide a model against which to compare the amphibian decline data from Neotropical areas.

Immediate action is needed to initiate qualitative and, wherever possible, quantitative monitoring of amphibian populations in the relatively few areas of Africa where baseline information exists. Whenever possible, amphibian survey and monitoring should be incorporated into the ongoing baseline inventory activities that are currently under way at forested sites across the continent. Reptile and amphibian communities have the potential to be informative bio-indicators, as the composition of these communities responds to even moderate levels of habitat disturbance. Quantitative surveys can also be applied to cross-tropical comparisons and are important in theoretical aspects of community ecology and evolution (Janzen 1976). Special attention should be focused on studies that test the ability of various species and communities of amphibians to serve as indicators or measures of the effects of different forest and wetland management regimes.

TARGETED MONITORING AND ECOLOGICAL STUD-
IES OF EXPLOITED AND ENDANGERED SPECIES

Monitoring programs need to be established over a broad area to directly address the status of species that are known to be exploited or threatened. Species of immediate concern include: goliath frogs (*Conraua goliath*), forest tortoises (*Kinixys*), monitor lizards (*Varanus*), and the various genera of pythons and chameleons. The IUCN Crocodile Specialist Group (IUCN/SSC 1992) considers west and central Africa a high-priority area for baseline surveys of crocodile populations.

To even begin to cogently address conservation and management concerns of exploited reptile and amphibian species, basic ecological and natural history information is urgently needed. Little or nothing is known of the biology of even the most easily recognized species occurring in the forest zone. Whereas more information exists for some wide-ranging species in the savanna and more arid portions of their range, knowledge of their biology in mesic forests is lacking.

INCREASED EMPHASIS ON SYSTEMATICS AND
INTEGRATING SYSTEMATICS AND CONSERVATION

Taxonomic treatments and revisions are necessary for many groups to more adequately assess diversity and to clarify distributions and species boundaries. Phylogenetic hypotheses are also important for biogeographic analyses. Amphibians in particular may prove informative in unraveling the history of anthropogenic and climatic changes in the African forests. We recognize that scientific collecting must be conducted in a responsible manner, balancing the need for scientific knowledge with conservation, political, and cultural sensitivities. However, we are alarmed at the recent trend, particularly in the Neotropics, that discourages and often prohibits scientific collection, while allowing wholesale destruction of forested ecosystems. Campbell and Frost (1993: 53–56)

and Greene and Losos (1988) provide excellent discussions of various aspects of this dichotomy. As systematists become more engaged in the conservation process, they bring a unique phylogenetic perspective on conserving genetic diversity that can be incorporated into conservation priority-setting exercises and triage (Vane-Wright et al. 1991).

IDENTIFICATION AND TRAINING
OF NATIONALS IN HERPETOLOGY

It will be difficult to generate and sustain interest in herpetological research and conservation over the long term without the participation and leadership of national researchers and wildlife managers. The current state of spotty geographic and historical herpetological exploration is almost exclusively the result of intermittent expatriate research. National professionals are better placed to ensure the continuation of research programs and conservation projects, and to foster future national involvement.

Certainly, any long-term monitoring programs will have to be run by national staff. It is essential that more attention be given to providing educational and training opportunities to a wide range of wildlife professionals, from park wardens and ecologists to undergraduate and postgraduate students, and that faculty at African universities be given opportunities to develop and maintain their standing in the international academic community. International funding institutions and universities in developed countries should channel grants and scholarships to herpetology and other disciplines in which there is little current national expertise. The ultimate goal, however, must be not only to provide professional opportunities for individual Africans, but also to assist and strengthen African institutions, so that they may assume an increasingly important role in providing advanced and specialized training within Africa.

## STUDIES OF HUNTING AND SUSTAINABLE USE

Hunting studies that measure the sustainability of offtake need to include amphibians and reptiles. Because of their lower energy requirement, these species are able to maintain high biomass levels and may be candidates for sustainable management programs (Klemens and Thorbjarnarson 1995). Yet, because certain reptiles are usually long lived and have low fecundity, they are also vulnerable to overexploitation. In the case of many turtles, even incidental pressure, such as that caused by hunting, elevated mortality, or collection, may have dramatic and devastating effects on populations (Klemens 1989; Congdon et al. 1993).

In Africa, megavertebrates have garnered much of the conservation and research attention, as well as the majority of the public's interest. Reptiles and amphibians have been the beneficiaries of reserves created for Africa's great mammalian wildlife spectacles. Now, as biologists and the general public gain a more sophisticated appreciation of entire ecosystems and the value of unique habitats, we can begin to expand the concept and practice of conservation to encompass a much fuller array of the total diversity and functions of these systems—a diversity of which amphibians and reptiles are integral components. While we have little doubt that research and conservation of amphibian and reptile diversity (and that of other lesser-known groups) in tropical Africa will continue to be piggy-backed on the charismatic megavertebrates, the stage is set for a new era of herpetological discovery on the continent.

## Postscript

In August 1997, a special symposium entitled "Africa—The Neglected Continent" was convened at the Third International Herpetological Congress in Prague, at which many of the concepts and actions proposed in this chapter were presented. One outcome of the congress was the reconstitution of IUCN's African Amphibian and Reptile Specialist Group. In its reorganization, the Group recognized the dual challenges of gathering basic information and integrating those data with conservation activities. In so doing, the group incorporated as its goals many of the challenges put forth in this chapter.

ACKNOWLEDGMENTS

We would like to thank the editors of this book for the opportunity to present this information and their helpful criticism during the compilation of the manuscript. DPL thanks J. A. Campbell and Bryan Curran for their comments on earlier drafts. MWK acknowledges financial support for herpetological field research in Tanzania provided by the Bay Foundation, as well as support for biodiversity research and capacity building received from the John D. and Catherine T. MacArthur Foundation. Field work in Tanzania was facilitated through the offices of the Tanzanian Commission for Science and Technology (COSTECH), Tanzania National Parks (TANAPA), Serengeti Wildlife Research Institute (SWRI), and the University of Dar es Salaam. Darrel Frost kindly made available a copy of his working manuscript "Amphibian Species of the World." MWK acknowledges the contributions to various aspects of this study made by L. S. Ford, K. M. Howell, P. D. Moehlman, A. M. Nikundiwe, J. C. Poynton, and J. Rasmussen, and thanks Nicole Silton Klemens for her assistance in compiling the data presented in figure 17.1.

REFERENCES

Amiet, J.-L. 1972. Les *Cardioglossa* camerounaises. *Sci. et Natur.* 114:11–24.

———. "1977" [1978]. Les *Astylosternus* du Cameroun (*Amphibia anura, Astylosterninae*). *Ann. Fac. Sci. Yaoundé* 23–24:99–227.

———. 1978. Les amphibiens Anoures de la région de Mamfé (Cameroun). *Ann. Fac. Sci. Yaoundé* 25:189–219.

———. 1980. Révision du genre *Leptodactylodon* Andersson (*Amphibia, Anura, Astylosterninae*). *Ann. Fac. Sci. Yaoundé* 27:69–224.

Andersson, L. G. 1903. Neue Batrachier aus Kamerun von den Herren Dr. Y. Sjöstedt und Dr. S. Junger gesammelt. *Verh. Zool. Bot. Ges. Wein.* 53:141–145.

———. 1905. Batrachians from Cameroon collected by Dr. Y. Sjöstedt in the years 1890–1892. *Ark. Zool.* 2:1–29.

———. 1907. Verzeichnis einer Batrachier-Sammlung von Bibundi bei Kamerun des Naturhistorischen Museums zu Wiesbaden. *Jahrb. Nassau. Ver. Naturk.* 60:228–245.

———. 1909. Über einige der *Hylambates*-Formen Kameruns. *Jahrb. Nassau. Ver. Naturk.* 62:103–110.

Barbour, T., and A. Loveridge. 1928. A comparative study of the herpetological faunae of the Uluguru and Usambara Mountains, Tanganyika Territory with descriptions of new species. *Memoirs Mus. Comp. Zool. Harvard* 50(2):87–265, plates 1–4.

Broadley, D. G., and K. M. Howell. 1991. A checklist of the reptiles of Tanzania, with synoptic keys. *Syntarsus* 1:1–70.

Butler, J. A., and E. Shitu. 1985. Uses of some reptiles by the Yoruba people of Nigeria. *Herpetol. Rev.* 16(1):15–16.

Campbell, J. A., and D. R. Frost. 1993. Anguid lizards of the genus *Abronia:* Revisionary notes, descriptions of four new species, a phylogenetic analysis, and key. *Bull. Am. Mus. Nat. Hist.* 216:1–121.

Carcasson, R. H. 1964. A preliminary survey of the zoogeography of African butterflies. *E. African Wild Life J.* 2:122–157.

Congdon, J. D., A. E. Dunham, and R. C. Van Loben Sels. 1993. Delayed sexual maturity and demographics of Blanding's turtles (*Emydoidea blandingii*): Implications for conservation and management of long-lived organisms. *Conserv. Biol.* 7(4):826–833.

De La Riva, I. 1994. Anfibios anuros del Parque Nacional de Monte Alén, Río Muni, Guinea Ecuatorial. *Rev. Esp. Herp.* 8:123–139.

Dubois, A. "1986" [1987]. Miscellanea taxinomica batrachologica (I). *Alytes* 5(1–2):7–95.

Frost, D. R. 1996. Amphibian species of the world, revised and updated with complete synonymies. American Museum of Natural History.

Fuchs, V. E. 1939. The Lake Rukwa Expedition, 1938. *Geographical J.* 94(5):368–387 and 1 map.

Gartshore, M. E. 1984. The status of the montane herpetofauna of the Cameroon highlands. Pages 204–240 in S. N. Stuart, ed. *Conservation of Cameroon Montane Forests.* International Council for Bird Preservation, London.

Gentry, A. H. 1988. Changes in plant community diversity and floristic composition on environmental and geographic gradients. *Annals of the Missouri Botanical Garden* 75:1–34.

Grandison, A. G. C. 1981. Morphology and phylogenetic position of the West African *Didynamipus sjostedti* Andersson, 1903 (Anura Bufonidae). *Monit. Zool. Ital. (N. S.) Suppl.* 15(11):187–215.

Greene, H. W., and J. B. Losos. 1988. Systematics, natural history, and conservation: Field biologists must fight a public-image problem. *BioSci.* 38(7):458–462.

Hawthorne, W. D. 1993. East African coastal forest botany. Pages 57–99 in J. C. Lovett and S. K. Wasser, eds. *Biogeography and Ecology of the Rain Forests of Eastern Africa.* Cambridge University Press, Cambridge.

Hoffman, R. L. 1993. Biogeography of East African montane forest millipedes. Pages 103–114 in J. C. Lovett and S. K. Wasser, eds. *Biogeography and Ecology of the Rain Forests of Eastern Africa.* Cambridge University Press, Cambridge.

Howell, K. M. 1993. Herpetofauna of the eastern African forests. Pages 173–201 in J. C. Lovett and S. K. Wasser, eds. *Biogeography and Ecology of the Rain Forests of Eastern Africa.* Cambridge University Press, Cambridge.

Hughes, B., and D. H. Barry. 1969. The snakes of Ghana: A checklist and key. *Bull. I.F.A.N.* 31(3):1004–1041.

IUCN/SSC. 1992. *Crocodiles: An Action Plan for Their Conservation.* J. Thorbjarnarson, comp. H. Messel, F. W. King, and J. P. Ross, eds. IUCN. Gland, Switzerland.

Janzen, D. H. 1976. The depression of reptile biomass by large herbivores. *Am. Nat.* 110(973):371–400.

Joger, U. 1990. The herpetofauna of the Central African Republic, with description of a new species of *Rhinotyphlops* (Serpentes: Typhlopidae). Pages 85–102 in G. Peters and R. Hutterer, eds. *Vertebrates in the Tropics.* Museum Alexander Koenig, Bonn.

Kingdon, J. 1990. *Island Africa: The Evolution of Africa's Rare Animals and Plants.* Collins and Sons, London.

Klemens, M. W. 1989. The methodology of conservation. Pages 1–4 in I. R. Swingland and M. W. Klemens, eds. *The Conservation Biology of Tortoises.* Occas. Pap. IUCN/SSC 5. IUCN, Gland, Switzerland.

———. 1998. The male nuptial characteristics of *Arthroleptides martiensseni* Neiden, an endemic torrent frog from Tanzania's Eastern Arc Mountains. *Herp. J.* 8(1):35–40.

Klemens, M. W., and D. Moll. 1995. An assessment of the effects of commercial exploitation on the pancake tortoise *Malacochersus tornieri* in Tanzania. *Chelonian Conserv. and Biol.* 1(3):197–206.

Klemens, M. W., A. M. Nikundiwe, and L. S. Ford. 1997. Amphibian and reptile biodiversity of the Udzungwa Mountains National Park: Some preliminary findings. Pages 31–32 in M. Nummelin, ed. *Abstracts: International Conference on the Eastern Arc Mountains, 1–5 December 1997, Morogoro, Tanzania*.

Klemens, M. W., and J. B. Thorbjarnarson. 1995. Reptiles as a food resource. *Biodiversity and Conserv.* 4(3):281–298.

Knoepffler, L. P. 1966. Faune du Gabon (amphibiens et reptiles). *Biologia Gabonica* 2(1):3–24.

———. 1974. Faune du Gabon (amphibiens et reptiles), 2: Crocodiles, chéloniens, et sauriens de l'Ogoué-Ivindo et du Wolen N'Tem. *Vie et Milieu* 24(1) sér. C:111–128.

———. N.d. Liste des vertébrés du Bassin de l'Ivindo (République Gabonaise)—poissons exceptés—amphibiens pp. 2–4; reptiles pp. 5–8. Mimeographed list extracted from various publications of the "Mission biologique au Gabon (1963–1969), Laboratoire de Primatologie et d'Ecologie Equatoriale (1970–1978), and the Institut de Biologie Tropicale (1979)." Copy on deposit in the library of the Herpetology Department, American Museum of Natural History, New York.

Laurent, R. F. 1954. Reptiles et batraciens de la région de Dundo (Angola) (deuxième note). *Publ. Culturais da Companhia de Diamantes de Angola* 23:35–84.

———. 1956. Contribution à l'herpétologie de la région des Grands Lacs de l'Afrique Centrale, I: Généralités; II: Chéloniens; III: Ophidiens. *Ann. Mus. Roy. Congo Belge.* Zool. sér. 8 vols. 48:1–390.

———. 1964. Reptiles et amphibiens de l'Angola (troisième contribution). *Publ. Culturais da Companhia de Diamantes de Angola* 67:11–165.

———. 1972. Amphibiens. *Exploration du Parc National des Virunga* 2d sér. 22:1–125.

———. 1973. A parallel survey of equatorial amphibians and reptiles in Africa and South America. Pages 259–265 in B. J. Meggers, E. S. Ayensu, and W. D. Duckworth, eds. *Tropical Forest Ecosystems in Africa and South America: A Comparative Review*. Smithsonian Institution Press, Washington, D.C.

Lawson, D. P. 1993. The reptiles and amphibians of the Korup National Park Project, Cameroon. *Herp. Nat. Hist.* 1(2):27–90.

Loveridge, A. 1932. New reptiles and amphibians from Tanganyika Territory and Kenya Colony. *Bull. Mus. Comp. Zool. Harvard* 72(10):375–387.

———. 1933a. Reports on the scientific results of an expedition to the southwestern highlands of Tanganyika

Territory, 1: Introduction and zoogeography. *Bull. Mus. Comp. Zool. Harvard* 75(1):1–43, plates 1–3.

———. 1933b. Reports on the scientific results of an expedition to the southwestern highlands of Tanganyika Territory, 7: Herpetology. *Bull. Mus. Comp. Zool. Harvard* 74(7):197–416, plates 1–3.

———. 1935. Scientific results of an expedition to rain forest regions in eastern Africa, 1: New reptiles and amphibians from east Africa. *Bull. Mus. Comp Zool. Harvard* 79(1):3–19.

———. 1936a. Scientific results of an expedition to rain forest regions in eastern Africa, 5: Reptiles. *Bull. Mus. Comp. Zool. Harvard* 79(5):209–337, plates 1–9.

———. 1936b. Scientific results of an expedition to rain forest regions in eastern Africa, 7: Amphibians. *Bull. Mus. Comp. Zool. Harvard* 79(7):369–430, plates 1–3.

———. 1942a. Scientific results of a fourth expedition to forested areas in east and central Africa, 4: Reptiles. *Bull. Mus. Comp. Zool. Harvard* 91(4):237–373, plates 1–6.

———. 1942b. Scientific results of a fourth expedition to forested areas in east and central Africa, 5: Amphibians. *Bull. Mus. Comp. Zool. Harvard* 91(5):377–436, plates 1–4.

———. 1944. Scientific results of a fourth expedition to forested areas in east and central Africa, 6: Itinerary and comments. *Bull. Mus. Comp. Zool. Harvard* 94(5):191–214, plates 1–4.

———. 1957. Check list of the reptiles and amphibians of east Africa (Uganda; Kenya; Tanganyika; Zanzibar). *Bull. Mus. Comp. Zool. Harvard* 117(2):i–xxxvi, 153–362.

Lovett, J. C. 1993a. Climatic history and forest distribution in eastern Africa. Pages 23–29 in J. C. Lovett and S. K. Wasser, eds. *Biogeography and Ecology of the Rain Forests of Eastern Africa*. Cambridge University Press, Cambridge.

———. 1993b. Eastern Arc moist forest flora. Pages 33–55 in J. C. Lovett and S. K. Wasser, eds. *Biogeography and Ecology of the Rain Forests of Eastern Africa*. Cambridge University Press, Cambridge.

Mertens, R. 1937. Eine neue, tiergeographisch bemerkenswerte Eidechse aus Kamerun. *Senckenbergiana* 19:381–384.

———. 1938. Herpetologische Ergebnisse einer Reise nach Kamerun. *Abhand. Senck. Naturf. Ges.* 442:1–52, plates 1–10.

Mitchell, J. C., J. A. Wicknick, and C. D. Anthony. 1996. Effects of timber harvesting practices on peaks of otter salamander (*Plethodon hubrichti*) populations. *Amphib. Rep. Conserv.* 1(1):15–19.

Moehlman, P. D., M. W. Klemens, and A. M. Nikundiwe. 1995. Preliminary report on the biodiversity of Tarangire National Park, Tanzania. Report by the Wildlife Conservation Society (Bronx, N.Y.) to Tanzania National Parks, TANAPA (Arusha).

Moreau, R. E. 1963. Vicissitudes of the African biomes in the Late Pleistocene. *Proc. Zool. Soc. London* 141:395–421.

———. 1969. Climatic changes and the distribution of forest vertebrates in west Africa. *J. Zool., London* 158:39–61.

Nieden, E. 1908. Die Amphibienfauna von Kamerun. *Mitt. Zool. Mus. Berlin* 3:489–518.

———. 1910. *Die Reptilien (außer den Schlangen) und Amphibien.* Fauna deutsch. Kolonien. Kamerun, Berlin.

Noble, G. K. 1924. Contributions to the herpetology of the Belgian Congo based on the collection of the American Museum Congo Expedition, 1909–1915. *Bull. Am. Mus. Nat. Hist.* 49(2):147–347.

Noss, A. J. 1995. Duikers, cables, and nets: A cultural ecology of hunting in a central African forest. Ph.D. dissertation, University of Florida, Gainesville.

Pannell, C. M., and F. White. 1988. Patterns of speciation in Africa, Madagascar, and the tropical Far East: Regional faunas and cryptic evolution in vertebrate-dispersed plants. *Monographs in Systematic Botany from the Missouri Botanical Gardens* 25:639–659.

Parker, H. W. 1936. The amphibians of the Mamfe Division, Cameroons, 1: Zoogeography and systematics. *Proc. Zool. Soc. London* 1936:135–163, plate 1.

Perret, J.-L. 1966. Les amphibiens du Cameroun. *Zool. Jahrb. Abt. Syst. Oko. Geog.* 93:289–460.

———. 1977. Les *Hylarana* (Amphibiens, Ranidés) du Cameroun. *Revue Suisse Zool.* 84(4):841–868, plates 1–5.

Petranka, J. W., M. E. Eldridge, and K. E. Haley. 1993. Effects of timber harvesting on southern Appalachian salamanders. *Conserv. Biol.* 7(2):363–370.

Pitman, C. R. S. 1974. *A guide to the snakes of Uganda.* Rev. ed. Wheldon and Wesley, Codicote, England.

Poynton, J. C. 1998. Tanzanian bufonid diversity: Preliminary findings. *Herp. J.* 8(1):3–6.

Poynton, J. C., and D. G. Broadley. 1985a. Amphibia Zambesiaca, 1: Scolecomorphidae, Pipidae, Microhylidae, Hemisidae, Arthroleptidae. *Ann. Natal Mus.* 26(2):503–553.

———. 1985b. Amphibia Zambesiaca, 2: Ranidae. *Ann. Natal Mus.* 27(1):115–181.

———. 1987. Amphibia Zambesiaca, 3: Rhacophoridae and Hyperoliidae. *Ann. Natal Mus.* 28(1):161–229.

———. 1988. Amphibia Zambesiaca, 4: Bufonidae. *Ann. Natal Mus.* 29(2):447–490.

———. 1991. Amphibia Zambesiaca, 5: Zoogeography. *Ann. Natal Mus.* 32:221–277.

Richards, P. W. 1973. Africa, the "Odd man out." Pages 21–26 in B. J. Meggers, E. S. Ayensu, and W. D. Duckworth, eds. *Tropical Forest Ecosystems of Africa and South America: A Comparative Review.* Smithsonian Institution Press, Washington, D.C.

Robinson, J. G., and K. H. Redford. 1991. Sustainable harvest of Neotropical forest animals. In J. G. Robinson and K. H. Redford, eds. *Neotropical Wildlife Use and Conservation.* University of Chicago Press, Chicago.

———. 1994. Measuring the sustainability of hunting in tropical forests. *Oryx* 28(4):249–256.

Roux-Estève, R., and G. R. De Witte. 1975. Serpents. *Exploration du Parc National des Virunga,* 2d sér. 24:1–121.

Sanderson, I. T. 1936. The amphibians of the Mamfe division, Cameroons, 2: Ecology of the frogs. *Proc. Zool. Soc. London* 1936:165–208.

Scharff, N. 1993. The Linyphiid spider fauna (Araneae: Linyphiidae) of mountain forests in the Eastern Arc mountains. Pages 115–132 in J. C. Lovett and S. K. Wasser, eds. *Biogeography and Ecology of the Rain Forests of Eastern Africa.* Cambridge University Press, Cambridge.

Schiøtz, A. 1963. The amphibians of Nigeria. *Vidensk. Medd. Dansk Naturh. Foren.* 125:1–92.

———. 1967. The treefrogs (Rhacophoridae) of west Africa. *Spolia zool. Mus. Haun.* 25:1–346.

———. 1975. *The Treefrogs of Eastern Africa.* Steenstrupia, Copenhagen.

Schmidt, K. P. 1919. Contributions to the herpetology of the Belgian Congo based on the collection of the American Museum Congo Expedition, part 1: Turtles, crocodiles, lizards, and chameleons. *Bull. Am. Mus. Nat. Hist.* 39:385–624.

———. 1923. Contributions to the herpetology of the Belgian Congo based on the collection of the American Museum Congo Expedition, part 2: Snakes. *Bull. Am. Mus. Nat. Hist.* 49:1–146.

Schmidt, K. P., and R. F. Inger. 1959. Amphibians, exclusive of the genera *Afrixalus* and *Hyperolius. Explor. Parc Nat. Upemba. Mission G. R. De Witte.* 56:1–264.

Sjöstedt, Y. 1897. Reptilien aus Kamerun West-Afrika. *Bih. Svenska Vet. Akad. Handl.* 23/Afd.IV(2):1–36, plates 1–3.

Sternfeld, R. 1908. Die Schlangenfauna von Kamerun. *Mitt. Zool. Mus. Berlin* 3:397–432.

———. 1917. Reptilia und Amphibia. *Ergebnisse der zweiten deutschen Zentral-Afrika- Expedition 1910–1911, 1: Zoologie* (11):407–509.

Stewart, M. M. 1967. *Amphibians of Malawi.* State University of New York Press, Albany.

Stucki-Stern, M. C. 1979. *Snake report 721.* Herpeto-Verlag, Teuffenthal, Switzerland.

Uthmoller, W. 1942. Die Schlangen Ostafrikas in ihrem Lebensraum...der Vulkane Kilmandscharo, Meru, Hanang, der grossen ostafrikanischen Bruchstufe und des Usambaragebirges. *Archiv für Naturgeschichte Leipzig (NF)* 10:1–70.

Vane-Wright, R. I., C. J. Humphries, and P. H. Williams. 1991. What to protect? Systematics and the agony of choice. *Biological Conservation* 55:235–254.

Vitt, L. J., and G. R. Colli. 1994. Geographical ecology of a neotropical lizard: *Ameiva ameiva* (Teiidae) in Brazil. *Canadian Journal of Zoology* 72:1986–2008.

Vitt, L. J., P. A. Zani, and C. M. Lima. 1997. Heliotherms in tropical rain forest: The ecology of *Kentropyx calcarata* (Teiidae) and *Mabuya nigropunctata* (Scincidae) in the Curuá-Una of Brazil. *Journal of Tropical Ecology* 13:199–220.

Werner, F. 1897. Über Reptilien und Batrachier aus Togoland, Kamerun und Tunis, 1. *Verh. Zool. Bot. Ges. Wien* 47:395–408.

———. 1898. Über Reptilien und Batrachier aus Togoland, Kamerun und Tunis, 2. *Verh. Zool. Bot. Ges. Wien* 48:191–231.

———. 1899. Über Reptilien und Batrachier aus Togoland, Kamerun und Deutsch-Neuguinea. *Verh. Zool. Bot. Ges. Wien* 49:132–157.

# IV
## Humans and the Forest

# Overview of Part IV

Lisa Naughton-Treves

D**URING THE 1990S**, there was a profound shift in international conservation strategies. Instead of preserving pristine nature in parks, many conservationists now aim to integrate local economic needs with goals involving biodiversity. Driving this shift in conservation strategies are several factors. For one, conservationists now realize that most parks and reserves are too small to conserve the ecosystem processes that sustain biodiversity. They are also belatedly aware of the value of indigenous knowledge in managing ecosystems, and they recognize that entirely excluding people from land and resources within parks is often impractical or unethical. As we enter the twenty-first century, the "local community" occupies center stage as a promising steward of biodiversity. But just who is the local community? And how well do local environmental agendas match global ones?

The authors of the following six chapters delve into these difficult questions by exploring the subsistence and economic strategies of forest peoples in five African countries. There is marked variation among the chapters in terms of social and environmental context (for example, human population densities range from 1.1 to >200 people/km², for study sites in Gabon and Uganda, respectively). Some chapters focus on communities neighboring national parks, and others examine residents of vast tracts of forest. But all six chapters document profound changes in forest use owing to recent socioeconomic change. For example, the authors observe increased hunting intensity for the bushmeat trade, widespread adoption of sedentary farming practices, and growing local employment with commercial enterprises, including logging companies and tourism operators. In all cases, cultural norms and local conventions governing forest use and ownership have been transformed or erased by the growing pressure from regional and international markets, as well as by government intervention and the influx of immigrants. Whereas all the authors of these chapters observe similar threats to rain forest conservation, they each propose different conservation solutions. Taken together, the chapters in this section stress the need for tailoring popular models for conservation to local contexts.

In chapter 18, Andrew J. Noss provides a richly detailed description of the subsistence and economic strategies of one of the most ancient forest-dwelling peoples, the Aka people of the Central African Republic. Aka livelihoods are now enmeshed with regional and global markets. Unfortunately, they continue to participate in cash economies only as subordinates to other immigrant groups. Noss questions the viability of Aka traditional forest management for achieving sustainability, partly because their management traditions rely on mobility and assume resource abundance, but also because Aka forest use today depends on their interactions with other social groups and macro-economic forces. Chapter 19 offers an intriguing contrast. Alden Almquist considers the ritualized forest-use practices and territoriality of the Apagibeti of north-central DR Congo a vital tool for conservation. Chapter 20 concerns an equally remote site in northeastern Gabon. Here, Sally A. Lahm describes the market hunting practices of resident communities. Despite low human population density (1.1 individuals/km²), large-bodied mammals have been largely eliminated in the vicinity of villages, and hunters now rely on ever more remote hunting camps. In Chapter 21, Richard B. Peterson describes the social and environmental impact of immigration on the Ituri Forest frontier. Peterson also reveals the ethnic and economic heterogeneity within local populations and the resulting conflicts over land and resources. Nonetheless, Peterson emphatically calls for an indigenous, non-Western philosophy of conservation. Chapters 22 and 23 shift the focus to forest management within agricultural landscapes. Chapter 22 explains patterns of crop loss to wildlife among sedentary farmers neighboring Kibale National Park, Uganda, and offers management suggestions. In Chapter 23, Yaa Ntiamoa-Baidu describes traditional and introduced protected-area-system strategies for biodiversity conservation in Ghana.

The six chapters in this section confirm the importance of local participation in conservation efforts. Local citizens continue to be dependent on forest resources, and they strongly influence conservation outcomes. Africa's rain forests will not survive without local participation and support for conservation. But the authors also reveal profound challenges to community-based conservation efforts. No matter how remote the site, no author described a socially cohesive, isolated community managing the forest according to locally evolved conventions. This suggests that instead of searching for local community as an idealized alternative to the state, conservationists must address the political and economic factors that favor sustainable forest use. For example, without secure land tenure, few individuals will invest in long-term management practices. Fostering a sound process for local stakeholders and grassroots organizations to negotiate forest access rights and management responsibilities is a difficult and highly political action, but it is an essential first step. After all, the rain forest is much more than simply a "resource," it may be a core element in a people's sociopolitical identity and survival.

# CHAPTER 18

# Conservation, Development, and "the Forest People"

## The Aka of the Central African Republic

Andrew J. Noss

African "Pygmies" have long been of interest to outsiders—explorers, anthropologists, and conservationists—because of their "hunter-gatherer" lifestyle and their presumed status as the first inhabitants of central African rain forests. Today several distinct Pygmy groups exist (figure 18.1): the Aka/BaAka/BiAka of Congo-Brazzaville and the Central African Republic (CAR), the Baka and BaKola/Gyeli of Cameroon, the BaBongo/Akoa of Congo-Brazzaville and Gabon, the Mbuti/Efe of Congo-Kinshasa, and the BaCwa/Twa of Congo-Kinshasa and Rwanda (Bahuchet 1991a). One of the largest and best-studied groups is the Aka. This chapter begins with a review of historical and anthropological perspectives on the Aka in CAR. Figure 18.2 presents principal research sites where information has been gathered on the Aka: Bagandou, Bayanga, Mongoumba, and Ndele. Following sections discuss the economic and social history of the Bayanga region, describe variation among Aka settlements today, and focus in depth on the village of Mossapoula. A concluding section examines the conservation implications of current Aka economic and subsistence patterns.

## Historical and Anthropological Perspectives

The Aka currently number about 10,000 across 100,000 km of southern CAR and northern Congo (Bahuchet and Thomas 1986; Bahuchet 1991a). The Aka were probably the first inhabitants of central African forests, living from hunting and gathering (Bahuchet et al. 1991). But agricultural peoples speaking Bantu, Sudanic, and Ubanguian languages began moving into forests as long ago as 2,000 B.P. (Vansina 1985a, 1985b, 1986). The Aka language has evolved over time from a language of borrowed Bantu origin into its own distinct and autonomous language (Bahuchet and Guillaume 1982; Bahuchet 1990). The Aka shared their forest knowledge in guiding other ethnic groups in migrations into or across the forest, and patterns of mutual relations developed between the Aka "hunter-gatherers" and "farmers or fishers" of other ethnic groups (Arom and Thomas 1974; Delobeau 1978; Demesse 1978; Bahuchet 1985a, 1985c). Across their range, and over time as successive migrations occurred, Aka groups established these relations with farming and fishing villages of nineteen other ethnic groups: the Banda/Yangere, Isongo, Kaka, Monzombo, Mbati, Ngbaka, Ngando/Bangandu, Ngoundi, Pande, Sangha-Sangha, and others (Bahuchet and Guillaume

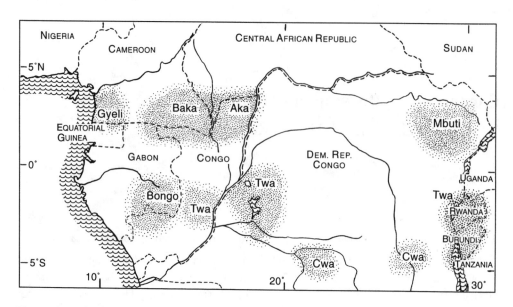

Figure 18.1. Distribution of African Pygmies. Adapted from Bahuchet 1985b, 1991a.

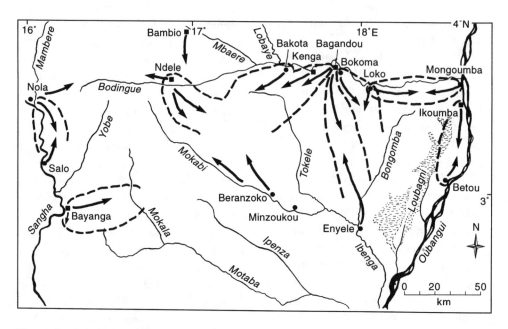

Figure 18.2. Principal research sites in the Central African Republic: Bagandou, Bayanga, Mongoumba, and Ndele. Adapted from Bahuchet and Thomas 1986.

1982; Bahuchet and Thomas 1986; Bahuchet 1991a).

Anthropologists describe a traditional "symbiosis" model for inter-ethnic relations wherein the Aka exchanged meat and other forest products for iron, salt, and agricultural products (Demesse 1978; Bahuchet 1979b, 1989; Bahuchet and Thomas 1986; Delobeau 1989). Perhaps neither group needed the other to survive, but they specialized and traded for mutual convenience (Hewlett 1977). Or perhaps neither group was self-sufficient and both could benefit from trade (Vansina 1985b). In fact, the dichotomy is overly simplistic, as the Aka have long practiced farming, while farmers and fishers also have long hunted and gathered (Vansina 1985a, 1985b, 1986; Bahuchet et al. 1991). By settling in forest clearings and along rivers, the immigrant peoples may have actually forced the Aka into the forest and led them to specialize increasingly in forest products (Bahuchet and Guillaume 1979, 1982).

The traditional system of mutual reliance and exchange was significantly modified by colonialism, beginning with the Atlantic trade in slaves and ivory. Fleeing the slave trade and later forced-labor programs, other ethnic groups like the Ngbaka in CAR hid themselves in the forest, where they depended to a greater degree on Aka assistance (Thomas 1963; Bahuchet 1991a). Much of the ivory was produced by Aka hunters on behalf of their trading partners, who increasingly sought to control, and to "own," the Aka. With the decline of elephant populations, French concession companies beginning in the 1890s demanded taxes in wild rubber, *Landolphia owariensis*. Tax collection focused on the more sedentary agricultural and fishing communities. Increasingly burdened with rubber collection and forced to labor in logging, mining, and plantation activities, other ethnic groups depended more and more on the Aka for meat.

In the 1920s the demand for duiker (*Cephalo-phus* spp.) skins in Europe boomed, and this impetus, together with the demand for meat from their neighbors, probably encouraged the Aka to adopt nets from their neighbors at this time. Net hunting specializes in the small duikers, whereas the traditional Aka spear hunting focused on larger game. Net hunting requires the cooperation of large numbers of people, up to 100 hunters at a time, and thus produces larger settlements: 40–60 adults, as opposed to the smaller settlements of 15–25 adults that existed previously. As agriculture expanded after 1950, the Aka became increasingly important as a source of labor (Bahuchet and Guillaume 1982; Bahuchet 1989; Hewlett 1991).

Across their range, the Aka present today a variety of socioeconomic patterns and relationships with different ethnic groups. Delobeau (1978, 1989) describes in detail the social and economic interdependence between the Monzombo and the Aka in such activities as fishing, house construction, cooking, and farming. Through relatively strong alliances between lineages, the Monzombo also assist the Aka with bride-price payments, funerals, and court cases. Some researchers characterize the changes from traditional symbiotic trade between two distinct groups to increasing participation by the Aka in village activities (including farming and wage labor) as change toward a single society with two castes, the Aka being the underclass or lower caste (Delobeau 1978, 1989; Motte 1982). A constraint to the Aka settling permanently into this system is the seasonal demand for labor and the growing urban demand for meat (Berry et al. 1986).

Some researchers therefore describe a more or less rigid seasonal activity cycle for Aka groups. In the Bangandou area, Bahuchet (1988, 1990) and Hudson (1990) describe seasons based on natural resource abundance (figure 18.3): an August–September caterpillar season, when large numbers of caterpillars fall to the ground before pupating and can be collected for

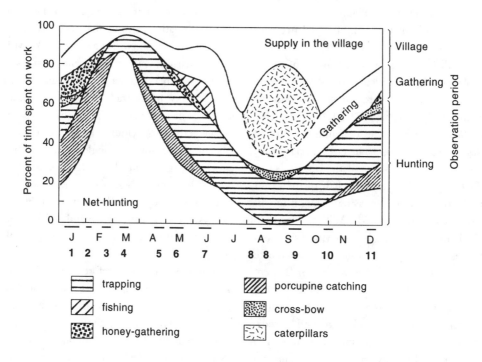

**Aka calendar before arrival of white people**

| Season | | Month | Dwelling | Gathering | | | | Hunting | | | Task group |
|---|---|---|---|---|---|---|---|---|---|---|---|
| Aka dry season | Last rains | Nov | Mixed camp, 25-30 people | Honey-locating | | | | Bow (monkeys), net (porcupine) | | | Man alone or father and son, and women and girls together |
| | | Dec | | | | | | | | | |
| | Dryness | Jan | Gathered camps, 70-100 people | | Yams | | Oleaginous kernels, Iringia | | Large framework hunting (duikers) | | Everybody: men, women, and children |
| | | Feb | | | | | | | | | |
| | First rains | Mar | Mixed camp, 25-30 people | Honey-collecting | | | | Individual: bow (monkeys), framework net (porcupine) | | | Man alone or father and son, and women and girls together or conjugal family |
| | Less frequent showers | Apr | | | | | | | | | |
| | | May | | | | | | | | | |
| | | Jun | | | | | | | | | Men together Women together |
| Aka rainy season | Heavy rains | Jul | Male camp Female camp | | | Mushrooms | Caterpillars | Spear: red hog, bongo, elephant, gorilla, chimpanzee | | | |
| | | Aug | Mixed camp | | | | | | | | Everybody or conjugal family |
| | | Sep | | | | | | | | | |
| | | Oct | Male camp Female camp | | | | | | | | Men together Women together |

Figure 18.3. Aka seasonal activity calendar: (*top*) Aka pre-colonial activity calendar from the late 1800s; (*bottom*) time devoted to food-getting activities by calendar month. Adapted from Bahuchet 1988.

consumption and for sale, and a honey season during the December–February dry season. In the long dry season several small Aka groups join together for net hunting, or to work in villagers' fields (Bahuchet 1979a, 1979b; Motte 1982). Some Aka establish temporary forest camps that move every two months, spending half the year or more in the forest and the other half near a village of another ethnic group. Others establish semi-permanent forest camps that are occupied for 6–8 months each year for 10 years or more, planting crops of their own nearby (Hudson 1990).

In Bangandou, Bahuchet (1985b, 1990) describes Aka territories shared by a regional band of roughly 100 individuals comprising several residential groups of about 30 individuals each. These territories of approximately 400 km² are long and narrow trapezoids following 100 km trails "owned" by particular Ngando lineages and extending south into Congo-Brazzaville (figure 18.4). Long-term visits permit access to resources in other bands' territories (Bahuchet 1991b). In this area territories are more or less exclusive with respect to other Aka bands, but the Ngando also have territories in the forest along the same trails, and use any territory or resources as they pass through (Bahuchet 1979b). Where the Aka have relations with the Ngbaka, the latter "own" the land and allow the Aka to hunt on it (Arom and Thomas 1974). Although the spe-

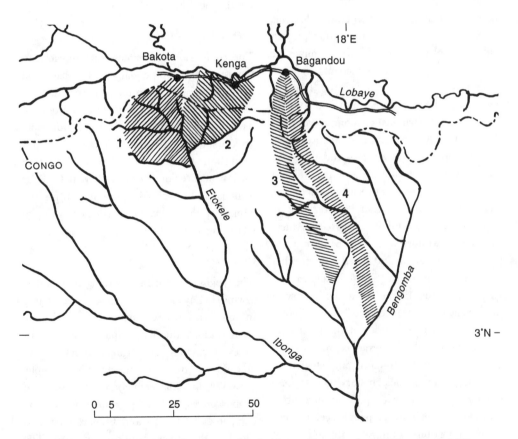

Figure 18.4. Aka resource exploitation ranges, showing four regional bands: (1) Bakota band, (2) Kenga band, (3) Bagandou band of the Bombolongo lineage, and (4) Bagandou band of the Bonzombi lineage. Adapted from Bahuchet 1979b.

cific manifestations vary, the Aka are always subservient to, and generally exploited by, their "partners" of other ethnic groups.

## Economic and Social History of the Bayanga Region

Various economic forces have influenced and continue to influence Aka life in the Bayanga region, the southwestern corner of CAR. Economic agents have been much more important than political actors in driving regional ecological and socioeconomic change. These actors have been principally foreign—concession companies, commercial societies, logging companies, and conservation organizations—and focus primarily on the exploitation of natural resources. Current economic activities with links outside the region include coffee plantations, logging, diamond mining, and a conservation project. Only coffee plantations and diamond mines are owned and operated by local residents.

Precolonial inhabitants of the Bayanga region were the Mbimu, Ngundi, and Sangha-Sangha peoples, who fished along the Sangha river; and the Aka, who hunted the forests east of the Sangha into what is today Congo (Demesse 1978). Historical accounts suggest that, long before the first Europeans visited the area in the 1890s, the Sangha river was an important trade conduit for slaves to the Americas; ivory, wild rubber, and palm nuts to Europe; and manioc and dried fish to other parts of central Africa (Sautter 1966; Harms 1981; Kalck 1993). The economic history of the Bayanga region since 1900 is one of boom-and-bust cycles, as a series of foreign enterprises have brought temporary economic prosperity followed by collapse and stagnation (Kretsinger 1993). Social history reflects these economic patterns, with population centers growing and declining as people seeking economic opportunities migrate into and out of the region.

### THE CONCESSION SYSTEM

French concession companies after 1898 used forced labor, a head tax that could be paid in forest products, and trade to generate most of their profits from the extraction of ivory and wild rubber (Serre 1960; Coquery-Vidrovitch 1972; Zoctizoum 1983; Colchester 1993). Under German administration (1911–1918) the Compagnie Forestière Sangha-Oubangui used Bayanga as a base for extracting ivory, wild rubber, animal skins, and palm nuts (Carroll, n.d.). During this period the Aka were the principal forest product extractors, turning over ivory, duiker skins, rubber, and palm nuts to "patrons" of other ethnic groups (Ballif 1954; Demesse 1957; Sautter 1966; Bahuchet and Guillaume 1982; Hewlett 1991).

### COFFEE PLANTATIONS

The concession companies were replaced in the 1930s by commercial societies based on agricultural production and coffee production, but which continued to extract forest products. Two societies were established in Lindjombo, 30 km south of Bayanga. Coffee plantations drew the first Gbaya and Mbimu immigrants to the Bayanga region from elsewhere in CAR, as well as additional Aka migrants from the Mokala watershed 50 km to the east in Congo (Kretsinger 1993; Giles-Vernick 1995).

### LOGGING

The Yugoslav company Slovenia-Bois built a sawmill in Bayanga and began selective logging in 1972 with a maximum of 600 employees. Immigration induced by employment opportunities increased Bayanga's population from several dozen to approximately 5,000 people. However, poor management, changes in world timber prices, and competition from logging companies in Congo and Cameroon caused Slovenia-Bois to go bankrupt in 1986. The abrupt closing left all employees jobless, with several months of back pay outstanding. Many stayed in Bayanga, waiting for their back pay and for logging to resume, but the town's population declined to less than 2,000 (Telesis 1991).

In 1993 a French company began logging operations in Bayanga, eventually renaming itself Sylvicole de Bayanga. Working the original Slovenia concession east and north of Bayanga until August 1994, Sylvicole de Bayanga obtained a new logging concession south and west of Bayanga in September 1994. The new concession preempted an effort by the German government and the World Wildlife Fund (WWF) to buy out logging rights in the region in order to consolidate conservation efforts. In 1995 logging employed about 120 men. Two dozen of these were Aka, mostly from Mossapoula and Yandoumbe, working as prospectors and laborers. Several of the Aka moved with their families to camps near the new logging site. In 1997 the CAR government suspended the company's activities on the grounds of mismanagement, then revoked the concession, and WWF is again trying to buy out logging rights.

DIAMOND MINING

Diamonds have surpassed coffee and timber as CAR's most important export (World Bank 1990). The majority of diamonds are supplied by an estimated 50,000 artisanal producers who excavate pits several meters across and several meters deep in streambeds (Zoctizoum 1983). Exploitation areas have shifted gradually south toward Bayanga. In 1994 a concentration of more than 2,000 people occupied a mining area on the northern boundary of the reserve itself near Yobe, only 20 km from Bayanga, benefiting from Sylvicole de Bayanga's logging roads. Most miners are immigrants from outside the Bayanga region, but local residents increasingly participate as well, as miners, porters, or merchants supplying food and commodities.

**Integrated Conservation and Development**

The Dzanga-Sangha integrated conservation and development project (figure 18.5) was established with the creation of the Dzanga-Ndoki National Park and the Dzanga-Sangha Special Reserve in 1990 by the government of CAR. The project is headquartered in Bayanga and is administered by the World Wildlife Fund-US. The project has multiple conservation objectives: to protect important populations of forest elephants, western lowland gorillas, and other wildlife; to preserve a significant portion of CAR's remaining rain forest; and to maintain the cultural integrity of the Aka (Carroll 1986; Doungoubé 1990).

The project includes the following conservation components: the Dzanga-Ndoki National Park, comprising two core protected areas totaling 1,220 km²; anti-poaching measures; tourism promotion to generate revenues; and conservation education. In order to reinforce its conservation objectives, the project also incorporates rural development components: most important, a mobile health-care unit that regularly visits Aka settlements, as well as employment for Aka men as park guides and day laborers. The project permits certain subsistence and income-generating activities within the 3,360 km² Dzanga-Sangha Special Reserve: farming in a kilometer-wide strip along the main Lindjombo-Salo road, and subsistence hunting and gathering using traditional harvest methods (Carroll 1986; Doungoubé 1990). An association of sixty representatives from all settlements in the project area was formed in 1994 to identify and implement community development activities—for example, agroforestry, crop protection from elephants, and health care—using the 40% of park visitor fees allocated to community development.

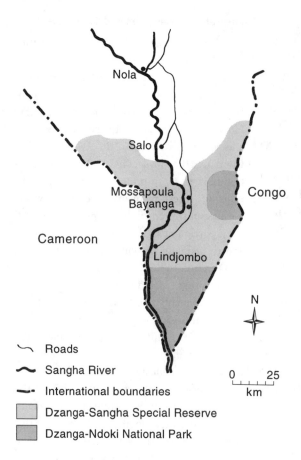

Roads

Sangha River

International boundaries

Dzanga-Sangha Special Reserve

Dzanga-Ndoki National Park

N

0      25
km

Figure 18.5. Dzanga-Ndoki National Park and Dzanga-Sangha Special Reserve. Adapted from Doungoubé 1990, courtesy of the World Bank, 1990.

## Aka Settlements in the Bayanga Region

A brief description of Aka settlements in the Bayanga region (figure 18.6; table 18.1) reveals a wide range of economic strategies and patterns of inter-ethnic relations, according to the economic opportunities and constraints each settlement faces, and resulting from the economic activities described above.

### BELAMBOKE, MONASSAO, AND NGUENGUELI

Established by French missionaries in 1973 and 1975 to foster socioeconomic development for the Aka (Koulaninga 1987), these villages drew Aka migrants from as far as Bambio, 120 km to the northeast. Belamboke is the largest of the Aka settlements and an important medical center. The Aka here are also the best educated in

the region: several individuals operate a community store, work in the hospital, and lead mass. As Monassao has grown, some residents have moved a short distance south to form the village of Nguengueli.

The mission goal of "accompaniment" supports house construction, manioc cultivation and processing, primary education and adult literacy programs, and health services, and encourages social and political awareness through voter registration and equal treatment of the Aka (Koulaninga 1987). Although the president of CAR declared them "citizens" in 1989, most Aka residents are not allowed to vote because they lack birth certificates or identification cards (Kretsinger 1993). In 1994, with some encouragement from the missionar-

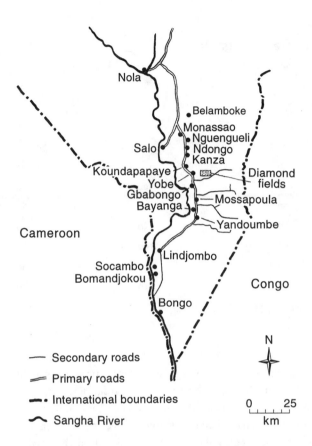

Figure 18.6. Population centers in the Bayanga region. From Noss 1995.

ies, the Monassao Aka went on strike against outside employers and merchants on two occasions: once for higher porter wages, and once for higher manioc prices. Both strikes were successful. This may be the first example of Aka political mobilization.

Although the villages are located in a savanna, residents continue to hunt (using both nets and snares) in nearby forests. In 1994, net and crossbow hunters in forest camps were supplying meat to the diamond fields near Yobe and further north. Some groups remained in forest camps for as many as four months in 1994, and may range well into Congo. Most of those in forest camps return to the villages at the start of the school year for the children to attend classes.

These villagers also produce manioc for consumption and sale, and work as day labors in neighboring villages and towns. Finally, many individuals now work in the nearby diamond fields, mining their own claims or serving as porters for other miners and merchants. In general, the Aka seek to emulate their neighbors in these ways in order to "be like villagers" and to be recognized as citizens (Bahuchet 1991a).

## MOSSAPOULA AND YANDOUMBE

Before Slovenia-Bois opened in 1971, the Mossapoula Aka lived next to the Kaka fishing village of Mbinjo, which is now part of Bayanga. By the early 1980s most of the Aka had moved to the airstrip and worked for the logging company. They moved 5 km north of

Table 18.1

## Estimated Population Figures
## for Aka-Inhabited Settlements in the Bayanga Region

| SETTLEMENT | AKA | OTHER | TOTAL |
|---|---|---|---|
| Belamboke[a] | 850 | | 850 |
| Monassao[a] | 750 | | 750 |
| Nguengueli[a] | 600 | | 600 |
| Mossapoula[a] | 320 | 50 | 370 |
| Yandoumbé[b] | 320 | | 320 |
| Lindjombo[c] | 220 | 500 | 720 |
| Koundapapaye[b] | 80 | | 80 |
| Gbabongo[b] | 60 | 150 | 210 |
| Socambo[b] | 60 | 100 | 160 |
| Bomandjokou[b] | 40 | 150 | 190 |
| Yobé[b] | 40 | 10 | 50 |

*Sources:* Data from (a) Noss 1995; (b) Kretsinger 1993; (c) Giles-Vernick 1995.

Bayanga to their present location at Mossapoula in the mid-1980s, citing conflicts with neighbors of other ethnic groups, insufficient land for manioc fields, and the lack of a nearby water source. Some of their Kaka patrons moved with them (Kretsinger 1993).

Originally living next to the Sangha-Sangha in Bayanga, the Yandoumbe Aka moved to the Kenie river next to the Slovenia-Bois sawmill in the mid-1970s. At this site the Aka suffered tremendously from disease, alcoholism, and abuse from Bayanga creditors. The American musicologist Louis Sarno, who had been living with the Aka since 1986, persuaded them in 1990 to move to the current village site 3 km south of Bayanga (Kretsinger 1993; Sarno 1993). In 1994, Yandoumbe split into three sections, with two groups moving 1 km closer to Bayanga. Bayanga farmers have also begun to plant fields south of Yandoumbe, in order to escape elephant depredations closer to Bayanga. They hire the Aka to work in their fields, but increasingly compete with them for

land, renewing the conflicts that the Aka sought to escape by moving away from Bayanga.

Most Mossapoula and Yandoumbe Aka have their own manioc fields, at times even selling manioc to Bayanga residents in a reversal of traditional trade patterns. Some men are formally employed by the logging company, the project, and researchers. Both men and women also work as day laborers in Bayanga. Men at times also make raffia roof shingles and tap raffia wine for sale. Every evening a group of 2–30 Bayanga market women visit Mossapoula and Yandoumbe to sell such "town" products as bread, manioc, alcohol, peanut butter, and other goods. In return the Aka sell "forest" products: meat, koko leaves (*Gnetum buckholzianum*), payo nuts (*Irvingia excelsa*), mushrooms, and caterpillars.

Net hunting is practiced year-round on day hunts from Mossapoula and Yandoumbe, and seasonally from forest hunting camps. At least three groups from each village moved to forest camps for several weeks in 1994. Several men in each village increasingly hunt with cable snares, purchased with income from formal employment. This hunting method is more complementary to formal employment than is net hunting.

## LINDJOMBO

Following the decline of coffee plantations and the closure of the logging company across the Sangha River in Cameroon, Lindjombo today is a town in decline (Giles-Vernick 1995). In the 1950s, the Aka worked in coffee plantations and cultivated manioc farms. Demesse (1957, 1980) reports that at that time they did not net hunt, and they entered the forest only to hunt elephants. Many Aka today have small fields but are not self-sufficient in manioc and often work in their neighbors' coffee plantations or manioc fields. The Lindjombo Aka established several separate net hunting and spear hunting camps in 1994, with one group staying in the forest for over four months.

## SMALLER SETTLEMENTS

Residents of Bomandjokou and Socambo have moved south from Lindjombo in order to fish and farm, occasionally accompanied by some of their Aka "clients" (Kretsinger 1993). These Aka net hunt infrequently, and did not establish forest camps in 1994. Few have fields, and most work for their "patrons."

The Gbabongo Aka also depend on their patrons for manioc, obtained in exchange for work in fields, in diamond mines, and on snare lines. These Aka plant few fields of their own and net hunt infrequently. Although they did not make forest camps in 1994, the previous year they formed joint forest camps with Mossapoula groups. Highly dependent on their neighbors, they are exploited and even physically abused, and some have moved to other Aka villages (Kretsinger 1993).

The Aka in Koundapapaye left Yobe in 1990 to create an Aka-only village, and they are self-sufficient in manioc. However, they continue to work intermittently as day laborers in order to earn cash. Farmers from Bayanga have recently moved to the Koundapapaye area as well, saying that elephant damage to their fields near Bayanga is too severe. They now occupy land both north and south of the Aka village, in effect recapturing Aka labor and re-creating competition for land as well as ethnic conflicts that the Aka originally fled in establishing Koundapapaye. The Aka hunt with nets and with snares, but just on day trips from the village, not from forest camps. When Slovenia-Bois was logging the surrounding area in the 1970s, several of the men worked for the company.

Two small Aka settlements near Yobe were established in 1994. Residents say that they do not have time to work in the diamond fields or net hunt because they must protect their fields from elephants.

## Mossapoula Today

A more detailed examination of individual activity patterns and mobility in the village of Mossapoula provides additional insight into Aka life today, in particular the relations between economic opportunities and forest resource exploitation patterns (see Noss 1995, 1997b, 1998a, 1998b). The Mossapoula Aka do not depend solely on hunting and gathering. In fact, the large size and relative permanence of Mossapoula results from the availability in Bayanga of opportunities for formal labor and day labor. Agriculture is also becoming increasingly important.

### ACTIVITY DISTRIBUTION

For all Aka adults and adolescents in Mossapoula, principal daily activities were recorded for 90 days of observation during the period from December 1993 to November 1994: 126 women and girls, 113 men and boys, and 19,656 total person-days. With respect to individual activity patterns, a subsample of 113 women and girls and 99 men and boys was derived, whose activities were tracked for at least 60 days. Those excluded at this level were added to or removed from the list in the middle of the observation period. Table 18.2 presents individual activity patterns: how individuals occupy themselves over time. Table 18.3 presents community activity patterns: how the community occupies itself each day. Individuals switch among activities from day to day, while households and the community as a whole pursue several activities simultaneously. The pattern of mean values for individual activities is virtually identical to the pattern for community activities.

Activities are categorized in the following fashion: formal labor, day labor, net hunt, for-

est camp, and absent from Mossapoula. An additional catch-all category that accounts for 40–45% of total workdays includes other activities in or near Mossapoula: field work, child care, fishing, gathering, tapping raffia wine, and making raffia roof shingles.

Formal labor includes employment with researchers, with a World Bank forestry survey team, and with a filmmaker. Only the filmmaker employs women. In addition, the project hires several men to work at the tourist center in Bayanga and to guide tourists. The logging company hires only men, but when logging operations shifted south of Lindjombo in 1994, a number of Mossapoula families moved to a camp near the work site. Formal labor is the single most important activity for men, occupying

Table 18.2

## Mossapoula Individual Activity Distribution

| WOMEN (N=113) | RANGE[a] | AVERAGE[a] | SD (±) | AVERAGE DAYS/YEAR |
|---|---|---|---|---|
| Mossapoula | 5.6–100.0 | 45.5 | 19.0 | 166 |
| Net hunt | 0.0–43.3 | 18.1 | 11.0 | 66 |
| Day labor | 0.0–33.3 | 11.3 | 7.6 | 41 |
| Formal labor | 0.0–44.4 | 5.5 | 11.9 | 20 |
| Forest camp | 0.0–20.0 | 3.2 | 5.4 | 12 |
| Absent | 0.0–94.4 | 14.4 | 23.4 | 53 |
| MEN (N = 99) | | | | |
| Mossapoula | 1.1–95.6 | 40.2 | 22.4 | 147 |
| Formal labor | 0.0–93.3 | 24.2 | 22.0 | 88 |
| Net hunt | 0.0–32.2 | 11.6 | 8.7 | 42 |
| Day labor | 0.0–66.7 | 7.0 | 10.1 | 26 |
| Forest camp | 0.0–20.0 | 3.2 | 5.0 | 12 |
| Absent | 0.0–94.4 | 11.6 | 23.4 | 42 |

*Sources:* Noss 1997b.

*Notes:* N = 90 observation days. Total days per year are less than 365 days because not all individuals were on the records list for the full 90-day observation period. Differences in means between women and men are statistically significant for the following categories: Mossapoula ($t = 1.84$, $P < 0.1$), net hunt ($t = 4.73$, $P < 0.01$), day labor ($t = 3.50$, $P < 0.01$), and formal labor ($t = -7.82$, $P < 0.01$). Differences in means for forest camps ($t = 0.04$) and absent ($t = 0.86$) are not significant.

a   Figures are given in percentage of days per individual.

Table 18.3

## Mossapoula Community Activity Distribution

| WOMEN (N=113) | RANGE[a] | AVERAGE[a] | SD (±) | AVERAGE DAYS/YEAR |
|---|---|---|---|---|
| Mossapoula | 14.3–75.4 | 42.5 | 15.8 | 54 |
| Net hunt | 0.0–46.0 | 16.7 | 16.0 | 21 |
| Day labor | 0.0–42.9 | 10.9 | 9.8 | 14 |
| Formal labor | 0.0–16.7 | 5.3 | 6.0 | 7 |
| Forest camp | 0.0–19.8 | 3.1 | 4.4 | 4 |
| Absent | 4.0–23.8 | 13.3 | 5.4 | 17 |
| MEN (N=116) | | | | |
| Mossapoula | 16.4–59.5 | 37.8 | 10.4 | 43 |
| Net hunt | 0.0–35.3 | 10.4 | 10.6 | 12 |
| Day labor | 3.4–16.4 | 10.4 | 3.4 | 12 |
| Formal labor | 0.0–25.9 | 6.2 | 4.5 | 7 |
| Forest camp | 0.0–19.8 | 2.9 | 4.3 | 3 |
| Absent | 11.2–32.8 | 20.9 | 4.9 | 24 |

*Source:* Noss 1997b.

*Notes:* N = 90 observation days. Total persons per day are less than 126 and 116, respectively, because not all individuals were on the records list for the full 90-day observation period. Differences in means between women and men are statistically significant for the following categories: Mossapoula ($t = 2.36$, $P < 0.05$), net hun ($t = 3.13$, $P < 0.01$), formal labor ($t = -19.10$, $P < 0.01$), day labor ($t = 4.09$, $P < 0.01$), and absent ($t = 4.33$, $P < 0.01$). Differences in means are not significant only in the case of forest camp ($t = 0.37$).

a    Figures are given in percentage of persons per day.

24% of their workdays, but only 5% of workdays for women.

Day labor includes work in Bayanga, Gbabongo, or Mossapoula. Women most often work in manioc fields or coffee plantations, while men may clear fields and cut palm nuts, raffia leaves, rattan, or construction poles. Gun hunting and work in the diamond fields are also classified as day labor. During the observation period, four different men were hired to hunt with a Bayanga resident's gun. Only one individual, a Sangha-Sangha man married to an Aka woman and living in Mossapoula, spent any significant time in the diamond fields. Other men and women made only short visits. Day labor is more important for women than for

men: 11% versus 7% of workdays, respectively.

Net hunting is the most important activity for women, at 18% of workdays, and the second most important activity for men, at 11%. The average daily returns per net hunter in FCFA (570 FCFA equal about $US1) from meat alone equal the minimum logging company wage of 200 FCFA per day, but are well below the official minimum wage of 650 FCFA per day that can be earned from formal employment with the project. In addition, net hunts do not take place every day, and Mossapoula residents can maximize their income by pursuing formal employment. However, formal employment opportunities for the Aka are limited and temporary. Individuals drop in and out of the formal employment system, engaging in hunting activities in between intermittent formal employment.

New opportunities do not entirely replace net hunting: virtually all individuals continue to practice net hunting at least occasionally and maintain the skills in their repertoire. Should the alternative opportunities evaporate, individuals may participate more in net hunting. But currently women emphasize such food-producing activities as net hunting and day labor, while men emphasize cash-earning formal employment. Net hunting continues to be practiced because it provides an income-earning and subsistence opportunity for Mossapoula residents who cannot find formal employment, and supplies returns in addition to meat. But several Mossapoula residents are shifting to snare hunting, which permits more intensive exploitation of multiple resources: individuals can harvest meat and earn wages at the same time.

MOBILITY

Mossapoula residential and occupational mobility takes place along two axes: a north-south axis between Mossapoula and other villages, and an east-west axis between Mossapoula and forest hunting camps. Mossapoula residents are absent 11–14% of the time, visiting other settlements from Bomandjokou in the south to Belamboke in the north. People change residences because of conflicts with their neighbors, in order to visit family members, and to pursue such economic opportunities as logging, diamond mining, and farming. Occupational mobility focuses on wage-labor opportunities in other settlements, especially Bayanga, or on forest exploitation: hunting, gathering, fishing, logging, conservation, research, and diamond mining. Mossapoula residents extend their exploitation ranges, for hunting as well as for other economic opportunities, by moving to other settlements in the region where they have

relatives. This north-south mobility between villages far exceeds the importance of east-west or village-forest mobility.

In 1993, more than half the Mossapoula Aka moved into forest camps in February and March, the middle of the dry season. There were as many as three separate camps, frequented by market women from Bayanga and Gbabongo coming to buy meat and sell manioc. The camps shifted every couple of weeks, moving gradually further away from Mossapoula. The group that stayed the longest in the forest made its last camp roughly 25 km east of Mossapoula, and finally returned to the village in early December, after ten months in the forest. During the later months, no market women were visiting their camp with manioc, and they were subsisting principally on wild yams, honey, and meat.

In 1994, forest camps were relatively small and short lived, consisting of 30–60 persons and lasting no more than two months during two main periods: February–March and July–September. More than 60% of Mossapoula residents never moved to forest camps. On average, Mossapoula residents spent only 3% of their time in forest camps, less than twelve days of the year (SD ± 19.2).

Forest camp residents supplied themselves with manioc by sending people back to Mossapoula, Bayanga, or Gbabongo to sell meat and buy manioc. One camp also obtained manioc from three Bayanga market women who came to buy meat: the women stayed for three days, until their manioc was finished. During five days when the market women were present, 77% of marketable meat from hunting was traded, compared to less than 20% of meat from Mossapoula day hunts that is sold. Lacking the alternative subsistence and economic opportunities available to Mossapoula residents, forest camp residents depend much more on trading or selling meat.

The principal reason cited for not moving to

forest camps, and for not staying longer in the forest, was that people had to clear and plant manioc fields. Rather than "meat hunger" as an incentive to hunt, they more frequently cite "manioc hunger" as a reason for not hunting or for not establishing forest camps. Other individuals said that they preferred to stay in the village where they could find salaried work—they complained they could not earn money living in the forest. Two out of five men in one camp were employed at the same time by a researcher: they spent two weeks working for her and then two weeks in the forest hunting camp. Such mobility would not have been easy if the hunting camp were further from Mossapoula, so establishing a camp near Mossapoula was a way to combine two activities that were important to

them. Forest camps are generally family groups led by experienced men 40–50 years old. Because several of these men were employed by the logging company in 1994, they stayed in Mossapoula, as did their family members who might have followed them into the forest.

The camps in 1994 were no more than 10 km from Mossapoula, thus on the edge of the range exploited by day hunts out of Mossapoula (figure 18.7). The forest camping time ends with the beginning of the dry season in December, when people return to Mossapoula to plant fields. Therefore, those who left the village as late as July did not have time to get very far from Mossapoula, as camps move away only gradually. One group that might otherwise have stayed longer returned to the village because

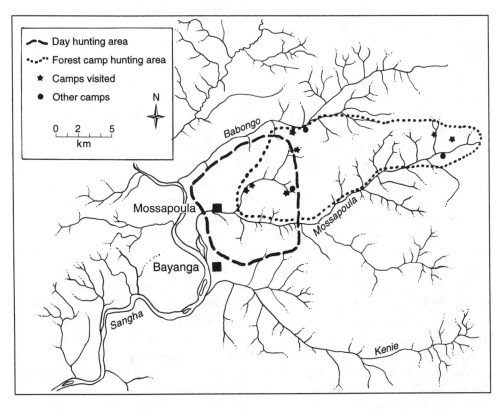

Figure 18.7. Mossapoula net-hunting range. Forest camps move frequently, and "other camps" refers to temporary sites previously occupied by groups visited. From Noss 1995.

people there heard that the project guards were looking for them to chase them out of the National Park, where camping and hunting are not permitted.

## Conservation Implications of Aka Resource Exploitation

The history of subsistence and economic strategies described above is important because it determines current patterns of natural resource exploitation. For at least a century the Aka have participated in local and extra-local trading networks. Relations with non-Pygmy peoples exert a greater influence on the Aka use of space than do resource constraints. The Aka cannot be understood in isolation, and Aka activities today depend on the situation of neighboring groups that include immigrant populations and international organizations (Bahuchet 1979b, 1991b; Bahuchet and Guillaume 1982; Kretsinger 1993). These groups influence the resource base available to the Aka, presenting opportunities and constraints for their activities (Pederson and Wæhle 1988). Nor can development or conservation activities address Aka concerns separate from those of their neighbors. Together they must be considered as an integrated economic and social system (Bailey et al. 1990).

The case of elephant crop depredations reveals some of these relationships, as well as the complexities associated with conservation and development efforts. Under park protection, elephants have become increasingly bold and frequent crop raiders, particularly in Bayanga, Gbabongo, and Mossapoula. Crop damage in Mossapoula has discouraged some Aka from farming and may prolong their dependence on such alternatives as hunting or day labor for Bayanga residents. Koundapa-paye and Yandoumbe were created by Aka groups seeking self-sufficiency in manioc production and independence from exploitative relationships with their neighbors. But Bayanga farmers are now beginning to farm near both these villages, where elephant damage is perceived to be lower, thus renewing land and resource competition with Aka residents and recapturing their labor.

The Aka today pursue a wide variety of economic and subsistence strategies that may include hunting and gathering, day labor for villagers, specialist meat production for meat traders or loggers, permanent or seasonal wage labor, and cultivating their own fields or laboring in others' fields. Variation occurs at several levels and at very small geographic and temporal scales: among neighboring settlements, within groups that divide time between sequential options, and within settlements or groups that subdivide such that families or individuals pursue alternative strategies simultaneously (Hudson 1990). This variation results from a suite of socioeconomic influences: immigration of other ethnic groups, competition with other hunters, population growth, agriculture, diamond mining, coffee plantations, logging, tourism, missions, government centers, roads, and commerce (Demesse 1978; Bahuchet and Guillaume 1979; Pederson and Wæhle 1988; Bahuchet 1991a). Socioeconomic changes entail ecological changes as well, by exploiting natural resources and converting natural ecosystems to settlements and farms.

Current subsistence and economic activities reflect adaptation to the opportunities and constraints that these changes present. For example, logging provides formal employment directly to the Aka, while stimulating markets for meat and other forest products. But at the same time, Bahuchet (1979a, 1985b) suggests that caterpillar outbreaks are relatively small and irregular in the Bayanga region because logging selectively harvests trees that are important caterpillar food plants, in particular *Entandrophragma cylindricum, E. utile,* and *Triplochiton scleroxylon.* Although immigrant miners and loggers of other ethnic groups represent markets for forest products and provide day-labor opportunities for Aka residents, these groups also compete with the Aka by directly harvesting forest products, in particular bushmeat, themselves. Finally, the creation of national parks provides employment to the Aka in tourism-related activities, in addition to health and development resources. But parks simultaneously restrict Aka hunting activities, in particular by virtually eliminating the Mossapoula forest camping range. Furthermore, buffer zones like the Dzanga-Sangha Special Reserve, which allow the Aka to hunt using "traditional" nets while prohibiting cable snares (the principal hunting method for all other ethnic groups), are strongly resented by the other groups.

The Aka are socially and economically subordinated by other ethnic groups. In general, the Aka respond only to changes imposed by others: if others want to hunt, farm, mine, log, or protect areas that the Aka currently occupy or exploit, they do so. In the past, Aka mobility—particularly their ability to move into the forest for extended periods—was an important means to escape excessive exploitation and to avoid conflict. As hunting ranges decline, as hunting declines in importance relative to other activities, and as Aka demand for goods and services increases, town-forest mobility also declines. Town-town mobility is much more important, as the Aka move to take advantage of economic opportunities.

Many Aka seek to increase their independence from other ethnic groups in new ways: by establishing separate settlements, by cultivating manioc, by pursuing formal employment opportunities, and by producing directly for markets—for example, making raffia roof shingles or mining diamonds. However, by competing for the same resources, they foster resentment from other ethnic groups. Rising human population densities and competition for resources will induce further socioeconomic change in the Bayanga region.

The direction of change is difficult to predict, as in the past economic booms and opportunities have been temporary, producing a pattern of change that is more cyclical than linear. Most Aka individuals, even those with formal employment, intermittently hunt and gather, thereby maintaining these skills in their repertoire should other opportunities fail. Diversification of activities at the individual and community levels is therefore a risk-reducing strategy. For some individuals, engagement in formal employment precludes other activities. But formal employment opportunities are limited, and other members of the household or community continue to hunt and gather. Formal employment may also be intermittent, and over time individuals must support themselves by practicing other activities as well. Thus even the savanna villages of Belamboke, Monassao, and Nguengueli continue to net hunt in neighboring forests. Other "formal" or "day" employment, for example logging and diamond mining, need not preclude hunting and gathering: those directly employed can simultaneously hunt with cable snares, and other Aka groups can establish camps to supply miners and loggers with meat and other forest products. The forest remains important as one source of economic and subsistence returns.

In the Bayanga region, each Aka settlement exploits a hunting and gathering range surrounding the settlement. Occasionally neighboring settlements will collaborate in net hunting and in forest camping, as did Gbabongo and Mossapoula residents in 1993. But local residents of other ethnic groups ignore any Aka land or resource usufruct rights, hunting or extracting other forest resources anywhere they choose. In legal terms, all land and forest resources belong to the state, and no community or individual owns anything beyond land occupied by dwellings or crops. The Aka have never had a land- or resource-use area separate from that of their neighbors, nor is there any political mobilization in CAR for independent communal land or resource tenure. Natural resources remain open-access resources, or are awarded by the government in exploitation concessions to foreign organizations, and local residents have no incentive to conserve them.

Furthermore, it is extremely difficult to identify the original inhabitants of the Bayanga region who may have a claim as "original" land owners, as virtually the entire current population, Aka and other, results from immigration for economic opportunities in recent decades. Ethnic diversity, compounded by recent immigration, complicates efforts to build community unity and support for conservation efforts (Noss 1995, 1997a, 1998a).

Patterns of resource use were sustainable in the past because of low population densities and mobile settlements. The same patterns, such as net hunting by Mossapoula residents, are probably not sustainable today, with growing populations and more permanent settlements. Most hunting and gathering from Mossapoula now takes place within a 110 km² range accessible on day trips. This range is limited by the Sangha River to the west, the Dzanga-Sangha National Park to the east, and other settlements to the north and south. From traditional exploitation ranges of 4 km² per person described above, Mossapoula residents today have about one-tenth this area available to them: 2.7 persons per km². Even within this reduced range they must compete with Bayanga residents who also hunt and gather. The intensified pressure by net hunting alone within a reduced range is unsustainable for two of the four primary game species, namely, the duikers *Cephalophus monticola* and *C. dorsalis* (Noss 1995, 1998a, 1998b).

But like their neighbors, the Aka appear to be more concerned with short-term economic advancement than with long-term sustainable use of resources. When one type of resource or opportunity is exhausted, they will switch their attention to other resources or opportunities that will become available. They adopt practices or technologies and engage in activities that provide higher economic and subsistence returns, greater independence from exploitative relationships, and security by diversification that keeps other options open.

Bird-David (1992a, 1992b) argues that resource-exploitation behavior of "hunter-gatherers" like the Aka results from deep cultural attitudes: they "procure" resources from the environment, taking whatever they see, and trusting the environment to share its resources with them according to their needs, assuming abundance rather than scarcity. The "environment" for the Aka includes not only the resources of the natural environment but also the neighboring ethnic groups and international organizations of the socioeconomic environment. An important characteristic of their opportunistic foraging is the relative lack of management costs or investments: resources are harvested or "mined" rather than managed. Natural resources and land are not owned until they are harvested or occupied, respectively. Although they may result in more sustainable long-term exploitation, the adoption of resource management practices of any kind imposes additional short-term costs (Alvard 1994). The historical pattern has been to exploit resources or economic opportunities while they last, then move on to another area or opportunity. These resources or opportunities may recover at a later date, through the actions of nature, God, or markets, and can then be exploited anew.

Conservation efforts in this context face several challenges. The Aka and their neighbors use forest resources opportunistically, not within a long-term sustained management system, unless the "fallow" periods resulting from mobile populations and intermittent exploitation are considered to be a management system. But natural resource exploitation is an important component of the mixed subsistence and economic system in which all forest residents engage, with the relative importance of each component varying temporally and geographically. Development programs in forests represent economic opportunities that attract immigrants, thereby increasing pressure on natural resources. Conservation is resented because it reduces the range of opportunities. Conservation programs are judged in the same way as other international companies and projects that local people have experienced—as temporary booms to be exploited for the short-term employment benefits and services they provide, rather than as possibilities for long-term forest conservation. In this way, it is very difficult for a conservation project to compete with a logging company.

## ACKNOWLEDGMENTS

The author's field research was assisted by a grant from the Joint Committee on African Studies of the Social Sciences Research Council and the American Council of Learned Societies with funds provided by the Ford, Mellon, and Rockefeller Foundations. Additional funds were provided by two Grants-in-Aid of Research from Sigma Xi, The Scientific Research Society; and by the World Wildlife Fund under the United States Department of Agriculture Agreement No. 93-G-155.

## REFERENCES

Alvard, M. S. 1994. Conservation by native peoples: Prey choice in a depleted habitat. *Human Nature* 5:127–154.

Arom, S., and J. M. C. Thomas. 1974. *Les Mimbo, Génies du Piégeage et le Monde Surnaturel des Ngbaka-Ma'bo (République Centrafricaine)*. SELAF, Paris.

Bahuchet, S. 1979a. Utilisation de l'espace forestier par les Pygmées Aka, chasseurs-cueilleurs d'Afrique centrale. *Information sur les Science Sociales* 18:999–1019.

———. 1985a. Circulation et échanges en Afrique tropicale: Relations entre chasseurs-cueilleurs Pygmées et agriculteurs de forêt de Centrafrique. *Revista de Prehistoria* 6:86–97.

———. 1985b. *Les Pygmées Aka et la Forêt Centrafricaine: Ethnologie Écologique*. SELAF, Paris.

———. 1985c. Linéaments d'une histoire humaine de la forêt du bassin congolais. *Mémoires du Muséum National d'Histoire Naturelle, Série A Zoologie* 132:297–315.

———. 1988. Food supply uncertainty among the Aka Pygmies (Lobaye, Central African Republic). Pages 118–149 in I. de Garine and G. A. Harrison, eds. *Coping with Uncertainty in Food Supply*. Clarendon Press, Oxford.

———. 1989. Les pygmées Aka et Baka: Contribution de l'ethnolinguistique à l'histoire des populations forestières d'Afrique centrale. Doctoral thesis, Université René Descartes, Paris.

———. 1990. Food sharing among the Pygmies of Central Africa. *African Study Monographs* 11:27–53.

———. 1991a. Les Pygmées d'aujourd'hui en Afrique centrale. *Journal des Africanistes* 61:5–35.

———. 1991b. Spatial mobility and access to resources among the African Pygmies. Pages 205–257 in M. S. Casimir and A. Rao, eds. *Mobility and Territoriality: Social and Spatial Boundaries among Foragers, Fishers, Pastoralists, and Peripatetics*. Berg, New York.

Bahuchet, S., ed. 1979b. *Pygmées de Centrafrique: Etudes Ethnologiques, Historiques, et Linguistiques sur les Ba.Mbenga (aka/baka) du Nord-ouest du Bassin Congolais*. SELAF, Paris.

Bahuchet, S., and H. Guillaume. 1979. Relations entre chasseurs-collecteurs pygmées et agriculteurs de la forêt du nord-ouest du bassin congolais. Pages 109–139 in S. Bahuchet, ed. *Pygmées de Centrafrique: Etudes Ethnologiques, Historiques, et Linguistiques sur les Ba.Mbenga (aka/baka) du Nord-ouest du Bassin Congolais*. SELAF, Paris.

———. 1982. Aka-farmer relations in the northwest Congo basin. Pages 189–211 in E. Leacock and R. Lee, eds. *Politics and History in Band Societies*. Cambridge University Press, Cambridge.

Bahuchet, S., D. McKey, and I. de Garine. 1991. Wild yams revisited: Is independence from agriculture possible for rain forest hunter-gatherers? *Human Ecology* 19:213–243.

Bahuchet, S., and J. M. Thomas. 1986. Linguistique et histoire des Pygmées de l'ouest du bassin congolais. *Sprache und Geschichte in Afrika* 7:73–103.

Bailey, R. C., S. Bahuchet, and B. Hewlett. 1990. Development in the Central African rainforest: Concern for forest peoples. Pages 260–269 in K. Cleaver, M. Munasinghe, M. Dyson, N. Egli, A. Peuker, and F. Wencélius, eds. *Conservation of West and Central African Rainforests*. World Bank, Washington, D.C.

Ballif, N. 1954. *Dancers of God*. Sidgwick and Jackson, London.

Berry, J. W., J. M. H. van de Koppel, C. Sénéchal, R. C. Annis, S. Bahuchet, L. L. Cavalli-Sforza, and H. A. Witkin. 1986. *On the Edge of the Forest: Cultural Adaptation and Cognitive Development in Central Africa*. Swets North America, Berwyn, Ill..

Bird-David, N. 1992a. Beyond the "hunting and gathering mode of subsistence": Culture-sensitive observations on the Nayaka and other modern hunter-gatherers. *Man* 27:19–44.

———. 1992b. Beyond the "original affluent society": A culturalist reformulation. *Current Anthropology* 33:25–47.

Carroll, R. W. 1986. The creation, development, protection, and management of the Dzanga-Sangha Dense Forest Sanctuary and the Dzanga-Ndoki National Park in southwestern Central African Republic. Report. Yale University, New Haven.

———. N.d. Central African Republic background information. World Wildlife Fund, Bangui, Central African Republic.

Colchester, M. 1993. Slave and enclave: Towards a political ecology of equatorial Africa. *Ecologist* 23:166–173.

Coquery-Vidrovitch, C. 1972. *Le Congo au Temps des Grandes Compagnies Concessionaires, 1898–1930*. Mouton, Paris.

Delobeau, J. M. 1978. Yamonzombo et Yandenga: Histoire des relations entre les gens du fleuve et les gens de la forêt (XIXè–XXè siècle). *Cahiers Congolais d'Anthropologie et d'Histoire (Brazzaville)* 2:43–55.

———. 1989. *Yamonzombo et Yandenga: Les Relations entre les Villages Monzombo et les Campements Pygmées Aka dans la Sous-préfecture de Mongoumba (Centrafrique)*. SELAF, Paris.

Demesse, L. 1957. *A la Recherche des Premiers-Ages: Les Babinga*. Pierre Amiot, Paris.

———. 1978. *Changements Techno-économiques et Sociaux chez les Pygmées BaBinga (nord Congo et sud Centrafrique)*. SELAF, Paris.

———. 1980. *Techniques et Economie des Pygmées Babinga*. Institut d'Ethnologie, Paris.

Doungoubé, G. 1990. Central African Republic: The Dzanga-Sangha Dense Forest Reserve. Pages 75–79 in A. Kiss, ed. *Living with Wildlife: Wildlife Resource Management with Local Participation in Africa*. World Bank, Washington, D.C.

Giles-Vernick, T. 1995. "A dead people": Migrants, land and history in the rain forests of central Africa. Ph.D. dissertation, Johns Hopkins University, Baltimore.

Harms, R. W. 1981. *River of Wealth, River of Sorrow*. Yale University Press, New Haven.

Hart, J. A. 1979. Nomadic hunters and cultivators: A study of subsistence interdependence in the Ituri Forest of Zaire. Master's thesis, Michigan State University, Lansing.

Hewlett, B. 1977. Notes on the Mbuti and Aka Pygmies of Central Africa. Master's thesis, California State University, Chico.

———. 1991. *Intimate Fathers: The Nature and Context of Aka Pygmy Paternal Infant Care*. University of Michigan, Ann Arbor.

Hudson, J. L. 1990. Advancing methods in zooarchaeology: An ethnoarchaeological study among the Aka. Ph.D. dissertation, University of California, Santa Barbara.

Kalck, P. 1993. *Central African Republic*. Clio Press, Oxford.

Koulaninga, A. 1987. L'éducation chez les pygmées de Centrafrique. Doctoral thesis, Université de Paris V René Descartes, Paris.

Kretsinger, A. 1993. *Some Development Options for the Ba Aka Within the Dzanga-Sangha Dense Forest Reserve*. World Wildlife Fund, Bangui, Central African Republic.

Motte, E. 1982. *Les Plantes chez les Pygmées Aka et les Monzombo de la Lobaye (Centrafrique): Contribution à Une Etude Ethnobotanique Comparative chez des Chasseurs-cueilleurs et des Pêcheurs-cultivateurs dans un Même Milieu Végétal*. SELAF, Paris.

Noss, A. J. 1995. Duikers, cables, and nets: The cultural ecology of hunting in a Central African forest. Ph.D. dissertation, University of Florida, Gainesville.

———. 1997a. Challenges to integrated conservation and development or community-based conservation in Central Africa. *Oryx* 31(3):180–188.

———. 1997b. The economic importance of communal net hunting among the Ba Aka of the Central African Republic. *Human Ecology* 25(1):71–89.

———. 1998a. Cable snares and nets in the Central African Republic. In J. G. Robinson and E. Bennett, eds. *Hunting for Sustainability in Tropical Forests*. Columbia University Press, New York.

———. 1998b. The impacts of BaAka net hunting on rainforest wildlife. *Biological Conservation* 86(2):161–167.

————. 1998c. Cable snares and bushmeat markets in a central African forest. *Environmental Conservation* 25(3): 228–233.

Pederson, J., and E. Wæhle. 1988. The complexities of residential organization among the Efe (Mbuti) and Bamgombi (Baka): A critical view of the notion of flux in hunter-gatherer societies. Pages 75–90 in T. Ingold, D. Riches, and J. Woodburn, eds. *Hunters and Gatherers, 1: History, Evolution, and Social Change.* St. Martin's Press, New York.

Sarno, L. 1993. *Song from the Forest: My Life among the Ba-Benjellé Pygmies.* Houghton Mifflin, Boston.

Sautter, G. 1966. *De l'Atlantique au Fleuve Congo: Une Géographie du Sous-peuplement.* Mouton, Paris.

Serre, J. 1960. *Histoire de la République Centrafricaine.* Ecole Nationale d'Administration, Bangui, Central African Republic.

Telesis. 1991. *Sustainable Economic Development Options for the Dzanga-Sangha Reserve, Central African Republic.* Telesis USA for World Wildlife Fund, Providence.

Thomas, J. 1963. *Les Ngbaka de la Lobaye: Le Dépeuplement Rural chez une Population Forestière de la République Centrafricaine.* Mouton, Paris.

Vansina, J. 1985a. Esquisse historique de l'agriculture en milieu forestier (Afrique Equatoriale). *Muntu* 2:5–34.

————. 1985b. L'homme, les forêts et le passé en Afrique. *Annales Economies, Sociétés, Civilisations* 40:1307–1334.

————. 1986. Do Pygmies have a history? *Sprache und Geschichte in Afrika* 7:431–445.

World Bank. 1990. *Projet d'aménagement de ressources naturelles.* World Bank, Washington, D.C.

Zoctizoum, Y. 1983. *Histoire de la Centrafrique.* Harmattan, Paris.

# CHAPTER 19

# Horticulture and Hunting in the Congo Basin

A Case from Central Africa (DR Congo)

Alden Almquist

Efforts to preserve rain forest fauna outside reserves are unlikely to succeed without a clear understanding of resident human populations and of the impact of their resource extraction practices and strategies on the forest habitat. To further that understanding, it may be useful to examine the patterns of adaptation of a single rain forest community of hunters and horticulturalists, namely, the Apagibeti of north-central DR Congo. Local patterns of resource usage in both horticulture and hunting, of technological changes, and of parallel changes in social organization and relations of production will be the analytic focus of this brief social history. Finally, some implications for wildlife conservation programs will be suggested.

The Monveda Apagibeti live north of the Dua River in a rain forest plateau environment. Numbering no more than four thousand, they live in the second least densely populated of DR Congo's eight regions, with fewer than 8 persons/km². They are distributed among some thirteen villages, which are grouped by the state into three separate chiefdoms. The villages are joined by a long forest path that is bisected by river and marshes at each end, sharply limiting access by outsiders.

Their isolation from the DR Congo road network is anomalous. Historically, Belgian colonial officials obliged forest-dwelling groups to relocate along roads for purposes of easier supervision, taxation, enforcement of compulsory cash-crop cultivation and road maintenance, and provision of medical and educational services. Apagibeti success in forming good relations with colonial authorities spared them relocation, and their patterns of interaction and adaptation to the forest environment did not suffer the disruption visited on most Congolese rain forest communities.

Geographical isolation does not imply social isolation. A patrilineal, virilocal people, living between the more numerous Ngbandi to the north, Budja to the east, and Ngombe and Bale to the south, Apagibeti appear to have historically sought "outside" allies and resources. In precolonial times, trade, warfare, intermarriage, and use of the kinship tie of the mother's brother produced an intermixture of peoples along the Dua River. Warriors by tradition, some Apagibeti probably served as recruits in Congo Free State forces; many later saw service in colonial-era armies fighting the Germans in west Africa during World War I. Catholic mission evangelists established churches in most Pagibeti villages for periods of outside employment in plantations and towns.[1] In one village surveyed, some 65% of all resident males had "outside" work experience.

## Horticulture

### CROPS AND PLANTING STRATEGIES

Apagibeti practice *swidden* (or shifting) horti-culture. They fall into Miracle's (1967) wide-spread long-fallow category of land use. Mira-cle, in his classic survey of agriculture in the Congo Basin, distinguished four major subsys-tems based on the degree to which restoration of soil fertility is left to nature: (1) classic trop-ical long-fallow systems, the most frequently reported; (2) ash-fertilizer-dependent long-fallow systems, found in the southeastern cor-ner of the basin; (3) compost-dependent long-fallow systems, which were found only in savanna areas; and (4) short-fallow systems relying on animal manures or irrigation, which were restricted to the eastern rim of the basin. Swidden cultivators frequently planted special fields or gardens in addition to their large fields, and it was not uncommon for them to grow thirty, or more, different crops. Swidden agri-culture was, in fact, so complex and varied in its manifestations that Miracle concluded that "development programs which are suited to one tribal economy may well fail in another."

Apagibeti plant as many as thirty crops, the three most widely planted being plantains, manioc, and maize, with peanuts next in impor-tance. Crop rotation and intercropping are practiced, with plantains or maize being planted first and manioc planted last. Peanuts are occasionally also planted first, or they may be planted together with plantains.

In addition, small quantities of varied food crops are planted in selected niches, as on the edges of large fields or the paths leading to them, or near the homestead. These include three varieties of bananas, sweet manioc, pep-pers, several varieties of root tubers, several varieties of squash, papaya, limes, raffia palms, oranges, guavas, pineapples, tangerines, man-gos, avocados, coconut, and kola nut. Whereas cotton was formerly planted, cash crops are not currently cultivated except by a few farmers who grow surplus coffee, which they carry out and sell at the nearest road market.

New cultigens are continually being intro-duced, occasionally with significant impact. The planting of large numbers of raffia palms after World War II, for example, effectively ended the local brewing of plantain beer in favor of the felling and tapping of palm trees for palm wine production. The status of the palm also shifted from that of a private to a public good, following a typical pattern for newly introduced tree crops. As with land, trees are not owned. When trees are new and scarce, however, usufruct rights belong to the planting individual and his household. Only when established and plentiful do the trees become a public good, available to all comers.

Use of the ethnographic present obscures the shifting dynamic of property rights in tree crops. A 1920 ethnography would note that whereas papaya trees could be harvested by any-one, all oil palm, pitch, and two species of bark cloth trees would be known and respected as "belonging" to particular households. A 1980 ethnography would have to class all of the just-mentioned trees as free or public goods. Scarce tangerine, mango, coconut, avocado, and lime trees, however, would have to be entered on the list as new "private usufruct" tree crops.

### DIVISION OF LABOR

Formally, fields are allocated by chiefs. Produc-tion is by household, although the men's tasks of cutting and clearing are frequently done in work groups of lineage males. After two or three years of use, land is left fallow. After five to fifteen years it may again be cleared for reuse. Some fields' produce is not for the general household but is disposed of by the woman cul-tivator as she sees fit.

Although cultivated crops form the mainstay of the local diet, they are accorded far less status than food gained from hunting and trapping. Men often spend more time in the forest hunting and trapping than they do in the village. With the first June rains, men, together with a few boys, leave the village one by one for their forest shelters, leaving the remainder of the horticultural tasks of planting, weeding, and harvesting to the women. The horticultural tool kit is simple, consisting largely of machetes, hoes, axes, and digging sticks.

The five months from January through May are those of maximum activity in the village. Tasks are divided by sex (table 19.1). Children assist adults in their tasks and frequently haul burdens, deliver goods, and, most visibly, continually dispatch messages between village adults. They have specific roles in some rites, notably hunting rituals and funerals, and they are the preferred players in such daily social dramas as the collection of debts and the circulation and sale of meat at dusk.

Table 19.1

## Pagibeti Division of Labor

| MEN'S TASKS | WOMEN'S TASKS |
| --- | --- |
| Cutting, clearing, and burning of fields | Planting, weeding, and harvesting of fields |
| Clearing and widening of village-garden trails and of the broad intervillage trail | Daily acquisition and transport of water, firewood, garden produce to homestead |
| Repair or construction of walls and roofs of village houses, men's houses, chiefs' houses, and garden houses | Clay-water mixing and daubing of walls and floors of village houses and chiefs' houses |
| Repair or production of machetes, axes, hoes, knives, arrowheads, and spears by local forgers | Pottery production |
| Repair of hunting nets, guns, flashlights, and snares; twine production for net and snare repair | Daily food preparation including the pounding, leaching, sun drying, and preparation of manioc |
| Construction of curved slat-beds and chairs | Daily child care |
| Weaving of mats and baskets | |
| Cutting, drying, shaping, and stitching of animal-hide quivers, sheaths, and hats | |
| Harvesting of palm-nut bunches and of sap from *ebeke* (pitch) trees and fuel for night lamps and night-hunting torches | Production of cooking oil, soap, and body lotion from palm oil and palm-nut kernels |
| Organization of village-wide communal net hunts and of late dry-season termite hunts | Organization and execution of group fishing by damming and poisoning streams; participation in communal net hunts |
| Organization and execution of court cases and of hunting and other rites | Execution of first-birth rites |
| Palm beer brewing | |

## SOCIAL RELATIONS

The introduction of new cultigens has had marked social consequences. An example is the postwar Belgian colonial government program of enforced cotton cultivation, which, however resented, had the additional unintended consequence of contributing to an alteration in the balance of power between the sexes. Men claim that women became too free or, in other words, that the authority of men over women declined. The brief period of cotton cultivation enriched women planters and made them more independent from their husbands for purchases of clothing and other items. Discontented wives and daughters could and sometimes did flee their husbands' or fathers' villages for their home villages or for the urban centers of Businga or Lisala. Changes favoring women in the arenas of ritual practice, children's naming practices, and bride-price distribution followed.

## Hunting

Notwithstanding the overwhelming dependence of Apagibeti on cultivated food, both sexes maintain a strong preference for meat. It is not at all unusual to be accosted during the dry season by a villager with a dire complaint of hunger. Among Apagibeti, to say "I have not eaten today" means "I have not eaten meat today." Although rainy-season hunting is a predominantly male activity, both sexes work toward the preparation of food stores for the "entry," as the disappearance of men into the forest is termed; women and children may travel out to the forest shelters if the hunters and trappers fail to return promptly with the first fruits of the hunt to end the "time of hunger" of the late dry season.

## SHIFTING HUNTING PRACTICES

For hunting purposes, usufruct rights apply to forest land, and these are allocated politically by chiefdom. Each of the three Pagibeti chief- doms has its own bounded section of forest within which its component villages, lineages, and households maintain a network of trails, shelters, and bridges for rainy-season hunting and trapping of forest game. Individuals from one village may hunt the forests of another village only if specifically invited to do so by local residents.

Each small hunting or trapping group of two or three men maintains two or three rainy-season forest shelters. They rotate their residence in these shelters from year to year so as to "let the animals rest" or, alternatively, "let the animals give birth." Forests, then, are managed somewhat like fields, with one section worked one year and another the next. Forests are even left "fallow" during the dry season. With rare exceptions, no hunting is conducted during this time except in the immediate environs of the village. Hunters' rationale for this practice is the high demand for dry-season labor and the need to "let the animals rest" or "let the animals come back" after their disturbance by, or flight, from the intensive hunting of the rainy season.

Forests are highly differentiated environments. Apagibeti distinguish between three different kinds of forest, namely, white forest, black forest, and vine forest, depending on the prevalence of specific species of trees in each. And within this overarching tripartite division, Apagibeti distinguish a multiplicity of other domains. There is forest with or without *knogo* (a leaf crucial to shelter roofing), with or without water (meaning not only an absence of streams but also a lack of any of the half-dozen species of water-bearing vines or trees on which hunters depend and in whose vicinity long-term forest shelters may be established), with or without *woke* (shallow forest lakes, fed and emptied by streams, whose edges serve as a habitat for bongo and other species sought by hunters), with or without *wume* (grassy clearings favored by buffalo and other game), and so on.

Two categories of game are distinguished for hunting purposes: "animals of the above" and "animals of the below." Distinctive hunting techniques and technologies are used in each domain. Small mesh nets and bow and poisoned arrows are used in hunting "animals of the above," primarily monkeys and chimpanzees. Spears, hunting dogs, small and large mesh hunting nets, traps and snares, and locally fabricated guns are used in hunting "animals of the below."

Here, again, use of the ethnographic present obscures the dynamism of Pagibeti practice; new technologies are adopted and adapted with the same alacrity as are new cultigens. The local manufacture and use of guns adapted to firing short spears instead of slugs has entirely displaced the older log-fell technique in elephant hunting. Since they were introduced in the 1960s, miners' headlamps have largely displaced the pitch-torch previously used for illumination in communal night net hunts, and made night hunting by individuals possible. Night hunts with miners' headlamps is now the one practice whose adoption the Apagibeti commonly cite as overly destructive of local game.

The most dramatic technological innovation was arguably the widespread adoption of wire snares during the post-Independence period. These supplemented or supplanted traditional snares made locally of twine, and all hunters and trappers now use them to some extent. Their durability from season to season and their greater effectiveness in holding snared wildlife are cited by hunters as reasons for their adoption. An individual hunter may set as few as 30 or as many as 160 snares in the vicinity of his rainy-season forest shelter. He then walks the circuit of snares and dispatches any game caught in them with a spear or machete. A wide variety of game is caught in this way, including duiker, other species of antelope, chevrotain, iguana, and pig. The wire snare has also influenced village social rela-

tions. Adopting wire snares enabled children to maintain a circuit of rainy-season snares and thus increase their income and relative independence.

Other hunters are specialists who hunt predominantly buffalo, elephant, or monkey. To specialize means to acquire not only the tools needed but also the "medicine" useful in making the kill. Getting the right medicine is a private and individual matter. As a local aphorism has it, *nto na nto na eboke te ngake* (to each person, his own leaf [or medicine]). Neighboring ethnic groups are visited by the hunter in order to acquire the knowledge and the medicine needed to succeed; one goes to the Ngombe for monkey-hunting medicine, to the Ngbandi for buffalo-hunting medicine, and to the Buans for elephant-hunting medicine. Those who hunt buffalo and bongo antelope with hunting dogs and spears generally have medicine that they feed to their dogs to make them "angry" and "hot" to hunt. At least one medicine, *pomoli*, is strictly proscribed. Its use empowers the user to, in effect, exchange the lives of his agnatic kin for the lives of forest game. Men have been brought to trial in Businga for suspicion of its use when unusually large numbers of kills in the forest coincide with unusually frequent deaths in the successful hunter's lineage.

Typically, men remain in their rainy-season forest shelters until supplies of garden food, salt, or palm oil are depleted, or until they have killed and smoke-dried a large quantity of game. At that point they either send a child back to the village to get women to bring out more supplies and carry back the smoke-dried meat, or they shoulder the burden of meat themselves and walk it out through the forest to sell at the markets of neighboring Ngbandi, Ngombe, or Budja ethnic groups, a trip of at least two to three "men's" days from most forest shelters. The forest also provides seasonal foraging opportunities for mushrooms, fruits, grubs, snails, honey, and termites.

Apagibeti have no intervillage markets among themselves but do market produce locally. Dried meat ported out and sold at road markets, together with the sale of the skins of bongo and buffalo, and occasionally leopard skins, okapi skins, or ivory, constitute the principal Pagibeti trade goods. Income earned is used to purchase those few items not produced in the village: salt, gunpowder, matches, steel wire, iron rod (for slugs), nylon line, batteries, metal pots and pans, and clothing. Remaining funds are spent primarily on taxes, medical care, bride-prices, school fees, and such key events as New Year's feasts, first-birth rites, and funerals.

## CONCEPTUALIZING THE FOREST

The forest and its game have much more than merely economic significance. The forest serves at once as a resource, a refuge, and an element of collective identity. During routine discussion in the village about hunting, I was spontaneously told on several occasions, "The forest saves us." This attitude sharply contrasts with that found in studies of forest farmers who trade their produce for forest products from hunter-gatherers (Turnbull 1968) and in a large number of structuralist studies of farmers and pastoralists; these have presented an understanding of the bush as a wild and threatening place contrasting with the secure zone of the village and its fields or corrals (Willis 1968).

For Apagibeti, the reverse is often the case. Apagibeti treat the forest as a refuge. When the state census taker or tax collector makes his rounds, some Apagibeti flee to the forest. One unregistered villager answered my warnings about the imminent arrival of the census taker or tax collector with the phrase "The forest is large." Older men exhausted by the civic demands of playing the elder during long dry-season court cases sometimes flee alone to their unoccupied forest shelter for relief from village social pressures. And in ritual it is the forest

that must be "protected" from threats emanating from the village, not the other way around.

## OPENING THE FOREST

Dry-season communal hunts vividly display and reproduce Apagibeti understandings regarding the forest. Villagers say that the forest will freely provide them with animals so long as they keep it "open" or "awake." One form of the communal hunt is even named the *paye leka* (forest wakening), which is performed following the burial of a deceased villager. The death of any villager "closes the forest" to hunters; no more game will be given up until the forest is reawakened or opened. (For more details on this and other forest awakening rites, see Almquist 1991.)

To reopen the forest after a death, the village gathers at dawn—men, women, and children alike. They prepare weapons, foodstuffs, hunting nets, and dogs for a two- to three-day hunt in the adjoining forest. First the assembled nets, spears, knives, and machetes are piled together in the village center and a water-saliva mixture containing the saliva of everyone in the village (that is, everyone who is of age, everyone whose "eyes are open") is sprinkled over the stacked hunting tools in a rite called *soseye ngi* (the washing of the village). This action is necessary in order to wash away the effect of both the recent death and any anger secretly harbored inside any of the hunt participants; unexorcised, such death or anger could close the forest.

Next, a special red leaf is retrieved; the leaf was rubbed with spittle from the deceased and placed under the *tolo*, or ancestral spirit shrine, just before burial. The morning of the hunt, the leaf is removed and borne into the forest to be cast into the first circle of strung nets.

Nets may be strung or "raised" from five to twelve times in the course of a day, depending on hunting fortunes and the size of the hunting group. A strict protocol governs conduct in the forest camps. At evening camp, men may discuss that day's or a previous day's hunt, or they and the camp's women and children may tell *mato*, or animal tales, to one another. The talk is important and powerful; talking of animals brings dreams of animals, and this means killing those animals on the following day. As talk appropriate to the forest, *mato* are rarely told in the village, and never before sunset (not even when the anthropologist asks for a special late-afternoon taping session). Conversely, youths talking about village affairs at the forest camp are frequently rebuked for speaking "village talk" deemed inappropriate to the hunting camp.

Other practices may be needed to awaken the forest. Although private medicine belts, customarily worn on private hunts, are banned, hunters whose nets are not attracting game may legitimately ask a neighbor to engage in reciprocal rubbing of skin with a particular leaf (*daleye*). Alternatively, they may take a leafy branch and beat the forest floor with it until they cut through the earth (*nya zonga*). In either case they address, while acting, the names of all those animals they wish to step into their nets.

If, after two or three raisings of the nets, no game is killed, a diagnosis of misfortune will be made using one of two oracles, the antelope-sending oracle (*ntiseye mboko*) or the bursting-pot oracle (*mbolongo*). In the first a hunter is selected by his peers to conduct the oracle. The hunter is selected on the basis of the degree to which he possesses a "good back" (*ngonga enza*), or good fortune in his enterprises, particularly in the hunt. The elected hunter then addresses the forest while beating the ground with a branch, reciting the reason for the hunt, telling its history, and citing the individual names of those animals the hunters hope to kill. He concludes, "Antelope, if we are to kill much, die male." After he repeats this several times, the hunt begins. If the first antelope caught in the strung circle of nets is male, the subsequent day's hunting should be good.

The bursting-pot oracle is more elaborate and is employed only when the assembled hunters have repeatedly strung their nets in a circle, loosed the dogs and beaters into the "head" of the circle, flushed the hidden game out in the open and into the nets, and yet found the catch insufficient. A twenty-odd animal kill is considered good; after two or three days of desultory kills, *mbolongo* is suggested.

Once again, a man chosen by his fellow hunters for his good back is selected. He goes out in the evening and cuts red wood scrapings from the *mbolongo* tree. The next morning at dawn he combines these with water and places the mixture in a small clay pot over a fire. While he works, the assembled hunters may dance a dance miming the movements of the animals they hope to kill. The oracle operator covers the top of the pot with a large *ngongo* leaf, sealing it tightly by wrapping a string around the lip. He then draws three or four trails radiating out from the fire center like spokes from the center of a wheel. As he draws each line he addresses the oracle, citing the name of a specific cause of misfortune and asking that it be identified and thus exorcised: "*Mbolongo*, I cut like this, I go, I cook you. May the forest open, may the forest boil over all at once. May we kill many animals. . . . If it be a pregnancy, spill here; if it be *mangodo*, spill here; if it be a death, spill here. . . . I cut like this, I go, I cook, may the forest open, may we kill animals." When the pot boils long enough, its leaf cover splits open and a water-foam-*mbolongo* mixture bursts out onto one of the three or four radiating spokes, indicating which of the causes of misfortune was responsible and exorcising it in a single stroke.

Some of the water-*mbolongo* mixture is then taken from the pot and carried in a leaf cone either by the oracle operator or by a child to the site of each new raising of the nets during the day. The remaining mixture is taken to the path where the operator squats facing the forest, his back to the village; he then hurls the remainder of the mixture over his shoulder in the direction of the village, stands up, and walks toward the forest without looking back.

All the perceived threats to opening the forest come from the village. An unknown death, an unknown pregnancy, or the unknown eating of a first kill by a *mangodo*, "he whose top teeth erupted first," will close the forest. Hunters who violate the ethic of the communal hunt by secretly harboring anger toward their peers or by using witchcraft or private medicines can also prevent the forest from "boiling over" with game. Whereas forest spirits do exist, they restrict their agency to the tangling up of traps and similar acts; unlike village spirit agents, they do not close the forest and they cannot kill.

If the forest is opened and the kill is good, the animals are collected and given to the huntmaster at the end of the day. He tallies them, displays them publicly to assembled hunters, and oversees the butchering and distribution of the day's kill. All participants receive something, even if their nets trapped nothing. The bulk of the kill is taken back to the village to feed villagers and funeral guests. In the most common variant of the communal hunt, that conducted out of dry-season hunger rather than after a burial, rules for meat distribution differ. In such hunts, the head of each animal is given to the first to arrive at the scene of the kill, the intestines to the second, and the limbs to the animal porter. Half the animal goes to the owner of the net that caught it and only the other half is turned over to the huntmaster for distribution to the group.

Additionally, the killing of any animal said to be of the "fathers," or ancestors, is an occasion for displaying and reproducing the principle of birth order that structures so much of daily interaction. Animals of the fathers or of the *tolo* (ancestral spirit shrine) include all spotted cats, turtles, and several species with striped or spotted fur. Whenever those are killed, they must be turned over to the oldest-lineage male for butchering and consumption. He in turn places some portions under the *tolo* so that the ancestors might eat. Only if the fathers, living and dead, are so placated will the forest continue to yield its bounty.

A lineage may also have the forest selectively closed to its members if its members kill or eat the animal that is proscribed to it. Pagibeti hunters generally honor these proscriptions; they are pragmatists, however, and while sparing the proscribed animal that they encounter while hunting, they freely share the information of where it was spotted to hunters who are not of their lineage, a courtesy reciprocated by those hunters when a complementary opportunity presents itself.

## PEOPLE OF THE LAND

One of the identity labels that Apagibeti apply to themselves is that of "people of the land," as opposed to their southern Bale neighbors, the "people of the water." The identity has practical significance. If an elephant is wounded on land but dies in the water, then the people of the land get to claim it. If it is wounded in the water but dies on land, then the people of the water have rights to it.

The identity also has ideological significance. Apagibeti state that ever since Belgian colonial officers forced the river people to plant cotton on riverbank land, the river people have acquired a taste for the game animals that they found on the land. Throughout the 1960s and 1970s, armed groups of Bale hunters penetrated deep into Pagibeti forests, hunting with snares and guns alike. As their inroads increased, the Pagibeti fought back. Knowing that either killing or the drawing of blood could bring state intervention, the Apagibeti fought in a controlled manner by destroying Bale traps, burning Bale forest shelters, and surrounding and beating Bale hunters, stripping them of all weapons, food, and, in some cases, clothing. The Bale nonetheless continued to come back, and by 1980 the fighting had largely ceased. Mutual destruction and raiding of snares was the last ongoing form of the conflict at that date. A phrase repeated time and again in relation to the Bale, and usually in a tone of disgust, was "Do fish come out on the land and walk?" While obviously fighting to preserve their forest game, they were simultaneously fighting to preserve an important part of their identity. Yet Apagibeti are keenly aware of the limitations of their power. When army paratroopers from the three-day-distant military camp at Kota-Koli enter the forest to machine-gun elephants for ivory, no formal denunciations are made and no resistance is offered.

## Implications for Conservation Initiatives

Some have argued that a more productive horticulture could serve as an alternative to reliance on forest game for basic food resources. The argument is unpersuasive. First, as thirty years of Belgian agricultural research concluded, "slash and burn" in conditions of low population density is an effective means of farming tropical soils. Research cited by Moran (1982) came to the same conclusion. While low in absolute productivity, the diversity of crops and cropping sequences ensures a fairly reliable and varied source of foodstuffs.

Apagibeti are aware of their privileged nutritional status relative to neighboring groups; one stated bluntly, "In the matter of eating we are chiefs." In 1979 a visiting medical team from the regional community development center of IMELOKO took serological and height and weight data from Pagibeti village children and effectively confirmed local perceptions. Apagibeti children had better scores than any of those tested from villages at the nine other IMELOKO dispensaries, all of which were sited along roads.

Second, from the local perspective, the possession of sufficient food calories, proteins, and nutrients is not the issue. Apagibeti like meat and declare themselves hungry after a full meal of plantains and peanuts or other plant foods. A more "fruitful" approach might be to encourage the expansion of existing practices of raising chickens and goats. Free-ranging chickens have long been raised for sale, gifts, and consumption on ceremonial occasions and during times of "hunger." Goats are a more recent introduction and are raised almost exclusively for sale.

Third, any reform of local practice (including the above) is at best a Band-Aid; the principal threat to the forest and its fauna comes from the outside. Apagibeti have shown their willingness to take up arms in defense of their hunting grounds against poachers from neighboring groups, but they cite the DR Congo state's declaration that the land belongs to all DR Congo citizens as one reason for abandoning the fight. Conservationists could usefully promote land tenure reform, which would legally vest rights in land and resources in local communities.

Finally, and fundamentally, conservationists could reverse the old missionary model by asking what we can learn from Africans rather than what we can teach them. Models of positive African ecological practices and ideas do exist (see, for example, Biodiversity Support Program 1993; Richards 1985; and Warren et al. 1989). In the case of the Apagibeti, a program might build on such existing conservation practices as the allocation of hunting rights to specific social units in specific areas of forest, the alternation of trapping areas within the forest from one year to the next, and the ban on individual hunting during the dry season.

Indigenous ideas may be particularly useful in such a program if we are not too proud or too rushed to use them. The West's abstract lessons concerning the dependence of human populations on the biosphere appear redundant to a group that identifies itself as a "people of the land," which models a key rite of passage, first-birth rites, on the behavior of animals in their environment (Almquist 1985), and whose hunting rituals explicitly link the well-being of the forest with the well-being of the villagers. The idea, for example, that forest productivity is directly dependent on proper human conduct and human agency is just one of the many implications of the *mbolongo* oracle and of the *soseye ngi*, or village washing. The concept of *pomoli*, of a natural balance between human and animal populations and of the fatal consequences for humans of extracting too much from the forest, could profitably serve as one model for an indigenous ecological ethic in broader African rain forest conservation efforts.

Also, at least within DR Congo, an approach that uses non-governmental agencies would appear to have the best chance of success. As Michael Schatzberg and others have richly documented, the state is not merely inefficient but is actively oppressive and greatly feared by the citizenry. Where the state cannot be avoided, attempts to revitalize local institutions as part of a state-sponsored initiative might prove desirable. One of the infrequent development successes reported by Arnould, a fisheries project in Niger, based its success in part on the revitalization of the role of fisher professional associations led by the fishmaster and watermaster. The success cost a government ministry official some of his power over vehicles and other resources, however, and the long-term success of the effort does not look bright.

REFERENCES

Almquist, A. 1985. Symbolic consensus and ritual practice: An ethnography of differentiation in Pagibeti rites of passage. Ph.D. dissertation, Indiana University, Bloomington.

———. 1991. Divination and the hunt in Pagibeti ideology. Pages 101–111 in P. Peek, ed. *African Divination Systems*. Indiana University Press, Bloomington.

Biodiversity Support Program. 1993. *African Biodiversity: Foundation for the Future*. World Wildlife Fund, Washington, D.C.

Guenther, M. 1988. Animals in bushman thought, myth and art. Pages 192–202 in T. Ingold, D. Riches, and J. Woodburn, eds. *Hunters and Gatherers*. Berg Publishers, Oxford.

Miracle, M. 1967. *Agriculture in the Congo Basin: Tradition and Change in African Rural Economies*. University of Wisconsin Press, Madison.

Moran, E. 1982. *Human Adaptability: An Introduction to Ecological Anthropology*. Westview Press, Boulder, Colo.

Richards, P. 1985. *Indigenous Agricultural Revolution: Ecology and Food Production in West Africa*. Westview Press, Boulder, Colo.

Rigby, P. 1968. Gogo rituals of purification. In E. R. Leach, ed. *Dialectic in Practical Religion*. Cambridge University Press, Cambridge.

Turnbull, C. 1968. *The Forest People*. Simon and Schuster, New York.

Warren, D. M., L. J. Slikkerveer, and S. O. Titilola, eds. 1989. *Indigenous Knowledge Systems: Implications for Agriculture and International Development*. Studies in Technology and Social Change, Publ. No. 11. Technology and Social Change Program, Iowa State University, Ames.

Willis, R. G. 1968. Changes in mystical concepts and practices among the Fipa. *Ethnology* 7(2):139–157.

NOTE

1. Whereas the term *Apagibeti* refers to the people, *Pagibeti* is used to identify inanimate objects or cultural practices pertaining to the Apagibeti.

CHAPTER 20

# Hunting and Wildlife in Northeastern Gabon
## Why Conservation Should Extend Beyond Protected Areas

Sally A. Lahm

Like many other African nations, Gabon depends heavily on the sale of natural resources from tropical forests for revenue at both national and local levels. Many residents of rural communities sell wood products, bushmeat, and fish because they have no other sources of income. A number of factors have precipitated changes in attitudes and resource-use practices of rural and urban residents in Gabon (Lahm 1993a). These include permanent settlement; the replacement of spears, crossbows, and nets with shotguns and cable snares; the abandonment of traditional beliefs that helped regulate use of natural resources; and participation in the modern cash economy. What does this sale of natural resources imply for long-term conservation prospects? To better manage resources, we must understand factors underlying people's use of timber and wildlife, as well as the resulting ecological impacts.

Gabon's low human population density and large expanses of forest make it an important sanctuary for the African tropical forest biological community (see, e.g., McShane and McShane-Caluzi 1990). However, our lack of knowledge of the intricacies of the tropical forest ecosystem, combined with current eco-nomic exigencies, should warn us against implying, as some researchers have (e.g., Adams and McShane 1992), that a laissez-faire attitude should be promoted toward conservation and wildlife management in this country. Studies indicate that rural human settlement patterns and commercial resource exploitation in Gabon have long-lasting and significant effects on wildlife and habitat (Barnes et al. 1991; Lahm 1993a, 1994; White 1994a; White and Tutin, this volume). This chapter discusses why Gabonese villagers sell bushmeat, and the impact of combined subsistence and commercial hunting on wildlife. Conservation implications are drawn from these and other human activities in the forest ecosystem.

Gabon straddles the equator on the west central African coast, with a total surface area of 267,667 km². According to 1994 data from the Ministère de la Planification du Gabon (MDP), the human population density (3.8/km²) and annual growth rate (2%) are low compared to many other African countries. Between 50% and 60% of Gabon's people live in urban areas. As a result of a government-enforced village regroupment policy, most citizens live along major roads. Gabon is thus characterized by narrow bands of human habitation along roads and some waterways, and extensive areas of uninhabited terrain in the forest interior.

The major sources of revenue in Gabon are oil, manganese, uranium, and timber. Since the decrease in oil prices in 1986 and the devaluation of the CFA franc in 1994, logging, urban unemployment, and the dependency of rural residents on the sale of crops, bushmeat, and other forest products for income have increased greatly. Logging companies and entrepreneurs are steadily advancing into previously unlogged areas, and coastal forests are being relogged at ever increasing turnover rates (Tutin 1992; Lahm 1992, 1993a).

The forest of Gabon is part of the Guineo-Congolian phytogeographic region, which stretches from Guinea to eastern DR Congo (White 1983). In western Gabon, the vegetation type is the classic "dense humid evergreen forest of low to medium altitude" (Aubréville 1948; Letouzey 1968). The evergreen characteristic decreases gradually with distance from the coast, culminating in the northeast in a forest of transition toward the semi-deciduous vegetation of Congo and Cameroon (Caballe 1978). Caballe (1983) estimated that 15% of the country is composed of swamps, mangroves, steppe, and savanna, and at least 75–80% is forested. The true extent and status of the Gabonese forest are unknown and will remain so until better remote sensing data are available. More than 130 species of mammals have been identified throughout Gabon, including 20 species of primates, 16 species of artiodactyls, and at least 30 species of bats (IPN 1978; Emmons et al. 1983).

### The Field Study on the Impact of Hunting

In 1988 a field study of local hunting practices and ecological impacts was launched at the Institut de Recherche en Ecologie Tropicale near Makokou in northeastern Gabon ($0°34'$ N, $12°52'$ E; 516 m altitude). This region has the lowest human population density in the coun-try, 1.1 persons/km² (MDP data, 1994), and is continuously forested.

Ecologically similar study sites were selected based on forest structure and habitat types. One village on each of the three roads radiating from Makokou was chosen for collection of socioeconomic and biological data. The three villages were between 30 and 42 km from Makokou, and each had been established in its present location for at least twenty years.

Data on village subsistence and income-generating activities were gathered through a questionnaire and informal conversations with local residents. Game harvested by village hunters and trappers was recorded opportunistically in the three communities on all non-flying animals weighing more than 1 kg.

A comparison of the relative numbers and distribution of mammal species in forest sites exposed to different levels of hunting pressure was made by establishing sixteen transects of 5 km in length: two within 5 km of each village; two in a less-hunted intermediate zone 7–10 km from each village; and four in a non-hunted remote area at least 100 km from Makokou. Twelve surveys (six day, six night) were completed on each transect between March 1988 and December 1990 using standardized sheets for direct observations and indirect evidence of animal presence (dung, tracks, vocalizations, feeding signs). Survey data collected were analyzed following Whitesides et al. (1988). Data from surveys on each set of transects (village, intermediate, remote) were combined. There was no significant variation among sites (Kruskal-Wallis one-way ANOVA; $P > 0.05$).

IMPORTANCE OF BUSHMEAT
IN THE LOCAL ECONOMY

The decline in the national economy has resulted in high unemployment and a drastic reduction in the flow of cash and material goods from working family members to their relatives in villages. Rural exodus continues as young villagers move to urban areas in search of work and a better life. Few achieve their goals, and most eventually return to their villages.

With limited alternative sources of income, many rural families rely heavily on the sale of bushmeat and fish for cash. Men hunt day and night with shotguns and set snares of wire cable. Of ninety households sampled, 68% sold bushmeat and 46% sold fish regularly. As elsewhere in the tropics, there is now little difference between subsistence and commercial hunting in villages (Redford 1992). If a hunter searching for game for his family kills a large animal or more than one, he is likely to sell part of the meat.

Villagers have also become increasingly dependent on the use of hunting camps to find game. Now, hunters and trappers sell bushmeat acquired during either short subsistence-oriented searches near their villages or longer expeditions at forest camps. According to village elders, hunting camps were formerly used only to supply large quantities of meat for important rituals and ceremonies, and smoked and fresh meat to village families in times of seasonal scarcity.

Results from this study reveal that 73% of forty-seven hunters regularly sold from one-half to two-thirds of their gain from short hunts in village environs (table 20.1). Only 6%

of hunters sold none. Of twenty-five men with hunting camps, 84% sold one-half to two-thirds of their meat.

GAME TAKEN BY VILLAGERS

Artiodactyls, particularly duikers, bear the greatest hunting pressure (table 20.2). Of a sample of 254 animals killed by village hunters and trappers, artiodactyls represented 57.5% of the total. The smallest species, the blue duiker, accounted for 65.1% of all artiodactyls killed. The remaining ungulates included four other species of duiker, the bush pig, the chevrotain, and the sitatunga.

The second most numerically important prey group, primates, represented 18.5% of the sample. The majority were guenons. Others included the mandrill, the grey-cheeked mangabey, the black and white colobus, the chimpanzee, and the gorilla. The majority of rodents captured were brush-tailed porcupines.

Artiodactyls may be more vulnerable to exploitation than other animals because they are exposed to both day and night hunting as well as trapping. Primates are rarely caught in traps and are usually hunted only during the daytime.

Reptiles (dwarf crocodile, monitor lizard), artiodactyls, and porcupines provide the most popular bushmeat and can be easily sold. Most of these species in the sample were sold rather

Table 20.1

## Percentage of Game Sold
## by Hunters from Village Day Hunts and Hunting Camps

| AMOUNT SOLD | VILLAGE (N = 47) NUMBER[a] | % TOTAL[b] | AMOUNT SOLD | CAMP (N = 25) NUMBER[a] | % TOTAL[b] |
|---|---|---|---|---|---|
| None | 3 | 6.0 | None | 2 | 8.0 |
| $\frac{1}{3}$ | 10 | 21.0 | $\frac{1}{3}$ | 2 | 8.0 |
| $\frac{1}{2}$ | 5 | 11.0 | $\frac{1}{2}$ | 9 | 36.0 |
| $\frac{2}{3}$ | 29 | 62.0 | $\frac{2}{3}$ | 12 | 48.0 |

a   The number of hunters who sold each amount.

b   The percentage of hunters who sold each amount.

Table 20.2

## Game Taken by Hunters and Trappers, and Percentages of Prey Groups Sold

| PREY GROUP (N=254) | NO. KILLED | % KILLED | % SOLD |
|---|---|---|---|
| Reptiles | 5 | 2.0 | 100 |
| Pangolins | 9 | 3.5 | 11.0 |
| Carnivores | 12 | 4.7 | 67.0 |
| Rodents | 35 | 13.8 | 63.0 |
| Primates | 47 | 18.5 | 40.0 |
| Artiodactyls | 146 | 57.5 | 66.0 |

*Note:* "Number killed" refers to the number of animals killed in each prey group. "Percent killed" refers to the percentage of the 254 animals killed that were in each prey group. "Percent sold" refers to the percentage of the kill in each prey group that was sold.

than eaten in the villages, despite an expressed preference by villagers for the meat of these animals (Lahm 1993b).

Most of the carnivores captured, including civets, genets, mongooses, and leopards, were also sold, probably owing to a strong local taboo against the consumption of animals with spotted coats (Lahm 1993b). These species may be valuable for their skins, which are used in rituals. Hunters often kill taboo species or other animals that they normally would not eat because they can usually sell the meat or skin to someone else.

DIFFERENTIAL RESPONSES TO HUNTING

Hunting and trapping have the greatest impact on populations of large-bodied animals. Table 20.3 demonstrates that populations of all game species decreased considerably at hunted sites compared to non-hunted sites, but the greatest reduction occurred for species of greatest body size. The body weight of each species was estimated at 75% of the average adult weight to allow for differences in population structure (Feer 1993).

Results for the blue duiker were inconclusive at non-hunted sites, possibly because of poor observation conditions, but the density of this species at intermediate sites was significantly higher than in village transects. The relatively small decrease noted for the heaviest species, the bay duiker, may reflect its apparent ability to survive near villages as a result of such particular eco-ethological characteristics as nocturnality and association with habitat types similar to secondary regenerating vegetation (Feer 1988). Survey results for abundance of larger mammals indicate similar trends (table 20.4).

When body weight of game species and percentage reduction of these species in hunted areas are compared, the data indicate that hunters are not harvesting the heavier game species in proportion to their natural densities (figure 20.1). As large game is eliminated near villages, the zone of exploitation slowly expands, and local smaller species experience increased pressure from hunting and trapping. When these populations are also severely reduced, hunting and trapping zones extend further into the forest.

In order to educate, feed, and clothe their families, village residents in Gabon sell what they or the forest produce: crops, basketry, wood, bushmeat, and fish. If wildlife is affected in this manner in a region with 1.1 persons/km$^2$, what are the effects of hunting where human

Table 20.3

## Differences in Density Between Hunted and None-Hunted Sites, and Percent Reduction of Game Species

| SPECIES | HUNTED DENSITY (NO/KM²) | NON-HUNTED DENSITY (NO/KM²) | % REDUCTION | BODY WEIGHT (KG) |
|---|---|---|---|---|
| Moustached guenon | 12.5 | 22.0 | 43 | 3.0 |
| Crowned guenon | 11.1 | 19.8 | 44 | 3.4 |
| Blue duiker | 30.4 | 53.0[a] | 43 | 3.7 |
| Spot-nosed guenon | 21.9 | 80.2 | 73 | 3.8 |
| Grey-cheeked mangabey | 2.5 | 51.2 | 95 | 5.8 |
| Black and white colobus | 0.8 | 6.8 | 88 | 6.4 |
| Peter's duiker | 0.6 | 6.7 | 91 | 15.0 |
| Bay duiker | 2.5 | 5.8 | 57 | 16.5 |

a   Data from intermediate transects.

*Sources:* Weights from Emmons et al. 1983; Feer 1988; Lahm 1993a.

Table 20.4

## Differences in Abundance Between Hunted and Non-Hunted Sites and Percentage Reduction of Large Game Species

| SPECIES | HUNTED (NO./KM²) | NON-HUNTED (NO./KM²) | % REDUCTION | BODY WEIGHT (KG) |
|---|---|---|---|---|
| Chevrotain | 0.2 | 0.9 | 78 | 7.5 |
| Yellow-backed duiker | 0.0 | 0.3 | 100 | 26.8 |
| Chimpanzee | 0.3 | 3.6 | 92 | 31.8 |
| Bush pig | 3.6 | 17.0 | 79 | 33.5 |
| Sitatunga | 0.05 | 0.3 | 83 | 40.2 |
| Gorilla | 0.0 | 2.4 | 100 | 60.3 |

*Sources:* Weights from Haltenorth and Diller 1988; Emmons et al. 1983; Lahm 1993a; Barnes and Lahm, unpublished data.

population density is higher? In some parts of Gabon where there are more people, a more extensive network of roads, and greater access to urban markets by car or train, there is evidence of local extermination, not just reduction, of most species of larger mammals within 10 km or more of rural towns and villages (Lahm 1994).

As in Gabon, Hart (1978) reported that Mbuti Pygmy hunters in DR Congo sold or traded 47–76% of their game meat, even when harvests were low. Local duiker populations decreased significantly after the Mbuti began to participate in market hunting. Elsewhere in DR Congo (Colyn et al. 1987), and in northern Congo (Robineau 1971), hunters sold most of the meat of the larger and more popular species such as ungulates in town, and ate or sold monkeys and smaller animals in villages.

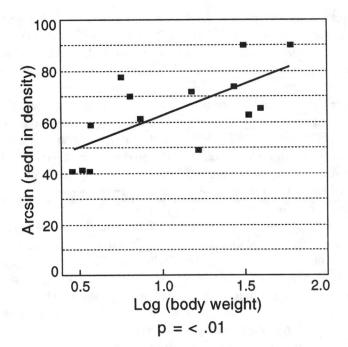

Figure 20.1. Effect of body weight on population reduction of hunted game.

## CHANGING HUNTING STRATEGIES

As hunting pressure on local wildlife populations intensifies and animals are eliminated or adopt specific behavioral or ecological strategies to avoid human contact, hunting zones expand and the importance of night hunting and hunting camps increases. Many species are vulnerable to night hunting, particularly ungulates that may hide in the forest undergrowth, because hunters can locate them by eye reflection. In Gabon, estimated densities for blue duikers observed near villages during daytime surveys were 4.3/km² compared to 30.4/km² for night surveys on the same transects (Lahm 1993a). Payne (1992) reported that blue duikers constituted 41.2% of night kills, compared with 3.2% of day kills, of village hunters in southern Cameroon.

Hunting camps established in lightly hunted forest at considerable distances from villages are frequently the only sites where hunters now have good possibilities of finding large and numerous prey. Daily yields from camp hunts in Cameroon and Gabon were at least twice as high as those from daytime hunts near villages (Infield 1988; Lahm 1993a).

## EFFECTS OF COMBINED DISTURBANCE

Villagers exploit their local forests intensively through hunting, trapping, gathering, fishing, woodcutting, and agriculture. Consequently, these combined activities probably cause changes and interruptions in the behavior, diet, activity rhythm, and reproductive cycle of both game and non-game species through continual disturbance and removal of sources of food and shelter. Mammals that thrive near villages are usually characterized by small body size, nocturnal activity pattern, and cryptic behavior, which facilitate avoidance of humans.

Populations of larger game species may decline from a combination of natural mortality and the direct and indirect effects of hunting. Trapping and other human activities even cause abortions and elevated mortality of adults and offspring, as reported by Geist (1971). Movement patterns and access to food sources and conspecifics may also be affected.

## Consequences of Uncontrolled Hunting

Little is known about the life cycles and causes of natural fluctuations in populations of African forest animals or how they respond to human exploitation, even for some of the most common game species. If populations of the blue duiker, the most abundant and widespread ungulate in Gabon (Lahm 1994), are experiencing significant decline in numbers where village hunting and trapping are intensive, we should clearly be concerned about the status of all wildlife in this country. The data from Gabon suggest that the observed reductions in animal populations near villages occurred within the 20–30 years since the communities were permanently established. If subsistence hunting alone were practiced, would animal numbers be higher? A long-term study of hunting by villagers in Neotropical forest showed that hunters there also preferred game of large body size and that even subsistence hunting resulted in the extinction of game species (Vickers 1991).

The situation in Gabon is complex because social, economic, and technological changes are involved. Traditional hunting and trapping practices were selective, using different tools and techniques for specific prey types. Now, shotgun hunting is concentrated on larger mammals and birds, and cable snares capture prey ranging in size from porcupines to apes. Because populations of most larger game species have been severely reduced near villages, much hunting and trapping occurs in the deep forest.

In the past, most bushmeat was probably acquired by "garden hunting" in the vicinity of village plantations (Linares 1976; Dove 1993). This supplied meat for the family and protection for crops. Although most mammals living near villages today are small and nocturnal, larger, more mobile species also adopt survival strategies to enable them to avoid human contact. For example, elephants and bush pigs typically raid crops only at night and withdraw into the forest at daybreak. Mandrills dig up manioc tubers in the early morning and late afternoon when village women are not in their plantations (Lahm 1993a).

Crop raiding by animals is a serious problem in many parts of Gabon. One of the factors influencing the severity of crop raiding in any given area is the lack of protection of crops by village men, many of whom either seem to be disinterested in traditional male activities or prefer deep-forest hunting and trapping. Of nearly three thousand families interviewed throughout the country, 36% said that their crops were unprotected against animal depredations (Lahm 1994). Thus, even in regions where few large mammals exist close to villages, plantations are often overrun by such smaller species as cane rats.

The data presented here show that small remnant populations of some larger mammals do exist near villages in some areas and that the more abundant and often destructive small mammals are underused as sources of meat. An experimental program of community-controlled hunting and trapping in selected villages in Gabon is suggested to determine if wildlife could be managed on a semi-captive basis to allow for the recovery of game species that are now experiencing heavy hunting pressure.

The promotion of reduced hunting far from villages and the use of traditional crop protection measures could be combined with such agroforestry projects as proposed by Hladik (1986) and Redford et al. (1992). These steps could serve to attract wildlife, provide a stable supply of meat from plantation perimeters, decrease hunting time and pressure on crucial "reservoir" animal populations in deep forest, and give alternative income to villagers. Of vital importance also is the enforcement of already-established legislation concerning hunting quotas, gun ownership, and commercial exploitation of wildlife, all of which remain largely uncontrolled in Gabon.

### IMPACT ON FOREST REGENERATION

The loss of these species may have consequences beyond a diminished food supply for villagers and impoverishment of local fauna. Many of the most intensively hunted game species play a key role in animal and plant population regulation and forest regeneration through predation, herbivory, trampling, seed destruction and dispersal, and pollination (Redford 1992). White (1994b) estimated that 70% of plant species in the Lopé Reserve in central Gabon are dispersed by animals. In the Ivory Coast, some plant species did not regenerate where elephants had been eliminated (Alexandre 1978). Are the structure and composition of heavily hunted forests different from those containing an intact animal community? This issue needs to be addressed.

### HUNTING AT LOGGING CAMPS

What are the effects on wildlife populations of the establishment and expansion of logging operations? White (1994a) found chimpanzees to be negatively affected by logging practices in a wildlife reserve in Gabon even though there was no hunting. Elsewhere in Gabon, subsistence and commercial hunting are widespread and largely uncontrolled in logging concessions, with greater negative impacts.

Increased crop raiding by elephants and other large mammals in villages near abandoned logging camps may be influenced by intensive local hunting during logging operations (Lahm 1994). When camps are vacated, hunting pressure subsides and animals may move into the surrounding terrain, eventually arriving in fields of neighboring villages.

Logging operations are often located in remote regions where human-wildlife contact is rare. These areas and the extensive uninhabited forest tracts beyond village hunting zones serve as "reservoirs" for intact animal communities. In northern Congo, there was a tremendous increase in commercial hunting with the arrival of logging camps (Wilkie et al. 1992). Similar consequences are to be expected in Gabon with the acceleration of logging and the progressive movement of village and professional commercial hunters into remote areas.

Would forestry companies be willing to forbid or limit hunting in their concessions if it were shown that hunting activities of their employees and of commercial hunters using logging roads are detrimental to wildlife? Who would enforce the regulations? People engaged in conservation in African forests must increasingly deal with these issues, trying to work with, rather than against, those who exploit the forest.

### RURAL DEVELOPMENT

Another important topic concerns alternative sources of both income and protein for rural people who live by hunting or inhabit areas where most game has been eliminated. If rural development programs provide jobs for local people, it is possible that they will use the income to buy guns, cable snares, and chainsaws, thus hastening the demise of wildlife. Furthermore, if agricultural development is successful in providing income and decreasing rural exodus, wildlife populations may decline as more forest is cleared.

The difficulties in implementing sustainable hunting in tropical forest, even in areas of low human population density, are numerous and would require the concomitant development of alternative income-generating activities for villagers. Perhaps game farming would be a feasible means of providing meat for sale and local consumption, thus relieving hunting pressure on wildlife. Species characterized by high density and reproductive rates have the best potential for sustainable use, but there are few tropical forest game animals that meet these criteria. Therefore, most forest species can probably sustain only minimal harvest (Robinson 1993; Feer 1993).

Advocates of conservation must confront these types of problems as well as cultural differences in attitudes. Other cultures may have ideas about biodiversity and conservation that are quite different from Western concepts (Redford and Stearman 1993). People in the tropics want a better standard of living. They often have no means of earning income other than exploiting natural resources, and are unlikely to give up short-term benefits to the family for long-term community gains.

The potential for conflict between hunters and wildlife conservationists in Gabon is great. Most natural resources occur outside parks and reserves. There may be no future for protected areas if uncontrolled use of biological resources is widespread elsewhere. Buffer zones and park boundaries can easily be overrun by people in need of food, fuel, farmland, and income. We do not know if rural exodus will offset the current level of hunting, but there will probably always be a high demand for bushmeat in towns and cities.

Wildlife and timber are indeed abundant in Gabon, but so are logging operations, oil companies, bushmeat vendors, hunters, and trappers. Considering the current economic situation in Gabon, the creation and implementation of appropriate management plans and professional supervision of forest resource exploitation should be priorities.

Government personnel responsible for natural resources need to be involved in a variety of activities in addition to protected-area manage-ment, including surveillance of commercial resource exploitation (oil exploration and production, mining, fishing, logging, hunting) and collaborative interaction with rural communities. Village residents possess a wealth of knowledge concerning local flora and fauna. In the interest of effective management of resources, government agents and villagers could benefit from a positive working relationship.

Conservation efforts in African forests should focus not only on creating protected areas, but also on helping people manage natural resources outside parks and reserves by finding the answers to questions such as those asked in this chapter. The challenge is to find solutions appropriate for local conditions and to include the often-neglected rural citizens in the decision-making process.

ACKNOWLEDGMENTS

Funding for research was provided by the Wildlife Conservation Society and by World Wildlife Fund-US. I thank Paul Posso and CENAREST for use of the facilities at the Institut de Recherche en Ecologie Tropicale at Makokou and the people of numerous villages for technical and logistical assistance during the study. Richard Barnes, John Robinson, and Lee White commented on the manuscript.

REFERENCES

Adams, J., and T. U. McShane. 1992. *The Myth of Wild Africa*. Norton, New York.

Alexandre, D. -Y. 1978. Le rôle disseminateur des éléphants en forêt de Tai, Côte-d'Ivoire. *La Terre et la Vie* 32:47–71.

Aubréville, A. 1948. Etude sur les forêts de l'Afrique equatoriale. *Bois et Forêts des Tropiques* 2:24–49.

Barnes, R. F. W. 1990. Deforestation trends in tropical Africa. *African Journal of Ecology* 28:161–173.

Barnes, R. F. W., K. L. Barnes, M. P. T. Alers, and A. Blom. 1991. Man determines the distribution of elephants in the rain forests of northeastern Gabon. *African Journal of Ecology* 29:54–63.

Caballe, G. 1978. Essai sur la géographie forestière du Gabon. *Adansonia* 217(4):425–440.

———. 1983. Vegetation. Pages 34–37 in J. Barret, ed. *Géographie et Cartographie du Gabon*. EDICEF, Paris.

Colyn, M., A. Dudu, and M. ma Mbelele. 1987. Données sur l'exploition du "petit et moyen gibier" des forêts ombrophiles du Zaïre. Pages 109–146 in B. des Clers, ed. *Wildlife Management in Sub-Saharan Africa*. Fondation International pour la Sauvegarde du Gibier, Paris.

Dove, M. R. 1993. The responses of Dayak and bearded pig to mast-fruiting in Kalimantan: An analysis of nature-culture analogies. Pages 113–137 in C. M. Hladik, A. Hladik, O. F. Linares, H. Pagezy, A. Semple, and M. Hadley, eds. *Tropical Forests, People and Food*. UNESCO and the Parthenon Group, Paris.

Emmons, L. H., A. Gautier-Hion, and G. Dubost. 1983. Community structure of the frugivorous-folivorous forest mammals of Gabon. *Journal of Zoology, London* 199:209–222.

Feer, F. 1988. *Stratégies écologiques de deux espèces de bovidés sympatriques de la forêt sempervirente Africaine (Cephalophus callipygus et C. dorsalis): Influence du rhythme d'activité*. Doctoral thesis, Université Pierre et Marie Curie, Paris.

———. 1993. The potential for sustainable hunting and rearing of game in tropical forests. Pages 691–709 in C. M. Hladik, A. Hladik, O. F. Linares, H. Pagezy, A. Semple, and M. Hadley, eds. *Tropical Forests, People and Food*. UNESCO and the Parthenon Publishing Group, Paris.

Geist, V. 1971. A behavioral approach to the management of wild ungulates. Pages 413–424 in E. D. Duffey and A. S. Watt, eds. *The Scientific Management of Animal and Plant Communities for Conservation*. Blackwell Scientific Publications, Oxford.

Hart, J. A. 1978. From subsistence to market: A case study of the Mbuti net hunters. *Human Ecology* 6:325–353.

Haltenorth, T., and H. Diller. 1988. *Collins Field Guide to the Mammals of Africa Including Madagascar*. S. Greene Press, Lexington, Mass.

Hladik, C. M. 1986. Le gibier disponible dans la forêt dense naturelle et les possibilités d'augmenter sa biomasse par une stratégie agroforestière. Pages 229–233 in M. Maldague, A. Hladik, and P. Posso, eds. *Agroforesterie en Zônes Forestières Humides d'Afrique*. UNESCO, Paris.

Infield, M. 1988. Hunting, trapping and fishing in villages within and on the periphery of the Korup National Park. WWF Publication 3206/A9.6.

IPN (Institut Pédagogique National). 1978. *Les Mammifères de la Forêt Gabonaise*. Ministère de l'Education Nationale, Libreville, Gabon.

Lahm, S. A. 1992. Human activity and animal populations at Remboue III drilling site, Abanga block. Report, British Petrole Gabon.

———. 1993a. Ecology and economics of human/wildlife interaction in northeastern Gabon. Ph.D. dissertation, New York University, New York.

———. 1993b. Utilization of forest resources and local variation of wildlife populations in northeastern Gabon. Pages 213–227 in C. M. Hladik, A. Hladik, O. F. Linares, H. Pagezy, A. Semple, and M. Hadley, eds. *Tropical Forests, People and Food*. UNESCO and the Parthenon Press, Paris.

———. 1994. The impact of elephants and other wildlife on agriculture in Gabon. Report to WWF-International and the Environment and Development Group.

Letouzey, R. 1968. *Etude Phytogéographique du Cameroun*. Encyclopédie Biologique 69. Editions Paul Lechevalier, Paris, and MDP Ministère de la Planification, Libreville, Gabon.

Linares, O. F. 1976. "Garden hunting" in the American tropics. *Human Ecology* 4:331–349.

McShane, T. O., and E. McShane-Caluzi. 1990. Conservation before the crisis: A strategy for conservation in Gabon. Report, WWF-US.

Payne, J. C. 1992. A field study of techniques for estimating densities of duikers in Korup National Park, Cameroon. Master's thesis, University of Florida, Gainesville.

Redford, K. H. 1992. The empty forest. *Bioscience* 24(6):412–422.

Redford, K. H., B. Klein, and C. Murcia. 1992. Incorporation of game animals into small scale agroforestry systems in the Neotropics. Pages 333–359 in K. H. Redford and C. Padoch, eds. *Conservation of Neotropical Forests: Working from Traditional Resource Use*. Columbia University Press, New York.

Redford, K. H., and Stearman, A. M. 1993. Forest-dwelling native Amazonians and the conservation of biodiversity: Interests in common or in collision. *Conservation Biology* 7(2):248–256.

Robineau, C. 1971. *Evolution Economique et Sociale en Afrique Centrale: L'Exemple de Souanke*. ORSTOM, Paris.

Robinson, J. G. 1993. The limits to caring: Sustainable living and the loss of biodiversity. *Conservation Biology* 7(1):20–28.

Tutin, C. 1992. Gabon. Pages 168–174 in J. A. Sayer, C. S. Harcourt, and N. M. Collins, eds. *Conservation Atlas of Tropical Forests: Africa*. Macmillan, London.

Vickers, W. T. 1991. Hunting yields and game composition over ten years in an Amazon Indian territory. Pages 53–81 in J. G. Robinson and K. H. Redford, eds. *Neotropical Wildlife Use and Conservation.*University of Chicago Press, Chicago.

White, F. 1983. *The Vegetation of Africa*. UNESCO, Paris.

White, L. J. T. 1994a. The effects of commercial mechanised logging on forest structure and composition on a transect in the Lopé Reserve, Gabon. *Journal of Tropical Ecology* 10:313–322.

————. 1994b. Patterns of fruit-fall phenology in the Lopé Reserve, Gabon. *Journal of Tropical Ecology* 10:289–312.

Whitesides, G. H., J. F. Oates, S. Green, and R. P. Kluberdanz. 1988. Estimating primate densities from transects in a west African rain forest: A comparison of techniques. *Journal of Animal Ecology* 57:345–367.

Wilkie, D. S., J. G. Sidle, and G. C. Boundzanga. 1992. Mechanized logging, market hunting, and a bank loan in Congo. *Conservation Biology* 6(4):570–581.

# CHAPTER 21

# Conservation—For Whom?

## A Study of Immigration onto DR Congo's Ituri Forest Frontier

Richard B. Peterson

*In fact, however, every ecological question that involves human activity is not only an ecological question, but also both a moral question and a political question.... The ecological dimension of the problem will be resolved only as part of a comprehensive resolution of the whole problem, including its moral and political dimensions.* (Spooner 1987:660)

*To blame colonizing peasants for uprooting tribal peoples and burning the rainforest is tantamount to blaming soldiers for causing wars.... To understand the colonists' role in deforestation, one must ask why these families enter the rainforest in the first place.* (Nations and Komer 1983:14)

*But conservation in the African context should not be seen merely in terms of nature conservation; it is also an important part of the reform process in land tenure systems, land-use patterns, and land management practices.* (Areola 1987:277)

In the spring of 1989, I spent six months living and working with indigenous and immigrant farmers in three villages located along the main trans-Ituri road. I was trying to gain a better understanding of the nature, causes, and impacts of the spontaneous human immigration that appeared to be bringing so many social and ecological changes to the forest.

All the farming peoples currently living in the Ituri Forest immigrated to the area at some point in time, but certain ethnic groups have lived there much longer than others. In the villages of Badengaido, Tobola, and Eringeti, these groups are the Mbo, Ndaka, Bira, Lese, and Mbuba, who first began moving into these areas of the forest at least several centuries ago. In contrast, nearly all the immigrants of this study have moved to these villages within the past thirty years. The great majority of immigrant households have settled only since 1979.

Approximately one-fifth of the immigrant households I worked with in the spring of 1989 had been on the frontier for only one year or less. Although it is difficult to separate these villages into distinct groups of immigrant and indigenous households, one can differentiate relatively clearly between those households that have a long history of residence in the particular natural, cultural, and political setting of this part of the forest, and those that don't. One can also distinguish immigrant households according to the ethnic affiliation of the head.

In this chapter, I consider immigrant households as those in which the head is affiliated with an ethnic group that has historically been centered outside this particular forest milieu. This does not imply that all immigrants coming to the Ituri share more characteristics than any one immigrant group may share with the Ituri's indigenous inhabitants. Indeed, it is useful to make a distinction between those immigrants coming from western forest environments, similar to the Ituri, and those whose home is the eastern savannas. For example, the Budu and Bali are relative newcomers to the village of Badengaido and differ in many respects from the indigenous Mbo and Ndaka. However, they also have a long history of dwelling in the forest—in this case, that to the north and west of the village. In contrast, the Nande, a major immigrant group found in all three villages, have much less experience with forest environments. They are more accustomed to the highland savannas lying to the east of the Ituri, from which they originate. In my analysis I do make some comparisons between those households that have the strongest historical connections to the three villages and those that have come more recently from other locales. But these comparisons are necessarily qualified by the knowledge that immigrant groups are not all alike, but come from different environments.

During the course of interviews, immigrant farmers would often ask me what the final goal of my research was. This was a very difficult question to answer because I was unable to promise that the research would lead to any solutions to the difficulties they faced: the back-breaking toil of clearing trees, the distance from the closest health care (three to four hours' walk away), animals ravaging their gardens, high prices for commercial goods and low prices for their produce, poor schools for their children, harassment by petty state officials—the list goes on (table 21.1).

At times, immigrants would respond to my questions in ways that were truly humbling and sparked my own questioning of the proper role of research, conservation, and development. For example, Gaspar, an immigrant farmer, was clearly uneasy when I began asking and writing down the names of his household members. When I got to the subject of crops, he clammed up and exclaimed, "Don't trouble me with all your questions. What good is this all going to do for me? You know life here is one endless crisis. What are you going to do to help? You *mazungu* (white people) come and study us Africans and say they live this way, they plant this and that, they fish or hunt or sell their goods, but how does that ever bring any good back to us? Don't trouble me."

To slow the loss of primary forest and lessen the disruption of indigenous societies, both of which result from immigration, as well as to ease the crises that immigrants themselves face, both research and action are needed. Actions that address these issues have a greater chance of success when they are preceded by time spent trying to understand more fully the exact nature, causes, and dynamics of immigration. Through the research summarized here (table 21.2), I attempted to gain that understanding—to find out who these immigrants are and why they are coming, how they live and use land and forests, how they interact with those who have lived in the forest for a much longer time, how they acquire access to the forest's resources, and how their movement into the forest is linked to broader political and economic forces within Congolese society.

The conservation efforts under way in the Ituri will affect the future of the forest, the

Table 21.1

## Major Difficulties Perceived by Villagers

| DIFFICULTY | % OF TOTAL HOUSEHOLDS RESPONDING[a] (NUMBER OF HOUSEHOLDS) | % OF TOTAL HOUSEHOLDS GIVING ANSWER AS FIRST RESPONSE[a] (NUMBER OF HOUSEHOLDS) |
|---|---|---|
| Lack of medical care/sickness | 53 (69) | 21 (27) |
| Difficulties obtaining food and/or palm oil | 41 (54) | 18 (22) |
| Labor shortage | 12 (16) | 7 (9) |
| Difficulty obtaining money | 11 (14) | 6 (8) |
| Difficulty obtaining land/clearing forest | 11 (14) | 5 (6) |
| Difficulties of gold prospecting | 10 (13) | 8 (10) |
| Animals ravaging gardens | 10 (13) | 7 (9) |
| Poor access to markets/lack of transportation | 8 (11) | 2 (2) |
| Harassment by the state | 8 (10) | 5 (6) |
| Ethnic, family, and social discrimination, discord, and jealousies | 7 (9) | 4 (5) |
| Poor schools | 7 (9) | 2 (3) |
| Unequal exchange/high prices in stores | 6 (8) | 2 (3) |
| No problems | 5 (7) | 5 (7) |
| Poor housing | 4 (5) | 2 (3) |
| Lack of garden tools | 2 (3) | 1 (1) |
| Few religious converts | 2 (2) | 2 (2) |
| Lack of clothes | 2 (2) | 1 (1) |
| Poor drinking water | 2 (2) | — |
| Lack of credit for small farmers | 1 (1) | 1 (1) |
| Mbuti stealing from gardens | 1 (1) | 1 (1) |
| Corruption | 1 (1) | 1 (1) |
| Drunkenness of indigenous people | 1 (1) | 1 (1) |
| Currency devaluation | 1 (1) | — |
| Park guards jailing people for shooting animals in gardens | 1 (1) | — |

a   Multiple responses were tallied; therefore, percentages do not sum to 100.

Ituri's indigenous peoples, and its more recent inhabitants, immigrant farmers. Several organizations are currently working to address the crucial issues facing both the forest and the peoples of the Ituri. Most pertinent is the work of the Institut Congolais pour la Conservation de la Nature (ICCN), the Wildlife Conservation Society (WCS), and the Gilman Investment Company (GIC), a private American firm involved in the capture and breeding of okapi. The World Wide Fund for Nature (WWF) has also been involved in the Ituri conservation efforts. Since the mid-1980s these groups have been cooperating to create a multi-use conservation area out of a portion of the Ituri. In May 1992, the government of what was then Zaire

Table 21.2

## Indigenous and Immigrant Peoples in Three Ituri Villages

| VILLAGE | STAGE OF NANDE IMMIGRATION[a] | DISTANCE FROM NANDE HOMELAND (KM) | % OF HOUSEHOLDS NANDE | MAJOR INDIGENOUS GROUPS | MAJOR IMMIGRANT GROUPS | MAJOR LAND USES | LAND TENURE SYSTEM PRACTICED |
|---|---|---|---|---|---|---|---|
| Badengaido | Early | ±470 | 7 | Mbo, Ndaka, Mbuti | Budu, Bali | Shifting cultivation for food production; 27% of households produce coffee on permanent fields, 3–5 ha in size; significant gold propecting; minor hand-sawn timber production; Mbuti traditional hunting and gathering | Open access; little payment to and control by indigenous people for and over their land |
| Tobola | Intermediate | ±345 | 62 | Bira, Lese, Mbuti | Nande | Shifting cultivation for food production; 71% of households produce coffee on permanent fields, 3–5 ha in size; minor gold prospecting; minor hand-sawn timber production; Mbuti traditional hunting and gathering | Open access; little payment to indigenous people for their land; immigrant access to land facilitated by immigrants holding local-level political positions |
| Eringeti | Advanced | ±135 | 81 | Mbuba, Mbuti | Nande | No or short-fallow food cultivation; significant coffee production on a large scale (km²); post-boom gold prospecting; significant mechanized timber production; Mbuti hunting superseded by Mbuti gathering construction materials and engaging in agriculture | Unofficial land markets; buying and selling of both primary forest and fallowed land; political and economic control in the hands of immigrants |

a   Stages are determined on the basis of the level of settlement by Nande, a major immigrant group from the North Kivu region.

officially gazetted the 13,000 km² Okapi Wildlife Reserve. A Reserve Management Plan, written by a WWF consultant, was completed and accepted by the ICCN. The WWF also initiated a conservation education program that provided training, materials, and methods to primary and secondary school teachers from the area for teaching children in surrounding village schools about conservation. More recently, the WCS has established the Centre de Formation et Recherche en Conservation Forestière (CEFRECOF) at Epulu, a research center set up to train Congolese students in field ecology and conservation management. As of June 2000, these efforts are in a state of transition, as the various projects at Epulu and many people worldwide anticipate and hope for a peaceful settlement to Congo's current political crisis and civil war.

One purpose of my study is to inform the conservation and development agencies that are working in the Ituri Forest about who these immigrants are, and to help those responsible for creating policies on immigration to view immigrants as actual people and not just as threats to the forest. At the same time, the study is an appeal for a more concerted effort on the part of conservation and development organizations to recognize and support indigenous peoples' claims to the lands and forests of the Ituri.

## Major Findings from the Ituri Case Study

- Although most immigrants come from two major population centers on the northern and eastern edges of the forest, the total array of the ethnic and geographic origins of immigrants is extremely widespread (figure 21.1). Not only does this indicate the strength of the Ituri's pull, it also makes the Ituri increasingly ethnically diverse, to the point where the proportion of indigenous populations is decreasing while that of immigrants is rising.

- The abundance and easy acquisition of land, the lure of liberalized gold prospecting, and the presence of other family members already settled on the frontier are major reasons why immigrants are coming to the Ituri. Added to these, urban unemployment, high population densities, a dearth of arable land, and an unequal distribution of land all push people to leave their home areas located outside the forest.

- Each of the three major reasons for immigrating (land, family, and gold) is linked to more fundamental and contingent causes of migration rooted in Belgian colonial policy. Colonial recruitment of Nande (a major immigrant group from North Kivu Region) for forest-based mines and plantations, and political and economic favoritism toward the agricultural Nande over indigenous forest people, laid the basis for migration into and conversion of the forest. Today these policies have also resulted in the political, economic, and demographic marginalization of forest groups compared with an immigrant majority.

- The land- and forest-use patterns of the Ituri's contemporary immigrants differ from the practices of indigenous farmers and are generally less environmentally sustainable. Immigrants practice both more extensive and more intensive agriculture leading to greater forest clearing and more rapid soil depletion. They also place more emphasis on cash-crop production that results in permanent conversion of forest to field. Finally, immigrants make less use of and possess less knowledge about forest products than do local forest people. Cleared agricultural land is, in general, more valuable to immigrant farmers than harvests of naturally occurring forest resources.

- The more extensive and intensive agricultural practices of immigrants require

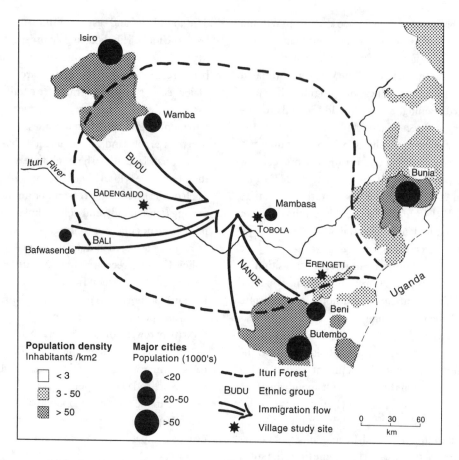

Figure 21.1. Major immigration flows into the Ituri Forest. Adapted from Hart 1979.

greater labor input, which they supply in three primary ways. Formally organized cooperative labor groups allow immigrants to increase the efficiency and extent of garden clearing and planting. Encouraging other family members to settle on the frontier and provide labor also increases the flow of immigrants and the rate of forest clearing. Finally, the Mbuti (indigenous hunter-gatherers) provide cheap labor on a day-to-day wage basis, leading to fundamental changes in the villager-Mbuti exchange relationship that are often detrimental to Mbuti culture and quality of life.

• The rigid and more narrowly defined wage labor system that immigrants employ, their negligible demand for the forest products the Mbuti have to offer, and the deterioration of the forest resource base that results from immigration, all combine to move the Mbuti into an increasingly dependent, rather than interdependent, position vis-à-vis villagers. Although the Mbuti do have some choice in how they respond and adapt to these new relationships and situations, their adaptability is dependent on the existence of primary forest. Forest survival in turn depends primarily on the degree to which indigenous rights to and uses of the forest can be strengthened and defended.

- Land tenure systems in the Ituri become more formal and institutionalized as immigration and population density increase. In areas of low population density and an abundance of land, immigrants are acquiring land under an "open-access" arrangement in which indigenous people receive very little payment for their land. Closer to the source area of immigration, the growth of unofficial land markets gives indigenous farmers greater payment for their land, but also encourages many to sell off most of their land rights. Where immigration is the most advanced, as in the southern Ituri, political and economic control by immigrants facilitates the means by which they increasingly gain tenure over lands once held by indigenous groups under customary tenure arrangements. Because the Mbuti have only indirect claims to forest areas through their farmer patrons, their rights to the forest are the least recognized and the most endangered.

- Immigrants tend to migrate to the Ituri and not elsewhere because there, unlike in other areas of the region, the position of indigenous land chiefs is relatively weak. Furthermore, forest tenure systems are generally not sufficiently structured and formalized so as to control immigrants' tenurial innovations that increase the extent of individualized land holding. The reasons for these forest tenure features are many and complex. Paramount is that the shifting settlement patterns engendered by a forest agroecology have discouraged the permanent establishment of formal tenurial and political institutions. Institutional malleability may have served as a means of coping in a difficult, isolated, and yet ever-changing sociopolitical and natural milieu, but as far as the more formal tenure systems brought in by immigrants are concerned, indigenous tenure flexibility translates into tenure collapse.

- Nonetheless, rather than being culprits, most immigrants are themselves victims of disruptive tenurial and social changes in their home areas. The growing individualization of tenure and the breakdown of social controls on land transactions have allowed a wealthy entrepreneurial class to accumulate land, dispossessing rural small holders. With few options left them, these farmers often end up migrating into socially and environmentally sensitive areas like the Ituri. Tenure change and social and environmental disruption within the forest is inextricably linked to processes of land-based capital accumulation and an unequal distribution of land taking place outside the forest.

## Comparisons to Other Tropical Forest Frontiers

The Ituri case is at once both unique and typical in regard to the ways immigrant farmers contribute to social and ecological change in the world's tropical forests. Part of its uniqueness lies in the Ituri's relative degree of isolation in comparison to west African rain forests and other forests within the Congo Basin. Until recently, the indigenous farmers and hunter-gatherers of the Ituri have had less contact than most other forest peoples with such large-scale commercial enterprises as plantations and logging and mining operations. In Cameroon, Congo, and the Central African Republic, logging, mining, and plantation agriculture have contributed to a greater commercialization of the farmer and hunter-gatherer exchange system. These areas are also marked by increased levels of wage labor, sedentarization, and adoption of agriculture among hunter-gatherers (Bailey et al. 1990; Bahuchet 1991). Until recently, the Ituri's relative isolation has also limited the amount of contact the Ituri's indigenous peoples have had with new groups of immigrant farmers who do not hold the same traditions of exchange. Now the degree of contact is rapidly increasing, as the commoditization of land in areas bordering the Ituri, combined with the lure of liberalized gold prospecting and the relative abundance of land on the frontier, bring new people into the forest.

Immigration into the Ituri also differs from that into other central African forests by the extent to which capital accumulation and land dispossession outside the forest act as primary causes of immigration. The Ituri is bordered in the east by highlands ideal for coffee growing and cattle ranching. This heightens the role that unequal land distribution, as opposed to simple population growth (the more commonly stated cause of African environmental degradation), plays in deforestation. Study of the Ituri case thus yields better understanding of the ways environmental destruction on the frontier is not so much "a problem arising from the relationship between people and the environment," but "an outcome of relationships among people that determine how the environment is to be utilized" (Painter 1988:1).

In contrast, such features as the lure of gold, the abundance of land, and the unequal distribution of land in forest-bordering areas make the Ituri case typical of many tropical forest frontiers in other regions of the world. In general terms, what is happening in the Ituri is similar to what began happening in the Brazilian Amazon more than thirty years ago. Yet unlike the Amazonian situation, human immigration into the Ituri is completely spontaneous, with little involvement by the government. The absence of government-planned colonization projects does not mean, however, that the Ituri's immigration is any less destructive.

## Implications for Conservation and Meeting Human Needs in the Ituri

To maintain the subsistence and cultural base of the Ituri's indigenous people, it is crucial that intact primary forest continue to exist. Beyond the biological reasons for conserving the Ituri's diverse flora and fauna lie social and cultural reasons. The integrity of the forest ecosystem supports the complex interdependencies by which its indigenous human inhabitants survive. Without intact primary forest, life for the Ituri's indigenous peoples will most likely be characterized by increased poverty and marginalization. Already there is ample evidence of the ways in which primary forest loss and forest resource depletion are contributing to growing urbanization and the breakdown of indigenous cultural and socioeconomic systems. In Mambasa (one of the Ituri's largest towns), for example, the Mbuti have become a primary cheap labor pool, and young Mbuti are raised in environments bereft of the cultural, social, economic, and metaphysical benefits that the forest and the "enforested" community provide.

Thus there is an urgent need for legal protection for a part of the Ituri Forest. Conservation efforts have succeeded in establishing the Okapi Wildlife Reserve in the heart of the Ituri region. It is possible that a portion of the reserve will eventually become a national park (Hall 1990; Hart and Hart 1990). However, it is important that such efforts go beyond simply preserving a biological laboratory, to include support for Mbuti and indigenous forest cultivators' rights to the subsistence use of the forest. As yet, the degree to which the reserve will recognize and support indigenous peoples' land rights and allow for their traditional use of the forest remains unclear. A program to elicit local people's participation in developing the reserve's socioeconomic policies has been carried out by a WCS team of Congolese and American social scientists. It is hoped that such efforts will provide some clarification regarding the reserve's policies vis-à-vis indigenous peoples and their use of forest resources.

Those working to establish any sort of protected area in the Ituri must recognize that even though many areas of the forest may be sparsely inhabited, they are not without some sort of tenurial claim. As a report to the World Bank on development and indigenous peoples in central African forests explains, "For the purposes of planning the development or protection of any area of land, it should be assumed *a priori* that any forest is occupied by some sort of person, or some clan, lineage, or group. Even if no overt signs of occupation... are evident, the land is most likely occupied intermittently and exploited by people whose lifestyles depend upon frequent movement" (Bailey et al. 1990:26).

Furthermore, beyond the simple recognition of indigenous land rights, conservation agencies may actually enhance their efforts to conserve the forest by allowing indigenous people to continue their subsistence use of forest resources. This would be one way of gaining support for conservation among the Ituri's indigenous peoples. By defending and supporting these indigenous use rights, conservation agencies could help keep more unsustainable practices (such as logging and plantation agriculture) from taking hold.

The low-intensity resource use and mobile settlement patterns of indigenous forest peoples limit the effect they have on the forest (Bailey et al. 1990:27). The Mbuti, for example, use the forest in ways that rarely lead to an overexploitation of resources. Their knowledge of the forest and their forest-based skills have much to contribute to the design of models of sustainable forest management. They, however, have the least political power of any group, and virtually no official rights to the land and forests they rely on. If anything, the state would prefer to see the Mbuti leaving the forest, settling along the road, and becoming active agriculturalists. As immigration increases, the depletion of forest resources and the decreasing proximity of primary forest move the Mbuti into an increasingly perilous and powerless position, with few alternatives other than selling their labor as a means of earning a living. Conservation projects that attempt to benefit the Mbuti by offering them employment are not sufficient. What is needed is to guarantee the protection of primary forest in such a way as to allow the Mbuti to maintain the forest-based subsistence they depend on, both materially and culturally. Therefore, political efforts to strengthen Mbuti and indigenous farmers' rights to and subsistence use of the forest need to be made part of an Ituri conservation plan.

One possibility that requires further investigation is group registration of land on concessionary terms. Theoretically, it would be legal to register land areas once held under customary tenure in the name of a group, rather than an individual or corporation. In the Democratic Republic of Congo (DR Congo), concessionary rights can be granted to a *collectivité* (mid-level political administrative division corresponding most closely to ethnic and clan distinctions), which is considered a legal personality (Salacuse 1985:43). It may be legally possible for an entire village, and perhaps an entire ethnic group represented by a collectivité, to receive concessionary title to their ancestral lands. This would strengthen indigenous claims to land and might inhibit outsiders' encroachment on indigenous lands. Because the Mbuti are not represented by their own collectivité, however, it is vital that they be guaranteed joint concessionary rights with the indigenous farming ethnic group with whom they are associated. This would resemble the customary manner by which the Mbuti have maintained their access to areas of forest over many years and would assert the fact that the Mbuti and indigenous forest farmers must be considered together as one integrated economic and social system (Bailey et al. 1990:6).

Because current trends favor the individualization of tenure and governments lack precedent for granting concessions to indigenous groups, application for such a concession would require support from a broader community than just the village or collectivité. The international conservation groups, funding agencies, and branches of the Congolese government interested in keeping the Ituri Forest intact may be in a position to offer that support. This procedure may seem idealistic given the realities of Congolese politics, but it is not without workable models from other tropical forests of the world. One example is the Brazilian government's creation of Extractive Reserves, which preserve indigenous peoples' use rights to the forest and prohibit encroachment by outsiders (Hecht and Cockburn 1989). Numerous federations of indigenous peoples in Central and South America have gained international support for the defense of their lands against encroachment by outsiders, whether they be colonists or multinational logging enterprises. Why should the less well known but equally endangered groups of indigenous peoples in central African forests not receive similar types of support?

Just as important as defending indigenous land rights is avoiding immigration policies that "blame the victim." Although immigrants do generate certain social and ecological changes that can significantly disrupt indigenous forest societies, most immigrants are themselves facing severe crises. While current conservation actions in the Ituri are trying to benefit the local people, distinguishing between those who should benefit and those who shouldn't on the basis of whether they are indigenous or immigrant is ethically controvertible given the general state of crisis and rural poverty. A better approach is to guarantee that the benefits the park may offer local people will not be appropriated and turned to the benefit of one group over another.

Policies that blame the victim can too easily overlook some of the potential benefits that immigrants may bring to forest societies. These may include greater access to health care, education, transportation, credit, and commercial goods; greater opportunities for marketing agricultural products; and increased economic activity in general. However, the word *potential* must be underlined. When increased opportunities and benefits enter into a society already marked by a skewed distribution of political and economic power, the strong often benefit much more than the weak. As immigration progresses, these positions of strength and weakness are increasingly occupied by immigrants and indigenous peoples, respectively. When this inequality is coupled with the social and ecological disruptions that immigration carries in tow, indigenous people end up playing a losing game.

Furthermore, immigration policies must be tempered by an analysis that places the changes wrought by immigration within the broader context of the economic, social, and political structures fueling immigration in its areas of origin. Deforestation on the Ituri frontier cannot be viewed simply as an isolated and local-level phenomenon. To properly understand the processes by which land in the Ituri changes hands, forest-based farmer and Mbuti societies are disrupted, and environmental degradation is aggravated, it is necessary to view these local developments "in relation to a 'macrosystem' with its characteristic economic and sociopolitical relations" (Schoepf and Schoepf 1987:27).

From this point of view we find that the immigration of farmers into primary forest is as symptomatic as the destruction of the forest itself—symptomatic of more fundamental causes of environmental degradation that are rooted in the unjust relations between different groups in Congolese society. Both immigration and forest destruction stem in part from a growing tendency throughout much of Congo toward individualization of land tenure, whether through manipulation of the legal concessionary system or through the spontaneous creation of unofficial land markets. These mechanisms increase the level of social inequality within rural Congolese society and favor a growing inequality in distribution of land. The resulting land scarcity deepens the poverty of rural producers and fosters increased migration into such socially and environmentally sensitive areas as the Ituri. Therefore, the deforestation and social disruption the Ituri's immigrants engender needs to be seen as a desperate, last-choice response to deeper causal factors of environmental degradation. Unless efforts to create areas of strict protection are coupled with initiatives directed at the root causes of forest destruction, the long-term success of conservation efforts will be jeopardized.

Thus the key to eliminate the adverse impacts of immigration lies in efforts undertaken in the source areas of immigration. Reforms that reverse the trend toward increasingly unequal distribution of land in areas outside of the forest are going to be crucial for alleviating immigration into such fragile areas as the Ituri. A similar conclusion is reached by a United Nations Center for Human Settlement (UNCHS) study on spontaneous settlement: "The judicious use of remaining land reserves must be part of a general land policy, and pressures for their settlement must be reduced through the effective implementation of land reforms in those existing rural areas where land distribution is skewed in favor of large landowners, leaving large portions of the rural population landless and unable to earn a living on the land" (UNCHS 1986:29).

Such reforms require concerted and committed political action that many conservationists may feel is beyond their scope. But larger and more powerful groups, which are also becoming interested in the conservation of the Ituri, may be able to effect such changes. Any sort of tenure reform, however, needs to avoid policies of individualization and titling alone, which have often led to concentration of land holding by an elite through manipulation of the titling mechanism.

If small holders' customary claims to land can be reinforced, they will more likely adopt various sustainable agricultural intensification technologies that can lessen the need to migrate (Cleaver 1990:4). A growing body of both theoretical and empirical knowledge is being gleaned from experiences with agroforestry, sustainable agriculture (e.g., managed fallow, crop and field rotation, organic fertilizer, soil conservation, intercropping), diversification of the household economy (e.g., such alternative income-generating activities as fish farming, small animal husbandry, sustainable gathering and marketing of forest products), extractive reserves, and other types of eco-development in the humid tropics (Posey et al. 1984; Clay 1987; Lugo et al. 1988). Such conservation and development initiatives merit further study to devise implementation methods for the particular cultural and environmental contexts of the areas bordering the Ituri. In addition, efforts to improve the land-use practices of people remaining on already deforested lands directly adjacent to protected areas, and in the reserve buffer zones, are crucial to slowing the rate of movement into areas of intact primary forest.

One of the most difficult questions confronting conservation planning for the Ituri concerns the immigrants who have already settled within the reserve. Most, if not all, of these immigrants have settled along the main trans-Ituri road and reside in the proposed reserve buffer zones. Relocating these people outside the reserve holds extreme economic and social costs that make it an unfeasible option. Current buffer-zone policy aims to incorporate and bring benefits to buffer-zone communities by encouraging them to participate in the capture of okapi within their adjacent forests (Hall 1989:10). The ICCN, in cooperation with the GIC (formerly Animals in Motion), maintains a captive breeding population of okapi at the reserve headquarters in Epulu for sale to Western zoos (one okapi can earn as much as US $250,000). Through their participation, local villagers are to receive the benefits of employment. The village community may also receive remuneration from the reserve for the okapi captured in their forests in the form of improvements on the village school, dispensary, or local chief's house. As a WWF report explains, "The theory is that if people benefit from this economically lucrative activity, then there will be incentive to protect forest along the road" (Hall 1990:3).

This policy, however, largely bypasses the immigrant members of these communities. In fact, it is a purposeful means to benefit local villagers specifically, given that they are the only ones with the skills necessary for the capture of okapi. Although I agree that those with the longest-standing claims to the forest should be the first to benefit, given the state of the economy in DR Congo, it is important not to forget that both immigrant and indigenous farmers in the buffer zones are living under conditions of poverty and crisis. This is especially true in comparison to the lives, including my own, of the people pushing most fervently for conservation. Immigrants cannot be ignored.

Without the provision of alternative land-use practices, immigrant farmers living in buffer zones will continue clearing new forests at an accelerated rate. Some conservationists feel that any conservation or development initiatives taken within buffer zones will only increase forest destruction by attracting more people to the area. But this is open to question because, as this study shows, the reasons people immigrate to the forest are much more fundamental. Population growth will likely continue in the area with or without attempts to promote more sustainable land use within buffer zones.

The other option, to implement no alternatives and let immigrant communities within buffer zones "fizzle" as their garden clearing eventually reaches zone limits, holds both practical and ethical complications. The problem with a buffer-zone policy that simply sets limits to forest cutting (roughly 5 to 10 km on either side of the road), without trying to change land use and implement alternatives, is that soon the 5 to 10 km boundary will be reached. It will then have to be controlled by reserve guards, which will be difficult and will end up leading to nothing but conflict and animosity between the reserve and local people. The reserve, being part of the state, will in the end win, and people will move off somewhere else to search for their life, with little or no help from the reserve.

A policy limiting tree cutting to a narrow buffer zone is nothing more than a policy of slow suffocation and less justifiable than intentional relocation with the provision of compensation to people for what they have lost. Such a policy goes down easy if one views people as social phenomena in a generalized, objectified manner that leads to large-scale analyses and diagnoses of the social trends of human use of the forest. But if one sees people as individuals, each of subjective and unique quality, and listens to the realities and personalized stories of crisis and coping, such a policy seems very cruel and insensitive, and in the end doomed. It is a difference between a policy that sees people as the problem and one that attempts to understand where people are coming from.

This study recommends that actions to mitigate the adverse impacts of immigration proceed along two lines: first, they should seek to stem the flow of new people moving into the forest. This can be done in several ways. One is to make it more difficult for immigrants to receive access to land by strengthening and supporting the indigenous tenurial claims to the lands and forests of the Ituri. Another and extremely important way is to work in the source areas of immigration to reverse an unequal distribution of land while promoting alternative sustainable land-use practices among would-be immigrants so that migration into the forest is not their only option.

A second line of action is to work with those immigrants who have already settled in the Ituri to develop alternative income-earning and food-producing possibilities that improve land use and take pressure off the forest. Assertions that such measures would cause population growth to rise and encourage more immigration into the forest are as yet unfounded. When combined with political efforts to control and limit immigration, the benefits of such measures would likely outweigh the risks of exacerbating forest destruction.

One benefit would be to gain more support and understanding for conservation among the Ituri's immigrants, more perhaps than Western-style conservation education programs can instill. An interview with a Nande immigrant farmer illustrates this point well. In response to my argument for the need of a conservation education program to teach people the value of the forest, Kahindo expressed the truth that unless such education efforts are linked with practical, tangible efforts to improve the quality of life through more efficient, less destructive land-use practices, "the teaching will be useless, go in one ear and out the other." Conservation education based on some theoretical abstract reasons for concern, such as loss of species or the greenhouse effect, will carry little weight and have a small chance of success in this context of desperate attempts just to keep food on the table.

It is important that an Ituri conservation plan include human beings as well as high biodiversity, exotic and endemic species, and intact ecosystems as considerations in approaching the goal of forest conservation. The human beings living in these forests are part of its ecosystem and need to be incorporated into decisions about what happens to the forest. This includes all people living in the area, not only indigenous populations but immigrants as well. The challenge is not how to limit these human communities, but instead how to work with them to promote more sustainable land-use practices that improve their living conditions while limiting environmental destruction. Locking them out in order to preserve nature will not in the long run keep the forest intact, because such a policy discourages rather than encourages their support of conservation's goals. As recent research from around the world shows (see, e.g., Sibanda and Omwega 1996; Western 1997), conservation as preservation no longer works.

In conclusion, of the many pressures confronting Africa's rain forests, the immigration of shifting cultivators is the one with the most human face. Unlike the large-scale lumber and gold concessionaires, most immigrants' stories are those of people simply struggling for a means to survive under a political and economic system that benefits only a minority. Those who would conserve areas of intact forest within the Ituri must recognize that blame for local-level environmental degradation can rarely be attributed to the people living within those areas. Rather, it is issues of tenure and the politico-economic structures and policies outside these areas that are the most important to address. Efforts to limit immigration must be applied to the roots of the problem rather than to the symptoms alone.

Within indigenous forest communities, indigenous people must retain the freedom and power to choose their own path. There is a need to trust indigenous peoples' abilities to choose ways that are not going to destroy the resource base on which they depend. After all, is it not the people who have lived in the forest for hundreds of years who have the greatest knowledge and experience of it? Respecting that knowledge is an important first step in the search for workable strategies to guarantee the forest's future.

REFERENCES

Areola, O. 1987. The political reality of conservation in Nigeria. Pages 277–293 in R. Anderson and R. Grove, eds. *Conservation in Africa: People, Politics, and Practice.* Cambridge University Press, Cambridge.

Bahuchet, S. 1991. Les pygmées d'aujourd'hui en Afrique Centrale. *Journal des Africanistes* 61(1):5–36.

Bailey, R., S. Bahuchet, and B. Hewlett. 1990. Development in central African rain forest: Concerns for tribal forest peoples. Report prepared for the World Bank.

Clay, J. 1987. *Indigenous Peoples and Tropical Forests: Models of Land Use and Management from Latin America.* Cultural Survival, Cambridge, Mass.

Cleaver, K. 1990. The population, agriculture, environment nexus in central and western Africa. Paper presented to the "Conférence sur la Conservation et l'Utilisation Rationelle de la Fôret Dense d'Afrique Centrale et de l'Ouest." Abidjan, Ivory Coast, November 5–9.

Hall, J. 1989. Rapport préliminaire sur les limites du futur Parc National de l'Okapi. Report to the Institut Zaïrois pour la Conservation de la Nature.

———. 1990. Conservation politics in Zaire: The protection of the Ituri Forest. *WWF Reports* (April–May):3–4.

Hart, J. A. 1979. Nomadic hunters and village cultivators: A study of subsistence interdependence in the Ituri Forest of Zaire. Master's thesis, Michigan State University, Lansing.

Hart, J. A., and T. B. Hart. 1990. Program for training and bio-monitoring in the Ituri Forest of Zaire. Project proposal submitted to USAID, March.

Hecht, S., and A. Cockburn. 1989. *The Fate of the Forest: Developers, Destroyers and Defenders of the Amazon.* Verso, London.

Lugo, A., J. R. Clark, and R. D. Child, eds. 1988. *Ecological Development in the Humid Tropics.* Winrock International Institute for Agricultural Development, Morrilton, Alaska.

Nations, J. D., and D. I. Komer. 1983. Rainforests and the Hamburger Society. *Environment* 25(3):12–25.

Painter, M. 1988. Comanagement with whom? Conservation and development in Latin America. Paper presented to the symposium "Culture: The Missing Component in Conservation and Development." Washington, D.C., April 8–9.

Peterson, R. B. 1991. To search for life: A study of spontaneous immigration, settlement, and land use on Zaire's Ituri Forest frontier. Master's thesis, University of Wisconsin, Madison.

Posey, D. A., J. Frechione, J. Eddins, L. F. DaSilva, D. Myers, D. Case, and P. MacBeath. 1984. Ethnoecology as applied anthropology in Amazonian development. *Human Organization* 43(2):95–107.

Salacuse, J. W. 1985. The national land law system of Zaire. Report to the University of Wisconsin Land Tenure Center and USAID/Kinshasa.

Schoepf, B. G., and C. Schoepf. 1987. Food crisis and agrarian change in the eastern highlands of Zaire. *Urban Anthropology* 16(1):5–37.

Sibanda, B., and A. Omwega. 1996. Some reflections on conservation, sustainable development and equitable sharing of benefits from wildlife in Africa. *Southern Africa Journal of Wildlife Research* 26(4):175–181.

Spooner, B. 1987. Insiders and outsiders in Baluchistan: Western and indigenous perspectives on ecology and development. Pages 58–68 in P. D. Little and M. Horowitz, eds. *Lands at Risk in the Third World: Local-Level Perspectives.* Westview Press, Boulder, Colo.

United Nations Centre for Human Settlement (UNCHS-HABITAT). 1986. *Spontaneous Settlement Formation in Rural Regions, Volume 1: Conceptual Framework and Guidelines for Action.* UNCHS, Nairobi.

CHAPTER 22

# Farmers, Wildlife, and the Forest Fringe

Lisa Naughton-Treves

The role of local communities in the conservation of African forests and wildlife occupies center stage in the eyes of a growing number of scholars and practitioners (Anderson 1987; Kiss 1990; Adams and McShane 1992; Western 1994). This new focus derives in part from failed attempts by centralized African governments to protect forests by making them entirely off limits to rural populations. The call for community-based conservation is further bolstered by case studies documenting the intricacies and resilience of indigenous management of biological resources (Rocheleau et al. 1993; see Almquist, this volume). Amid the celebration of local communities as rightful stewards of biological resources, the terms "local" and "community" are seldom carefully defined. Too often it is assumed that people inhabiting remote areas are socially cohesive and will manage forests or wildlife for collective long-term benefit. Field experience indicates, however, that residents of biologically rich areas are not necessarily socially cohesive or ethnically homogeneous (Peterson, this volume), and their land management goals may be markedly different from the environmental goals of the national or global citizenry. A subsistence farmer is likely to place higher priority on maximizing food security than on protecting charismatic megafauna. The potential con-

flict between local and global environmental concerns is starkly revealed where conservationists attempt to maintain large or potentially dangerous animals in forest habitat adjacent to smallholder farmers (KIFCON 1992; Newmark et al. 1994; Mubalama 1996; Tchamba 1996). At such sites, local communities may be hostile to conservation programs.

Ugandan governmental authorities and non-governmental conservation organizations share the goal of making forest and wildlife conservation locally viable (Mugisha 1996). Promoting community-based conservation in Uganda is especially challenging given the context of rapid population growth (3.4% per year between 1990 and 1995), densely settled rural landscapes (at $10^3$ individuals/km², Uganda is the sixth most densely populated country in sub-Saharan Africa), and declining forest resources (approximately 1% of total forest area was lost per year between 1981 and 1990) (World Resources Institute 1996). The forests of Uganda have been largely converted to agriculture during the past half-century (Hamilton 1984). Today, closed-canopy forests cover approximately 3% of Uganda's land surface and are located predominantly in western Uganda (Howard 1991).

The forests and wildlife remaining in Uganda's parks and reserves are becoming increasingly isolated in agricultural landscapes, as elsewhere in east Africa (see Newmark 1996). Farmers who settle on the edge of forest parks and reserves are vulnerable to crop loss to wildlife and on occasion risk their lives defending their crops from large, dangerous animals. Such losses and perceptions of vulnerability inhibit local support for wildlife conservation. If people are physically hurt or killed by wildlife, the political implications of human-wildlife conflict rapidly escalate. In neighboring Kenya, such conflicts have led the Kenya Wildlife Service (KWS) to fence off parts of several forest reserves, including the Aberdares, Mount Kenya, and Shimba Hills (Booth et al. 1992; Mwathe 1992; D. Western, pers. comm.).

Many conservation biologists object that fencing further insularizes forests that are already inadequate in size to maintain ecosystem processes (Western 1994; Newmark 1996). Social analysts meanwhile harshly criticize attempts to isolate forests from the rural poor or to erect barriers between people and wildlife (Adams and McShane 1992). How, then, can forests and wildlife be maintained amid densely settled rural landscapes? The dilemma is profound. To elevate local tolerance for wildlife requires technical, social, and economic interventions and a thorough understanding of the dynamics of human-wildlife relations in a development context.

This chapter offers a close examination of the relationship between smallholder agriculturalists and forest wildlife on the edge of Kibale National Park, Uganda. The fundamental questions at Kibale are similar to those at other sites: How much damage are wildlife actually causing? Who absorbs the cost of conserving wildlife? How do farmers cope with the risk of crop loss? How can local costs be balanced by benefits? To address these questions, I draw on two years of monitoring crop loss to wildlife, testing coping strategies of farmers,

and surveying local attitudes toward conservation. This case study illuminates the challenges and opportunities facing conservationists working to save forests and wildlife in densely settled human landscapes.

## Kibale National Park and Its Neighbors

Long before Kibale National Park was established in 1948, the Batoro of western Uganda shared a predominantly forested landscape with wild animals of extraordinary diversity and density (Taylor 1962; Hamilton 1981). Batoro farmers traditionally attempted to balance crop loss to mammals with bushmeat gains by trapping or hunting animals in and around their fields, a commonplace practice among African forest peoples (Koch 1968; Vansina 1990). Other traditional coping strategies included planting widely dispersed fields, guarding, intercropping, and rotational planting (Naughton-Treves 1996). The relationship between the Batoro and surrounding wildlife in the precolonial period was not necessarily harmonious. Crop damage by wildlife, particularly by elephants, made some arable land uninhabitable for sedentary farmers (Osmaston 1959).

At the beginning of the twentieth century, the linked strategies of farming and hunting were decoupled when British colonial authorities prohibited "native hunting" and declared all wild animals sole property of the Crown (Graham 1973). The colonial Ugandan Game Department thereafter protected Royal Game within parks for elite hunters, and elsewhere initiated militaristic campaigns to eradicate problem animals, such as elephants, hippopotamuses, and leopards (Graham 1973). In Uganda, as elsewhere in east Africa, these vermin control programs targeted the steadily expanding agricultural frontier (Brooks and Buss 1962; Graham 1973; Booth et al. 1992; Ville 1995). The lucrative ivory trade undoubtedly fueled this "pogrom of Uganda's elephants" (Graham 1973:176). During subsequent decades, the combined impact of colonial control operations, the ivory trade, deforestation, and civil war served to extirpate large animals from much of western Uganda and isolate surviving wildlife in protected areas (Struhsaker 1987).

Today most Batoro farmers have no contact with large wild animals unless they live near Kibale National Park (figure 22.1). Kibale is a 766 km² forest remnant much celebrated for its exceptional diversity and density of primates, and considered a possible Pleistocene forest refugium (Struhsaker 1981a, 1981b). Among the assemblage of wildlife present at Kibale are species notorious for raiding crops, such as olive baboons (*Papio cynocephalus anubis*), redtail monkeys (*Cercopithecus ascanius*), elephants (*Loxodonta africana*), and bush pigs (*Potamochoerus* sp.).

Currently, 54% of land within 1.0 km of Kibale's boundary is used in smallholder agriculture (Mugisha 1994). Agriculturalists in the area belong to two predominant ethnic groups, the indigenous Batoro and the immigrant Bakiga, who began arriving in the area during the 1950s (Turyahikayo-Rugyema 1974). The Batoro chiefs at the time allocated land to immigrants on the outskirts of their settlements, hoping to buffer Batoro farmers from crop damage by wildlife (Kabera, in Aluma et al. 1989). Today, both groups plant a mixture of more than thirty species of subsistence and cash crops; bananas (*Musa* spp.), maize (*Zea mays*), beans (*Phaseolus vulgaris*), yams (*Dioscorea* spp.), and cassava (*Manihot esculenta* ) cover the greatest area. Planting regimes are influenced by ethnicity: Batoro favor brewing bananas and sweet potatoes, and are likely to leave more of their farm in fallow, while Bakiga farmers plant more

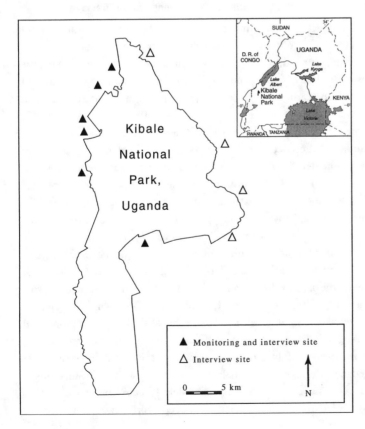

Figure 22.1. Kibale National Park, Uganda, with wildlife crop damage monitoring and interview sites indicated.

groundnuts, sorghum, and maize, and use intensive cultivation strategies (Naughton-Treves 1996). In both groups, women generally assume responsibility for cultivating food crops, while men tend such cash crops as brewing bananas, which they distill for local sale. Farm sizes are small (1.4 ha, on average), and population density is high (272 individuals per km²). Within this diverse farming system, various animals forage on both subsistence and cash crops, resulting in much frustration and resentment against the park among local cultivators (Aluma et al. 1989; Edmunds 1994).

### Extent of Crop Loss to Wildlife (and Livestock)

Weekly monitoring over two years in ninety-six farms revealed that more than 90% of crop-raiding events occurred within 160 m of the edge of Kibale. This distance varied by animal: it was smallest for elephants (126 m) and greatest for bush pigs (245 m) and palm civets (290 m) (see Naughton-Treves 1998 for details on research methods). The frequency and amount of damage varied greatly among wildlife species. For example, every week, redtail monkeys caused damage to many farms around Kibale. Whereas each event was small (about 16 m² on average), the sum of their damage over two years at all study sites matched or exceeded the damage caused by two larger animals (chimpanzees and bush pigs). The pattern of elephant damage was quite different. Most farms around Kibale suffered no damage by elephant. When elephants visited a single farm, however, the amount of damage was unpredictable and on occasion destroyed an entire farm's production. Elephant damage summed at the park level over two years was exceeded only by baboon damage. Moreover, elephant events were highly localized. Thus, at the park level, redtail monkey damage appears to be relatively evenly spread among farms on the edge of the forest, while elephant damage appears clumped.

The monitoring data also revealed that livestock (goats and cattle, principally) were the second most frequent source of crop damage by animals over 3 kg. In fact, livestock damage accounted for 16% of the total area of crops damaged over the study period. This unexpected result is discussed below in the context of farmers' perceptions of crop damage.

Across ninety-six farms within 500 m of the forest boundary, banana fields suffered on average 4% damage by area, cassava fields 7%, and maize fields 6% (Naughton-Treves 1998). However, average values mask the extent of variation in damage. For example, more than half the monitored banana fields suffered no damage, but one-quarter of those damaged lost more than 400 m². Moreover, certain villages suffered more intense damage than others (for example, Nyabubale lost five times more crops than did Nkingo).

### Farmers' Perceptions of Crop Damage

In spite of considerable variation in amounts of damage, farmers were nearly unanimous and highly vociferous about the severity of the problem. The heightened local perceptions of crop damage to wildlife can be partly explained by the variation in the frequency and magnitude of crop loss. Whereas crop losses generally averaged less than 10%, 7% of farmers lost more than 50% of their planted maize and cassava. Local perception of risk apparently reflects extreme damage events, not average losses.

Perceptions oriented toward extreme events may also explain why farmers complained less about redtail monkeys and small animals than about large animals, such as elephants. Small animals will not destroy an entire field in a single raid; their damage is self-limiting (Burton et al. 1978). Eighty of the ninety-six monitored farms suffered regular damage by redtails, yet only ten farmers described redtails as the worst animals (that is, 12.5% of those losing crops to redtails considered them the worst pests). Baboons damaged crops on sixty farms and were named worst by thirty-four farmers (56.7%). Similarly, bush pigs damaged fifty-one farms and were named worst by twenty-seven farmers (52.9%). Finally, elephants damaged crops on only eight farms, yet fourteen farmers reported them as the worst animal (175%). Elephant damage is potentially catastrophic, as an entire farm can be lost to elephants in a single night. Elephants foraging outside a national park may therefore create a "subsistence crisis" (Scott 1976). Local farmers are also well aware of the physical danger of elephants. Only elephant damage caused people to abandon their farms. In sum, the potential extreme damage by a large animal shapes people's perceptions more strongly than do persistent small losses.

Although the majority of damage occurred to bananas, farmers perceived maize and sweet potatoes to be the most vulnerable crops. Maize and sweet potato fields are preferred crops of baboons and bush pigs. On occasion, these animals completely destroyed fields of maize and sweet potatoes, whereas no banana field was ever entirely consumed, not even by elephants. This is due in part to a dramatic difference in field sizes (for example, maize fields were, on average, about 1% the size of brewing banana fields). Again, the potential for total loss may shape local perception more than frequent small losses. Another key difference is that maize and sweet potatoes have relatively short growing seasons in which an entire field of plants ripens simultaneously. If damage occurs any time after the first few weeks of growth, a farmer must wait until the next season to replant (Goldman 1986). Thus an entire season's maize production is at risk when baboons enter a field near harvest time. Finally, losing maize and sweet potatoes is especially significant as they are both staple food crops. A loss of staple crops creates immediate food insecurity (Porter 1979; Hill 1997).

Personal farming strategies interact with sociocultural factors to influence perceptions of crop vulnerability. The amount of labor invested by an individual farmer in a specific crop influenced his or her ranking of crop vulnerability. For example, men complained more frequently about banana damage, and women about cassava damage. Similarly, Batoro farmers cultivate more brewing bananas and sweet potatoes than do Bakiga, and they focus disproportionately on damage to these two crops.

## Who Is Most Vulnerable to Crop Loss to Wildlife?

Proximity to the forest is the most powerful overall predictor of crop damage by wildlife at Kibale. Land immediately on the forest edge is generally less expensive than land elsewhere (about 20% less, in part because of crop raiding risk and the fact that such land requires clearing before planting). During interviews and mapping exercises with communities, it became apparent that the immigrant Bakiga are disproportionately represented on the forest edge, and farms on the edge are more likely to be purchased (rather than inherited) than farms elsewhere. These patterns of landholding correspond to the historical practice of Batoro chiefs allocating land to immigrants on the outskirts of their settlements, in essence buffering Batoro farmers from crop damage by wildlife (Kabera, in Aluma et al. 1989). A similar settlement pattern was observed at Budongo forest in northwestern Uganda, where immigrants from what was Zaire were disproportionately represented adjacent to the forest edge (K. Hill, pers. comm.). Immigrants outnumber indigenous peoples around other protected areas and hence suffer greater crop losses (see, e.g., Milton and Binney 1980; N. Salafsky, unpublished data; S. Kanuhi, KWS, pers. comm.).

The disproportionate presence of Bakiga immigrants on the edge of Kibale is predicted by environmental risk theory. Often the most vulnerable people live in the most vulnerable regions (Susman et al. 1983). Minority communities in particular may reside in marginal areas and have lower coping capacities than other community members (Dow 1992). Yet, the elevated risk of crop loss to wildlife that comes with living on the edge of Kibale National Park may be counterbalanced with benefits. Interview data revealed that land near the forest edge is readily available and less expensive, so the Bakiga tend to own larger farms there ($\chi^2 = 44$, $P < 0.0001$, $N = 109$). Living next to the forest also allows ready access to various resources, in particular fuelwood (Naughton-Treves 1996). A parallel example is found at Royal Chitawan National Park in Nepal, where immigrants ("hill people") outnumber traditional groups on the park's edge (Milton and Binney 1980). These immigrants suffer the most crop damage from wildlife (rhinoceros, chital, and wild pigs), but they are also gaining economic power over traditional residents (Milton and Binney 1980). These economic gains may ultimately enable immigrants to cope better with crop losses.

## How Do Farmers Cope with the Risk of Crop Loss?

Farmers residing on the forest edge vary in their capacity to cope with crop losses to wildlife. By far the most frequently identified defensive strategy was guarding (78%, N = 145). In multivariate tests, there was no negative association found between guarding investment and crop loss (Naughton-Treves 1998). Given that farmers guard more in high-risk settings and that this study measured guarding effort only on a weekly basis, the true effect of guarding remains unknown. Other reported defensive methods included fencing (3.4%), trenches (1.4%), trapping (1.4%), and poisoning wildlife (0.7%). Farmers may have been reluctant to admit to killing wildlife, as it is illegal. Traps or poison were encountered in 15% of the monitored farms (N = 96).

Each farmer's coping strategies are constrained and facilitated by complex factors operating at different levels. For example, the owner of a small farm has little leeway for the location of planted crops of different palatability to wildlife. She is not able to plant a buffer of a banana plantation between maize fields and the forest edge, a planting strategy observed on most of the large farms around Kibale. Affluent owners of large farms also occasionally use pasture to separate their food crops from the forest edge. These options are not available to a subsistence farmer with only a small plot and minimal capital. Affluent farmers may also employ others (often others' children) to guard their fields, whereas poorer farmers must either face the risk of crop loss with no guard posted, or sacrifice other opportunities, such as schooling, to leave a child vigilant over crops. Ultimately, these passive defense measures (guarding or planting buffers) were considered costly by most farmers and only partially effective (Naughton-Treves 1998).

Although farmers around Kibale were not observed working communally to defend their crops, an individual's vulnerability was influenced by her neighbors' activities. For example, multivariate tests revealed that an individual living in a village where several individuals hunted in their fields suffered lower levels of damage to bush pigs, even if she herself did not hunt (Naughton-Treves 1998). Ultimately, an individual's best defense against losing crops to wildlife was to have a neighbor's crops between her and the forest. Today, large landholders (including Bakiga farmers) take advantage of this defense by leasing plots to other farmers immediately along the forest boundary. Researchers elsewhere in Africa have also noted that a densely settled band of farms forms the best barrier to wildlife incursions deep into agricultural lands (Bell 1984a; Hawkes 1991; Hill 1997).

Farmers universally complained that their

traditional defensive options were curtailed by the government; specifically, they could not kill raiding wildlife or "chase them deep into the forest." Currently, farmers cannot legally hunt or trap wildlife without formal permission from the Game Department. During meetings and interviews, a fundamental consensus emerged that local defensive options are constrained by government ownership of wildlife. This view of wildlife as state property caused considerable resentment (see also Booth et al. 1992). Many farmers bitterly referred to elephants as *ebitungwa bya gavumenti* or *ente za gavumenti* (the government's livestock or cattle). This comment reflects the widespread belief that the government is responsible for the problem and should either help guard, cull animals, or build a fence (for a U.S. parallel see Conover 1988; see also Lewis 1988).

The issue of proprietorship explains, in part, why people did not complain about livestock damage. When a goat or cow damages crops, customary law dictates that the owner of the animal must compensate the victim. In fact, some individuals capture a raiding goat or cow and hold it as ransom until the owner pays for damage. Farmers repeatedly drew the analogy of the government being a bad neighbor, allowing its "livestock" to damage other people's crops. Seldom did they mention that many farmers frequently graze their cattle and goats illegally within the boundaries of the park. Obviously, the traditional local social contracts regarding grazing rights and restitutions for animal crop damage are not operational between farmers and Kibale National Park.

## Management Recommendations

The results of this study indicate that the majority of crop damage caused by wildlife species (>3 kg) is confined to a narrow band of farms lying within 160 m of the forest boundary. This tight relationship between damage and distance to the forest has been reported at other sites where parks are surrounded by dense agricultural settlement, such as Aberdares National Park, Kenya, where the majority of damage occurred within 400 m of the forest edge (S. Kanuhi, pers. comm.). Within the zone of greatest crop damage surrounding Kibale, baboons, bush pigs, and redtail monkeys were ubiquitous crop raiders, whereas elephants caused infrequent, localized, and large-scale damage. The clumped distribution of crop damage by elephants parallels observations from other sites (Bell 1984b; Hawkes 1991; Damiba and Ables 1993; Lahm 1995; Languy 1996; Mubalama 1996; Tchamba 1996). The fact that elephants damaged crops only at certain villages becomes especially interesting given the history of elephant activity at these sites. Game department archives dating to 1951 reveal frequent elephant damage at key sites (such as Sebitoli). Elephant foraging patterns are likely to be shaped by a complex set of factors operating at a large scale (such as forest disturbance by logging in the north of Kibale, and a history of heavy poaching in the south).

The considerable variation in crop damage observed between crops, wildlife species, and villages is typical for crop raiding at other parks in Africa (Bell 1984b; Hawkes 1991; Damiba and Ables 1993; Languy 1996). This variation makes universal management prescriptions difficult, and also hinders efforts to compare the magnitude of the problem at different parks. In Cameroon, for example, elephants damaged <10% to 30% of planted crops around a national park (Tchamba 1996), depending on the type of crop and the village sampled. In Niger, farmers neighboring a park lost only 0.2% of their crops to elephants, yet certain individual farmers lost 100% (Taylor, in Tchamba 1996). Average crop loss to wildlife at a site in Malawi was estimated at 17%, but again, individual farmers suffered from 0% to 100% (Bell 1984a). Whereas elephants damaged 0.5% to 1.5% of crops on average per year around a faunal reserve in Gabon, some villages suffered losses of 8% (Languy 1996).

Confounding comparisons between sites is the different sampling regime of each study. In some studies, particularly where human population density was low and animals migrate regionally, crop damage was frequently recorded several kilometers from the park boundary, and damage levels were estimated over extensive areas (Tchamba 1996). Studies at densely populated sites (e.g., >200 persons/km²) typically base damage estimates on areas within 1 km of the forest boundary (CARE 1994; S. Kanuhi, pers. comm.).

Farmers around Kibale can reduce their risk of crop loss to bush pigs by hunting, and can lower their vulnerability to other animals by planting less palatable crops or pasture near the forest boundary. Some farmers have managed to increase the size of their farms at the forest boundary by buying more land. This may reduce risk, as large farms are less likely to lose an entire season's production due to a single incursion by wildlife. Other farmers have moved away from the forest, or have rented out the most vulnerable parts of their farms. In general, people are poorly equipped to defend their crops from elephants. Most vulnerable are those individuals growing food crops on small farms (<1 ha), immediately on the forest edge, and at villages where elephants are common.

The current conflict between farmers and wildlife at the edge of Kibale Forest echoes the traditional pattern of conflict at African agricultural frontiers. In historic times, agricultural frontiers were dynamic. At some sites, shifts in animal distributions or movements forced farmers to abandon cultivation at the forest edge because of heavy crop losses or danger (Osmaston 1959; J. Vansina, pers. comm.). More recently, as problem animals were eradicated, a major constraint on settled agriculture was removed and the frontier expanded (Graham 1973). Today, park edges mark a perma-

nent boundary where wildlife habitat meets agriculture or other intensive land uses. In sum, human-wildlife conflict itself is not new. What is novel about today's problem is that the government is attempting to maintain substantial wildlife populations adjacent to densely settled agriculture, while also limiting farmers' traditional range of strategies for coping with risk.

From a national or international perspective, crop losses of 4–7% in a narrow band of farmland appear to be a trivial price for maintaining endangered wildlife and forest habitat. In fact, the zone of heaviest crop loss (approximately 200 m beyond the forest boundary) could be considered 3,000 ha of "extra" habitat for wildlife at Kibale (area calculated by Mugisha [1994]). But living within this extra habitat are four thousand frustrated farmers who protest vehemently against the use of their land as a park for grazing wild animals. The human-wildlife conflict at the edge of Kibale serves as a vivid example of one of the greatest dilemmas in contemporary conservation: balancing global environmental goals with local residents' concerns. The global mandate to conserve wildlife imposes risk on Kibale's neighbors, who vary in their capacity to cope with crop losses. Although most local farmers recognized the purpose of Kibale National Park and acknowledged obtaining benefits, such as fuelwood, they were intolerant of the risk of crop loss to wildlife, particularly elephants.

The managers of Kibale National Park are increasingly aware that local support is important for long-term forest protection. Kibale National Park staff who field complaints about wildlife crop damage may be better equipped to define appropriate management responses if they understand the nature of local perceptions regarding crop raiding:

• Perceptions apparently reflect past or potential severe events more than steady, small losses

- People perceive larger animals as the worst pests, even though actual damage from small animals is often worse
- Personal investment in a crop or farming strategy will influence an individual's perception of risk
- Present conditions (such as wildlife belonging to the government) limit people's coping strategies
- At Kibale, as at many other sites, local residents perceive large and potentially life-threatening animals, such as elephants, to be the responsibility of the wildlife authorities (Booth et al. 1992; Hawkes 1991)

The government of Uganda is placed in the difficult position of protecting wildlife while guarding the property and lives of its citizenry. Management efforts to minimize human-wildlife conflict around protected areas fall into two general categories: (1) raising public tolerance for wildlife, and (2) deterring wildlife from entering croplands. These approaches will be discussed below in the context of Kibale.

## Raising Public Tolerance for Wildlife

A variety of strategies have been used around African parks and reserves to increase public tolerance for wildlife, ranging from direct monetary compensation, employment opportunities, and revenue sharing to campaigns to increase awareness of the value of wild animals to the community.

### MONETARY COMPENSATION

Schemes involving monetary compensation for crop damage have been attempted on several occasions in eastern and southern Africa, with poor results (Booth et al. 1992). Such schemes were nearly always abandoned after it was shown that they were open to abuse or blatant corruption, were cumbersome and expensive to administer, and reduced motives for self-defense among farmers (Bell 1984a; Booth et al.

1992). Around parks and reserves in Zimbabwe today, the government offers monetary compensation only in cases where human life has been lost to wildlife (J. Murombedzi, Project CAMPFIRE, pers. comm.).

Farmers around Kibale still retain hope of receiving compensation for crop loss to wildlife. Offering direct compensation to all households suffering damage from bush pigs or primates obviously would not be feasible because hundreds of households are affected on a regular basis. Victims of elephant damage deserve special consideration, however. They are relatively few in number and have minimal capacity to defend their crops. A plan to directly compensate farmers suffering elephant damage is likely to have the same problems as other compensation plans. However, in areas of persistent elephant raiding, it may be appropriate to run an experimental "insurance" plan designed and administered at the village level, and possibly based on traditional systems of restitution for livestock damage. Villages suffering repeated crop loss to elephants also ought to have priority access to jobs associated with the park or to tourism revenue-sharing programs. Alternatively, funds should be made available to purchase the farms of individuals who have long suffered elephant damage and who wish to sell their land. Such an offer, however, would have to be made very carefully so as not to raise fears of evictions, or to create inflated land values on the forest edge.

### JOBS AND REVENUE SHARING

Increasing economic benefits from wildlife for neighboring populations is a broadly recognized means to elevate local support for conservation (Harcourt et al. 1986; Western 1989; CARE 1994). The greatest successes have come from projects where considerable economic revenue is directly tied to wildlife survival (Winer 1996). Around a game ranch in Burkina Faso, local people are reportedly glad that the ranch and its elephants are so close because they receive such benefits as jobs, meat from culling operations, fishing on the ranch, and the opportunity to sell local crafts to tourists (Damiba and Ables 1993).

Managers at Kibale hope to improve tolerance for crop loss by channeling nature tourism revenues to local communities through village-level councils (Mugisha 1996). These pilot efforts are patterned after community-based game-ranching projects in southern Africa, primarily in savanna settings and on sparsely inhabited communal lands adjacent to parks (Lewis 1988; Kiss 1990; Winer 1996). Similar success at Kibale National Park will be difficult because of high human population densities and a prohibition on hunting. It is more appropriate to examine revenue-sharing programs around forested parks in the nearby region (Weber 1995). Like Bwindi National Park to the south, Kibale draws significant numbers of tourists eager to view chimpanzees and other primates, thus there is potential for significant tourism revenue.

### ENVIRONMENTAL EDUCATION AND LOCAL RESOURCE USE

Most respondents were aware of the conservation purpose of Kibale, partly because of an active environmental education campaign (Nganzi 1994). However, the belief of half the respondents that benefits of Kibale accrue primarily to the Ugandan government reveals a degree of local alienation from the park (N = 145). Most people also believed that local people benefit from the park (84%) but identified the illegal activities of collecting fuelwood (44%) and building materials (28%) as the most important benefits.

Permitting resource use inside national parks is a highly controversial subject. Several conservationists argue that unless local communities directly benefit by obtaining resources from the

park, they will always resent the presence of the park (Western and Pearl 1989). Yet determining the appropriate resources and defining sustainable harvest rates is difficult, as is controlling the actual rates of harvest (Ludwig et al. 1993). Pilot efforts in this direction are ongoing at Kibale related to the harvest of wild coffee, medicinal plants, and honey (J. Kasenene, Makerere University Biological Field Station, pers. comm.). Meanwhile, local residents extract timber products (firewood, fuelwood for charcoal, and construction material), hunt, and graze their livestock in the park illicitly. Legalizing these uses would not necessarily elevate public tolerance of wildlife, and should not be considered until there is better definition of who is allowed access to park resources and with what rights. Opening resource use to the general public at this time would inevitably lead to overexploitation of resources (Ostrom 1990). The long-term goal must be to match local rights to resources with local responsibility for conservation.

### Deterrence of Wildlife

Currently, farmers around Kibale attempt to deter wildlife from entering their fields by guarding, lighting fires, keeping dogs, and hunting. These methods are only partially effective, and in some cases exact high social costs. Continual guarding has far-reaching ramifications, particularly for children and adolescents, whose education may suffer as they stay at home to guard fields, and who may lose sleep or suffer greater exposure to such diseases as malaria through nocturnal or crepuscular guarding.

Further research is needed to quantify the benefits of crop guarding. Farmers' reports, my own observations of wildlife on the edge of the forest, and published accounts (Mohnot 1971; Horrocks and Hunte 1986) suggest that animals will flee or avoid intensely guarded fields.

## DISTURBANCE SHOOTING

On occasion, the Ugandan Park Service sends a guard to frighten crop-raiding elephants back to the forest by shooting in the air. This is effective only on a temporary basis. Elephants elsewhere have become habituated to this exercise (Booth et al. 1992). In Malawi, for example, plots guarded by an above-average density of hunter units (who shot to scare, not kill, elephants) showed no significant reduction in crop damage rates (Bell 1984a).

## BUFFER CROP

Planting less palatable crops near the forest boundary is a logical management option because crop damage is highly correlated with proximity to the forest edge. Thus, while a buffer of unpalatable cultivars providing little concealment is unlikely to stop wildlife incursions altogether, it may reduce the damage. The current list of buffer crops being considered by managers at Kibale includes soybeans, sunflowers, tobacco, tea, and Mauritius thorn.

Preliminary trials were conducted using soybeans as a buffer. During the first season in which soybeans were planted, minimal damage occurred. By the third season, however, bush pigs damaged a substantial portion of the crop (Naughton-Treves, unpubl. data). Ultimately, soybeans were no more effective a buffer than were beans, a crop grown traditionally in the area.

To be successful as a buffer, a cultivar must be profitable, unpalatable, and planted over a large enough area to reduce the attractiveness of crops beyond. Most farmers around Kibale own small landholdings and do not cooperate with neighbors with regard to crop selection, maintenance, or timing. For example, one farmer at Kibale planted tea immediately on the forest boundary and planted maize just beyond, approximately 100 m from the forest. His tea plantation was small, however (about 5,000 m²), and wildlife simply traveled through his neighbor's fallow land and then fed in his maize field. Given the small landholdings on the edge of Kibale, a buffer is a viable option only if neighbors collaborate in their planting.

Tea is a popular buffer crop in highland Africa, as it is not consumed by any wildlife species (Reuling et al. 1992; Kanuhi 1995). The Nyayo tea zone around the Aberdares National Park, Kenya, offers lessons. First, a great deal of natural forest was lost when this tea zone was planted within the boundaries of the park. Second, the tea zone has not served as an absolute barrier. Where it is 50 m wide, elephants cross the tea without hesitation (Reuling et al. 1992). Elsewhere, they use fallow patches within the plantation to cross the larger areas of tea (S. Kanuhi, pers. comm.). In sum, a tea buffer is unlikely to deter elephants unless it is planted continuously and extensively. Such a planting regime is beyond the scope of an individual farmer, and would require collective or corporate ownership. Another great drawback of tea is the tremendous volume of fuelwood required for processing the leaves. For every 3 ha of tea planted, 1 ha of fuelwood is required (S. Kluge, Toro Tea manager, pers. comm.).

## ELECTRIC FENCES

According to Adams and McShane (1992:58), "a fenced park in Africa is a symbol of failure." Yet in 1992, KWS authorities decided that there was "no option but to erect barriers if substantial wildlife populations, particularly elephant, are to survive in rural areas of Kenya" (in Booth et al. 1992:33). They argued that their decision was based on the failure of a variety of other strategies marshaled to reduce the human-animal conflict.

Electric fences have been used effectively at many sites in southern and eastern Africa to control crop-raiding mammals. The cost of building these fences runs from U.S.$500 to U.S.$1,500 per km, and the annual maintenance costs are estimated at 30% of the initial capital costs (Booth et al. 1992). Electric fences may be even more expensive to maintain in moist forest environments (Blair and Noor 1979). Experience elsewhere indicates that the minimum voltage required is 3,000 V, and that it is necessary to kill elephants that persistently break the fence. The literature indicates that constructing fences is futile without the support of local communities to assist in fence maintenance (Booth et al. 1992).

Beyond the substantial labor and capital costs of fencing, a long-term environmental cost arises from the insularization of natural habitat. Permanently confining wildlife to a patch of natural habitat results in a loss of genetic diversity and concomitant vulnerability to disease, habitat degradation if populations grow, and disruption of natural disturbance regimes. Such a conservation strategy resembles zoo management more than ecosystem management.

A profound dilemma now exists, in that wildlife at many sites are confined to pockets of habitat in a matrix of densely settled agriculture. From an ecological perspective, habitat must be expanded to conserve large animals, yet fences are the immediate political answer to a desperate human situation. Fencing should be considered only as a last resort, when other strategies have failed. Ideally the expenditures on a fence at a given protected area should be matched by funds toward establishing wildlife corridors to neighboring natural habitats. In 1993 the Ugandan National Park Service evicted thousands of settlers from the game corridor connecting Kibale Forest with Queen Elizabeth National Park. This corridor may eventually allow wildlife to re-establish regional patterns of movement; however, considerable controversy and protest persist over the eviction of settlers.

## CULLING

Killing problem animals has been a standard method of control used throughout Africa. Yet unless populations are eradicated, control shooting has proved to have little influence in reducing crop damage (Bell 1984a; Balakrishnan and Ndhlovu 1992; Booth et al. 1992; Tchamba 1996). Yet sharing game meat with local residents has increased public support for conservation in several cases.

Bush pigs apparently thrive in disturbed habitats and at forest-agricultural margins (Vercammen et al. 1993). Their meat is highly valued around Kibale, suggesting the option to legalize hunting of bush pigs by farmers on their own property. Considering the heavy toll of snares on such endangered species as chimpanzees at Kibale (Ghiglieri 1984), the technology and locale of bush pig hunting would have to be controlled so as not to include other species and not to occur within the park. Without even rudimentary data on population dynamics and movement patterns of elephants, elephant culling is not an appropriate option.

Ultimately, no single management strategy will stop crop loss to wildlife. In fact, the goal of management should not be strictly to control the problem, but rather to raise general tolerance of wildlife among the farmers, enhance their methods of defense, and lessen the impact of severe losses by elephants.

Given the context of human–wildlife conflict, state ownership of wildlife, and high human population density, fostering community-based wildlife conservation around Kibale is especially challenging. The Uganda National Park Service has taken the first steps toward building long-term local support and responsibility for Kibale by promoting village conservation councils along the forest edge (Mugisha 1996; A. Rwetsiba, UNP, pers. comm.). It is hoped that these councils will eventually negotiate resource access issues, receive and allocate tourism revenue, and help guard Kibale from resource use by outsiders. Environmental education programs should focus on supporting these councils and their decisions. Outreach and extension programs should also take into account that the closest neighbors of Kibale have distinct attributes in terms of ethnicity, land tenure, and wealth.

Certain conditions are promising for building legitimate local support for Kibale. First, crop loss is confined to a relatively narrow band of farmers, so managers have a target group for their efforts to share benefits, including growing tourism revenue. Second, at the park level, elephant damage at Kibale is relatively low compared to other forest sites, and no one has been killed by elephants (in contrast to such sites as Aberdares National Park or Shimba Hills in Kenya; see Mwathe 1992). Third, the current government of Uganda supports environmental protection and local participation, and park staff at Kibale consider both a priority (Mugisha 1996). Finally, the system of local governance (local councils) promoted by the national government is functioning relatively well in western Uganda, allowing park managers the opportunity to negotiate with legitimate local institutions regarding rights and responsibilities to Kibale National Park.

Local support is crucial to the long-term conservation of Kibale National Park. Ultimately, however, maintaining elephants and other large, far-ranging wildlife species in a densely settled agricultural landscape will continue to require management interventions by institutions beyond the level of the individual or village. This study demonstrated that most large wildlife species rely on the national park as a refuge of natural habitat and do not travel far into densely settled agricultural land. Assuming that local-level environmental management is uniformly better than anything organized at a national or regional scale is

romantic and not a useful perspective (Ostrom and Wertime 1995). The survival of Kibale Forest and its wildlife will depend on the management decisions of both local and national institutions, as well as the outcome of negotiations between local citizens and conservation authorities.

ACKNOWLEDGMENTS

This study was approved by the Ugandan National Research Council and received generous support from the Wildlife Conservation Society, the Fulbright-Hays Program, the Makerere University Biological Field Station, and the American Association of University Women. P. Baguma, F. Mugurusi, and P. Kagoro provided first-rate field assistance. A MacArthur Foundation postdoctoral fellowship at CIS, Princeton University, supported the write-up and analysis. C. Chapman, K. Redford, and A. Treves offered invaluable advice and critical comments.

REFERENCES

Adams, J. S., and T. O. McShane. 1992. *The Myth of Wild Africa*. Norton, New York.

Aluma, J., C. Drennon, J. Kigula, S. W. Lawry, E. S. K. Muwanga-Zake, and J. Were. 1989. *Settlement in Forest Reserves, Game Reserves, and National Parks in Uganda*. Makerere Institute of Social Research, Kampala, Uganda.

Anderson, D. 1987. Managing the forest: The conservation history of Lembus, Kenya, 1904–63. In D. Anderson and R. Grove, eds. *Conservation in Africa*. Cambridge University Press, Cambridge.

Balakrishnan, M., and D. E. Ndhlovu. 1992. Wildlife utilization and local people: A case-study in Upper Lupande Game Management Area, Zambia. *Environmental Conservation* 19(2):135–144.

Bell, R. H. V. 1984a. The man-animal interface: An assessment of crop damage and wildlife control. In R. H. V. Bell and E. McShane-Caluzi, eds. *Conservation and Wildlife Management in Africa*. U.S. Peace Corps Office of Training and Program Support, Malawi.

———. 1984b. Monitoring of public attitudes. In R. H. V. Bell and E. McShane-Caluzi, eds. *Conservation and Wildlife Management in Africa*. U.S. Peace Corps Office of Training and Program Support, Malawi.

Blair, J. R., and N. M. Noor. 1979. Initial report on feasibility of preventive measures. Paper read at "Incompatible Neighbors: Workshop on Elephant Damage," Trolak, Perak, Malaysia.

Booth, V. R., N. N. Kipuri, and J. M. Zonneveld. 1992. *Environmental Impact of the Proposed Fencing Programme in Kenya: Elephant and Community Wildlife Programme.* Kenya Wildlife Service, Nairobi.

Brooks, A. C., and I. O. Buss. 1962. Past and present status of the elephant in Uganda. *Journal of Wildlife Management* 26(1):38–50.

Burton, I., R. W. Kates, and G. F. White. 1978. *The Environment as Hazard.* Oxford University Press, New York.

CARE. 1994. The people of Bwindi Impenetrable National Park. CARE, Kabale, Uganda.

Conover, M. R. 1988. Effect of grazing by Canada geese on the winter growth of rye. *Journal of Wildlife Management* 52:76–80.

Damiba, T. E., and E. D. Ables. 1993. Promising future for an elephant population: A case study in Burkina Faso. *Oryx* 27(2):97–103.

Dow, K. 1992. Exploring differences in our common future(s): The meaning of vulnerability to global environmental change. *Geoforum* 23:417–436.

Edmunds, D. 1994. Report on preliminary findings from a study of local resource management institutions near the Kibale Forest National Park. MISR, Land Access Program, Kampala, Uganda.

Ghiglieri, M. P. 1984. *The Chimpanzees of Kibale Forest.* Columbia University Press, New York.

Ghiglieri, M. P., T. M. Butynski, T. T. Struhsaker, L. Leland, S. J. Wallis, and P. Waser. 1982. Bush pig (*Potamochoerus porcus*) polychromatism and ecology in Kibale Forest, Uganda. *African Journal of Ecology* 20:231–236.

Goldman, A. C. 1986. Pest hazards and pest management by small scale farmers in Kenya. Ph.D. dissertation, Clark University.

Graham, A. D. 1973. *The Gardeners of Eden.* Allen and Unwin, London.

Grubb, P. 1993. The afrotropical suids, Phacochoerus, Hylochoerus, and Potamochoerus: Taxonomy and description. In W. L. R. Oliver, ed. *Pigs, Peccaries and Hippos.* IUCN, Gland, Switzerland.

Hamilton, A. C. 1981. The quaternary history of African forests: Its relevance to conservation. *African Journal of Ecology* 19:1–6.

———. 1984. *Deforestation in Uganda.* Oxford University Press, Nairobi, Kenya.

Harcourt, A. H., H. Pennington, and A. W. Weber. 1986. Public attitudes to wildlife and conservation in the Third World. *Oryx* 20(3):152–154.

Hawkes, R. K. 1991. Crop and livestock losses to wild animals in the Bulilimamangwe Natural Resources Management Project Area. Center for Applied Social Sciences, University of Zimbabwe.

Hill, C. M. 1997. Crop-raiding by wild vertebrates: The farmers' perspective in an agricultural community in western Uganda. *International Journal of Pest Management* 43(1):77–84.

Horrocks, J. A., and W. Hunte. 1986. Sentinel behaviour in vervet monkeys: Who sees whom first? *Animal Behaviour* 34(5):1566–1567.

Howard, P. C. 1991. *Nature Conservation in Uganda's Tropical Forest Reserves.* IUCN, Gland, Switzerland.

Ingham, K. 1978. *The Kingdom of Toro in Uganda.* Methuen, London.

Kanuhi, S. 1995. Elephant crop raiding around Aberdares National Park. Master's thesis, Moi University, Nairobi, Kenya.

KIFCON. 1992. Report on a survey to assess the damage caused by wild animals to farm-households adjacent to the Aberdares Forest Reserve. Kenya Indigenous Forest Conservation Project, Nairobi, Kenya.

Kiss, A. 1990. Living with wildlife: Wildlife resource management and local participation in Africa. World Bank Technical Paper Number 130/Africa Technical Department Series, Washington, D.C.

Koch, H. 1968. *Magie et Chasse dans la Fôret Camerounaise.* Berger-Lerrault, Paris.

Lahm, S. A. 1995. Survey of crop raiding animals. *Gnusletter* 14(2–3):22–23.

Languy, M. 1996. Suivi et atténuation de l'impact des éléphants et autres mammifères sauvages sur l'agriculture au Gabon. WWF-Programme pour le Gabon, DFC-Direction de la Faune et de la Chasse, Libreville, Gabon.

Lewis, D. M. 1988. Survey of perceptions toward wildlife for two village communities with different exposure to a wildlife conservation project. Zambia National Parks and Wildlife Services, Lupande, Zambia.

Ludwig, D., R. Hilborn, and C. Walters. 1993. Uncertainty, resource exploitation, and conservation: Lessons from history. *Science* 260(17):36.

Milton, J. P., and G. A. Binney. 1980. Ecological planning in the Nepalese Terai Threshold. International Center for Environmental Renewal.

Mohnot, S. M. 1971. Ecology and behaviour of the Hanuman langur, *Presbytis entellus,* invading fields, gardens and orchards around Jodhpur, western India. *Tropical Ecology* 12(2):237–249.

Mubalama, L. 1996. An assessment of crop damage by large mammals in the Réserve de Faune à Okapis in the Ituri Forest, Zaire—with special emphasis on elephants. Institut Zaïrois pour la Conservation de la Nature, Canterbury, England.

Mugisha, A. R. 1996. *Collaborative Management: The Ugandan Experiences.* World Conservation Congress, Montreal.

Mugisha, S. 1994. Land cover/use around Kibale National Park. MUIENR, RS/GIS Lab, Kampala, Uganda.

Mwathe, K. M. 1992. A preliminary report on elephant crop damage in areas bordering Shimba Hills National Reserve. Kenya Wildlife Service Elephant Program.

Naughton-Treves, L. 1996. Uneasy neighbors: Wildlife and farmers around Kibale National Park, Uganda. Ph.D. dissertation, University of Florida, Gainesville.

———. 1998. Predicting patterns of crop damage by wildlife around Kibale National Park, Uganda. *Conservation Biology* 12(1):1257–1269.

Newmark, W. D. 1996. Insularization of Tanzanian parks and the local extinction of large mammals. *Conservation Biology* 10(6):1549–1556.

Newmark, W. D., D. N. Manyanza, D.-G. M. Gamassa, and H. I. Sariko. 1994. The conflict between wildlife and local people living adjacent to protected areas in Tanzania: Human density as a predictor. *Conservation Biology* 8(1):249–255.

Nganzi, P. B. 1994. The socio-economic impact of Kibale Tree-Planting Project and Conservation Education Project on the local population. Makerere International Institute of Environmental Education, Kampala, Uganda. Report.

Osmaston, H. A. 1959. *Working Plan for the Kibale and Itwara Central Forest Reserves, Toro District, Western Province, Uganda.* Forest Department, Uganda Protectorate, Entebbe.

Ostrom, E. 1990. *Governing the Commons.* Cambridge University Press, Cambridge.

Ostrom, E., and M. B. Wertime. 1995. International Forestry Resources and Institutions (IFRI) research strategy. Workshop in political theory and policy analysis, Indiana University, Bloomington.

Porter, P. W. 1979. *Food and Development in the Semi-Arid Zone of East Africa.* Foreign and Comparative Studies/African Series 32. Maxwell School of Citizenship and Public Affairs, Syracuse University, Syracuse, N.Y.

Reuling, M., U. Matiru, S. Njumbi, and M. Litoroh. 1992. Mount Kenya and Mount Elgon forest elephant survey. Kenya Wildlife Service/EC, Nairobi, Kenya. Report.

Rocheleau, D., M. Jama, and B. Wamalwa-Muragori. 1993. Gender, ecology, and agroforestry: Science and survival in Kathama. In B. Thomas-Slayter and D. Rocheleau, eds. *Gender, Environment, and Development in Kenya: A Grassroots Perspective.* Lynne Rienner, Boulder.

Scott, J. C. 1976. *The Moral Economy of the Peasant.* Yale University Press, New Haven.

Struhsaker, T. T. 1981a. Forest and primate conservation in East Africa. *African Journal of Ecology* 19:99–114.

———. 1981b. Vocalizations, phylogeny and palaeogeography of red colobus monkeys (*Colobus badius*). *African Journal of Ecology* 19:265–283.

———. 1987. Forestry issues and conservation in Uganda. *Biological Conservation* 39:209–234.

Susman, P., P. O'Keefe, and B. Wisner. 1983. Global disasters, a radical interpretation. In K. Hewitt, ed. *Interpretations of Calamity.* Allen and Unwin, Boston.

Taylor, B. K. 1962. The western lacustrine Bantu. In D. Forde, ed. *Ethnographic Survey of Africa.* Sidney Press, London.

Tchamba, M. N. 1996. History and present status of the human/elephant conflict in the Waza-Logone region, Cameroon, West Africa. *Biological Conservation* 75:35–41.

Turyahikayo-Rugyema, B. 1974. The history of the Bakiga in southwestern Uganda and northern Rwanda, 1500–1930. Ph.D. Dissertation, University of Michigan, Ann Arbor.

Vansina, J. 1990. *Paths in the Rainforests.* University of Wisconsin Press, Madison.

Vercammen, P., A. H. W. Seydack, and W. L. R. Oliver. 1993. The bush pigs (*Potamochoerus porcus* and *P. larvatus*). In W. L. R. Oliver, ed. *Pigs, Peccaries and Hippos.* IUCN, Gland, Switzerland.

Ville, J.-L. 1995. Man and elephant in the Tsavo area of Kenya: An anthropological perspective. *Pachyderm* 20:69–72.

Weber, W. 1995. Monitoring awareness and attitude in conservation education: The Mountain Gorilla Project in Rwanda. Pages 28–48 in S. Jacobson, ed. *Conserving*

*Wildlife: International Education and Communication Approaches.* Columbia University Press, New York.

Western, D. 1989. Conservation without parks: Wildlife in the rural landscape. In D. Western and M. Pearl, eds. *Conservation for the Twenty-first Century.* Oxford University Press, New York.

———. 1994. Linking conservation and community aspirations. In D. Western and R. M. Wright, eds. *Natural Connections.* Island Press, Washington, D.C.

Western, D., and M. Pearl, eds. 1989. *Conservation for the Twenty-first Century.* Oxford University Press, New York.

Winer, Nicholas. 1996. Collaborative management and community rights in the Chobe Enclave of N. Botswana. Paper read at IUCN Workshop on Collaborative Management, Montreal.

World Resources Institute. 1996. *World Resources, 1996–97.* Oxford University Press, New York.

# CHAPTER 23

# Indigenous Versus Introduced Biodiversity Conservation Strategies

## The Case of Protected Area Systems in Ghana

Yaa Ntiamoa-Baidu

Many countries in Africa have undergone rapid population growth. Unfortunately, this growth has not been matched by an equally rapid rate of development of appropriate technologies and management strategies to meet the increasing demands for resources. Concurrently, the traditional strategies for biodiversity conservation that have existed in African communities have gradually eroded. The result is extensive habitat destruction, degradation, and severe depletion of wildlife, with serious consequences for biodiversity conservation on the continent. Biodiversity is defined here as the variety of life forms, measured in terms of biomes, ecosystems, species, and genetic varieties and the interactions among them.

In the beginning of the twentieth century, concern over rapid destruction of forests and dwindling wild animal populations in Ghana and other African countries led colonial administrators to introduce protected area systems based on Western knowledge and values. Introduced protected area systems in Ghana comprise forest reserves and wildlife conservation areas (national parks, game production reserves, wildlife sanctuaries, and strict nature reserves). Control of these areas is vested in central government. A policy of externally enforced exclusion is pursued, and no serious attempts are made to involve the local communities in the management of the protected areas. This situation, naturally, generates antagonism and often results in conflicts between local communities and wildlife or forestry officers.

Sacred groves in Ghana and other west African countries provide a good example of indigenous protected area systems within rural communities. Ntiamoa-Baidu (1991a) identified three categories of indigenous strategies that advertently or inadvertently conserve biodiversity in Ghana. These are strategies that:

1. protect particular ecosystems or habitats (such as sacred groves, royal burial grounds, sacred rivers);
2. protect particular animal or plant species (such as totem and tabooed species); and
3. regulate exploitation of natural resources (such as closed seasons for harvesting or hunting).

These strategies are often enshrined in religious or cultural beliefs and superstitions and enforced by taboos. The taboos have no legal backing but have been strong enough in the past to make people obey the regulations.

This chapter analyzes indigenous and introduced protected area systems in Ghana, threats to the systems, and factors causing those threats. It then proposes mechanisms through which the two systems could be integrated to evolve a structure that promotes, empowers, and enables—rather than alienates—local communities in the conservation of their biological resources.

## Indigenous Protected Area Systems

In the past, small patches of forest were set aside, usually close to settlements, as sacred lands that were strictly protected by customary laws. Such areas still exist in rural Ghana and are known as *Abosompow/Asoneyeso* (shrine), *Mpanyinpow* (ancestral forests), and *Nsamanpow* (burial grounds); collectively they are known as sacred or fetish groves. A number of sacred groves have been destroyed as a result of urban and infrastructure development, but many still survive.

Several categories of groves exist (Dickson 1969; EPC 1976; Dwomoh 1990). Many are small (less than 1 ha), often comprising an object (such as a tree, stone, or rock) considered to be a god and its immediate surroundings. One example is the Malshegu sacred grove near Tamale in northern Ghana (Dorm-Adzobu et al. 1991; Ntiamoa-Baidu et al. 1992). Such small areas may be insignificant in terms of biodiversity conservation.

More commonly, the patch of forest in which the royals of a particular village were buried was protected because of respect for the dead and the belief that ancestral spirits live there. Entry into such forests was prohibited, and only a limited class of people (such as members of the royal family, village elders, and clan heads) were allowed access for burial purposes.

Many rivers and streams that provided the main source of drinking water for a village were also considered sacred. The surrounding forest lands were protected in the belief that the spirit of the river resided in the forest. Taboos associated with such sites included prohibition of cultivation of forest lands on the river banks, prohibition of use of fisheries resources within the river, and restrictions on access to the river on certain days. These taboos prevented defilement of the river. Although protection of the forests along riverbanks was based on religious and cultural beliefs, it served as river corridor management.

Often, patches of forests were protected because they supported sacred, totem, or tabooed species that were believed to have special spiritual or cultural values and associations. Many clans in Ghana have a wild animal or plant species as their symbol. For example, the leopard (*Panthera pardus*) is the symbol of the Bretuo clan of the Akan people (Azanwule clan in Nzema); that of the Oyoko (Alonwoba) clan is the raffia palm (*Raffia hookeri*). Traditionally, such species were strictly protected. In some cases, even touching the species was forbidden. The Boabeng-Fiema Monkey Sanctuary is an example of a grove protected because the forest supports black and white colobus monkeys (*Colobus polykomos* and *Cercopithecus mona*), considered sacred by the people of Boabeng and Fiema villages (Akowuah et al. 1975; Ntiamoa-Baidu 1987; Fargey 1991; Ntiamoa-Baidu et al. 1992).

Some sacred forests have their origins in historical events. For example, the Pinkwae grove (near Katamanso), a forest of 1.2 km², was the battleground of a war between the people of Katamanso and the Ashantis in 1826. It is believed to be the abode of the spirits of ancestors who died in the war and of the Aflye god whose powers enabled the Katamanso people to defeat the Ashantis (Lieberman 1979; Dorm-Adzobu 1990; Ntiamoa-Baidu et al. 1992). The Asantemanso grove (near Esumegya) is believed to contain the cave from which the seven clans of the Ashanti tribe originated (Ntiamoa-Baidu et al. 1992).

The total number of sacred groves in Ghana is unknown. A survey by the Ghana Forestry Commission returned a figure of 1,904 groves, of which 79.1% were in the south. Whereas many groves are too small to be of biological significance, a number have potential for biodiversity conservation.

Sacred groves are controlled by traditional authority (usually the fetish priest in charge of the god of the grove, the chief of the village, and heads of relevant clans). The responsibility for protection of the grove is vested in the entire community, but a select group of people or a family has the duty to enforce the rules. The conservation strategy, which is one of preservation, is enshrined in taboos and numerous cultural and religious rites and is maintained through reverence for the gods and ancestral spirits.

The Nkodurom grove, approximately 5 km$^2$ of primary forest that has been preserved by traditional beliefs for about three hundred years, provides an example of the mechanisms protecting sacred groves. The grove and its associated shrine, Anokye Kumaam, are located on the outskirts of Paakoso village (6°45' N, 1°30' W), 3 km east of Fumesua, off the Accra-Kumasi road. Paakoso derives its name from the main stream in the village, Pekoo (literally translated as "interested in war"). *Nkodurom* means "a source of war medicine," or a place where such medicines are prepared. The importance of the grove is linked with the war between the Ashantis and Denkyeras during the reign of the King Osei Tutu, the founder of the Asante empire. It is believed that the spiritual powers of his fetish priest, Okomfo Anokye, and herbal preparations from the Nkodurom sacred grove gave the Ashantis victory in war.

The final authority, ownership, and control of Nkodurom lies with the Asantehene (King of the Ashantis). This authority is exercised through the chief of Paakoso, representing the Agona clan (the royal family of the village). The high priest of the fetish (Anokye Boabeduro) is selected from the Asene clan (the clan to which Okomfo Anokye belonged). On appointment, the high priest remains in office for life and leads the people in the performance of all purification rites at the grove.

The taboos governing the grove include prohibition of:

- all forms of use, including farming, hunting, and collection of any plant material;
- access, except by traditional authorities for the performance of customary rites, or persons authorized after performance of certain purification rites and pouring of libation;
- access to all persons on Thursdays (believed to be Okomfo Anokye's day of rest, when he communed with the gods and spirits); and
- access to menstruating women.

The grove was held in such high esteem in the past that if an offender for any crime in any part of the Asante Kingdom sought refuge in Paakoso village, he could not be punished. The people of Paakoso strongly believe that the presence of the grove and the fetish protect them from various ills, such as outbreaks of epidemics and infant and maternal mortality.

Ensuring that the regulations governing the grove are strictly adhered to is the responsibility of the descendants of the Atabriso family. These traditional guards regularly patrol the periphery of the grove and arrest intruders, who are sent to the Paakoso chief for the necessary customary sanctions. The sanctions, which are exacted for the purpose of pacifying and purifying the gods and spirits, vary depending on the gravity of the offense. However, they usually consist of a cash fine, several bottles of schnapps, and a sheep for sacrifice to the gods. A culprit who refuses to honor the summons of the Paakoso chief or to accept and pay the penalty imposed upon him is sent to the Asantehene's court, where he is given a much heavier punishment.

The grove was not demarcated, and the rules governing access and usage were unwritten and had no legal backing. They were strictly observed, however, and most people believed that something dreadful would happen to them if they disregarded any of the rules or refused to offer sacrifices to pacify the gods and purify themselves. The traditional guards received no remuneration for their work but considered it their honorable duty.

Sacred groves have survived purely because of the strong traditional beliefs upheld by the local people, and the spiritual, religious, and cultural attachments to the groves. The major virtue of this strong culture-based practice is that it encourages community participation in natural resource conservation and sustains positive awareness of nature and the linkages between man and nature.

## Introduced Protected Area Systems

### FOREST RESERVES

The Forestry Department was established in 1909, but forest reserves (areas legally constituted for permanent forestry production) were not established until 1927, when the Forest Ordinance was passed. There are now more than 280 forest reserves in Ghana covering approximately 23,729 km². Although the primary purpose of most forest reserves in Ghana presently appears to be timber production, the maintenance of environmental and ecological stability was a major objective in their establishment in the early days of forest management. Thus, reserves were established along the forest-savanna borders to prevent the advancement of savanna vegetation into the forest zone, hilly areas were reserved to protect the headwaters of major rivers and prevent erosion, and reserves were scattered throughout the forest zone to maintain hydrological and climatic conditions.

### WILDLIFE CONSERVATION AREAS

The Department of Wildlife, formerly the Game Branch of the Forestry Department, was established in 1965 with the responsibility of managing Ghana's wildlife resources both within and outside conservation areas. The first six wildlife conservation areas were legally established in 1971 (Legislative Instrument 701). In 1974, the Ghana Wildlife Conservation Policy was adopted. The policy aims at reserving representative assemblages of Ghana's fauna and flora. There are currently sixteen legally constituted and two proposed wildlife conservation areas (one strict nature reserve, seven national parks, six game production reserves, and four wildlife sanctuaries), covering 14,333 km².

Most of the wildlife reserves within the forest zone were former forest reserves or parts of forest reserves. Bia National Park and Game Production Reserve, for example, were converted to their present status from the Bia Tributaries South Forest Reserve.

### WILDLIFE CONSERVATION AREAS AND FOREST RESERVES

Normally, wildlife conservation areas and forest reserves are clearly demarcated with boundary pillars, and staff are paid by the government to maintain the boundaries, enforce the laws governing use and access, and carry out prescribed management activities. Each reserve has a warden or conservator, or an officer of junior rank in charge of the routine administration of the reserve, who is assisted by an appropriate number of support staff.

The annual expenditures (comprising administrative costs, personal emoluments, and travel, protection, and management costs) for two typical forest districts (Kumasi West and Takoradi) in 1991 were 47.340 million and 9.596 million Cedis, respectively (approximately 950 Cedis = U.S.$1). The revenues accruing from timber and other forest products were 36.590 and 16.771 million Cedis, respectively.

Hunting, capturing, or destruction of any animal, and the collection or destruction of any plants, are legally prohibited in all wildlife conservation areas. The law gives the chief wildlife officer authority to grant permission for collection of flora and fauna from reserves, but this authority has so far been exercised only for scientific collections.

The forestry policy is more flexible regarding use of forest reserves. Production reserves may be given out as timber concessions and are logged under conditions prescribed by the department. In some cases, the original landowners retain rights for hunting and collection of certain forest products from the reserves.

A deficiency in the protection strategy of both the Forestry Department and the Department of Wildlife is the lack of involvement of local communities. Both departments pursue a policy of externally enforced exclusion of local people. Protective measures invariably marginalize the people, and there has been little or no attempt to encourage local community participation in the management of the protected areas. Also, little effort has been put into educating communities around the reserves about the aims and objectives of reserving a particular area and of the benefits that such communities stand to gain from the protection activities (Ntiamoa-Baidu 1991b).

## Effectiveness of the Two Systems for Biodiversity Conservation

### RESOURCES PROTECTED

Ghana's total land area is estimated to be 23.9 million ha, of which about 15% is under protection (wildlife and forest reserves). Identifiable ecological zones in the country include rain for-

est (3%), moist forest (31%), interior savanna (57%), coastal savanna (5%), and the Volta Uke (4%). It is estimated that 70% of the original 8.22 million ha of closed forest in the country has been destroyed (International Institute for Environment and Development 1992).

Introduced protected areas within the seven forest types found in Ghana cover approximately 1.79 million ha (table 23.1; Hall and Swaine 1981). This figure is equivalent to as much as 73% of the forests remaining in Ghana (based on the estimate that only 30% of the country's original 8.2 million ha of closed forest is left). The limited data on animal populations in the country indicate that the majority of the populations of all taxa, excepting rodents and invertebrates, are in protected areas.

The total number of sacred groves in Ghana, and the area they cover, is unknown, and the biodiversity of many groves has not been studied. There is, however, some evidence of their botanical value. For example, Hall and Swaine (1981) found in sacred groves the only surviving specimens of the inner-zone subtype of the dry semi-deciduous forest as well as the southern marginal forest. In many areas, sacred groves constitute the only remnant forest amid severely degraded forest lands and farmlands. Admittedly, the small sizes of the groves make them far less important than the introduced protected-area systems in terms of biodiversity conservation. The groves, however, form a matrix of biotic islands with potential for conservation of remnant communities of flora and fauna.

### THREATS AND ENCROACHMENT ON PROTECTED AREAS

Deforestation is one of the key environmental problems facing Ghana. Protected areas in forest zones suffer increasing pressure from the demands of agriculture and forest products. Illegal farming is a problem that forestry and wildlife officers constantly have to contend with; for example, approximately 50% of Kogyae, the only strict nature reserve, is reported to have been devastated by commercial yam farming.

While illegal felling of trees continues to be a major problem in forest reserves (the 1991 annual report of the Kumasi West Forest district documents ten cases of illegal timber extraction), wildlife populations in all reserves are under constant threat from illegal hunting. Illegal collection of other forest products, like chewing sticks, poles, cane, and snails, is also a source of regular conflict between local people and wildlife or forestry officers (Ntiamoa-Baidu 1991b). The survival of sacred groves is threatened by the erosion of the traditional beliefs that have sustained the system. A number of sacred groves have been gradually encroached on by surrounding farms, and a number have already been lost to development projects (EPC 1976; Ntiamoa-Baidu et al. 1992). The breakdown of beliefs can be attrib-

uted to Western-type education and religion, immigration of people who may have no respect for local traditions, and the lack of modern legislation to reinforce traditional rules (Ntiamoa-Baidu 1990, 1991a, 1991b).

## The Way Ahead

LOCAL PEOPLE'S PERCEPTION
OF FORESTS AND WILDLIFE

The importance of forests and wildlife to west Africans is well documented. Forest products and wild animals provide valuable sources of income, food, building materials, and many household tools (Asibey 1974; Sale 1981; Ntiamoa-Baidu 1987; Falconer 1992). In Ghana, the timber industry is the third most important foreign exchange earner, after cocoa and gold, accounting for 4.5% of the total gross national product (International Institute for Environment and Development 1992).

A study of the perceptions and values of forests and wildlife to four villages around for-

Table 23.1

## Area of Forest and Wildlife Reserves in the Various Forest Types

|       | AREA OF SUBTYPE ENTIRELY INCLUDED IN THE RESERVES | AREA PARTLY IN RESERVES | TOTAL RESERVED AREA IN TYPE | TOTAL AREA OF TYPE | % OF TYPE RESERVED |
|-------|------|------|------|------|------|
| WE    | 1,375 | 536 | 1,910 | 6,570 | 29 |
| ME    | 4,400 | 1,131 | 5,531 | 17,770 | 31 |
| UE    | 0 | 292 | 292 | 292 | 100 |
| MSSE  | 1,859 | 632 | 2,491 | 18,460 | 13 |
| MSNW  | 3,567 | 474 | 4,041 | 14,430 | 28 |
| DS12  | 136 | 557 | 693 | 8,630 | 823 |
| DSF2  | 2,592 | 290 | 2,882 | 12,810 | 4 |
| SM    | 62 | 31 | 93 | 2,360 | 29 |
| SO    | 1 | 5 | 6 | 20 | 22 |
| Total | 13,992 | 3,948 | 17,939 | 81,342 | |

*Source:* Adapted from Hall and Swaine 1981.

*Notes:* All figures are given in km. Forest types are: WE = wet evergreen; ME = moist evergreen; UE = upland evergreen; MSNW = moist semi-deciduous northwest subtype; MSSE = moist semi-deciduous southeast subtype; DSFZ = dry semi-deciduous fire-zone subtype; DSIZ = dry semi-deciduous inner-zone subtype; SM = southern marginal; SO = southeast outlier.

est or wildlife reserves in western Ghana recorded a wide range of values including food resource, income, and maintenance of ecological stability (tables 23.2 and 23.3). Several species of wild animals were reported to be symbols of the people's cultural identity, and a number of animal and plant species were used in medicines for a variety of diseases (Ntiamoa-Baidu 1991b). More than 90% of the respondents considered establishment of protected areas by central government to be useful and contended that there would be no primary forests left if the government had not established the reserves.

Therefore, there is no doubt that wildlife resources are of immense importance to the rural Ghanaian and that people appreciate both the need for protection and the role of protected areas. Despite this, communities living around reserves continue to show antagonism to the protection activities of government agencies. Relationships between the people in the communities studied in western Ghana and wildlife staff were particularly strained.

The question that emerges is: Why do people continue to destroy forests and overexploit forest resources that are so vital to their own survival, yet remain antagonistic to a system that seeks to conserve those resources? Six main factors can be identified as militating against protected area management, underlining the continued pressure on forest resources and the antagonism and conflicts between local people and conservation authorities:

1. Poverty (the lack of basic necessities of life)
2. Exclusion of local communities from the management of protected areas and the resources they contain, which the people consider to be theirs by birthright
3. Ignorance and lack of comprehension of the limited rate of renewability of natural resources, coupled with the fact that popu-

Table 23.2

## Value of Wild Animals to Communities Living in the Vicinity of Forest or Wildlife Reserves in Western Ghana

| VALUE | SCORE | % TOTAL |
|---|---|---|
| Source of food | 257 | 53 |
| Direct income generation | 92 | 19 |
| Cultural | 48 | 10 |
| Recreational/aesthetic | 38 | 8 |
| Maintenance of environmental/ecological stability | 17 | 4 |
| Medicinal | 17 | 4 |
| Educational/Scientific | 12 | 2 |
| Total | 481 | 100 |

*Source:* Ntiamoa-Baidu 1991b.

Table 23.3

## Value of Forests to Communities Living in the Vicinity of Forest or Wildlife Reserves in Western Ghana

| VALUE | SCORE | % TOTAL |
|---|---|---|
| Maintenance of environmental/ecological stability | 173 | 27 |
| Source of timber | 171 | 27 |
| Source of land for cultivation | 118 | 18 |
| Source of food | 76 | 12 |
| Source of household utilities | 57 | 9 |
| Medicinal | 26 | 4 |
| Direct income generation | 17 | 3 |
| Cultural | 1 | 0 |
| Total | 639 | 100 |

*Source:* Ntiamoa-Baidu 1991b.

lations of most wild animal species are so
low that the current rate of exploitation is
unsustainable

4. Low public awareness of general conservation issues and regulations
5. Misunderstanding of the policies and
   functions of the government conservation
   agencies
6. Poor public relations on the part of the
   conservation officers

These basic issues must be addressed in
evolving a sustainable protected area system
that is acceptable to the people.

## WHO BENEFITS FROM PROTECTED AREAS?

In the Western world, resources conserved are
mainly those for which people have no immediate need. In Africa, people are being asked to
conserve resources that they depend on for their
everyday needs. Protective efforts are purported
to be in the interest of the people, but who really
benefits from forest and wildlife reserves? Is it
the local person who has lost his land to conservation and has neither access to the resources of
his land to support his traditional way of life nor
access to modern developments to improve his
quality of life? Is it the forestry or wildlife officer
whose livelihood depends on the maintenance of
conservation departments? Is it the government, which requires foreign exchange earnings to provide amenities, most of which is consumed by the city or town dweller? Or is it the
Western world, where natural resources have
been destroyed in the process of development,
and where forest products are required to
maintain the overconsumptive way of life, but
also where conservation of tropical forests in
the interest of global environmental health and
biodiversity conservation are vehemently
advocated?

Protected areas in Ghana and Africa as a
whole will survive only if they provide benefits
that are substantial enough for the local com-

munities. There is, therefore, an urgent need to
address the question of equitable distribution
of the benefits from protected areas. Alternatives for the local communities and development options that will increase protected area
benefits must be found. These might include
development of appropriate agricultural technologies and facilities to increase productivity
through increasing yield per unit area rather
than increasing acreage of arable land; promotion of agricultural practices that encourage
forest conservation; and farming of favorite or
useful wild animal or plant species, such as
grasscutters (*Thryonomys swindetianus*) and
chewing sticks.

## INVOLVING LOCAL PEOPLE
## IN PROTECTED AREA MANAGEMENT

It is clear that the externally enforced exclusion
of local communities from reserves and alienation of the people from the management of
protected areas do not promote good relations
and do not encourage local support for conservation. Mechanisms for the participation of
local communities in the management of natural resources in their area must be explored and
instituted. This will help to eliminate the feeling of reserves being for "the government" and
the general antipathy toward such reserves.

Areas that local communities can be involved
in include law enforcement and education. A
simple representation of relevant local groups
on site management committees and advisory
boards of reserves can greatly facilitate communication of project aims and objectives and
activities to the people, promote a participatory
feeling, and eliminate unnecessary misunder-
standings (Ntiamoa-Baidu 1991c).

Another area worth exploring is whether
some protected areas that are currently closed
to any form of use can sustain some measure of
it, at least at the local subsistence levels. Protected area management policies and strategies
must be reassessed for the evolution of an inte-

grated approach that incorporates local needs and promotes efficient resource use. Collection of minor forest products like snails and medicinal plants that are currently prohibited in wildlife reserves, and protection of species that are considered pests, are some of the issues that require review.

## COORDINATION OF CONSERVATION EFFORTS

To maximize biodiversity conservation both within and outside protected areas, there is urgent need to coordinate conservation efforts and integrate traditional knowledge and practices in modern conservation strategies. Little attempt has been made beyond the local level to explore and develop the potential of sacred groves for biodiversity conservation. Ntiamoa-Baidu et al. (1992) outline a management strategy for Ghana's sacred groves that advocates nationwide inventory of the groves and the biological resources they contain, legislation to reinforce the traditional regulations regarding use and access, and provision of resources to improve local people's capability to manage their groves. It is hoped that the requisite financial resources can be found for the adoption of this strategy to ensure the survival of the groves.

In recent years, forest reserves in Ghana found to contain significant populations of wild animals have been converted to wildlife conservation areas and placed under the administration of the Department of Wildlife on the grounds that the Forestry Department's policy does not adequately protect wildlife. In theory, this may seem laudable. In practical terms, it is difficult to envisage any real gains in terms of protection efficiency in view of the inadequate staffing levels of the Department of Wildlife. Moreover, the greater part of the intact forests remaining in the country (and most likely the best wildlife habitats) are within forest reserves under the administration of the Forestry Department. Under such circumstances it would be more productive to evolve a system that incorporates wildlife management into forest management practices in all forest reserves, rather than to shift a few reserves to the administration of another department that is struggling to maintain the few areas already under its control.

## EDUCATION AND PUBLIC AWARENESS

Four of the six factors identified as militating against protected area management in Ghana are related to lack of education, public awareness, and understanding of conservation issues. As the *World Conservation Strategy* clearly acknowledges, it is virtually impossible to expect people to change their attitudes toward forests and wildlife when they do not understand the conservation issues at stake. It is encouraging that several organizations in Ghana—including non-governmental organizations like the Ghana Wildlife Society (GWS) and the Green Forum—are working to increase public awareness of wildlife and environmental issues in the country. The role of education and public awareness programs in wildlife conservation cannot be overemphasized. It is anticipated that funding agencies will continue to recognize and support this important component in the quest for sustainable use of natural resources in Africa.

### ACKNOWLEDGMENTS

An earlier version of this chapter appeared in *African Biodiversity Series*, no. 1 (May 1995). Produced by the Biodiversity Support Program, c/o World Wildlife Fund, 1250 24th Street, N.W., Washington, D.C. 20037. I am grateful to Kwasi Kese, deputy chief conservator of forests, for direction and assistance during the data collection and for his support and encouragement. My thanks go to B. Y. Ofori-Frimpong, principal game warden, Accra; James Addy, district forestry officer, Kumasi; and Stephen A. Adu, district forestry officer, Takoradi, for their time and assistance during the data collection. The efforts of my assistants, S. K. Nyame and Alfred A. Nuoh, in the data collection and analysis are gratefully acknowledged. I am also grateful for the warm welcome and help of the chiefs and elders of Paakoso and to Baffour Domfeh-Gyeabour, Nsumankwaahene, Asantehene's chief fetish priest, for his time and for directing us to various groves in Ashanti. Funding for this chapter was provided by the Biodiversity Support Program, a consortium of the Nature Conservancy, the World Resources Institute, and the World Wildlife Fund funded by the U.S. Agency for International Development.

REFERENCES

Akowuah, D. K., K. Rice, A. Merz, and V. A. Sackey. 1975. The children of the gods. *Journal of the Ghana Wildlife Society* 1(2):19–22.

Asibey, E. O. A. 1974. Wildlife as a source of protein in Africa south of the Sahara. *Biological Conservation* 6:32–39.

Dickson, K. B. 1969. *Historical Geography of Ghana.* Cambridge University Press, London.

Dorm-Adzobu, C. 1990. Pre-assessment of Pinkwae Forest at Katamanso in the Tema District. Report.

Dorm-Adzobu, C., O. Ampadu-Agyei, and P. G. Veit. 1991. *Religious Beliefs and Environmental Protection: The Malshegu Sacred Grove in Northern Ghana.* WRI, Washington, D.C., and Acts Press, Africa Centre for Technology Studies, Kenya.

Dwomoh, D. 1990. Forest conservation: The contribution of sacred groves (a study of Sekyere West District, Ashanti Region). Bachelor's thesis, University of Ghana, Legon.

EPC (Environmental Protection Council). 1976. Traditional approaches to conservation: Sacred groves in Ghana. Report prepared for the Environmental Protection Council, Accra, Ghana.

Falconer, J. 1992. People's uses and trade in nontimber forest products in Southern Ghana: A pilot study. ODA Report.

Fargey, P. J. 1991. Assessment of the conservation status of the Boabeng Fiema Monkey Sanctuary. Report submitted to the Flora and Fauna Preservation Society.

Hall, J. B., and M. D. Swaine. 1981. *Distribution and Ecology of Vascular Plants in a Tropical Rain Forest: Forest Vegetation in Ghana.* W. Junk, The Hague, The Netherlands.

International Institute for Environment and Development. 1992. Environmental synopsis of Ghana. Overseas Development Administration. Report.

Lieberman, D. D. 1979. Dynamics of forest and thicket vegetation on the Accra plains, Ghana. Ph.D. dissertation, University of Ghana.

Ntiamoa-Baidu, Y. 1987. West African wildlife: A resource in jeopardy. *Unasylva* 39(2):27–35.

——.1991a. Conservation of coastal lagoons in Ghana: The traditional approach. *Landscape and Urban Planning* 20(1–3):41–46.

——. 1991b. Local perceptions and value of wildlife to communities living in the vicinity of Forest National Parks in Western Ghana. In *Protected Area Development in Southwest Ghana,* Report by Environment Development Group, Oxford.

——. 1991c. Coastal wetlands conservation: The Save the Seashore Birds Project. Pages 91–95 in A. Kiss, ed. *Living with Wildlife.* World Bank Technical Paper 130.

Ntiamoa-Baidu, Y., L. J. Gyamfl-Fenteng, and D. Abbiw. 1992. Management strategy for sacred groves in Ghana. Report prepared for the World Bank and the Environmental Protection Council.

Sale, J. B. 1981. The importance and value of wild plants and animals in Africa. IUCN, Gland, Switzerland.

# V
# Applied Research and Management

# Overview of Part V

William Weber

WHEREAS AFRICAN RAIN FORESTS have received less research attention than other areas, this attention deficit is even greater when it comes to applied research. In response, this section presents eight chapters discussing research that is explicitly intended to inform and influence conservation management decisions at particular sites. The chapters cover a broad geographic range, from Ghanaian lowland forests in west Africa, across the Congo Basin to its eastern rim in the Afromontane forests of Uganda and Rwanda, to the island world of Madagascar.

Topically, the selection is equally diverse. Five chapters approach management from a primarily biological perspective, two describe the applications of social science, and one takes a more interdisciplinary angle. The science itself ranges from detailed floristic and faunal inventories to participatory social science and a historical review. Parks and protected areas provide the management focus for several chapters, but multiple use and human impacts are issues of concern to all. Four chapters address the commercial forestry sector. Finally, it should be noted that certain chapters from part IV—notably those on hunting, forest farmers, and immigration—could easily have fit within this section, given their strongly applied focus.

In chapter 24, "Designing a New National Park in Madagascar," Claire Kremen et al. describe the rare situation where biologists are able to conduct surveys to inform decisions on boundaries *before* the establishment of a new protected area. They describe criteria, methods, and limits of their multi-taxa approach, concluding that butterflies provide the greatest sensitivity to habitat differences within their rain forest ecosystem. Ultimately, their efforts were central to the creation of the 2,100 km² Masoala National Park. In chapter 25, Caroline Tutin and Amy Vedder look at a single charismatic focal species in their review of gorilla conservation and research in central Africa. This comprehensive chapter looks at the relations between many kinds of research and conservation action in several different countries. According to the authors, action is most effective when informed by studies that document the specific ecological needs of each of the three gorilla subspecies. Multidisciplinary research further helps to clarify the variety of threats that confront these different populations. Logging, hunting, and habitat loss pose vastly different challenges to lowland (as opposed to mountain) gorillas, and thus affect the selection of management strategies from ecotourism to controls on road development. Even within the lowland set of countries, Tutin and Vedder argue that different human population densities greatly alter the time frame for action.

Chapters 26 through 29 describe applied research of direct relevance to the future of both conservation and commercial forestry in the African rain forest, from west Africa to the east African highlands. The first two contributions discuss the impacts of logging on wildlife, whereas the others consider the effects of certain practices on tree and forest regeneration. In chapter 26, Lee White and Caroline Tutin report that chimpanzee populations in Gabon react much more negatively to logging than do gorillas. According to their thoughtful assessment, chimpanzees are far more traumatized and take much longer to return to logged areas for mostly social and behavioral reasons. This analysis leads to recommendations for change in certain timber management practices and raises questions for further research.

Andrew Plumptre targets a broader array of taxa in chapter 27, "The Effects of Habitat Change Due to Selective Logging on the Fauna of Forests in Africa," with particular attention to monodominant forests in Uganda and DR Congo. In both areas, Plumptre describes commercial forestry practices, including selective arboricide, that have the unusual effect of increasing tree species richness, as they actively promote a transition from monodominance to more diverse forests, for entirely commercial reasons. In the process, certain wildlife species are favored, including many of the more charismatic frugivorous primates and birds. Management recommendations are made for those species more dependent on monodominance.

John Kasenene draws on his own extensive experience in Uganda's Kibale Forest to consider the effect of gap size on natural forest regeneration. In chapter 28, he looks at the variable use histories of comparable forest compartments and concludes that commercially valuable hardwoods regenerate far better in smaller natural gaps. The larger clearings created by most logging operations, according to Kasenene, facilitate an increase in seed and seedling predation by rodents and more intensive browsing of saplings by elephants and other large herbivores. This finding raises serious questions about the prospects for truly sustainable logging at any scale beyond artisanal pit-sawing.

Chapter 29 takes us into the highly regimented realm of forestry management in Ghana. While acknowledging past mistakes in both the colonial and post-independence periods, William Hawthorne argues that modern Ghanaian forestry should be a model for other African nations. This model is technical, based on extensive sampling, rarity, and value ratings for individual tree species, and a collective diversity value for stands. The result of all these surveys and analyses is a finely tuned array of management regimes that will challenge Ghana's foresters and political leaders alike.

Bryan Curran and Richard Tshombe, trained social scientists who are also conservation practitioners, are the authors of chapter 30, "Integrating Local

Communities into the Management of Protected Areas." This chapter is rich in qualitative information derived from their extensive work with local populations in northeastern DR Congo and southeastern Cameroon. For DR Congo's Ituri Forest, they describe conflicts between communities and conservation agents over official decisions made without their input, and failure to keep promises made before the conservationists' work began. In Cameroon, however, Curran and Tshombe found that the local community had much greater opportunity for input before any final decisions were made on reserve design or function. This, in turn, created more options for collaborative solutions. In most instances the authors walk the narrow line between information and advocacy, although they openly cross that line when they feel that support for local human interests is required. Indeed, under certain circumstances they challenge the idea that areas should be completely protected.

The final chapter in this section is an anomaly. Although it is more history than social science, and although it focuses on a different type of habitat, "Politics, Negotiation, and Conservation" is extremely relevant to this section: its subject is the decades-long process of striking, maintaining, and adapting an agreement between conservationists and local populations. Authors Alison Richard and Robert Dewar were part of a team that set out to establish the Beza Mahalafy Reserve in southern Madagascar in the 1970s. They describe the compromises and conflicts that have characterized their initial agreement with local people and a still-evolving program of buffer-zone management. The identification of significant conflicts between local and regional political interests is all the more telling when one extrapolates from the restricted area around Beza Mahalafy to the immense scale—ecological and political—of most of the Congo Basin initiatives considered in this volume.

These chapters, in addition to others from earlier sections, demonstrate the potential for good information to be applied to improved conservation management. The information can be biological, behavioral, socioeconomic, or experiential; preferably it will be multidisciplinary. Good managers are desperate for information that is relevant to their concerns, and researchers are generally pleased to find an existing management structure into which they can feed their results. It is a combination that should encourage more applied research, closer collaboration between conservation scientists and practitioners, and a better future for the African rain forest.

# CHAPTER 24

# Designing a New National Park in Madagascar

## The Use of Biodiversity Data

Claire Kremen, David Lees, Vincent Razafimahatratra, and Heritiana Raharitsimba

Most parks and reserves around the world were designed without the benefit of prior biodiversity surveys: rare indeed are the opportunities to collect and analyze biodiversity data for planning protected areas (Scott et al. 1987; Kremen 1994). In designing the Masoala National Park in Madagascar, however, we were fortunate to have both the time and the funds to carry out biodiversity surveys first.

Inventories can provide a significant basis for making ecological or biogeographic comparisons. Increasingly, authors are calling for biological inventories to be conducted with attention to ecological sampling design and appropriate choice of target groups (Coddington et al. 1991; Di Castri et al. 1992; Kremen et al. 1993; Kremen 1994; Colwell and Coddington 1994), to allow the application of inventory results to conservation, among other reasons. By necessity, however, inventory programs usually must compromise between sampling the maximum number of habitats and microhabitats in a region and sampling replicates of habitats over space and time to avoid overlooking rare or seasonally limited species. Despite this caveat, biological inventory data provide an essential informational component for making sound conservation decisions (Scott et al.

1987). The increasing use of such data in conservation planning represents a significant advancement over previous attempts to delineate protected areas without reference to detailed surveys.

The Masoala Peninsula is a land mass of some 4,300 km² in northeastern Madagascar (figure 24.1). Formerly entirely covered with pristine rain forest, the peninsula still has 3,000 km² of primary forest, making it the largest single block of low- to mid-elevation humid evergreen forest left in Madagascar. The principal threats to this area are slash-and-burn agriculture and logging. The design of the national park, established as a priority for conservation in Madagascar's national Environmental Action Plan (World Bank 1988), was entrusted to a consortium of international non-governmental organizations—CARE International Madagascar, the Wildlife Conservation Society, and the Peregrine Fund—which are implementing the Masoala integrated conservation and development program (ICDP). CARE is the lead organization in the consortium assuring the development component, whereas the Wildlife Conservation Society is responsible for the conservation component, including park design. The consortium acts under the authority and guidance of the Malagasy Direction des Eaux et Forêts and the Association National pour la Gestion des Aires Protégées. The consortium's ICDP proposal, funded by the United States Agency for International Development, allowed for fifteen months to complete various studies that would allow the park plan to be based on a body of scientific data and ecological and economic principles.

Early in the process of this work, we established a set of general guidelines for reserve design (table 24.1) and a flowchart methodology by which to apply these criteria (figure 24.2). Applying the methodology required

socioeconomic information on human settlement and land-use patterns; biological information on the distributions of vegetation cover, habitat types, and biodiversity; and studies assessing the feasibility of a variety of projects in sustainable agriculture, forestry, ecotourism, and non-timber forest resources.

The results of these studies were used to define a park of 210,000 ha, with a forestry management zone outside the park of 50,000 ha (figure 24.3). To summarize briefly, the proposed park limits would protect the full spectrum of habitat types and biodiversity found on the peninsula in multiple representatives,

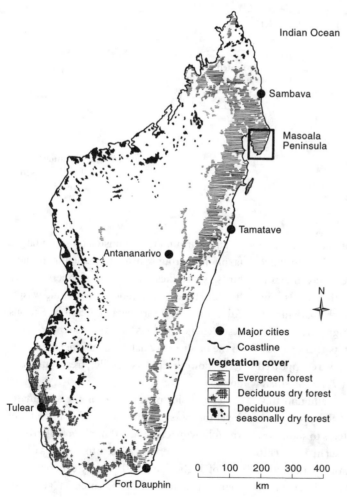

Figure 24.1. Madagascar, showing the Masoala Penninsula and vegetation cover.

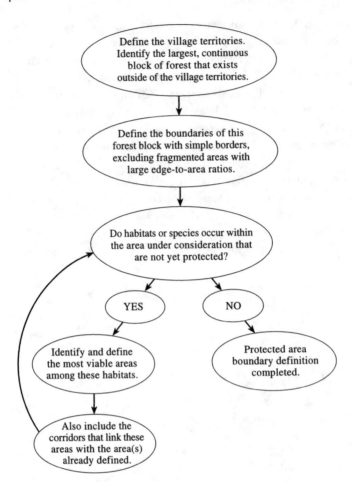

Figure 24.2. Boundary delineation flowchart used to guide reserve design.

focusing on the protection of the largest and most pristine habitat blocks. To minimize social conflicts, the park borders were drawn so that most landholders retained their lands. The large forested area left outside the park (about 100,000 ha) was deemed too fragmented to protect with certainty. Because of its greater accessibility, however, a portion of this area appeared best suited as a management zone for small-scale community-based forestry and non-timber forest projects (Kremen et al. 1995).

The purpose of this chapter is to show how the biodiversity inventory was set up in order to aid in establishing and evaluating park design. A systematic inventory was carried out at nine localities around the peninsula on seven target taxonomic groups: primates, small mammals, birds, butterflies (including skippers), tiger beetles, scarab beetles, and ichneumonid wasps. Additional inventory data gathered by various researchers at various localities also contributed to the biodiversity database and included ants, ferns, flowering plants, reptiles, freshwater fish, and bats.

The choice of the taxa for the systematic inventory was influenced by at least several of the following criteria, although not all groups met all the criteria: (1) desire to sample a diversity of higher taxa with widely different ecological characteristics (Kremen et al. 1993;

Table 24.1

# Guideline for Park Delimitation

| GUIDELINE | CONSERVATION PRINCIPLES |
|---|---|
| The largest continuous area consisting of natural habitats must be protected. | Larger areas are easier to manage because they have greater ecological viability. Larger areas may also protect highly mobile or migratory species and allow zones of contact between different habitat types. |
| The area should contain several representative examples (at a minimum) of the existing habitat types. | Protection of different habitat types protects a larger percentage of biodiversity and maintains ecological interactions between habitat types, leading to long-term viability. Several examples of each habitat type should be protected to guard against environmental catastrophes, of either natural or anthropogenic origin. |
| The area to be protected should consist largely of primary or relatively undisturbed habitats. | By protecting primary habitats, we protect the biological diversity that is most at risk, and maintain examples of threatened natural ecosystems. |
| Special consideration must be given to rare or threatened habitats; rare, threatened, or endangered species; and locally endemic species The size of the area protected must be large enough to maintain viable populations of these species. | By giving special consideration, we maintain the most endangered elements of biodiversity. |
| If there is a choice between two areas, that which will protect the maximum number of species of interest should be chosen. | Such choices are invoked to protect the maximum biodiversity. |
| The headwaters of rivers should be included. | Protecting headwaters minimizes erosion and siltation downstream, and maintains water quality and aquatic habitats for a variety of terrestrial and aquatic species. |
| The park limits should be simple (circular rather than lobed). | Simple limits minimize the edge-to-area ratio, thereby reducing edge effects and management problems. |
| The limits should be easy to see and to respect, following geographic features when possible. | Simple limits facilitate patrolling and allow local people to have full knowledge of park boundaries. |
| Corridors should be protected which link natural habitats. The corridor protected must be wide enough to encourage animal movement, or it must include a zone where forest regeneration can be actively encouraged. | Corridors maintain landscape-level connectivity, allowing gene flow between neighboring populations of wild species and permitting migrations and recolonizations that are often necessary owing to environmental change. |
| Human settlements must not exist within the park. | The exclusion of human settlements from the protected area will minimize conflict between humans and wildlife. |
| There should be points of access for ecotourism and patrolling. | Tourism must be encouraged as a source of revenue both for park management and for local populations living in the park's peripheral zone. |

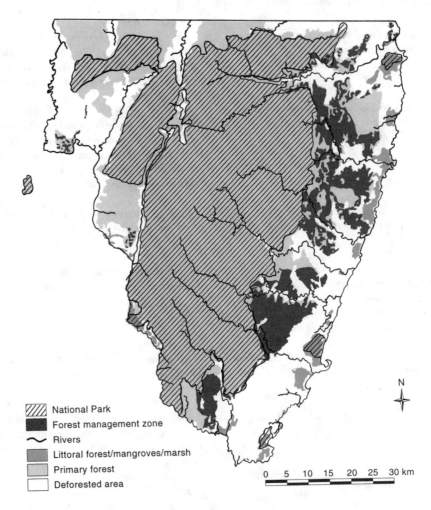

Figure 24.3. The proposed boundaries of the national park and its forestry management zone.

Kremen 1994); (2) presence of in-country specialists to conduct the inventory; (3) ease and
replicability of sampling the taxon; and (4)
ability to discriminate between or identify
species within the taxon. Madagascar, as
opposed to many African nations, has no large
mammal species whose area requirements,
once known, might inform the conservation
planning process as "umbrella" indicators (but
see Merenlender et al. 1998; Kremen et al.
1999). Therefore, in contrast to some other

biological surveys conducted for conservation
purposes, this inventory was concerned with
the distributions of multiple taxonomic
assemblages (that is, a geographic characterization of biodiversity) rather than with the
distribution of individuals from one or several
umbrella species. Geographical variation in
the distributions of species assemblages across
the landscape then provided a basis for deciding the minimal set of regions that must be
protected. Here, we present the detailed data

from butterflies and skippers only, because information on the other groups has been presented elsewhere (Andriamampianina 1995; Razafimahatratra and Andriamampianina 1995; Razafindrakoto 1995; Thorstrom and Watson 1997; Sterling and Rakotoarison 1998; Kremen et al. 1999).

## Designing the Biodiversity Inventory

Initial decisions regarding park boundaries were based primarily on landscape-level elements—that is, the distribution of primary vegetation, habitat types, and human settlement patterns. Thus a primary goal of the biodiversity inventories was to provide data serving as an assessment tool at a finer scale to ensure that the proposed limits of the park would protect the biodiversity of the area. The sampling localities for the biodiversity inventory were selected with this conservation planning goal in mind.

We therefore chose the most significant and evident environmental gradients to establish the sampling localities so as to include the widest range of environmental and habitat diversity within the survey. By so doing, we could provide the most stringent test of whether the proposed park would protect the biodiversity of the area. In essence, our null hypothesis was that no differences occur in the floral and faunal species composition along these gradients. These gradients included:

- The east-west gradient. A mountain chain of 600–1,200 m runs north to south, separating the peninsula into two zones (figure 24.4). The western, bay side of the peninsula is characterized by extremely steep slopes and annual rainfall exceeding 6 m per year (unpublished data from Andranobe field station, 1991–1994). The eastern, ocean side has gentler relief, higher temperatures, and lower rainfall (an average of 2.75 m per year from the Cap Est

weather station) as well as higher wind speeds (Andriamampianina 1995). Rainfall is thought to influence the distribution of vegetation in Madagascar (Gentry 1993; Abraham et al. 1996).

- The elevational gradient. Forested habitats exist on the peninsula from sea level to 1,200 m. Species distributions are often limited by elevational characteristics because of associated edaphic or microclimatic conditions (Whittaker 1975). The full elevational gradient was therefore investigated.
- The north-south gradient. In analogy to island systems, one might expect a "peninsular effect," with certain species dropping out or being replaced with increasing distance from the "mainland" (Ricklefs 1973).
- The soils gradient. A geologic map showing parent rock type allowed us to identify three soil categories: lateritic soils on granite parent rock, soils on basalt parent rock, and sandy coastal soils (Hottin and Liandrat 1964). Soil type may vary based on factors other than parent rock; however, no soil classification information was available to guide the design of the inventory. These parent rock types do show distinct differences in vegetation community composition and structure (Dupuy and Moat 1997). Whereas the canopy of humid evergreen forest on lateritic soils on Masoala is typified by *Canarium* (Buseraceae), *Diospyros* (Ebenaceae), *Symphonia* (Clusiaceae), *Tambourissa* (Monimiaceae), *Weinmania* (Cunoniaceae), and *Anthostema* and *Uapaca* (Euphorbiaceae), littoral forests on sandy soils are characterized by *Symphonia* and *Calophyllum* (Clusiaceae), *Terminalia* (Combretaceae), *Hernandia* (Hernandiaceae), *Brexia* (Saxifragaceaea), *Anthostema* and *Uapaca* (Euphorbiaceae), *Intsia* (Fabaceae), *Sarcolaena* and *Leptolaena* (Sarcolaenaceae), and *Faucherea* (Sapotaceae)

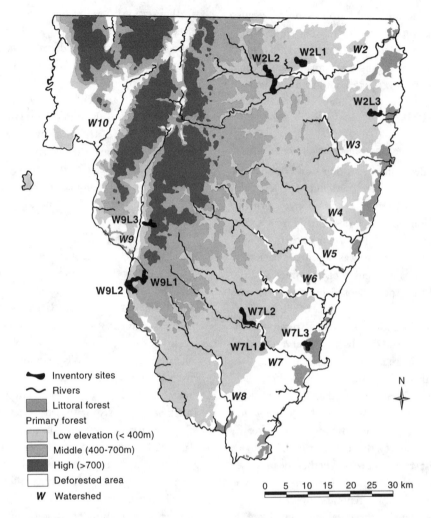

Figure 24.4. Bio-inventory sites and elevational zones on the Masoala Peninsula. *W* indicates the major watersheds (*W*2–*W*10), and bio-inventory sites are named within each major watershed (W2L1, W2L2, W2L3, etc.).

(G. Rahajasoa, G. Schatz, P. Lowry, and R. Rabevohitra, unpublished data).

- Degree of threat to forests from slash-and-burn farming. Slash-and-burn farming had already decimated a portion of the peninsula's forests (figure 24.5), and the area adjacent to this zone was at high risk of deforestation because of the presence of numerous villages and human preference

for regions of low elevation and low slope for farming (Kremen et al. 1999). Therefore, it was important to know whether the highly threatened forests differed significantly in species composition from more remote areas in order to determine the importance of including these threatened areas within the park.

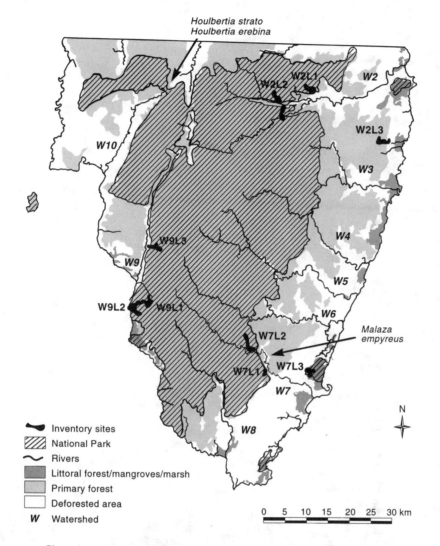

Figure 24.5. Sites where three rare butterfly species were found, but only outside the park boundaries to date. *Houlbertia strato* and *H. erebina* occur in disturbed habitats and are probably not at risk. *Malaza empyreus* occurs over a wide geographic range but at extremely low abundance. It is likely to occur in similar habitat within the park but has not yet been detected.

## BIODIVERSITY SAMPLING LOCALITIES

Nine localities were studied around the peninsula (see figure 24.4; table 24.2). All the localities were situated in primary forests, although some existed adjacent to deforested areas or were almost completely surrounded by deforestation (see table 24.2). At each locality, between six and ten sampling sites were identified along preexisting or new trails. These sites, and the trails themselves, were selected to sample the full elevational range at each locality as well as microhabitat differences within localities (for example, riparian, ridge, slope, bamboo, and flood-plain forest).

Table 24.2

## Characterization of Biodiversity Sampling Localities

| LOCALITY CODE | LOCALITY NAME | NORTH-SOUTH | EAST-WEST | ELEVATIONAL RANGE (M) | PARENT ROCK TYPE | HABITAT TYPE |
|---|---|---|---|---|---|---|
| W2L1 | Sarahandrano[a,b] | N | E | 50–430 | granitic/quartz | rain forest |
| W2L2 | Antsamanarana River[c] | N | E | 50–540 | granitic | rain forest |
| W2L3 | Ambery River[a] | N | E | 10–100 | basalt | littoral forest |
| W7L1 | Manosona River[a,b] | S | E | 20–50 | granitic | rain forest |
| W7L2 | Antafononana River[c] | S | E | 40–330 | granitic | rain forest |
| W7L3 | Andranomainty[a] | S | E | 15 | old sand dunes | littoral forest |
| W9L2 | Andranobe Field Station[c] | — | W | 0–600 | granitic | rain forest |
| W9L1 | Be Dinta Base Camp[c] | — | W | 550–700 | granitic | rain forest |
| W9L3 | Ambohitsitondroina Ambanizana | — | W | 600–1,100 | granitic | rain forest and ericaceous forest |

a  These localities were adjacent to, or nearly surrounded by, deforested areas.

b  Peripheral localities.

c  Interior localities.

BUTTERFLY AND SKIPPER INVENTORY

Butterfly inventories were conducted using three methodologies: trapping with fruit bait, stationary observations, and netting or observing on transect walks. Incidental observations were also made at each inventory locality, as well as in other localities around the peninsula. Methods for trapping and stationary observations are described in Kremen 1994. Traps were set and stationary observations conducted at the pre-identified sampling sites. Transects were walked, and all butterflies within 5 m of the transect were collected and identified, or identified on the wing. The total amount of time spent walking transects was also noted. No transect-walking data were taken at W9L2 and W9L3.

DATA ANALYSIS
AND BIOGEOGRAPHIC COMPARISONS

Each data set (trapping, observations, and transect walks) was combined to develop a presence-absence site by species matrix. Note that the unit of analysis is the sampling site, not the sampling locality. The sampling localities represented in figure 24.4 and table 24.2 simply represent a convenient way of referring to a group of sampling sites. The following comparisons were drawn from the sampling-site data.

- East-west. For sampling sites under 500 m only, all eastern sampling sites were compared to all western sites. Only sites under 500 m were considered because all the east coast sites are lower than 500 m. Otherwise, the east-west comparison could be confounded by differences owing to elevation, given that only west coast localities include areas above 500 m.
- Elevation. Sites at each locality were grouped according to elevation, with three classes: 0–400 m (lowland), 400–700 m (mid-elevation), and over 700 m (montane). These three classes generally conform to changes in vegetative structure and compo-

sition with elevation (Lowry et al. 1997).
- North-south. For localities on the eastern side only, northern sites were compared with southern sites (N = W2L1, W2L2, and W2L3; S = W7L1, W7L2, and W7L3). West coast sites were not included because they all occurred midway in latitude between the other two areas, and a west-east difference could therefore be confounded with a north-south difference.
- Parent rock type. For elevations less than 250 meters for east coast areas only, sites representing the three different parent rock types—granite, basalt, and ancient sand dunes—were compared. Only sites under 250 m on the east coast were included because two of the parent rock categories, basalt and ancient sand dunes, occurred only under such conditions.
- Degree of threat. Sites in the highly threatened area of the eastern forests (adjacent to the deforested zone; see figure 24.5) were compared with sites remote from threat. Sites were restricted to low- and mid-elevation forests on lateritic soils.

In drawing comparisons using presence-absence data (see below), incidental observations were used to "disqualify" a comparison based on the results obtained from the sampling sites. For example, if a species was found in eastern sites only but not in western sites, an incidental observation on the west coast would nonetheless disqualify this species as an "eastern-limited species." Conversely, however, species for which only incidental observations were available were not examined at all in drawing comparisons. Treating incidental observations in this manner is the most conservative way for testing the null hypothesis that no differences occur along the gradients studied (in other words, the use of incidental observations makes it more difficult to falsify the null hypothesis).

## ASSESSING RARITY

Following Rabinowitz et al. (1986), who recognized that a spectrum of rarity states can be defined based on geographic range, habitat breadth, and relative abundance, we classified the rarity of each butterfly species found on Masoala according to the codes given in table 24.3. Classes one through four were considered rare. Rarity defined in the manner of Rabinowitz et al. (1986) is necessarily qualitative and somewhat subjective (Gaston 1994). Here, we considered species to be narrow in geographic range if they were strictly limited to a portion of the eastern rain forest region (such as the Bay of Antongil region or the northern rain forests). Species known to occur throughout the eastern rain forests or in both eastern and western habitats were considered to have a wide geographic range. Species were considered to have narrow habitat breadth if they were strictly limited to lowland, mid-elevation or high-elevation forests; or if they were limited to open areas; or if they were found only within a particular vegetation type. Any species known to occur in more than one of these categories was considered to have broad habitat breadth. The assessment of abundance introduced the greatest degree of subjectivity and simply could not be assessed for many species. Species were classed as abundant if they were known to be found commonly anywhere within their range and as scarce if they were relatively uncommon throughout their range.

A more quantitative method for assessing rarity is to calculate the three-dimensional range size of each species (Gaston 1994). We calculated three-dimensional range sizes for certain species for which data were available (the mycalesine satyrines) by multiplying altitudinal ranges by the number of $1°$ grid squares occupied by the species in Madagascar (Lees et al. 1999). To compare rarity assessments given by the two methods, we then logarithmically scaled the three-dimensional range sizes obtained to a scale of one to seven.

## LIMITATIONS OF THESE DATA

Systematic sampling biases potentially occur in this data set because of the following constraints: (1) certain habitat types were oversampled while others were undersampled relative to their abundance on the peninsula (habitat abundance was estimated using Geographic Information Systems analysis); for example, lowland forest habitat was oversampled by about seven sampling sites, while mid- and high-elevation forest habitats were undersampled by about two to four sampling sites, respectively; (2) some localities were sampled during the early part of the rainy season (November–December), while others were sampled during the late part (January–February); (3) localities W9L2 and W9L3 were sampled in 1991, whereas all other localities were sampled during the 1993–1994 field season; (4) no path-walking data exist for W9L2 and W9L3. Such limitations typify the sampling biases of inventory data because adequate per-

Table 24.3

## Rarity Classification Scheme

| GEOGRAPHIC RANGE | WIDE | WIDE | NARROW | NARROW |
|---|---|---|---|---|
| HABITAT BREADTH | BROAD | NARROW | BROAD | NARROW |
| Relative abundance: somewhere abundant | 7 | 6 | 5 | 2 |
| Relative abundance: everywhere scarce | 4 | 3 | 2 | 1 |

*Source*: Data from Rabinowitz et al. 1986.

*Note:* Classes 1 through 4 are considered here to be rare.

sonnel, funds, and time rarely exist to allow the inventory to be conducted with sufficient replicates in space and time.

## Butterfly Distributional Patterns and Conservation Implications

### SPECIES BY LOCALITY MATRICES

Appendix 24.1 summarizes the data from all three sampling methods as a presence-absence matrix by locality. The incidental observations are also noted. All species endemic to Madagascar are indicated, and the habitat types in which species are most often found (primary forest, secondary forest, and degraded habitats) are noted.

### DIVERSITY, RARITY, AND ENDEMISM

This butterfly inventory of Masoala yielded 135 butterfly species (see appendix 24.1). The species diversity at any given locality varied between 25 and 67 species, with locality W2L1 the most diverse, followed by W9L2. The diversity of the Masoala inventory was similar to that at the Ranomafana National Park area in southeastern Madagascar, where a comparable investment of effort in butterfly inventories yielded approximately 120 species (Kremen 1992, 1994, and unpublished data). An estimated 63% of the species known from the Masoala Peninsula are also found in the Ranomafana region.

Ninety of the Masoala butterfly species are endemic to Madagascar (67.2%). Among these endemic species, the Masoala list includes five new species, currently known only from the Masoala Peninsula (*Masoura* 53 and *Saribia* sp. nov.) or the peninsula plus neighboring areas (*Henotesia* 14a, *Henotesia* 16, *Henotesia* 20). One other species has not yet been definitively identified, and may prove to be a new species limited to Masoala (here provisionally labeled *Strabena* 7), although it may also be a variant of *Strabena batesii*. Another species, *Houlbertia erebennis*,

was formerly more widespread but is now known only from Masoala and the surrounding region. Thus between five and seven butterfly species appear to be locally endemic to the peninsula and neighboring areas (Lees 1997). This region (figure 24.6) seems to be an important area of endemism in Madagascar, for both butterflies and other groups (Dransfield and Beentje 1995). Narrow endemic species merit particular attention as conservation indicators.

Table 24.4 shows the 22 butterfly species found on Masoala that are deemed rare according to the Rabinowitz classification system (see table 24.3). The total number of rare species found at each locality is given in appendix 24.1. These rare species also merit special attention as conservation indicators.

As a test of the Rabinowitz classification scheme, we compared the rarity assessments given by this scheme with those provided by the three-dimensional range-size measure (Gaston 1994) for the subset of butterfly species for which both measures could be calculated (the mycalesine satyrines). The two schemes exhibited a good correspondence, increasing our confidence in the rarity assessments conducted for all the butterflies using the Rabinowitz method. Whereas species often differed slightly in the values assigned by the different methods, no species classified as rare by the Rabinowitz method (that is, with a rarity code less than 4) received a value above 4.45 using the three-dimensional range-size method.

## Regional Comparisons Within Masoala Using Butterflies as Indicators

### EAST-WEST COMPARISON

Ten species of butterflies were found exclusively in the eastern inventory sites, and two species were found only in the western sites (table 24.5). However, of these twelve species, seven were observed less than ten times through the three sampling methods. Because

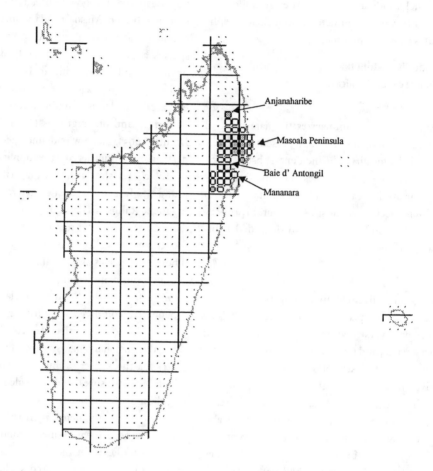

Figure 24.6. Quarter-degree grid cells where five to seven locally endemic butterfly species co-occur (see also table 24.4).

of these low frequencies, it is difficult to state definitively whether these seven species are limited to one region or not, although these findings can be taken as suggestive. Species-richness patterns mirrored uniqueness, with 82 species in the east and 55 in the west (not including incidentally observed species).

Among the four more abundant species, *Henotesia* 14a, a new species known only from the Bay of Antongil region, was found abundantly at five of the six eastern study sites (N=105) but never in the western study sites.

*Henotesia bicristata,* a species limited to the northeastern rain forest and the Sambirano region of northwest Madagascar (historically known from Ile St. Marie), was also found only in the eastern region. *Galerga hyposticta* was limited to the west and *Gnophodes betsimena* was limited to the east; both these species are widely distributed at low to moderate densities elsewhere in Madagascar. *Masoura* 53 (a new species endemic to Masoala [Lees 1997]) also appeared to be limited to its eastern region.

Certain species showed trends toward

Table 24.4

# Rare Species of Masoala

| SPECIES | GEOGRAPHIC RANGE[a] | HABITAT BREADTH[b] | RELATIVE ABUNDANCE[c] | RABINOWITZ RARITY CODE[d] | HABITAT | LOCAL ENDEMISM |
|---|---|---|---|---|---|---|
| *Houlbertia erebennis* | N | N | S | 1 | lowland rain forest | Baie d'Antongil |
| *Saribia* sp. nov. | N | N | S | 1 | lowland rain forest | Masoala Peninsula |
| *Acraea sambavae* | N | ? | S | 1 or 2 | rain forest | central eastern forests |
| *Henotesia 20* | N | N | ? | 1 or 2 | montane forest | Masoala to Anjanaharibe |
| *Henotesia 14a* | N | N | A | 2 | rain forest | Eastern Masoala Mananara |
| *Henotesia 16* | N | N | A | 2 | rain forest | Mananara to Anjanaharibe |
| *Henotesia sabus* | N | N | A | 2 | montane forest | Antongil to Anjanaharibe |
| *Houlbertia erebina* | N | N | A | 2 | disturbed forests | northeastern forests |
| *Houlbertia strato* | N | N | A | 2 | disturbed forests | northeastern forests |
| *Masoura 53* | N | B | S | 2 | rain forest | eastern Masoala Peninsula |
| *Charaxes andranodorus* | W | N | S | 3 | rain forest | |
| *Charaxes cowani* | W | N | S | 3 | rain forest | |
| *Cymothoe lambertoni* | W | N | S | 3 | rain forest | |
| *Gnophodes betsimena* | W | N | S | 3 | rain forest | |
| *Hypolimnas anthedon drucei* | W | N | S | 3 | rain forest | |
| *Hypolimnas dexithea* | W | N | S | 3 | rain forest | |
| *Malaza empyreus* | W | N | S | 3 | rain forest | |
| *Papilio mangoura* | W | N | S | 3 | rain forest | |
| *Sallya hopensis* | W | N | S | 3 | rain forest | |

Table 24.4 (*continued*)

| SPECIES | GEOGRAPHIC RANGE[a] | HABITAT BREADTH[b] | RELATIVE ABUNDANCE[c] | RABINOWITZ RARITY CODE[d] | HABITAT | LOCAL ENDEMISM |
|---|---|---|---|---|---|---|
| *Sallya madagascariensis* | W | N | S | 3 | rain forest | |
| *Saribia decaryi* | W | N | S | 3 | rain forest | |
| *Acraea hova* | W | B | S | 4 | rain forest | |

a  N=narrow; W=wide.
b  N=narrow; B=broad.
c  S=scarce; A=abundant; ?=unknown.
d  For rarity codes, see table 24.3.

Table 24.5

# Species Richness and Uniqueness Across Environmental Gradients

| ENVIRONMENTAL GRADIENT | | | SPECIES RICHNESS | | | UNIQUE SPECIES[a] | | | RARE SPECIES | | |
|---|---|---|---|---|---|---|---|---|---|---|---|
| ZONE 1 | ZONE 2 | ZONE 3 | ZONE 1 | ZONE 2 | ZONE 3 | ZONE 1 | ZONE 2 | ZONE 3 | ZONE 1 | ZONE 2 | ZONE 3 |
| East | West | — | 82 | 55 | — | 10 (4) | 2 (1) | — | 5 | 0 | — |
| Low | Mid | High | 89 | 62 | 32 | 32 (6) | 1 (1) | 6 (1) | 3 | 1 | 2 |
| North | South | — | 75 | 61 | — | 20 (20) | 4 (1) | — | 6 | 1 | — |
| Granite | Basalt | Sand | 75 | 52 | 24 | 16 (5) | 2 (0) | 1 (1) | 2 | 1 | 0 |
| Periphery | Interior | — | 72 | 80 | — | 3 (0) | 11 (3) | — | 0 | 3 | — |

a  The subset of unique species with more than ten observations is given in parentheses.

greater abundance in one region. *Appias sabina confusa, Melanitis leda,* and *Mylothris phileris* all had higher relative abundances in the west, while *Leptosia alcesta, Henotesia uniformis,* and *Salamis anacardii duprei* were more frequently observed in the eastern region. All these species are also broadly distributed throughout the eastern rain forests of Madagascar.

In summary, the butterfly inventory shows that regional differences exist between the fauna of the eastern and western regions of the Masoala Peninsula. We can confidently assert that at least one new butterfly species (*Henotesia* 14a) is limited on Masoala to its eastern region. *Masoura* 53 may also be limited to the east. Conservation decision making should take into account the available data on species that are rare, newly discovered, potentially endemic, or limited to threatened habitat types, so as to ensure that the ensemble of protected areas fully represents associated biodiversity.

ELEVATIONAL COMPARISON

In the course of this study, thirty-two species were recorded uniquely in the lowland zone, one uniquely from the mid-elevation zone, and six from the high-elevation zone only (see table 24.5). Species richness declined at high altitude. Of the species recorded only in the lowland zone, it was noted that more than half are commonly found in degraded areas and were recorded at low frequency in this study (table 24.6). Most of the degraded habitat on the Masoala Peninsula also occurs in the lowlands. Because the study sites themselves were located in primary forests, it is therefore logical to conclude that the species found only at lowland sites that are common to degraded areas had invaded primary forest habitats from adjacent disturbed areas. These species were consequently not considered to be restricted to lowland primary forest habitat (in fact, the authors have observed many of these species in disturbed habitats at higher elevations elsewhere in Madagascar). Rather, their presence in the lowland primary forest sites reflects the distribution of degraded habitats on Masoala.

Among the fourteen other species found only at the lowland inventory sites, only four were observed more than ten times (*Charaxes cacuthis, Salamis anteva, Saribia perroti,* and *Saribia tepahi;* see table 24.5). All these species are widely distributed in the eastern rain forests, where they have also been observed at higher elevations (for example, at Ranomafana National Park in southeastern Madagascar). Of the infrequently noted species recorded only in the low-elevation zone, four are potentially of conservation interest: *Charaxes andara,* a species that is extremely rare in the eastern rain forests of Madagascar, although abundant in western dry forest (F. Hawkins, pers. comm.; Jenkins 1990), *Houlbertia erebennis,* a species now known only from the lowland forests of Masoala and Mananara (also recorded historically from Ile St. Marie), *Hypolimnas dexithea,* well distributed in eastern rain forests but at low densities, and *Saribia* sp. nov., a new species known only from Masoala.

*Acraea sambavae,* the only species limited to

Table 24.6

## The Relationship Between Habitat Degradation and Elevation for Species Limited to Elevational Zones

| ELEVATIONAL ZONE | NUMBER OF LIMITED SPECIES | NUMBER TOLERATING DISTURBED CONDITIONS |
|---|---|---|
| Low | 32 | 18 |
| Mid | 1 | 0 |
| High | 6 | 0 |

the mid-elevation zone, was observed ten times. While known from other eastern rain forests, this species is thought to be scarce throughout its range (Jenkins 1990; Pierre 1994).

For five of the six species limited to high-elevation forests, the number of individuals recorded was below ten. Despite the paucity of the data, available evidence suggests that these species are limited to the high-elevation forest type. *Henotesia exocellata*, observed only once during this study, is always associated with elevations greater than 1,000 m at all other localities in Madagascar (Lees et al. 1999). The other four species were frequently noted through incidental observations at the higher elevations. They were never observed incidentally at lower elevations. In general, the low sampling counts for the high-elevation species resulted from the poor weather encountered at the high-elevation sites, and from the fact that fewer high-elevation sites were sampled during the study relative to the amount of high-elevation habitat occurring on the peninsula. Considering these factors, a surprising number of species were found to be limited to the high-elevation sites. Species of high conservation interest from the high-elevation zone include *Henotesia* 20 and *Henotesia sabas*, both known only from Masoala and the neighboring reserve to the north, Anjanaharibe (see figure 24.6), and *Strabena* 7.

Certain species showed elevational trends in their abundance distributions. Seven species were much more abundant in lowland forests, falling off in abundance with increasing elevation. Of these seven species, *Acraea ranavalona*, *Henotesia strigula*, and *Eurema floricola* are commonly found in primary and secondary forests, and their enhanced abundance in the lowland areas may be in part related to their ability to tolerate adjacent disturbed habitats. *Henotesia* 14a (endemic to the peninsula) and *Henotesia* 16 (endemic to the larger region shown in figure 24.6) showed strong preferences for lowland forest habitat. *Henotesia*

*undulans* showed its peak abundance in mid-elevation forests, and *Mylothris phileris* was most abundant in the high-elevation forests.

In summary, the butterfly inventory shows that faunal differences occur between the low-, mid- and high-elevation forest habitats of the Masoala Peninsula. In addition, the low and high elevational bands appear to be characterized by several butterfly species that are potentially endemic to the Masoala and neighboring areas (low: *Henotesia* 14a, *Henotesia* 16, *Houlbertia erebennis*, and *Saribia* sp. nov.; high: *Henotesia sabas*, *Henotesia* 20, and *Strabena* 7).

NORTH-SOUTH COMPARISON

Twenty species were recorded only in the north, and four were recorded only in the south, with species-richness patterns paralleling uniqueness (see table 24.5). Again, only a few of these species (six) were recorded in sufficient numbers to draw conclusions. Three *Acraea* species were found in the north but not in the south, including the rare species *Acraea sambavae*. *Graphium endochus* and *Junonia goudoti* were also observed only in the north. *Charaxes cowani* and *Sallya madagascariensis*, both rare species, were found only in the south. *Charaxes cowani*, though listed as known only from a forest 32 km south of Ambositra (Jenkins 1990), is in fact widespread in the eastern rain forests of Madagascar. Its status with respect to *Charaxes antambalou* needs investigation. All these species are found throughout the eastern rain forest region. Another rare species apparently limited to the north was *Cymothoe lambertoni*, an extremely rare species with a disjunct distribution in Madagascar that was previously known from northern sites (Montagne d'Ambre, Tsaratanana) and from one collection record in the southeast (Viette 1971; Lees 1997). *Henotesia undulans* was found in greater abundance in the north than the south, whereas *Charaxes cacuthis* showed the opposite pattern.

In general, the north-south comparison supported the prediction of a "peninsular effect"—certain species found in the north of the peninsula drop out as one proceeds south. Most of these species, however, were also found in the western sites. Supporting the concept of the peninsular effect, a recent inventory of the northern edge of the peninsula (along W10, see figure 24.5) added seven species to the butterfly list. Several species found just north of the peninsula along the coastal road to Antalaha (such as *Colotis mananhari*) were never found on the peninsula. In fact, many other species and even genera common to the eastern rain forest biome are strikingly absent from Masoala, among them the genera *Hovala, Perrotia,* and *Miraja* (Hesperiidae); *Heteropsis* and *Admiratio* (Satyrinae); *Lepidochrysops* (Lycaenidae); and many *Strabena* species (Satyrinae) (Lees 1997; Lees et al. 1999).

SOIL TYPE COMPARISON

Sixteen species were found only on granite-based soils, of which five species were observed more than ten times (see table 24.5). However, given that the majority of sampling sites—and indeed, the majority of the peninsula—are on granite-based soils, the presence of species found only on granite is more likely to be due to sampling error than to true differences (in other words, these species exist on the other soil types but were not detected there). Only two species were found to be "unique" to basalt and one "unique" to sandy soils. However, these three species were also found on granite-based soils above 250 m (recall that granite sites above 250 m were excluded from the comparison because sandy and basalt soils do not occur above this elevation). Thus no species were found to be exclusively limited to the basalt or sandy soil types. The lack of species unique to the rarer soil types is surprising, and suggests that soil diversity does not tend to enhance the butterfly diversity on the peninsula. In fact, the forests found on basalt and sandy soils were

generally depauperate compared to those found on granite-based soils; however, this finding may have been caused by the fragmented nature of these forests (see figure 24.5).

DEGREE OF THREAT

Only three species were observed uniquely in the highly threatened peripheral zone, and none of these species were observed more than ten times. All these species were widespread species, and two were associated with disturbed or secondary forest conditions. Thus no species of special conservation concern appeared to be limited to this zone. In contrast, the interior region of forest had eleven species not found in the peripheral zone, of which three (*Acraea sambavae, Houlbertia erebennis,* and *Henotesia sabas*) were considered rare. Most of the other species unique to the interior zone were forest-restricted species.

COMPARISON WITH INVENTORY DATA FROM OTHER INSECT GROUPS

Regional trends observed in butterfly distributions were mirrored by those found for cicindelid and scarab distributions, although regional differences were much more strongly marked in these other groups (Razafimahatratra and Andriamampianina 1995; Andriamampianina 1995). As with the butterfly comparisons, cicindelids had a greater number of unique species in the lowlands, in the east, in the north, and on granite soils. The cicindelid community was also depauperate on both basalt and sandy soils. The scarab species distribution showed identical trends, except for the east-west comparison, in which the greatest number of unique species was found in the west. For both taxa, regions with higher uniqueness always had higher species richness. For each regional comparison for each taxonomic group, a high percentage of species (between 43% and 90%) was limited to one zone. Scarab and cicindelid species distributions thus showed marked species turnover along the environmental gradients investigated.

## Conservation Implications

This butterfly inventory showed that in order to protect the full complement of rare and locally endemic butterfly species on the Masoala Peninsula, it is necessary to include examples of eastern, western, northern, southern, lowland, mid-elevation, and high-elevation forests of this area. In contrast, the peripheral zone of the eastern forest did not add rare or locally endemic species. These results, along with inventory data from other taxa, showed that the rain forests of the Masoala Peninsula are not uniform in floral and faunal distributions and aided (1) in justifying the need for a large park encompassing the distinct regions of Masoala, and (2) defining the area that could be left out of the park as a buffer zone without omitting sensitive species.

Specifically, table 24.7 and figure 24.7 show how the inclusion of various sampling localities within the park added the butterfly species of greatest concern to the list of species that will be protected. Using the principle of complementarity (Vane-Wright et al. 1991; Pressey et al. 1993; Colwell and Coddington 1994), sampling localities were added sequentially based on maximizing the cumulative number of rare species thereby protected (see table 24.7). These sampling localities comprised three bio-inventory localities inside the proposed park boundaries (W9L2, W2L2, and W9L3).

All but three of the rare species would be protected by the three localities within the park. Of the three remaining species, two (*Houlbertia erebina* and *H. strato*) occur in disturbed habitats in the northern part of the peninsula (see figure 24.5 and table 24.7). Habitat for these two species is probably not at risk even if it is not included within the park boundaries. The third species, *Malaza empyreus*, a widespread but extremely scarce species, was observed only once in an area just outside the proposed park boundaries (see figure 24.5 and table 24.7). A large area of similar habitat is included within the park boundaries, and future inventories may locate the species elsewhere.

By then examining the full list of species that would be protected by including these

Table 24.7

## Rare Species as Indicators of Conservation Importance

| LOCALITY | RARE SPECIES ADDED | CUMULATIVE RARE SPECIES PROTECTED | CUMULATIVE PRIMARY FOREST SPECIES (NON-RARE) PROTECTED | CUMULATIVE SPECIES PROTECTED |
|---|---|---|---|---|
| **Inside Park Boundaries** | | | | |
| W9L2, Andranobe, western low- and mid-elevation forests | +12 species | 12 | 25 | 65 |
| W2L2, Antsamanarana, northeastern low- and mid-elevation forests | +16 species | 18 | 29 | 86 |
| W9L3, Ambohitsitondroina Ambanizana, western high-elevation forests | +1 species | 19 | 33 | 91 |
| **Outside Park Boundaries** | | | | |
| W10, north of forest corridor | +2 species | — | — | — |
| north of W7L1, eastern lowland forest | +1 species | — | — | — |

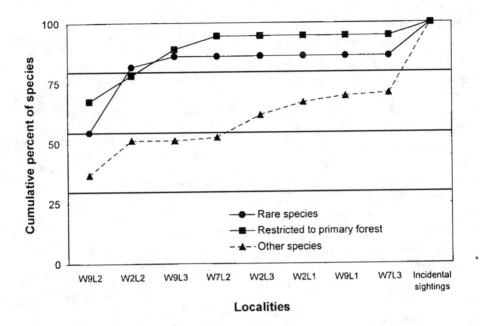

Figure 24.7. Evaluating the Masoala National Park design: conservation of key localities for butterfly rarity, endemism, and diversity.

three sites within the park, it was determined that the majority of the peninsular butterfly species would receive protection (91 of 135 species; see table 24.7). Of the remaining 44 species, only 2 occur exclusively in primary forests, and the rest are exclusive to, or also occupy, degraded or secondary forest habitats. These species are not at risk of extinction.

Fifty-two of the 55 species restricted to primary forests would be protected at one of the three sites (see table 24.7). Thus, using the rare and locally endemic species as indicators worked well in this case to predict the areas of conservation importance: these three localities harbored 95% of the butterfly biodiversity of concern. The entire park would protect all but one of the forest-restricted butterflies of the Masoala Peninsula (*Malaza empyreus;* see figure 24.5). Comparisons with other taxonomic groups showed that the rare, vulnerable, or

endangered species of birds and mammals would also be protected.

Interestingly, the bird and primate data did not differentiate environmental gradients on Masoala to the same extent that the insect data did (Kremen et al. 1999). Although differences in species composition between inventory localities were certainly evident, few birds or primates were found to be limited to any one biogeographic zone within the Masoala Peninsula. Evidently, many bird and primate species, being either more wide ranging or more catholic in their tastes, do not respond to these environmental differences at the same spatial scale as many insect species. The insect taxa, therefore, proved to be the most informative indicators for examining patterns in the distribution of biodiversity on the spatial scale required for the Masoala National Park planning problem. The bird and primate inventories, however, led to the discovery of several potential flagship species with great conservation value, such as the serpent eagle, *Eutriorchis astur* (Thorstrom et al. 1995), and hence to an evaluation of the park plan on the basis of these species' area requirements (Kremen et al. 1995; Merenlender et al. 1998), a more traditional approach (Shaffer 1981).

Should this approach to evaluating park design be applied in African rain forests? On the one hand, studies of the area and ecological requirements of the large mammal fauna of the African rain forests can inform the conservation planning process in an efficient manner by providing umbrella indicators. Large-scale censuses that focus on one to several large mammal species can be conducted so as to cover large areas relatively rapidly. In contrast, the Masoala inventory could sample only about 0.5% of the entire forested area, since it was necessary to stay at each locality for approximately two to three weeks to sample the diversity of target groups adequately.

On the other hand, by focusing on the large mammal fauna alone, African conservationists may risk missing fine-scaled geographic variation in the distribution of biological diversity (Kremen et al. 1993), as this chapter demonstrates. Gradients of environmental variation ought to be investigated using plant, insect, reptile, or amphibian indicator assemblages that respond to fine-scaled geographic variation, to ensure that areas of local endemism and unique communities and habitats are not excluded from protected area networks. Ideally, a variety of taxonomic assemblages should be studied because distributional patterns are not always congruent between groups (Prendergast et al 1993; Humphries et al. 1995).

ACKNOWLEDGMENTS

This research was conducted under the authority of the Masoala Project, implemented by CARE International, the Wildlife Conservation Society, and the Peregrine Fund under the supervision of Madagascar's Direction des Eaux et Forêts and Association National pour la Gestion des Aires Protégées. Funding was provided by the USAID-SAVEM project, the Biodiversity Support Program, the MacArthur Foundation, the Wildlife Conservation Society, and the National Geographic Society. The Stanford Center for Conservation Biology developed the Masoala Geographic Information Systems, which produced the illustrations in this chapter.

REFERENCES

Abraham, J., R. Benja, M. Randrianasolo, J. U. Ganzhorn, V. Jeannoda, and E. G. Leigh. 1996. Tree diversity on small plots in Madagascar: A preliminary review. *Revue d'Ecologie (Terre et Vie)* 51:93–116.

Andriamampianina, L. 1995. Contribution à l'étude de la repartition et de la structure des peuplements des insectes Scarabeidae (Coleoptères) de la Presqu'ile de Masoala. DEA thesis, Faculté de Sciences, Université d'Antananarivo, Madagascar.

Coddington, J. A., C. E. Griswold, D. S. Dávila, E. Peñaranda, and S. F. Larcher. 1991. Designing and testing sampling protocols to estimate biodiversity in tropical ecosystems. Page 1048 in E. C. Dudley, ed. *The Unity of Evolutionary Biology: Proceedings of the Fourth International Congress of Systematic and Evolutionary Biology.* Dioscorides Press, Portland.

Colwell, R. K, and J. A. Coddington. 1994. Estimating terrestrial biodiversity through extrapolation. *Philosophical Transactions of the Royal Society of London B* 345:101–118.

di Castri, F., J. R. Vernhes, and T. Younés. 1992. A proposal for an international network on inventorying and monitoring of biodiversity. *Biology International* 27:1–27.

Dransfield, J., and H. Beentje. 1995. *The Palms of Madagascar.* Royal Botanical Gardens, Kew, and the Palm Society, London.

Du Puy, D. J., and J. Moat. 1996. A refined classification of the primary vegetation of Madagascar based on the underlying geology: Using GIS to map its distribution and to assess its conservation status. Pages 205–218 in W. R. Lourenc, ed. *Biogéographie de Madagascar.* ORSTOM, Paris.

Gaston, K. J. 1994. *Rarity.* Chapman and Hall, London.

Gentry, A. H. 1993. Diversity and floristic composition of lowland tropical forest in Africa and South America. Pages 500–547 in P. Goldblatt, ed. *Biological Relationships Between Africa and South America*. Yale University Press, New Haven.

Hottin, G., and E. Liandrat. 1964. Géologie de la presqu'île de Masoala: Travaux du bureau géologique no. 118. Service Géologique, Antananarivo, Madagascar.

Humphries, C. J., P. H. Williams, and R. I. Vane-Wright. 1995. Measuring biodiversity value for conservation. *Annual Review of Ecology and Systematics* 26:93–111.

Jenkins, M. D. 1990. *Madagascar: Profil de L'Environnement*. UICN-L'Alliance Mondiale pour la Nature, Gland, Switzerland.

Kremen, C. 1992. Assessing the indicator properties of species assemblages for natural areas monitoring. *Ecological Applications* 2:203–217.

———. 1994. Biological inventory using target taxa: A case study of the butterflies of Madagascar. *Ecological Applications* 4:407–422.

Kremen, C., R. Colwell, T. L. Erwin, D. D. Murphy, R. F. Noss, and S. Muttulingam. 1993. Arthropod assemblages: Their use as indicators for biological inventory and monitoring programs. *Conservation Biology* 8:796–808.

Kremen, C., J. Rakotomalala, V. Razafimahatratra, O. Rakotobe, P. Guillery, A. Zakandraina, and Jaomanana. 1995. *Proposition des limites du Parc National Masoala*. CARE International, Wildlife Conservation Society, and Peregrine Fund, Antananarivo, Madagascar.

Kremen, C., V. Razafimahatratra, R. P. Guillery, J. Rakotomalala, A. Weiss, and J. Ratsisompatrarivo. 1999. On scientific grounds: Designing the Masoala National Park in Madagascar. *Conservation Biology* 13:1055–1068.

Lees, D. C. 1997. Systematics and biogeography of Madagascar mycalesine butterflies and applications to conservation evaluation. Ph.D. thesis, University of London, London.

Lees, D. C., C. Kremen, and L. Andriamampianiana. 1999. A null model for species richness gradients: Bounded range overlap of butterflies and other rainforest endemics in Madagascar. *Biological Journal of the Linnean Society* 67:529–584.

Lowry, P. P., G. E. Schatz, and P. B. Phillipson. 1997. The classification of natural and anthropogenic vegetation in Madagascar. Pages 93–123 in S. M. Goodman and B. D. Patterson, eds. *Natural change and human impact in Madagascar*. Smithsonian Institution, Washington, D.C.

Merenlender, A., C. Kremen, and M. Rakotondratsima. 1998. Monitoring impacts of natural resource extraction on lemurs on the Masoala Peninsula, Madagascar. *Conservation Ecology* 2: www.consecol.org/Journal/vol2/iss2/art5/.

Pierre, J. 1994. Les Acraeidae des Comores et de Madagascar (Lepidoptères Rhopalocères). *L'Entomologiste* 48:351–364.

Prendergast, J. R., R. M. Quinn, J. H. Lawton, B. C. Eversham, and D. W. Gibbons. 1993. Rare species, the coincidence of diversity hotspots and conservation strategies. *Nature* 365:335–337.

Pressey, R. L., C. J. Humphries, C. R. Margules, R. I. Vane-Wright, and P. H. Williams. 1993. Beyond opportunism: Key principles for systematic reserve selection. *Trends in Ecology and Evolution* 8:124–128.

Rabinowitz, D., S. Cairns, and T. Dillon. 1986. Seven forms of rarity and their frequency in the flora of the British Isles. Pages 182–204 in M. E. Soule, ed. *Conservation Biology: The Science of Scarcity and Diversity*. Sinauer Associates, Cambridge, Mass.

Rakotindraibe, F., and F. Raharimalala. 1994. *Contribution à L'Etude de la Flore des Aires Protégées de Madagascar: Les Ptéridophytes de la Presqu'île Masoala*. Revue du Centre National de la Recherche sur l'Environnement, Antananarivo, Madagascar.

Razafimahatratra, V., and L. Andriamampianina. 1995. Inventaire des deux groupes d'insectes Coleoptères: Scarabeidae et Cicindelidae de la presqu'île Masoala et comparaisons biogéographiques. Annexe V.4. In C. Kremen, ed. *Proposition des Limites du Parc National Masoala*. Care International Madagascar, the Wildlife Conservation Society, and the Peregrine Fund, Antananarivo, Madagascar.

Razafindrakoto, Y. 1995. Contribution à l'étude des rongeurs et insectivores de la presqu'île de Masoala. DEA thesis, Université d'Antananarivo, Madagascar.

Ricklefs, R. E. 1973. *Ecology*. W.H. Freeman, New York.

Scott, J. M., B. Csuti, J. D. Jacobi, and J. E. Estes. 1987. A geographic approach to protecting future biological diversity. *BioScience* 37:782–788.

Shaffer, M. L. 1981. Minimum population sizes for species conservation. *BioScience* 31:131–134.

Sterling, E. J., and N. Rakotoarison. 1998. Rapid assessment of primate species richness and density on the Masoala Peninsula, eastern Madagascar. *Folio Primatologica* 69:109–116.

Thorstrom, R., and R. T. Watson. 1997. Avian inventory
and key species of the Masoala Peninsula, Madagascar.
*Bird Conservation International* 7:99–115.

Thorstrom, R., R. T. Watson, B. Damary, F. Toto, M.
Baba, and V. Baba. 1995. Repeated sightings and first cap-
ture of a live Madagascar serpent-eagle *Eutriorchis astur*.
*Bulletin of the British Ornithologist's Club* 115:40–45.

Vane-Wright, R. I., C. J. Humphries, and P. H. Williams.
1991. What to protect? Systematics and the agony of
choice. *Biological Conservation* 55:235–254.

Viette, P. 1971. Note sur *Cymothoe lambertoni* Ch.
Oberthur. *Bulletin de la Société Entomologique de France*
76:32–34.

Whittaker, R. H. 1975. *Communities and Ecosystems*.
Macmillan, New York.

World Bank, United States Agency for International
Development, Cooperation Suisse, UNESCO, United
Nations Development Program, and Worldwide Fund for
Nature. 1988. *Madagascar Environmental Action Plan*.
World Bank, Washington, D.C.

## Species by Locality Matrix

| SPECIES | ENDEMISM[a] | HABITAT TYPE[b] | W2L1[c] | W2L2[c] | W2L3[c] | W7L1[c] | W7L2[c] | W7L3[c] | W9L1[c] | W9L2[c] | W9L3[c] | INCIDENTAL SIGHTINGS[d] |
|---|---|---|---|---|---|---|---|---|---|---|---|---|
| Acleros leucopyga | e | p | i | | | | x | x | | i | | W7 |
| Acraea dammii | e | p | x,i | x | | | x | | x | x | | W7, W9, NM |
| Acraea encedon | e | d | | | | | | | | | | W10 |
| Acraea eponina | e | d | | | | | | | | | | W7 |
| **Acraea hova** | **e** | **p** | x | x | x,i | | x | | x,i | x | x | **NM** |
| Acraea igati | e | p | x,i | | | | | | i | x,i | x | W2, NM |
| Acraea mahela | e | d | | | | | | | | | | W2 |
| Acraea masamba | e | p | x,i | x | | | | | x,i | x | x | W10, NM |
| Acraea ranavalona | e | p/s/d | x,i | x | x,i | x | x | x | x,i | x | | W2, W3, W7, W9, W10, NM |
| **Acraea sambavae** | **e** | **p** | | x,l | | | | | x | l | | W2 |
| Acraea zitja | e | d | | | x | | | | | | | W2, W3 |
| Appias epaphia | e | d | | | | | | | | | | W2, W3 |
| Appias sabina confusa | e | p | x,i | | x,i | x | x | | x,i | x,i | x | W2, W3, W7, W9, W10, NM |
| Artitropa boseae | e | d | | | | | | | | | | W2 |
| Aterica rabena | e | p/s | x,i | | x | x | x | x | x | x | x,i | W2, W7, W9, W10, NM |
| Azanus sitalces | e | d | | | | x | | x | | | | W2 |
| Belenois helcida | e | p | x | | x | x | x | | i | x,i | x | W2, W7, W10, NM |
| Borbo borbonica | e | d | x | x | | | | | | | | W2, W7 |
| Byblia anvatara | e | d | x | | | | | | | | | W2, W3, W7, W9 |
| Cacyreus darius | e | p/s | | | | | | | x | | | W2, W9, W10 |
| Catopsilia thauruma | e | p/s | x | x | | | | | | | | W3, W10 |
| Celaenorrhinus humbloti | e | p | | | | | | | | | | W9, W10 |
| Charaxes analava | e | p | x | | x | x | x | | x | x | x | |
| Charaxes andara | e | p | | x,i | | | | | | x,i | | W9 |

APPENDIX 24.1 (continued)

| SPECIES | ENDEMISM[a] | HABITAT TYPE[b] | W2L1[c] | W2L2[c] | W2L3[c] | W7L1[c] | W7L2[c] | W7L3[c] | W9L1[c] | W9L2[c] | W9L3[c] | INCIDENTAL SIGHTINGS[d] |
|---|---|---|---|---|---|---|---|---|---|---|---|---|
| **Charaxes andranodorus** | **e** | **p** | x | x | x | x | x | x | x | x | x | **W9** |
| Charaxes antambalou | e | p | x,i | x | x | x | x | x |  | . | x,i | W10 |
| Charaxes cacuthis | e | p | x | x | x | x | x | x |  | x | x | W9 |
| **Charaxes cowani** | **e** | **p** |  |  |  | x | x | x |  | x | x |  |
| Charaxes zoolina | e | p | x,i | x | x | x | x |  |  | x |  | W9 |
| Coeliades fervida | e | p/s/d | x | x | x | x |  |  |  |  |  | W2, W9, W10 |
| Coeliades forestan | e | p/s/d |  |  |  | i |  |  |  | i |  | W2, W3, W7 |
| Coeliades rama | e | p/s/d |  |  |  |  |  |  | x |  |  | W2 |
| Coeliades ramanatek | e | p/s/d |  |  | x |  |  |  |  |  |  | W2, W3, W7, W10, NM |
| **Cymothoe lambertoni** | **e** | **p** |  | x | x,i |  |  |  |  |  |  |  |
| Cyrestis camillus elegans |  | p/s | x,i |  | x |  | x |  |  |  |  | W3, W7, W9, W10 |
| Danaus chrysippus |  | d |  |  | i |  |  |  |  |  |  | W2, W3, W7, W10, W9 |
| Deudorix antalus |  | s |  |  |  |  |  |  |  |  |  | W7 |
| Eagris sabadius |  | p/s |  | x,i |  |  |  |  | x |  |  | W2, W3, NM |
| Eicochrysops hippocrates |  | s/d |  | i |  |  |  |  |  | i | i | W2, W7 |
| Eicochrysops sanguigutta |  | p/s | x | i |  |  |  |  |  | i | i | NM |
| Euchrysops malathana |  | s/d |  |  |  |  |  |  |  |  |  | W2, W3 |
| Eurema brigitta pulchella |  | d |  |  | i |  |  |  |  |  |  | W2, W3, W7, W10 |
| Eurema desjardinsi |  | d |  |  |  |  |  |  |  |  |  | W7 |
| Eurema floricola |  | p/s | x | x | x | x | x,i | x | x | x,i |  | W2, W7, W10, NM |
| Eurytela dryope |  | s/d |  | x | x |  | x |  |  |  |  | W2 |
| Eurytela narinda |  | p/s | x | x | x | x | x | x |  | x |  |  |
| Euxanthe madagascariensis |  | p | x | x | x | x | x |  |  | x |  | W7, W9, NM |
| Fulda coroller | e | p/s/d | x | x,i |  |  |  |  | x,i | x |  | W7, W2, W3, W9, W10 |
| Fulda rhadama | e | s/d | x |  | x |  |  |  |  |  |  | W9, W7, W3, W10, W2 |

| SPECIES | ENDEMISM[a] | HABITAT TYPE[b] | W2L1[c] | W2L2[c] | W2L3[c] | W7L1[c] | W7L2[c] | W7L3[c] | W9L1[c] | W9L2[c] | W9L3[c] | INCIDENTAL SIGHTINGS[d] |
|---|---|---|---|---|---|---|---|---|---|---|---|---|
| Galerga hyposticta | e | p | | | | | | | x, i | i | | W10 |
| **Gnophodes betsimena** | **e** | **p** | **x** | **x, i** | **x, i** | | x | | | | | |
| Graphium cyrnus | e | p/s | x, i | x | x | x | x | x | x, i | x | | W7, W9, W10, NM |
| Graphium endochus | e | p | x | x | | | | | x, i | x | x | W9, W10, NM |
| Graphium evombar | e | p | x | x | x | x | | | x | x, i | x, i | W7 |
| **Henotesia 14a** | **e** | **p/s** | **x, i** | **x, i** | **x, i** | **x** | **x** | | **x, i** | **x** | | **W2, W7** |
| **Henotesia 16** | **e** | **p** | **x, i** | **x, i** | **x, i** | **x** | **x, i** | | **x, i** | **x, i** | **x** | **W9, W10** |
| **Henotesia 20** | **e** | **p** | | | | | | | | | **x, i** | **W10, high elev.** |
| Henotesia ankaratra | e | s/d | | | x, i | | | | | | | W2 |
| Henotesia bicristata | e | p | | x, i | | | | x | | | | W2 (littoral forest) |
| Henotesia exocellata | e | p | | | | | | | | | x, i | |
| Henotesia iboina | e | p/s | x, i | x, i | | x | x, i | x | x | x, i | x | W7, W9, W10 |
| Henotesia laeta | e | s/d | | | | | | | | | | W10 |
| Henotesia maeva | e | p/s | x, i | x | x, i | x | x | x | | x | | W7, W9, W10, NM, W2, W3 |
| Henotesia narcissus fraterna | e | p/s/d | x | x | x | x | x | x | | | | W7, W9, W10, NM, W2, W3 |
| Henotesia obscura | e | p | x, i | x, i | x, i | x | x | | x, i | | | W2, W7 |
| Henotesia parva | e | p | | | | | | | | | | W10, high elev. |
| Henotesia pauper | e | p/s | x, i | x, i | x, i | x | x | x | x | x, i | x, i | W2, W7, W9, W10, NM |
| **Henotesia sabas** | **e** | **p** | | | | **x** | **x** | | **x** | **i** | **x, i** | **W10** |
| Henotesia strigula | e | s/d | x, i | x, i | x | x | x | | | x | | |
| Henotesia subsimilis | e | p/s | | x | | x | x | | x | x, i | | W7, W2, W9 |
| Henotesia undulans | e | p | x, i | x, i | | x | x | | x | | | W10 |
| Henotesia undulosa | e | p | | | | | | | | x, i | | |
| Henotesia uniformis | e | p/s | x, i | x, i | x, i | x | x | x | x | x, i | x | W9 |
| Henotesia vola | e | p | i | i | x, i | x | x | x | x | x, i | | |

APPENDIX 24.1 (*continued*)

| SPECIES | ENDEMISM[a] | HABITAT TYPE[b] | W2L1[c] | W2L2[c] | W2L3[c] | W7L1[c] | W7L2[c] | W7L3[c] | W9L1[c] | W9L2[c] | W9L3[c] | INCIDENTAL SIGHTINGS[d] |
|---|---|---|---|---|---|---|---|---|---|---|---|---|
| **Houlbertia erebennis** | **e** | **p** | | **x, l** | | | | | | | | **W10** |
| **Houlbertia erebina** | **e** | **s/d** | | | | | | | | | | **W10** |
| Houlbertia passandava | e | p | x, i | x, i | x, i | x | x, i | | x, i | x, i | | W2, W7, W10 |
| **Houlbertia strato** | **e** | **s/d** | | | | | | | | | | **W10** |
| **Hypolimnas anthedon drucei** | **e** | **p/s** | **x, l** | **x** | **l** | | | | | **l** | | **W2, W10** |
| **Hypolimnas dexithea** | **e** | **p** | **x** | **x** | **l** | | | | | | | **W9, W2** |
| Iolaus argentarius | e | p | x, i | | | | x | | | | | |
| Iolaus ceres | e | p | | | | | | | | i | | |
| Iolaus maryra | e | p/s | | i | x | | | | | i | | W2 |
| Iolaus mermeros | e | s/d | | | | | | x | | | | W7, W9 |
| Junonia andremiaja | e | s/d | x | x | | | | x | x | | | W2, W9, W10 |
| Junonia eurodoce | e | p/s | x | | | x | x | x | x | x | | W7, W9, W10 |
| Junonia goudoti | e | p/s/d | x | | x | | x | x | | x, i | x | W7, W10, W2 |
| Junonia oenone epiclelia | | s/d | x | x | | x | | | | | | W3, W7, W9, W10, W2 |
| Junonia rhadama | e | s/d | | | | | | | | | | W10 |
| Lampides boeticus | | s/d | | | | | | | | | | W2 |
| Leptomyrina phidias | | d | i | | | | | | | | | W2, W3 |
| Leptosia alcesta | e | p/s | x, i | x | x, i | x | x | x | x | x, i | | W2, W3, W7 |
| Leptosia nupta | | p/s | x, i | | | | | | | | | W9, W10, NM |
| Leptotes rabefaner | e | p/s/d | x | | | | x | | | i | | W2, W3, W7, W10 |
| **Malaza empyreus** | **e** | **p** | | | | | | | | | | **W7, near W7L1** |
| Masoura ankoma | e | p | | x | | | | | x | | | |
| **Masoura 53** | **e** | **p** | | **x** | **x** | | | **x** | | | | |
| Masoura antahala | e | s/d | x | x | x | | x | | x | x, i | | W10 |
| Melanitis leda | e | s | x | | | | | | | | | W3, NM, W2 |

| SPECIES | ENDEMISM[a] | HABITAT TYPE[b] | W2L1[c] | W2L2[c] | W2L3[c] | W7L1[c] | W7L2[c] | W7L3[c] | W9L1[c] | W9L2[c] | W9L3[c] | INCIDENTAL SIGHTINGS[d] |
|---|---|---|---|---|---|---|---|---|---|---|---|---|
| *Mylothris phileris* | e | p/s | x | x | i | x | | | x,i | x,i | x | W7, W10, NM, W2 |
| *Neptidopsis fulgurata* | | p/s | | | | | | | | | | NM |
| *Neptis kikideli* | e | p/s | x,i | x | x | x | x | x | | x,i | x,i | W10, NM, W2 |
| *Neptis saclava* | | p/s | | | x | x | | | | x,i | | W2 |
| *Papilio dardanus* | e | p/s/d | x,i | x | x,i | x | x | | | x | | W7, W9, W10, W2 |
| *Papilio delalandei* | e | p | x,i | x,i | x | x | x,i | | x | | x | W7, W9, W10, NM, W2 |
| *Papilio demodocus* | | d | x | | x | x | | x | | | | W7, W3, W10, W2 |
| *Papilio epiphorbas* | | p/s/d | x | | | x | | | | | | W3, W7, W9, W10, NM, W2 |
| **Papilio mangoura** | **e** | **p** | | | | | | | | - | | |
| *Papilio oribazus* | e | p | x | | | x | | | | x | x | W9, W10 |
| *Pardopsis punctatissima* | e | d | | | | | | | | | | W2 |
| *Parnara naso poutieri* | | d | | | | | | | | | | W2 |
| *Phalanta madagascariensis* | e | p | x,i | x,i | x,i | x | x | x | x | x | | W2, W9, W10 |
| *Phalanta phalantha* | | d | | | | | | | | | | W2, W3, W7 |
| *Ploetzia amygdalis* | e | p | | | | | i | | | | | |
| *Pseudacrea imerina* | e | p | x | x | x,i | x | x,i | | x | x,i | x,i | W7, W10, NM |
| *Pseudacrea lucretia* | | p/s | x,i | x | x | x | | | x | x | | W7, W9 |
| *Rysops scintilla* | e | p/s/d | | | | | | | | | | W2, W3 |
| *Salamis anacardii duprei* | e | p/s | x,i | x | x | x | x | | | | | W7, W9, W10 |
| *Salamis anteva* | e | p | x | x,i | x | x | | | x,i | x,i | | W10, W2 |
| **Sallya howensis** | **e** | **p** | | | | | | | | - | | |
| **Sallya madagascariensis** | **e** | **p** | | | | x | | x | | **x** | | **W9, NM** |
| **Saribia decaryi** | **e** | **p** | | | | | | | | - | | |
| **Saribia sp. nov.** | **e** | **p** | **x,i** | **x** | | | x | | | - | | |
| *Saribia perroti* | e | p | x,i | | x,i | x,i | x,i | | | i | | W9, W2 |

APPENDIX 24.1 (*continued*)

| SPECIES | ENDEMISM[a] | HABITAT TYPE[b] | W2L1[c] | W2L2[c] | W2L3[c] | W7L1[c] | W7L2[c] | W7L3[c] | W9L1[c] | W9L2[c] | W9L3[c] | INCIDENTAL SIGHTINGS[d] |
|---|---|---|---|---|---|---|---|---|---|---|---|---|
| *Saribia tepahi* | e | p | x | x,i | x,i | x,i | x | | | i | | W9 |
| *Spalgis tintinga* | e | s | | | | | | | | | | W2, W3 |
| *Strabena ? batesii* (7) | e | p | | | | | | | | | x,i | |
| *Strabena ? perroti* (6) | e | p | | | | | | | | | x,i | |
| *Strabena andriana* | e | p/s/d | | | | x | x | | x,i | i | | |
| *Strabena corynetes* | e | s/d | | | | | | | | | | W10 |
| *Strabena smithii* | e | p/s/d | x,i | | | x,i | x,i | | | i | | |
| *Strabena tamatave* | e | s/d | | | | | | | | | | W2 |
| *Tagiades insularis* | e | p/s | | x | i | | x | | | | | W2, W9, NM |
| *Zizeeria knysna* | e | p/s/d | | | | x | | | | i | | W2, W7 |
| *Zizina antanossa* | | d | | | | | | | | | | W3, W7 |
| Total (135) | 91 | | 67 | 64 | 57 | 48 | 51 | 24 | 40 | 65 | 25 | |
| Rare species (22) | | | 8 | 11 | 8 | 5 | 7 | 4 | 5 | 12 | 5 | |

*Notes:* Data for rare species are given in boldface type.

a   e = endemic at the species level.

b   p = primary forest; s = secondary forest; d = degraded habitat.

c   x = present in sample; i = incidental sighting.

d   Codes W2–W10 refer to the watershed in which the species was observed (see figure 24.3).

CHAPTER 25

# Gorilla Conservation and Research in Central Africa

## A Diversity of Approaches and Problems

Caroline E. G. Tutin and Amy Vedder

The African rain forest contains an outstanding diversity and abundance of large mammals, which is unmatched in any other forest system of the world. Within this rich assemblage is found an unparalleled array of primate species, including three of the world's four great apes. Given their relatedness to our own species, the variety and importance of ecological roles they play, and their ability to garner support as flagship species, primates attract both scientific and conservation attention. This is certainly true for the gorilla *(Gorilla gorilla)*, the largest living primate.

Although constituting a single species, gorillas demonstrate a significant degree of morphological and genetic variation, occupy greatly different habitats, and play varied ecological roles. In addition, gorillas are confronted with an array of threats, which vary both spatially and temporally. As a result, a variety of approaches to conservation have been developed. Both biological and socioeconomic research has proven vital to these processes, identifying site- and population-specific factors important to conservation of the forest and its fauna, and providing recommendations for appropriate actions to address conservation problems and opportunities. Gorillas are one of the best-studied animal species in the wild, and some populations figure among the most endangered. They are not a typical primate, but research and conservation efforts over the past decades provide an informative set of case studies that exemplify the variety of problems faced in African forest and fauna conservation, illustrating a range of possible approaches, many of which have broad relevance.

### Overview of *Gorilla gorilla*

The fragmented nature of gorilla distribution (see below) is reflected by significant levels of genetic variation (Ruvolo et al. 1994; Garner and Ryder 1996), even though gorillas belong to a single species. Although commonly considered to consist of three subspecies *(G. gorilla gorilla, G. g. graueri, and G. g. beringei)*, taxonomists continue to debate these distinctions as well as finer classifications such as demes (Groves 1970, 1971; Sarmiento et al. 1995, 1996). The species *Gorilla gorilla* indisputably varies both among and within regions, and this variation calls for protection of multiple and diverse sites, which in turn benefits the conservation of a range of habitats in the central African forests.

Gorilla distribution within central Africa is

extensive, but a gap of about 750 km exists between the western populations found in Congo, Gabon, Cameroon, Equatorial Guinea, Central African Republic, Cabinda, and Nigeria, and the eastern populations located in DR Congo, Uganda, and Rwanda (figure 25.1). This large-scale geographical disjunction is found in a number of other plant and animal species, and has been considered as evidence for large-scale changes in forest cover during climatic upheavals of the late Pleistocene. Within the geographical range of the species, gorilla distribution is often fragmented, with many geographically isolated populations. Some of this fragmentation is ancient and likely to be related to vegetation changes owing to past changes in climate, some is due to ecological factors that define suitable habitat, and some is due to human activities.

Forests inhabited by gorillas are difficult to characterize. They vary widely in altitude (from sea level to 4,000 m), degree of maturity (from colonizing forest to mature forest), extent of canopy cover (from closed-canopy rain forest to open montane vegetation), and plant species richness (from diverse forests of the Guineo-Congolian region to species-poor montane forests). In all these habitats gorillas co-occur with a diverse array of other species. For these reasons, gorillas are well suited to act as an effective umbrella species for a range of African forest habitats.

Gorillas are considered generalist herbivores, with high caloric requirements because of their large body size. Diet varies with habitat but always includes staple leaf and stem-pith foods, usually from herbaceous plants. In all but montane habitats, fruit figures prominently and introduces large seasonal variations in diet, as most plants fruit at particular times of year. Small quantities of insects are eaten regularly, either accidentally (ingested with herbaceous foods) or deliberately (predation

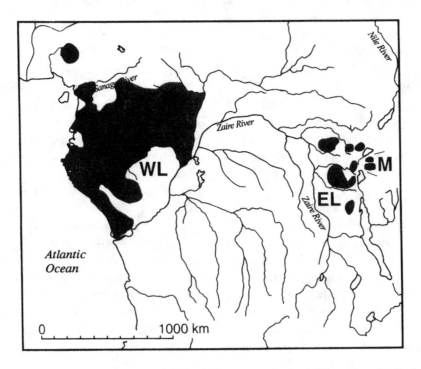

Figure 25.1. Gorilla distribution in central Africa. WL = western lowland; EL = eastern lowland; M = mountain. Adapted from Kingdon 1997.

on ants or termites). Gorillas eat succulent fruit when it is available, but all populations depend on an abundant supply of plant foliage, either permanently or when fruit is scarce. The habitat requirements of gorillas are substantial, and thus population densities are generally low. Gorilla groups show nomadic movements within their large home ranges in response to seasonal shifts in food availability.

Whereas the size of home ranges varies, the minimum habitat area required for a small but potentially viable population of gorillas (for example, those of the Virunga Volcanoes or Bwindi Forest) is on the order of 400 km². Such an area is sufficient to maintain viable populations of most sympatric forest fauna, which reinforces the value of gorillas as an umbrella species.

Gorillas play important ecological roles within their habitats, including the maintenance of forest structure and the dispersal of seeds. Forests that lose these animals will most certainly change in botanical composition and structure, implying a greater conservation loss than that of the single species.

It is generally accepted that gorillas live in relatively stable family groups, ranging in size from a few individuals to thirty or forty family members. Individuals remain with their natal group until they approach maturity. At this point most males (12–15 years old) leave solitarily, later forming their own family groups, whereas many females (6–10 years old) join solitary males or existing families. Smaller numbers of both males and females have been observed to remain in their natal groups. Strong social bonds develop among family members, which are illustrated in the form of cooperation during within-family conflicts as well as group defense when the family is threatened.

Threats faced by gorillas take many forms, from loss of habitat to direct hunting. In the past, collection for museums and capture of live infants for zoos took a heavy toll, but national and international laws and conventions now control most trade. Habitat loss through conversion of forests to agricultural land has brought gorillas to the brink of extinction in areas with dense human population. The most critically affected are remaining populations in Nigeria and those in Rwanda and Uganda, and pressure is mounting in parts of eastern DR Congo. In less densely populated west-central Africa (Congo, Gabon, Cameroon, Equatorial Guinea, and Central African Republic), hunting for meat is usually the greatest threat to gorillas. As a large and imposing primate capable of attacking humans if threatened, gorillas are perceived as dangerous by most humans who share their range, and the gorillas' image is not helped by a tendency to raid plantations in some areas. To protect gorilla populations effectively, the support and understanding of local human populations are essential, but obtaining them presents a conservation challenge because of potential conflicts between the interests of humans and those of gorillas.

In this chapter we review current research and conservation activities focusing on the three subspecies of gorilla (*Gorilla g. beringei*, the mountain gorilla; *G. g. gorilla*, the western lowland gorilla; and *G. g. graueri*, the eastern lowland gorilla) and examine the diversity of problems presented and progress found on the ground. Investment, in both scientific research and conservation, has been high, and although gorillas are neither a typical primate nor a typical mammal, efforts to study and protect wild gorillas in central Africa over the past two decades illustrate the evolution of conservation philosophy and practice. An analysis of progress is pertinent to conservation initiatives across tropical Africa and helps define the problems that must be addressed in the coming decades.

## Mountain Gorillas

Mountain gorillas (*G. g. beringei*) were the first gorillas to receive significant scientific attention

(Donisthorpe 1958; Schaller 1960), and they have continued to be the subject of numerous intensive and long-term field studies. This scientific knowledge has been extensively translated to the public through various media, and resultant conservation activities have concentrated first and most visibly around these animals. Therefore, mountain gorillas have in many ways become the standard-bearer for their species. However, these gorillas are in fact the least numerous of their conspecifics, are ecologically distinct from their western relatives, and face the most severe threats to their existence (Weber and Vedder 1983). Such distinctions provide the context for an important—though specialized—approach to conservation.

The subspecies *G. g. beringei* is generally considered to consist of two populations found in the Western Rift highlands: one in the Virunga Volcanoes, at the intersection of the international borders of DR Congo, Uganda, and Rwanda; and the second in the Bwindi Forest, 30 km to the north in Uganda (questions on the taxonomic status of the Bwindi population have been recently raised: see Sarmiento et al. 1995). Of these two, the Virunga population is better known and more closely monitored, and has been the first focus of field conservation efforts. For these reasons, the Virunga gorillas provide a particularly instructive example of research and conservation efforts, and will be the focus of this section.

In the late 1950s, George Schaller conducted a survey of gorillas in eastern DR Congo (Schaller and Emlen 1963) and a groundbreaking study on the ecology and behavior of mountain gorillas in the Virunga Volcanoes (Schaller 1960). This work was expanded on by Dian Fossey and numerous other researchers, who established the first longitudinal field studies of non-provisioned apes (Fossey 1974). Given the surprising ease of habituating individual gorillas—and entire families—to the presence of researchers, sci-entists have been able to observe gorilla behavior directly and to compile a picture of mountain gorilla life in remarkable detail. As a result of this unique opportunity, investigations following Schaller's concentrated first on social behavior. Gorillas were observed to develop strong social bonds and consistent relationships between individuals (Watts 1992, 1994), form stable family structures, establish new families via emigration of mature males and females, and demonstrate very little agonistic behavior, although they eventually showed some incidence of infanticide (Watts 1989). Long-term studies provided all-too-rare data on demographic parameters in the wild, via the compilation of birth and death statistics on a pool of 142 animals, each individually known. This information was effectively translated via television and print media to the eager American and European public. In these faraway places, mountain gorillas became intimately known, many by name, as "gentle giants" living in close-knit families.

Subsequent investigations began to address ecological questions, unattended since Schaller's first work. Virunga gorillas were found to be generalist herbivores, consuming a large volume of foods, chiefly leaves and stems (Vedder 1984; Watts 1984). In areas of open woodland with associated herbaceous patches, food resources are abundant and are associated with short, somewhat predictable day ranges and consistent activity budgets (Vedder 1984; Watts 1984). Nevertheless, Virunga gorillas show preference for rare food items, which may serve to balance nutrients in an otherwise fibrous, low-quality diet (Vedder 1989). Understanding of gorilla movements and resource requirements fed directly into conservation planning.

Extraordinary motivation for gorilla conservation has been evident since the earliest encounters in the Virungas. Stimulated by the persistent efforts of Carl Akeley, originally in the region as a collector of gorilla specimens for the American Museum of Natural History, the Virunga mountains were gazetted in 1925 as the first national park in Africa, "to make the world safe for mountain gorillas." George Schaller's early surveys pointed out the limited size of the Virunga population (approximately 400–500 individuals), but concerns were raised further as the population was censused in the early 1970s, showing a drastic decline to an estimated 275 animals (Harcourt and Groom 1972). This realization prompted scattered efforts to reduce human and cattle incursions into the park in Rwanda, and also a multidisciplinary investigation of the causes for this decline and the larger socioeconomic context in which mountain gorilla conservation occurred (Weber and Vedder 1983). Direct ecological studies of the gorillas indicated no resource shortage that could account for the population crash, and instead suggested that a combination of prior war (in DR Congo) and continued low-scale poaching were the most likely explanation of these dramatic results (Weber and Vedder 1983). In the broader context, however, the real challenge to conservation seemed to lie in an overall lack of recognition of the value of the gorillas or their habitat to the people and nations that were responsible for their continued survival. Given other more pressing priorities for assuring basic human needs in an impoverished region, there existed little understanding of the gorillas, their rarity, the ecological services provided by their forest habitat, and the economic importance (or potential) of each to the people of the region (Weber 1987).

Building from a strong base of information on the biology of the Virunga gorillas, their population trends, and the attitudes of local people and national authorities, the Mountain Gorilla Project was initiated in Rwanda in 1979 (Vedder and Weber 1990). The program quickly received both American and European support, from a public far better informed and with far greater financial means than the local

population. The Mountain Gorilla Project was designed to approach conservation from multiple angles: to improve enforcement for the long-standing national park, to generate revenue—and thereby national and local interest—for the park, and to increase the Rwandan public's awareness about conservation of the gorillas and their forest. These actions were intended to have immediate, intermediate, and long-term conservation impacts, respectively. As a result of these efforts, park patrols were improved and encroachment into the reserve was minimized. A public awareness campaign was initiated and, in the first two years, reached more than 100,000 people in schools and communities in the area bordering the national park, presenting issues of gorilla conservation as well as the benefits of protecting the watershed within gorilla habitat.

Most visibly, a tourism program was established to generate interest in—via revenue for—conservation. This program took advantage of the charisma of gorillas (which drew in tourists), the ease of habituation and observation of gorillas in such an open habitat, the intimate knowledge of mountain gorilla behavior gained from past research, and the predictability and accessibility of Virunga gorilla home ranges. Development of the program was dependent on the following conditions: the willingness of the Rwandan government to adopt such an approach; the existence of a good road and communications infrastructure; and a healthy climate favorable for visitors. Given conservation objectives, tourism was limited to small, carefully guided groups visiting one of a select number of gorilla families, with strict codes of conduct designed to reduce the potential for negative impacts.

Through the efforts of many governmental officers, conservationists, and scientists, continued conservation efforts have been supported to date in Rwanda, although certainly not without adaptation over the years since the inception of the Mountain Gorilla Project. As a result, dramatic change was documented up to the time of the civil war in 1990. Poaching of gorillas had been reduced to zero since 1984, tourism became the third-ranked source of foreign exchange for the nation, and recognition of gorillas and the national park was positive and widespread on both local and national levels (Weber 1993). The government of Rwanda rejected a proposal to develop cattle ranches in the park, hired additional park personnel to ensure its protection, and incorporated an additional 5 km² of land into the national park. Despite tangible successes, more progress was desired in distributing benefits from tourism more locally, resisting pressure to increase tourist visitation numbers, protecting gorillas from potential disease transmission from visitors and guides, and maintaining support for long-term conservation awareness activities.

The best measure of conservation success, however, lies in population trends. The Virunga gorillas have been carefully monitored since the early 1970s, demonstrating strong signs of success (table 25.1). Having dropped to a low of an estimated 252 individuals in 1981, the population has increased to a high of 324 in 1989, the date of the most recent census. At the same time, the proportion of immature animals increased, and the average gorilla group size grew. Allowing for some lag time in conservation impact, it

is reasonable to relate the upturn in gorilla numbers to increased conservation efforts, starting with the inception of the Mountain Gorilla Project in 1979. The ultimate test, however, has taken place since this last count: a four-year civil war that tore the nation of Rwanda apart, resulting in the loss of an estimated 800,000 lives. Insecurity in the region during the subsequent seven-year period has not allowed an updated census; however, only seventeen gorillas are known to have been killed during this chaotic time. Remarkably, in the midst of the civil war, government and rebels pledged to protect the national park and its gorillas, and although a thorough assessment has not yet been conducted, little direct impact of the war has been detected in the portion of the park inhabited by gorillas. It is truly extraordinary that the conservation of endangered gorillas has achieved a profile high enough to protect them from the atrocities of catastrophic war.

Following initial success in the Mountain Gorilla Project, similar conservation programs were initiated over the borders in DR Congo and Uganda, including protection, tourism, and education elements, focusing first on the Virunga gorillas and later the Bwindi population (Aveling and Aveling 1989; Butynski and Kalina 1993). With adaptations to specific sites, specific opportunities, and specific funding constraints, the approach has been extended to

Table 25.1

## Virunga Mountain Gorilla Census Results

| YEAR | NUMBER | SOURCE |
| --- | --- | --- |
| 1959 | 400–500[a] | Schaller 1963 |
| 1971–1973 | 274 ± 5% | Harcourt and Groom 1972 |
| 1976–1978 | 268 ± 5% | Weber and Vedder 1983 |
| 1981 | 254 ± 5% | Harcout et al. 1983 |
| 1986 | 293 ± 5% | Vedder 1989 |
| 1989 | 324 ± 5% | Sholley 1990 |

a   This figure is an estimate.

benefit other wildlife and forests in the region: in Nyungwe, Rwanda; Tongo, DR Congo; and Kibale, Uganda (Weber 1993).

In sum, the conservation of Virunga gorillas grew from an in-depth understanding of the animals, the product of detailed research. The resultant intimate picture of mountain gorilla behavior and ecology and the socioeconomic factors working for and against conservation provided the basis of information for threat assessments, protection plans, careful tourism design, and analysis of population trends. It also stimulated international concern for, and dedication to conservation of, these charismatic animals. Intensive research has been followed by intensive management of the population. Aside from the period of war and major unrest in Rwanda, the Virunga gorillas have been protected by one of the highest deployments (by density) of national park guards in the world (1 guard/2 km²), approximately one-third of the gorilla families have been monitored on a near-daily basis as focal tourist or research groups, and veterinary services have been available for injured or ill individuals since 1985. Conservation has therefore become an extremely high-input and high-profile process for this small, fragile population of gorillas. For its continued survival, this input must continue well into the future, and must maintain an uncompromising, protectionist approach.

## Western Lowland Gorillas

By 1980, a general fear existed that western gorilla populations had declined alarmingly as a result of subsistence hunting and the capture of youngsters. Remaining wild populations of the subspecies *G. g. gorilla* had been estimated as numbering 9,000 individuals for the IUCN Red Data Book (Gartlan 1980). From 1980 to 1983, Tutin and Fernandez undertook a nationwide census of lowland gorillas (and sympatric chimpanzees) in Gabon. The results of the census showed that Gabon alone har-

bored a population of about 35,000 gorillas (Tutin and Fernandez 1984). The explanation of this unexpected finding was clear. Previous estimates of population sizes had been based on the assumption that gorillas were restricted to secondary, regenerating forests, where the herbs on which they were believed to depend for their food were common. Such habitats, created by human activities, particularly slash-and-burn agriculture, cover a minority of the geographical range of gorillas in west-central Africa's rain forests, as human population density in the region is low. The Gabon census showed that although gorillas do occur at high population densities in areas of secondary forest not subjected to hunting, they also occur at lower densities in primary and seasonally flooded forest habitats. Censuses in the Central African Republic (Carroll 1988; Fay 1989) and in Congo (Fay and Agnagna 1992) confirmed and extended these findings of ecological flexibility and revealed that western lowland gorillas not only occupy such unlikely habitats as swamp forests but live there at high population densities (Fay et al. 1989; Blake et al. 1995).

Whereas Gabon is the only habitat country to have been systematically surveyed for gorilla populations, many surveys and censuses have been conducted in large areas of conservation interest in the Central African Republic, Congo, Cameroon, and Equatorial Guinea. Gorillas' presence has been confirmed throughout the central part of their range with no major fragmentation of populations. In addition, gorillas had been thought to be extinct in Nigeria (Gartlan 1980), but surveys in 1986 and 1988 showed that they survive in three small isolated populations near the border with Cameroon (Harris et al. 1987; Harcourt et al. 1989). These gorillas are, and probably have been for many thousands of years, separated by 450 km from the nearest gorilla populations in Cameroon, and their taxonomic status remains unknown. If gorillas survive in Cabinda at the south of the species' range, they are likely to be few. The recent censuses and surveys mean that in the space of twenty years, estimates of remaining wild populations of western lowland gorillas have increased by an order of magnitude, from 9,000 to 100,000 (Harcourt 1995).

The surveys and censuses of the 1980s were followed by the establishment of field research stations, first at Lopé in central Gabon and later at Dzanga-Sangha in the southern Central African Republic and at Nouabale-Ndoki in northern Congo. From all these long-term studies a similar picture of western lowland gorilla ecology has emerged. In these tropical lowland rain forest habitats gorillas have diverse diets and eat a wide array of fruit as well as such folivorous foods as leaves and bark. For example, at Lopé 220 foods have been recorded for gorillas (compared to about 50–75 for mountain gorillas [Watts 1984]), and 44% of these are fruit (Williamson et al. 1990; Tutin and Fernandez 1993a). Western lowland gorillas are relatively arboreal, harvesting 76% of their foods in trees and building 35% of their night nests above the ground (Tutin et al. 1995). Day ranges and home ranges are relatively large, groups tend to be smaller, and intergroup interactions are less aggressive in lowland than in mountain gorillas—all adaptations that appear to be linked to their frugivory (Tutin et al. 1992; Remis 1994; Goldsmith 1995; Tutin 1996). Another result of their frugivory is that important variations in diet occur both within and between years, as most tree species produce fruit seasonally and crops last for a limited time. When fruit is scarce, gorillas turn to a largely folivorous diet, eating herbs and the bark of some tree species (Rogers et al. 1988, 1994). Aquatic plants are eaten in large quantities, and gorillas at Dzanga-Sangha and Nouabale-Ndoki make regular use of open swamps (called *bais*), spending hours up to their knees or even chests in water, feeding on abundant pith and leaves (Williamson et al. 1988; Fay et al. 1989; Mitani

1992; Nishihara 1992; Kuroda et al. 1996).

In contrast to mountain gorillas, western lowland gorillas have proven difficult to habituate to human presence (Tutin and Fernandez 1991), and details of social behavior remain scarce. Visibility is limited within tropical forests, but at Nouabale-Ndoki daily monitoring of the large Mbeli bai began in 1994, and invaluable data on social behavior and group structure are beginning to emerge (Olejniczak 1994). Some authors have suggested that gorillas in tropical forest habitats show fission-fusion social organization similar to that of chimpanzees and unlike that of mountain gorillas (Jones and Sabater Pi 1971; Mitani 1992; Remis 1994; Goldsmith 1995), while others (Tutin 1996; Fay 1997; C. Olejniczak, pers. comm.) doubt that a fundamental difference exists in social organization between mountain and lowland gorillas. A definitive answer to this and other outstanding questions about western lowland gorilla social structure could come from "fingerprinting" of DNA from hairs shed in night nests (Wickings 1993; Morin et al. 1994). The reliable identification of individual gorillas from DNA extracted from hairs will permit precise description of group structure and allow definition of home ranges and monitoring of demographic changes within local communities.

The frugivory of western gorillas has important consequences for forest dynamics, as these large-bodied, wide-ranging apes swallow seeds of most of their fruit foods and the seeds pass through the gut to be deposited intact in feces. Research has shown that not only are gorillas high-quality dispersers of seeds but also that gorilla nest sites (where about 50% of feces are deposited) provide the best chance of survival for seedlings of some tree species (Voysey et al. 1999a, 1999b). Gorillas swallow larger seeds than do chimpanzees (Tutin and Fernandez 1994), and an endemic tree species at Lopé, *Cola lizae,* depends entirely on gorillas for dispersal of its large seeds (Tutin et al. 1991).

The elusive nature of western lowland gorillas has meant that field researchers have developed a number of indirect methods to study their ecology, several of which permit intersite comparisons. Feeding ecology is generally similar at all sites, with the proportion of fruit in the diet increasing with plant species diversity, but some cultural differences exist: for example, the fruit of several species eaten by gorillas at Nouabale-Ndoki is not eaten by Lopé gorillas even though the trees are common (Moutsambote et al. 1994; Tutin et al. 1994). Similarly, all gorillas in lowland tropical forests eat insects, but although rates of consumption are similar, with around 30% of feces containing insect remains, different prey are eaten at different sites (Tutin and Fernandez 1983, 1992; Carroll 1986). The possibility of local differences in social organization cannot be ruled out.

It will take many more years before western lowland gorillas reveal all their secrets, but the long years of "detective work" in central African rain forests have contributed greatly to our understanding of the gorillas' habitat. The emphasis has been on ecology, with faunal inventories, population censuses, and community studies combined with botanical prospection, vegetation description, and monitoring. The result, over time, is a set of unique databases (e.g., White et al. 1993, 1995; Moutsambote et al. 1994; Tutin et al. 1994; White 1994). Data collected during these long-term studies have contributed to advances in understanding of other aspects of tropical forest ecology, for example, relationships between climatic variables and flowering in some tree species (Tutin and Fernandez 1993b; Tutin and White, in press) and the impact of past climate change on present-day vegetation (Oslisly et al. 1996; White et al. 1998).

The past fifteen years of research in the forests of central Africa have shown that healthy populations of western lowland gorillas remain through much of the central part of the subspecies range in southern Cameroon, southeast Central African Republic, Equatorial Guinea, Gabon, and Congo. Population densities vary from about 0.2 to 10 individuals per km², depending on habitat and human activities (hunting, agriculture, and selective logging). It is clear that *Gorilla g. gorilla* is not endangered in the wild. The reasons for the relatively healthy status of *G.g. gorilla* compared to the other two subspecies are (1) the greater area of habitat and (2) the generally low density of human populations in the tropical rain forests of west-central Africa. Forest clearance for agriculture is a pressing problem in some areas, notably Nigeria, southern Congo, and parts of Equatorial Guinea, but deforestation rates in Gabon, northern Congo, and most of the gorillas' range in Cameroon and Central African Republic are low.

The remaining gorillas in Nigeria occur in three isolated populations, and total numbers are estimated at 75–150 individuals (Harcourt et al. 1989; Oates et al. 1990; Fay 1997). Many of the conservation approaches developed with mountain gorillas are appropriate for all critically endangered populations, but what of the bulk of western lowland gorillas? Does the sheer weight of their numbers mean that they are not a conservation priority?

The surveys and censuses of apes and subsequent censuses of forest elephants (Barnes et al. 1995) drew attention to the tremendous potential the forests of west-central Africa offer for ecosystem conservation (Tutin and Fernandez 1987; McShane 1990). Large areas of forest remain intact and harbor as yet unquantified levels of species diversity. Unfortunately, it is difficult to convince people of the importance of protecting areas of pristine forest before they become threatened. Gorilla researchers Richard Carroll and Mike Fay campaigned successfully for the creation of the large Dzanga-Sangha National Park, the Central African Republic's first rain forest park, and Mike Fay went on to be instrumental in the gazetting of the Nouablale-Ndoki National Park in northern Congo. Gorillas, as a charismatic flagship species, played a major role in these success stories, but the creation of the two new parks will protect the diverse array of animal and plant species that share the habitat. In 1988, Lopé Faunal Reserve (no national parks exist in Gabon) was chosen as a field site for a regional conservation project (ECOFAC) largely because of the ongoing ecological research there. However, other proposals for the creation of new protected areas in the region (e.g., Gartlan 1990; Hecketsweiler 1990; Wilks 1990; Fa 1991) have, on the whole, been ignored. Existing protected areas cover 4–7% of the countries concerned, but only a minority of these already scanty zones receive effective management and protection. In Gabon, for example, logging is permitted in protected areas, and efforts to exclude logging from a core area of 1,670 km² within the Lopé Reserve succeeded only late in 1996.

Changing economic conditions, particularly the 1994 devaluation of the local currency (the CFA franc), have increased pressure on forests throughout the region. The advance of selective logging into areas believed to be naturally protected by their inaccessibility has occurred with astonishing speed. While selective logging per se does not have a lasting negative impact on gorilla populations (see White and Tutin, this volume), associated commercial hunting is posing a major new threat to some gorilla populations in southern Cameroon and northern Congo (Pearce 1995; Wilkie et al. 1992). The expansion of agricultural activities that often follows roads opened in tropical forests by logging operations occurs infrequently in this region because human population density is generally low.

Hunting therefore remains the major threat

currently facing western lowland gorillas, and whereas no quantitative data exist, there are reasons to believe that gorillas are particularly vulnerable to hunting. Although hunters may not target them specifically, the fear that gorillas engender means that they are often shot in perceived self-defense. When confronted by a threat, adult male gorillas will approach and display, making them easy targets for hunters. If the leader male is killed, the fate of his family is precarious, as unweaned infants are likely to be victims of infanticide (Fossey 1984; Watts 1989). Sometimes live infants are obtained as a by-product of this hunting, and in places where a market for pets is perceived to exist, capture of infants is more deliberate. This has a devastating impact, as adults, particularly silverback males and mothers, are killed trying to defend infants, which leads to social disruption and loss of breeding females. The pet trade is one of the threats facing all the great apes (a price of their high "cuddle factor"), and efforts to stop illegal capture of young apes have involved the creation of orphanages or sanctuaries for confiscated youngsters. The death rate for all young apes is high, but compared to chimpanzees and orangutans, even fewer gorilla orphans survive the trauma of capture and captivity, and it seems that the species is psychologically fragile. In 1987 an orphanage was established in Congo to rehabilitate young gorillas that had been confiscated from poachers, and in 1995–96 a first group of these gorillas was introduced into the Lefini Reserve, a protected area containing no wild gorillas.

Gorillas are protected by national laws throughout central Africa, but enforcement is generally weak. The encouraging upward revision of the numbers of wild western lowland gorilla might have led to renewed collecting by zoos, but fortunately this has not happened. Growing conservation awareness within the zoo community, along with improved husbandry (construction of large enclosures and group living) and management (establishing the Studbook list of all captive gorillas and a system of breeding loans), has led to a marked increase in captive breeding. Today, the captive population of gorillas, which is made up almost exclusively of the western subspecies, is considered self-sustaining.

Only a minority of wild western lowland gorillas live within protected areas, and although they are generally better off than those in habitats without protected status, their future is not assured. The success of tourism with mountain gorillas has led to ecotourism being identified as a promising way to integrate conservation and sustainable rural development, but many pitfalls exist (Boo 1990; Weber 1993). To date, western lowland gorilla groups have remained shy even after many years of effective protection and researchers' presence. Gorillas in tropical rain forest habitats have large home ranges and long day ranges and, when eating fruit, are difficult to track, as they leave little trail. Thus it is difficult to locate a gorilla group on a regular basis, and this, combined with limited visibility within the forest habitat, creates unfavorable conditions for habituation (Tutin and Fernandez 1991).

There are exceptions: the behavior of groups toward humans varies, depending largely on the response of the silverback male, some of which are more tolerant than others. At Odzala National Park (Congo), a gorilla group living in a forest close to villages where tradition forbids hunting of apes, accepted observers quickly (M. Bermejo, pers. comm.). The ease of viewing provided by the open swamps at Nouable-Ndoki, Dzanga-Sangha, and Odzala holds potential for tourism without the necessity of habituation. But is ecotourism based on western lowland gorillas a promising or viable proposition? It is already clear that the ideal conditions for ecotourism (assurance of finding habituated groups and clear visibility), which have been created with mountain goril-

las in Rwanda, DR Congo, and Uganda, are unlikely ever to be achieved in the rain forests of central Africa (Weber 1993). At Lopé, a community-based program of ecotourism was initiated by ECOFAC in 1995, but gorillas took a back seat, and the marketing was based on a "tropical rain forest experience." The projected revenues are modest, and it is unlikely that tourism, while it may create jobs and generate a sustainable economic input locally and nationally, will ever figure among the top industries of the country. In Gabon, as in neighboring countries, ecotourism alone will not provide a credible economic argument for rain forest protection in the foreseeable future, but it has proved to be a promising way of involving local people in conservation. New approaches to gorilla tourism development are planned at Dzanga-Sangha (A. Blom, pers. comm.) and at Lopé (E. deMerode, pers. comm.), driven by the growing demand for environmentally friendly adventure tourism and the magnetic attraction of gorillas, but a solution to the "Homophobia" of these apes needs to be found if the expectations of visitors (which will inevitably be based on media contact with mountain gorillas) are not to be disappointed.

Whereas direct economic arguments for the need to protect gorillas within tropical rain forests are elusive, ecological arguments are obvious. The large body size and "sloppy" fruit-eating habits of gorillas (Tutin and Fernandez 1994) make them seed dispersers par excellence, and aspects of their behavior (such as their choice of nest sites and their wide range) contribute to the demonstrated quality of gorilla dispersal in terms of germination success, seedling survival, and growth—at least for the first three years (Tutin et al. 1991; Voysey et al. 1999a, 1999b). The loss of gorillas from forests in central Africa will lead to long-term changes in plant species composition, as has been demonstrated for forests in Ivory

Coast from which elephants have been exterminated (Alexandre 1980). The management of tropical forests for long-term sustainability of timber production depends on maintaining the links in the chain of regeneration provided by frugivores. This type of ecological argument for protecting rain forests is, alone, unlikely to hold much weight with planners and economists, and it is gorillas' power to sway public opinion and generate funds for conservation that is needed. Research continues to define the gorillas' role in tropical ecosystems, and as a flagship species with wide popular appeal, gorillas can serve as ambassadors to ensure the protection of the myriad plants and animals with which they coexist.

### Eastern Lowland Gorillas

Eastern lowland gorillas (*G. g. graueri*), though first surveyed by scientists and first visited by tourists, remain the least known of the gorilla subspecies. Even after cursory investigation, however, it is clear that the subspecies shows a degree of variation taxonomically (Sarmiento et al. 1996) and ecologically (Goodall and Groves 1977; Yamagiwa and Mwanza 1994) that renders them difficult to characterize as a single group. Even the common name, eastern *lowland* gorilla, is misleading given their presence in high montane forest as well as lowlands, and for that reason they are referred to here as Grauer's gorillas.

From the earliest surveys, the population and distribution of Grauer's gorillas were known to be limited. Schaller estimated the subspecies to number between 5,000 and 15,000 individuals, found in fragmented patches in eastern DR Congo (Emlen and Schaller 1960). Although no new data were collected, the population was assumed to have declined significantly by 1980 because of increased human pressure (Gartlan 1980), but a more recent comprehensive field survey estimated current numbers to be between 7,600

and 22,400 (Hall et al. 1998). Grauer's gorilla distribution spans an altitudinal gradient from 400 m to 3,000 m, comprising a wide variety of habitats from dense lowland rain forest to the open woodlands and bamboo forests of montane systems. The fragmentation of the habitat may be due to several very different factors: gorilla immigration from the east, and a consequent divergence caused by the dense network of rivers in the region (Groves 1970), as well as past and current human pressures from agriculture and hunting. Although the Grauer's gorilla population is considered to be far larger than the mountain gorilla population, this "island-like" nature of Grauer's distribution has left the subspecies vulnerable to some of the same problems encountered by their mountain relatives because of small population size.

All information about the biology of Grauer's gorillas is derived from a single population, located in the Kahuzi-Biega National Park. Gorilla social structure and behavior resemble that found in the other two subspecies, showing stable family groups, with group sizes intermediate between those of the mountain and western lowland subspecies. The park comprises both low- and high-altitude regions, and thus the gorillas found in the park demonstrate great ecological variability. Grauer's gorillas feed largely on fruit where and when available, and on herbaceous foods where these dominate (Yamagiwa 1994). Their home ranges and daily ranges vary greatly depending on time of year and location, correlated with availability of food resources, especially bamboo shoots and fruit (Goodall 1979; Yamagiwa 1994).

Given diversity in habitat, human economic activities, access to markets, and human population densities and culture, threats to Grauer's gorillas differ across their range. In the eastern portions of their distribution, competition for land is greatest, resulting in loss of forest habitat to agriculture and livestock grazing. This is particularly true in the highland sector of

Kahuzi-Biega National Park, as well as in the Itombwe region to the south and Tshiaberimu to the north. Further west, hunting becomes a more important factor, as small farms are interspersed within the forest and local people hunt gorillas for food and to deter them from (or punish them for) raiding crops. Most recently, tremendous pressure has been placed on the forest and its wildlife from two civil wars and refugee movements. The full impact of this pressure is not yet known.

To date, conservation actions have focused almost entirely on the Kahuzi-Biega population of Grauer's gorillas. The original sector of the national park was established in 1970, a result of lobbying by Adrian DeSchriver, who had earlier guided large game hunting safaris in the region. DeSchriver hired poachers as park employees, prohibiting them from hunting, and in 1973 initiated the first gorilla tourism program to provide income for the park. In 1975, park boundaries were extended to their current positions. As is true for the Virunga Volcanoes, the area is one of scenic beauty, comfortable climate, and good accessibility, all of which resulted in a growing tourist clientele. However, as tourist pressures grew and a conservation ethic focusing on the gorillas dissipated, the system degenerated as visits (involving high frequency, large numbers of people, and a negative impact on high-altitude vegetation) became disturbing to the animals. As the quality of the tourist experience declined, and better-controlled programs began in the Virungas for mountain gorillas (in Rwanda and DR Congo), tourist numbers and benefits dropped off in Kahuzi-Biega. Although management and visitation systems improved significantly in the 1990s with development agency support, disturbance caused by the presence of Rwandan refugee camps and the following wars in 1996 and 1998–99 reduced tourist activities and consequent benefits to an insignificant level. Nevertheless, conservation efforts on the part of park

staff and technical assistants have continued. The degree of negative impact was effectively limited until 1999, when lack of security resulted in the deaths of at least sixty gorillas (G. Debonnet, pers. comm.).

Outside the Kahuzi-Biega area, gorilla conservation faces even greater challenges. Local people and authorities are most concerned with development options for a human population that is growing both in numbers and in expectations. Given increased conversion of forest lands for agricultural use, and the likelihood that large-scale investors will be interested in further developing both mining and forestry activities in the region once the political situation in DR Congo settles, conflict between people and gorillas will continue to accelerate. Meanwhile, local forces are not favorable to the establishment of new protected areas.

In this context, conservation must focus on effective management of existing parks and reserves. In particular, Kahuzi-Biega National Park remains a vital reserve for Grauer's gorillas, most recently reported to harbor approximately 75% of the total population (Hall et al. 1998). Yet most of these animals are found in the lowland sector of the park, where impacts of the war have certainly been great, and where management, tourism, and protection activities have always been the weakest. In this region, a significant new push for strong conservation must be made.

Beyond Kahuzi-Biega, innovative education and protection programs should be enacted which reinforce the values of the gorillas' forest habitat to local communities and help to inform people of real costs and benefits of living with gorillas.

Fifteen years ago, mountain gorillas were on the brink of extinction, and populations of the eastern and western lowland subspecies were feared to be severely threatened. Today, an overview of the status of gorillas in central

Africa is more optimistic. Research and conservation investment in gorillas has intensified over the past decade, and up-to-date information is available from all habitat countries except Cabinda. However, there is no room for complacency, and continued investment and innovation are required to cope with the great diversity of habitats, threats, and human contexts associated with wild gorilla populations.

Research has brought to light the unexpected ecological flexibility of gorillas, while national, regional, and local censuses have identified priority populations, providing the baseline data essential to conservation planning. Priority populations include both those that are most imminently threatened and those that can be effectively protected before threats arrive. Translating research findings into conservation action is not an automatic process, but the charismatic appeal of gorillas has helped raise awareness and funds.

Although no simple solutions exist because of the diversity of situations, some basic common approaches can be identified. No gorilla population can be protected if its continued existence is (or is perceived as) an obstacle to the well-being of humans sharing the ecosystem. Conflicts of interest arise when a desire exists to convert gorilla habitat into agricultural land, when traditional hunting occurs in gorilla habitat, or when gorillas raid plantations. These problems have all intensified as human populations continue to grow and firearms become widespread. The enforcement of laws that protect either areas of forest or individual species is essential, but the "stick" will be more effective if accompanied by a "carrot."

The most successful positive incentive for the protection of gorillas is ecotourism. The example from the Virunga Volcanoes in Rwanda shows that gorilla viewing can generate substantial revenues, making the gorillas a precious natural resource that can be "exploited" in a sustainable way. Analysis of ecotourism proj-

ects with African apes shows that great care must be taken in their planning and management (Weber 1993), but the Rwandan example has been transposed successfully to sites in DR Congo and Uganda. Tourism depends on political stability, but the indisputable economic value of mountain gorillas has been a major factor in their having been relatively unscathed during the catastrophic horrors of the past five years in Rwanda and eastern DR Congo. This value, however, has not proven powerful enough to resist years of sustained war and insecurity in the Kahuzi-Biega region, where Grauer's gorillas have been significantly affected. In any case, gorilla ecotourism requires reliable viewing of habituated individuals, and such conditions have not so far been achieved for western lowland gorillas, although attempts are being made in the Central African Republic, the Congo Republic, and Gabon. Other types of ecotourism, which market the chance of seeing unhabituated gorillas as part of a "tropical rain forest experience," are promising ways of generating sustainable revenues for local communities but will make less economic impact at a national level.

Essential ingredients of all conservation efforts are education and training. To this end, researchers can play an active role both in providing input to community- or school-based conservation education programs and in training local people at all levels, from tourist guides and park guards to doctoral students and wildlife managers. Sustained education programs do change attitudes, and children are particularly open to new concepts, especially if these are firmly grounded in their local context (Weber 1995). Education and training are long-term undertakings, but they are fundamental elements of eventual success. Furthermore, it is important to communicate research results in a language that will be understood by local people, politicians, and wildlife managers: 90% of gorillas live in countries where French is the official language, but more than 90% of scientific and popular publications about gorillas are in English. Dedicated nationals can achieve things impossible for outsiders, as shown by the gorilla trackers at Karisoke and the personnel of DR Congo's National Parks Service, who continue to protect wildlife and parks in extraordinarily difficult circumstances (Hart et al. 1996). Encouraging local individuals and organizations to become involved in every aspect of conservation is essential, but particularly so in the case of conservation education.

It is clear that an adaptive and flexible approach to conservation is needed, as threats can rapidly change in nature or degree. Political instability is difficult to predict, as is the impact of economic change. If present economic conditions prevail, all non-protected rain forests in west-central Africa will be subjected to selective logging. Although several new rain forest national parks have been created, the total area of protected gorilla habitat is still lamentably small. Selective logging disturbs gorillas, but in the absence of hunting, population densities return to normal levels after 3–5 years, and light logging may actually increase the carrying capacity of the habitat by increasing the amount of herbaceous vegetation (see Oates 1996; White and Tutin, this volume). Presently, however, hunting occurs on all logging concessions at least to satisfy the needs of the workforce and often at a commercial scale, with meat being sent for sale in towns (Wilkie et al. 1992; Pearce 1995). Aspects of gorilla behavior and social organization make the species particularly vulnerable to hunting, and if a market exists for captured infants, hunting is even more devastating. Whereas international trade in live gorillas has been stopped by laws and conventions, great care must still be taken to educate people in habitat countries about the negative impact of the pet trade.

Looking to the future, it is clear that the severely threatened isolated populations in Rwanda, Uganda, DR Congo, and Nigeria (possibly also in Cabinda) will continue to depend on sustained multi-faceted conservation programs for their survival. The gorillas of Nigeria pose the greatest problem because their numbers are so low. These gorillas are understandably shy around humans, as hunting still occurs in their habitat (Harcourt et al. 1989; Oates et al. 1990), so ecotourism does not seem a likely solution. Ongoing research will provide essential baseline data and may suggest novel approaches to this critical situation.

The large populations of gorillas that inhabit the rain forests of west-central Africa and parts of eastern DR Congo present an enormous conservation challenge. Current protected areas are inadequate both in terms of their total area and in terms of the institutional capacities of the countries of the region to enforce existing legislation (see, e.g., Hart and Hall 1996). Major sustained expenditure is needed, but this will not come from the hard-pressed national budgets of the habitat countries. Foreign aid has earmarked large amounts of money for forest conservation in central Africa, but lack of coordination and the absence of overall strategies have hindered progress to date, and it has been argued that the trend to see sustainable development as the only route to conservation creates as many problems as it solves (see, e.g., Oates 1996). The establishment of trust funds to assure a regular income to maintain particular protected areas is a possible solution, but will donors have sufficient confidence to invest for the future given the political instability and widespread corruption that exist in the region?

In developed countries, such large, potentially dangerous animals as bears, wolves, and large carnivores survive only within strictly demarcated parks and reserves. This is already the case in Rwanda and Uganda, and the only hope of gorillas' persistence in Nigeria is through the extension of the Cross River National Park. If this is the inevitable future of gorillas, chimpanzees, and elephants through-out Africa, we must accept that 80–90% of present-day populations are doomed. If we are not prepared to envisage such a future scenario, conservation action should focus on the major threats to gorilla populations, which differ from country to country. In densely populated countries (Cameroon, eastern DR Congo, and Equatorial Guinea) deforestation is the major threat, and at current rates, forests and gorillas will have disappeared from all but protected areas in less than a hundred years (Barnes 1990; Harcourt 1995). In countries with low human population densities (Gabon, Congo, and Central African Republic), hunting is the major threat to gorillas. Deforestation in these countries is not an immediate threat. Selective logging will occur but should not adversely affect gorillas if associated hunting is stopped.

There is growing awareness that timber harvesting from remaining tropical forests should be carried out sustainably, driven by consumer pressure in the West. The Forestry Stewardship Council has developed a certification system that defines and imposes criteria of sustainability such that consumers can identify "green" timber and wood products. The present criteria focus on forestry issues and pay scant attention to the ecological importance of animals as essential links in the natural process of regeneration, even though 70–80% of plants in tropical forests depend on animals for seed dispersal (White 1994). To obtain the green label, forestry companies are obliged to "protect biodiversity and ecological processes" (Upton and Bass 1995), but specific certification contracts do not require a ban on, or even regulation of, hunting (see, e.g., Upton 1996). Whereas efforts to promote sustainability of tropical timber harvesting are laudable, the present process has been developed with minimal input from ecologists, and pressure should be increased to stop the high levels of hunting facilitated in remote forest areas by logging companies (Pearce 1995).

Ultimately, all the problems of protecting gorillas and their habitats are related to human population growth. The various present-day scenarios across the gorillas' range in Africa are part of a dynamic continuum. The persistence of a viable population of mountain gorillas in the Virunga Volcanoes (Rwanda, DR Congo, and Uganda), which are surrounded by areas with the highest human population densities of the continent and affected by civil unrest of catastrophic proportions, is testimony to what can be achieved by multi-faceted conservation and research. Yet if we refuse to accept as inevitable the fact that in 100–200 years' time all remaining wild gorillas will be confined to similar small, isolated islands of protected forest, then we must continue to seek innovative, appropriate ways that allow gorillas and humans to coexist within the tropical forests of west-central Africa.

REFERENCES

Alexandre, D.Y. 1980. Caractère saisonnier de la fructification dans une forêt hygrophile de Côte d'Ivoire. *Revue Ecologie* 34:335–350.

Aveling, C., and R. Aveling. 1989. Gorilla conservation in Zaire. *Oryx* 23:64–70.

Barnes, R. F. W. 1990. Deforestation trends in tropical Africa. *African Journal of Ecology* 28:161–173.

Barnes, R. F. W., A. Blom, M. P. T. Alers, and K. L. Barnes. 1995. An estimate of the numbers of forest elephants in Gabon. *Journal of Tropical Ecology* 11:27–37.

Baulleu, E. E. 1980. The Centre International de Recherche Médicale de Franceville (CIRMF) and the symposium on the great apes of Africa. *Journal of Reproduction and Fertility Supplement* 28:1–2.

Blake, S., E. Rogers, M. J. Fay, M. Ngangoue, and G. Ebeke. 1995. Swamp gorillas in northern Congo. *African Journal of Ecology* 33:285–290.

Boo, E. 1990. *Ecotourism: The Potentials and Pitfalls.* World Wildlife Fund, Washington, D.C.

Butynski, T. M., and J. Kalina. 1993. Three new mountain national parks for Uganda. *Oryx* 27:214–224.

Carroll, R. W. 1986. Status of the lowland gorilla and other wildlife in the Dzanga Sangha region of southwestern Central African Republic. *Primate Conservation* 7:38–41.

———. 1988. Relative density, range extension and conservation potential of the lowland gorilla in the Dzanga-Sangha region of southwestern Central African Republic. *Mammalia* 52:309–323.

Donisthorpe, J. 1958. A pilot study of the mountain gorilla (*Gorilla gorilla beringei*) in South-West Uganda, February to September 1957. *South African Journal of Science* 54:195–217.

Emlen, J. T., and G. Schaller. 1960. Distribution and status of the mountain gorilla (*Gorilla gorilla berengei*)—1959. *Zoologica* 45:41–52.

Fa, J. E. 1991. Conservación de los ecosistemas forestales de Guinea Ecuatorial. In M. Collins and J. Sayer, eds. *Bibliothèque de la Conservation de l'UICN*. IUCN, Gland, Switzerland.

Fay, J. M. 1989. Partial completion of a census of the western lowland gorilla (*Gorilla gorilla gorilla*) in southwestern Central African Republic. *Mammalia* 53:202–215.

———. 1997. The ecology, social organization, populations, habitat and history of the western lowland gorilla (*Gorilla gorilla gorilla* Savage and Wyman 1847). Ph.D. dissertation, Washington University, St. Louis.

Fay, J. M., and M. Agnagna. 1992. Census of gorillas in northern Republic of Congo. *American Journal of Primatology* 27:275–284.

Fay, J. M., M. Agnagna, J. Moore, and R. Oko. 1989. Gorillas in the Likouala swamp forests of north central Congo. *International Journal of Primatology* 10:477–486.

Fossey, D. 1974. Observations on the home range of one group of mountain gorillas (*Gorilla gorilla beringei*). *Animal Behaviour* 22:568–581.

———. 1984. Infanticide in mountain gorillas (*Gorilla gorilla beringei*) with comparative notes on chimpanzees. In G. Hausfater and S. B. Hrdy., eds. *Infanticide: Comparative and Evolutionary Perspectives*. Aldine, New York.

Garner, K. J., and O. A. Ryder. 1996. Mitochondrial DNA diversity in gorillas. *Molecular Phylogenetics and Evolution* 6(1):39–48.

Gartlan, S. 1980. Western gorillas. Manuscript.

———. 1990. La conservation des écosystèmes forestiers du Cameroun. In M. Collins and J. Sayer, eds. *Bibliothèque de la Conservation de l'UICN*. IUCN, Gland, Switzerland

Goldsmith, M. L. 1995. Ranging and grouping patterns of western lowland gorillas (*Gorilla g. gorilla*) in the Central African Republic. *Gorilla Conservation News* 9:5–6.

Goodall, A. G., and C. P. Groves. 1977. The conservation of eastern gorillas. Pages 599–637 in P. Rainer and G. H. Bournes, eds. *Primate Conservation*. Academic Press, London.

Groves, C. P. 1970. Population systematics of the gorilla. *Journal of Zoology* 161:287–300.

———. 1971. Distribution and place of origin of the gorilla. *Man* 6:44–51.

Hall, J. S., K. Saltonstall, B.-I. Inogwabini, and I. Omari. 1998. Distribution, abundance and conservation status of Grauer's gorilla. *Oryx* 32:122–130.

Harcourt, A. H. 1995. Population viability estimates: Theory and practice for a wild gorilla population. *Conservation Biology* 9:134–142.

Harcourt, A. H., and A. F. G. Groom. 1972. Gorilla census. *Oryx* 11:355–363.

Harcourt, A. H., J. Kineman, G. Campbell, I. Redmond, C. Aveling, and M. Condiotti. 1983. Conservation and the Virunga gorilla population. *African Journal of Ecology* 21:139–142.

Harcourt, A. H., K. J. Stewart, and I. M. Inahoro. 1989. Nigeria's gorillas: A survey and recommendations. *Primate Conservation* 10:73–76.

Harris, D., J. M. Fay, and N. Macdonald. 1987. Report of gorillas from Nigeria. *Primate Conservation* 8:40.

Hart, J. A., and J. S. Hall. 1996. Status of eastern Zaire's forest parks and reserves. *Conservation Biology* 10:316–327.

Hart, T. A., J. A. Hart, and J. S. Hall. 1996. Conservation in the declining nation state of Zaire. *Conservation Biology* 10:685–686.

Hecketsweiler, P. 1990. La conservation des écosystèmes forestiers du Congo. In M. Collins and J. Sayer, eds. *Bibliothèque de la Conservation de l'UICN*. UICN, Gland, Switzerland.

Jones, C., and J. Sabater Pi. 1971. Comparative ecology of *Gorilla gorilla* (Savage and Wyman) and *Pan troglodytes* (Blumenbach) in Rio Muni, west Africa. *Bibliotheca Primatologica* 13:1–96.

Kingdon, J. 1997. *The Kingdon Field Guide to African Mammals*. Academic Press, London.

Kuroda, S., T. Nishihara, S. Suzuki, and R. A. Oko. 1996. Sympatric chimpanzees and gorillas in the Ndoki Forest, Congo. Pages 71–81 in W. C. McGrew, L. F. Marchant, and T. Nishida, eds. *Great Ape Societies*. Cambridge University Press, Cambridge.

McShane, O. 1990. Conservation before the crisis—an opportunity in Gabon. *Oryx* 24:9–14.

Mitani, M. 1992. Preliminary results of the studies of wild western lowland gorillas and other sympatric diurnal primates in the Ndoki Forest, northern Congo. Pages 215–244 in N. Itoigawa, Y. Sugiyama, G. P. Sackett, and R. K. R. Thompson, eds. *Topics in Primatology, Volume 2: Behavior, Ecology and Conservation*. University of Tokyo Press, Tokyo.

Morin, P. A., J. J. Moore, R. Vhakraborty, L. Jin, J. Goodall, and D. S. Woodruff. 1994. Kin selection, social structure, gene flow, and the evolution of chimpanzees. *Science* 265:1193–1201.

Moutsambote, J. M., T. Yumoto, M. Mitani, T. Nishihara, S. Suzuki, and S. Kuroda. 1994. Vegetation list of plant species identified in the Nouabalé-Ndoki Forest, Congo. *Tropics* 3:277–293.

Nishihara, T. 1992. A preliminary report on the feeding habits of western lowland gorillas (*Gorilla gorilla gorilla*) in the Ndoki Forest, northern Congo. Pages 225–240 in N. Itolgawa, Y. Sugiyama, G. P. Sackett, and R. K. R. Thompson, eds. *Topics in Primatology, Volume 2: Behavior, Ecology and Conservation*. University of Tokyo Press, Tokyo.

Oates, J. F. 1996. Habitat alteration, hunting and the conservation of folivorous primates in African forests. *Australian Journal of Ecology* 21:1–9.

Oates, J .F., D. White, E. L. Gadsby, and P. O. Bisong. 1990. Cross River National Park (Okwangwo Division), feasibility study, appendix 1: Conservation of gorillas and other species. Report.

Olejniczak, C. 1994. Report on a pilot study of western lowland gorillas at Mbeli Bai, Nouabale-Ndoki Reserve, northern Congo. *Gorilla Conservation News* 8:9–11.

Oslisly, R., B. Peyrot, S. Abdessadok, and L. J. T. White. 1996. Le site de Lopé 2: Un indicateur de transition écosystémique *ca* 10,000 BP dans la moyenne vallée de l'Ogooué (Gabon). *C. R. Acad. Sc. Paris, Sér. IIa* 323:933–939.

Pearce, J. 1995. *Slaughter of the Apes*. World Society for the Protection of Animals.

Remis, M. J. 1994. Feeding ecology and positional behavior of western lowland gorillas (*Gorilla gorilla gorilla*) in the Central African Republic. Ph.D. dissertation, Yale University, New Haven.

Rogers, M. E., C. E. G. Tutin, R. J. Parnell, B. C. Voysey, E. A. Williamson, and M. Fernandez. 1994. Seasonal feeding on bark by gorillas: An unexpected keystone food? Pages 27–43 in J. R. Anderson, J. J. Roeder, and N. Herrenschmidt, eds. *Current Primatology, Volume 1: Ecology and Evolution*. B. Thierry. Université Louis Pasteur, Strasbourg, France.

Rogers, M. E., E. A. Williamson, C. E. G. Tutin, and M. Fernandez. 1988. Effects of the dry season on gorilla diet in Gabon. *Primate Report* 19:29–34.

Ruvolo, M., D. Pan, S. Zehr, T. Goldberg, T. Dosptell, and M. von Dornum. 1994. Gene trees and hominid phylogeny. *Proceedings of the National Academy of Science, USA* 91:8900–8904.

Sarmiento, E. E., T. M. Butynski, and J. Kalina. 1995. Taxonomic status of the gorillas of the Bwindi-Impenetrable Forest, Uganda. *Primate Conservation* 16:40–43.

———. 1996. Gorillas of Bwindi-Impenetrable Forest and the Virunga volcanoes: Taxonomic implications of morphological and ecological differences. *American Journal of Primatology* 40:1–21.

Schaller, G. B. 1960. The conservation of gorillas in the Virunga Volcanoes. *Current Anth.* 1(4):331.

———. 1963. *The mountain gorilla*. University of Chicago Press, Chicago.

Schaller, G. B., and J. T. Emlen. 1963. Observations on the behavior of the mountain gorilla. In F. C. Howell, ed. *African Ecology and Human Evolution*. Viking Fund Publications in Anthropology.

Sholley, C. 1990. *Census of the Virunga gorilla population, 1989*. African Wildlife Foundation.

Tutin, C. E. G. 1996. Ranging and social structure of lowland gorillas in the Lopé Reserve, Gabon. Pages 58–70 in W. C. McGrew, L. F. Marchant, and T. Nishida, eds. *Great Ape Societies*. Cambridge University Press, Cambridge.

Tutin, C. E. G., and M. Fernandez. 1983. Gorillas feeding on termites in Gabon, West Africa. *Journal of Mammalogy* 64:511–513.

———. 1984. Nationwide census of gorilla (*Gorilla g. gorilla*) and chimpanzee (*Pan t. troglodytes*) populations in Gabon. *American Journal of Primatology* 6:313–336.

———. 1987. Gabon: A fragile sanctuary. *Primate Conservation* 8:160–161.

———. 1991. Responses of wild chimpanzees and gorillas to the arrival of primatologists: Behaviour observed during habituation. Pages 187–197 in H. O. Box, ed. *Primate Responses to Environmental Change*. Chapman and Hall, London.

———. 1992. Insect-eating by sympatric lowland gorillas (*Gorilla g. gorilla*) and chimpanzees (*Pan t. troglodytes*) in the Lopé Reserve, Gabon. *American Journal of Primatology* 28:29–40.

———. 1993a. Composition of the diet of chimpanzees and comparisons with that of sympatric lowland gorillas in the Lopé Reserve, Gabon. *American Journal of Primatology* 30:195–211.

———. 1993b. Relationships between minimum temperature and fruit production in some tropical forest trees in Gabon. *Journal of Tropical Ecology* 9:241–248.

———. 1994. Comparison of food processing by sympatric apes in the Lopé Reserve, Gabon. Pages 29–36 in B. Thierry, J. R. Anderson, J. J. Roeder, and N. Herenschmidt, eds. *Current Primatology, Volume 1: Ecology and Evolution*. Université Louis Pasteur, Strasbourg, France.

Tutin, C. E. G., M. Fernandez, M. E. Rogers, and E. A. Williamson. 1992. A preliminary analysis of the social structure of lowland gorillas in the Lopé Reserve, Gabon. Pages 245–254 in N. Itoigawa, Y. Sugiyama, G. P. Sackett, and R. K. R. Thompson, eds. *Topics in Primatology*. University of Tokyo Press, Tokyo.

Tutin, C. E. G., R. J. Parnell, L. J. T. White, and M. Fernandez. 1995. Nest building by lowland gorillas in the Lopé Reserve, Gabon: Environmental influences and implications for censusing. *International Journal of Primatology* 16:53–76.

Tutin, C. E. G., and L. J. T. White. In press. Primates, phenology and frugivory: Present, past and future patterns in the Lopé Reserve, Gabon. In D. M. Newbery, H. H. T. Prins, and N. Brown, eds. *Dynamics of Populations and Communities in the Tropics*. Blackwell, Oxford.

Tutin, C. E. G., L. J. T. White, E. A. Williamson, M. Fernandez, and G. McPherson. 1994. List of plant species identified in the northern part of the Lopé Reserve, Gabon. *Tropics* 3:249–276.

Tutin, C. E. G., E. A. Williamson, M. E. Rogers, and M. Fernandez. 1991. A case study of a plant-animal relationship: *Cola lizae* and lowland gorillas in the Lopé Reserve, Gabon. *Journal of Tropical Ecology* 7:181–199.

Upton, C. 1996. Summary of assessment report: Gabonese forestry operations of ISOROY which are managed by Leroy Gabon, Gabon, Central Africa. Qualifor Report, November.

Upton, C., and S. Bass. 1995. *The Forest Certification Handbook*. Earthscan, London.

Vedder, A. L. 1984. Movement patterns of a group of free-ranging mountain gorillas (*Gorilla gorilla beringei*)

and their relation to food availability. *American Journal of Primatology* 7(2):73–88.

———. 1989. Feeding ecology and conservation of the mountain gorilla (*Gorilla gorilla beringei*). Ph.D. dissertation, University of Wisconsin, Madison.

Vedder, A. L., and W. Weber. 1990. The mountain gorilla project. In A. Kiss, ed. *Living with Wildlife: Wildlife Resource Management with Local Participation in Africa*. World Bank, Washington, D.C.

Voysey, B. C., K. E. McDonald, M. E. Rogers, C. E. G. Tutin, and R. J. Parnell. 1999a. Gorillas and seed dispersal in the Lopé reserve, Gabon, I: Gorilla acquisition by trees. *Journal of Tropical Ecology* 15:23–38.

———. 1999b. Gorillas and seed dispersal in the Lopé reserve, Gabon, II: Survival and growth of seedlings. *Journal of Tropical Ecology* 15:39–60.

Watts, D. P. 1984. Composition and variablility of mountain gorilla diets in the central Virungas. *American Journal of Primatology* 7:323–356.

———. 1989. Infanticide in mountain gorillas: New cases and a reconsideration of the evidence. *Ethology* 81:1–18.

———. 1992. Social relationships of immigrant and resident female mountain gorillas, 1: Male-female relationships. *American Journal of Primatology* 28:159–181.

———. 1994. Social relationships of immigrant and resident female mountain gorillas, 2: Relatedness, residence, and relationships between females. *American Journal of Primatology* 32:13–30.

Weber, W. 1993. Primate conservation and ecotourism in Africa. In C. S. Potter, J. L. Cohen, and D. Janczewski, eds. *Perspectives on Biodiversity: Case Studies of Genetic Resource Conservation and Development*. AAAS Press, Washington, D.C.

———. 1995. Monitoring awareness and attitude in conservation education: The mountain gorilla project in Rwanda. Pages 28–48 in S. K. Jacobsen, ed. *Conserving Wildlife: International Education and Communication Approaches*. Columbia University Press, New York.

Weber, W., and A. L. Vedder. 1983. Population dynamics of the Virunga gorillas: 1959–1978. *Biological Conservation* 26:341–366.

White, L. J. T. 1994. Biomass of rain forest mammals in the Lopé Reserve, Gabon. *Journal of Animal Ecology* 63:499–512.

White, L. J. T., R. Oslisly, K. A. Abernethy, and J. Maley. 1998. *Aucoumea klaineana:* A Holocene success story now

in decline? In *Proceedings of ECOFIT Symposium.*
ORSTOM, Paris.

White, L. J. T., M. E. R. Rogers, C. E. G. Tutin, E. A.
Williamson, and M. Fernandez. 1995. Herbaceous vege-
tation in different forest types in the Lopé Reserve,
Gabon: Implications for keystone food availability.
*African Journal of Ecology* 33:124–141.

White, L. J. T., C. E. G. Tutin, and M. Fernandez. 1993.
Group composition and diet of forest elephants, *Lox-
odonta africana cyclotis*, Matschle 1900, in the Lopé
Reserve, Gabon. *African Journal of Ecology* 31:181–199.

Wickings, E. J. 1993. Hypervariable single and multi-
locus DNA polymorphisms for genetic typing of non-
human primates. *Primates* 34:323–331.

Wilkie, D. S., J. G. Sidle, and G. C. Boundzanga. 1992.
Mechanized logging, market hunting, and a bank loan in
Congo. *Conservation Biology* 6:570–580.

Wilks, C. 1990. La conservation des écosystèmes
forestiers du Gabon. In M. Collins and J. Sayer, eds. *Bibliothèque de la Conservation de l'UICN.* UICN, Gland,
Switzerland.

Williamson, E. A., C. E. G. Tutin, and M. Fernandez.
1988. Western lowland gorillas feeding in streams and on
savannas. *Primate Report* 19:29–34.

Williamson, E. A., C. E. G. Tutin, M. E. Rogers, and M.
Fernandez. 1990. Composition of the diet of lowland
gorillas at Lopé in Gabon. *American Journal of Primatology* 21:265–277.

Yamagiwa, J. 1994. Present situation and conservation of
mountain and eastern lowland gorillas. *Reichorai Kenkyul
Primate Res.* (Japanese with English summary)
10(3):347–362.

Yamagiwa, J., and N. Mwanza. 1994. Day-journey length
and daily diet of solitary male gorillas in lowland and
highland habitats. *International Journal of Primatology*
15(2):207–224.

# CHAPTER 26

# Why Chimpanzees and Gorillas Respond Differently to Logging

## A Cautionary Tale from Gabon

Lee J. T. White and Caroline E. G. Tutin

Commercial mechanized logging of tropical rain forest typically results in the destruction of about 50% of all trees present before exploitation, although damage varies greatly between sites, ranging from about 5% to more than 80% (see Johns 1992). Resulting changes in vegetation structure and composition are reflected by changes in animal communities. The exploitation of tropical timbers contributes significantly to the economies of many developing countries, but if this source of revenue is to be maintained, it is vital that logging operations be conducted on a sustainable basis. One cannot consider forest exploitation as sustainable unless the integrity of its ecological processes is maintained (Wyatt-Smith 1987). Plant-animal interactions in the tropical rain forest are complex and as yet poorly understood (see, e.g., Price et al. 1991). Given our current lack of knowledge of, for example, the role of animal seed dispersers in forest regeneration, or the effects of vegetation disturbance on insect pollinators, it is impossible to predict the long-term effects of logging. In practice, there is rarely even an attempt to monitor the initial changes after logging.

Primates are perhaps the best studied order of mammals, but our understanding of how primate communities respond to vegetation change following logging is poor. Johns and Skorupa (1987) reviewed the responses of rain forest primates to habitat disturbance and identified body size and percentage frugivory as useful predictors of how species respond to logging (large-bodied frugivores being most susceptible). However, there have been few community-level studies in Africa. Those that are available (Martin and Asibey 1979; Skorupa 1986; Howard 1991; Plumptre and Reynolds 1994) show that even closely related species may respond very differently. For example, in Kibale, Uganda, *Colobus guereza* increases in density in logged forest, while *Colobus badius* becomes significantly less abundant. In addition, responses of any given species can differ markedly between sites that are geographically close (for example, the findings of Skorupa 1986; Howard 1991; and Plumptre and Reynolds 1994, in Kibale, Kalinzu, and Budongo, Uganda, demonstrate opposite responses for several species). Johns (1989) illustrated this point well for several species of primate in different sites in Tekam Forest Reserve, Peninsular Malaysia.

Among the African apes, the chimpanzee

(*Pan troglodytes*) would appear to be more eco-logically adaptable than the gorilla (*Gorilla gorilla*) and the pygmy chimpanzee (*Pan panis-cus*). The geographical range of chimpanzees is much wider (Osman-Hill 1969), and they occur in a variety of habitats, including lowland trop-ical rain forest, Afromontane forest, and arid savannas, in which the only forest cover is in narrow galleries. In these savanna areas a single chimpanzee community may range over hun-dreds of square kilometers (Baldwin et al. 1982). They are serious crop raiders in some areas (Sugiyama 1981) and fashion a variety of tools to enable them to overcome diverse prac-tical problems in their daily routine (McGrew 1992). Nevertheless, studies in both east and west Africa have shown that chimpanzee densi-ties decline following logging (Tutin and Fer-nandez 1984; Skorupa 1986) while gorillas seem little affected (Tutin and Fernandez 1984).

Recent studies of the western lowland gorilla (*G. g. gorilla*) (Williamson et al. 1990; Tutin et al. 1991; Tutin and Fernandez 1993) and the eastern lowland gorilla (*G. g. graueri*) (Yamagiwa et al. 1994) have shown that both races consume large quantities of fruit in low-land tropical rain forest habitats. This evidence contrasts with early studies in montane and secondary habitats, which led to the classifica-tion of gorillas as folivores (e.g., Schaller 1963; Fossey and Harcourt 1977). In lowland tropical rain forest in Lopé, Gabon, the diets of gorillas and chimpanzees are quantitatively similar and contain many species of fruits, about 80% of which are consumed by both species (Tutin et al. 1991; Tutin and Fernandez 1993). However, their diets diverge during the annual dry sea-son, when fruit availability is low. Gorillas sub-sist almost entirely on vegetative foods, while chimpanzees maintain some of the fruit com-ponent of the diet and show social adjustment, reducing intraspecific competition for scarce fruit foods by foraging alone or in small groups (Tutin et al. 1991). It seems that subtle physio-logical differences allow gorillas to subsist on an essentially foliverous diet, while chimpanzees have to maintain a minimum fruit intake. This is supported by the fact that gorillas thrive in old secondary vegetation where chimpanzee densities are low (Tutin and Fernandez 1984) and in montane habitats where chimpanzees are absent (e.g., Schaller 1963).

It is possible that gorillas' adaptability and chimpanzees' dependence on fruit is the reason for their different responses to logging. Logging damage results in decreased densities of trees, which provide the majority of fruits consumed by chimpanzees and gorillas (Williamson et al. 1990; Tutin and Fernandez 1993), and hence in decreased food availability. However, gorillas might benefit from lush secondary vegetative growth in logged forests. Commercially impor-tant tree species in Africa tend to be important food trees for primates (see, e.g., Struhsaker 1975; Martin and Asibey 1979). If loggers cut species whose fruit are keystone foods for chim-panzees, even low-intensity exploitation could have serious effects on densities (Leighton and Leighton 1983). Hunting levels in central Africa often increase dramatically during and after logging (Dowsett-Lemaire 1991; Wilkie et al. 1992). Densities of both gorillas and chim-panzees decrease significantly in response to hunting, however (Tutin and Fernandez 1984), so it is unlikely that this can explain their differ-ent reactions to logging.

Tutin and Fernandez (1984) undertook a nationwide survey of gorilla and chimpanzee densities in Gabon and found sizable popula-tions of both species in much of the country. However, the completion of a railway cutting across Gabon from Libreville to Franceville in 1987 opened up large tracts of the interior of the country to logging (Tutin and Fernandez 1987; White 1994a). The aims of this study were to assess the damage levels due to logging and to investigate the responses of sympatric gorillas and chimpanzees to these changes. The study was based in the Lopé Reserve, central Gabon, for several reasons: a field study of gorillas and chimpanzees was started there in 1983 and is ongoing, and hence there is a wealth of background information (e.g., Tutin and Fernandez 1993); the reserve supports large populations of both species (White 1994b); logging had been under way in some parts of the reserve for up to twenty-five years, so long-term effects of logging could be analyzed; and finally, the reserve is effectively protected from poaching, providing the opportunity to evalu-

ate the effects of logging in the absence of this complicating factor.

Gabon straddles the equator along the west coast of Africa ($3°$ N–$3°$ S, $8°$ E–$15°$ E) and covers 267,667 km². The natural vegetation of 85% of the country is lowland rain forest (Caballé 1983), although the present extent of forest cover is estimated at 200,000 km², or 75% (Myers 1991). Research was carried out in the Lopé Reserve (5,320 km²), the largest of five legally protected areas in Gabon (figure 26.1). About 2,500 km² of the reserve have been

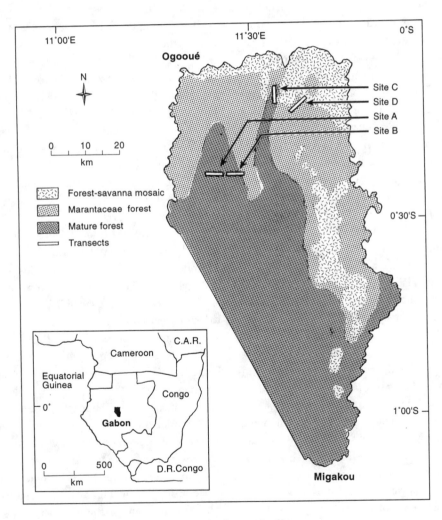

Figure 26.1. Location of the Lopé Reserve and the four study sites.

selectively logged at low intensity (1–2 trees per hectare), principally for one tree species, *Aucoumea klaineana*. The effects of logging on vegetation in the Lopé have been described in detail by White (1994a). Mean annual rainfall at the Station d'Etude des Gorilles et Chimpanzées (SEGC) is 1,503 mm (1984–1997). There are considerable interannual variations in the amount and distribution of rainfall: consistently low rainfall in July and August constitutes a major dry season, which is a feature of the annual cycle, although its onset and its duration vary between years, such that it may commence in June and extend into September; December to February are generally relatively dry months, but no distinct second dry season can be defined. The major dry season is a period of fruit scarcity (Tutin et al. 1991; White 1994b), when frugivorous mammals change their dietary habits or behavioral patterns in order to cope with low food availability (see, e.g., Tutin et al. 1991).

Two major forest types occur within the reserve. The first type, Marantaceae Forest, is dominated by species of Bursuraceae, Sterculiaceae, Mimosaceae, Ochnaceae, and Caesalpiniaceae, including a recently described species, *Cola lizae* (Sterculiaceae), which makes up about a quarter of all individuals above 10 cm dbh (Hallé 1987). Dominant species by basal area are (in decreasing order) *Aucoumea klaineana* (Burseraceae), *Cola lizae*, *Lophira alata* (Ochnaceae), *Pentaclethra macrophylla* (Mimosaceae), and *Dacryodes buettneri* (Burseraceae). Abundant ground vegetation is dominated by *Haumania liebrechtsiana* and *Megaphrynium* spp. (Marantaceae) and *Aframomum sericeum* (Zingiberaceae), which, combined, reach stem densities of 80,000 ha$^{-1}$ (Rogers and Williamson 1987) and form thickets rising up to heights of 10 m, creating vine towers around trees. Canopy cover is about 85% (see White, this volume, chapter 11).

The second forest type, Mature or Closed-Canopy Forest, has higher species diversity and different species composition than Marantaceae Forest. Dominant families are Burseraceae, Caesalpiniaceae, Olacaceae, Myristicaceae, and Euphorbiaceae, and the most important species in terms of basal area are *Aucoumea klaineana*, *Dacryodes buettneri*, *Scyphocephalium ocochoa* (Myristicaceae), *Santiria trimera* (Burseraceae), and *Coula edulis* (Olacaceae). Ground vegetation is sparse and dominated by shrubs in the family Rubiaceae. Combined Marantaceae and Zingiberaceae stem counts are below 10,000 ha$^{-1}$. Canopy cover is about 95%.

Work was undertaken in four study sites in the northern half of the reserve, one (the SEGC main study area) in Marantaceae Forest and three in Closed-Canopy Forest. White (1992) and Tutin et al. (1994) describe vegetation in these sites in detail. The four sites had been logged at different times in the past. A single line-transect 5 km in length was cut crossing the major drainage features in each site, starting from a randomly chosen point on a baseline. Table 26.1 presents summary descriptions of the four sites.

Damage levels in Gabon due to logging are typically about 10% (Wilks 1990). Lopé is no exception. Site A was logged during this study, and overall damage levels were about 5% (considering loss of trees ≥10 cm dbh, basal area, or canopy cover). However, only about half of this site was logged. The remainder was left unexploited because of steep terrain. Most of Sites B and C had been exploited, so lower stocking densities in these sites probably reflected higher damage levels (about 10%) due to more thorough logging. Tree species characteristic of secondary vegetation (e.g., *Anthocleista* spp., *Macaranga* spp., *Pauridiantha* spp., *Psorospermum* spp., *Xylopia* spp.) had become established in areas opened up by logging but were not common in any of the sites, being mostly restricted to roadsides (White 1994a).

# Table 26.1
## Details of Four Study Sites

| SITE | FOREST TYPE | TIME SINCE LOGGING (YEARS) | CANOPY COVER (%)[a] | D | BA | M | Z | R | S |
|---|---|---|---|---|---|---|---|---|---|
| A¹ | CCF | unlogged | 95 | 464 | 39.3 | 6,640 | 250 | — | — |
| A² | CCF | 0–1 | 90 | 442 | 36.8 | — | — | 0.7 | 5.0 |
| B | CCF | 3–5 | 93 | 381 | 33.2 | 10,140 | 1,730 | 1.3 | 7.5 |
| C | CCF | c. 15 | 94 | 388 | 40.0 | 9,560 | 390 | 1.8 | 7.2 |
| D | MF | c. 25 | 84 | 305 | 28.6 | 48,870 | 4,270 | 2.0 | — |

*Source:* For further details, see Tutin et al. 1994; White 1992, 1994; White et al. 1995.

*Notes:* A¹ = site A before logging; A² = site A after logging; D = density of trees and lianes ≥ 10 cm dbh (individuals ha⁻¹) as determined on a transect of 5 km x 5 m; BA = corresponding basal area of trees and lianes ≥ 10 cm dbh (m² ha⁻¹); M = stem density of Marantaceae from a sample of 1,000 quadrats of 1 m² at 5 m intervals along the transects (no repeat was undertaken after logging); Z = stem density of Zingiberaceae from a sample of 1,000 quadrats of 1 m² at 5 m intervals along the transects (no repeat was undertaken after logging); R = percentage of site covered by visible roads; S = percentage of site covered by visible skidder trails (in site D, trails could not be detected with certainty); CCF = Closed-Canopy Forest; MF = Marantaceae Forest.

a Canopy cover was assessed using a point quadrat method (Grieg-Smith 1983), which involved looking up vertically through a clinometer and scoring "+" or "−" depending on whether the sight fell on vegetation. This process was repeated 1,000 times (at 5 m intervals along each transect). The percentage of points that hit vegetation is given to represent canopy cover.

Herbaceous plants in the Marantaceae and Zingiberaceae families are keystone foods (*sensu* Terborgh 1986) for both gorillas and chimpanzees in the long dry season in Lopé (Tutin et al. 1991). Densities of these plants were somewhat elevated in logged forest, particularly close to old roads and in large gaps, such as those caused by logging (see White et al. 1995). However, densities in logged Closed Canopy Forest did not approach those found naturally in Marantaceae Forest (see also White, this volume, chapter 11).

Wildlife censuses were conducted along each transect. Standard primate census techniques (Whitesides et al. 1988) were employed to record sightings of apes. Transects were walked once or twice per month between February 1989 and June 1991, giving 21, 19, 35, 31, and 43 repetitions in sites A (before and after logging), B, C, and D, respectively. In addition, ape nest counts were undertaken monthly. For each gorilla and chimpanzee nest site sighted, the location along the transect, age, and the distance to all other nests sighted from the transect were noted. Maps of each nest site were made as scale drawings assigning a number to each nest, and detailed data were taken for all nests (including those not visible from the transect). The distance to the nest group center was determined by calculating the average distance of all nests from the transect, scoring one side positive and the other side negative if nests occurred on both sides. Both these data sets were analyzed following Whitesides et al. (1988), using perpendicular distances to the first individual sighted (and correcting for group spread) or to the nest group center to calculate strip widths (see White 1994b for more details).

Table 26.2 gives mean encounter rates for gorillas and chimpanzees, and nest sites for both species. There were no statistically significant differences in sighting distances between sites (Kruskal-Wallis one-way ANOVAs, $P > 0.05$ in all cases—Siegel and Castellan 1988). Group

and nest sighting frequencies fitted a Poisson distribution, so a one-way ANOVA with Tukey HSD multiple comparisons was used to test for differences between sites, using square-root transformed data (Sokal and Rohlf 1981; see White 1994b). There were statistical differences ($P < 0.05$, two-tailed) in encounter rates for some sites). Gorilla nest counts showed that Site D had higher gorilla densities than the other sites, but there was no clear response to logging. For chimpanzees, Site A had high rates of observations and nest site encounters. Observation rates for Site A were statistically higher than for Sites B, C, and D. Nest encounters were statistically higher before logging in Site A, and Site B, logged three years before this study, had particularly low chimpanzee densities.

Figure 26.2 shows a graphical representation of chimpanzee nest data. The cumulative mean encounter rate is calculated after each census on each transect and plotted against the number of censuses. After initial noise due to small sample sizes, data for Sites B, C, and D settle close to their final overall averages, as do the data for Site A after logging. However, Site A had high nest group encounter rates in the first few months of the study, with ten, nine, and eight fresh nest groups in the first three months, respectively. These were concentrated in the last 2 km of the transect, in the area later exploited and closest to logging activities at the time. In all the other sites combined, only one census with ten fresh nest groups occurred, and otherwise counts did not exceed five and were rarely above three. During these first months of work in Site A, large, excited groups of chimpanzees were frequently encountered, or heard pant hooting, screaming, and drumming. After the fifth month of the study, logging was under way about 1 km from the transect, and the number of nests (and individuals seen and heard) decreased. After logging, nest encounter rate was stable but low, and comparable to Site B.

Table 26.2

## Mean Encounter Rates for Gorillas, Chimpanzees, and Their Nests

| SPECIES | SITE A$^1$ | | SITE A$^2$ | | SITE B | | SITE C | | SITE D | |
|---|---|---|---|---|---|---|---|---|---|---|
| | MER | N | MER | N | MER | N | MER | N | MER | N |
| *Gorilla gorilla* | | | | | | | | | | |
| Sightings | 0.20 | 4 | 0.17 | 3 | 0.11 | 4 | 0.13 | 4 | 0.21 | 9 |
| Nest sites | 0.44 | 4 | 0.25 | 3 | 0.58 | 14 | $^D$0.35 | 7 | 1.04 | 27 |
| Density (km$^{-2}$) | 0.4 | | 0.3 | | 0.5 | | 0.3 | | 1.0 | |
| *Pan troglodytes* | | | | | | | | | | |
| Sightings | 0.45 | 9 | 0.17 | 3 | $^A$P | 0 | $^A$0.03 | 1 | 0.14 | 6 |
| Nest sites | 4.56 | 41 | $^{A,D}$1.00 | 12 | $^{A,C,D}$0.75 | 18 | 1.90 | 38 | 2.81 | 73 |
| Density (km$^{-2}$) | 1.1 | | 0.2 | | 0.2 | | 0.5 | | 0.7 | |

*Notes:* A$^1$ = site A before logging; A$^2$ = site A after logging; MER = mean encounter rate per census; N = number of observations; P = present but not seen. Superscript letters indicate statistical differences between sites: superscript A indicates a statistical difference between this site and site A; superscript B indicates a statistical difference between this site and site B; and so on.

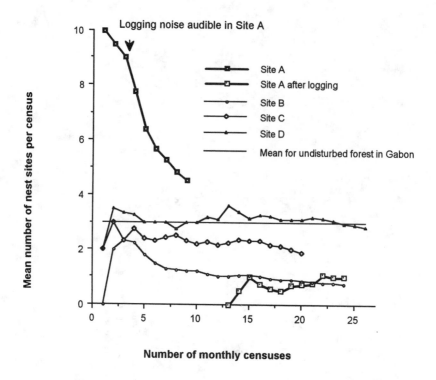

Figure 26.2. Mean encounter rates for chimpanzee nest sites (the curve for Site A after logging is shown continuing in time from censuses conducted before exploitation began).

Levels of logging damage in this study were low. The study was undertaken in a reserve, and there was no hunting in any of the study sites. White (1994c) studied patterns of fruit production in Lopé and found no evidence for differences in the amount of fruit available in the four sites included in this study. Logging concentrated on *Aucoumea klaineana*, which is not a food species for apes. Several species that produce fruits eaten by chimpanzees or gorillas were cut, but none of these provide keystone foods in the dry season, when community fruiting is at a minimum. Therefore, one would predict that logging in Lopé was unlikely to have had serious ecological effects on apes. In addition, logging caused a small increase in density of some species of Marantaceae and Zingiberaceae, keystone foods in the dry season.

Logging in Site A, however, resulted in an immediate decrease in chimpanzee numbers, and there was no sign of recovery during the first year after logging. When this study began, densities were high and the chimpanzees seemed unusually excited, calling and drumming on tree buttresses frequently. Pre-logging nest counts correspond to a density of 1.1 chimpanzees/km² (White 1994b), but densities in the first four months of the study were higher, suggesting perhaps that chimpanzees had been displaced by the logging to the north. One has to be cautious about extrapolating the effects of logging from studies in several sites logged at different times (Johns 1989). However, Sites A, B, and C were located in the same forest type and had similar plant species composition. Chimpanzee nest counts in Site A after logging,

Site B (logged 3–5 years previously), and Site C (logged 10–15 years previously) correspond to densities of 0.2, 0.2, and 0.5 individuals/km², respectively (White 1994b). In Site D, chimpanzee density was 0.7 individuals/km², which is the mean chimpanzee density in undisturbed rain forest in Gabon (Tutin and Fernandez 1984). This suggests that the recovery of chimpanzee densities after logging is a slow process, taking 15–25 years.

Given that damage levels are low and food trees are not selectively extracted, why should there be such a marked and long-term decrease in densities of an ecologically adaptable species like the chimpanzee? The abruptness of the response suggests that the reason is not an ecological one. If it were, one would expect a gradual decrease in condition, a lowering of reproductive rate or a population crash at some later date, when any resulting reduction in food availability became critical during a period of fruit scarcity (see Foster 1982). Thus, it seems likely that the explanation is a social one.

Chimpanzees live in stable communities with large home ranges, which partially overlap. Territorial behavior is shown, notably patrolling and defense of boundaries. Encounters between groups can be violent and can result in mortality, particularly among males and infants (Goodall et al. 1979; Nishida et al. 1985; Goodall 1986). Long-term studies at Gombe and Mahale in Tanzania have documented (once in each site) extreme intercommunity aggression, during which individuals from a larger community systematically tracked and killed members of another, resulting in its extinction. During these periods of conflict there were violent group attacks by members of one community on both males and females of the other, which resulted in fatal injuries. Whenever females enter the territory of a "foreign" community, they are generally subject to attacks by both resident males and resident females, which may result in the death and cannibalism of their infants (Goodall 1986).

Humans and chimpanzees have coexisted in Lopé for at least 60,000 years (Tutin and Oslisly 1995). There is evidence from Cameroon that Stone Age humans consumed (and presumably hunted) chimpanzees (de Maret et al. 1987). Chimpanzees in Lopé have remained shy despite eleven continuous years of effort to habituate them (Tutin and Fernandez 1990), suggesting that they actively avoid humans if possible, although there have been no human residents in the immediate vicinity for a century or more.

It seems likely that the noise and disturbance caused by loggers make chimpanzees leave the immediate area. In Site A, logging operations advanced on a continuous front 5–10 km wide from north to south. Noise from active areas can be heard from 5 km or more away and is likely to cause stress and disruption within a community of chimpanzees with no previous experience of such disturbance. It is not unlikely that the advancing chaos could displace entire chimpanzee communities.

The high chimpanzee density observed in Site A at the beginning of this study could reflect such a displacement, and the general excitement might have been the result of two communities coming into contact. The displacement of an entire community of chimpanzees would probably never occur naturally. However, observations from such long-term field sites as Gombe and Mahale suggest that it could result in violent conflict, which could cause high mortality and the eventual disappearance of one group, likely the smaller. This rather disturbing scenario would explain the changes caused by logging. Chimpanzees may respect boundaries between their own and neighboring communities over a number of years and tend to show tense, nervous behavior when in peripheral areas of their range. Conspecifics from neighboring social groups who intrude into the home range are aggressively expelled by the residents. Therefore, one would expect surviving chimpanzee communities to be cautious about exploring the empty areas left in the wake of logging (see Goodall 1986:526).

Gorillas showed no consistent pattern of density change in sites logged at different times. Site D had significantly higher gorilla densities than the other sites, and this was because it was in a different forest type with an abundance of food plants in the families Marantaceae and Zingiberaceae (White 1994b). Any increases in density of these herbs due to logging were minimal compared to natural differences between different forest types (see table 26.1). One would predict that recently logged forest might be able to support slightly higher gorilla densities given the small increase in keystone herb foods, but the loss of up to 10% of trees that provide fruits in the dry season may well counteract this. Chimpanzees consume increased quantities of these herbs during periods of fruit scarcity (Tutin et al. 1991), so they might be expected to benefit similarly. Logging certainly does not create the type of secondary vegetation found on old village sites and which supports high densities of gorillas (Tutin and Fernandez 1984). Indeed, it seems that any increases in herb densities are short lived and would probably disappear before gorillas, which have a slow reproductive rate (Harcourt et al. 1980), had time to respond.

Hence, there seems to be no ecological reasons to expect increased gorilla numbers in logged forest in Lopé. But why do they not show the same response as chimpanzees? There are perhaps two reasons why gorillas are not subject to the same social stress during logging. First, home ranges of adjacent groups overlap almost entirely, and intergroup encounters rarely involve aggression (Tutin et al. 1992), so gorilla groups will be better able to avoid active logging. Second, the social organization of gorillas is based around a dominant adult male (silverback) who takes on the role of protector (see,

e.g., Schaller 1963). Therefore, female and immature group members are likely to be less stressed during logging than are chimpanzees, whose fission-fusion organization (Goodall 1968; Nishida 1968) means that individuals need to be more independent, and hence that females with infants will be particularly vulnerable to logging, since they will be attacked if displaced into the home range of another community. There are parallels between the chimpanzee and the orangutan (*Pongo pygmaeus*), an essentially solitary species, which was heavily hunted during the Stone Age by humans (this continues today in some places) and which also moves out of areas being logged. However, this ape lacks strict territories and should be able to adapt socially to these displacements (van Schaik 1992 and pers. comm.).

Where hunting occurs alongside or after logging, its effect needs to be considered. It could be argued that gorillas will be more vulnerable to hunting than chimpanzees because the death of the silverback, the most likely target for hunters, could result in the deaths of all unweaned group members due to infanticide (Watts 1990) (although the intimidating presence of the silverback may have made them less susceptible to attack than chimpanzees). Yet chimpanzees respond to imitations of duiker calls by hunters, advancing toward the source of the call, where they are easily shot. Tutin and Fernandez's more general survey results (1984), which showed chimpanzee densities to be lower in logged forests while gorillas are little affected, included areas of intense hunting pressure. Neither species was specifically targeted by hunters, so the differences detected in logged and unlogged forests are likely to have been due to logging per se and not to hunting.

The chimpanzee is an intelligent species able to survive in a wide variety of habitats. One would predict that they could adapt easily to habitat changes due to logging. Despite low-intensity logging, which leaves keystone food species essentially untouched, densities of chimpanzees declined significantly immediately after exploitation and remained low for many years. The explanation for this appears to be social, rather than ecological. This emphasizes the complexity of the tropical rain forest ecosystem. To understand the ecological impact of logging, we need to improve our knowledge of the ecological interactions within tropical rain forests. Furthermore, this study demonstrates that in order to understand the effects of logging on large mammals, knowledge of their social behavior is required. The only means of acquiring such knowledge is to undertake long-term multidisciplinary field studies.

This chapter should be taken as both a warning and a pointer. It stresses our lack of ability to predict the effects of logging in tropical forests. We have no firm explanation for the consistent and long-lasting drop in chimpanzee numbers in Lopé and in other logged forests across Africa, only rather sinister theories. In a system where 70–90% of plant species are dependent on animals (particularly larger mammals and birds) for dispersal of their seeds (see, e.g., White 1994c), one cannot assess the sustainability of a logging operation without monitoring changes in densities of these species. Chimpanzees alone are major seed dispersers (Tutin and Fernandez 1993). Future studies should aim to assess whether other highly territorial animal species are similarly susceptible to logging, as may be the case for gibbons (J. E. Bennett, pers. comm.). In addition, it is imperative that further research be undertaken to determine the mechanisms that shape the response of chimpanzees to logging, given that most of their habitat will be logged in the near future (see, e.g., Johns and Skorupa 1987).

Plumptre and Reynolds (1994) found no evidence of negative effects of logging on chimpanzees in Budongo, Uganda. This forest, covering 428 km², has been logged selectively for more than sixty years but maintains a high chimpanzee density. Logging has occurred at different times in a series of compartments each of about 4–10 km² (calculated from a figure in Plumptre and Reynolds 1994), smaller than the home range of chimpanzee communities in Budongo, estimated at about 15–20 km² (Reynolds 1965). In these circumstances, chimpanzees would have the option of moving to a different part of their home range during logging without being forced to trespass into the territory of another community. This type of logging is on a very different scale from the mechanized exploitation under way in much of central Africa, where concessions cover hundreds or thousands of square kilometers.

Several studies have shown that better planning of extraction programs can substantially reduce the environmental damage caused and can significantly reduce financial costs of exploitation (see, e.g., Buenaflor and Tiki 1989; Marn 1982, cited in Johns 1989). It is a great advantage to build advance roads during the dry season to facilitate exploitation during the rains. It is possible that this could be combined with different logging patterns, which minimize the negative effects on wildlife. For the moment, however, our knowledge of these changes is minimal, as has been stressed elsewhere (see Johns 1992). We are not in a position to make firm recommendations to logging companies or government agencies, beyond such commonsense measures as minimizing damage and limiting hunting.

In the past it was unheard of for logging companies in Gabon to undertake environmental impact studies. The same was true elsewhere in central Africa, unless such a study was a prerequisite for obtaining a development bank loan (Wilkie et al. 1992). In this respect, logging companies are somewhat behind the oil industry, which began to undertake impact studies about ten years ago (Basquin et al. 1991; White 1994a). There are signs that some logging companies operating in the tropical rain forests of central Africa are beginning to shoulder their responsibilities and to think about how to minimize their impact on the environment. Unless this trend extends to the entire industry in the near future, the elusive goal of sustainable exploitation of tropical forests will remain just that. Two factors working strongly against any changes in logging practice are the short duration of logging permits and poor supervision of logging operations by government officials in much of Africa. Any new awareness among loggers needs to be driven by the government departments responsible for overseeing logging activities. Until this is achieved, it is unlikely that recommendations aimed at reducing the environmental effects of logging will be converted into action. Therefore, a priority for conservation and development organizations should be further investigation of: (1) the ecological effects of logging; (2) possible means of minimizing negative effects; and (3) training of government personnel to supervise and monitor logging operations.

ACKNOWLEDGMENTS

Lee J. T. White thanks the Wildlife Conservation Society, the Leverhulme Trust, the Royal Society (London), the Conder Conservation Trust, the Richard Brown Scholarship (University of Edinburgh), the Conrad Zweig Trust, the Institute of Cell, Animal and Population Biology (University of Edinburgh), and the Société Forestier du Gabon (SOFORGA) for financial and logistical support. Caroline E. G. Tutin thanks L. S. B. Leakey Foundation, the World Wide Fund for Nature, and especially the Centre International de Recherches Médicales de Franceville for financial support. Alphonse Mackanga-Missandzou, Joseph Maroga-Mbina, and the Direction de la Faune kindly granted permission to work in Lopé. We thank Gordon McPherson, Chris Wilks, Frank White, and David Harris for help with botanical identifications. Kate Abernethy, Liz Bennett, and Carel van Schaik commented on the manuscript.

REFERENCES

Baldwin, P. J., W. C. McGrew, and C. E. G. Tutin. 1982. Wide-ranging chimpanzees at Mt. Assirik, Senegal. *International Journal of Primatology* 3:367–385.

Basquin, P., G. Van Beek, P. Christy, B. Clist, R. Guicherit, S. Lahm, A. Moungazi, J. Reitsma, H. Waardenburg, L. White, and C. Wilks. 1991. *Maguelou: An Environmental Study of the Ofoubou Area*. Dupont.

Buenaflor, V., and T. Tiki. 1989. *Forest Management Research and Development Papua New Guinea*. UNDP/FAO, Lae, Papua New Guinea.

Caballé, G. 1983. Végétation. Pages 34–37 in Barret, J., ed. *Géographie et cartographie du Gabon: Atlas illustré*. EDICEF, Paris.

Dowsett-Lemaire, F. 1991. Exploitation du bois dans le bassin de Kouilou au Congo: Flore et faune du bassin du Kouilou (Congo). *Turaco Research Report* 4:53–56.

Fossey, D., and A. H. Harcourt. 1977. Feeding ecology of free-ranging mountain gorilla (*Gorilla gorilla beringei*). Pages 415–447 In T. Clutton-Brock, ed. *Primate Ecology*. Academic Press, London.

Foster, R. B. 1982. Famine on Barro Colorado Island. Pages 201–212 in E. G. Leigh, A. S. Rand, and D. M. Windsor, eds. *The Ecology of a Tropical Forest: Seasonal Rhythms and Long Term Changes*. Smithsonian Institution Press, Washington, D.C.

Goodall, J. v. L. 1968. The behaviour of free-living chimpanzees in the Gombe Stream Reserve. *Animal Behavior Monographs* 1:161–311.

———. 1986. *The Chimpanzees of Gombe*. Harvard University Press, Cambridge.

Goodall, J. v. L., A. Bandora, E. Bergmann, B. Busse, H. Matama, E. Mpongo, A. Pierce, and D. Riss. 1979. Intercommunity interactions in the chimpanzee population of the Gombe National Park. Pages 13–53 in D. Hamburg and E. McCown, eds. *The Great Apes*. Benjamin and Cummings, Menlo Park, Calif.

Grieg-Smith, P. 1983. *Quantitative Plant Ecology*. Studies in ecology, vol. 9. University of California Press, Berkeley.

Hallé, N. 1987. *Cola lizae* N. Hallé (Sterculiaceae), nouvelle espèce du Moyen Ogooué (Gabon). *Adansonia*, série 4, 9(3):229–237.

Harcourt, A. H., D. Fossey, K. S. Stewart, and D. P. Watts. 1980. Reproduction in wild gorillas and some comparisons with chimpanzees. *Journal of Reproduction and Fertility*, supplement 28:59–70.

Howard, P. C. 1991. *Nature Conservation in Uganda's Tropical Forest Reserves*. IUCN, Gland, Switzerland.

Johns, A. D. 1989. *Timber, the Environment and Wildlife in Malaysian Rain Forests.* Institute of South-east Asian Biology, University of Aberdeen, Scotland.

———. 1992. Species conservation in managed tropical forests. Pages 15–53 in J. A. Sayer and T. C. Whitmore, eds. *Realistic Strategies for Tropical Forest Conservation.* IUCN, Gland, Switzerland.

Johns, A. D., and J. P. Skorupa. 1987. Responses of rain forest primates to habitat disturbance: A review. *International Journal of Primatology* 8:157–191.

Leighton, M., and D. R. Leighton. 1983. Vertebrate responses to fruiting seasonality within a Bornean rain forest. Pages 181–196 in S. L. Sutton, T. C. Whitmore, and A. C. Chadwick, eds. *Tropical Rain Forest: Ecology and Management.* Blackwell Scientific, Oxford.

Maret, P. de, B. Clist, and W. Van Neer. 1987. Résultats des premières fouilles dans les abris sous roche de Shum Laka et d'Abeke au nord ouest du Cameroun. *L'Anthropologie* 91:559–584.

Marn, H. M. 1982. The planning and design of the forest harvesting and log transport operation in the mixed dipterocarp forest of Sarawak. UNDP/FAO Field Document MAL/76/008, no. 17.

Martin, C., and E. O. A. Asibey. 1979. Effect of timber exploitation on primate populations and distribution in the Bia rain forest area of Ghana. Paper presented to the Seventh IPS Congress, Bangalore, India.

McGrew, W. C. 1992. *Chimpanzee Material Culture: Implications for Human Evolution.* Cambridge University Press, Cambridge.

Myers, N. 1991. Tropical forests: Present status and future outlook. *Climatic Change* 19:3–32.

Nishida, T. 1968. The social group of wild chimpanzees in the Mahale Mountains. *Primates* 7:141–204.

Nishida, T., M. Hiraiwa-Hasegawa, T. Hasegawa, and Y. Takahata. 1985. Group extinction and female transfer in wild chimpanzees in the Mahale National Park, Tanzania. *Z. Tierpsychol.* 67:284–301.

Osman-Hill, W. C. 1969. The nomenclature, taxonomy and distribution of chimpanzees. Pages 22–49 in *The Chimpanzee,* vol. 1. Karger, Basel, Switzerland.

Plumptre, A. J., and V. Reynolds. 1994. The impact of selective logging on primate populations in the Budongo Forest Reserve, Uganda. *Journal of Applied Ecology* 31:631–641.

Price, P. W., T. M. Lewinsholme, G. W. Fernandes, and W. W. Benson. 1991. *Plant-Animal Interactions: Evolutionary Ecology in Tropical and Temperate Regions.* John Wiley and Sons, New York.

Reynolds, V. 1965. *Budongo: A Forest and Its Chimpanzees.* Natural History Press, New York.

Rogers, M. E., and E. A. Williamson. 1987. Density of herbaceous plants eaten by gorillas in Gabon: Some preliminary data. *Biotropica* 19:278–281.

Schaller, G. B. 1963. *The Mountain Gorilla: Ecology and Behavior.* University of Chicago Press, Chicago.

Siegel, S., and Castellan, N. J. 1988. *Nonparametric Statistics for the Behavioral Sciences.* McGraw-Hill, New York.

Skorupa, J. P. 1986. Responses of rainforest primates to selective logging in Kibale Forest, Uganda: A summary report. Pages 57–70 in K. Benirschke, ed. *Primates: The Road to Self-Sustaining Populations.* Springer-Verlag, New York.

Sokal, R. R., and F. J. Rohlf. 1981. *Biometry.* 2d ed. W. H. Freeman, New York.

Struhsaker, T. T. 1975. *The Red Colobus Monkey.* University of Chicago Press, Chicago.

Sugiyama, Y. 1981. Observations on the population dynamics and behaviour of wild chimpanzees at Bossou, Guinea, in 1979–80. *Primates* 26:435–444.

Terborgh, J. 1986. Keystone plant resources in the tropical forest. In M. E. Soulé, ed. *Conservation Biology: The Science of Scarcity and Diversity.* Sinauer Associates, Sunderland, Mass.

Tutin, C. E. G., and M. Fernandez. 1984. Nationwide census of gorilla (*Gorilla g. gorilla*) and chimpanzee (*Pan t. troglodytes*) populations in Gabon. *American Journal of Primatology* 6:313–336.

———. 1987. Gabon: A fragile sanctuary. *Primate Conservation* 8:160–161.

———. 1990. Responses of wild chimpanzees and gorillas to the arrival of primatologists: Behaviour observed during habituation. Pages 187–197 in H. O. Box, ed. *Primate Responses to Environmental Change.* Chapman and Hall, London.

———. 1993. Composition of the diet of chimpanzees and comparisons with that of sympatric lowland gorillas in the Lopé Reserve, Gabon. *American Journal of Primatology* 30:195–211.

Tutin, C. E. G., M. Fernandez, M. E. Rogers, E. A. Williamson, and W. C. McGrew. 1991. Foraging profiles of sympatric lowland gorillas and chimpanzees in the Lopé Reserve, Gabon. *Philosophical Transactions of the Royal Society, London B* 334:179–186.

Tutin, C. E. G., and R. Oslisly. 1995. Homo, Pan and Gorilla: Co-existence over 60,000 years at Lopé in central Gabon. *Journal of Human Ecology* 28: 597–602.

Tutin, C. E. G., L. J. T. White, E. A. Williamson, M. Fernandez, and G. McPherson. 1994. List of plant species identified in the northern part of the Lopé Reserve, Gabon. *TROPICS* 3:249–276.

van Schaik, C. 1992. Orangutans as indicators of biomass and diversity of wildlife and human disturbance in Sumatra. Presentation and p. 82 in *Abstracts of the XIVth Congress of the International Primatological Society*. Strasbourg, France.

Watts, D. P. 1990. Ecology of gorillas and its relation to female transfer in mountain gorillas. *International Journal of Primatology* 11:21–45.

White, L. J. T. 1992. Vegetation history and logging damage: Effects on rain forest mammals in the Lopé Reserve, Gabon. Ph.D. dissertation, University of Edinburgh.

———. 1994a. The effects of commercial mechanised logging on forest structure and composition on a transect in the Lopé Reserve, Gabon. *Journal of Tropical Ecology* 10:313–322.

———. 1994b. Biomass of rain forest mammals in the Lopé Reserve, Gabon. *Journal of Animal Ecology* 63:499–512.

———. 1994c. Patterns of fruit-fall phenology in the Lopé Reserve, Gabon. *Journal of Tropical Ecology* 10:289–312.

White, L. J. T., M. E. R. Rogers, C. E. G. Tutin, E. A. Williamson, and M. Fernandez. 1995. Herbaceous vegetation in different forest types in the Lopé Reserve, Gabon: Implications for keystone food availability. *African Journal of Ecology* 33:124–141.

Whitesides, G. H., J. F. Oates, S. M. Green, and R. P. Kluberdanz. 1988. Estimating primate densities from transects in a west African rain forest: A comparison of techniques. *Journal of Animal Ecology* 57:345–367.

Whitmore, T. C. 1984. *Tropical Rain Forests of the Far East*. 2d ed. Clarendon Press, Oxford.

Wilkie, D. S., J. G. Sidle, and G. C. Boundzanga. 1992. Mechanized logging, market hunting, and a bank loan in Congo. *Conservation Biology* 6:570–580.

Wilks, C. 1990. *La Conservation des Ecosystèmes Forestiers du Gabon*. IUCN, Gland.

Williamson, E. A., C. E. G. Tutin, M. E. Rogers, and M. Fernandez. 1990. Composition of the diet of lowland gorillas at Lopé in Gabon. *American Journal of Primatology* 21:265–277.

Wyatt-Smith, J. 1987. The management of tropical moist forest for the sustained production of timber: Some issues. IUCN/IIED Tropical Forestry Policy Paper 4.

Yamagiwa, J., N. Mwanza, T. Yumoto, and T. Maruhashi. 1994. Seasonal changes in the composition of the diet of eastern lowland gorillas. *Primates* 35:1–14.

# CHAPTER 27

# The Effects of Habitat Change
# Due to Selective Logging on the Fauna of Forests in Africa

Andrew J. Plumptre

Forest disturbance through natural or human-induced events alters the structure and composition of forests, affecting regeneration patterns and the availability of food for animals. Selective logging as a means of obtaining a sustainable yield of timber is promoted as a compromise between destroying a forest completely and leaving it intact (Whitmore and Sayer 1992). Most countries where tropical forests occur cannot afford to put money into protected areas unless they can generate some income from these areas. It will never be possible to finance total protection for the huge area of tropical forest in the central African basin. Selective logging, however, could provide one way of generating income without completely destroying the forest. If logging is to be truly sustainable, it is important to know how the fauna and flora of these forests are affected by selective logging in order to identify which animals are likely to be at risk and how logging practices could be improved so that this risk is lessened.

It is estimated that about 40% of the central African forest basin consists of monodominant forest dominated by a small number of species of trees (Connell and Lowman 1989; Hart et al. 1989). In certain areas, such as the Ituri Forest in eastern Congo where *Gilbertiodendron dewevrei* is monodominant, one species may form more than 80% of the standing trees (Hart 1985; Hart et al. 1989). Most of these dominant species belong to the Leguminosae and therefore bear bean-like fruits. The fruiting of these trees often follows a masting pattern, where all the trees fruit at one time, and consequently at other times of year there is often a dearth of fruit.

Between 1991 and the present, the Budongo Forest Project in western Uganda has been undertaking research specifically aimed at answering questions concerning the effects of selective logging. The Budongo Forest Reserve contains monodominant forest, dominated by *Cynometra alexandri*, which is thought to steadily replace other forest types (Eggeling 1947a) to form a climax community. Several researchers have carried out studies of the effects of selective logging on different faunal groups in the reserve. This chapter summarizes the findings of this work and develops a general model that predicts how different taxa and guilds within taxa will change in abundance following disturbance. This model was then tested in the Ituri Forest to determine how applicable it is to other sites in Africa.

Table 27.1
## Logging History of the Study Compartments

| COMPARTMENT | DATE LOGGED | VOLUME OF TIMBER REMOVED (M³/HA) | ARBORICIDE TREATMENT[a] |
|---|---|---|---|
| N15 | unlogged | none | none |
| KP11–13 (KP) | unlogged | none | none |
| B4 | 1941–42 | 34.8 | 1957 |
| N3 | 1947–52 | 80.0 | 1960–61 |
| N11 | 1960 | 26.5 | 1959–60 |
| W21 | 1963–64 | 36.1 | 1963–64 |
| B1[b] | 1935 | 19.9 | 1958 |
| B1[b] | 1983–86 | 21.5 | none |
| K4 | 1988–92 | 20.7 | none |

a   Arboricide was applied to all trees considered unprofitable for harvesting, particularly *Cynometra alexandri*.

b   Compartment B1 has been harvested twice, and both sets of figures are given.

## The Budongo Forest

The Budongo Forest Reserve (1°37'–2°00' N, 31°22'–31°46' E) covers 793 km² of moist semi-deciduous forest and grassland, 428 km² of which are forested. The reserve lies at a mean altitude of 1,100 m (range 950–1,200 m) and was gazetted by the British Colonial Administration between 1932 and 1939 (Howard 1991). Timber was extracted at low levels from about 1910 until the first foresters were posted to the forest in the early 1930s (Paterson 1991). It was always intended that the forest be harvested sustainably, and logging operations were detailed in working plans from 1933 (Harris 1933; Eggeling 1947b; Trenaman et al. 1956; Philip 1964). Eggeling (1947a) classified the vegetation into four forest types: *Cynometra* forest, mixed forest, colonizing woodland, and swamp forest. The first three types, he argued, followed an ecological succession from colonizing woodland to mixed forest to *Cynometra* forest, with colonizing-mixed and *Cynometra*-mixed ecotones. Details about the species composition of these forest types can be found in Eggeling 1947a and Synnott 1985. The most

valuable timber species in the forest, *Lovoa trichilioides, Milicia excelsa, Khaya anthotheca, Entandrophragma angolense, E. cylindricum,* and *E. utile,* tended to occur in the mixed forest and became excluded, he concluded, by the slower growing *Cynometra alexandri.*

Initially, it was planned to remove all the old timber trees of more than 1.3 m diameter at breast height (dbh) followed eighty years later by felling of smaller trees with the aim of creating a two-stage uniform crop of trees that would be felled at forty-year intervals (polycyclic felling). During the 1950s research showed that damage to the forest due to the logging operations and the slow growth rates of the trees meant that polycyclic felling would lead to lower yields than felling on longer cycles (Dawkins 1958). Therefore the logging plans were changed to monocyclic felling at eighty-year intervals (Philip 1964). Felling limits of timber trees were reduced to 85 cm dbh. Offtake of timber varied between 10 m³/ha and 80 m³/ha in different compartments of the forest. Replanting of mahoganies (*Khaya* and *Entandrophragma*) was carried out in logged areas

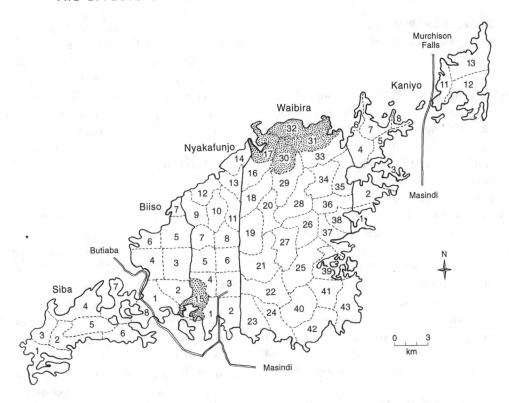

Figure 27.1. A map of Budongo Forest Reserve, showing the position of the seventy compartments. In table 27.1 and in the text, compartments are identified by a letter-number combination in which the letter indicates a region of the forest: S (Siba), B (Biiso), N (Nyakafunjo), W (Waibira), K (Kaniyo), and KP. Solid lines separate these regions, whereas dotted lines separate compartments within a region. The stippled areas are protected areas that have never been logged or treated with arboricide.

during the 1940s and early 1950s to encourage the regeneration of these species; however, research showed that natural regeneration was just as successful, and replanting ceased by the mid-1950s (Trenaman et al. 1956). It was found that opening up the canopy during the logging operation aided the growth of the mahoganies (Trenaman et al. 1956). Consequently, during the 1950s and 1960s, arboricide (2,4,5-T and 2,4-D in 1:2 proportions and mixed with diesel) was applied to trees that were not marketable ("weedy species"), particularly *Cynometra*, to

open up the canopy, reduce the extent of mono-dominant forest, and encourage the spread of the mixed forest (Philip 1964). This treatment ceased in the 1970s when more trees became marketable and it became difficult to import the chemicals (Synnott 1985).

Research was carried out in eight of the seventy compartments in the forest reserve (figure 27.1), two unlogged areas and six logged areas that were selected to represent different harvesting histories since 1940. The histories of these eight compartments are given in table 27.1.

## Studies of the Effects of Logging on the Fauna in Budongo

VEGETATION

Determining the effects of the forest management activities on the vegetation is not easy because many of the major changes took place some time ago. However, there are two sources of information that allow us to determine how Budongo has changed over time: aerial photographs and permanent plots.

In 1951 and in 1990, sets of aerial photographs were made of the whole forest, which allowed the main habitat types to be mapped before and after arboricide treatment (Plumptre 1996). These show that the extent of *Cynometra*-dominated forest was reduced greatly (loss of 103 km², or 24% of the forest), while mixed forest expanded (increase of 176 km², or 41% of the forest). Colonizing forest was also reduced during this period as the forest matured and the colonizing forest became mixed forest.

In addition, Eggeling (1947a) established eleven permanent sample plots during the mid-1930s which have been monitored over the years and were recently evaluated (Sheil 1996). These plots show that Eggeling's hypothesis of a succession from colonizing forest to *Cynometra* forest via mixed forest does hold in general,

although there seems to be some cycling as trees mature and die. Studies of an additional plot that was established in the 1950s to evaluate arboricide treatment also show that where arboricide was applied there is a reduction in *Cynometra alexandri* and an increase of trees that produce small fleshy fruits, such as *Celtis durandii*, *Croton macrostachys*, and *Maesopsis eminii* (Rukundo 1997).

PRIMATES

Five primate species occur in Budongo Forest (Plumptre and Reynolds 1994; Fairgrieve 1995): blue monkey (*Cercopithecus mitis stuhlmannii*), redtail monkey (*C. ascanius schmidti*), black and white colobus (*Colobus guereza occidentalis*), olive baboon (*Papio anubis*), and chimpanzee (*Pan troglodytes schweinfurthi*). Primate densities were determined by censusing using line transect methods (Buckland et al. 1993; Plumptre and Reynolds 1994, 1996). Whereas there was variation found in densities between compartments in the forest, when the densities were calculated for logged and unlogged compartments it was found that the densities of *C. mitis*, *C. ascanius*, and *C. guereza* were significantly higher in logged compartments. The densities of *P. anubis* and *P. troglodytes* did not differ significantly between compartments (table 27.2).

Table 27.2

## Densities of Primate Species in Logged and Unlogged Forest

| SPECIES | UNLOGGED (NO./KM²) | LOGGED (NO./KM²) | Z-TEST |
|---|---|---|---|
| *Cercopithecus mitis* | 15.6 | 58.2 | *** |
| *Cercopithecus ascanius* | 8.3 | 46.4 | *** |
| *Colobus guereza* | 27.0 | 44.2 | *** |
| *Papio anubis* | 14.0 | 11.0 | ns |
| *Pan troglodytes* (sighting) | 3.2 | 2.8 | ns |
| *Pan troglodytes* (daily nest production) | 1.7 | 1.1 | ns |

*Note:* The results of Z-tests between these two estimates are given ($* = P < 0.05$, $** = P < 0.01$, $*** = P < 0.001$). Density estimates from line transects are positively skewed, and 95% confidence intervals are calculated assuming a log-normal distribution and are hence not symmetrical about the mean (Buckland et al. 1993).

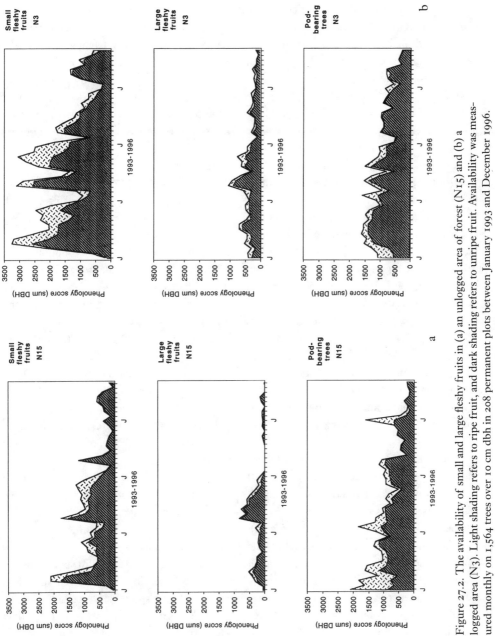

Figure 27.2. The availability of small and large fleshy fruits in (a) an unlogged area of forest (N15) and (b) a logged area (N3). Light shading refers to ripe fruit, and dark shading refers to unripe fruit. Availability was measured monthly on 1,564 trees over 10 cm dbh in 208 permanent plots between January 1993 and December 1996.

Long-term studies of the diets of four of these primates and the effect of predation by the crowned eagle *Stephanoaetus coronatus* on habituated groups showed that food supply was the main factor determining the densities of primates in Budongo forest. *C. mitis* and *C. ascanius* both had strong positive correlations between primate density and the percentage of ripe and percentage of fleshy fruit in their diet. *C. guereza* density was strongly correlated with the density of *Celtis durandii,* a tree whose leaves and fruits were preferred food items for this monkey in all eight study areas. *P. troglodytes* density was strongly correlated with density of trees bearing large fleshy fruits, a food item forming a large percentage of their diet (Plumptre and Reynolds, in prep.). Predators might have had an effect on the primate densities in Budongo; however, the predation risk from crowned eagles measured on seventeen habituated groups was about 0.001%, which was less than the risk of death faced by a driver in a passenger car (0.006%) in the Netherlands (Prins 1996). Just as drivers tend to think that this level of risk is inconsequential, it is probable that the monkeys do not alter their behavior greatly. The only other potential predator was chimpanzees, but again the amounts of meat or bones found in more than six hundred fecal samples examined was very small (about 1%).

Studies of the phenology of trees in two compartments (one logged and the other unlogged) over four years show that the availability of small fleshy fruits in logged forest (cut forty years ago) was greater than in unlogged forest (figure 27.2), while large fleshy fruit availability was about the same. Consequently, the removal of *Cynometra* monodominance in Budongo through arboricide treatment increased the availability of trees bearing small fleshy fruits, which benefited three of the monkeys. Chimpanzees and baboons tend to range over large areas (one habituated chimpanzee community in Budongo uses at least 10 km²) and hence use logged and unlogged areas as

part of their home range in Budongo. This is probably why no difference was found in their densities between logged and unlogged areas.

BIRDS

Birds were censused using point counts in five study compartments (Compartments N15, KP11–13, B1, N3, and W21; see table 27.1, figure 27.1) chosen to reflect different periods of logging recovery (Owiunji 1996; Owiunji and Plumptre 1998). In each compartment, five 2 km transects were established running north-south or east-west at random intervals along a baseline. At every 100 m along each transect a point was marked to allow repeated counts to be made at each point. The distance from the observer to the bird seen was measured using a range finder, and these distances were analyzed to determine densities for the common species following Buckland et al. (1993). Bird species were assigned a guild specialization following observations and consultations with ornithological experts in Kenya and Uganda.

Table 27.3 gives the densities for the twenty-two most commonly recorded species in logged and unlogged compartments, and their forest specialization. Most species did not show significant density differences between logged and unlogged forest. It was expected that those species known to be forest specialists, which are not found outside fairly mature forest, would be the species to be affected by the forest management in Budongo; those that can live in disturbed or fragmented forest would be more abundant. However, where differences in density occurred, not all species affected negatively by logging were forest specialist species, and not all those benefiting were forest specialists, either. For some species it was possible to calculate densities in each compartment. For yellow-whiskered greenbul (*Andropadus latirostris*), red-bellied paradise flycatcher (*Terpsiphone rufiventer*), and rufous thrush (*Neocossyphus fraseri*) there was a significant positive correla-

Table 27.3

## Comparison of Bird Densities in Logged and Unlogged Forest

| SPECIES[a] | UNLOGGED (NO./KM²) | LOGGED (NO./KM²) | Z-TESTS | HABITAT CATEGORY[b] |
|---|---|---|---|---|
| Red-bellied paradise flycatcher | 123.1 | 23.7 | *** | FF |
| Rufous thrush | 57.7 | 20.9 | *** | FF |
| Yellow-whiskered greenbul | 73.7 | 28.7 | *** | FF |
| Green hylia | 110.6 | 82.2 | * | F |
| Red-chested cuckoo | 18.1 | 4.0 | * | F |
| Red-tailed greenbul | 60.7 | 57.1 | ns | FF |
| Western black-headed oriole | 39.0 | 18.7 | ns | F |
| White-throated greenbul | 114.8 | 83.2 | ns | FF |
| Tambourine dove | 47.6 | 39.1 | ns | F |
| Bristlebill | 16.6 | 12.0 | ns | FF |
| Crested malimbe | 31.8 | 29.8 | ns | F |
| Blue-breasted kingfisher | 9.6 | 4.3 | ns | FF |
| Nicator | 27.6 | 28.9 | ns | F |
| Fire-crested alethe | 16.8 | 20.8 | ns | FF |
| White-thighed hornbill | 45.6 | 53.6 | ns | FF |
| Yellow-crested woodpecker | 13.1 | 30.7 | ns | FF |
| Spotted greenbul | 59.1 | 76.6 | ns | FF |
| Olive-green camarotera | 31.3 | 51.5 | ns | FF |
| Olive sunbird | 151.9 | 168.3 | ns | FF |
| Brown-crowned eremomela | 21.7 | 58.4 | * | F |
| Chestnut wattle-eye | 37.3 | 118.2 | *** | FF |
| Little greenbul | 94.5 | 205.4 | *** | F |

a   All twenty-two species were sighted more than twenty-five times.

b   FF = forest specialist; F = forest generalist.

*   $P < 0.05$; *** $P < 0.001$; ns = not significant.

tion between density and the percentage of *Cynometra* and *Cynometra*-mixed forest (the most mature forest types). This supports the hypothesis that higher densities in unlogged compartments can be explained by changes in forest composition for these three species.

The differences in the mean percentage of different guilds of birds from the point count data from each transect in logged and unlogged forest were tested using a T-test. Table 27.4 gives the differences between guilds of birds in logged and unlogged forest. Correlations were found between the number of birds in each guild and measures of forest structure, and the positive correlations support the logged-unlogged comparisons (table 27.3). Open understory (low VDEN), higher percentages of canopy cover, and *Cynometra* forest are characteristic of

Table 27.4

## Mean Percentage of Birds in Each Guild for Logged and Unlogged Compartments with Results of T-Tests Between the Two Columns

| GUILD | UNLOGGED FOREST (%) | LOGGED FOREST (%) | T-TEST | POSITIVELY CORRELATED | NEGATIVELY CORRELATED |
|---|---|---|---|---|---|
| Frugivore | 3.5 | 10.6 | * | VDEN, Mahogany | C, %COV |
| Frugivore-insectivore | 22.1 | 22.6 | ns | | |
| Gramnivore-insectivore | 1.7 | 4.0 | ns | | |
| Leaf-gleaning insectivore | 39.5 | 34.4 | * | %COV | |
| Bark-gleaning insectivore | 1.0 | 2.2 | * | | %COV |
| Terrestrial feeding insectivore | 5.8 | 7.1 | ns | LD, M | C |
| Sallying insectivore | 21.2 | 10.4 | * | %COV, C | S, VDEN |
| Nectarivore | 4.7 | 6.9 | * | | %COV |
| Omnivore | 0.5 | 1.6 | * | LD | CL,C |
| Raptor | 0.0 | 0.2 | ns | | |

*Notes:* Spearman rank correlations between measures of forest type and forest structure are listed where they were significant ($P<0.05$). LD = Liana density; VDEN = vegetation density at ground level; %COV = % canopy cover; C = % *Cynometra* forest; M = % mixed forest; CL = % colonizing forest; S = % swamp forest; Mahogany = density of mahogany trees.

\* $P<0.05$; ns = not significant.

unlogged forest in Budongo, while mixed forest, higher vegetation density at ground level, and lower percentages of canopy cover are found in selectively logged forest.

### SMALL MAMMALS

Five of the study compartments (N15, KP11–13, B1, N3, and W21) used for the primate and bird studies were selected for the study of small mammals (Musamali 1996). Live trapping studies of rats and shrews with Sherman traps and pitfall traps were conducted between April and August 1993 and between October 1994 and March 1995.

Twelve small mammal species were recorded during the ten-month study. Species diversity was low compared with other tropical forests, given the large trapping effort and number of sites studied. Two species formed the bulk of captures: *Praomys jacksoni* (60.3%) and

*Hylomyscus stella* (28.0%). One species, *Malacomys longipes*, was trapped only in swamp forest along streams where water was constantly available. All other species were caught in logged and unlogged areas. Higher total numbers of rodents were caught in unlogged forest. T-tests were made between the mean number of captures of the two most abundantly trapped rodents in logged and unlogged forest. Both had significantly higher capture rates in unlogged forest (*P. jacksoni*: $T = 6.82$, $df = 116$, $P < 0.001$; *H. stella*: $T = 2.44$, $df = 116$, $P < 0.05$). For *P. jacksoni*, which had a large sample size (>50), these differences were maintained during both wet and dry seasons. There were strong correlations between capture frequency for these two rodents and the density of *Cynometra alexandri* on the trap site, supporting the finding of higher numbers in undisturbed forest.

Studies in Kibale Forest in Uganda showed that although rodent diversity may be higher in logged forest, densities of *P. jacksoni* and *H. stella* (the two most abundant species in Budongo also) were higher in unlogged forest (Isabirye-Basuta and Kasenene 1987; Lwanga 1994). These two species feed on seeds and insects and may rely on the *Cynometra* seeds that appear in the December–February dry season in Budongo Forest. In one year in Budongo when the *Cynometra* failed to fruit, there were much lower capture rates of these rodents (A. Stanford, pers. comm.).

AMPHIBIANS

An undergraduate expedition from Oxford University Zoology Department came to Budongo in June–August 1996 to survey amphibian populations in Budongo. Methods included ad lib sampling of different areas, attempting to build up a species list for the forest, and also the intensive searching of 8×8 m quadrats between the ground and a height of 1.5 m. Eighty quadrat searches were undertaken in two adjacent compartments (B1 and N15; see table 27.1), forty in unlogged forest and forty in logged forest. All frogs found were collected and preserved for identification at Makerere University and the Natural History Museum in London.

Twenty-two species that have been identified (three remain to be identified) were found in both N15 and B1 in the quadrats. However, alpha diversity seems to be higher in B1, with all twenty-two found after sampling eleven quadrats, compared with twenty-two quadrats in N15. Apart from two species, most frogs occurred at both sites. *Ptychadena* sp. was found only in N15, and *Leptopelis christyi* was found only in B1. Two species, *Arthroleptis adolfifriderici* and *A. poecilonotus*, dominated the numbers of individuals caught at both sites (more than 90% at both sites). Table 27.5 summarizes the counts from the plots.

Amphibians are thought to be particularly sensitive to changes in microhabitat (Johns 1985; Clarke 1989). The findings from this short study, however, indicate that in Budongo most species have survived the two logging events that have occurred in the logged compartment.

ARTHROPODS

Thomas Wagner spent from June to August 1994 fogging the canopies of selected trees with pyrethrum insecticide, collecting arthropods. Because many arthropod species are thought to be specific to certain tree species, he collected from the same tree species in logged and unlogged forest. Eight trees each of *C. alexandri*, *Trichilia rubescens*, and *Rinorea ardisiaeflora* were sampled in an unlogged site, and eight of each in a logged site. In addition, eight *Rinorea ardisiaeflora* and eight *Teclea nobilis* were sampled in swamp forest. Arthropods were sorted into orders, and in the case of beetles they were morphotyped under each family.

Table 27.5

## Number of Frogs Found, Density, Shannon-Wiener Diversity, and Evenness for Unlogged and Logged Forest

|  | UNLOGGED (N15) | LOGGED (B1) |
|---|---|---|
| Total no. frogs caught in 2,560 m² | 173 | 151 |
| Density (no./64 m² plot) | 4.3±0.5 | 3.8±0.6 |
| Diversity H' | 0.40 | 0.43 |
| Evenness | 0.47 | 0.51 |

A total of 130,000 arthropods was collected from the sixty-four trees that were sampled; of this total, 30,000 were beetles. The species accumulation curves for beetles were still rising steeply after eight trees had been sampled at each site, and the curve still rose steeply even if all sixty-four trees were combined. Analysis of relative biomass showed that ants and Lepidoptera dominated the biomass of arthropods on all the trees.

Phytophagous beetles are thought to be specialized on certain tree species (T. Wagner, pers. comm.), and many are monophagous. When the number of Chrysomelid beetles was compared with the number in the families Lagriinae and Alleculinae from the Budongo data set, it was found that for all trees in unlogged forest, the ratio of Chrysomelids to the Alleculinae and Lagriinae was high, while in logged forest the ratio of Chrysomelids declined.

Although these results are preliminary, and further collections have recently been made in the dry season, they indicate that proportions of most orders of arthropods (apart from the phytophagous beetles) did not differ greatly between logged and unlogged forest. However, the data indicate that phytophagous beetles are probably less dependent on one species of tree than they often are in Europe because of the much larger number of tree species in the tropics.

## A Model of the Effects of Logging on Wildlife

One of the main objectives of the research of the Budongo Forest Project was to develop a general model that is capable of predicting the responses of wildlife to selective logging in monodominant forest. Figure 27.3 gives a schematic diagram of the model and looks at two possible scenarios: (1) if logging or forest management activities (such as arboricide treatment) remove the monodominant tree species and (2) if the monodominant species is not removed. This model assumes that no human settlement takes place, and no hunting

occurs, in the logged forest. The model focuses on trees, birds, and primates because in the other studies there were no correlations made between animal abundance and measures of habitat structure or composition, so it is more difficult to predict how these populations might change.

### MONODOMINANT SPECIES NOT REMOVED

If the monodominant species is not removed, then gap sizes created by logging are necessarily going to be small and harvesting intensity will be light. Consequently, the impact on the forest will not be high and may not be very different from the natural pattern of tree falls. Hart (1995) showed that in monodominant *Gilbertiodendron dewevrei* forest, seedlings of this species could survive for many years (49% survived for ten years) in the understory waiting for a gap to grow into. Regeneration in gaps is likely to consist of many seedlings and saplings of the monodominant species, and the forest composition is unlikely to change greatly. Return to mature monodominant forest structure will probably be less than the time for one generation of the monodominant species. Consequently it is predicted that animal populations will not change very much.

### MONODOMINANT SPECIES REMOVED

If the monodominant species is removed—either by arboricide treatment, as occurred in Budongo during the 1950s and 1960s, or by harvesting it for timber, as one sawmill currently does in Budongo—then much larger gaps will be created, and the composition of the forest will change. Larger gaps will encourage quicker-growing light-dependent species, which will compete more effectively with the seedlings of the monodominant species in the gap. Consequently, a disturbed mixed forest will be created. Many of the colonizing trees produce small fleshy fruits, so the availability of food for frugivores will increase.

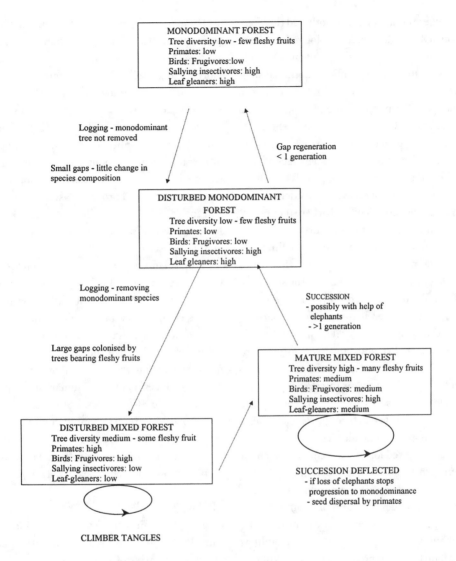

Figure 27.3. A model predicting the changes in primate and bird communities following disturbance in monodominant forests in Africa. See text for a detailed explanation.

The creation of large gaps will lead to dense foliage in the forest understory, which will have effects on the bird species that respond to forest structure. Sallying insectivores and leaf-gleaning insectivores were positively correlated with dense cover in the canopy (see table 27.4). Consequently, their populations will drop soon after logging opens the canopy. By contrast, frugivores and nectarivores will increase in number because they seem to prefer a denser understory vegetation and a more open canopy.

Climbers increase in number in disturbed forest and can take over large gaps in the forest. There are some large areas in Budongo that have not shown any recovery between 1992 and

1996 where climbers have smothered the vegetation. Other more recently logged sites are already regenerating, and saplings are higher in these sites than in the climber-smothered sites. Foresters used to practice climber cutting in Budongo (Philip 1964) to encourage natural regeneration. Climbers have been shown to significantly reduce the growth rates of trees in Budongo (Plumptre, unpublished data) if they get into the canopy. Consequently, climbers can slow down forest recovery considerably and will act as a negative feedback where they occur, maintaining the forest in a disturbed state.

The forest understory will remain fairly dense for many years. Even forty-five years after logging in Budongo Forest, the structure is dissimilar to mature forest (Plumptre 1996). Gradually, however, if there is no further logging, the forest will develop into a mature mixed forest. Primates and frugivorous bird numbers will remain relatively high in comparison with monodominant forest. Sallying and leaf-gleaning insectivores will become more common as forest structure approaches a more closed canopy and open understory.

Mature mixed forest may remain as such for several generations or even indefinitely, depending on what the causes of monodominance are. It has been suggested that monodominance in Budongo is a deflected succession caused by elephants selectively feeding on young saplings of certain trees. It is possible that the loss of elephants due to poaching that occurred in the 1980s could lead to the maintenance of mature mixed forest. The importance of primates, and to some extent frugivorous birds, in seed dispersal has been shown in Budongo (Plumptre, unpublished data). These animals slow the progression to monodominant forest by dispersing and promoting the germination of seeds of trees bearing fleshy fruits. Thus elephants and frugivores act as a negative feedback on the progression to monodominance.

If monodominance is due to superior competitive abilities of the monodominant tree species, then the mature mixed forest will gradually change back to monodominant forest, provided no further logging takes place. This will probably take several generations. As growth rates of trees in Budongo are 0.2–0.7 cm dbh per year (Sheil 1996), a tree of 70 cm dbh will have taken about 140 years to reach that girth. *Cynometra* trees can reach more than 1 m in diameter and probably last between 150 and 200 years before they die of old age. Therefore, recovery of monodominant forest will be on this timescale.

Tropical forests, however, are not stable systems. Hart et al. (1997) showed that for the past 4,000 years the Ituri Forest has changed in species composition, often quite dramatically. *Gilbertiodendron dewevrei*, which now is monodominant over much of the southern Ituri, was not found in this area earlier than about 500 years ago. Trees that were abundant in the past are now rare. Consequently, although this model predicts a return to monodominant forest following heavy logging disturbance, the forest may vary in composition by that time, and even the monodominant species may differ.

## Testing Some Aspects of the Model

One of the main criticisms of the findings from Budongo Forest is that the results are not representative of other forests. Budongo lies at the northeastern extreme of a line of forests that occurs in western Uganda and is probably a remnant of the eastern limit of the last forest expansion when the climate was wetter and warmer (Kingdon 1990). Consequently, Budongo may have a tendency to contain those species that were able to colonize new areas rapidly. In addition, Budongo can be thought of as an island of forest among grassland and cultivation. Edge effects on species composition could be important in determining what exists there today. Therefore, mature forest in Budongo may contain species that are ideally adapted to disturbance.

A test of this model's applicability to other monodominant forests was made in monodominant *Gilbertiodendron* forest. Research was carried out in the Ituri Forest in eastern DR Congo on the bird communities in monodominant, mature mixed, and secondary forest. Birds are good subjects to study because they have many guilds and therefore can be used to test the model. Primate studies had already been made in this area (Thomas 1991; J. Hart, unpublished data) from which comparisons of primate responses could be made.

The Ituri Forest is the name given to about 70,000 km² of tropical lowland forest in northeastern DR Congo which occurs in the watershed of the Ituri River. Three main forest types are found in the Ituri: monodominant forest composed of *Gilbertiodendron dewevrei*, which can comprise over 75% of the individuals (Hart 1985); mixed forest; and secondary forest formed by shifting cultivation along the sides of the roads in the forest (Wilkie 1989). Although the disturbance from shifting cultivation is not exactly the same as that of selective logging, it still leads to similar structural changes to the forest (denser understory and more open canopy).

Table 27.6 shows the percentage of captures in the three sites. The secondary forest is significantly different from the two primary sites for four guild categories: frugivore-insectivores ($F = 14.94$, $P < 0.001$; secondary > mixed and monodominant), ground-feeding insectivores ($F = 7.73$, $P < 0.01$; mixed and monodominant > secondary), nectarivores ($F = 6.3$, $P < 0.01$; secondary > mixed and monodominant), and gramnivores ($F = 7.09$, $P < 0.01$; monodominant > secondary and mixed).

If frugivores and frugivore-insectivores are combined into one category, there is a significantly higher percentage of these species in secondary forest ($F = 14.94$, $P < 0.001$). The high value for gramnivores in monodominant forest is attributable to Grant's bluebill (*Spermophaga poliogenys*), which may be more of a frugivore in this forest. Sallying insectivore catches did not differ significantly because of low sample sizes. However, if the total numbers caught are examined (monodominant = 23, mixed = 18, and secondary = 8), it can be seen that there is a tendency for more sallying insectivores in mature forest.

Table 27.6

## Percentage of Captures in the Three Forest Types in the Ituri Forest

| GUILD[a] | MONODOMINANT FOREST | MIXED FOREST | SECONDARY FOREST |
|---|---|---|---|
| Frugivore | 1.0 | 0.9 | 0.8 |
| Frugivore-insectivore | 8.2 | 10.7 | 26.8 |
| Insectivore[b] | 55.3 | 53.2 | 24.9 |
| Sallying[b] | 8.3 | 4.6 | 1.3 |
| Gramnivore | 4.5 | 0.3 | 0.8 |
| Nectarivore | 23.3 | 27.7 | 40.8 |
| Omnivore | 7.5 | 7.2 | 5.6 |
| Raptor | 0.2 | — | 0.3 |

a    Guilds are not as detailed as in Budongo Forest because the biology of many of the species is not as well known.

b    Sallying insectivores are given as a subset of insectivores.

The results suggest that shifting cultivation as it is currently practiced in the Ituri Forest causes marked changes to the bird communities that live in this forest region. The overlap in species composition between the secondary forest and the two primary forest sites was much lower than the overlap between the two primary forest sites. This was despite the fact that the mixed forest site has a fair amount of disturbance from wind and contains a fairly broken canopy.

Examining only guilds, however, masks the different responses to disturbance among species. Species that seem to be particularly affected by disturbance include some ground-feeding insectivores such as ground thrushes (*Zoothera*). Four ground thrushes were caught in monodominant forest and one at both of the other sites, although the thrush in the secondary forest site was caught in a strip of mature riverine forest between patches of secondary forest. During a major study of the birds in the forests of Uganda by the Uganda Forest Department in which 14,216 birds were caught during 1,210,978 meter-net-hours (mnh), only three ground thrushes were captured: one black-eared ground thrush and two gray ground thrushes. After capturing 3,993 birds during 158,508 mnh in Budongo Forest, we have caught only one black-eared ground thrush, which was in undisturbed forest. Consequently, the Ituri Forest seems to be particularly rich for these ground-feeding insectivore thrush species. At all the sites where the individual thrushes were captured, the understory was very open with a short herb layer of Marantaceae and Zingiberaceae up to a height of about 40 cm. These species are likely therefore to suffer from forest disturbance, and this may be why they are uncommon in the "island" forests of Uganda.

Brown-chested alethes (*Alethe poliocephala*), a species that follows ants and feeds on insects flushed up by the ants, also seem to require undisturbed forest, and this is also true in Budongo Forest. However, fire-crested alethes (*Alethe diademata*), which also follow ants, seem to be able to adapt and show no difference in densities between primary and secondary forest in both forests. In contrast, many frugivore-insectivores, such as greenbuls, seem to do well in disturbed forest, and some species in the Ituri were found only in the disturbed secondary forest.

The effects of disturbance in both forests produced similar results in terms of community changes. Certain sallying and leaf-gleaning insectivores were at lower numbers in disturbed forest, while frugivores or frugivore-insectivores, mainly greenbuls, were at higher numbers. Most species of primates, which are also frugivorous, increased in density in mixed and secondary forest in the Ituri (Thomas 1991; J. Hart, pers. comm.). Only the owl-faced monkey (*Cercopithecus hamlynii*) may prefer monodominant forest, a possible consequence of leopard predation (J. Hart, pers. comm.) because there are fewer duikers in monodominant forest that would attract the leopards. This primate forages on the ground a lot and therefore faces a greater risk from leopard predation than other primates.

Consequently, the predictions of the model seem to hold for the Ituri Forest. The few other studies of the effects of logging on birds in Africa show similar responses to disturbance. Allport et al. (1989) showed that very disturbed forest (farm bush) in Sierra Leone had more frugivorous and gramnivorous birds and fewer leaf-gleaning and ground-feeding insectivores. Logged forest was similar in bird community composition to primary forest, but it is not clear how heavy the logging was. Dranzoa (1995) found that the biomasses of frugivore-insectivores and gramnivores were at higher density, and the biomasses of sallying, bark-gleaning, and ground-feeding insectivores were at lower density, in logged forest in Kibale Forest Reserve, Uganda. Arboreal foliage gleaners decreased in logged forest also, but understory foliage gleaners increased. Neither of these studies was in monodominant forest, however, so it appears that disturbance of mature mixed forest can cause similar changes in bird community composition. Primate responses in Kibale were variable, however, and the model does not appear to apply to these animals when logging occurs in mixed forest (Skorupa 1988). Further work on rodents, amphibians, and arthropods is required in order to see how these species are affected by forest disturbance elsewhere in Africa.

### Conservation and Use of Monodominant Forest

Forest management leading to the removal of monodominance in these forests does benefit many species, particularly charismatic frugivorous species such as many primates and some birds. Consequently it does appear that you can enjoy the benefits of both logging and species conservation, provided that the damage to the forest is limited so that regeneration can occur relatively quickly. There is a danger that monodominant forests will be considered less diverse and hence less valuable for conservation. Yet studies of tree species richness in the Ituri are actually showing that monodominant forest can be as species rich as mixed forest, although this is true only when a large area is sampled. What differs is species richness over small areas. The findings from Budongo and the Ituri Forest that such charismatic animals as primates benefit from removal of the monodominance may lead some people to argue that these forests should be converted to mixed forest. However, these studies also found that certain species—particularly the rare ground thrushes, other ground-feeding or leaf gleaning insectivores, and sallying insectivores—occur more frequently in monodominant forest or where there is a very open understory. Consequently, these less diverse forest areas can be essential for the conservation of some species.

The reason for the rarity of ground thrushes in Uganda is probably that most forests are effectively small islands, and the forest is affected by the ecotone between the forest and savanna or cultivation. Therefore, if these birds are to be protected, it will be important to have large areas set aside for them to maintain their populations. However, they do appear to be able to live in corridors of mature forest that are linked to larger blocks.

Logging operations that aim to harvest monodominant forest on a sustainable-yield basis could put into place certain practices that should benefit those species that depend on characteristics of monodominant or mature forest:

- Leave certain areas of the concession (or an adjacent protected area) untouched or do not remove the monodominance of the forest so that animal populations dependent on this can survive.
- Use directional felling so that the gaps created are minimized. This will not only help wildlife but also help the regeneration of the forest for the subsequent crop.
- Leave corridors along streams or rivers in the forest (preferably 30 m or more on either side).
- Leave mature seed trees to provide seed for natural regeneration. Plumptre (1995) found that the diameter at which the average *Khaya* starts to produce fruit is 50 cm, and for *Entandrophragma* the diameter is 70 cm. Many of these mahoganies are harvested below these diameters, and consequently there is little seed produced following logging.
- Prevent hunting in the concession by logging company employees and people following the logging roads. Little has been said about the effects of hunting combined with logging, because this chapter concentrates on the effect of disturbance to the forest. Hunting, however, usually poses the greatest threat to animal populations, particu-

larly large-bodied species, and logging combined with hunting is currently threatening most of central Africa's large mammals.

These practices would serve to: maintain tracts of monodominant forest, upon which certain species depend; maintain corridors; reduce the degree of disturbance at logging sites; and promote adequate regeneration, thereby conserving wildlife and promoting the possibility of conservation.

## REFERENCES

Allport, G., M. Ausden, P. V. Hayman, P. Robertson, and P. Wood. 1989. The conservation of the birds of Gola Forest, Sierra Leone. ICBP Study Report No. 38.

Buckland, S. T., D. R. Anderson, K. P. Burnham, and J. L. Laake. 1993. *Distance Sampling: Estimating Abundance of Biological Populations*. Chapman and Hall, London.

Clarke, B. T. 1989. Real vs apparent distributions of dwarf amphibians: *Bufo lindneri* Mertens 1955: A case in point. *Amphibia-Reptilia* 10:297–306.

Connell, J. H., and M. D. Lowman. 1989. Low-diversity tropical rainforests: Some possible mechanisms for their existence. *American Naturalist* 134:88–119.

Dawkins, C. 1958. The management of tropical high forest with special reference to Uganda. Imperial Forestry Institute Paper 34, Oxford University.

Dranzoa, C. 1995. Bird populations of primary and logged forests in the Kibale Forest National Park, Uganda. Ph.D. dissertation, Makerere University, Kampala, Uganda.

Eggeling, W. 1947a. Observations on the ecology of the Budongo rainforest, Uganda. *Journal of Ecology* 34:20–87.

———. 1947b. *Working Plan for the Budongo and Siba Forests, First Revision, Period 1945–1954*. Government Printer, Entebbe, Uganda.

Fairgrieve, C. 1995. The comparative ecology of blue monkeys (*C. mitis stuhlmannii*) in logged and unlogged forest, Budongo Forest Reserve, Uganda. Ph.D. dissertation, Edinburgh University, Scotland.

Harris, C. M. 1933. *Working Plan Report for the Bunyoro Forests, Uganda*. Government Printer, Entebbe, Uganda.

Hart, T. B. 1985. The ecology of a single-species-dominant forest and a mixed forest in Zaire, Africa. Ph.D. dissertation, Michigan State University, East Lansing.

———. 1995. Seed, seedling and sub-canopy survival in monodominant and mixed forests of the Ituri Forest, Africa. *Journal of Tropical Ecology* 11:443–459.

Hart, T. B., J. A. Hart, and P. G. Murphy. 1989. Monodominant and species-rich forests of the humid tropics: Causes for their co-occurrence. *American Naturalist* 133:613–633.

Hart, T. B., J. A. Hart, R. Deschamps, M. Fournier, and M. Ataholo. 1997. Changes in forest composition over the last 4,000 years in the Ituri basin, Zaire. In L. J. G. Van der Maesen, ed. *The Biodiversity of African Plants*. Proceedings of the Fourteenth AETFAT Congress, Wageningen, Netherlands. Academic Press, London.

Howard, P. C. 1991. *Nature Conservation in Uganda's Tropical Forest Reserves*. IUCN, Gland, Switzerland.

Isabirye-Basuta, G., and J. M. Kasenene. 1987. Small rodent population in selectively felled and mature tracts of Kibale Forest, Uganda. *Biotropica* 19:260–266.

Johns, A. D. 1985. Selective logging and wildlife conservation in tropical rainforest: Problems and recommendations. *Biological Conservation* 31:335–375.

Kingdon, J. 1990. *Island Africa*. Collins, London.

Lwanga, J. S. 1994. The role of seed and seedling predators and browsers in the regeneration of two forest canopy species (*Mimusops bagshawei* and *Strombosia scheffleri*) in Kibale Forest Reserve, Uganda. Ph.D. dissertation, University of Florida, Gainesville.

Musamali, P. B. 1996. The ecology, diversity and relative abundance of small mammals in primary and disturbed forest compartments of Budongo Forest Reserve. Master's thesis, Makerere University, Kampala, Uganda.

Owiunji, I. 1996. The long-term effects of forest management on the bird community of Budongo Forest Reserve, Uganda. Master's thesis, Makerere University, Kampala, Uganda.

Owiunji, I., and A. J. Plumptre. 1998. Bird communities in logged and unlogged compartments of Budongo Forest, Uganda. *Forest Ecology and Management* 108:115–126.

Paterson, J. D. 1991. The ecology and history of Uganda's Budongo Forest. *Forest Conservation and History* 35:179–187.

Philip, M. S. 1964. *Working Plan for Budongo Central Forest Reserve (1964–74), Third Revision*. Uganda Government Printer, Entebbe, Uganda.

Plumptre, A. J. 1995. The importance of "seed trees" for the natural regeneration of selectively logged tropical forest. *Commonwealth Forestry Review* 74:253–258.

————. 1996. Changes following sixty years of selective timber harvesting in the Budongo Forest Reserve, Uganda. *Forest Ecology and Management* 89:101–113.

Plumptre, A. J., and V. Reynolds. 1994. The impact of selective logging on the primate populations in the Budongo Forest Reserve, Uganda. *Journal of Applied Ecology* 31:631–641.

————. 1996. Censusing chimpanzees in the Budongo forest. *International Journal of Primatology* 17:85–99.

Prins, H. H. T. 1996. *Ecology and Behaviour of the African Buffalo: Social Inequality and Decision Making*. Chapman and Hall, London.

Rukundo, T. 1997. The long-term effect of canopy treatment on tree diversity in the Budongo Forest: An evaluation of a 25 hectare permanent plot study. Master's thesis, Makerere University, Kampala, Uganda.

Sheil, D. 1996. The ecology of long term change in a Ugandan rainforest. Ph.D. dissertation, Oxford University.

Skorupa, J. P. 1988. The effect of selective timber harvesting on rain forest primates in Kibale Forest, Uganda. Ph.D. dissertation, University of California, Davis.

Synnott, T. J. 1985. A checklist of the flora of Budongo Forest Reserve, Uganda with notes on ecology and phenology. CFI Occasional Paper 27. Oxford Forestry Institute, Oxford.

Thomas, S. C. 1991. Population densities and patterns of habitat use among anthropoid primates of the Ituri Forest, Zaire. *Biotropica* 23:68–83.

Trenaman, K. W., H. C. Dawkins, and C. Swabey. 1956. *Working Plan for Budongo, Siba and Kitigo Central Forest Reserves, Second Revision, 1955–1964*. Government Printer, Entebbe, Uganda.

Whitmore, T. C., and J. A. Sayer. 1992. *Tropical Deforestation and Species Extinction*. Chapman and Hall, London.

Wilkie, D. S. 1989. Human settlement and forest composition within the proposed Okapi rainforest reserve in northeastern Zaire: Creating the first thematic map of the Ituri forest using digital analyses of Landsat TM imagery. Final report to WWF, WCS, and NSF. WWF project 3249.

# CHAPTER 28

# Lost Logging

## Problems of Tree Regeneration
## in Forest Gaps in Kibale Forest, Uganda

John Kasenene

Commercial mechanized logging typically results in destruction of about 50% of the original forest stand (see, e.g., Ewel and Conde 1976; Johns 1992). Timber trees are killed directly during the felling process. As they fall, they crush smaller trees and saplings below, opening up light gaps in the forest. Other trees are destroyed during transport of logs to loading areas, as well as during construction of access roads and camps. In the vast majority of tropical rain forests, little attention is paid to minimizing damage levels, and the only form of forest management is to leave the area undisturbed for a number of years, with the assumption that natural regeneration will be sufficient to support further logging in the future (Poore et al. 1990). Following disruption by logging, recovery depends on survival of already established saplings and seedlings; regeneration from soil seed banks and seed rain from surviving mature trees; recolonization from outside the disturbance; or coppicing of damaged trees (see, e.g., Young 1985; Marquis et al. 1986; Kasenene 1987; Uhl et al. 1988). However, seedlings and saplings of shade-tolerant tree species do not usually survive in extensive gaps,

such as those caused by logging, and seedling establishment and survival of these species tends to be inversely related to gap size (Richards 1964; Whitmore 1975; Swaine and Whitmore 1988).

In Uganda, mechanized selective logging opens up the forest canopy by more than 50% (Kingston 1967). The northern third of the Kibale Forest, Uganda, was selectively logged between 1967 and 1969, creating numerous large gaps in the canopy (Struhsaker 1975; Kasenene 1984, 1987; Kasenene and Murphy 1991; Skorupa 1988). In the remaining intact natural forest, small to medium gaps occur due to natural tree falls or tree death (Kasenene 1991). Twenty years after logging, regeneration of trees seemed not to have occurred in areas damaged by logging, particularly in large gaps (see, e.g., Kasenene 1984).

In this chapter I shall explain why regeneration of trees has been adversely affected, and I shall discuss implications for management of tropical forests for sustainable timber production. Because logging was undertaken more than thirty years ago, it was not possible to assess regeneration immediately following exploitation, but an assessment of why regeneration does not presently occur can shed some light on the fundamental problems.

## Background

The Kibale Forest (560 km²) in midwestern Uganda is located on a plateau, which descends from 1,590 m in the north to 1,110 m in the south (Kingston 1967; Wing and Buss 1970; figure 28.1). Rainfall is low to moderate (c. 1,700 mm/year) and is spread quite evenly through the year, although March–May and September–November are generally wetter than other months (Struhsaker 1975). Temperatures are low, with mean minima and maxima of 12.7°C and 25.5°C, respectively, and an annual average of 20.5°C. About 60% of the

forest, now a national park, is medium-altitude tropical moist forest. The natural high forest is composed predominantly of evergreen trees, but some deciduous species also occur. In undisturbed areas, trees rise to over 50 m and undergrowth is sparse, composed of shade-tolerant herbs and shrubs. The remaining 40% of the park is a mosaic of grassland, woodland-thicket, and colonizing vegetation, thought to have been the result of past human habitation in the area (Kingston 1967).

Kibale National Park was, until 1994, classified as a Central Forest Reserve and was divided into a number of compartments for management purposes. Up to 1969, selective logging affected some northern and central parts of the forest (Struhsaker 1975). This resulted in large

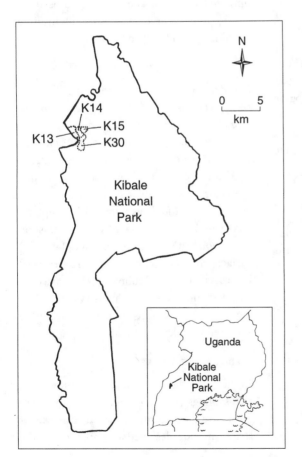

Figure 28.1. Map of the Kibale National Park showing the major vegetation types and physical features, and Uganda (inset) showing the approximate location of the Kibale National Park.

forest gaps, which allowed much sunlight into the forest understory, resulting in the formation of a dense tangle of herbs, shrubs, and climbers, dominated by plants in the Acanthaceae family (predominantly *Acanthus* spp.). A diversity of animals resident in Kibale, including elephants, bushbuck, duikers, and rodents, frequent the dense ground-story vegetation in logging gaps.

This study was undertaken in four compartments with different logging histories. Compartments K-13 and K-15 were heavily damaged during logging (50% loss of canopy cover and basal area), and K-13 was also subject to arboricide treatment to encourage regeneration of commercial species (only data on animal incursions and sapling or pole densities were collected for K-13); compartment K-14 was lightly damaged (25% loss of canopy cover and basal area); and compartment K-30 was in unexploited mature forest. Skorupa (1988) gives detailed management histories and vegetation descriptions for the compartments and discusses the effects of logging damage on primates.

All gaps in each compartment were mapped. Forty study gaps were then randomly selected and marked in each forest compartment. The gap was delimited by a combination of two methods: (1) visual estimation of the gap edges based on the distribution of such light-demanding plant species as *Brillantaisia nitens, Mimulopsis solmsii,* and *Trema guineensis;* (2) light-meter readings were taken in the gap and within surrounding forest, and the point at which readings declined rapidly to forest levels was taken as the edge of the gap. Once the boundary of the gap had been mapped, a scale drawing was made on graph paper to calculate the surface area.

## Some Sources of Plant Recruitment in Forest Gaps

### SOIL SEED BANKS

In order to assess the composition of soil seed banks in K-14, K-15, and K-30 (data were not collected in K-13), samples of the top 10 cm of surface soil with decomposed litter were collected from 1 m² plots in 28 forest gaps (G) and in closed forest adjacent to the gaps (F). The soils were transferred to nursery beds that were protected by screens against contamination and seed or seedling predation. The topsoil was spread thinly over subsoil removed from below the 10 cm depth of each sample and was watered every evening, except on rainy days. Two replicate trials were undertaken, each lasting seven months. Each month any plant sprouts were uprooted, counted, and categorized as herb, shrub, vine, or tree seedlings, and the tree seedlings were identified to species level.

### SEED RAIN AND EXISTING GAP VEGETATION

In order to examine the contribution of buried seeds, seed rain, and extant gap vegetation on tree regeneration in the same three compartments, two plant regeneration plots, each 4 m², were established in 20 randomly selected gaps in each study area. Each quadrant was divided into four 1 m² subplots, treated as follows: existing vegetation was removed in subplot 1 (Q1); subplot 2 (Q2) was screened after existing vegetation had been removed; in subplot 3 (Q3) surface soil and vegetation were removed before screening; and subplot 4 (Q4) was a control left intact. These plots were visited monthly for 20 months, and any sprouts were collected and identified as above.

### HERBIVORY AND GAP REGENERATION

In the 40 marked gaps in all four study areas, tree saplings (ranging from 1.5 cm to 10 cm dbh) were counted and identified on a series of strips 5 m wide, designed to sample 40% of the surface area of each gap. All saplings and pole stems counted were classified by growth form as either normal, coppicing, or stem sprouts. In addition, the number of times stems had been snapped and then coppiced was recorded. Each gap was then examined each month to assess animal incursions and intensity of browsing on tree seedlings, saplings, and poles. Each instance of debarking, defoliation, pushover, snapping, or uprooting was recorded.

Figure 28.2 shows the distribution of gap sizes in the four study areas. Mean size of natural gaps in unlogged forest (K-30) was 256 m² (range 100–663 m²). In lightly logged forest (K-14), mean gap size was 467 m² (range 75–1,800 m²), whereas in the two heavily logged areas, K-13 and K-15, it was 938 m² (range 227–3,313 m²) and 1,307 m² (range 73–7,100 m²), respectively. There were statistically significant differences between all sites, with the exception of K-13 and K-15 (Simulta-

neous Test Procedure of Govindarajulu and Dwass 1979).

SOIL SEEDLING BANKS

Table 28.1 shows the summary of plant life forms, numbers of species, and density of plants sprouted from the forest soil (F) and forest gap soil (G) from compartments K-14, K-15, and K-30. Seedlings of non-woody species dominated, particularly in logged forest samples. The percentages of tree seedlings sprouting in forest soil were 10.2%, 9.3%, and 16% for K-14, K-15, and K-30, respectively, whereas seedlings of herbaceous plants constituted 80.4%, 71.2%, and 53.1%, respectively. As far as trees were concerned, forest soils consistently had greater numbers of species and higher densities of seedlings than did gap soils (Mann-Whitney U-test—Sokal and Rohlf 1981; $P < 0.05$).

The seeds of commercially valuable or

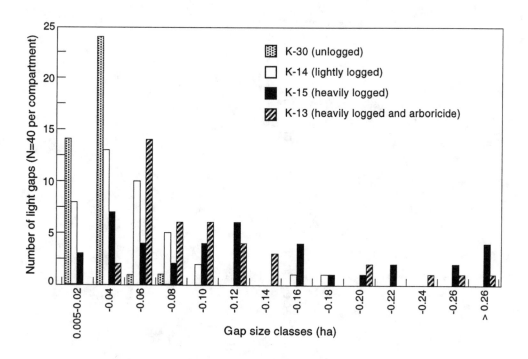

Figure 28.2. Distribution of gap sizes in four study sites, Kibale Forest, Uganda.

Table 28.1
## Mean Species Richness and Density of Plant Life Forms

| PLANT LIFE FORMS | MEAN NUMBER OF PLANT SPECIES | | | | | | MEAN DENSITY OF PLANTS (NO./M²) | | | | | |
| --- | --- | --- | --- | --- | --- | --- | --- | --- | --- | --- | --- | --- |
| | K-14 (N=28) | | K-15 (N=28) | | K-30 (N=28) | | K-14 (N=28) | | K-15 (N=28) | | K-30 (N=28) | |
| | F | G | F | G | F | G | F | G | F | G | F | G |
| Tree seedlings | 2.3 | 1.3 | 3.0 | 1.3 | 3.4 | 1.6 | 43.2 | 1.6 | 33.7 | 2.3 | 57.3 | 15.1 |
| Shrub species | 1.0 | 2.0 | 2.4 | 3.0 | 3.9 | 1.4 | 7.9 | 26.9 | 12.6 | 48.4 | 17.7 | 20.9 |
| Herbaceous species | 18.0 | 6.4 | 23.0 | 8.9 | 11.3 | 5.0 | 256.4 | 478.7 | 243.3 | 709.0 | 125.9 | 270.0 |
| Vines and climbers | 1.1 | 10.8 | 3.9 | 1.4 | 2.7 | 1.0 | 3.6 | 4.6 | 9.6 | 7.3 | 4.9 | 4.6 |
| Mean total | 22.4 | 10.8 | 32.3 | 14.6 | 21.3 | 9.0 | 310.2 | 512.2 | 299.2 | 767.0 | 205.8 | 310.6 |

Note: Plants were germinated in 1 m² plots of the top 10 cm of forest soil (F) and gap soil (G) from each of the three compartments (K=14, K=15, and K=30) and were grown in a screened nursery.

canopy tree species were rare in both forest and gap soil seed banks. Seedlings of two valuable species did germinate: low densities of *Celtis africana* occurred in both closed forest and forest gap soils, and *Cordia millenii* occurred in soils from uncut forest. The remaining tree species that germinated were of no commercial value (table 28.2).

## SEED RAIN AND EXISTING GAP VEGETATION

The number of plant species and their densities observed in the plant regeneration plots in situ were significantly lower than those germinated under screened nursery conditions from forest or gap soils (see tables 28.1 and 28.3). This suggests either that there was high seed or seedling predation in natural conditions or that a greater proportion of seeds fail to germinate under forest gap conditions. The plant regeneration plots with vegetation removed (Q1), vegetation removed and screened (Q2), and vegetation and surface soils removed (Q3) had similar numbers of plant species germinating in all the study compartments; all were higher than the control plots (Q4) (F = 4.10, 4.42, and 7.06 respectively, all $P < 0.05$, Duncan's small test). The density of herbaceous sprouts in all treated plant regeneration plots was high compared to that of tree, shrub, or climber sprouts (all F > 20.93, $P < 0.005$, table 28.3). The mean density of germinating herbaceous vegetation was highest in Q1 and Q2, intermediate in Q3, and lowest in Q4.

These observations demonstrate that the existing gap vegetation had a strong suppressive influence on the germination of newly dispersed or buried seeds. In K-15, no newly germinated or established tree seedling survived in Q4 subplots six months after sampling began. By the end of the study (twenty months), all subplots in K-15 had lost any newly recruited tree seedlings, while subplots in other compartments (K-14 and K-30) had lost 80% of the tree seedlings. Thus tree seedling mortality was higher in the large gaps found in heavily logged forest than in gaps in lightly cut or uncut mature forest.

## GAP TREE REGENERATION AND HERBIVORY

There were significant negative correlations (*P* ≤ 0.01 in each case) between gap size and density of saplings and poles in all compartments except K-13 (figure 28.3). There was also a highly significant difference between mean tree sapling and pole densities in uncut, lightly cut, and heavily cut compartments (Simultaneous Test Procedure of Govindarajulu and Dwass 1979—$P < 0.001$) with densities in uncut forest ranking highest, showing that regeneration was retarded in larger gaps typical of logged forest. The proportion of tree saplings and poles exhibiting normal growth differed significantly between the study sites, with K-30 having the greatest number and proportion of well-formed individuals ($F = 32.36, P < 0.001$) (table 28.4). Saplings and poles of trees in large gaps in logged forest tended to be coppiced or broken more frequently. Damage was principally due to elephant browsing. The frequency of elephant browsing in monitored plots was higher in logged forest (K-14 and K-15) than in uncut mature forest (*P* < 0.005, Wilcoxon's test). Between 70% and 90% of the saplings and poles of the canopy tree species *Newtonia buchananii, Lovoa swynnertonii, Aningeria altissima, Blighia unijugata, Antiaris toxicara,* and *Funtumia africana* were snapped or coppiced in logged forest. In unlogged forest 51% had signs of breaks or coppicing, suggesting that animal browsing impact is more pronounced in larger forest gaps created by logging than in small or medium gaps caused by natural tree mortality. A recent study of elephant movements in Kibale has confirmed their preference for logged forest (Buhanga, pers. comm.).

Table 28.2

## Mean Density and Frequency of Occurrence of Tree Seedlings

| TREE SPECIES | UNDISTURBED FOREST SOIL | | | | | | FOREST GAP SOIL | | | | | |
|---|---|---|---|---|---|---|---|---|---|---|---|---|
| | K-14 | | K-15 | | K-30 | | K-14 | | K-15 | | K-30 | |
| | D | F | D | F | D | F | D | F | D | F | D | F |
| *Celtis durandii* | 22.8 | 1.0 | 6.2 | 1.0 | 12.2 | 0.8 | 1.3 | 0.3 | 0.2 | 0.5 | 3.0 | 0.8 |
| *Celtis africana* | 2.4 | 0.2 | — | — | 2.4 | 0.6 | 0.4 | 0.2 | 0.5 | 0.2 | 0.2 | 0.1 |
| *Trema guineensis* | 16.8 | 0.4 | 25.0 | 1.0 | 35.2 | 1.0 | — | — | 1.0 | 0.1 | 11.5 | 0.8 |
| *Teclea nobilis* | 0.2 | 0.2 | — | — | 4.0 | 0.8 | 0.4 | 0.1 | — | — | — | — |
| *Neoboutonia sp.* | — | — | 1.0 | 0.3 | — | — | — | — | 0.6 | 0.3 | — | — |
| *Chaetacme aristata* | — | — | 1.0 | 0.1 | — | — | — | — | — | — | — | — |
| *Clausena anisata* | — | — | 0.8 | 0.1 | — | — | — | — | — | — | 0.4 | 0.2 |
| *Diospyros abyssinica* | — | — | — | — | — | — | 0.2 | 0.1 | — | — | — | — |
| *Cyphomandra betacea* | — | — | 0.6 | 0.2 | 1.0 | 0.1 | — | — | — | — | — | — |
| *Cordia millenii* | — | — | — | — | 0.6 | 0.1 | — | — | — | — | — | — |
| *Maesa lanceolata* | — | — | — | — | 0.9 | 0.4 | — | — | — | — | — | — |
| Unidentified | 1.1 | 10.8 | 3.9 | 1.4 | 2.7 | 1.0 | 3.6 | 4.6 | 9.6 | 7.3 | 4.9 | 4.6 |
| Total number of species | 4 | | 6 | | 9 | | 4 | | 4 | | 4 | |
| Mean density total | 42.2 | | 33.8 | | 57.7 | | 2.3 | | 2.3 | | 15.1 | |

*Note:* D = mean density (no./m²); F = frequency of occurrence (100% = 1.0). Tree seedlings were germinated from 1 m² plots (n = 28 duplicates for each compartment) of the top 10 cm of forest soil and forest gap soil tended in a screened nursery.

Table 28.3

## Mean Species Richness and Density of Plants Newly Germinated in Manipulated Gap Plots

| QUADRANT | Q1 | | | Q2 | | | Q3 | | | Q4 | | |
|---|---|---|---|---|---|---|---|---|---|---|---|---|
| COMPARTMENT | K-14 | K-15 | K-30 | K-14 | K-15 | K-30 | K-14 | K-15 | K-30 | K-14 | K-15 | K-30 |
| Species richness (species/m²) | | | | | | | | | | | | |
| Tree seedlings | 1.1 | 0.4 | 1.4 | 1.1 | 0.5 | 1.3 | 0.5 | 0.2 | 0.7 | 0.1 | 0.0 | 0.2 |
| Shrub species | 1.1 | 1.1 | 1.2 | 1.0 | 0.7 | 0.7 | 0.8 | 0.8 | 0.5 | 0.1 | 0.1 | 0.0 |
| Herbaceous plant | 3.1 | 3.2 | 4.2 | 4.1 | 2.4 | 4.0 | 3.5 | 3.3 | 2.4 | 0.2 | 0.6 | 0.7 |
| Vines and climbers | 0.5 | 0.9 | 0.9 | 0.7 | 1.0 | 1.5 | 1.0 | 0.7 | 0.4 | 0.1 | 0.1 | 0.2 |
| Mean total | 5.8 | 5.5 | 7.7 | 6.9 | 4.6 | 7.5 | 5.8 | 5.0 | 4.0 | 0.5 | 0.8 | 1.1 |
| Density (no./m²) | | | | | | | | | | | | |
| Tree seedlings | 14.3 | 11.7 | 10.5 | 8.7 | 31.1 | 6.7 | 10.4 | 20.5 | 9.6 | 0.3 | 0.0 | 1.5 |
| Shrub species | 25.1 | 3.5 | 13.5 | 5.2 | 4.9 | 2.7 | 19.0 | 21.8 | 5.5 | 0.1 | 0.2 | 0.0 |
| Herbaceous plants | 165.1 | 236.0 | 102.2 | 223.0 | 276.9 | 99.0 | 39.1 | 104.2 | 16.1 | 0.9 | 11.0 | 1.9 |
| Vines and climbers | 18.0 | 6.5 | 3.2 | 0.7 | 9.0 | 4.8 | 1.1 | 2.9 | 10.4 | 0.2 | 0.1 | 0.3 |
| Mean total | 206.3 | 257.7 | 122.4 | 237.6 | 321.0 | 113.2 | 69.6 | 149.4 | 31.6 | 1.5 | 11.3 | 3.7 |

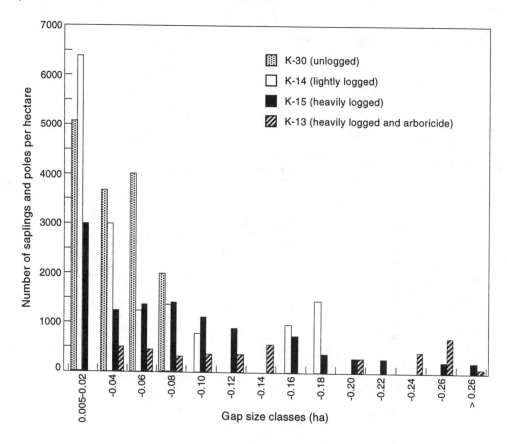

Figure 28.3. Differences between mean tree sapling and pole densities in compartments with different logging histories.

This study has shown that twenty years after logging in Kibale Forest, Uganda, regeneration of commercially valuable trees (and other canopy species) in large gaps remains poor. There are a number of reasons for this. First, trees are poorly represented in soil seed banks. Second, tree seedlings tend to be smothered by the rapid formation of a tangle of herbs, shrubs, and climbers. Third, logging results in larger gaps than those due to natural tree deaths, and these large gaps are regularly frequented by elephants, which break or otherwise damage a high proportion of seedlings and saplings. Finally, logging in Kibale results in increased density and species diversity of rodents, which are major seed and seedling predators and thus have strong negative effects on regeneration and survival of seedlings, particularly those of commercial species (Kasenene 1984). The resulting poor forest regeneration is mirrored by the failure of the primate community to recover to pre-logging densities more than twenty years after logging (Struhsaker 1975; Skorupa 1988; Howard 1991).

A number of studies have focused on the influence of large mammals on the regeneration of tropical forests (Eggling 1947; Wing and Buss 1970; Short 1983; Hart 1986; Merz 1986). Eggeling (1947) observed that elephants, where they are common, can be a lim-

Table 28.4

## Mean Densities of Canopy Tree Saplings and Poles and Percentage Species Populations Regenerating Normally, as Coppices, or as Stem Sprouts

| STUDY PLOT[a] | MEAN DENSITY (NO./HA/SP.) | NORMAL GROWTH[b] (NO./HA/SP.) | COPPICES[b] (NO./HA/SP.) | STEM SPROUTS[b] (NO./HA/SP.) | MEAN NO. OF BREAKS AND COPPICES PER STEM |
|---|---|---|---|---|---|
| K-13 | 24.7 | 1.9 (7.7) | 19.9 (80.6) | 2.9 (11.7) | 1.5 |
| K-14 | 53.7 | 15.0 (28.2) | 26.7 (50.2) | 11.5 (21.6) | 1.1 |
| K-15 | 15.7 | 2.0 (12.7) | 9.5 (60.5) | 4.2 (26.8) | 1.8 |
| K-30 | 71.5 | 29.9 (41.8) | 32.2 (45.0) | 10.4 (14.5) | 0.8 |

a   Plots K-14, K-15, and K-30 are forest gaps, whereas K-13 is heavily cut and treated forest.

b   Numbers in parentheses indicate percentages of species populations.

iting factor in the natural development of vegetation. In the Ituri Forest of DR Congo, large forest gaps were frequented by okapi, and no marked seedlings survived two years of observation (Hart 1986). The studies of Short (1983) and Merz (1986) in west African forests show that large mammals, including elephants, prefer secondary forest and large forest gaps to undisturbed mature forests. This perhaps fits with observations that gap size had little effect on regeneration in a site in Malaysia without large mammal species (Brown and Whitmore 1992; Kennedy and Swaine 1992).

In logged areas or large forest gaps at Kibale, tree regeneration is impeded through selective browsing of young trees, leading to the perpetuation of a dense tangle of herbs (see Kortlandt 1984). The intensity of forest use by elephants increases with the degree of modification or disturbance of primary forest (Kasenene 1987). This implies that the impact of elephants on regeneration in forest gaps could be controlled if the methods of tropical forest exploitation created small to medium forest gaps akin to natural tree fall.

This study shows that we cannot simply sit back and wait for tropical forests to recover after mechanized selective logging. It suggests that loggers should aim to create small to medium forest gaps akin to natural tree falls

(<600–800 m²), where the impact of animal disturbance will be low and regeneration of primary forest species may occur. Careless logging resulting in unnaturally large gaps is likely to cause severe ecological problems for many forest plants, although it may favor elephants. Human intervention by way of seed introduction, enrichment planting, and protection might be necessary in degraded tropical forest areas where regeneration of canopy trees is impeded and inadequate, yet where the objective is for timber production. Although one cannot generalize from Kibale to all African tropical forests, this study should be taken as a warning. Minimizing damage levels is likely to be of major importance if tropical forests are to regenerate successfully following logging, and it is critical that the regeneration process be monitored closely in order to identify potential problems. In Kibale there has been little regeneration in thirty years following logging. If the same is true of other Ugandan forests, there is little hope for renewable forest management in the country.

ACKNOWLEDGMENTS

The National Council for Research and Technology, Office of the President, and the Uganda Forest Department gave me permission to conduct research in Kibale Forest. I am indebted to Peter Murphy and Tom Struhsaker for their support during the study and criticism of earlier manuscripts. The Wildlife Conservation Society, the African Wildlife Foundation, and the National Science Foundation (Grant No. Int-8411306) are recognized for financial and material support for field research at Kibale Forest and post-graduate course work and dissertation writing at Michigan State University, East Lansing.

REFERENCES

Brown, D., and T. C. Whitmore. 1992. Do dipterocarp seedlings really partition rain forest gaps? *Philosophical Transactions of the Royal Society of London B* 335:369–378.

Eggeling, W. J. 1947. Observations on the ecology of the Budongo rain forest, Uganda. *Journal of Ecology* 34:20–87.

Ewel, J., and L. Conde. 1976. Potential ecological impact of increased intensity of tropical forest utilization. USDA Forest Service, Madison, Wis.

Govindarajulu, Z., and M. Dwass. 1979. Simultaneous test for randomness and location of symmetry in one-sample case. In *Proceedings of the Second Prague Symposium on Asymptotic Statistics*. North-Holland, Amsterdam.

Harrop, J. 1962. Soils. In *Atlas of Uganda*. Uganda Department of Lands and Surveys, Entebbe.

Hart, T. B. 1986. The ecology of single-species dominant forest and a mixed forest in Zaire, Africa. Ph.D. dissertation, Michigan State University, East Lansing.

Howard, P. C. 1991. *Nature Conservation in Uganda's Tropical Forest Reserves*. IUCN, Cambridge, England.

Johns, A. D. 1992. Species conservation in managed tropical forests. Pages 15–53 in J. A. Sayer and T. C. Whitmore, eds. *Realistic Strategies for Tropical Forest Conservation*. IUCN, Gland, Switzerland.

Kasenene, J. M. 1984. The influence of selective logging on rodent populations and the regeneration of selected tree species in the Kibale Forest, Uganda. *Tropical Ecology* 25:178–195.

———. 1987. The influence of mechanized selective logging, logging intensity and gap size on the regeneration of a tropical moist forest in Kibale Forest, Uganda. Ph.D. dissertation, Michigan State University, East Lansing.

Kasenene, J. M., and P. G. Murphy. 1991. Post-logging tree mortality and major branch losses in Kibale Forest, Uganda. *Forest Ecology and Management* 46:295–307.

Kennedy, D. N., and M. D. Swaine. 1992. Germination and growth of colonizing species in artificial gaps of different sizes in dipterocarp rain-forest. *Philosophical Transactions of the Royal Society of London B* 335:357–366.

Kingston, B. 1967. *Working Plan for the Kibale and Itwara Central Forest Reserves*. Uganda Forestry Department, Entebbe.

Kortlandt, A. 1984. Vegetation research and the "bulldozer" herbivores of Africa. Pages 205–226 in A. C. Chadwick and S. L. Sutton, eds. *Tropical Rain-Forest: The Leeds Symposium*. Blackwell, Oxford.

Marquis, R. J., H. J. Young, and H. E. Braker. 1986. The influence of understory vegetation cover on germination and seedling establishment in a tropical lowland wet forest. *Biotropica* 18:273–278.

Merz, G. 1986. Counting elephants (*Loxodonta africana cyclotis*) in tropical rain forests with particular reference to the Tai National Park, Ivory Coast. *African Journal of Ecology* 24:61–68.

Poore, D., P. Burgess, J. Palmer, and T. Synott. 1990. *No Timber Without Trees: Sustainability in the Tropical Forest*. Earthscan, London.

Richards, P. W. 1964. *The Tropical Rain Forest: An Ecological Study*. Cambridge University Press, Cambridge.

Short, J. C. 1983. Density and seasonal movements of forest elephant (*Loxodonta africana cyclotis*, Matschie) in Bia National Park, Ghana. *African Journal of Ecology* 21:175–184.

Skorupa, J. P. 1988. The effects of selective timber harvesting on rain-forest primates in Kibale Forest, Uganda. Ph.D. dissertation, University of California, Davis.

Sokal, R. R., and F. J. Rohlf. 1981. *Biometry*. 2d ed. W. H. Freeman, New York.

Struhsaker, T. T. 1975. *The Red Colobus Monkey*. University of Chicago Press, Chicago.

Swaine, M. D., and T. C. Whitmore. 1988. On the definition of ecological species groups in tropical rain forests. *Vegetatio* 75:81–86.

Uhl, C., K. Clark, N. Dexxeo, and P. Maquirino. 1988. Vegetation dynamics in Amazonian treefall gaps. *Ecology* 69:751–763.

Whitmore, T. C. 1975. *Tropical Rain Forests of the Far East*. Clarendon Press, Oxford.

Wing, L. D., and I. O. Buss. 1970. *Elephants and Forests*. Wildlife Monographs 19.

Young, K. R. 1985. Deeply buried seeds in a tropical wet forest in Costa Rica. *Biotropica* 17:336–338.

# CHAPTER 29

# Forest Conservation in Ghana

## Forestry, Dragons, Genetic Heat

W. D. Hawthorne

The forestry situation in Ghana is very dynamic. Between the time when this study was first presented and the publication of this book, a number of highly relevant papers were published. This chapter has been left largely unaltered from its initial form, as it has been referred to in some of the subsequent papers and provided the first airing of some of the concepts discussed. Readers should interpret use of the present tense accordingly, and more current information may be found in the following subsequent publications: Hawthorne and Abu Juam 1995; Parren and de Graaf 1995; and Hawthorne 1996.

### Historical Perspective

In some forestry circles, forest management in Ghana is hailed as a model case. Yet supposed paragons invite keen criticism, and one does not have to try hard in Ghana to find examples of bad forestry practice. Efforts are being made to change these practices. This chapter provides some background to forest conservation in Ghana, a summary of the types of data and analyses being used by the Forestry Department (FD) to inform conservation decisions, and mention of solutions implemented or recommended. It has been written during a key

phase of development of policy in the conservation bodies, so few of the conclusions are as yet set in stone.

The brief history of forestry in Ghana is one of an inspired first phase, a period of decline, and a period of fitful activity and partial recovery from 1985 until the present. Relics of all phases are found in the status and management of Ghana's forests today. Positive trends in forest conservation are working against a recent history of devastation, much of which was officially sanctioned.

Deforestation worried the authorities even at the beginning of the twentieth century and led to the formation of the FD in 1909 (Hall 1987; Hawthorne 1990b; Gordon 1992). Most forest reserves were established in the first half of the century in hilly or swampy areas and in watersheds, to protect the climate, soils, and water supply and to prevent the savanna from spreading. A secondary aim of the reserves was to circumscribe and protect stocks of timber trees for careful management by the authorities and to protect supplies of non-timber forest produce for the villagers surrounding the forests (Foggie 1957, 1962; Foggie and Piasecki 1962). Although conservation of biodiversity in toto was not a stated reason for reservation, as much forest as was politically feasible at the time was put under reservation. This amounted to about 20% of the potential forest zone (the authorities were aiming for 25%), including a broad spectrum of forest types and the most sensitive landscapes. Today these reserves hold virtually all that is left of the forest.

Forest conservation in Ghana is today mostly part of the general forest management activity of the FD, with two important exceptions. First, small unreserved but variously protected forests (often sacred groves) make a significant contribution to the area of certain types of dry forest and are the main refuge of a few rare species, even though these forests constitute just a small portion of the national forest area. Second, the Game and Wildlife Department (GWD) was established in 1965 and has complete responsibility for the forest in Nini-Suhien and Bia South National Parks, and partial responsibility for Ankasa and Bia South Game Production Reserves (Asibey 1972; Asibey and Owusu 1982). Some areas listed as forested national parks (e.g., by Gordon 1992) include insignificant areas of forest, being mostly savanna. Kakum was recently established as a "special-case" national park, with encouragement of some eco-tourism.

National parks have protection from logging and (in theory) from hunting and other local uses. Game production reserves (GPR) are supposedly managed sensitively, allowing exploitation in the context of special wildlife protection. In fact, no special care has been taken in Bia GPR, which has therefore been significantly disturbed, but there are no current plans to log Ankasa GPR, which has been reasonably protected, at least from a botanical point of view.

Pressure to exploit reserves has grown, and now virtually all have been logged at some time. Starting in the 1950s, increasing areas of reserved forest were exploited under the "selection system," with log yields calculated as a con-

servative proportion of mature and not-quite-mature trees, with reference to stock maps showing all large trees in the forest (Baidoe 1970, 1976; Asabere 1987). Protection Working Circle areas (PWCs) were set aside within productive forest reserves, usually because they were either not loggable or too poor in commercial species. Other parts of forest reserves were designated for conversion or research (or, most often, both), but these were usually vaguely and temporarily defined management units. Some PWCs were to be protected "in perpetuity," but even in these, district officers had discretionary powers to permit logging or other activities. With no national framework for the protection in the first place, this led to the increasing disturbance of PWCs, particularly during the FD's decline. On paper, PWCs supposedly cover about 1/3 of the forest reserves, but less than half this area supports undisturbed forest today.

The period of decline, from the mid-1960s to the early 1980s, applied to the whole economy, not only to the management of natural resources. Although the economic collapse toward the end of this period probably helped protect some forests from logging, overexploitation had previously been rife, and legal and illegal farms proliferated inside some reserves (Martin 1991). Large areas of forest reserve—sometimes even of healthy forest—were lost to farms through the encouragement of *taungya*, whereby farmers were allowed to farm supposedly poor forest land if they planted commercial species of trees before shifting sites. With generally limited support and control, this practice amounted to little more than legalized deforestation. A few mono- (or oligo-) cultures of often exotic species were created, but the scheme seems largely to have been a failure even on its own terms, and it has now been stopped. Managers need to be wary of similar schemes reappearing within a Trojan horse of, say, community forestry, which in other guises might be expected to play a growing and positive role, even inside reserves, in forest harvesting and restoration.

Forest managers were to blame during this phase for another serious error: the introduction of "salvage logging" of unlimited "overmature" trees, which in effect represented the dissolution of the original, much more gentle selection system. Admittedly, this had been introduced as a temporary measure, but the resulting excessive logging, compounded by unofficial exploitation, weakened the forests in a way that made them more susceptible to severe degradation by fire, particularly following the fires of 1983 (Hawthorne 1993, 1994). The unlimited removal of supposedly overmature trees has been stopped since about 1989; this change would ideally be pushed a step further so that the oldest and largest trees, which are often heart-rotted, receive complete protection on a tree-by-tree basis (within the context of fine-grained protection discussed below). Attempts were made in the 1960s to define new reserves in the then rather remote and untouched western region, although clearly the arrangements were not satisfactory to the local chiefs (landlords) or land users, because today these reserves (such as Bia Tawya, Manzan, and Trans-Bia) have suffered most seriously from "illegal" farming and support little forest. Mass "disobedience" leading to complete deforestation has also occurred in the Volta region reserves (such as the Togo plateau) because of the high demand for farmland, associated partly with movement of dispossessed people before the damming of the Volta Lake in 1966.

By the early 1980s, Ghana's once shining suit of forestry armor had developed two decades of rust. We are now in the third phase, during which fresh components are being grafted onto whichever practices from the past have not been discarded. Years of careful polishing are needed to restore those mechanisms deemed worthy, and years of adjustment may be needed while new components—for instance, greater consideration for biodiversity and greater community involvement—are integrated. Non-governmental organizations are likely to play an increasing role in local forest use, particularly if the FD proves too slow to provide attractive initiatives. Meanwhile, the rough coexists with the smooth: good personnel are discouraged by poor incentives; satellite imagery and other technology is brought to a halt by power cuts and maintenance problems; healthy forest reserves exist alongside overlogged and burned forests. Since 1989 the felling cycle has been lengthened from twenty-five to forty years, certain species can be felled only under special permit, and yield allocation has been tightened. But the timber companies, many with "direct lines" to people in power, are becoming hungry. So although there is cause for optimism, there is also no shortage of past and recent management mistakes from which to learn and threats of which to be wary.

### The Forest Vegetation and Applied Research

The botanical layout of Ghana's forests is among the best known in the tropics (Hall and Swaine 1981; Hawthorne 1990b). The ordination and classification of forests made in the 1970s by Hall and Swaine (1976, 1981; see forest types in figure 29.1) provides an unusually robust foundation for all work on growth, yield, and biodiversity studies, without which subsequent work would be harder and of less value.

Forest inventory and management projects (now just one forest inventory and management project), supported within the FD by outside agencies since 1983, have tried to fill gaps in technical knowledge and biometrical data needed for forest management. This work has included a 1/400 sample of trees down to 5 cm dbh in all remaining productive forest (Wong 1989; Hawthorne 1995); a network of Perma-

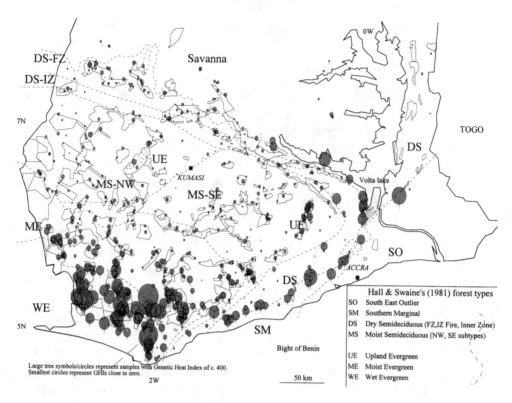

Figure 29.1. Map of the forest zone in southern Ghana, showing forest reserves within solid lines and forest types between dashed lines (according to Hall and Swaine 1981; see key). Tree symbols (shaded circles with trunk-like lines) designate botanic survey samples of the plant community, sized according to Genetic Heat Index (GHI). A hotspot area, with generally elevated GHIs, is clearly visible in the southwestern corner of the country. Adapted from Hall and Swaine 1981.

nent Sample plots (Bird and Sackey 1991); growth/yield models (Alder 1990); a field guide to trees (Hawthorne 1990a); ecological studies of the effects of logging and fire (Hawthorne 1993, 1994); a study of the market in non-timber forest products (NTFPs) (Falconer 1992); and selected inventories of NTFPs.

In the national Temporary Sample Plot (TSP) tree inventory, which is used mainly to assess the standing populations of timber and other common trees, rare tree species are represented very unreliably. However, certain patterns revealed by the TSP data have strong implications for biological conservation. For instance, timber trees are of conservation con-

cern in spite of their commonness, and a national picture of their population structure is therefore vital for careful scientific management; moreover, by classifying commoner trees into guilds, trends in the response of the forest to various types of disturbance can be monitored (Hawthorne 1993, 1994, 1995). There is little point in striving to restore or sustain a healthy forest without some index of whether this has been achieved. Balance of guilds provides one "handle" on the plant community, and this type of information is available on a grid for all forested reserves remaining in the country, albeit only as a 0.25% sample (see Hawthorne 1996).

Additionally, 600 Permanent Sample plots throughout the forest zone have been established (or retained) to provide information on stand dynamics, but these will not bear their main crop of data for several years (Bird and Sackey 1991).

Some information required for careful management of biodiversity has had to come from more specialized research. One such subproject is a review of the PWCs within the reserved forest, combined with a botanical survey of all the forests. The aim is to appraise and revise the framework for the maintenance of plant genetic diversity (Hawthorne and Abu Juam 1995). Components of this survey will be reviewed below.

The FD unfortunately has no current capability for directly managing the animal life in forest reserves on a sound scientific basis. The protected area review deals only with the vegetation as a whole and with plant species in particular. A closer link with GWD is a priority for sound management, and there are signs that a complete faunal survey, with recommendations for management, may soon be under way. In the meantime, animals in forest reserves will have to be protected "on the backs" of the plants.

## The Botanical Survey and Protected Areas Review: Methods and Tools

In January 1990, a small team under the Forest Resources Management Project was set up to reevaluate the Protection Working Circle. The field surveys fell under two overlapping headings, the PWC survey and the botanical survey. Sometimes both surveys were performed by the same team.

The PWC survey provided data relating to the history and management of forest reserves, particularly regarding which areas were supposed to have been protected, and whether or not they had been. The botanical survey built on Hall and Swaine's botanical survey data, which led to their classification of forest types (Hall and

Swaine 1981). The aim was to build a database of the distribution of forest vascular plants in Ghana to enhance our picture of the botanical wealth of the forests, allowing the forests to be classified with greater resolution. This is useful in ensuring that all areas of Ghana's forest community are adequately protected, because there is substantial floristic variation within each of Hall and Swaine's zones. The distribution of rare plants, rather than more widespread ones, was a matter of prime concern, and the floristic "anatomy" of individual reserves was examined, whereas Hall and Swaine's survey rarely had more than one plot in a reserve and was mainly concerned with the commoner species.

### METHOD FOR THE BOTANICAL SURVEY

Readers are referred to Hawthorne and Abu Juam 1995 for a summary of the botanical survey field methods. Plotless and measured (1/16 ha) samples were enumerated in a variety of forest types and conditions. Most samples were of 50–150 species. Much of the work in this survey involved collection, curating, and identification of specimens, which provided a permanent and precise record of much of the team's findings. All PWC and botanical survey data are stored on computer. A summary of these data can be exported to a map-displaying program called FROGGIE (see Hawthorne and Abu Juam 1995), on which species distributions, tree stem densities, and conservation priority scores can be viewed over a map of forest reserves.

### FROGGIE

Forest Reserves of Ghana: Graphical Information Exhibitor (FROGGIE) is a simple map-based database designed to provide access to a wide variety of forest reserve and botanical data (see Hawthorne 1995), including densities of chosen size classes of tree species or guild for any or all of the forest reserves; the known distribution of all forest species to the nearest 100 m; Genetic Heat Indexes (see below), Pioneer indexes, and other calculated properties of samples; and overall forest condition, textual notes, or landscape attributes.

The jungle of information on species in plots and forests is indigestible without ways of summarizing it. For biodiversity purposes, the following scheme has been devised. FROG-GIE is in effect an interactive appendix to Hawthorne and Abu Juam 1995. The generalization and evolution of FROGGIE as a tool have subsequently been pursued as a research project at the Oxford Forestry Institute.

### Star Rating of Conservation Value

All Ghana's forest species have been scored for conservation priority, with most consideration given to rarity. A classification giving similar significance to rarity and perceived threat (like IUCN's red list criteria) has not been adopted, because we are more directly concerned with management of forest fragments rather than individual species, and the threat factor (which is, on such a local level, very fickle anyway) is better considered independently of the botanical interest of the forest.

"Weedy" species, or savanna species found only at the margin of forests, score nothing. The classification focuses on the conservation of those species in Ghana's forests that would not flourish if all forest were to be removed. The remaining species are rated Black Star, Gold Star, Blue Star, or Green Star, in order of decreasing conservation priority. Fairly common or widespread species that are heavily exploited or that show great economic potential are rated as Red Star species, requiring individual attention. Where exploited species have an otherwise high conservation rating, they are allocated Black or Gold Stars as appropriate.

The allocation of Green to Black Stars to species has been made through consideration of a variety of factors concerning the ecology, taxonomy, and distribution at all levels. These considerations are summarized in table 29.1 and appendix 29.1. The primary, but not the only, factor used in the classification is rarity, both in Ghana and internationally. Only species rare on both scales can score Black Stars, although rarity on either scale can contribute to a Gold (or Blue) Star rating.

The scores for species can be combined into scores for communities, to highlight "genetic hotspots," but there are various ways to do this. One could just list or count the number of Black or Gold Star species found in a reserve. This gives a moderately coherent pattern, with neighboring reserves having similar numbers, but more weight can be given to stars of higher value for a more sensitive index, and the con-

Table 29.1

## Considerations for Star Ratings

| STAR | NUMBER OF SPECIES | | NUMBER OF DEGREE SQUARES OCCUPIED | NO. OF TEN-DEGREE SQUARES OCCUPIED | CHOSEN WEIGHT |
| | TOTAL | SAMPLED | | | |
|---|---|---|---|---|---|
| Black | 52 | 5 | 1.6 ± 0.49 | 1.2 ± 0.4 | 27 |
| Gold | 201 | 16 | 7.8 ± 3.76 | 2.88 ± 0.86 | 9 |
| Blue | 361 | 35 | 24.5 ± 12.58 | 7.34 ± 4.08 | 3 |
| Red | 78 | 5 | 39.6 ± 15.9 | 11.2 ± 3.76 | 2 |
| Green | 969 | 43 | 69.2 ± 49.76 | 17.02 ± 10.45 | 1 |
| (Weeds) | 273 | 0 | >100 | >20 | 0 |
| (Savanna species) | 265 | 9 | 75.3 ± 54.3 | 19.8 ± 10.04 | 0 |

centration of rare species in the sample (rather than a total) can be calculated, for samples showing a fair cross-section of the community. This gives a weighted index, as follows.

### CALCULATION OF GENETIC HEAT INDEX

$$GHI = \frac{(BK \times BKweight) + (GD \times GDweight) + (BU \times BUweight) + (RD \times RDweight) \times 100}{BK + GD + BU + RD + GN}$$

where BK = number of Black Star species, GD = number of Gold Star species, BU = number of Blue Star species, GN = number of Green Star species, RD = number of Red Star species, BKweight = weight for Black Star species (as explained below), GDweight = weight for Gold Star species, and so on. Species for which details are unknown are excluded from these sums entirely.

The main feature of this Genetic Heat Index (GHI) is that it reflects the concentration of rare species in the community rather than, for instance, the number per unit area. If there were only "rare" and "common" species, it would express the percentage of rare species in the sample, but the index responds to the various classes of rarity. An important consideration, therefore, is the weight given to the different stars.

The weights for the star categories have been calculated from the inverse of the areas of earth occupied by the average for the species in each category. "Higher-starred" species occupy smaller areas on earth and are allocated a proportionately higher score. Taken to its extreme, each species would have its own "inverse area occupied" score, but this would be impractical to calculate even crudely. Rather, a subsample has been taken of species with their geographical ranges within Africa published (few rated species extend beyond Africa) in the series *Distributiones Plantarum Africanarum*. In this series of maps, the known distribution of species is plotted on a degree-square matrix. Obviously, this is a sample, not a complete set,

but it does provide a fair index of the range occupied by each species.

This degree-square sample was compared also with the number of ten-degree squares occupied, which is likely to be a more stable figure but is less subtle. The results of this analysis are shown in table 29.1.

The weights for the stars used in calculation of GHI were chosen by:

1. allocating 1 to Green Star species (the standard forest species of no special rarity value);
2. at both one-degree-square and ten-degree-square resolutions;
3. choosing a round number between the two values for each star. In the end, the values 1–3–9–27 were chosen because they form a geometric series. Note that Red Star species are scored here without consideration for their commercial value or degree of threat due to exploitation (see below).

The GHI is, then, a weighted mean rarity value score for a sample of species. It is not a diversity index—more a "bioquality" index. Diversity scores can be monitored separately. Indeed, the southern dry types of forest are species poor but have a high GHI; there are few species, many of which are rare, so the forest is rich in individuals of rare species. The result, as it turns out, usually makes sense as a conservation priority score to botanists with knowledge of Ghana, although the system does give a rather low rating to the communities dominated by plants rare in Ghana but widespread in Africa. This particularly affects the Upland Evergreen forest type (Hall et al. 1973), which is anyway recommended for protection in toto because it is a forest type of extremely limited extent, and on steep-sided hills. Further work with the entire database will provide a similar index based entirely on rarity in Ghana for comparison, and some species may change their star category as we learn more about them, but there is no intention to alter the star rating substantially. Greater precision or refinement in such an index would be superfluous, as the linkage between GHI and any potential conservationist's action is so loose.

It can be noted that the GHI or a similar index, with a star or equivalent rating tailored to local circumstances, could allow direct comparison of the "genetic heat" of communities across the continent. It is necessary only to alter the weighting according to the definition of the categories used (although a weight of 1 would have to be attached to species equivalent to the Green Stars here). Where time allowed, each species could carry its own weight.

## Reddish Star
### (Economically Threatened) Species

Reddish Star is the generic term for common species threatened by exploitation (Reddish Stars include not only Red Stars but also Scarlet and Pink Stars, as explained below). Commercial extinction (market collapse) of Reddish Star species is more likely than biological extinction. The most serious biological threat is the possibility of reduction of genetic quality as a consequence of removal of favored individuals from the population, as has been described for the "proper" mahogany (*Swietenia mahagani*), of which wild populations have supposedly been reduced through centuries of exploitation to low-branched trees (Styles and Khosla 1976). Protection measures in operation (discussed below) should prevent this effect in Ghana.

Most of the forest species most threatened through overexploitation are trees. Species are allocated a Reddish Star only if exploitation now affects or potentially affects a significant percentage of all plants in Ghana. Certain non-tree species whose populations ought to be monitored closely include species of canes and Marantaceae, for which there is a heavy demand in at least some areas (Falconer 1992). A few of these (e.g., *Thaumatococcus danielii*, a source of sweetener) have been allocated Reddish stars, but the background data are sparse. The situation is slightly more clear for non-timber trees than it is for non-trees. There is an immense and diverse market in certain chew-stick *Garcinia* spp. (Adu-Tutu et al. 1979; Falconer 1992). The trees are not particularly abundant. *Garcinia* spp. are shade tolerant and quite probably slow growing. A specific research project is needed to investigate the feasibility of "farming," or otherwise redressing the supply-demand imbalance for these species.

We have a much clearer picture of the significance of exploitation of trees for timber. Ultimately, Permanent Sample plot data and the "GHAFOSIM" yield modeling program (Alder 1990) may prove highly useful in this, but even rough-and-ready calculations of the type in appendix 29.1 are revealing.

Appendix 29.1 shows the number of trees counted during all inventories whose results were available at the time of writing. All Reddish Star trees, and all tree species exported in 1989, are included, as are a few non-timber common Green Star tree species (for comparison). The total number of actual trees measured, rather than a calculated density, is listed. The total number of individual trees standing in Ghana's forest reserves can be estimated at around 500–750 times the tabulated values, for trees of 30 cm dbh or greater.

Appendix 29.1 also shows the number of trees logged in one year (1989), according to log measurement certificates issued for that year (no doubt a slight underestimate). The number of trees logged is divided by the average population density for each species, to give a figure expressing the area required to produce the logs that were extracted in 1989 for each species. This has been done for two lower size limits, because different trees have different girth limits for logging, and because for the trees with larger girth limits, the >60 cm size range gives some indication of the potential recruitment as well.

Put simply, the higher the number in the last two columns of the appendix, the more serious was the effect of logging on that species in 1989. There are several sources of error in these data, so the precise values are not particularly meaningful. Rather surprisingly, *Funtumia elastica* is at the top of the table, probably because of sampling error, sporadic demand, and exploitation at sizes even less than 60 cm dbh. Otherwise, the trends accord with expectations reasonably well.

Based on considerations such as these, Reddish Star species can be split into three subcategories, or shades of red (see appendix 29.1):

1. Pink Star species, which are significantly exploited, but not yet so much as to cause concern for their commercial future. Pink species will probably become more exploited soon; indeed, it is Ghana's intention to promote many of the minor timbers to take the pressure off the more desirable ones.

2. Red Star species, for which current rates of exploitation may well present a significant danger of commercial damage or genetic erosion over, say, a ten-year period. Managers should consider restricting the practices of loggers with regard to these. They should certainly be among the species for which individuals are selected for retention in each logged compartment.

3. Scarlet Star species are under more imminent threat than Red Stars and face priorities for serious control.

Red Star species are subject to continual revision as markets and populations change. Since this chapter was first prepared Wong (1994) has updated them, referring to an earlier draft of this chapter as a starting point. The greatest priority for careful protection, of course, must go to a few species which are both exploited and rare. These have been allocated Black to Blue Stars as appropriate. Two species that deserve careful attention are *Sclerosperma mannii* (rare miniature palm used for roofing in the wet evergreen zone) and *Talbotiella gentii* (Black Star, endemic tree cut for charcoal) (Swaine and Hall 1981). *Talbotiella gentii*, especially, is so rare and so exploited that it would be worth mounting a special project just to protect it and to provide alternative fuel supplies.

Economic indexes analogous to the GHI can be calculated according to specific interests. For general purposes, an economic index is defined as the percentage of Reddish Star species in the sample, weighted according to shade of red. A tree-specific index has proved more revealing, expressed as percentage of Red Star tree stems in a tree sample. Economic Index (EI) is used informally below simply to refer to the unweighted percentage of Red Star species, unless "EI (stems)" is specified.

## The State of the Forest and Reasons for Its Erosion

Although satellite pictures show a remarkable conformity between the theoretical reserve boundaries and the actual forest cover, the quality of the "forest" within the boundaries is quite variable, with conditions worsening with increasing dryness (see Hawthorne and Abu Juam 1995; a disproportionately high amount of good forest is located in the proposed Hill Sanctuaries, described below). Negative influences on forest quality have come from both outside and inside the FD's sphere of influence. Onslaught by factors largely beyond the managers' control—such as drought-cum-fire, the allocation by superior powers of forest reserves for mining, and illegal farming—have been important. Inasmuch as they have not always been completely beyond the control of the Ministry of Natural Resources as a whole, there is some potential to reduce their influence, particularly as the ministry becomes increasingly aware of the value of keeping the remaining reserves as unmined natural forest.

The single most important unplanned destructive factor has been fire. Even if all exploitation in Ghana's forests were to stop today, fire would remain for many years an immense threat against which managers seem hopelessly impotent. It is Ghana's dragon (and St. George is nowhere to be seen). Even though occasional fires may have occurred throughout history (Hall and Swaine 1981), fire has recently presented a serious challenge to the long-term existence of semi-deciduous forest in Ghana (the threat is insignificant in wetter forests). Since 1982, fire has significantly altered the structure and composition of 30% of the remaining semi-deciduous forest; at least 4 million m$^3$ of high-quality timber (Redwoods and Iroko) have been lost (Hawthorne 1994).

Fire is enhanced by a variety of factors, including drought, past fires, logging or other disturbance, and proximity to savanna. The damage inflicted to the forests in the 1970s almost certainly encouraged the widespread fire damage after the drought in the 1980s. The problem is particularly serious because burned forest burns more readily than unburned forest, and management has no experience of restoration of burned forest (other than by conversion to teak or a similar plantation). It is hoped that research and indigenous-species trials will begin soon on firebreaks and techniques for the restoration of Asukese Forest Reserve (discussed in Hawthorne 1994), but conclusive results may be decades away. Any management prescriptions in the meantime are based on guesswork.

The reduction of disturbance in fire-prone areas would seem an obvious essential step in tackling the problem. Unfortunately, advocating widespread complete protection in the semi-deciduous forest zones is unrealistic, as these are traditionally the most productive forests and are generally low on the priority list in terms of GHI. However, yield allocation is being reduced in fire-prone areas, and it may be possible to prevent tree felling in areas with a burned understory. Recommendations have also been made that the pattern of logging be altered in fire-prone forest, with collars of undisturbed, protected forest surrounding logged areas.

Past official mistakes include those actions that had undesirable, unplanned repercussions, and those that had a planned effect that is now seen as undesirable. Fire damage can be seen as partly attributable to mistakes of the first kind, which can be reduced by research and education. Within the broad timescales that forest managers have to consider, mistakes of the second kind are also easy to make. For instance, unmarketable trees in one decade, poisoned during intensive silvicultural improvement operations, have become desirable in the next.

Although the FD is generally committed to the maintenance of forest within reserves, there is much dispute about how the ideal forest should be defined. At one extreme, for a significant proportion of foresters and consumers, plantations of exotics are not only a fair substitute, but even a marked improvement on natural forest. Plantations have an expanding role in Ghana, at least to take the pressure off the natural forests, and in some cases forest reserves can benefit from planting. However, it has not only been monotonous, weed-dominated, low-value, degraded parts of forest reserves that have been converted to monocultures, although there has been some limited success with this. Rather, natural forest in fair to good condition has been clear-cut to make way for plantations. Plantation, even where the understory is well developed, has a GHI lower than that of natural forest nearby. Subri, the biggest reserve in Ghana, is much used for non-timber forest produce (Falconer 1992) and has a fairly high GHI (coming 17th out of 220 in the league table of average reserve GHI). Partly because it was difficult to log and poor in commercial species, it is being partially converted to an industrial *Gmelina* plantation for pulpwood. This will probably be seen by an increasing number of observers, not least by villagers around the reserve, as one of Ghana's big mistakes of the 1980s and 1990s.

Most of the mess in Ghana's forests has come from factors more within the FD's control: excessive yield allocation (with concomitant fire damage), taungya, local corruption, and similar lack of local control have been the main reasons for forest degradation. However, some degree of excuse can be found in the facts of economic deterioration (improving in the 1990s): poor incentives for workers (incentives being reviewed); unawareness of the (fire) hazards of overexploitation and taungya; and misguided policies (which may be declining). To the extent that these factors are soluble, there is hope.

A significant part of the problem has derived from imprecise, contradictory, or confusing guidelines or, in other contexts, "loopholes." Maintenance of biodiversity is a recent goal. Quite apart from the many good reasons for conservation of biodiversity as a primary management aim, a dogma that includes its maintenance can be seen as the only sensible way to respond to the problem of ever-shifting goals in forest management.

## The Distribution of Genetic Heat and Its Implications

The methods used for sampling the Genetic Heat Index (GHI) and Economic Index (EI) are mentioned above. What are the implications of the results of this survey, as regards management to encourage biodiversity?

Forest samples from the same area are fairly coherent in terms of GHI, in spite of variation in history of disturbance (see figure 29.1). Extensive areas are "cool" (low GHI) all over, with sometimes rapid gradients into areas that are usually hot, except in spots where heavy disturbance has occurred recently. In this part of the Guinean region, the pattern of high GHI is bimodal, with the driest and the wettest forests scoring similarly, and standing out as opposed peaks, or genetic hotspots. Forest managers ought to encourage activities in sympathy with this pattern, with production mostly conducted in the intermediate (Moist Evergreen and Moist Semideciduous) "cool" zones. Because of the mode of definition of these hotspots, they express significance on an international level and should justify international (financial and other) assistance in their conservation, particularly as their protection may represent a cost to Ghana that it is unlikely to recoup quickly.

Given that fire risk is enhanced by logging, and that the dry forests are the most threatened (for this and other reasons), there is a strong priority for protection from logging—or rather, a need for a more extensive network of protected areas—in these driest forests. All Southern Marginal and South East Outlier forests are being recommended for complete protection, but the total area is small. Most of this forest zone remains as scattered, often unreserved hilltop and sacred forest fragments, and their conservation would probably best be managed entirely at the local level with help from a coordinating agency.

At the other GHI pinnacle, the threats to the wettest (Wet Evergreen and wetter Moist Evergreen) forest reserves have to a great extent stemmed directly from official policy, with some parts of some reserves (Subri, Neung North and South) being currently converted and other areas (Neung North, potentially Cape Three Points) being mined. The current GHI framework justifies the cessation of all this officially sanctioned deforestation in these areas. Official damage should be reduced and exploitation dampened, and in addition, the hotspot forest zones should receive a disproportionate area of special protected areas (discussed below).

In spite of national trends, the GHI does show some local variation, with important implications for managers. In Cape Three Points Forest Reserve (the reserve with the 10th highest GHI out of 220), which approaches the sea front and is situated across a steep rainfall gradient, there is a wide variation in GHI, with extremely high scores in the hilly areas to the north. In this case, any logging should be restricted to the south, and special protection is recommended for the hills. Individual landscape units have been recommended for protection elsewhere, where they form a definable block.

### Pattern of Disturbance

For a small area of forest, the response of the GHI to various types of forest usage is of prime interest. Exploitation brings a temporary, local disaster to the forest community. The early months on recently abandoned logging trails are associated with a GHI and EI of close to zero in the associated pioneer-dominated plant community, but these can soon pick up to the same level or a higher one than those in the surrounding forest. Within the first five years of the developing community, the EI (stems) is often very healthy: some timber trees are pioneers, and most are non-pioneer light demanders and tend to make a good showing in secondary forest, provided that there are adequate timber trees close to the disturbed area (Hawthorne 1994). Similarly, our few samples analyzed to date show that log-skidding trails, and other patches of fine-grained disturbance in intimate contact with the surrounding community, recover ("reheat") quickly in plant-genetic terms.

The forest structure may need centuries to recover if ancient trees have been felled, and this may have significant repercussions on the animal community. Nevertheless, the area affected directly by logging in Ghana exceeds 25% only in extreme cases, and the repercus-sions of recently reduced yield prescriptions should be that at least 80% of a logged forest compartment remains barely affected, with a retention of many of the largest trees. Such disturbance must have negative effects on the pre-existing system, but the accrued benefits, not least by justifying to a poor country the value of maintaining natural forest, outweigh them.

In contrast, heavier disturbance, especially where the topsoil is lost—for example, from log loading bays, main access roads, and farms—or, in the extreme, where complete deforestation is spread over many hectares, recovers much less reliably even when the disturbance, which tends to persist in such areas, finally desists. Secondary forest far away from an undisturbed forest patch probably recovers GHI and EI more slowly than a similar area closer to good forest (see Hawthorne 1996). So the effects of disturbance should be minimized by preventing the formation of large gaps. As the early forest managers realized and as one can easily imagine, small wounds that are well dispersed in space and time ("fine-grained" wounds) are rapidly healed. Large-grained wounds are not. Such observations have led to numerous recommendations that are gradually being adopted (or re-adopted) in Ghana.

### Pattern of Protection

Clearly, the pattern or granularity of disturbance contributes enormously to the effects and chances of recovery from a given quota (measured, perhaps, as number of trees removed) of disturbance. Limited dispersibility of plants and the greater maintenance of the less-extreme forest environment must explain the benefits of ensuring the proximity of undisturbed forest to disturbed forest. The implication of this is that "fine-grained protection" should be applied wherever forest is being disturbed.

## FINE-GRAINED PROTECTION

Fine-grained protection concerns the protection of (1) selected individual plants, especially trees, in a variety of circumstances; (2) small clusters of trees and their understory in otherwise deforested areas; and (3) all forest over sensitive parts of the landscape that are too small or impermanent to be catalogued on the national level. The aims of fine-grained protection are:

- Structural or environmental: to prevent erosion or deterioration of physical resources; to maintain a significant amount of the canopy at all times, and a complete canopy cover on banks of rivers and streams, lakes and swamps, and steep slopes.
- Botanical: to maintain within the matrix of disturbed areas the population densities of seed-bearing trees at a high enough level to ensure good regeneration and to reduce genetic erosion, particularly for those species that are most exploited.
- Biological: to provide continuity through time of those habitats represented only in or among old trees or undisturbed sites, and to maintain these as refuges within otherwise inhospitable terrain.

Protection must be carried out on the fine-grained, even tree-by-tree level throughout all forest reserves during all practices. The existence of protected areas in no way removes the burden of conservation in other parts of forest reserves. If readers think that such fussy management is a tall order in tropical forests, they should note that many aspects of such a system function already in Ghana, and were working in the 1950s. The problem is to extend and standardize this essential level of protection.

In the case of logging, needs for fine-grained protection have devolved to recommendations that the number of trees felled per compartment be limited (a maximum of 200 stems per compartment of c. 140 ha) and that more standard "good logging practices" always be adopted (Baur 1964). Logging in Ghana has for some time been operating under these rules. There is room for improvement in the treatment (royalty rate, girth limits, retention factors) of each exploited species according to its own ecological merits, but steps in this direction have been taken.

## SPECIES-SPECIFIC CONTROLS

Certain controls should apply to individual species, if they are heavily exploited in proportion to their abundance (Red Star species), and such controls are included as fine-grained protection. The main thrust of protection of Red Star species is at the tree-by-tree level in each compartment; this is (bar a few important subtleties) already taken care of in the yield allocation procedure. For a logged forest to regenerate and remain healthy and viable economically, the nearest good stocks of timber species could not be in a protected area ten kilometers away. Fine individuals of Red Star species should be selected by forest officers for protection before logging. Very old (so-called overmature) trees, which are often rotten in the middle but which have huge crowns and often seem to fruit profusely, should be particularly protected (not least for the animals and epiphytes they support), as these take longest to replace. There is a lot to be said in favor of the Forestry Department somehow making especially favored individuals unloggable (perhaps by shooting metal fragments or a dye into them), with a clear notice to that effect attached. Fireproofing such trees would be beneficial in some areas as well. There is some way to go in the institutionalizing of such tree-by-tree protection.

All Black Star species require special protection, but few of these are exploited, and in most cases they will be adequately protected without species-level protection on the ground. Only a few NTFPs seem to require intervention to prevent overexploitation or genetic erosion. The control of hunting in forest reserves is a matter of urgency; the FD has traditionally not taken a very active or scientific interest in this aspect, and there is vast scope for improvement.

Where plantations are being made, fine-grained protection requires that trees, particularly those other than Green Star species, not be felled and that even pockets of a few meters of healthy forest be left where they are found, to become embedded in the plantation as havens for wildlife. Of course, these rules would prevent the destruction of healthy natural forest to make way for plantation. There is in Ghana no shortage of reserved savanna, illegal farmland, and savagely burned areas for planters to work on. Where there is a mosaic, planters are encouraged to plant in the degraded pockets only, and to forgo their preferred orthogonal blocks. Again, such practices are only beginning to be implemented.

### LARGE-GRAINED PROTECTION: THE PWCS REVAMPED

Fine-grained protection is vital to the long-term health of the forests but is inadequate without the protection of larger blocks—which brings us to a reappraisal of the PWCs and the justification for their existence. The following points outline five ways in which fine-grained protection would be inadequate on its own to maintain Ghana's forest ecosystem (there are probably others as well). Large-grained protection would provide:

1. A second line of defense. Larger, constantly protected areas are more likely to be positively defendable from fire, logging, or other violations.
2. A functioning ecosystem. Many components of the ecosystem may not be able to operate in a broken mosaic of small forest islands.
3. Community tokens and provenance reserves. There is a need for provenance reserves where a population of economic species is protected completely from "creaming" in an undisturbed environment, and where reference samples of the community as a whole can be protected.
4. A minimization of edge effects. Small pockets of forest have a high proportion of edge, and consequently will tend to develop a "weedy" understory, as they are more susceptible to treefall, fire invasion, desiccation, and illumination. Protected areas are more resistant to fire.
5. Complete protection of special parts of the landscape. Steeply sloping, hilly, or other environmentally sensitive land often covers large areas and needs protecting as a whole. This includes most of the protected, unlogged areas existing already. Many fetish groves are located within forest reserves and require special protection from disturbance.

Large-grained protection—the revitalized PWC—includes a network of areas large enough individually to be mappable, manageable, and worth demarcating and guarding, and to possess the other attributes mentioned above. A compartment-sized area (c. 140 ha) is the most convenient minimum size, given that finer-grained protection should already be operative within logged compartments. A larger size is usually highly desirable and would obviously be recommended by zoologists for larger mammals at least. Some fetish groves and other special sites are smaller, and may be best served by mapped smaller areas.

In these cases "protection" means complete and absolute exclusion of logging and any other usage that is detrimental to the canopy trees, including local debarking and use of canopy trees for firewood. Planting of any sort and the lighting of any fire should not be permitted in protected areas. No discretionary permits should be available at the district level—a main cause of the erosion of many PWCs in the past—although permits to fell an occasional tree without extraction vehicles could be considered by the planning branch or head office. Protection does not exclude local uses that do not affect the canopy, unless species are specifically protected or controlled, or unless the area has special protected status, and as long as the use is considered not to be detrimental to the natural balance of the forest, at the discretion of the district officer. Heavy removal of more than scattered trees or shrubs (>10%), or of more than scattered lianas, would qualify as detrimental to the natural balance. Digging for minerals and other such activities should, as a rule, be forbidden for similar reasons.

Special protected status includes protection from all disturbance and will normally apply to areas of high GHI amply surrounded by less strictly protected forest. The same rules would apply as for a national park. It can be noted that certain areas might be excluded from logging even if they are not permanently protected, on the basis of proposed amendments to the stock survey or yield allocation procedure (see Maginnis 1994). Compartments—even those with plentiful, mature timber trees—would be allocated no yield if their understory were, for instance, burned. Protected areas would include:

1. Hill (or difficult terrain) sanctuaries, defined as areas of predominantly 16° (30%) or steeper slope and enclosed plateaux or valleys, or extensive swampy areas, main rivers and their banks, and similarly inaccessible terrain. Slopes, swamps, and riverbanks of smaller extent are protected on a fine-grained level by existing logging rules. Hill sanctuaries cover 2,738 km², about 15% of the reserved area, with more than 20% of the remaining area of good or moderate forest. They are well represented in some zones (such as Upland Evergreen) and barely represented elsewhere (as in the Dry Semideciduous zone).

2. Sacred areas (with special protection if required). These are scattered, and of insignificant total area.

3. Genetic hotspots. These are areas, or samples of areas, with a high GHI. They would normally be specially protected. Outstanding areas include Ankasa GPR, northern parts of Cape Three Points, and many areas included under item 1 in the Wet Evergreen zone, which are therefore recommended for special protection. All the Southern Dry Forest remnants (all dry forest fragments included in Hall and Swaine's [1981] South East Outlier and Southern Marginal types) are also included.

4. General provenance reserves ensure that representative samples of every corner of the forest community be preserved in an undisturbed state. These should receive special protection. The Upland Evergreen zone should receive protection under heading 1 as well as this category.

5. Red Star provenance reserves. Areas supporting samples of economic trees not adequately covered by categories 1 and 2. The greatest need for these is for *Pericopsis alata* in the northwestern Moist Semideciduous zone, although such areas have not as yet been selected. Special protection would be advisable. Ignorance of the population genetics and intraspecific variation of such species requires that sites be chosen for protection with the assumption that much of the genetic variation within species follows patterns seen in the community as a whole.

Prime sites for such provenance reserve areas

have not yet been selected and will be chosen based on the following criteria: (a) limited previous disturbance; (b) richness in commercial species; (c) presence in a part of the spectrum of communities that is not well represented in other protected areas; (d) location in areas as distant as possible from other such sites, if this doesn't conflict with a–c; (e) insofar as possible, absence of threat from fire, farming, logging, and so on; (f) proximity to the middle of the reserve. Qualities a, b, e, and f will no doubt make many such areas rather inaccessible, but selected areas should not always be on hills or in swamps, as this will bias the sampled populations. Nevertheless, it seems inevitable that many areas for the protection of provenance reserves will have to be in such sites, because alternative sites are already too degraded.

Other tropical African countries, both those with similar or worse problems and those that can still afford to be more profligate with their forests, stand to benefit by taking more note of forestry research and development in Ghana. The continued actual protection, and a serious national commitment to protection of blocks of natural forest, represent a keystone of Ghana's success, whereas in neighboring countries declining reserved forest is symptomatic of limited commitment to conservation. The positive aspects of Ghana's forestry management mostly follow from this general level of national commitment. Ghana has also benefited from a sustained input of money from the ODA (U.K. Department of Overseas Development, now known as Department for International Development, or DFID) and other aid donors. This chapter has relied on inventory data that have been hundreds of person-years in the collection, and many other countries would not have seen the need for this research. In Ghana there has been a serious commitment at all levels to all aspects of forest protection, and it may perhaps be daunting to suggest that such commitment

cannot simply be learned. Nevertheless, there are some factors and facets from which observers might stand to benefit.

Databases of species—of their properties and qualities, however defined—are to a great extent country specific, but these data are almost essential to any coherent, scientifically guided conservation effort in tropical forest, even if only as a stimulus to further discussion among interest groups. To make most use of such information, however, species qualities need to be combined to form community scores, and for this, appropriate inventories have to be executed. Inventories alone are of some value but are rarely recent or extensive enough in the context of African forest conservation. Even where inventory data are extensive, the inventories tend to be performed by institutes or individuals that lack the most detailed species information. A coherent way of addressing this problem, and the accompanying implicit linkage of institutions or types of information and interest groups, represents one of the historic and current conservation achievements in Ghana that could be reinterpreted elsewhere.

The pattern of protected areas within a matrix of fine-grained protection is not new in Ghana; it is implicit in many of the working plans of the 1950s and 1960s. However, both levels of protection were allowed to slip, partly because of lack of firm statements as to their essential existence and their benefits, and partly because of outside pressure to dissolve them. Because the aims of previous protection were less clearly defined than today, the requirements were also less stringent, which made the policies pliable. Some effort has now been made to tighten up these policies. The continued development and practice of fine-grained protection in every meter of every forest reserve is a challenge, even to Ghana's foresters. To the extent that much of it is a formalization of commonsense forestry, there should be no problem, but problems arise in the gray or unfamiliar areas, such as degrees of tree retention in logged forest; the value of a particular natural forest pocket in a large plantation; how much disturbance to allow in an area of high genetic heat; and the means to prevent overexploitation of NTFPs. Should one recommend simple rules that can be religiously adhered to, or a flexible system of guidelines, which allow evolution of practice? If we accept that biodiversity, or rather "bioquality," is a safe item for managers to optimize, within the inevitable context of exploitation, we can also see that both useful eternal truths (protected areas) and more flexible formulas (fine-grained protection, species-specific priorities) are needed.

By attempting to enforce fine-grained protection by every forest officer and user, and by raising the profile of biodiversity in the list of forest reserve functions, it may be possible to prevent the sorts of official mistakes that have been so detrimental to the forests in the past. The task for the future is to convert the framework of fine-grained protection into practical rules of thumb, while providing for the evolution of such techniques and of the ways the forest is used.

Whereas some of the problems in Ghana's forests might be solved in such ways, the massive real and potential negative influence of evolving forces beyond the forester's control, particularly the economy and pressure via such powerful groups as miners, farmers, and loggers, must be acknowledged. If any of the advances of forest management in Ghana have any chance of working in the long term, Ghana and her forestry sector should be well rewarded for striving so hard toward conservation and sustainable use. Outside money should continue to flow in to help form the best schemes into practical solutions. If, when the dust settles, such approaches have been adopted, surely Ghana can expect to be one of the few tropical countries with a permit to sell its timber, and diverse other forest products, abroad. This, then, would provide an aspect of Ghanaian forestry that other African countries would do well to note, and would be a strong incentive for other African forestry departments to act in similarly positive ways.

REFERENCES

Adu-Tutu, M., Y. Afful, K. Asante-Appiah, D. Lieberman, J. B. Hall, and M. Elvin Lewis. 1979. Chewing stick usage in Southern Ghana. *Economic Botany* 33:320–328.

Alder, D. (Manas Systems Ltd.). 1990. Ghafosim: A projection system for natural forest growth and yield in Ghana. ODA-Ghana Forest Dept. report.

Asabere, P. K. 1987. Attempts at sustained yield management in the tropical high forests of Ghana. In F. Mergen and J. R. Vincent, eds. *Natural management of tropical moist forests.* Yale University Press, New Haven.

Asibey, E. O. A. 1972. Ghana's progress. *Oryx* 11:470–475.

Asibey, E. O. A., and J. G. K. Owusu. 1982. The case for high forest national parks in Ghana. *Environmental Conservation* 9(4):293–304.

Baidoe, J. F. 1970. The selection system as practiced in Ghana. *Community Forest Review* 49:159–165.

———. 1976. Yield regulation in the high forest in Ghana. *Ghana Forest Journal* 2:22–27.

Baur, G. N. 1964. *The ecological basis of rainforest management.* Forestry Commission, New South Wales.

Bird, N. M., and E. Sackey. 1991. The permanent sample plot programme. Internal report, Forestry Dept., Accra, Ghana.

Falconer, J. 1992. *Non-timber forest products in southern Ghana: A summary report.* ODA Forestry Series 2. Natural Resources Institute, Chatham, England.

Foggie, A. 1957. Forestry problems in the closed forest zone of Ghana. *Journal of West Africa Science Association* 3:141–147.

———. 1962. The role of forestry in the agricultural economy. Pages 229–235 in Wills, J. B., ed. *Agriculture and land use in Ghana.* Oxford University Press, London.

Foggie, A., and B. Piasecki. 1962. Timber, fuel and minor forest produce. Pages 236–251 in Wills, J. B., ed. *Agriculture and land use in Ghana*. Oxford University Press, London.

Gordon, D. M. 1992. Ghana. In J. A. Sugar, C. A. Harcourt, and N. M. Collins, eds. *The Conservation Atlas of Tropical Forests: Africa*. IUCN, Gland, Switzerland; World Conservation Monitoring Centre, Cambridge, England; Macmillan Press, London.

Hall, J. B. 1987. Conservation of forest in Ghana. *Universitas* 8:33–42 (University of Legon, Ghana).

Hall, J. B., J. Brookman-Amissah, D. Leston, and R. Dodoo. 1973. *Conservation and exploitation of Atewa range forest reserve*. Council of Scientific and Industrial Research, Accra, Ghana.

Hall, J. B., and M. D. Swaine. 1976. Classification and ecology of closed canopy forest in Ghana. *Journal of Ecology* 64:913–951.

———. 1981. Distribution and ecology of vascular plants in a tropical rain forest. In *Forest Vegetation in Ghana*. Geobotany 1. Junk, The Hague, The Netherlands.

Hawthorne, W. D. 1989. The regeneration of Ghana's forests. Interim report. ODA-Forestry Dept., Accra, Ghana.

———. 1990a. *Field guide to the forest trees of Ghana*. Ghana Forestry Series 1. Natural Resources Institute, Chatham, England.

———. 1990b. Knowledge of plant species in the forest zone of Ghana. Pages 177–185 in *Proceedings of the 12th plenary meeting of AETFAT*. Mitt. Inst. Allg. Bot. Hamburg.

———. 1993. *Forest regeneration after logging: Findings of a study in the Bia South Game Production Reserve, Ghana*. ODA Forestry Series 3. Natural Resources Insitute, Chatham, England.

———. 1994. *Fire damage and forest regeneration in Ghana*. ODA Forestry Series 4. Natural Resources Institute, Chatham, England.

———. 1995. Ecological profiles of Ghanaian forest trees. Tropical Forestry Papers 29. Oxford Forestry Institute, Oxford.

———. 1996. Holes and the sums of parts in Ghanaian forest: Regeneration, scale and sustainable use. *Proceedings of the Royal Society of Edinburgh* 104B:75–176.

Hawthorne, W. D., and M. Abu Juam. 1995. *Forest protection in Ghana*. IUCN, Gland, Switzerland.

Maginnis, S. 1994. Understory condition scoring of Ghanaian lowland tropical moist forest during stock survey: A technique for regulating the allowable cut in ecologically and structurally degraded production forest. *Forest Ecology and Management* 70:89–97.

Martin, C. 1978. Management plan for the Bia Wildlife Conservation Areas. Final report IUCN/WWF/Project 3206.

———. 1991. *The rain forests of west Africa*. Birkhauser, Berlin.

Parren, M. P. E., and N. R. de Graaf. 1995. *The quest for natural forest management in Ghana, Côte d'Ivoire and Liberia*. Tropenbos Series. Blackhuys, Leiden, Netherlands.

Styles, B. T., and P. K. Khosla. 1976. Cytology and reproductive biology of Meliaceae. In J. Burley and B. T. Styles, eds. *Tropical trees, variation, breeding and conservation*. Linnean Society, London.

Swaine, M. D., and J. B. Hall. 1981. The monospecific tropical forest of the Ghanaian endemic tree *Talbotiella gentii*. In H. Synge, ed. *Proceedings of the International Conference on Biological Aspects of Rare Plant Conservation*. Wiley, London.

Thompson, H. N. 1910. Gold Coast: Report on forests. *Colonial Reports Miscellaneous* 66:1–238.

Wong, J. L. G. 1994. Timber species classification and the assessment of exploitation problems. FIMP Discussion Paper 4. Ghana Forestry Department, Accra, Ghana.

Wong, J., ed. 1989. Forest inventory seminar proceedings, 29–30 March 1989, Accra. ODA/Ghana Forestry Department, Accra, Ghana.

## Star Ratings and Survey Criteria for Individual Species

| NAME (TRADE NAME) | STAR[a] | NUMBER OF TREES COUNTED IN 2,624 PLOTS (BY CM DBH SIZE CLASSES) | | | | | NUMBER OF TREES LOGGED IN 1989 | FOREST SUPPORTING 1989'S LOGGED TREES (KM²)[b] | |
|---|---|---|---|---|---|---|---|---|---|
| | | 5–10 | 10–30 | 30–60 | 60–90 | 90+ | | TREES >90 CM | TREES >60 CM |
| *Funtumia elastica* (funtum) | R | 279 | 715 | 523 | 1 | 0 | 636 | >Ghana | 15,900 |
| *Milicia excelsa/regia* (odum) | R | 20 | 80 | 522 | 292 | 284 | 16,017 | 1,480 | 730 |
| *Pericopsis elata* (afrormosi) | R | 2 | 8 | 22 | 33 | 24 | 1,438 | 1,580 | 663 |
| *Terminalia ivorensis* (emire) | R | 8 | 25 | 228 | 90 | 96 | 3,111 | 850 | 439 |
| *Tieghemella heckelii* (baku) | R | 25 | 110 | 142 | 67 | 62 | 2,123 | 900 | 432 |
| *Khaya ivorensis* (mahogany) | R | 135 | 186 | 488 | 359 | 193 | 8,581 | 1,166 | 408 |
| *Entandrophragma candollei* | R | 14 | 70 | 82 | 35 | 22 | 674 | 802 | 311 |
| *Entandrophragma utile* | R | 37 | 108 | 149 | 86 | 70 | 1,694 | 634 | 285 |
| *E. cylindricum* (sapele) | R | 75 | 228 | 396 | 222 | 173 | 3,968 | 602 | 264 |
| *Nauclea diderrichii* (kusia) | R | 22 | 37 | 127 | 111 | 58 | 1,503 | 608 | 233 |
| *Triplochiton sclerox* (wawa) | R | 146 | 488 | 3,190 | 2,785 | 1,846 | 39,948 | 568 | 226 |
| *Albizia ferruginea* | R | 12 | 54 | 243 | 132 | 89 | 1,798 | 530 | 214 |
| *Guibourtia ehie* (shedua) | R | 83 | 199 | 439 | 184 | 2 | 1,157 | 14,463 | 163 |
| *Pterygota macrocarpa* (kyere) | R | 123 | 386 | 1,355 | 738 | 143 | 5,474 | 1,004 | 163 |
| *Entandrophragma angolense* | R | 166 | 680 | 1,289 | 423 | 127 | 3,010 | 622 | 144 |
| *Lovoa trichilioides* | R | 52 | 99 | 157 | 84 | 36 | 600 | 438 | 131 |
| *Lophira alata* (kaku) | R | 9 | 27 | 59 | 78 | 85 | 733 | 226 | 118 |
| *Afzelia africana/*spp. (papao) | R | 21 | 54 | 432 | 137 | 40 | 774 | 509 | 115 |
| *Hallea* spp. (subaha) | R | 17 | 42 | 161 | 80 | 18 | 405 | 587 | 109 |
| *Rhodognaphalon brev.* (bombax) | R | 18 | 162 | 509 | 173 | 35 | 835 | 628 | 105 |
| *Guarea cedrata* | r | 389 | 613 | 741 | 206 | 34 | 945 | 727 | 103 |
| *Heritiera utilis* (Nyankom) | R | 66 | 168 | 568 | 192 | 25 | 777 | 818 | 94 |

APPENDIX 29.1 (continued)

| NAME (TRADE NAME) | STAR[a] | NUMBER OF TREES COUNTED IN 2,624 PLOTS (BY CM DBH SIZE CLASSES) | | | | | NUMBER OF TREES LOGGED IN 1989 | FOREST SUPPORTING 1989'S LOGGED TREES (KM²)[b] | |
|---|---|---|---|---|---|---|---|---|---|
| | | 5–10 | 10–30 | 30–60 | 60–90 | 90+ | | TREES >90 CM | TREES >60 CM |
| Antrocaryon micraster | R | 9 | 53 | 244 | 85 | 42 | 383 | 239 | 79 |
| Mansonia altissima | r | 84 | 164 | 958 | 214 | 10 | 637 | 1,676 | 75 |
| Pycnanthus angolensis (otie) | r | 207 | 486 | 2,109 | 1,080 | 107 | 2,967 | 727 | 66 |
| Erythrophleum ivorense | r | 12 | 15 | 74 | 64 | 53 | 273 | 135 | 61 |
| Aningeria spp. | r | 269 | 311 | 1,167 | 347 | 191 | 1,233 | 169 | 60 |
| Distemonanthus benthamianus | r | 71 | 243 | 366 | 133 | 45 | 326 | 191 | 48 |
| Antiaris toxicaria | r | 166 | 488 | 1,205 | 594 | 408 | 1,423 | 92 | 37 |
| Piptadeniastrum af. (Dahoma) | r | 185 | 777 | 2,308 | 1,326 | 683 | 2,561 | 98 | 33 |
| Nesogordonia papaverifera | r | 601 | 1,049 | 4,561 | 678 | 52 | 797 | 403 | 29 |
| Daniellia spp. (ogea/shedua) | r | 95 | 198 | 519 | 210 | 76 | 288 | 99 | 26 |
| Ceiba pentandra | gn | 186 | 305 | 784 | 386 | 937 | 880 | 25 | 17 |
| Millettia rhodantha | gn | 71 | 205 | 316 | 20 | 0 | 12 | 0 | 16 |
| Chrysophyllum spp. (akasa) | r | 423 | 1,058 | 2,147 | 513 | 47 | 274 | 153 | 13 |
| Mammea africana | r | 86 | 152 | 246 | 85 | 22 | 50 | 60 | 12 |
| Turraeanthus africanus | r | 170 | 612 | 3,846 | 627 | 53 | 293 | 145 | 11 |
| Albizia zygia | gn | 59 | 169 | 483 | 181 | 25 | 87 | 92 | 11 |
| Cynometra anania | r | 131 | 211 | 864 | 444 | 44 | 211 | 126 | 11 |
| Sterculia rhinopetala | r | 284 | 962 | 2,892 | 383 | 26 | 177 | 179 | 11 |
| Terminalia superba (ofram) | r | 138 | 184 | 719 | 1,027 | 410 | 606 | 39 | 11 |
| Amphimas pterocarpoides | gn | 269 | 415 | 382 | 191 | 59 | 102 | 45 | 11 |
| Albizia glaberrima | gn | 5 | 8 | 40 | 22 | 13 | 13 | 26 | 10 |
| Celtis mildbraedii | gn | 1,690 | 5,603 | 11,410 | 2,529 | 316 | 948 | 79 | 9 |
| Celtis spp. | gn | 229 | 927 | 1,571 | 386 | 47 | 0 | 0 | 0 |

| NAME (TRADE NAME) | STAR[a] | NUMBER OF TREES COUNTED IN 2,624 PLOTS (BY CM DBH SIZE CLASSES) | | | | | NUMBER OF TREES LOGGED IN 1989 | FOREST SUPPORTING 1989'S LOGGED TREES (KM²)[b] | |
|---|---|---|---|---|---|---|---|---|---|
| | | 5–10 | 10–30 | 30–60 | 60–90 | 90+ | | TREES >90 CM | TREES >60 CM |
| *Canarium schweinfurthii* | r | 31 | 37 | 135 | 57 | 32 | 27 | 22 | 8 |
| *Petersianthus macrocarpus* | gn | 59 | 388 | 2,096 | 736 | 204 | 270 | 35 | 8 |
| *Cylicodiscus gabunensis* | bu | 20 | 73 | 259 | 187 | 328 | 129 | 10 | 7 |
| *Celtis zenkeri* | gn | 215 | 761 | 2,308 | 495 | 89 | 83 | 24 | 4 |
| *Sterculia oblonga* | gn | 228 | 687 | 1,469 | 351 | 34 | 43 | 33 | 3 |
| *Anopyxis klaineana* | r | 11 | 28 | 82 | 85 | 32 | 8 | 7 | 2 |
| *Morus mesozygia* | gn | 84 | 253 | 310 | 63 | 5 | 3 | 16 | 1 |
| *Ongokea gore* | gn | 16 | 64 | 262 | 63 | 9 | 2 | 6 | 1 |
| *Parkia bicolor* | gn | 73 | 240 | 717 | 423 | 148 | 24 | 4 | 1 |
| *Khaya anthotheca* | R+ | 28 | 42 | 71 | 28 | 18 | 0 | 0 | 0 |
| *Khaya antho./grandif.* | R | 10 | 19 | 24 | 20 | 5 | 0 | 0 | 0 |
| *Guarea thompsonii* | R+ | 76 | 179 | 157 | 27 | 1 | 0 | 0 | 0 |
| *Celtis adolfi-friderici* | gn | 216 | 759 | 1,530 | 387 | 54 | 0 | 0 | 0 |
| *Holoptelea grandis* | gn | 29 | 85 | 262 | 92 | 28 | 1 | 1 | 0 |
| *Scottellia klaineana* | r | 755 | 1,293 | 1,472 | 154 | 15 | 0 | 0 | 0 |
| *Alstonia boonei* | gn | 132 | 240 | 945 | 542 | 311 | 0 | 0 | 0 |
| *Bombax buonopozense* | gn | 29 | 72 | 205 | 120 | 47 | 0 | 0 | 0 |
| *Bussea occidentalis* | gn | 76 | 283 | 977 | 126 | 14 | 0 | 0 | 0 |
| *Chrysophyllum perpulchrum* | gn | 140 | 280 | 1,047 | 288 | 32 | 0 | 0 | 0 |
| *Cola gigantea* | gn | 310 | 1,228 | 3,111 | 576 | 92 | 0 | 0 | 0 |
| *Cola nitida* | r+ | 484 | 1,589 | 987 | 13 | 0 | 0 | 0 | 0 |
| *Copaifera salikounda* | R | 98 | 112 | 46 | 14 | 16 | 0 | 0 | 0 |
| *Corynanthe pachyceras* | gn | 435 | 1,387 | 3,523 | 109 | 0 | 0 | 0 | 0 |

| NAME (TRADE NAME) | STAR[a] | NUMBER OF TREES COUNTED IN 2,624 PLOTS (BY CM DBH SIZE CLASSES) | | | | | NUMBER OF TREES LOGGED IN 1989 | FOREST SUPPORTING 1989'S LOGGED TREES (KM²)[b] | |
|---|---|---|---|---|---|---|---|---|---|
| | | 5–10 | 10–30 | 30–60 | 60–90 | 90+ | | TREES >90 CM | TREES >60 CM |
| *Dacryodes klaineana* | gn | 493 | 1,707 | 4,225 | 325 | 12 | 0 | 0 | 0 |
| *Dialium aubrevillei* | gn | 445 | 651 | 1,580 | 277 | 36 | 0 | 0 | 0 |
| *Garcinia afzelii* (nsoko) | R+ | 5 | 18 | 1 | 0 | 0 | 0 | 0 | 0 |
| *Garcinia epuntata* (nsoko) | R+ | 88 | 23 | 3 | 0 | 0 | 0 | 0 | 0 |
| *Garcinia kola* (tweapea) | R | 3 | 10 | 26 | 1 | 0 | 0 | 0 | 0 |
| *Garcinia* spp. | r | 6 | 7 | 0 | 0 | 0 | 0 | 0 | 0 |
| *Hannoa klaineana* | gn | 488 | 1,302 | 1,167 | 301 | 51 | 0 | 0 | 0 |
| *Hexalobus crispiflorus* | gn | 196 | 346 | 439 | 125 | 33 | 0 | 0 | 0 |
| *Klainedox gabonensis* | gn | 36 | 158 | 426 | 196 | 95 | 0 | 0 | 0 |
| *Lannea welwitschii* | gn | 73 | 262 | 758 | 216 | 16 | 0 | 0 | 0 |
| *Parinari excelsa* | gn | 77 | 165 | 429 | 141 | 78 | 0 | 0 | 0 |
| *Pentaclethra macrophylla* | gn | 46 | 158 | 659 | 179 | 48 | 0 | 0 | 0 |
| *Pterygota bequaertii* | R | 0 | 2 | 7 | 0 | 0 | 0 | 0 | 0 |
| *Ricinodendron heudelotii* | gn | 206 | 534 | 904 | 438 | 189 | 0 | 0 | 0 |
| *Treculia africana* | gn | 82 | 207 | 725 | 134 | 22 | 0 | 0 | 0 |
| *Trilepisium madagascariense* | gn | 87 | 352 | 2,255 | 213 | 15 | 0 | 0 | 0 |
| *Xylia evansii* | gn | 79 | 202 | 483 | 113 | 26 | 0 | 0 | 0 |
| *Berlinia* spp. | gn | 490 | 1,526 | 2,578 | 524 | 79 | 6 | 2 | 0 |
| *Blighia* spp. | gn | 581 | 1,074 | 1,508 | 280 | 49 | 0 | 0 | 0 |
| *in forest* | | 76 | 4 | 9 | 2 | 5 | 6 | | |

a   Stars are indicated as follows: R = Red Star; r = Pink Star; R+ = Scarlet Star; gn = Green Star; bu = Blue Star.

b   Figures indicate the area required to hold the number of logs extracted in 1989 (for two different size classes of trees).

CHAPTER 30

# Integrating Local Communities into the Management of Protected Areas

## Lessons from DR Congo and Cameroon

Bryan K. Curran and Richard Key Tshombe

Considerable attention has been focused in recent years on the rapid worldwide destruction of tropical forests and the concomitant loss of floral and faunal diversity. Although large expanses of tropical moist forest remain intact in central Africa, even these relatively remote areas are increasingly vulnerable to timber exploitation, mining, commercialization of forest products (especially bushmeat), large-scale agricultural and development projects, and growing levels of subsistence farming and hunting. In an effort to limit the damage and provide long-term protection to these forests, African governments, conservation organizations, and development agencies have initiated dozens of new protected area programs, with varying degrees of success.

As human populations continue to expand, the limits of the national park model for conservation are exposed. Complete protection is certainly justifiable in some cases, particularly where only small populations of rare or endangered species remain (for example, the mountain gorilla in Virunga National Park, Democratic Republic of Congo [DR Congo], and in the Volcanoes National Park in Rwanda, or the

northern white rhino in DR Congo's Garamba National Park). However, the creation of a national park does not automatically convey full protection, especially in those instances where the responsible national authorities are not capable of protecting the area because of inadequate training, staff, motivation, equipment, or financial means. Perhaps most important, parks often alienate local communities that have long-standing, legitimate claims to resources. As a result, the concept of buffer zones has arisen (Oldfield 1988; Sayer 1991), and new categories of protection have been proposed, including the biosphere reserve (Batisse 1982; UNESCO/UNEP 1984), natural biotic area, anthropological reserve (Stuart et al. 1990; IUCN 1992), and community forest reserves, where multiple uses have been advanced that allow local populations to exploit resources in a sustainable manner.

In order for multiple-use conservation strategies to be effective, it is vital that all stakeholders (local, regional, and national governments, conservation non-governmental organizations [NGOs], and local communities, among others) recognize the legitimacy of different values attributed to biological resources. In this sense, conservation projects must reconcile competing interests in order to succeed. Multiple-use conservation strategies require the participation of local communities in the design, implementation, management, and monitoring of projects. This participation has taken a variety of forms, but it focused initially on the premise that local populations must realize direct economic benefits that adequately offset the costs incurred from lost access to resources. As a result, the dual goals of biodiversity conservation and improving the quality of life of neighboring communities spawned Integrated Conservation and Development Projects, or ICDPs (Brown and Wycoff-Baird 1992; Wells et al. 1992). Although the theory and methodological approach of ICDPs have been broadly accepted by many within both the conservation and development communities, they must be considered experimental, and the factors that have limited their overall success require constant monitoring (Hough and Sherpa 1989; Brown and Wycoff-Baird 1992).

The idea that indigenous groups have lived in harmony with their environment for many generations has been advanced by those who favor greater control of the management of resources by local populations (Durning 1992; Hill and Press 1994; Kleymeyer 1994). However, the "ecologically noble savage" (Redford 1990) may never have existed. This purportedly holistic relationship between people and their environment is more likely the result of low human population densities, primitive technologies, and limited markets rather than conscious self-restraint. Humans have demonstrated time and again that as their numbers increase, as more "efficient" technology becomes available, or as economic opportunities arise, they tend to overexploit their resource base (Hart 1978; Hames 1979; Redford and Robinson 1991; Vickers 1991; Alvard 1993). Members of local communities rarely have adequate education or training, and

they lack appropriate resources to manage protected areas without external assistance. Yet local support is essential for long-term conservation goals, and indigenous populations themselves generally have interests in the sustenance of important resources. In some cases, indigenous peoples use sophisticated resource management strategies (Alcorn 1990; Zerner 1994; Poffenberger and McGean 1996). They must therefore be an integral part of the solutions to the problem of unsustainable resource use.

Too often, local populations are slated to receive the benefits of ICDPs without participating in project design or implementation (Chambers 1983; Cernea 1985). Another common error is the tendency of some projects to overemphasize rural development aspects at the expense of the core conservation goals (Kiss 1990). Without any explicit connection between the adoption of certain measures to protect biodiversity and resulting social or economic benefits, there may be no motivation for local populations to modify potentially destructive resource exploitation practices. And although there are certainly examples of conservation projects that generate significant foreign exchange for national governments and local communities (Weber 1987, 1994; Lewis et al. 1989; Vedder and Weber 1990; Metcalfe 1994), these are generally exceptions.

Those working to empower local peoples as stakeholders in the conservation of biodiversity have used a variety of methods to solicit information regarding the concerns and aspirations of communities living near protected areas (e.g., Chambers 1991). Originally designed to assist the international development community, these "rapid rural appraisals" or "participatory rural appraisals" (McCracken et al. 1988; National Environmental Secretariat et al. 1990; Schoonmaker-Freudenberger and Gueye 1990; Theis and Grady 1991; Bhatnagar and Williams 1992) represent an important step toward integrating indigenous peoples into the design of conservation projects, in that they explicitly rely on local participation for their success. However, caution should be exercised in determining how "rapid" these appraisals should be made; gaining the confidence of local communities in order to acquire legitimate and useful information is difficult in the short time frames mandated by many lender or donor agencies. Nonetheless, the real strength of these approaches lies in their flexibility and creativity, as well as in the wide variety of tools they employ: questionnaires, focal interviews, ranking exercises, group discussions, written and oral histories, games, surveys, diagrams, and resource maps can all be used for gathering baseline data, as well as for cross-checking information and monitoring progress of conservation and development objectives.

Integration of local communities into all phases of conservation projects requires a dedication to the principle that indigenous peoples are legitimate stakeholders, as well as a significant commitment of time and resources on the part of governments, international lenders and donor organizations, NGOs, and protected-area managers. This task will likely never be easy; even the initial step of identifying the "local community" is rarely straightforward, as considerable heterogeneity and stratification often exist within these communities, with each group having its own subsistence and economic concerns (Weber 1989). Nonetheless, when local populations are excluded from one phase or another of a development or conservation project, the results are nearly always the same—the project does not function as envisioned (see Cernea 1985). The most promising examples of conservation and resource management projects, and presumably those with the greatest potential for long-term success, are those in which local communities have been given a genuine participatory role (Breslin and Chapin 1984; Child 1984; Owen-Smith and Jacobsohn 1989).

In the following sections, we present case studies in which we have direct field experience working with local populations; many of the statements reflect our own observations. These examples illustrate some of the constraints inherent in this type of exercise, and demonstrate some of the lessons that can be applied toward successful collaboration and integration of local communities into long-term conservation of biodiversity. It should be noted that civil war in DR Congo and closure of the project in Cameroon have complicated field work at both sites in recent years.

## The Okapi Wildlife Reserve

### GEOPHYSICAL BACKGROUND

The Ituri Forest (figure 30.1), located in the northeastern corner of DR Congo, encloses more than 65,000 km² of tropical moist forest at the edge of the Congo Basin. The area is centered on the upper watershed of the Ituri River, 0–3° N and 27–30° E. The altitude ranges from 700 m to 1,000 m, and the annual rainfall averages about 1,900 mm, with a pronounced dry season from December to February. The Ituri is of particular interest in that it contains the greatest diversity of mammalian fauna of all DR Congo forests (Wilkie and Finn 1988). Although satellite imagery has demonstrated that more than 90% of the Ituri remains intact (Wilkie 1989), a number of pressures threaten the future of this rain forest. In addition to an ongoing civil war, these threats include gold prospecting, a growing commercialization of the bushmeat trade, small-scale timber extraction, charcoal production, and particularly immigration from areas well outside the forest (Hall 1990; Peterson 1990, 1991, this volume).

### THE PEOPLE OF THE ITURI FOREST

The Ituri Forest supports Bantu- and Sudanic-speaking shifting horticulturalists (approximately 140,000 people) and mobile hunter-gatherer populations (approximately 30,000–35,000) commonly referred to as Pygmies. Both groups

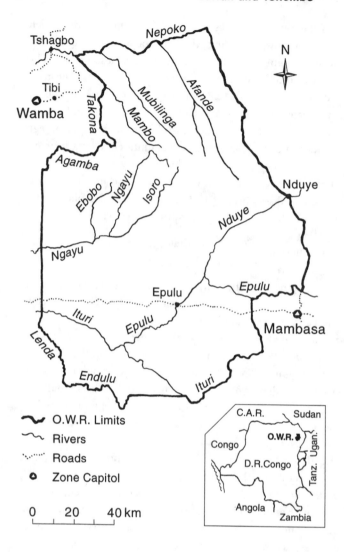

Figure 30.1. Okapi Wildlife Reserve (OWR), northeastern DR Congo.

depend on the forest and its resources. The Ituri is sparsely populated, with about one person per square kilometer in most areas. The Pygmies, known collectively as the BaMbuti, include the net-hunting Asua of the western Ituri, who are associated with the Budu, Ndaka, Beru, and Bali. The Efe, who are bow hunters distributed in the northern and eastern Ituri, are found near Mamvu and Lese gardeners. The Mbuti net hunters live in the central and southern Ituri in close association with Bila and Ndaka horticulturalists.

A symbiotic dependence between the Mbuti and their farming neighbors has existed for at least 2,000 years (Ehret and Posnansky 1982). The relationship includes an array of political, social, religious, and primarily economic ties that are passed down through clans over generations. The Pygmies provide such forest products as meat, honey, fruits, nuts, medicinal plants, and building materials, along with garden labor, in exchange for cultivated foods, metal, and cloth, among other items. These links extend to land tenure as well. Each horticultural clan holds ancestral rights, generally respected by neighboring farmers, to certain forest areas where shifting cultivation, hunting, fishing, and gathering of forest products are allowed. The Mbuti associated with a particular horticultural group also enjoy this exclusive access to an area of the forest in which to pursue their activities. In this manner, much of the land within the Ituri Forest is claimed by the indigenous peoples, although this system has faced mounting pressure in recent years.

Immigration into the Ituri by people seeking forest resources, arable land, and gold increased rapidly throughout the 1980s (Peterson 1990, 1991, this volume). According to Peterson, the immigrants are primarily Nande, who are abandoning the densely populated, eroded, and violently contested highland areas to the southeast of the Ituri. Other immigrant groups include the Budu and the Bali, who are entering the Ituri in large numbers from the north and west, respectively. Agricultural land on the forest frontier is often claimed simply by clearing it, with no compensation for the indigenous farmers. This loss of primary forest is exacerbated as immigrants often clear larger gardens for producing cash crops on a semipermanent basis. This frequently results in permanent conversion of forest to fields, and resulting soil degradation may ultimately prevent natural forest regeneration.

Associated with these increasing population densities are gradual changes in the traditional land tenure and exchange systems (Peterson 1990, 1991, this volume). Customary forest access claims have become more difficult to enforce, particularly as the immigrant horticulturalists such as the Nande begin to acquire political leverage. In addition, the symbiosis between the indigenous farmers and the Mbuti is becoming more tenuous, as the Pygmies tend to sell their labor to the higher-paying immigrant farmers and gold carriers, who have little use for more traditional exchange relationships

and are therefore less likely to assist the Mbuti beyond providing temporary employment. Large-scale commercial hunting has been discouraged since the creation of a protected area in 1992. As their hunting skills in the forest become underappreciated and job opportunities more scarce, many of the Mbuti are forced to clear gardens of their own to have access to cultivated foods, often on less productive land "given" to them by their horticultural neighbors. Not only are they unskilled gardeners (one Mbuti hunter noted, "Our nets are our 'gardens'"), but their land claims are the first to be endangered should disputes arise, which ultimately serves to marginalize them even further.

## THE OKAPI WILDLIFE RESERVE AND ITURI NATIONAL PARK

The okapi (*Okapia johnstoni*), a species of forest giraffe endemic to DR Congo, has a long-standing ritual significance among local cultures (Hart and Hart 1993). First observed by Westerners in the early twentieth century, the okapi has since been adopted as a national symbol by the Institut Congolais pour la Conservation de la Nature (ICCN), which established an okapi capture station (and nearby capture zones) in Epulu in the 1950s for the purpose of exporting these animals.

More recently, attention has shifted toward the creation of a faunal reserve and national park, which included the okapi capture zones, with Epulu as the administrative headquarters. Arrêté Ministériel No. 045/CM/ECN of May 2, 1992, officially gazetted the Okapi Wildlife Reserve (OWR; see figure 30.1), enclosing an area of almost 14,000 km². The creation of the planned Ituri National Park (4,850 km² as originally proposed), which would represent an integrally protected zone within the core of the OWR, requires a presidential decree. This has not been forthcoming, given the unrest within the country. The unique status of the okapi among the local communities may prove important to the conservation program in an area where approximately 23,000 Bantu and perhaps 8,000 BaMbuti currently reside and engage in their traditional farming, hunting, and gathering activities (Curran 1994). This is particularly true as more immigrants arrive in the Ituri, and as the long-standing respect accorded the okapi and the capture zones seems to be eroding.

## COMMUNITY PARTICIPATION
## IN THE OKAPI WILDLIFE RESERVE

The task of integrating the local population into an active role in the management of the OWR represents perhaps the most critical (and also the most difficult) problem facing the ICCN and conservation NGOs. Without dynamic community support for custody of the Ituri Forest by the ICCN, the prospects for successful long-term conservation will be limited, as local populations claim the forest within the reserve as communal property. Should these traditional land claims not be respected by reserve authorities, government officials, or conservation NGOs, protection will likely be a function only of the number of reserve guards. The OWR is simply too vast and logistically difficult to be effectively patrolled and managed without direct input and assistance from area residents.

A number of field missions have taken place, aimed at assessing the level of local participation and understanding of the goals of conservation in the Ituri (Curran 1991, 1992). Although early prospections for delimiting the OWR included preliminary contacts with local communities (Hall 1990), these populations did not actively participate in defining either the reserve limits or the objectives of the conservation project. Nonetheless, in 1991 most people living in the Ituri were supportive of the reserve, although its purpose was not well understood (Curran 1991). Instead, local communities gave their tentative approval as a

result of assurances by the visiting minister of the environment, who promised that their support for the OWR would lead to development projects (schools, health clinics, road repairs, and so on) that would improve their lives.

Subsequent work with local communities in the Ituri revealed a drastic change in attitude toward the OWR (Curran 1994). Beginning in August 1993, we began to explore how local communities might contribute more directly to the conservation efforts. This included the first detailed census of the human population living in and around the OWR, as well as village meetings, interviews, and surveys to provide critical socioeconomic information. This work, although both preliminary and qualitative in nature, revealed that many of the local communities were no longer supportive of the conservation project, and in some cases were vociferously opposed to the reserve. For instance, local communities found around the northern edges of the OWR refused to meet any longer with the ICCN. One particular problem was the lack of clear directives from the ICCN regarding reserve regulations. In addition, the continued economic and political deterioration in DR Congo had caused further impoverishment, which many people attributed to the presence of the reserve. Furthermore, the development projects promised by the government in 1991 had not been forthcoming, and these were now perceived as a disingenuous effort by the government to win local support for creation of the reserve. This combination of limited economic opportunities and broken promises had resulted in a deterioration of trust between local communities and the OWR management team, a situation further fueled by disorderly reserve guards, as well as by local government officials and missionaries promoting their own agendas. It was apparent that immediate measures were necessary in order to salvage some level of collaboration with local communities.

First, the level of contact with local communities and customary authorities around the OWR was dramatically increased. Meetings were arranged with local government officials, missionaries, and other potential stakeholders, who were provided with appropriate documents detailing the goals and objectives of conservation in the Ituri. An education team sponsored by the Gilman Investment Company (a private American company that assists the ICCN in management of the okapi capture station and other conservation-related activities) began to work closely with local communities. The team prepared a brochure that was translated into local languages and distributed around the Ituri as part of a much-expanded effort to provide critical information to local populations. New guards were hired from within local communities, and a comprehensive training course was designed and enacted to ensure their understanding of reserve regulations, with special attention to the rights of local communities. Perhaps most important, efforts to more fully integrate the local population into management of the reserve were revitalized.

Since 1985, much of the contact between the reserve management team, the DR Congo government, conservation NGOs, and local populations had occurred on the level of the Collectivité (a DR Congo geopolitical unit). Typically, the chief of the Collectivité was informed of the progress and problems regarding the OWR, on the premise that he would then disseminate this information to the population under his control. However, it was clear to our team that local communities had been poorly informed by their chiefs, which had exacerbated the distrust between residents, their representatives, and the ICCN. Village meetings that we convened suggested that a more effective tactic would be to consult directly with the population at the Groupement level (the geopolitical unit just below the Collectivité). We continued to involve the Col-

lectivité, but these administrative units are simply too large and inefficient to allow for effective collaboration. As such, the social science team of the Wildlife Conservation Society (WCS) and the ICCN decided to establish a Comité Permanent de Consultation Locale (CPCL) within every Groupement affected by the reserve. Although caution should be exercised in creating somewhat artificial structures with which local communities have no experience, we found no comparable traditional institution on which local representation could be built across such a large area.

In late 1993 CPCLs were initiated in thirty Groupements around the reserve, representing local communities from ten Collectivités. A total of thirty-three CPCLs were formed, as several Groupements were so large that it was necessary to create two CPCLs. Each CPCL consisted of a minimum of ten members (five Bantu men and two Bantu women, plus three Mbuti) who were selected locally (those Groupements with more than ten Localités were encouraged to designate additional members, such that all the Localités within or around the reserve were represented). A president and secretary were chosen by and from the members of each CPCL, and many of the responsibilities for the functioning of the CPCL rest with these two individuals. No salaries are paid to CPCL members, although they occasionally receive such small benefits as salt, soap, or T-shirts. The CPCLs have been encouraged to meet monthly to discuss conservation issues affecting their Groupement, and to submit written reports to the ICCN. The social science team, ICCN, and the reserve education team have met regularly with the CPCLs in order to discuss the importance and functions of the CPCL, present general OWR background information, provide materials for operation (paper, pens, envelopes), and so forth. Although this system of contact and conflict resolution may still not adequately represent the views of all stakeholders (Rupp 1994), some of the local communities have already begun to participate in the active management of the OWR. For example, these CPCLs are helping to establish zoning limits, assisting in anti-poaching and guard selection, helping in data collection, and defining restrictions on their own hunting and subsistence activities. In contrast, in certain areas of the OWR, continued distrust between local communities (especially the traditional chiefs, who are unhappy about the failure of the ICCN to follow through on development promises) and the reserve has led to complete cessation of CPCL activities. Whether local confidence and effective collaboration can be restored in these areas remains a serious issue for the ICCN if the CPCLs are to prove a sustainable mechanism for community involvement in conservation in the Ituri.

ISSUES FOR THE FUTURE

- *Immigration.* A significant problem for the future of the Okapi Wildlife Reserve is the control of immigration. As the 1994 census showed, more than 30,000 people live in the reserve, or within 15 km of its limits: nearly 10% of these residents immigrated to the Ituri in the two years after the Ministerial Arreté was signed in 1992 (Curran and Tshombe, unpublished census data). Therefore, certain restrictions on immigration must be adopted. Human population density remains low, yet unregulated immigration would increase the pressure on the resources of the forest to unsustainable levels. It is critical that the local communities assume an active role in restricting population increases, which will be more feasible if meaningful land tenure regulations can be enforced. The CPCLs and communities are participating in the creation of a system of zoning for subsistence and economic activities, which should give them greater control of their

land and provide a check on uncontrolled immigration.

- *Development opportunities.* In order to promote more active cooperation of the rural population around the OWR in the conservation of its valuable flora and fauna, some tangible benefits must be realized locally. If not, the unfulfilled government promises that enticed local chiefs to support the creation of the OWR will continue to plague the ICCN. Benefits might include generating employment within the reserve (particularly as new guards are added) and assistance with development projects, especially those that can be run with the help of active local NGOs. If the ICCN could help local communities to limit human–elephant conflicts, for instance, it is expected that certain economic benefits would accrue to residents, who could sell more agricultural goods. Without a good-faith gesture on the part of the reserve management team for development, it is likely that the local population will turn completely against the OWR. However, any development schemes must be small and carefully planned in direct collaboration with local communities; most important, they must be tied directly to conservation measures that will be respected and monitored by the local populations themselves. For instance, those areas where the CPCLs are functioning smoothly, and assisting in the control of poaching and population growth and the creation of effective zoning, should be given priority for initial development assistance.

- *Creation of the Ituri National Park.* According to DR Congo law (Ordonnance-Loi No. 69-041 of August 22, 1969), all human activities are prohibited within the borders of a national park. Such restrictions would be difficult (and perhaps inappropriate) to impose on a group of semi-nomadic hunter-gatherers like the BaMbuti. The population census indicated that certain places in the reserve are rarely used by the majority of the local population (Curran and Tshombe, unpublished data), and large mammal surveys also demonstrated that very little human activity occurs in certain parts of the OWR (J. Hart, unpublished data). In this sense, certain areas of the Ituri are already essentially protected. Clearly, this situation could change if the influx of outsiders were not effectively controlled. In that case, however, the entire reserve would be at risk, regardless of what types of regulations are applied.

Many people in the Ituri are convinced that the OWR is merely a smokescreen, a prelude to the creation of a larger national park, and they have been reluctant to support the reserve for that reason. It would be imprudent to gazette a totally protected area at present, endangering what progress has been made in integrating the local population into the management of the reserve. The ICCN has the authority already to control human activities within the OWR (MECN 1992:article 4). When the political climate permits renewed action, the ICCN should adopt a flexible approach to management which responds effectively to local conditions, including a system of zoning defined with community input. As long as these conditions are regularly monitored, there appears to be little need to legislate total protection at present.

## The Proposed Lobeke Protected Area

### GEOPHYSICAL BACKGROUND

The Lobeke region of southeastern Cameroon is located in the Eastern Province, within the Arrondissement of Moloundou (figure 30.2). The area includes part of the watershed of the Sangha River, 2.05–2.25° N and 15.40–15.45° E. The altitude ranges between 400 m and 700 m, and the annual rainfall averages about 1,800

mm, with a pronounced dry season from December to February. The Lobeke region is clearly of biological interest, as it contains some of the highest densities of forest elephants and western lowland gorillas in all Africa (Stromayer and Ekobo 1991), as well as large numbers of chimpanzees and other primates, leopards, and ten species of forest ungulates. In addition, the area maintains a high diversity of natural communities minimally disturbed by humans (Hall 1993). However, numerous outside commercial interests threaten the future of this forest. These include timber exploitation, safari hunting companies, still-active ivory markets, intensive poaching for bushmeat, exploitation of exotic animals (gorillas and gray parrots, for instance), and mining.

### THE PEOPLE OF THE LOBEKE REGION

In addition to their tremendous biological diversity, the forests of southeastern Cameroon surrounding Lobeke support Bangando horticulturalists, semi-sedentary forager-farmer Baka Pygmy populations, and small numbers of traders, all of whom rely, to varying degrees, on forest products for their livelihood. Except for the large concentrations of people around Moloundou (approximately 5,000 individuals), and to a lesser extent around logging camps and towns, the Lobeke region is sparsely populated (about 2 people/km²). These groups live almost exclusively along the main road running north-south from Yokadouma to Moloundou, and the villages nearest to Lobeke (between Koumela and Mbangoy I; see figure 30.2) include just over 5,000 people (WCS 1996). Smaller concentrations of Bakwele horticulturalists and other mostly non-indigenous people are found around the timber concessions in Kika, Bela, and Libongo, as well as along the roads leading to these areas. In addition, the forest is also exploited adjacent to the Ngoko and Sangha Rivers by groups from the Republic of Congo and Central African Republic.

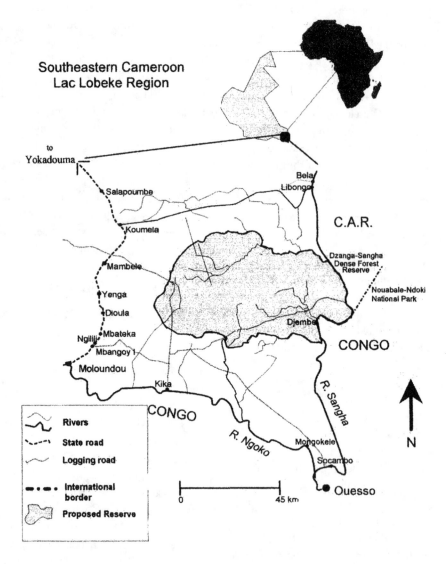

Figure 30.2. The Lobeke region of southeastern Cameroon. From Robinson and Bennett 2000, © 2000 Columbia University Press. Reprinted with the permission of the publisher.

As elsewhere in forested central Africa (see Ichikawa 1983; Kitanishi 1995), Lobeke's horticultural (Bangando) and hunter-gatherer (Baka) groups are linked by a network of political, social, religious, and primarily economic bonds (Bahuchet and Guillaume 1979; Bahuchet 1983). Each horticultural clan and associated Baka group asserts control over certain areas of the forest, where their subsistence activities are allowed. Although these land tenure claims overlap and are not formally demarcated, the forests of southeastern Cameroon are claimed by its indigenous communities according to local conventions. These local customs have recently begun to deteriorate with the changing social and economic forces that dominate the area.

At the urging of the Cameroonian government, and with assistance from missionaries and NGOs operating in the southeast, the Baka have become increasingly sedentary over the past three decades. Many are now living in permanent villages along the Yokadouma-Moloundou road, where they cultivate small plots of land. Modest numbers of Baka children are attending school, and there is greater access to health care. These factors, combined with the increased availability of unskilled, low-paying employment with the timber companies and safari outfits, have allowed many Baka to attain some degree of autonomy from their Bangando "patrons," who often egregiously exploit the Baka. However, this autonomy has also led to greater conflict with the Bangando, who resent the number of social service programs directed solely toward the Baka. These sociocultural problems are further exacerbated because of pressure exerted on the forest by foreign interests.

OUTSIDE PRESSURES ON THE FORESTS

- *Logging.* Southeastern Cameroon has been logged rather extensively for the past thirty years and with growing intensity since 1980. In 1995–96, five European-based timber companies were operating in the Lobeke area alone, including some concessions in the proposed conservation area. These groups export both raw timber (primarily mahoganies) and partially processed lumber. Although these products have been shipped in the past via the Sangha River to Brazzaville and then by train to Pointe-Noire, because of political turmoil in Congo timber is now transported by road to the Cameroonian port of Douala. In fact, timber companies operating in northwestern Congo have also begun to use the road to Douala, bringing

increased pressure on the forests of southeastern Cameroon. The Cameroonian government typically grants logging concessions for five years on a renewable basis, but timber companies have claimed that economic troubles and the decreasing availability of high-value trees may soon drive them to leave and move further north. In 1994 the government of Cameroon declined to grant a logging concession within the last remaining 40,000 ha of unexploited forest in the Lobeke region, in part because of the recommendations of a WCS multidisciplinary team (Curran 1993a; Hall 1993). However, more recent discussions have re-opened the prospect of additional exploitation in the area by "trading" a concession near Lobeke for the creation of an ecological corridor that would link the nearby proposed Boumba-Bek and Nki Reserves (M. van der Waal, pers. comm.). The social, economic, and ecological impacts of these logging operations are notable (Rodgers et al. 1991; Curran 1993a; Hall 1993). In a survey conducted among the local population in 1993, nearly 65% of respondents felt that timber companies posed the greatest threat to the forests of southeastern Cameroon (Curran, unpublished data). The most critical problem (according to 38% of respondents) is the easy access to forested areas created by opening logging roads. Poaching has become rampant, and the indigenous population has noted a significant decline in animal densities, necessitating longer forays into the forest for their traditional subsistence activities. According to Stromayer and Ekobo (1991), expatriate timber company employees estimate that the elephant populations have been reduced by one-half in southeastern Cameroon since 1980, coinciding roughly with the period of increased timber extraction.

Also, 98% of respondents to the 1993 survey noted the declining availability of tree species of value to the local population (especially for the collection of caterpillars [63%] and medicinal plants [22%]; Curran, unpublished data).

- *Poaching.* A well-organized and well-armed poaching network has moved into the Lobeke region, thanks in large measure to the timber companies' road system. Although some poaching includes the Bangando and Baka (the Baka in particular are sent into the forest with weapons at the bidding of third parties, often in search of ivory), a much greater impact is caused by outsiders. A census conducted in 1995–1996 found local villagers present in only one poaching camp among sixteen (WCS 1996). These outsiders include Cameroonians from other regions, who typically transport bushmeat for sale to timber company employees or to larger population centers elsewhere in the country. Also involved in the commercialization of bushmeat are hunters from Congo and Central African Republic, who can easily cross the unpatrolled international boundaries marked by the Ngoko and Sangha Rivers. These hunters typically ship the meat by pirogues toward Ouesso, Congo, from where it continues on to Brazzaville by air (Hennessey 1995). It is not uncommon for an individual poacher to set hundreds of wire snares at once, leading to local decimation and waste (due to rotting) of wildlife (WCS 1996). In addition to the commercial bushmeat operations, poaching is directed toward ivory collection and the capture of live animals (particularly young gorillas, chimpanzees, and African gray parrots), exotic skins, and other trophy items (bongo and leopard, for instance). The trade in African gray parrots (*Psittacus erithacus*) is particularly noteworthy, as individual permits for the capture of up to 1,000 birds have been issued annually by the Ministry of Environment and Forests (MINEF) in Yaoundé. Although quotas were reduced in 1995, interviews with parrot trappers indicated that fewer birds were being captured, suggesting that perhaps their populations have already been seriously affected (WCS 1996). And in 1997, the Convention on International Trade of Endangered Species (CITES) suspended all exports of parrots from Cameroon because of the lack of control. Trapping continues unabated, however; in fact, observation platforms constructed near important forest clearings in Lobeke have reportedly been taken over by parrot trappers, who are housing their birds there before transporting them to Douala (T. Smith, pers. comm.).

- *Sport hunting.* In 1996, three European-based safari companies were operating in the forests around Lobeke, and purportedly other groups are interested in the area as well (DeGeorges 1994). Safari companies generally hunt from December to June, serving wealthy foreign clients interested in trophy animals (primarily bongo, with buffalo, sitatunga, and elephant also of interest). Some of these companies have been hunting in the area for up to two decades, despite the uncertain legal status of their operations. Although they do receive government permission to hunt in this region, they have no clearly defined concessions. A 1993 proposal to create four large hunting concessions near Lobeke (effectively excluding the local population from much of what is considered their traditional forests) was tabled following the recommendations of earlier studies (Curran 1993a; Hall 1993). However, the issue of permanent legal hunting zones remains unresolved, and concessions continue to be granted on a provisional basis. Considerable friction exists between the indigenous population along the Yokadouma-Mouloundou road and the safari outfits. The local people resent the safari hunting practices and lack of communication. The safari hunters have supposedly never consulted with the local chiefs or the population in general. In fact, they are known to intimidate area residents by burning their hunting camps and possessions, and they have even threatened to shoot people found in the forest (WCS 1996).

## THE PROPOSED LOBEKE PROTECTED AREA

The outstanding biological diversity and importance of the Lobeke region have been recognized by the government of Cameroon and conservation organizations for many years. In 1987 the first proposal to create a reserve of approximately 400 km² was submitted (Harrison and Agland 1987). Further international attention was drawn to Lobeke when a proposed trinational conservation initiative with the Central African Republic (Dzanga-Sangha Dense Forest Reserve) and the Republic of Congo (Nouabale-Ndoki National Park) was drafted (Carroll and Weber 1991). Although Congo and the Central African Republic have officially gazetted their respective territories, the Cameroonian component continues to be studied by the government and interested NGOs.

The size of the proposed reserve has not been determined, and its management status is vague. Initial biological surveys of the forest recommended that the unexploited core area of 400 km² be expanded to encompass more than 4,000 km² in the "Lake Lobeke-Mongokele National Park," in order to ensure that additional areas received protected status (Stromayer and Ekobo 1991). Unfortunately, one of the zones proposed for protection was subsequently designated as a logging concession. In a succeeding report, an area of 2,125 km² was recommended as a wildlife reserve (Ekobo 1994). The forestry zoning plan accepted by the Cameroonian government includes proposals for the creation of a national park and a wildlife reserve (with an approximate area of 4,000 km²) in the southeastern corner of the country (Cote 1993). Should this plan be implemented, the forest surrounding Lobeke could become totally protected according to Cameroonian regulations. However, some form of multiple-use zone may well prove more appropriate, given the competing interests active in the area today. This could open the forest to timber exploitation and safari hunting, although these operations must be coordinated within the context of sustainable management of the entire area. The subsistence practices and economic concerns of the local population must also become an integral element in the conservation equation. The foreign revenue that Cameroon captures in the next few years from timber exploitation, and to a lesser extent from sport hunting, must be balanced with a farsighted scheme that integrally protects some of the forest, while reserving distinct areas for these other activities.

## COMMUNITY PARTICIPATION IN SOUTHEASTERN CAMEROON

During two periods (1992–1993 and 1995–1996) an interdisciplinary team worked with the local population as part of an overall program (Curran 1993; Hall 1993; WCS 1996) to create a protected area near Lobeke. The methodologies employed included focal interviews regarding historical and current perspectives on resource issues; written histories prepared by select area residents; group discussions on a variety of topics; seasonal activities calendars to better understand time budgets for forest-based subsistence and economic activities; and a detailed questionnaire of socioeconomic and conservation attitudes prepared and translated into the Bangando and Baka languages. These were administered by local residents trained in survey techniques. Finally, in an effort to promote greater involvement by all segments of the community, a video filmed by a local Catholic priest was shown in the largest villages along the Moloundou road. The footage allowed many individuals their first glimpse of forest wildlife in natural settings. The video was followed by discussions concerning conservation issues in general, problems directly relevant to the local population, and potential areas for protection in the Lobeke region. Village meetings were convened the following day (with Bangando and Baka discussions conducted separately, to ensure broad participation), focusing on how the Lobeke area could be protected without interfering with the traditional subsistence and economic activities of the local populations. In all instances, Bangando and Baka translators who also spoke French were employed, to ensure that the greatest possible interaction took place.

Foreign exploitation of forest resources has often undermined traditional local access to resources and led to conflicts. If management strategies for a protected area around Lobeke are poorly planned, they will certainly exacerbate these conflicts. However, if the local population participates in the resource management program and their lifestyles and beliefs are respected, there appears to be great potential for a real partnership in conservation in southeastern Cameroon. The indigenous residents have a basic understanding of the importance of conservation, and they often speak eloquently of the need to protect some part of the forest from the destruction currently occurring at the hands of outsiders. They report that they have never been consulted by the Cameroonian government, the timber companies, or the professional hunters, nor has any tangible benefit ever accrued locally from the exploitation of "their" forest. As such, there is interest in protecting the Lobeke region, not only to ensure that future generations of Bangando and Baka will have access to sufficient forest, but also as a backlash against exploitation by outsiders and the government. Many area residents have indicated that they would be willing to support total protection of a core area (even if their own activities were limited there) if an adequate amount of forest were also set aside for traditional subsistence and economic activities.

It is encouraging that the local communities recognize the benefits of conservation, regardless of their motivation. It would be naive to assume, however, that we have fulfilled the need to consult with the local population. Instead, we must build on this first step and continue to integrate the Bangando and Baka into any management plan for the Lobeke region. Toward that end, the WCS and the World Wide Fund for Nature (WWF) have sponsored research teams that explored both ecological and socioeconomic issues linked to the creation of this protected area (Curran 1993a; Hall 1993; Ekobo 1994; Elkan 1994; WCS 1996). More recently, a German development organization (GTZ) has continued to work with local communities (Davenport and Usongo 1998).

ISSUES FOR THE FUTURE

- *Delimitation of protected area boundaries.* As noted above, the size of the proposed conservation area (400–4,000 km²) and the management options (faunal reserve, national park, multiple-use zones, and so on) are still under discussion (Davenport and Usongo 1998). According to Cameroonian law, the creation of a national park or a wildlife reserve would strictly limit human activities in the area. Given that the local populations (especially the Baka) exploit even the core of the Lobeke region on an annual basis, especially for fish in the Lobeke River during the dry season (January–February), it will be diffi-

cult to prohibit these activities. Greater
attention must be given to the subsistence
needs of the Baka, who continue to use the
forest more intensively than the Bangando.
Management of the area will surely be
complicated by the presence of timber
concessions, safari hunting leases, and the
needs of the local populations. If an equitable and acceptable management plan for
the area is to be developed, it must integrate all stakeholders and include a number of different zones of protection. A
clause in the forestry regulations allows
some controlled exploitation in a wildlife
reserve, such that the Baka could hunt and
fish during the dry season. Another option
under Cameroonian law would be the creation of community forest reserves or community hunting reserves in conjunction
with the Lobeke conservation area, which
could bring about 5,000 km² under a sus-
tainable-use management plan. Any activity within this zone must be carefully
defined and monitored in collaboration
with local populations.

- *Economic opportunities.* The local economy
of southeastern Cameroon was until
recently largely dependent on cocoa, but
the collapse of world markets has forced
local communities to look elsewhere for
sources of income. The timber and safari
hunting industries are unlikely ever to
employ more than a few local people
(Rodgers et al. 1991; Curran 1993a). As is
often the case with permanently settled
human forest populations with few
income-earning opportunities (Redford
and Robinson 1991; Curran 1993b; Lahm
1993, this volume), people in the Lobeke
area have come to rely heavily on commercialization of bushmeat to generate income
(nearly 50% of respondents to the 1993
survey, including 61% of Baka males,
reported selling meat to raise money; Cur-

ran, unpublished data; WCS 1996). There
is little reason to expect that this activity is
sustainable over the long term; yet if no
viable alternatives exist, commercial hunting will continue (Fimbel et al. 2000).
Improvement in agricultural markets may
be possible, and a number of non-timber
forest products have also been noted by
local communities (honey and wild man-
gos, for instance). One particularly interesting option might be for the local communities to acquire the necessary permits
for the export of African gray parrots.
Exploitation of these birds is widespread
in Cameroon, especially in the southeast,
although minimal profits from this activity
reach local communities. It may be possible
to design a nesting system that would allow
the carefully controlled removal of young
birds in a manner that has limited effects
on overall parrot populations and significantly reduces the mortality of captured
birds, and yet at the same time assures that
profits remain with local people (C. Munn,
pers. comm.).

## What Lessons Have Been Learned?

The earth of the twentieth century is a human-
dominated landscape. If collaborative dialogue
is not established from the outset, there are very
few places left in the world where conservation
projects can be established without adverse
impacts on local people. This situation is certain
to become even more pronounced as human
populations expand. The conservation and
development literature is replete with calls for
consultation and collaboration with all stakeholders in project development, implementation, and management. Recognition that
indigenous communities have legitimate rights
to certain resources is an important step in the
right direction, as it is increasingly apparent
that forced protection is a short-term solution,
particularly in developing nations that rarely

have the financial resources to devote to biodiversity conservation. Identifying biological resources and defining the sustainable limits of their exploitation will be a major challenge for protected-area managers. It has often been claimed that the preservation of tropical forest is in the best interests of indigenous peoples who have traditionally exploited its resources. In reality, however, conservation measures frequently constrain subsistence and economic activities, and the long-term benefits from conservation—even restocking overhunted areas from core protection zones—are not readily apparent to local communities. If the legitimate concerns of local populations continue to be overlooked, protected-area projects will fail to achieve their objectives.

Each project must be taken case by case, as there are clearly no magic formulas for success. Yet conservationists must share lessons learned, and continue to search for innovative measures to ensure long-term protection of biodiversity. Experience in DR Congo's Ituri Forest and in the Lobeke region of southeastern Cameroon has made certain points worth reinforcing:

- Collaboration and participation of local communities are critical from the outset. Although numerous factors complicate successful conservation in the Ituri Forest, greater involvement of local peoples in the design of the Okapi Wildlife Reserve may well have served to limit current tensions between the management team and indigenous residents. Although there was early outreach to local communities (Hall 1990), subsequent difficulties may have resulted from the failure to correctly identify key stakeholders. Local chiefs, though an indisputable part of any bid to work with tribal peoples in Africa, often have very different agendas from the communities they represent. This is evidently true in the Ituri, and it is hoped that recent efforts to

involve a broader set of stakeholders will continue to pay dividends. In southeastern Cameroon, efforts to establish a protected area started on a promising note, in that collaboration began in the earliest phase of the project. The value of this type of collaboration was never more evident than during the proceedings of a meeting between local government officials and community representatives in February 1993. Government administrators explained in no uncertain terms that it was unlikely that the "uneducated peasant" Bangando and Baka really understood the objectives of a conservation effort. An elder Bangando man, clearly intimidated by speaking out before these officials yet determined to make a point, declared that it was the government that did not understand, and that the reconnaissance team was the first that ever bothered to talk to people. This led to a succession of local people declaring their support for a conservation project in the Lobeke area. Sitting down with local communities, listening to their concerns, and learning from their experiences are only the first steps, but clearly they are positive ones that can be used to gain momentum as the project moves into its design phase.

- The value of a conservation awareness program should not be underestimated. A properly designed program of conservation awareness is a critical component of the overall protected-area system, and there is little doubt that such programs can have an impact on resource use by local communities (Weber 1987, 1995; Wood and Wood 1987). Some of the conflict between the management team of the Okapi Wildlife Reserve and local communities stems from the lack of adequate communication. Initial community outreach programs in the Ituri concentrated on primary school programs and contact with "elites." Although it was expected that they would in turn pass along information to the poorer residents who make up the bulk of the local population, this rarely happened. Primary school education programs are certainly worthwhile exercises for future generations, but to be most effective, education approaches must target those who are exploiting the resources. The education team in the Ituri spent virtually no time explaining the reserve regulations to local communities, which resulted in considerable confusion regarding permitted activities within the protected area, and subsequently exacerbated the conflicts between the reserve management team and area residents. Initial steps to rectify this situation have been taken, with the presence of a new team of educators who are working closely with local communities to discuss issues most important to them. Without an adequate understanding of the regulations and how subsistence and economic activities affect their environment, there is little hope that local communities will modify their behavior and participate in management of protected areas. The development of a conservation extension program—which will include indigenous knowledge wherever appropriate—should be one of the priorities for the approach in southeastern Cameroon.

- ICDPs are not a panacea for success in conservation objectives. Local communities should benefit from conservation, particularly in those instances where people have lost their access to traditionally exploited resources (or had their access reduced). At the same time, conservation projects cannot be expected to solve all the social and economic woes of impoverished communities. Wherever possible, it should be stressed that conservation itself is the benefit (hence the education programs), and it should not be assumed that develop-

ment projects initiated by outside "experts" are the only viable solution. Local populations typically have a vested interest in protecting the resource base, and we should begin conservation projects by searching for those benefits that inspire and uplift local communities but do not carry high costs. The people of southeastern Cameroon made reasonable requests for "compensation" from the safari hunters (shared meat) and timber companies (access to scrap wood that is burned anyway) that would have cost very little; these requests were refused in both instances, intensifying conflict unnecessarily (Curran 1993a). It is true that in some instances, small-scale rural development projects may make a critical difference, but this should not be taken as a given.

- In those instances where development projects are deemed appropriate, they must be firmly linked to conservation objectives (Owen-Smith and Jacobsohn 1989). If there are no explicit connections between the adoption of certain measures to protect biodiversity and resulting social or economic benefits, there may be no motivation for local populations to modify potentially destructive resource exploitation practices. Written "contracts" that enable all partners to understand explicitly what is expected of them and that can be effectively monitored are an integral part of this connection. Given the attitude of local communities in the Ituri, the management team is left with little choice but to include some development projects, but it has been made clear that only those communities that actively participate in conservation measures will be eligible for development assistance. In southeastern Cameroon, the first aim should be to reduce conflicts between stakeholders, share benefits more equitably, and devise a system that will guarantee secure access to certain parts of the forest for the local communities (that is, a system that will restrict timber exploitation, safari hunting, and poaching). This system should include monitoring to ensure that local communities are not overexploiting their resources. Infrastructure- or income-generating projects should remain lower priorities for the present.

- Promises to local communities (other than promises to listen to them) should be avoided. It is critical not to raise the expectations of local populations unrealistically at the beginning of a conservation project. In the Ituri, promises of development assistance were made by a government official interested in securing written "permission" from traditional local authorities for the creation of the reserve. The failure of the government (and by extension, anyone now working in the reserve) to follow through on these promises has become a major point of contention for the local population. As a result, the conservation project will likely be forced to make some costly concessions on this point, even though very little has changed for local communities in terms of access to traditionally exploited resources because of reserve regulations. In the Lobeke region, it was continually stressed that the only promise was to listen to the local population, and to account for their concerns and aspirations in the overall protected-area management plan. The danger always exists that any actions by outsiders will be misinterpreted, but if care is taken and respect is shown to local communities, this problem can be avoided.
- A conservation project should realistically reflect local political, economic, and social realities. It is a mistake to assume that a national park will be better protected than multiple-use reserves, particularly in those cases where park authorities are not capable of ensuring complete protection. Rather, "an appropriate form of protected area should be designed for each situation, depending upon factors such as the social and economic needs of the local people, the particular needs of the species and ecosystems in each area, tourist and sport-hunting potential and research importance" (Stuart et al. 1990:14). Protected areas like the Okapi Wildlife Reserve in the Ituri and the proposed Lobeke reserve in southeastern Cameroon, which both include mobile hunter-gatherer communities with long-standing tenure, must make allowances for access to resources, which is incompatible with total protection. Protected-area managers must strive instead for an understanding with all stakeholders of the types of activities which will be permitted, as well as sanctions for illegal actions (for example, poaching or assisting outsiders in poaching, or extensive commercialization of bushmeat).

Integrating local populations into the management of protected areas is a protracted process, with the potential for many wrong turns and blind alleys. Yet if northern industrialized nations are truly dedicated to conservation of biodiversity, we must devise the methods and find the resources to make it happen. The lessons, both good and bad, that we learn in each case should allow the design and management of more and better conservation projects, which guarantee the long-term protection of floral and faunal diversity without negative impacts on local communities struggling for their own survival.

ACKNOWLEDGMENTS

We are grateful to colleagues at the Wildlife Conservation Society for the support and encouragement they have always offered, and for their dedication to finding ways to make conservation work with and for indigenous local communities. Thanks also to Amy Vedder, Bill Weber, Hilary Simons-Morland, Richard Peterson, David Wilkie, Matthew Hatchwell, and Lisa Naughton-Treves for comments on the manuscript of this chapter, and to Matthew Etter, Jodi Hilty, and Mika Boots for data entry and analysis. Terese and John Hart, James Powell, Cheryl and Rob Fimbel, Annette Lanjouw, Conrad Aveling, and Lee White have contributed many thoughtful ideas to the debate on the role of local communities in biodiversity conservation. In addition, our own field experiences have been supplemented by the shared interests and long discussions with numerous colleagues, including David Wilkie, Jefferson Hall, Robert Bailey, David Nzouango, Roger Fotso, Kambale Kisuki, and Stephanie Rupp. Many local peoples have welcomed us into their villages and gone to great lengths to make our work easier; we hope that in some small way our efforts will eventually repay their hospitality.

REFERENCES

Alcorn, J. 1990. Indigenous agroforestry strategies meeting farmer's needs. In A. Anderson, ed. *Alternatives to Deforestation: Steps Toward Sustainable Use of the Amazon Rain Forest*. Columbia University Press, New York.

Alvard, M. 1993. Testing the ecologically noble savage hypothesis: Interspecific prey choice by Piro hunters of Amazonian Peru. *Human Ecology* 21:355–387.

Bahuchet, S. 1985. *Les Pygmées Aka et la Forêt Centrafricaine*. SELAF, Paris.

Bahuchet, S., and H. Guillaume. 1979. Relations entre chasseurs-collecteurs pygmées et agriculteurs de la fôret du nord-ouest du bassin congolais. In Bahuchet, ed. *Pygmées de Centrafrique*. SELAF, Paris.

Batisse, M. 1982. The biosphere reserve: A tool for environmental conservation and management. *Environmental Conservation* 9(2):110–111.

Bhatnagar, B., and A. Williams., eds. 1992. *Participatory Development and the World Bank: Potential Directions for Change*. World Bank, Washington, D.C.

Breslin, P., and M. Chapin. 1984. Conservation Kunastyle. *Grassroots Development* 8(2):26–35.

Brown, M., and B. Wycoff-Baird. 1992. *Designing Integrated Conservation and Development Projects*. Biodiversity Support Program, Washington, D.C.

Carroll, R., and W. Weber. 1991. An integrated plan for regional forest conservation and management in southeastern Cameroon, southwestern Central African Republic and northern Congo. Proposal submitted to USAID by the World Wildlife Fund and the Wildlife Conservation Society.

Cernea, M., ed. 1985. *Putting People First: Sociological Variables in Rural Development*. A World Bank Publication. Oxford University Press, New York.

Chambers, R. 1983. *Rural Development: Putting the Last First*. Longman, New York.

———. 1991. Shortcut and participatory methods for gaining social information for projects. In M. Cernea, ed. *Putting People First: Sociological Variables in Rural Development*. A World Bank Publication. Oxford University Press, Washington, D.C.

Child, G. 1984. Managing wildlife for populations in Zimbabwe. In J. McNeely and K. Miller, eds. *National Parks, Conservation and Development: The Role of Protected Areas in Sustaining Society*. Smithsonian Institution Press, Washington, D.C.

Cote, S. 1993. Plan de zonage du Cameroun Forestier Meridional. Report prepared for the Cameroonian Ministry of the Environment and Forests.

Curran, B. 1991. The Okapi National Reserve: Its impact on the indigenous people of the area. Report prepared for the World Bank.

———. 1992. Appraisal of the Okapi Wildlife Reserve Management Plan. Report prepared for the World Bank.

———. 1993a. Preliminary assessment of issues affecting the human populations of the Lac Lobeke Region, southeastern Cameroon. Report prepared for the Wildlife Conservation Society, the World Bank, and the Biodiversity Support Program.

———. 1993b. Socio-economic assessment and management recommendations for Campo Reserve, southwestern Cameroon. Report prepared for the World Bank.

———. 1994. Preliminary results of socio-economic studies in the Okapi Wildlife Reserve, northeastern DR Congo. Report prepared for the Wildlife Conservation Society and the World Bank.

Davenport, T., and L. Usongo. 1998. Technical and financial progress report for the Lobeke Forest Project. Report prepared for WWF/GEF.

DeGeorges, A. 1994. Preliminary discussions leading to development of an elephant conservation program between the Cameroonian Ministry of Environment and Safari Club International. Report prepared for Safari Club International.

Durning, A. 1992. Guardians of the land: Indigenous peoples and the health of the earth. Worldwatch Paper 112. Worldwatch Institute, Washington, D.C.

Ehret, C., and M. Posnansky, eds. 1982. *The Archeological and Linguistic Reconstruction of African History*. University of California Press, Berkeley.

Ekobo, A. 1994. Elephants in the Lobeke Forest. Paper presented at the meeting of the African Elephant Specialist Group, Mombasa, Kenya.

Elkan, P. 1994. Preliminary results of a Bongo antelope survey in southeastern Cameroon. Report prepared for the Wildlife Conservation Society.

Fimbel, C., B. Curran, and L. Usongo. 2000. Enhancing the sustainability of duiker hunting through community participation and controlled access. In J. Robinson and L. Bennett, eds. *Sustainability of Hunting in Tropical Forests*. Columbia University Press, New York.

Hall, J. 1990. Conservation politics in DR Congo: The protection of the Ituri Forest. *WWF Reports* (April–May): 3–4.

———. 1993. Report on the strategic planning mission for the creation of a protected area in the Lobeke Region of southeastern Cameroon: Assessment of timber exploitation, safari hunting and preliminary vegetation analysis. Report prepared for the Wildlife Conservation Society, the World Bank, and the Biodiversity Support Program.

Hames, R. 1979. Comparison of the efficiencies of the shotgun and the bow in Neotropical forest hunting. *Human Ecology* 7:219–252.

Harrison, M., and P. Agland. 1987. Southeast Cameroon: A proposal for three new rainforest reserves. Report for the Cameroonian Secretary of State for Tourism.

Hart, J. 1978. From subsistence to market: A case study of the Mbuti net hunters. *Human Ecology* 6(3):325–353.

Hart, T., and J. Hart. 1993. Between sun and shadow. *Natural History* 101(11):28–35.

Hennessey, B. 1995. A study of the meat trade in Ouesso, Republic of Congo. Report prepared for GTZ and the Wildlife Conservation Society.

Hill, M., and A. Press. 1994. Kakadu National Park: An Australian experience in comanagement. In D. Western, R. Wright, and S. Strum, eds. *Natural Connections: Perspectives in Community-Based Conservation*. Island Press, Washington, D.C.

Hough, J., and M. Sherpa. 1989. Bottom up vs. basic needs: Integrating conservation and development in the Annapurna and Michiru Mountain Conservation Areas of Nepal and Malawi. *Ambio* 18(8):434–441.

IUCN. 1986. *Managing Protected Areas in the Tropics*. IUCN, Gland, Switzerland.

Kiss, A., ed. 1990. Living with wildlife: Wildlife resource management with local participation in Africa. World Bank Technical Paper No. 130, Washington, D.C.

Kitanishi, K. 1995. Seasonal food changes in the subsistence activities and food intake of the Aka hunter-gatherers in northeast Congo. *African Study Monographs* 16(2):71–118.

Kleymeyer, C. 1994. Cultural traditions and community-based management. In D. Western, R. Wright, and S. Strum, eds. *Natural Connections: Perspectives in Community-Based Conservation*. Island Press, Washington, D.C.

Lahm, S. 1993. Utilization and local variation of game resources in Gabon. In C. Hladik, A. Hladik, O. Linares, H. Pagezy, A. Semple, and M. Hadley, eds. *Tropical Forests, People and Food*. UNESCO, Paris; Parthenon Publishing, Carnforth, England.

Lewis, D., G. Kaweche, and A. Mwenya. 1989. Wildlife conservation outside protected areas: Lessons from an experiment in Zambia. *Conservation Biology* 4(2):171–180.

McCracken, J., J. Pretty, and G. Conway. 1988. *An Introduction to Rapid Rural Appraisal for Agricultural Development*. International Institute for Environment and Development, London.

MECN. 1992. Arrêté Ministériel No. 045/CM/ECN/92 du 02 Mai 1992 Portant Création et Délimitation d'une Réserve Naturelle Dénommée Réserve de Faune à Okapis. Government of Zaire, Kinshasa, DR Congo.

Metcalfe, S. 1994. The Zimbabwe Communal Areas Management Programme for Indigenous Resources (CAMPFIRE). In D. Western, R. Wright, and S. Strum, eds. *Natural Connections: Perspectives in Community-Based Conservation*. Island Press, Washington, D.C.

National Environmental Secretariat (Government of Kenya), Clark University, Egerton University, World Resources Institute, and Center for International Development and the Environment. 1990. *Participatory Rural Appraisal Handbook*. World Resources Institute, Washington, D.C.

Oldfield, S. 1988. *Buffer Zone Management in Tropical Moist Forests: Case Studies and Guidelines*. IUCN, Gland, Switzerland.

Owen-Smith, G., and M. Jacobsohn. 1989. Involving a local community in wildlife conservation: A pilot project at Purros, Southwestern Koakaland, SWA/Namibia. *Quagga* 27:21–28.

Peterson, R. 1990. Searching for life on Zaire's Ituri forest frontier. *Cultural Survival Quarterly* 14(4):56–62.

———. 1991. Whose forests? Land tenure dynamics on Zaire's Ituri forest frontier. Paper presented at the LTC Workshop on Tenure and the Management of Natural Resources in Subsaharan Africa, Rosslyn, Va.

Poffenberger, M., and B. McGean, eds. 1996. *Village Voices, Forest Choices: Joint Forest Management in India.* Oxford University Press, Delhi.

Redford, K. 1990. The ecologically noble savage. *Orion* 9:24–29.

Redford, K., and J. Robinson. 1991. Subsistence and commercial uses of wildlife in Latin America. In J. Robinson and K. Redford, eds. *Neotropical Wildlife Use and Conservation.* University of Chicago Press, Chicago.

Robinson, J. G., and E. L. Bennett. 2000. *Hunting for Sustainability in Tropical Forests.* Columbia University Press, New York.

Rodgers, J., J. Loung, B. Tarh, and V. Joiris. 1991. Economic, training and community needs assessment of the proposed Cameroon component of the Central African Regional Forest Conservation Initiative. Preliminary report to World Wildlife Fund and Wildlife Conservation International.

Rupp, S. 1994. Women's issues in conservation management of the Okapi Wildlife Reserve. Report prepared for Wildlife Conservation Society, ICCN, and CEFRECOF.

Sayer, J. 1991. *Rainforest Buffer Zones: Guidelines for Protected Area Managers.* IUCN, Gland, Switzerland.

Schoonmaker-Freudenberger, K., and B. Gueye. 1990. *RRA Notes to Accompany Introductory Training Module.* International Institute for Environment and Development, London.

Stromayer, K., and A. Ekobo. 1991. Biological survey of the Lake Lobeke Region, Southeastern Cameroon. Preliminary report to Wildlife Conservation International and the World Wildlife Fund.

Stuart, S., and R. Adams (with M. Jenkins). 1990. Biodiversity in sub-Saharan Africa and its islands: Conservation, management and sustainable use. Occasional Papers of the IUCN Species Survival Commission No. 6.

Theis, J., and H. Grady. 1991. *Participatory Rapid Appraisal for Community Development: A Training Manual Based on Experiences in the Middle East and North Africa.* International Institute for Environment and Development, London.

UNESCO/UNEP. 1984. *Conservation, Science and Society: Contributions to the First International Biosphere Reserve Congress, Minsk.* UNESCO, Paris.

Vedder, A., and W. Weber. 1990. Rwanda: Mountain gorilla project, Volcanoes National Park. In A. Kiss, ed. *Living with Wildlife: Wildlife Resource Management with Local Participation in Africa.* World Bank Technical Paper No. 130, Washington, D.C.

Vickers, W. 1991. Hunting yields and game composition over ten years in an Amazon Indian Territory. In J. Robinson and K. Redford, eds. *Neotropical Wildlife Use and Conservation.* University of Chicago Press, Chicago.

WCS. 1996. The Lobeke Forest, southeast Cameroun: Summary of activities (1988–1995). Report prepared for the Wildlife Conservation Society and the Government of Cameroun.

Weber, W. 1987. Sociological factors in the conservation of Afromontane forest reserves. Pages 205–229 in C. Marsh and R. Mittermeier, eds. *Primate Conservation in Tropical Rain Forest.* Alan R. Liss, New York.

———. 1989. Conservation and development on the Zaire-Nile Divide: An analysis of value conflicts and convergence in the management of Afromontane forests in Rwanda. Ph.D. dissertation, Institute for Environmental Studies, University of Wisconsin-Madison.

———. 1994. Ecotourism and primate conservation in African rain forests. Pages 129–150 in C. Potter, J. Cohen, and D. Janczewski, eds. *Perspectives on Biodiversity: Case Studies of Genetic Resource Conservation and Development.* AAAS Press, Washington, D.C.

———. 1995. Monitoring awareness and attitudinal factors in conservation education: The mountain gorilla project. Pages 28–48 in S. Jacobsen, ed. *Conserving Wildlife: International Education/Community Approaches.* Columbia University Press, New York.

Wells, M., K. Brandon, and L. Hannah. 1992. *People and Parks: Linking Protected Area Management with Local Communities.* World Bank, Washington, D.C.

Wilkie, D. 1989. Human settlement and forest composition within the proposed Okapi Rain Forest Reserve in northeastern Zaire: Creating the first thematic map of the Ituri Forest using digital analyses of Landsat TM imagery. Final report prepared for the World Wide Fund for Nature.

Wilkie, D., and J. Finn. 1988. A spatial model of land use and forest regeneration in the Ituri Forest of northeastern Zaire. *Ecological Modeling* 41:307–323.

Wood, D., and D. Wood. 1987. *How to Plan a Conservation Education Program*. International Institute for Environment and Development, Washington, D.C.

Zerner, C. 1994. Transforming customary law and coastal management practices in the Maluku Islands, Indonesia, 1870–1992. In D. Western, R. Wright, and S. Strum, eds. *Natural Connections: Perspectives in Community-Based Conservation*. Island Press, Washington, D.C.

# CHAPTER 31

# Politics, Negotiation, and Conservation

## A View from Madagascar

Alison F. Richard and Robert E. Dewar

This chapter is about the relationship between politics and economic development in the context of conservation and, in particular, about the implications of this relationship for planning projects aimed at protecting parks and reserves in Madagascar. We propose a framework to characterize the relationship and apply it to a case study from southern Madagascar. Although the setting for our case study is an area of dry forest, the issues that emerge are highly relevant to conservation efforts in other parts of the island too, including the rain forests of eastern Madagascar. Our focus is on the initiation of projects with external inputs, and not on the initiation of, or transition to, a self-sustaining effort. This partly reflects our experience and partly reflects our belief that it is unrealistic to expect conservation projects in Madagascar to become completely self-sustaining for the foreseeable future.

The past few years have seen a growing consensus that the protection of parks and reserves is most effectively pursued with the participation of people living around the protected areas, and that protection must go hand in hand with improvements in the quality of people's lives (MacKinnon et al. 1986; Weber 1987; Wright 1989; McNeely et al. 1990; Wells et al.

1992). The realization that successful protection commonly requires both that people participate and that they stop using, or reduce their reliance on, certain kinds of resources has generated growing support for integrated programs of protection and economic development. Many integrated programs take the form of targeting an area for total protection and delineating around it a buffer zone to be the site of development interventions. The rationale is explicit: protection will succeed only if the economies of neighboring areas provide alternatives that reduce or eliminate economic pressures on the protected zone.

One effect of this strategy has been to bring about a new emphasis on action and debate whereby conservation efforts in developing countries are most strongly linked to rural economic development among people distant from the national political center. Implicit in this approach is an important assumption: the interaction between national political agendas and conservation efforts is different in developed and developing countries. In developed nations, conservation goals are pursued primarily through political activity at the national level. Governments are pressured by conservation groups to protect areas by the enactment and enforcement of legal restrictions on activities antithetical to conservation goals. Political pressures on national governments force them to bear many of the direct and indirect costs of conservation.

Conservationists thus usefully focus their energy on identifying areas and species in need of protection, and on mobilizing public opinion as a way of encouraging governmental activity. In developing nations, these same tactics have proven less effective, if not therefore less important. A major difficulty is that the central governments of many developing nations are simply without sufficient resources to monitor, let alone restrict and control, the lives of people in distant rural or less populous areas. As a result, policy decisions at the national level generally have much less rapid and efficient consequences than in more tightly integrated economies (Deshmukh 1989; but see Hough and Sherpa 1989). Moreover, the economic activities that threaten areas of conservation priority are often the pursuit of life's necessities by impoverished populations without viable alternatives (Lewis 1988). Lacking resources, or the political and administrative strength to provide real options, central governments would be unrealistic, at the very least, to try to enforce protection and expect people to give up their current means of making a living.

Our thesis is that the removal of actions from the political center does not lessen the political content of these actions: local participation in conservation is as much a political process as national government participation, involving such issues as representation, authority, and legitimacy. In seeking local participation, then, a political strategy as well as an economic strategy is needed: in many conservation projects today, the locus of political activity is shifted, but not its importance or, indeed, its complexity. We argue that an extended period during which a bargain is struck and effective political structures are identified or established is likely to precede any real participation by local people in setting and achieving long-term conservation goals. This, in turn, has important implications for the initial planning of integrated conservation and development projects. Specifically, the role and goals of economic development in these projects need closer scrutiny, and the fact that, at the outset, development activities may have little to do with biodiversity and much to do with politics should not be viewed as evidence of the failure of the approach but rather as an integral part of it.

Within the dynamic political and economic framework that characterizes conservation efforts today, three kinds of development activities need to be envisaged and planned from the outset, the specific aims of which are different and distinct, and the timing and implementation of which are staggered:

1. *Striking a bargain.* Activities of this kind respond to the immediate, articulated needs of people living in the buffer zone, and not necessarily to any of the long-term pressures on the environment. The aim is to build reciprocity through a negotiated process of action and counteraction: we need you to stop hunting, burning, and collecting in this reserve—what do you need in order to stop? The "we" in this case are usually biologists, government officials, and other outsiders, including foreigners; the "you" are the local people, both as individuals and as represented in one or more ways.
2. *Direct substitution.* Building on the reciprocal relationships established in stage 1 to develop a more collaborative approach, a second phase of activities can be launched to halt inappropriate uses of the protected zone by establishing acceptable alternatives in the buffer zone.
3. *Anticipated substitution.* As knowledge of

the ecological and economic system increases and a working partnership between local and non-local participants emerges, the third and most difficult type of activity can be envisaged, aimed at pre-empting inappropriate uses of the protected area in the future by responding to anticipated economic pressures.

Political issues present themselves at every point along the way. For example, is the local community represented by its political or traditional leaders, men or women, elite or poor, landowners or sharecroppers, the powerful or the passionate? And on whose behalf speak the assortment of foreign and national academics, members of the international donor agencies, and government officials commonly seated on one side of the negotiating table together? Insofar as there are answers at all to these inherently political questions, they frequently owe more to historical accident than to considered strategy.

Our emphasis on the centrality of the political process at the local level carries implications for the delineation of buffer zones. A buffer zone is an area that receives special attention because it is adjacent to a protected area. The nature of this attention is quite variable (MacKinnon et al. 1986; Oldfield 1988), and how the idea of a buffer zone has been translated spatially and functionally has varied too, from a swathe of more or less arbitrary width to a region that is functionally coherent according to economic or ecological criteria. In reality, an evolving conception of the buffer zone is called for. At the outset, its operational definition is perforce narrowly determined by local political configurations. As stage 2 and stage 3 activities become possible, the criteria should broaden to encompass the wider network of economic interdependencies and needs that exert pressure on the protected area.

In the rest of this chapter, we illustrate and expand these ideas, using our experiences with the Beza Mahafaly Project in southern Madagascar over the past twenty-five years. We should emphasize that our view is retrospective, and that the framework proposed here is not the one in which the Beza Mahafaly Project goals and activities were articulated at the time but rather one that makes sense to us now, based on what we have learned.

## The Beza Mahafaly Project

The Beza Mahafaly Project began in 1975 when three field researchers (G. Ramanantsoa of Université de Madagascar, R. W. Sussman of Washington University, and A. F. Richard of Yale University) decided to work together to establish a nature reserve that would serve as a training ground for students at the School of Agronomy of the Université de Madagascar. The reserve would be a place where conservation was not only taught but put in practice to preserve critical, unprotected habitat. Selecting the location of the reserve fell to Ramanantsoa, who had the difficult task of finding an area that fulfilled two conditions: it should contain important natural resources of a type poorly preserved in existing reserves, and the local population currently using the forest to be designated as a reserve had to be willing to collaborate in the enterprise and—more to the point at the outset—to give up their use rights to the forest. The second condition was particularly difficult to meet, and only after many months of searching did Ramanantsoa find his way to the village of Beza Mahafaly, in the District of Beavoha. The district president found the proposed enterprise attractive, and he was able to convince his constituents of the advantages of establishing a tie with the university. Two parcels were duly set aside with the formal support of representatives from seven of the eight villages in the district: 100 ha of riverine *Tamarindus* forest, and 500 ha of nearby but non-contiguous xerophytic forest, dominated by Didiereaceae. The reserve is, thus, very small; only one other governmentally managed reserve is as small.

There are three phases in the history of the project. During the first phase, from 1977 to 1985, a small natural reserve and rough-and-ready facilities were established, guards were recruited locally, and a field school was developed for students from the School of Agronomy. Although the idea of undertaking or facilitating development activities in villages around the reserve was integral to the project's initial conception, in practice the idea did not become a reality during this first phase, for three reasons: first, the total budget was small ($120,500 provided by the World Wildlife Fund [WWF] over a seven-year period) and mandated for conservation and training. Second, the project was chronically understaffed at the management level. Finally, none of the project directors were development specialists, and thus they were forced into on-the-job training.

Constraints and problems notwithstanding, by 1985 the conservation and training components of the project were well established, and the first phase culminated with the formal inauguration of the reserve as a Réserve Spéciale (the first to be named for seventeen years) in November 1985. The most remarkable aspect of this event was that it was precipitated by the local political authorities through a petition to the central government for the establishment of a permanent reserve. Under the circumstances, this was an ill-deserved vote of confidence in the project, because the effect of the change in status was to make the forest's reserved status more permanent—with precious little to show in return from the standpoint of people locally.

The second phase of the project, from 1985 to 1992, was one of expansion and major capital expenditure, made possible by increased support from WWF and the addition of development assistance from the U.S. Agency for International Development (USAID). The most important events were the initiation of development activities intended to facilitate further and continued conservation. In addition, the larger budget made it possible to strengthen the training activities of the project, both within Madagascar and abroad. Camp facilities were now used not only by students from the School of Agronomy but also by government conservation agents attending seminars.

In 1992, the Beza Mahafaly project entered a third phase, with major development projects in abeyance and a new emphasis on the need to understand and respond to long-term pressures on the forests through efforts undertaken in collaboration with members of local communities. Issues of participation and representation resurfaced, as project leaders came to terms with the shortcomings of past arrangements. Support from WWF and USAID has continued into this phase, and in addition the Claiborne-Ortenberg Foundation has brought new resources and ideas to the task.

## Striking a Bargain

Support for the project from the local community was initially promoted by two factors: first, the forest was regarded as sacred—the nearest village bears the name Analafaly, meaning "at the sacred forest"—and the hunting of many animal species of the forest was forbidden, yet people from across the Onilahy River 10 km to the north had begun to penetrate the forest to hunt locally protected species, including lemurs and the land tortoise (*Geochelone radiata*). Help in keeping these outsiders away from the forest was certainly a factor in mobilizing local support. But the attractiveness of the proposed reserve was also discussed in economic and, more broadly, development terms.

Several immediate or future economic benefits were anticipated. The reserve could provide cash wages for villagers as forest guards, and the presence of Malagasy and foreign project staff on a permanent and temporary basis offered the prospect of some boost to the local cash economy. The district president was also explicit about the advantages to his community that would flow from interactions with the University of Madagascar. The district had fallen on hard times in the past two decades, in part because of the destruction of the intake to their irrigation network by a hurricane, and in part because of the deterioration of the road linking the community to the regional market town of Betioky, 35 km away. There was little chance for education and little contact with the outside world, and the reserve promised to bring such contact, albeit of a rather nonspecific sort.

The reserve was established, thus, through a process of negotiation between two groups with different but overlapping interests. The first development projects focused on needs perceived as high priorities by the local community: education, transportation, rice production, and water for household use. They were not designed to compensate directly for the cost of giving up the forest as current or future agricultural or grazing land, or as a source of lumber, honey, and other products.

A major hurdle to long-term development for several villages in the district was the lack of educational possibilities: there was no school within reasonable reach. The project supplied the materials and labor to construct a schoolhouse, the community supplied the teacher's house and furniture for the school, and the provincial government supplied the teacher.

The road from Beza Mahafaly to the market town of Betioky fell into increasing disrepair after 1974, and by 1979 it could be traveled only by oxcarts and four-wheel-drive vehicles. Transport costs for local products were extraordinarily high—as much as 30% of their market price. The project obtained the financing for road repairs, eventually enabling a Mercedes truck to make its tortuous way, except in the wettest weeks, from Betioky to the weekly market in Beavoha. This was one of the most expensive development interventions, but also one of the most sought-after and the most widely appreciated.

Rice is the most highly valued crop throughout Madagascar, and the south is no exception. The south is also the driest region of the island, and availability of water is a critical issue for cultivation of any kind. From 1950 to 1968, an irrigation canal carried water from the Onilahy River to 600 ha of rice fields belonging to villagers in the district. In 1968 a hurricane destroyed the intake, and a concurrent shift in the riverbed made the old intake unusable, even if rebuilt, without expensive new construction. With the loss of these fields, the district's economic output plummeted, and most families had to resort to rain-fed fields of uneven productivity. Help in making this canal operational again was the first and single most urgent request made of the project at its outset in 1977. In 1987, development agencies, provincial and national government agencies, and private contractors were finally all mobilized, and work started on the renovation, a substantial engineering project finally completed in a modified and more modest form in 1995.

The absence of clean, readily available water for household consumption is a chronic problem in much of southern Madagascar. In two villages in the District of Beavoha, drinking water was acquired during the dry season by digging temporary wells in the sandy riverbed to a depth of 2 m or more. Filtered by sand, this water is clear and uncontaminated but time consuming to obtain. In the wet season, water was sought directly from the river. Though easily obtained, the water is cloudy and polluted by cattle and goats and almost certainly a cause of gastrointestinal problems. The project provided wells for these villages in an effort to improve this situation.

## Direct and Anticipated Substitution

Finding satisfactory alternatives to exploitation of a protected area is inherently more difficult than striking the initial bargain, which provides the basis from which to launch such a search and buys the time needed to achieve it. Indeed, anticipating and heading off future pressures is a feat yet to be achieved by any project anywhere, to our knowledge. The very desirability of substitution as a total strategy is itself questionable (McNeely 1989). Since the reserve system was established in the 1920s, protection in Madagascar has been pursued almost exclusively through a strategy of total interdiction: in principle, even to set foot in a natural reserve is illegal for the people living around it, and only government authorities and scientists with permits are allowed entry. Arguably, this strategy may simply ensure that whatever marginal value the forest has locally, deriving from traditional uses of forest products, will eventually be eliminated by the progressive substitution of alternative sources of food, medicine, fuel, and construction materials.

Whereas the idea of extractive reserves deserves serious consideration, there are clearly many forms of exploitation of Madagascar's remaining forests that are not sustainable, and in these cases alternatives must urgently be sought. At Beza Mahafaly, at present, there are few direct pressures on the reserved forest because it has been well protected for the past fifteen years. Insofar as establishing a reserve disrupted a previous history of exploitation, the costs were borne by the local people who were obliged to seek other exploitable areas. In practice, however, population density was low, and there was sufficient forest nearby to meet and probably sustain local current needs for wild foods, medicines, deadwood for fuel, and pasture for herds.

The prognosis for the future is less positive. Forest cover is regarded by local farmers, almost certainly correctly, as a measure of soil fertility and adequate soil moisture, and thus forested areas are widely recognized as potentially valuable farming land (Barbour 1990). The Beza Mahafaly reserve, though prime agricultural land, for the moment is safe from conversion to fields by virtue of its legal status and the consensus of support. Probably a greater local threat in the short term is that of hamlets being established on the reserve's immediate borders, thereby bringing it inside the radius of significant environmental impact that surrounds a concentration of people and their domestic animals. Housing is relatively impermanent in rural southern Madagascar, and hamlets tend to gravitate to the vicinity of the active fields. The rain-fed agricultural fields found in the vicinity of the reserve need significant fallow periods, and the number of potential sites for fields is limited (Barbour 1990). There is a real likelihood that if a field management plan is not developed by the local farmers, forest will be cleared, new fields opened up, and new hamlets established right on the edges of the reserve.

For the medium term (5–10 years), then, the project needs to foster the development of a land management plan by people locally and, more generally, to encourage a relocation of agricultural activity away from the immediate borders of the reserve and other remaining forested areas. In this respect, the refurbishment of the irrigation canal holds particular value. The re-establishment of irrigated agriculture in the already-cleared rice fields should reduce the attractiveness of potential rain-fed fields, given the significant annual variability in rainfall. The location of these irrigated fields some kilometers away from the reserve would inevitably have a positive effect in terms of population settlement patterns.

Projections of future needs and pressures over the long term (10–20 years) are much more difficult to make, and the research and monitoring efforts essential to such projections should have been planned and started at the outset so that now, twenty-five years into the project, a considered response to these needs and pressures could be initiated. The reality, however, is that serious attention has only begun to be given to these issues (see, e.g., Green and Sussman 1990). One major factor of concern is our inability to predict local and regional population trends. Whereas Madagascar's national population continues to increase, there has long been significant migration from rural zones to urban centers, and this pattern has been most noticeable in the south (World Bank 1980:183). It is not yet clear whether the re-establishment of the irrigation canal will act to counter the regional trend and attract significant population movement to the District of Beavoha. However, three regionally significant future problems are evident:

1. continued cutting of standing forest, with an attendant increasing scarcity of naturally available wood for fuel and construction;
2. the failure of natural forest to re-establish on fallow or abandoned agriculture fields, for unknown reasons, although domestic animals are potentially important; and
3. deterioration of standing forest through grazing and browsing by the growing herds of cattle and goats.

Ironically, the very success of the project in achieving local goals may generate new external pressures: the continuing presence of forest, the development activities themselves, and the improved access through road repairs make the Beza area an emerging focus for entrepreneurial activities and, potentially, new settlers. The most effective ways of meeting these new challenges may be to ensure that the legal basis for protection is soundly established, and that the local community has a strong vested interest in the success of the project's goals.

The framework and case history presented in this chapter together serve to highlight three issues concerning projects integrating development and protected areas, which we believe should be considered as future projects are planned in Madagascar and elsewhere.

The first is the centrality of the political process in such projects, and the opportunities and dilemmas thereby created. The local political leaders in Beza Mahafaly are at the lowest salaried level in a national political hierarchy that groups districts (*firaisana*) into county-like units (*fivondronina*) and fivondronina into provinces (*faritany*); these same leaders are also the representatives or advocates of their constituencies to these more inclusive political units. They must act, thus, in two different capacities: enforcing the laws and regulations of a distant, poorly understood and distrusted central government in their political unit, and simultaneously trying to represent their constituencies' interests to these higher levels of government. In the second role, with an eye to their own political future, they enter negotiations with agendas determined in part by their web of local supporters. Their calculation of the expected benefits of an integrated project must be gauged across this entire constituency, and not merely in the immediate vicinity of a protected area. This is inevitable, given the inclination of politicians everywhere to service their political debts and please the greatest number of their constituents, but it can lead to problems of equity, with the villages bordering the reserve receiving poor compensation for their forgone resources. Inattention to the needs and contributions of those to whom no political debts are owed is also likely. Women are likely to be overlooked, for example, insofar as they are excluded from the formal political process (Carloni 1987).

These challenges have no simple resolution. However, recognition that they are significant may itself mean that more thought is given to representation and more care taken to ensure inclusiveness, so that more attention is paid to segments of the community not well represented by established political structures, such as women, or too small to have a strong voice in the political process, such as particular villages or even families.

At Beza Mahafaly, this dilemma surfaced repeatedly (see also Sussman et al. 1994). For example, when the negotiated development activities began to be realized, it became evident that the buffer zone for the reserve was being defined with respect to the formal political structure, rather than on economic or ecological grounds. Initiatives typically benefited the district as a whole as much or more than the villages adjacent to the reserve. The project's immediate neighbors voiced their grievances to project participants, and efforts were made to redress an inequity that was real as well as perceived. But the underlying problem remains: the need for village-level political support and engagement is partly at odds with the need for support at the more inclusive level of the district.

A second issue of concern is how to measure project progress. At Beza Mahafaly, the first development projects were not designed to compensate for the direct costs of giving up the forest for farming, as grazing land, or as a source of lumber and other products. They were designed to respond to immediate priorities of the local community: education, rice production, transportation, and water for household consumption. Diverse in scale and objectives, these activities had three features in common. First, they were all carried out in response to requests made by the political leadership of the community. Second, they did not reflect or fairly respond to differences in the sacrifices made by different segments of the community: villages most directly affected by the sequestering of land in the reserve did not necessarily benefit most from project activities.

Third, they did not bear directly on the maintenance of, or increase in, biodiversity in the area and thus could be said to have little or nothing to do with conservation. Indeed, the repair of the road could have an adverse effect on the conservation status of the area in the long run, by facilitating access by charcoalers and entrepreneurs in search of lumber for construction.

What, then, are the criteria by which to judge the success of the Beza Mahafaly project thus far? Certainly they must focus on the establishment of a productive and potentially enduring relationship between the local community and the newly arrived outsiders, be they Malagasy or foreign. We can find no generally applicable rules for measuring this, and in the absence of a baseline of information at the start of the project, there is in any case no basis for comparison. In a general way, however, we can point to three indications of success. One was the district's successful petition to the central government in 1985 to make the reserve a permanent Réserve Spéciale. This was perhaps the single most remarkable turn in the history of the project. The local political impetus of this move remains unclear, especially as its effect was to make the forest's reserved status more permanent at a time when there was as yet little progress in achieving the economic development goals. Possibly it served as a political gesture through which the district demonstrated its progressive qualities to a broader range of governmental authorities. Another indication of success was the local election of 1989 in which the successful candidates for district office campaigned on a platform of support for the project. This was notably true in the immediate vicinity of the reserve as well. A more recent indication, emerging since 1992, is that surrounding villages have begun organizing formally constituted conservation and development associations with the goal of playing a more structured and active role in the project.

A third general issue raised by the Beza

Mahafaly Project is one of scale. The Beza Mahafaly Project is small, arguably too small for useful lessons to be learned from it. The question has been and remains: Can this be made to work on a larger scale? We suggest that the answer to this question depends on the process of scaling up. If a large-scale project is simply an expanded small-scale project, based on the arguments presented in this essay we believe that the answer has to be no. Taking a more hierarchical view, however, a large-scale project can be recast as the integrated sum of many small-scale projects, and this leads us to ask a rather different question: How can small-scale projects, attuned to the scale of the local political systems with which they must necessarily work, be efficiently and effectively replicated over and over again?

Efforts to conserve protected areas cannot succeed in the long run without the active participation of people living around the areas to be conserved. This participation has to be permanent, and we have argued that if it can be achieved at all, it has to be through a flexible and responsive political process: it is a matter of negotiation between those who would conserve and those who have to carry on their lives doing without, or with less of, whatever it is that is being conserved. It is critical that project designers and managers negotiate their plans with local communities and their leaders. It follows that the buffer zone for development activities cannot be imposed simply by reference to economic or ecological criteria, but is itself a matter for political negotiation. On a more positive note, these negotiations can result in the development of a reinforcing relationship between a reserve and its neighbors, rather than the adversarial relationship that is often implied.

We believe that rather than focus attention purely on economic models of development, it is helpful to consider the process as encompassing three somewhat different activities car-

ried out in a shifting political landscape: striking a bargain, direct substitution, and anticipated substitution. All three kinds of activities should be incorporated in the design of a project from the outset. The success or failure of each must be measured in different ways, but all are undoubtedly critical to long-term success. The return on money and effort takes the longest to be realized in the case of anticipated substitution, but this does not reduce its importance. Conversely, the activities that result from the establishment of a relationship between the project and the local leadership, whatever form it takes, may seem less than immediately related to the conservation of biodiversity, but that, too, does not reduce their importance in achieving the long-term goals.

The Beza Mahafaly project began with the establishment of a relationship between the local population and a small group of outsiders. The primary contacts were with the established political leadership: district- and village-level elected officials. Without their support, there would have been no project. With their support alone, it is unlikely that conservation efforts around Beza Mahafaly can succeed in the long run.

ACKNOWLEDGMENTS

The Beza Mahafaly Project, which provided the starting point for this chapter, is a product of the ideas and efforts of many people, in particular Bob Sussman, Guy Ramanantsoa, Gilbert Ravelojaona, Pothin Rakotomanga, Andrianasolo Ranaivoson, Joseph Andriamampianina, and Sheila O'Connor. The project has received steady financial support from the World Wildlife Fund, particularly through the good offices of Tom Lovejoy, Russ Mittermeier, Rod Mast, and Barbara Wyckoff-Baird. Over the past five years, we have benefited from discussions with Art Ortenberg and Liz Claiborne, who have stopped us in our tracks more than once. Our thinking has been much influenced by working and talking with all these people. The views and conclusions presented here are our own, however, not those of the project as such, and they are not necessarily shared by all who have been involved. We thank the following people for comments on earlier versions of this chapter: Skip Barbour, Marc Edelman, Angelique Haugerud, Tom Lovejoy, Jim Murtaugh,

Hilary Simons Morland, Bill Weber, Mike Wells, and Barbara Wyckoff-Baird.

REFERENCES

Barbour, R. 1990. Technical report to the World Wildlife Fund.

Carloni, A. S. 1987. Women in development: A.I.D.'s experience, 1973–1985. Volume 1: Synthesis paper. Agency for International Development, Washington. D.C.

Deshmukh, I. 1989. On the limited role of biologists in biological conservation. *Conservation Biology* 3:321.

Green, G., and R. W. Sussman. 1990. Deforestation history of the eastern rain forests of Madagascar from satellite images. *Science* 248:212–215.

Hough, J. L., and M. N. Sherpa. 1989. Bottom up versus basic needs: Integrating conservation and development in the Annapurna and Michiru mountain conservation areas of Nepal and Malawi. *Ambio* 18:434–441.

Lewis, J. P., ed. 1988. Strengthening the poor: What have we learned? U.S.-Third World Policy Perspectives No. 10. Overseas Development Council, Washington, D.C.

MacKinnon, J., K. MacKinnon, G. Child, and J. Thorsell. 1986. *Managing Protected Areas in the Tropics*. IUCN, Gland, Switzerland.

McNeely, J. A. 1989. Protected areas and human ecology: How national parks can contribute to sustaining societies of the twenty-first century. Pages 150–157 in D. Western and M. C. Pearl., eds. *Conservation for the Twenty-First Century*. Oxford University Press, New York.

McNeely, J. A., K. R. Miller, W. V. Reid, R. A. Mittermeier, and T. B. Werner. 1990. *Conserving the World's Biological Diversity*. IUCN, Gland, Switzerland; WRI, CI, WWF-US, and the World Bank, Washington, D.C.

Oldfield, S. 1988. Buffer zone management in tropical moist forests. *IUCN Tropical Forest Paper* 5:1–49.

Sussman, R. W., G. Green, and L. K. Sussman. 1994. Satellite imagery, human ecology, anthropology, and deforestation in Madagascar. *Human Ecology* 22:333–354.

Weber, A. W. 1987. Socioecologic factors in the conservation of Afromontane forest reserves. Pages 205–229 in C. W. Marsh and R. A. Mittermeier, eds. *Primate Conservation in the Tropical Rain Forest*. A. R. Liss, New York.

Wells, M., and K. Brandon (with L. Hannah). 1992. *People and Parks: Linking Protected Area Management with Local Communities*. World Bank, WWF, and USAID, Washington, D.C.

World Bank. 1980. Madagascar: Recent economic developments and future prospects. World Bank, Washington, D.C.

Wright, M. 1989. Wildlands and human needs in Africa: The beauty of "imperfection." Paper presented at the 88th Meeting of the American Anthropological Association.

**VI**
**Conclusion**

# CHAPTER 32

# Trends and Challenges
# in Applied Research and Conservation

## A Practitioner's View from the African Rain Forest

William Weber and Amy Vedder

Applied interdisciplinary research holds great potential for enabling us to better understand the many knotty problems confronting the African rain forest. This information, in turn, should improve the design and implementation of projects that seek to integrate the interests of both conservation and development. Yet theory is far ahead of practice on both fronts, and it is not at all clear that we know how to close the gap.

This chapter traces the convergent evolution of trends in research and conservation over the past several decades: from scattered studies within a few narrow disciplines to a broader mix of more relevant subjects to a recent rise in multidisciplinary studies; from a colonial focus on parks and protection toward integrated multiple use to a more recent move toward community-based conservation; from a limited connection between science and conservation to a more explicit partnership that links research to the management needs of specific projects. It is not an academic review, but a practitioner's perspective based on twenty-five years of personal and professional experience.

## Roots in the Forest:
## Research and Conservation Before 1980

When we first arrived in Africa, in 1973, it was a savanna-centric world. The giants of research and conservation lived and worked on the east African plains, where their research focused almost entirely on the region's spectacular wildlife. Social and behavioral studies of these animals were most common, although some good ecological work was also being done. Conservation goals were straightforward: to create national parks and forcefully protect them against the incursions and depredations of local populations. The parks movement, launched in the United States in 1872, was transplanted to Africa with great enthusiasm by the colonial powers. By 1972, there were more than 250 national parks on African soil, more than 90% of them in savanna and semi-arid regions. A handful of parks provided minimal coverage of the rain forest realm.

The quest to date and describe our human origins, led by Louis and Mary Leakey, dominated social science in eastern Africa at that time. Living human cultures received much less attention, virtually none of it connected with conservation. Conservation was seen as the domain of the biologists, and most practitioners assumed the dual role of researcher and conservationist as a natural calling. Arguments among the region's *prima bwanas,* or leading men, over conservation practices took on the urgency of great ideological debates, and researchers lined up behind the competing paradigms in sometimes epic struggles. For nearly a decade, the battle raged over how to manage elephants as they adapted to life in Kenya's unnaturally small parks and reserves. On one side, the "dung beetle ecologists" favored natural selection and competition, while on the other side, the "shooters" called for drastic culling campaigns to reduce elephant numbers before they destroyed their own habitat. Thousands of elephants starved to death as the debate went on.

Life in the African forest zone was much lonelier, and generally less exciting, in the early 1970s. The colonial legacy was much less developed for both research and conservation. Scientific expeditions had been carried out in many areas, primarily for the purpose of botanical and zoological collections to be sent back to Brussels, Paris, and London. Ethnographic studies were also supported to provide descriptions of the diverse peoples under colonial rule. Information from the biological surveys was used to identify important areas for national parks in the Belgian Congo, yet almost all these reserves were in savanna and montane regions. Each of the colonial powers set aside forest reserves in rugged highland areas, but these were established for water catchment purposes, not to protect wildlife. At the same time, comprehensive bans on many forms of hunting in all forest areas caused potentially severe hardship for local populations. More commonly, they engendered disrespect and disregard for wildlife protection laws.

Few individuals committed themselves to long-term forest-based research during the colonial period before 1960, and even fewer linked their research to conservation. One early and notable exception was the naturalist Carl Akeley. Under contract to the American Museum of Natural History, Akeley went to the Virunga chain of volcanoes that separated Belgian and British holdings in east-central Africa to collect specimens of the recently discovered mountain gorilla. After killing four of these creatures in 1921, he dedicated himself to creating a reserve for the gorillas and initiating a long-term study of their behavior. Akeley succeeded in his first objective when Belgium agreed to establish Africa's first national park, the Albert Park, to protect the rare gorillas. His second goal was left unfulfilled, however, when he died of malaria soon after establishing a base camp in the Virungas. Although his wife, Mary, continued some of his work, it would be more than thirty years before George Schaller would complete the first definitive study of the mountain gorilla in the wild, in 1960.

It is worth noting that Akeley worked on the highland fringe of the African rain forest, from a base camp situated 3,000 m above sea level. In the late 1960s, other researchers would be attracted to these temperate highlands. Dian Fossey began her own studies of the mountain gorilla at the same Kabara camp used by Akeley and Schaller, then moved across the border in 1968 to establish the Karisoke Research Centre in Rwanda. Jane Goodall's pioneering studies of chimpanzees were conducted at Gombe Stream, on the mountain slopes overlooking Lake Tanganyika's eastern shore. Farther north in Uganda, Tom Struhsaker found the highland conditions of the Kibale Forest conducive to what would be a twenty-year study of the red colobus monkey. The three researchers had much in common: a focus on primates, with a particular interest in their behavior and social organization; a montane forest environment; and a conservation context in which poorly protected, very small reserves were completely surrounded by areas of intensive human settle-ment. This third condition resulted in escalat-ing conflicts with local populations, especially in the Virunga and Kibale Forests. Yet the research agendas of the 1960s and 1970s did not include any attention to local human concerns, and conservation was locked into the model of the "guarded fortress." The nascent industry of nature films and magazines of the early 1970s brought these scientists, their charismatic focal species, and their rain forest sites to the atten-tion of a global audience—and portrayed their single-handed struggles with local people as the essence of African conservation.

One notable exception to the highland model for forest research was the work of Colin Turnbull in the vast lowland Ituri Forest, of what was then the Belgian Congo. His research in the late 1950s and early 1960s was even more remarkable for its focus on the human species. In *The Forest People,* Turnbull described the hunter-gatherer existence of the Mbuti people as a complex, ancient tradition under threat from encroaching development and modern civilization. Turnbull may have overstated the traditional purity of the Mbuti, but he brought a singular way of life to the world's attention. He also highlighted the importance of the for-est itself in enabling the Mbuti to survive as a coherent culture. What he did not do was start a project to protect the Mbuti or their forest, as biologists working in the same Ituri Forest environment would later do in the 1980s. There remains a sharp distinction to this day between those who study animals and those who study people. Whereas social scientists often protest what is happening to their subjects yet remain unclear as to what could be done, biologists are quick to call for organized action in defense of their target species and habitats.

The 1970s saw a few more anthropologists and biologists scattered across the lowland African rain forest. Yet the worlds of research and conservation at that time remained con-strained and easily characterized. Research was more academic than applied, more biolog-ical than social, more behavioral than ecologi-cal, narrowly disciplinary, and generally unre-lated to conservation. Conservation efforts were concentrated in the montane forests, where traditional models of protection were under assault from the combined forces of poverty, expanding populations, and rising expectations in post-Independence Africa. There were no models for conservation in the lowland rain forest.

After trying to understand this world for two years as Peace Corps volunteers in eastern DR Congo, we decided to pursue a different approach to research and conservation. Our proposal, funded in 1977, sought to build on the work of Schaller and Fossey to understand and protect mountain gorillas in their Virunga forest refuge. Our research, however, was to be an integrated package of applied ecology and social science, intended to determine the basic needs of both the gorillas and local people. This research was also linked explicitly to conserva-tion needs, in that findings were to be applied to improve management. Finally, we were com-mitted to assuring these linkages and their application through direct personal involve-ment and training of others over a ten-year period between 1978 and 1988, and continued indirect involvement through an unbroken series of projects to this day.

The results of the Mountain Gorilla Project and several related initiatives in Rwanda and the surrounding region have been widely reported. They are commonly cited as suc-cesses, although their shortcomings are equally informative. Many see them as precursors to today's reigning models in integrated conser-vation and development, ecotourism, commu-nity-based conservation, and adaptive man-agement. To us, the process—an approach that called for interdisciplinary research, synthesis of information, application of findings to management, monitoring of effects, feedback, and adaptation—was inseparable from the results.

## The Paradigm Shift of the 1980s

Research and conservation in the African rain forest changed dramatically in the 1980s. In 1981, the *World Conservation Strategy* was published under the joint auspices of the International Union for the Conservation of Nature, the World Wildlife Fund, and the United Nations Development Program. Despite its grandiose title and overly simplistic prescriptions, the *Strategy* offered an appealing and timely synthesis that linked the worlds of conservation and development, and which continues to shape the debate over their relationship into the twenty-first century. The twin concepts around which all subsequent thought and action have evolved are: first, that economic development cannot be sustained in a state of environmental degradation; and second, that conservation cannot be sustained in a context of human impoverishment. They are attractive, if as yet unproven, corollaries.

The empirical foundation for the *World Conservation Strategy* came from a growing number of individuals in the conservation and development communities who saw their efforts stymied by recurrent problems beyond the traditional scope of their fields. In east Africa, the luster had dimmed from national parks that failed to generate tangible benefits for local populations. In the west African Sahel, agricultural development efforts were undermined by both drought and deforestation. From around the world, experience showed that fragile montane environments required special attention. Relatively little input came from rain forest areas, yet they would become prime testing grounds for emerging models of integrated conservation and development.

The 1980s saw an explosion in global attention to rain forests. An outpouring of scientific and popular books, articles, and films appeared

to inform us that the great biological diversity of the rain forest was matched only by our ignorance of its untold and unstudied millions of species. The potential loss of this newly discovered resource, whether in ethical, scientific, or economic terms, was dramatically highlighted by the first satellite images documenting rapid rates of deforestation in the Brazilian Amazon and presumed to be occurring elsewhere. Concern, too, was spreading for the fate of rain forest peoples, especially those deemed to be living traditional lifestyles. Anthropologists and indigenous leaders spoke eloquently of the essential lifeline between the health of the forest and the health of the forest people. The combination of exotic species and cultures proved irresistible, and the popular appeal was heightened by the reflected glow of celebrities attracted to the cause.

Out in the field, progress came more slowly. The first step was for researchers to come down from the mountains and get to know the lowland rain forest. Many of the contributors to this volume took that step at some point during the 1980s. The nature of the science being conducted also changed in some important ways. On the biological side, research on animal behavior and social organization was joined and soon surpassed by studies of wildlife ecology and the rain forest ecosystem. Many of these ecologists saw an explicit link between their activities and applied conservation. At the same time, the African forest's ethnic diversity attracted a growing number of anthropologists. Although they tended to concentrate on hunter-gatherers and other traditional societies, some social scientists were attracted to the dynamic conflicts faced by cultures, economies, and political structures in transition. A few explored subjects with an explicit link to conservation, although this was less common than among biologists.

The number of researchers was never great in any of these fields, and although they worked in isolation, most shared their African rain forest interests, passions, and circumstances with others. Many exchanged correspondence, and some exchanged site visits. Conferences offered an opportunity for discussion with a larger pool of colleagues, though generally within a limited set of disciplines. Universities provided the best forum for interdisciplinary discussion, and many started interest groups devoted to rain forests or the tropics. These attracted wide interest, but too few of the participants at that time had significant applied experience in the field. Not all field people were interested in applied or interdisciplinary work, either. Yet they were part of a core group that was rapidly expanding our understanding of the African forest, its wildlife, and its people throughout the 1980s. Indigenous African researchers stood out for their rarity within this group, both in the field and in academia.

Greater knowledge of native flora, fauna, and cultures did not directly result in global support to remedy the problems confronting the African forest. Instead, it was economic arguments of the potential utility of converting rain forest biota to pharmaceuticals, along with the growing specter of global warming due to deforestation, that finally forced the wealthy nations of the North to pay attention. By the end of the 1980s, private foundations and public agencies alike began to pour an unprecedented amount of money into the tropical forests of the world. Central and South America received most of the funding, but Africa and Asia would still receive more than anyone had dreamed possible even a few years before. The Roaring 1990s of rain forest funding had begun.

## Megabucks and Megadiversity

The promise of the *World Conservation Strategy* was its hopeful linkage between conservation and development. A decade of research in the 1980s seemed to indicate that the African rain forest, with its great biological wealth and profound human poverty, might be a good place to experiment with new models for integrated conservation and development. Several such experiments were initiated at the end of the decade, focusing on the most threatened sites: the relic coastal forests of Madagascar and the montane forests of east-central Africa. Ecotourism and limited subsistence uses were their prime development tools, while community dialogue and education promoted the cause of conservation. With moderate funding, these early initiatives achieved some significant results, often stemming the tide of forest conversion while generating discrete economic benefits to local populations. This modest success, however, served to justify a quantum leap in funding and a dramatic change in the nature of those projects funded.

The first changes came in second-generation projects in Madagascar and the Afromontane forest zone funded by the U.S. Agency for International Development, the European Community, and the World Bank. Generically referred to as Integrated Conservation and Development Projects (ICDPs), these enterprises saw their budgets increase five- to tenfold, often at the request of the donor agency, and multi-year, multi-million-dollar projects were common. These sums were not at all unusual for traditional development initiatives, but they were unheard of in conservation. The increased funding was intended to address the greater complexity of issues covered by an ICDP, especially those relating to human development and subsistence activities in and around the target forests—and to pay for the mostly Western experts brought in to design and manage these activities.

Yet while the complexity of projects increased almost exponentially, the related increase in understanding was linear, at best. Most of the imported advisers had developed their expertise in quite different contexts, either from forestry in the semi-arid Sahel or from more traditional agricultural settings. Very few had direct experience in rain forest environments, and on-the-job learning curves were steep. Many fell off the curves, so to speak. New knowledge from research was hard to come by, too. Most donor agencies viewed the "R" word as either irrelevant or an unaffordable luxury under conditions that called for action. Many conservationists agreed, however reluctantly, and research became a secondary priority at best.

At the end of the 1990s, these highland and island projects had recorded many successes. Lines were drawn and often strengthened around many of the relatively small reserves concerned, and a variety of development and negotiation activities seemed to have neutralized, if not lessened, the surrounding threats from encroachment. Yet we have extremely little empirical evidence—especially in published form—to tell us what did and didn't work, and for what reasons.

If the situation in the highlands and in Madagascar is uncertain, it is in many ways far more problematic in the African lowland rain forest. The threats are less immediate, or at least less readily evident, across much of a region that covers more than 2 million km²— more than half of which remains in a contiguous block within the Congo Basin. This extensive forest cover, however, tends to mask the myriad problems below. The lowland rain forest is the most diverse ecosystem on earth. While montane and island systems tend to have higher percentages of endemic species, the lowland forest has a vastly greater number of species, which interact in ways that make the overall ecosystem extremely complex. The human communities of the rain forest also represent a diverse range of cultures; economies that combine hunting, gathering, and farming; and connectivity with the modern world. Furthermore, these people and their activities are found dispersed throughout the forest, unlike

the more sedentary farming populations surrounding relic forests to the east and west of the lowland core.

As the decade of the 1990s began, our understanding of lowland rain forest systems lagged far behind their inherent complexity. This gap was especially apparent in the biological and ecological spheres. This may be because there were more anthropologists in the forest, or more likely because human subjects are better able to articulate their needs, values, beliefs, and actions. The small number of lowland forest ecologists also had a belated start on their complicated subject. Thus, although the human culture of the Mbuti might be wonderfully complex, we knew far more about how the Mbuti people lived in the Congo's Ituri Forest, at that time, than we did about the okapi and elephants that shared the same habitat.

This combination of complexity and ignorance did not stop international agencies and a growing number of private foundations from spending large amounts of money in the lowland African rain forest beginning in the early 1990s. ICDPs were deemed ready for export, and individual project budgets were at least as large as in the highlands, and often larger. Some projects had annual expenditures of more than U.S.$1 million. Overall, it is estimated that more than $70 million was earmarked for conservation across the central African forest region during the 1990s. This is far less than what was spent on development assistance, and insignificant in comparison with military expenditures for the same countries over a similar period. Still, it supported a small army of outside consultants—and created a windfall for formerly impoverished field biologists and social scientists. Although local communities did not receive as much total money as expatriate advisers, they were nonetheless awash in unprecedented levels of project-generated income and employment.

It is certainly true that this increased spending helped to create several new parks and limited-use zones across a region that had few such reserves. Yet problems have arisen that must be addressed. Many of these have to do with the initial project identification process. Target sites have been largely selected by foreign activists and donors, with little input to this critical activity from local communities or political structures. Selection criteria have also been strongly influenced by geopolitical considerations that favor certain countries and leave others begging for attention, regardless of conservation values. High biological diversity is generally weighted more heavily than other factors, yet diversity ratings are not entirely objective (many areas remain unsurveyed), and other sets of issues may be more important to the ICDP model. Finally, the mere presence of a funding spotlight often attracts other donors to the same site, like moths to a flame, leaving other would-be projects in the dark.

A more fundamental concern has been the continued lack of interest in, and support for, applied research and monitoring on the part of the international donor community. Biologists and social scientists have been employed in high numbers to staff ICDPs, and many of these individuals have significant training and prior research experience. This experience often comes from the particular project area. But once a project begins, most spend far more time and energy as advocates for either wildlife or local human communities, or as desk-bound administrators, than as practicing scientists in their respective fields. The result is that there is usually little baseline information about conditions before a given project is started, and even less information about how key variables—human welfare, animal numbers, ecosystem health—might change in response to project activities. This severely limits the potential for objective feedback and adaptive management, which is a stated goal of many of these projects.

## The Future

The past few decades have seen considerable progress in research and conservation in the African forest. Scientific endeavor has moved from the narrowly academic and compartmentalized to much more applied research in both the biological and social sciences. More multidisciplinary research is being conducted, and there is a stronger linkage between research and conservation. New models for conservation, from large-scale ICDPs to smaller community-based initiatives, provide more flexibility than traditional parks in addressing the needs of local human populations. Most striking of all is the greatly increased attention to the African rain forest itself, so long ignored outside of popular misrepresentations in word and image. Yet even greater progress is required today if these gains are to be consolidated and provide the foundation for a truly integrated, applied, and effective conservation science that is relevant to the future of the forest. Some promising lines of pursuit, and continued constraints and challenges, are outlined below.

If the first imperative in war is good intelligence—about comparative strengths, available resources, conditions in the field, and so on—then the battle to save the African rain forest and all its human and biological diversity also requires good intelligence. Conservation science must provide this information. Because human agency is the force that has the greatest impact on the state of the forest, we need especially good information from the social sciences, with particular attention to the following areas of study.

A fundamental problem for community-based conservation lies in the difference in scale between human social and political systems and natural ecosystems. This divergence is greatest in the lowland forest, where traditional community units are quite small and the forest is vast. Most rain forest reserves cover thousands of square kilometers; most communities cover no more than one hundred. In DR Congo's Okapi Wildlife Refuge, at least thirty distinct human communities representing several different cultures share a 5,000 km² reserve with the wildlife of the Ituri Forest. The multiple-use management plan for this reserve must eventually take into account the varied needs, interests, and attitudes of all of these groups. In Uganda's Kibale Forest, the views of one village that benefits as a gateway for tourists must be balanced with perspectives from other less favored communities around the forest. Expanded research on the subject of how such diverse and dispersed groups can cooperate, identify common goals, share costs and benefits of conservation, and reduce conflicts could provide information that is essential to sound management.

Conservationists are in the business of thinking about the long term. As we design more flexible reserves with allowances for traditional subsistence uses, we need to think about the terms "traditional" and "subsistence." If the local hunting tradition uses nets, are bows forbidden? Firearms? If a duiker is sold within a hunter's village, is that subsistence? What if it is sold at a market 20 km away, or smoked and transported by truck to a distant city? Who will make these decisions? If these societies are not to be locked up as living museum pieces, we urgently need to identify and refine mechanisms for change and adaptation that will last long beyond the average project cycle of 5–10 years.

The most widespread models for conservation in the African rain forest call for the participation of local populations in reserve design and management. But who speaks for the people? New methods that show some promise at detecting and interpreting this voice include role playing, community mapping exercises, and several forms of participatory assessment. With regard to decision making, questions remain as to the role of traditional leaders versus modern politicians. In many instances, non-governmental organizations are emerging that eclipse the power of both traditional and modern leaders, yet owe at least part of their standing to stakeholders from outside the country.

At the heart of the integrated conservation and development paradigm is a fundamental interest in biological conservation. If economic development is the primary interest, then efforts should concentrate on already settled and degraded land. Only an overriding concern for wildlife and ecosystem integrity should bring development to the world's most remote human communities and wildlands under this model.

A recent and welcome change in African rain forest research is the recognized need for a landscape perspective. The heterogeneity of the forest, the long-distance movements of animals, and the variety of human impacts all force the conservation biologist to look beyond core protected areas. This raises issues of connectivity between key ecosystems, and the need for greater sensitivity to the role that human–modified habitats must play in the overall forest landscape.

With multiple use the dominant model for management, researchers are beginning to study the long-term effects of hunting, logging, and mining on wildlife populations and ecosystems. This belated attention has already brought new perspectives to bear on management and caused a rethinking of such issues as the origins of species diversity and abundance. Selective logging appears to have lower impact than formerly thought, yet we know far too little about the long-term ecological sustainability of most extraction systems. Sustainable forms of hunting are even more elusive.

While management issues receive heightened attention, our overriding ignorance of the forest ecosystem also requires a renewed commitment to research in basic biology. Although a growing number of phenology and seed dispersal studies have increased our overall understanding of forest community ecology, research on other community interactions (such as predation) and on single-species ecology is still sorely needed. Donors, too, need to acquire the sophistication and patience to justify expanded, long-term support for this work.

Monitoring is essential to provide feedback for any program of adaptive management. Considerable thought and discussion are now aimed at determining which taxa, if any, might best serve as indicators of overall diversity, or as keystone, umbrella, or landscape species—and with what their status might be practically correlated. This question, and the related need for appropriate, efficient, and replicable field-monitoring methods, will be central to the ability of biological research to contribute to applied conservation and management.

Research in the African rain forest is much more multidisciplinary than ever before. This is especially true during the design phase of large projects, when the internal guidelines of agency funders require attention to multiple perspectives and interests. Research in the course of project implementation, however, is generally much less comprehensive and balanced, if it occurs at all. The solution to this problem requires that funders first see research as a source of valuable feedback on project progress and problems, as a tool for adaptive management. Once this need for information is acknowledged, the next step toward multidisciplinary work should be simpler, given that integrated projects require attention to both biological and socioeconomic issues.

Far more challenging is the leap from multidisciplinary to interdisciplinary research. In fact, interdisciplinary study and synthesis are extremely rare. The teams of specialists hired to design most ICDPs tend to produce multidisciplinary assessments, with information

compartmentalized into the different specialty areas. Synthesis and integration of this information, however, require a team leader to be able to take disparate studies or reports and integrate them in such a way as to form a new compound assessment that cannot be broken down into its component elements. In our experience, this ability is exceedingly rare. It remains to be seen whether the growing number of individuals now graduating from "interdisciplinary" university programs will prove more adept at this critical skill.

Fewer than twenty Africans with doctorates are actively working as conservation scientists across the Congo Basin. There are many reasons for this: colonial neglect of education in general, a lack of African institutions of higher learning, a dearth of financing for education abroad, and cultural biases that view field work as manual labor while equating success with suits and government offices. Whatever the causes, the situation must change. Local Africans play critical roles in almost every project as parabiologists, as education specialists, and in many other essential functions. Yet there is a great need for Africans—both male and female—to receive advanced academic training, to assume higher-level research positions in the biological and social sciences, and to use this training to help understand and resolve conservation problems. As this pool expands, there is a further need for the most qualified individuals to move into key leadership positions.

The intent of this chapter has been to provide an overview of trends in research and conservation during a fascinating and dynamic period of convergent change in both fields. More detailed and technical treatments of the issues raised here can and should be found by interested readers. Many large gaps remain in the published record, however. This is especially true for the applied-conservation side of the story. If there is a final message for the reader, it is that much more of this story should be published or otherwise disseminated, especially by those with relevant practical experience in the field. Too much of the public record is dominated by narrow academic tracts, untested theories, and sanitized, uncritical success stories intended to raise more funds than awareness.

The disconnection between advocacy and research, between theory and practice, is disturbing. It strongly implies that we have flown the model of integrated conservation and development through the past decade on autopilot, unable to detect the shifting winds and currents that might guide us toward alternative destinations. We are certainly engaged in a massive experiment with very high stakes and too little knowledge and control. There will be a price for failure, to be paid by local people, wildlife, and the forest itself. Their loss will be compounded if we do not learn from our experience. A functional science of conservation requires a more complete and accessible chronicle of successes and failures if it is to fulfill its potential and play a constructive role in the future of the African forest.

# Epilogue

## Conflict and Conservation in the African Rain Forest

Amy Vedder, Lisa Naughton-Treves, Andrew Plumptre, Leonard Mubalama,
Eugène Rutagarama, and William Weber

*We walked up the hill toward the church—a modest brick building with a bare metal cross
above the door. Arched windows helped give it more the air of a church, but almost all the
stained glass windows were broken. As we moved closer, a gaping hole appeared in the
back wall. The local curator led us inside and we carefully stepped over cloths draped on the
ground. As our eyes adjusted to the reduced light, we could see cloths everywhere—and
everywhere there were bones and skulls. These were the clothes of those who were
massacred and this was where they died. The church floor was piled deep with kangas, the
colored cloths that most Rwandan women wear as skirts. Shirts, trousers, and dresses were
still draped over what was left of their bodies. Kitchen pots and water gourds were mingled
with the human remains, along with red beans spilling from a sack, walking sticks, a
necklace, a rosary. The pile was so high we had to step across the benches that served as
pews. A single skull had been set on the altar, next to an open bible. Outside the church,
600 more skulls were stacked in tight rows, four to five deep, on a wattle stand which ran
the length of the annex. Skulls with jagged holes and long gashes; one with a knife still
stuck in the bone. Many children's skulls.... How can a nation, a people, ever recover from
such horror?—Amy Vedder, Ntarama, Rwanda, January 1999*

Over a hundred-day period in 1994, an estimated 800,000 people were slaughtered in
Rwanda. This killing episode, orchestrated by
the Hutu government against the minority
Tutsi, was the most concentrated in the history
of the world. Once known as the Switzerland of
Africa, Rwanda has now joined Nazi Germany
and Cambodia as one of history's most abhor-
rent case studies in genocide. Seven years later,
while an uneasy peace prevails within its borders, the aftershocks of the Rwandan cataclysm
continue to spread across the Congo Basin,

enveloping more and more nations in seemingly endless conflict at the heart of the African
continent.

War is no stranger to Africa. The armed
struggle for independence has merely given
way to more fratricidal combat. National borders that made sense only to remote colonial
powers now buckle under new stresses. At the
dawn of the twenty-first century, nearly a third
of sub-Saharan Africa's forty-two countries
are embroiled in international or civil war.
More than 1.7 million people have died in DR

Congo alone in the past two years. Millions of refugees have fled from their homelands and surged across porous borders into neighboring countries. In the final decade of the twentieth century, more than $U.S.400 billion were spent on warfare across the region.

Most of Africa's wars are internal, fought between a national army and guerrillas representing different regions or ethnic groups. However, many rebel groups—from Liberia and Sierra Leone to Congo and Uganda—have used bases in neighboring countries as staging areas for cross-border attacks and retreats. In the late 1990s, Rwanda and Uganda responded to such threats by sending troops to attack insurgents across their common border with the Democratic Republic of Congo. When this proved ineffective, they raised the ante and sought to overthrow DR Congo's legitimate, if unelected, government. Angola and Zimbabwe then sent troops to support the established Congolese government, for a combination of strategic and economic reasons. As this immense country is carved up by war, it has a destabilizing influence on the entire region. Locally, crops are burned, villages looted, roads mined, and civilians terrorized. Refugees flee in waves of tens of thousands. The costs in terms of human suffering are unfathomable, and the economic costs for many of the world's poorest nations are staggering. Politically, this conflict could ignite Africa's first continental war.

While we grieve the immense human costs of these wars, we also lament the serious environmental consequences of fighting. Soldiers, rebels, and refugees alike use wildlife and forests for sustenance and shelter. The social and environmental costs of war are amplified by contemporary ecological and socioeconomic conditions as well as technological change. Deforestation and forest fragmentation over recent decades have isolated wildlife populations, increasing the risk of species

extinction during war and its aftermath. The proliferation of land mines and vastly more destructive light weaponry has increased human death and suffering, and has made wildlife more vulnerable to poaching. Rural populations displaced by war are now more likely to turn to the bushmeat, fuelwood, or charcoal trade for economic sustenance. Finally, the unprecedented number of refugees has overwhelmed the capacity of international relief agencies to administer assistance, has exacerbated ethnic tensions, and has concentrated ecological damage in some of the region's most sensitive areas.

This epilogue is based on firsthand experience with efforts to sustain rain forest conservation during and after times of violent conflict. It is drawn primarily from the authors' experiences in Rwanda and DR Congo.

## A View from Rwanda

It has now been eleven years since Rwanda's civil war began, in late 1990, and seven years since its catastrophic genocide and overthrow of the former government. The ensuing path toward social, political, and economic recovery has been both slow and difficult. The conservation sector within Rwanda was not immune from the destruction, nor from the subsequent challenge of reclamation. Direct damage was immediate: the Tutsi-led Rwanda Patriotic Front (RPF) launched its effort to overthrow the Rwandan government with attacks from Uganda through the Volcanoes and Akagera National Parks. From 1991 through 1994, the two parks were intermittent battlegrounds where the combatants set fires, laid mines, stored weapons, and killed wildlife for food. Local populations also sought sanctuary and food within the reserves, especially the rain forest of the Volcanoes National Park. Spurred by the onset of the genocide in 1994, the RPF accelerated its attacks and quickly occupied more than half the country. The rout of the Rwandan army was slowed only by the arrival of French troops who allowed the soldiers and an influx of refugees to settle in, and increase pressure on, the Nyungwe Forest Reserve. There they engaged in widespread hunting, woodcutting, looting, and harassment of local populations before moving across the border into eastern DR Congo.

For all the horrors of this period in Rwanda, it can be argued that the environmental and political impacts were even more profound when this displaced refugee army settled in DR Congo. Refugee camps, supported by well-funded international relief agencies, were established on the edge of several national parks. Neighboring forests were depleted, wildlife populations were decimated, and legitimate local economies were overwhelmed by illicit trade in relief goods. Furthermore, the camps allowed the defeated Rwandan army to regroup, recruit new members, intimidate relief workers, politicize other refugees, and stage attacks back into Rwanda. Ultimately, the new Rwandan government sent its troops into DR Congo to close down the camps and again rout their adversaries. In the process, they triggered the overthrow of the corrupt Mobutu regime and launched a new wave of civil and regional war that continues to rage across the nation.

Back in Rwanda, some of the casualties of the fighting can be calculated. The natural savanna of the Akagera park has been largely converted to pasture for cattle, leaving little habitat for wildlife. The 400 km² Gishwati Forest has been completely cleared for farming. The Nyungwe and Virunga Forests, where conservation efforts were most active, have remained largely intact. Most remarkably, the flagship mountain gorilla population appears to have maintained its numbers somewhere around the still-precarious level of three hundred individuals. Humans who worked with the gorillas have not been so fortunate. Twelve Rwandan staff members from the Karisoke gorilla research station were killed, as were ten park guards; another thirty-five guards from

the Volcanoes park are presumed to have fled to DR Congo, where some joined the opposition; still others are in prison awaiting trial on charges of participation in the genocide. In Nyungwe, the chief warden was murdered soon after the formal fighting ended, and most of the senior staff fled to DR Congo. Several expatriate staff took great risks to protect Tutsi, but all eventually fled as the killing escalated, and only a few have returned. Rwanda's only doctoral-level field biologist has remained in England since the hostilities began. As a result, all field conservation and research operations lost key personnel, and the prospects for conservation of the country's extraordinary wildlife were considered very dim.

Yet despite the cataclysm and horror of war and the several years of chronic terrorism that followed, conservation and research did indeed continue. Rwandan staff who remained on the ground resumed work in the forest whenever security allowed, as well as during periods when their safety was very much at risk. In the Nyungwe Forest, these people were local junior staff, who organized their own work programs adapted to changing conservation needs. Research and conservation personnel joined together to reorient priority activities and to provide mutual support. As a result, over the past seven years some conservation efforts took place during all but one month in the Nyungwe Forest, and during all but thirteen months in Volcanoes National Park.

During these years non-governmental partners maintained, and often increased, their support for in-country staff and operations. Research and conservation funds were blended, made more flexible, and allocated to the most pressing needs. This support came from conservation and research NGOs from the United States and Europe. In stark contrast, bilateral and multilateral financial support for conservation from foreign governments and agencies was completely withdrawn.

## A View from Eastern Democratic Republic of Congo

By the early 1990s, the giant nation of Zaire was reeling from decades of institutional corruption, instability, and the general degradation of the state machinery, including the ability to manage the country's spectacular national parks. When the tidal wave of fleeing soldiers and refugees swept across its borders, it effectively ceded its sovereignty over the country's eastern frontier to the international relief agencies that moved in to manage, and mismanage, the situation. When that relief effort failed, the eastern half of the renamed DR Congo was plunged into a state of civil war, which continues to this day.

Conservation activities have been severely constrained and compromised by the conflicts in eastern DR Congo. In 1994, more than 1.2 million refugees lived in or next to the Virunga and Kahuzi-Biega National Parks. The presence of refugees, rebels, and uncontrolled Zairian government troops led to unprecedented poaching, forest clearing, and mining in the parks. Half the hippos in the Virunga National Park were killed, as were at least 150 of the park's 600 elephants. Equally devastating was the loss of wildlife habitat. The Virunga park lost nearly 115 km² of forest because of firewood collection by refugees and soldiers who sold the wood for profit in the Goma relief camp. Fully 90% of the refugees around Goma camped within walking distance of Virunga National Park, and up to 40,000 people entered the park every day. Between 410 and 770 tons of forest products were extracted from the park daily. The Mugunga camp, also on the edge of Virunga, had a similar devastating impact. In this case, Rwandan military personnel and Hutu militias sold wood and wildlife products from the park to refugees and traders.

The closure of the refugee camps in 1996 took great pressure off the parks. The "democratization" that followed the fall of Mobutu, however, was taken by many as a signal to enter and exploit reserves from which they had long been alienated and excluded. The result was another wave of woodcutting, clearing, gold mining, and hunting. The problem of hunting was exacerbated by the vastly increased number of weapons in circulation. Even traditional hunter-gatherers like the Mbuti began hunting with guns and selling bushmeat in northeastern DR Congo's Okapi Wildlife Reserve.

The core of Kahuzi-Biega National Park was invaded by Mayi-Mayi fighters, a group of Congolese dissidents partially supported by exiled Rwandan government forces. Kahuzi-Biega is home to approximately 70% of the eastern lowland gorilla population, which is endemic to eastern DR Congo. The Mayi-Mayi now control access to most of the park, where they are engaged in an unprecedented slaughter of wildlife. Outside observers report that most of the elephants and at least 25% of the gorillas have been killed in the highland sector of the park.

Throughout the troubles of the 1990s, local and international conservationists have struggled against high odds to maintain the integrity of DR Congo's parks and wildlife populations. Land mines are common along park borders, and anti-poaching patrols are subject to ambush if local people perceive guards to be overly aggressive. A uniform can be a death warrant for a guard mistakenly identified as a member of a rebel force or opposition ethnic group. In the Virunga park alone, which borders Rwanda's Volcanoes park, one assistant warden and thirty-five guards were killed between 1994 and 1999. Tensions between local people and park staff have led to looting, vandalism, arson, and the damage or destruction of reserve property.

Park protection is at an extremely low level. Guards are poorly equipped, morale is low, and many have resigned or abandoned their posts;

some have joined the various rebel groups competing for control in the region. Current government operating budgets for the parks are non-existent, and bilateral and multilateral funding has all been withdrawn. The limited funding that is available comes from a handful of international conservation organizations.

## Conclusions

Any war is unique, and the conservation lessons learned from times of conflict in Rwanda and DR Congo should not be overly generalized. However, a number of similarities have emerged from recent events in these two countries, and certain conclusions may be of broader value.

- The disruption of social and political organization leaves parks and protected areas at great risk. Laws and regulations break down, and national mandates for protection are likely to be disregarded. This vulnerability arises even in less traumatic transitions, including the process of "democratization."
- Parks and reserves, especially those along national borders, often become zones of military operations. Forested reserves, in particular, provide protection, shelter, and sustenance for local people and combatants alike. Some wildlife and forest resources are highly vulnerable to such exploitation.
- Dislocated populations and refugees may produce much greater impacts in neighboring host countries than at the immediate site of conflict. This was certainly true of Rwandan refugees in DR Congo, where relief agencies did not have the internal guidelines or staff capabilities to address ecological impacts of their operations.
- Reserve staff, especially those charged with security, may suffer enormously. In Rwanda and DR Congo, dozens were killed, most lost close family, and almost all were beaten, robbed, or threatened with

death. These individuals and their families have received no official compensation for their losses, and only modest support from non-governmental sources.

- International support from bilateral and multilateral government sources is likely to disappear during, and for long periods after, times of conflict. Non-governmental funding is more dependable and flexible, although amounts are generally much smaller. When bilateral and multilateral funding is renewed, conservation is given a low priority.
- Parks and reserves that depend on eco-tourism revenues may receive better protection during times of conflict. Eco-tourism, however, is highly sensitive to public perceptions of danger, and visitation, employment, and revenue levels may remain low long after a conflict has ended.
- Conservation in times of conflict is most effective at those sites where a long-term commitment had been demonstrated by a non-governmental, non-political organization. Such a commitment is successful to the extent that it builds a sense of purpose among staff, involves participation and responsibilities by local actors, and justifies confidence in continued postwar support. Sites without active community components or significant local staff fared much more poorly in both Rwanda and DR Congo.
- Current training by conservation organizations focuses on national leaders, and yet in Rwanda and DR Congo it was the junior staff that in all cases kept the conservation activities going as far as possible. This suggests a need for improved training, expanded responsibilities, and new career options for junior staff.
- Those staff who remained in Rwanda and DR Congo during the conflicts were motivated to continue working by their impression of commitment to a site from the con-

servation organization supporting them. That is, they believed that the NGOs would come back and that they would perhaps receive some compensation in the future. This was partly influenced by their perception of past treatment by their employers before the war.

- Quite a few staff in Rwanda and DR Congo continued working because they felt that they were doing something important for the country and the world—an attitude that indicates the importance of local education programs such as that of the Mountain Gorilla Project in Rwanda.
- An unanticipated result of civil disturbance in both countries is that nationals are now in positions of management authority. It is uncertain that this would have been the case otherwise. On the whole, the national staff have been highly responsible in their new positions.
- Conservationists must learn to lobby for support from an unfamiliar set of national and international players, including military bodies and relief agencies, which have very different objectives and agendas.

### From Horror to Hope

As we enter a new century, enduring peace remains elusive for DR Congo. Soldiers, mercenaries, and racketeers continue to terrorize local citizens and plunder the wildlife, minerals, and forests of this vast country. But we can draw hope for DR Congo and other war-torn African countries by looking at the case of Uganda.

Throughout the 1970s and much of the 1980s, Uganda was the prime example of political, economic, and social chaos in Africa. Hundreds of thousands were killed in its political pogroms and recurrent civil wars. The suffering of its people was unimaginable. More than a million refugees were scattered over the nation's interior and across its borders. Many fled into parks and reserves. During this period, the Ugandan government lost control of wildlife and parks entirely. Many guards on patrol in parks were outgunned and frequently killed by gangs and rogue soldiers. Other game guards poached wildlife in parks. Rhinoceros, cheetahs, African wild dogs, and roan antelope were completely eliminated in Uganda. Elephant herds were decimated; only young tuskless animals survived.

With the advent of peace in 1987, Ugandan civil society began to rebuild. Although the extreme northern and western regions still suffer violent conflict, elsewhere Ugandans are reconstructing their public and personal lives. One young woman grew up during wartime in Uganda's Arua District. Her father was tortured and killed, her brother died fighting the government of Milton Obote, and her mother went insane trying to comprehend the incomprehensible. As a child, she survived by hiding in the bush in neighboring Sudan. Years later, she has become one of Uganda's top wildlife biologists and the primary breadwinner for a large extended family.

As Uganda's civil society recovers, the country's wildlife and natural environment also prove resilient. The national government now publicly endorses conservation and, in many areas, promotes collaborative forest management with local communities. Elephant herds are growing quickly, and some Ugandan conservationists are lobbying to reintroduce wildlife species extirpated during the war. The healing process is fraught with tragic choices made in attempts to balance the needs of still fragile wildlife populations with urgent demands of the rural poor. Still, the experience of Uganda gives hope to those in neighboring DR Congo and Rwanda.

Africa has been through many ordeals, and has survived them all. Like the forest, its people have a remarkable capacity for recovery and renewal.

# Contributors

**Alden Almquist,** Literary Examiner, Library of Congress, Washington, D.C.

**Lauren J. Chapman,** Associate Professor, Department of Zoology, University of Florida, Gainesville, Florida

**Bryan K. Curran,** Africa Program Anthropologist, Director, Nouabale-Ndoki Project, Wildlife Conservation Society, Bronx, New York

**Robert E. Dewar,** Professor, Department of Anthropology, University of Connecticut, Storrs, Connecticut

**R. J. Dowsett,** Independent Researcher, Ganges, France

**F. Dowsett-Lemaire,** Independent Researcher, Ganges, France

**J. Michael Fay,** Africa Program, International Conservation, Wildlife Conservation Society, Bronx, New York

**Peter Grubb,** Researcher, London, England

**Alan Hamilton,** Plants Conservation Officer, Plants Unit, World Wildlife Fund, Panda House, Catteshall Lane, Godalming, Surrey, England

**John A. Hart,** Senior Conservation Zoologist, Africa Program, International Conservation, Wildlife Conservation Society, Bronx, New York

**Terese B. Hart,** Senior Conservation Ecologist, Africa Program, International Conservation, Wildlife Conservation Society, Bronx, New York

**W. D. Hawthorne,** Botanical Researcher, Department of Plant Sciences, University of Oxford, South Parks Road, Oxford, England

**Peter Howard,** Director, Africa Program, International Conservation, Wildlife Conservation Society, Bronx, New York

**John Kasenene,** Botany Department, Makerere University Biological Field Station at Kibale, Kampala, Uganda

**Michael W. Klemens,** Conservation Zoologist, Wildlife Conservation Society, Bronx, New York, and Research Associate in Herpetology, American Museum of Natural History, New York, New York

**Claire Kremen,** Assistant Professor, Department of Ecology and Evolutionary Biology, Princeton University, Princeton, New Jersey

**Sally A. Lahm,** Research Associate, Institut de Recherche en Ecologie Tropicale, Makokou, Gabon

**Nadine Laporte,** Research Scientist, Department of Geography, University of Maryland, College Park, Maryland

**Dwight P. Lawson,** Department of Biology, University of Texas at Arlington, Arlington, Texas

**David Lees,** Department of Entomology, Natural History Museum, Cromwell Road, South Kensington, England

**Daniel A. Livingstone,** James B. Duke Professor, Department of Zoology, Duke University, Durham, North Carolina

**Jean Maley,** Directeur de Recherche, Institut de Recherche pour le Développement (formerly ORSTOM), Paris, France; and Département Paléoenvironnements et Palynologie, Université de Montpellier-2, Montpellier, France

**Leonard K. Mubalama,** Senior Officer, Congolese Institute for Nature Conservation, Bukavu, DR Congo

**Lisa Naughton-Treves,** Assistant Professor, Department of Geography, University of Wisconsin—Madison, Madison, Wisconsin

**Andrew J. Noss,** Associate Conservationist, Latin America Program, International Conservation, Wildlife Conservation Society—Bolivia, Santa Cruz, Bolivia

**Yaa Ntiamoa-Baidu,** Director, Africa/Madagascar Region Program, World Wildlife Fund, WWF International, Gland, Switzerland

**Richard Oslisly,** Chargé de Recherche, Institut de Recherche pour le Développement (formerly ORSTOM), Laboratoire de Préhistoire du Muséum National d'Histoire Naturelle, Institut de Paléontologie Humaine, Paris, France

**Richard B. Peterson,** Assistant Professor of Environmental Studies, Antioch College, Yellow Springs, Ohio

**Andrew J. Plumptre,** Assistant Director, Africa Program, International Conservation, Wildlife Conservation Society, Bronx, New York

**Justina C. Ray,** Adjunct Professor, Faculty of Forestry, University of Toronto, Toronto, Ontario, Canada

**Heritiana Raharitsimba,** Wildlife Conservation Society Madagascar Program, Antananarivo, Madagascar

**Vincent Razafimahatratra** (deceased), Wildlife Conservation Society Madagascar Program, Antananarivo, Madagascar

**Alison F. Richard,** Provost of Yale University and Franklin Muzzy Crosby Professor of Anthropology, Department of Anthropology, Yale University, New Haven, Connecticut

**Eugène Rutagarama,** Director, Rwanda Office, International Gorilla Conservation Programme, Kigali, Rwanda

**J. Curt Stager, Professor,** Natural Resources Division, Paul Smith's College of the Adirondacks, Paul Smiths, New York

**David Taylor,** Associate Professor, Department of Geography, National University of Singapore, Kent Ridge, Singapore

**Richard Key Tshombe,** CEFRECOF Socio-Economic Unit Director, Africa Program, Wildlife Conservation Society, Ituri Forest, DR Congo

**Andrea K. Turkalo,** Associate Conservation Scientist, Africa Program, Wildlife Conservation Society, Bangui, Central African Republic

**Caroline E. G. Tutin,** Reader, Centre International de Recherches Médicales de Franceville, Gabon, and Department of Biological and Molecular Sciences, University of Stirling, Scotland

**Amy Vedder,** Director, Living Landscapes Program, International Conservation, Wildlife Conservation Society, Bronx, New York

**William Weber,** Director, North America Program, International Conservation, Wildlife Conservation Society, Bronx, New York

**David S. Wilkie,** Adjunct Associate Professor, Boston College, Waltham, Massachusetts

**Lee J. T. White,** Conservation Ecologist, Africa Program, International Conservation, Wildlife Conservation Society, Libreville, Gabon

# Index